Thomas Niederberger, Tobias Haller,
Helen Gambon, Madlen Kobi and Irina Wenk (eds)

The Open Cut

Action Anthropology
Aktionsethnologie

Volume/Band 2

LIT

THE OPEN CUT

Mining, Transnational Corporations and Local Populations

edited by

Thomas Niederberger, Tobias Haller,
Helen Gambon, Madlen Kobi
and Irina Wenk

LIT

Cover Photo: Mining in Cerro de Pasco, Peru; courtesy by Wikimedia, CC BY-SA 3.0.

Financial support:
Institute of Social Anthropology of the University of Bern,
Mittelbauvereinigung of the University of Bern,
Mercator Foundation,
IWGIA Group University of Zürich,
Burgergemeinde Bern,
and Tour de Lorraine Bern.

Bibliographic information published by the Deutsche Nationalbibliothek
The Deutsche Nationalbibliothek lists this publication in the Deutsche Nationalbibliografie; detailed bibliographic data are available on the Internet at http://dnb.d-nb.de.

ISBN 978-3-643-80151-7

A catalogue record for this book is available from the British Library

Klosbachstr. 107
CH-8032 Zürich
Tel. +41 (0) 44-251 75 05 Fax +41 (0) 44-251 75 06
E-Mail: zuerich@lit-verlag.ch http://www.lit-verlag.ch
Distribution:
In the UK: Global Book Marketing, e-mail: mo@centralbooks.com
In North America: International Specialized Book Services, e-mail: orders@isbs.com
In Germany: LIT Verlag Fresnostr. 2, D-48159 Münster
Tel. +49 (0) 2 51-620 32 22, Fax +49 (0) 2 51-922 60 99, E-mail: vertrieb@lit-verlag.de
In Austria: Medienlogistik Pichler-ÖBZ, e-mail: mlo@medien-logistik.at
e-books are available at www.litwebshop.de

Contents

Contributors

Editors Group

Thomas Niederberger (*1980) obtained his MA in Social Anthropology from the University of Zurich. His MA thesis dealt with community forest management in Jharkhand, India. His interest in mining conflicts started while travelling in the Argentinean Andes in 2010. He is the project coordinator for this book and a board member of *infoe CH.*

Prof. Dr. Tobias Haller (*1965) is Extraordinary Professor for Economic and Ecological Anthropology at the Institute of Social Anthropology, University of Bern. He is specialized in institutional theory, political ecology and common pool resource management, with fieldwork experience in Cameroon and Zambia, and currently engaged in an ongoing research on land grabbing. He is founder of *infoe CH,* editor/author of the infoe book on oil companies and indigenous peoples.

Helen Gambon (*1984) is a PhD student at the Centre for Development and Environment (CDE), University of Bern. In her dissertation, she analyses enabling and hindering factors for co-operation in bottom-up initiatives of institution building for the governance of natural resources based on fieldwork in the Indigenous Territory and Biosphere Reserve Pilón Lajas in the Bolivian Amazon. She is a board member of *infoe CH.*

Dr. des. Madlen Kobi (*1982) is a Postdoc at the Ethnographic Museum at the University of Zürich. She obtained her PhD in 2014 from the Institute of Social Anthropology, University of Bern. Her doctoral research focused on the construction, creation and contestation of cityscapes in Xinjiang Uyghur Autonomous Region, P. R. China. Her interest for the interaction of transnational companies and indigenous peoples stems from earlier engagement in lowland Ecuador, affected by oil exploitation. She is a board member of Incomindios and *infoe CH.*

Dr. des. Irina Wenk (*1976) is a lecturer at the Department of Social and Cultural Anthropology, University of Zurich. Her fields of specialization include Indigenous peoples' rights; mapping, titling and management of indigenous territories; natural resource conflicts; political ecology; colonial and post-colonial pacification processes, with a regional focus on the Philippines, Southeast Asia and Canada.

Authors

Amadea Tschannen (*1986) is an MA student at the Institute of Social Anthropology, University of Bern. In her MA thesis, she focuses on discourses of accountability of trans-nationally operating resource companies based in Switzerland. Her research analyses the role of the company in the transnational field of regulation and international law.

Angela Lindt (*1987) is an MA student at the Institute of Social Anthropology, University of Bern. Through investigations during an internship for the NGO 'Society for Threatened People' in 2012, she became interested in gold mining in Peru. For her MA thesis, she conducted fieldwork in Cajamarca, in the northern highland of Peru, focusing on social conflicts relating to Yanacocha's gold mine project Conga.

Annatina Foppa (*1984) obtained an MSc in philosophy of natural sciences at universities in Bern and Bremen and a CAS in science journalism. She worked as a freelancer for different Swiss Newspapers and is currently employed as specialist for radioactive waste disposal at the Swiss Federal Office of Energy.

Annette Wallimann (*1984) obtained her MA from the Institute of Social Anthropology, University of Bern in 2011. Her MA thesis, based on fieldwork in the Apurímac region in Peru, examined the large mining project Las Bambas in terms of social and economic impacts, the perceptions and strategies of different actors, and the ongoing development discourse dominated by the Swiss mining company Xstrata. She is currently working for WWF in Zürich.

Annick Vollmar (*1984) obtained her MA from the Institute of Social Anthropology, University of Bern. Her MA thesis focused on the impacts of micro-finance systems on livelihoods in Guatemala. After working as a junior research assistant for economic and political anthropology at the Institute, she did a Master of Advanced Studies in Development and Cooperation (NADEL) and is currently working as Regional Advisor on Market Systems Development for Helvetas.

Dr. Clare Oxby, Social Anthropologist (MA Cambridge, PhD SOAS London), spent a formative year living with Tuareg nomadic camel herders in Central Niger. She has taught Social Anthropology at the universities of Sussex, Oxford, Perugia and Bern, and has advised development agencies, notably the FAO, on livelihoods reconstruction following drought and other disasters. Current research on land use, as viewed from a pastoralist perspective, and notions of relatedness in migratory networks radiating from the Sahel.

Claudia Howald (*1986) lives in Quibdó/Colombia after having obtained her MA in Social Anthropology from the University of Neuchâtel. Currently she is coordinating the Instituto de Estudios del Pacífico and is part of the project "Observatorio Pacífico y Territorio" in Colombia. Her activities focus on mining, land and the local population in the Pacific Region.

Daniela Bucher (*1986) obtained an MSc in Geography from the University of Bern. For her thesis, she conducted fieldwork in Jabiru/Australia and focused on the impacts of uranium mining in the Kakadu Region, where she also volunteered at the office of the Gundjeihmi Aboriginal Corporation.

Eva Schober, (*1987) obtained her BA in Geography and Social Anthropology from the University of Bern in 2013. Her Bachelor's thesis focused on the enabling political, legal and biophysical conditions and impacts of land-grabbing in Ethiopia. Besides pursuing her MA studies, she currently engages for the association KonsumGlobal.

Germaine Françoise Spoerri (*1986) obtained an MA in Activism and Social Change from the School of Geography, University of Leeds (UK) and is currently a PhD student at the Department of Geography, University of Bern. The title of her ongoing research is: "Challenge Reappropriation: On negating neoliberal reterritorialisations of agricultural capital in post-2008 Greece".

Gregor Kaufeisen (*1988) finished his BA in Social Anthropology at the University of Bern with a thesis about the proposed gold mining project in Val Medel, Switzerland.

Günter Baitsch is a practising physician and a board member of the Swiss section of International Physicians for the Prevention of Nuclear War (see <www.ippnw.ch>).

Isabella Brom (*1988) obtained a BSc in Geography, a BA in Social Anthropology and an MA degree in Natural Hazard and Disaster Management with a focus on Geographic Information Science (GIS) at the Australian National University in Canberra. She is now working in the field of Geomatics and Management in Zürich.

Jacques Diezi, Professor (hon.) of the Department of Pharmacology and Toxicology of the University of Lausanne, is a board member of the Swiss section of International Physicians for the Prevention of Nuclear War (see <www.ippnw.ch>).

Joel Zumbühl, (*1990) obtained a BA and is currently an MA student in History and Social Anthropology at the University of Bern.

Karmen Ramírez Boscán (*1971) is a Wayuu indigenous woman from Colombia. Originally a graphic designer, she has in the last years worked as a defender of Human Rights in her home region affected by different mining activities. Currently living in Switzerland, she has funded the indigenous women's organization *Fuerza de Mujeres Wayuu* and has worked as a consultant for the ILO and for the Office of the High Commissioner for Human Rights (OHCHR) in Geneva.

Kathrin von Allmen (*1985) received her BA in Philosophy, Ecology and Biology in 2013 from the University of Bern. After an internship at the foundation for environmental information in Lucerne, she is currently completing her MA studies in Political, Legal and Economic Philosophy. In her master thesis she investigates the concept of collective responsibility in the field of global climate change.

Loretta Scherler (*1986) obtained her MA in Anthropology of Transnationalism and the State from the University of Bern. She has based her findings on field research in the mining region of the Northern Territory, Australia.

Matthieu Bolay (*1982) is a PhD student in Social Anthropology, University of Neuchâtel (*Laboratoire d'études transnationales et des processus sociaux*). His ongoing research is about the circulation of migrant workers and technology in the mining regions of West Africa (Guinea, Mali, Ivory Coast).

Maximilian Rau (*1991) obtained his BA degree in History and Social Anthropology from the University of Bern.

Dr. Roy Maconachie is Senior Lecturer in International Development at the University of Bath, UK. A human geographer by training, his research is concerned with the political economy of natural resource management with a specific interest in the socio-economic dimensions of small-scale mining in West Africa. His recent work in Sierra Leone explores how local livelihoods are being shaped by the increasing tension between artisanal mining and large-scale extraction, with a focus on how youth-led civil society groups are impacting the mining sector.

Dr. Saskia Walentowitz is a social anthropologist (Phd *École des Hautes Études en Sciences Sociales*, Paris), Senior Lecturer and Ambizione-Research Fellow at the Institute for Social Anthropology, University of Berne, from 2004 to 2015. She advised GTZ on their development projects and completed three years of ethnographic fieldwork in northern Niger. She works on the gendered dynamics of reproduction and kinship in Tuareg Berber and other Muslim contexts and participated in reproductive health and HIV-related research in Kenya. Her current studies and teachings promote a political anthropology of nature.

Image Credits

Additional image materials (to those produced by the chapters' authors) were provided by the following persons, institutions and organizations:

- Cover photo: Mining in Cerro de Pasco, Peru. 2009; courtesy by Wikimedia, <de.wikipedia.org/wiki/Cerro_de_Pasco>, Wikimedia commons: CC BY-SA 3.0.
- Covers of Part I, II, III and IV: details from the poster *The True Cost of Coal* of The Beehive Design Collective, <beehivecollective.org>, creative commons.
- Part I: Google Maps, DigitalGlobe, CNES/Spot Image/Astrium, Save Rosia Montana campaign, Wikimedia, Lottermoser (2007), and Notiwayuu.
- Part II: Community Administration Medel/Lucmagn, Terre des Hommes Suisse, Jacob Balzani Lööv, Meinrad Schade, Collectif Tchinaghen, World of Matter/Ursula Biemann, Stephan Suhner, Anna Leissing, Clothilde Laime, Dominic O'Brien, Karin Scheidegger, Bob Timonera, and Fastenopfer; Maps: Michael Curran.
- Part III: Meinrad Schade, Wikimedia, Annette Wallimann, Golda Fuentes, Alyansa Tigil Mina, Indymedia.org, Philipp Eyer, and Hambacher Forst campaign.
- Part IV: iconoclasistas, creative commons.

Acknowledgements

This book is the result of four years of inspiring collaborations, alliances and engagements between students, researchers, activists, and indigenous and civil society representatives from different domains and geographical backgrounds. As the editing group for *infoe CH*, we would like to thank in particular all the authors for their passionate engagement in the research and writing phase, as well as for their patience during the long editing process.

In these acknowledgements, we are only able to mention the most crucial contributors while we apologize for all those who have crossed our paths and are not mentioned by name, but have nevertheless inspired and pushed forward the realization of this project. This includes our families and friends.

In particular, we would like to thank the Institute of Social Anthropology at the University of Bern for hosting this project and providing generous technical and administrative support, as well as Stephan Rist and his team at the Centre for Development and Environment (CDE) of the University of Bern for the supervision of several theses, the provision of civil service assignments and their valuable advice in general.

Although most of the study was realized on a volunteer basis, it would not have been possible without financial support. The Institute of Social Anthropology at the University of Bern financed an assistant lecturer position for the initial research seminar, as well as for the supervision of related students' theses. The *Mittelbauvereinigung* of the University of Bern provided a project grant for the setup of infrastructure and the realization of several workshops and events. The Mercator Foundation's program "*engagier dich!*" contributed to the publication costs of the chapters written by Bachelor and Master students. Finally, the general costs for proof-reading, printing and layout were covered by contributions from the former IWGIA Group of the University of Zürich, the publication fund of the *Burgergemeinde Bern*, the *Tour de Lorraine* fund, and by generous private donations.

Financial contribution was an important pillar of the finalization of the project, but the study would not have been possible without the professional and engaged advice that experts from NGOs and research institutions shared with us. We would like to thank María Victoria Veracierto from the University of Buenos Aires for her insights into the socio-ecological conflicts in the Argentinean province of Jujuy, as well as Margit Huber and Jiyo Galindez for further inputs on the situation in Argentina; Katja Werthmann for contributing with her anthropological expert knowledge on small-scale mining in West Africa; Anna Leissing from the Swisspeace Center for Peacebuilding (KOFF) for her revisions and advice on the Guatemala case study; Eva Schmassmann from GfbV/Alliance Sud for inputs on the role of Swiss transnational companies in the extraction industries and the review of the chapter on Swiss gold refineries; Shalini Gera, who is a legal adviser for the PCSS cement workers' union, as well as Yvonne Zimmermann of Solifonds and Asgar Qadri for their contributions and critical inputs to the India case study; Daniel Hostettler from Fastenopfer for information on the Philippines case study, and Stephan Suhner from Arbeitsgruppe Schweiz-Kolumbien for Colombia; Amadea Tschannen and Stephy-Mathew Moozhiyil for reviewing the Comparison Part and the Zambia case study; Peter Diehl of WISE and Urs Ruegg for revising the chapter on health impacts; and Sandra Rüd-Probst, Annika Linkhorst, Jasmin Mertens and Nina Küng from the ETH Department of Environmental Systems Science for their comments on the chapter on the mining process and its environmental impacts.

For the linguistic improvement of our non-native English book contributions, we depended on the help of professional translators and proofreaders. German to English translations were contributed free of charge by Annabelle Piguet, Andrea Müller and Carmen Hidalgo Staub. The main editing, however, was done by Subashini Shriya and Satya Sivaraman from New Delhi, India. They were not only correcting our spelling and language mistakes, but at the same time engaged in human rights related campaigns in India and elsewhere. We would like to thank them for taking the time to work with us. Further, Dr. Michael Curran contributed the maps for the case studies, and a number of photographers, graphic designers, organizations and institutions agreed to publish their images free of charge or under creative commons licenses (see image credits above). We express our profound thanks to Ursula Grimm who dedicated many hours to the layout of this book, far below the usual market-rates.

Infoe CH received organizational support in the form of a civil service assignment from Urban Schwegler and Luciano Gagliardi; the *xiala* web-server collective from Zurich provided their expertise for the website and IT; the students Manuel Bär and Gregor Kaufeisen helped with fundraising; and Alexandra Ottinger facilitated the three-day workshop on action research and mining in Bern in November 2013.

Furthermore, we would like to thank the students who participated in the seminar at the Institute of Social Anthropology in 2011 and wrote essays on different case studies, some of which served as a first base for the chapters published in this book (as mentioned in the respective chapters): Anic Werder Picuasi Morán, Christina von Grünigen, Christoph Bürki, Joline Fehr, Jonas Jacob, Laura Rindlisbacher, Lucile Thalmann, Manuel Bär, Michelle Pfister, Rhea Belfanti, and Sara Ryser.

The first volume of the *Action Anthropology Series* on fossil fuels, published by infoe CH in English in 2007, only came into being through the passionate and relentless commitment of Max Mader, who sadly passed away in July 2011. It is in remembrance of Max that we started the ambitious venture of compiling a second volume and his inspiration accompanied the origination process of this project.

Last, but not least, we dedicate this book to all the individuals who daily engage in mining-related struggles for environmental and social justice, for greater democratic participation and for being heard in a capital-dominated economic system. If our publication can contribute even a small part to their struggles being more successful, the efforts taken were worth it.

The editors

The Open Cut
PART I
Introduction

PART I:

1. Introduction

Thomas Niederberger and Tobias Haller[1]

Around the year 2010, more and more reports about heavy mining-related conflicts started reaching Switzerland. Some of the companies involved had been based in this country since quite some time; others had recently moved their headquarters here, attracted by low taxes, political stability, a strong financial sector and a high quality lifestyle for those who could afford it. Besides some small NGOs and solidarity groups, few people had bothered about this development before, but now it has become increasingly difficult to ignore. The mining and commodity trading sectors had seen tremendous growth since the year 2000 and there was an obvious lack of knowledge about it—among the general public as well as in the academic spheres, including social anthropology and geography. Although there were some individual studies on local dynamics in mining areas, often provided by NGOs or press reports, we felt that there was a need for a comparative approach on mining impacts and strategies of local and corporate actors that would go beyond specific cases. The aim of the present book is quite simple, yet ambitious, at the same time: we want to gain a comparative overview on the social and environmental impacts of large-scale mining and provide empirical evidence on the way relationships around mining, transnational corporations and local populations are structured. Who are the actors in this field? What strategies drive their actions and interactions? Finally, it is also a way of relating ourselves to the extraction of the metals and minerals we often use as consumers, without thinking much about their origin.

In this introductory chapter, we first describe the writing process of this book. Second, we evaluate the relevance of mining as a driver for socio-environmental conflicts, and third, we introduce some of the relevant debates in social anthropology and related fields like human geography and political ecology. Finally, we explain our approach, the choice of case studies and the structure of the book.

1 Review by Madlen Kobi and Helen Gambon. The references for this chapter are included in those of Part III, page 503. Cover: detail from The True Cost of Coal, courtesy by The Beehive Design Collective, creative commons.

Fig. 1: Mopani Copper Mine, Mufulira, Zambia. Source: Google Maps / DigitalGlobe.

1.1 The Writing Process of this Book

For organizing this research, we decided to relaunch the Swiss section of the Institute for Ecology and Action Anthropology (*infoe CH*) at the University of Bern. *Infoe CH* had published a comparative book on fossil fuels extraction at the University of Zurich in 2000 (in English: Haller et al. 2007), which served as our model.

The title of this book—the Open Cut—refers to a technical term for large-scale mining "at open daylight" (also "open-pit" or "open-cast" mining), and metaphorically to the deep wounds that such mines typically cause to the cultural landscapes where they are situated.[2] The problems that arise from this point of departure—both theoretically and practically—are serious enough and do not need to be further emphasized. What normative implications are to be drawn from our findings is largely left to the reader. Nevertheless, if there is something like a "hope" that leads this investigation, it is perhaps the idea that mining conflicts can under certain circumstances lead to positive transformations and social innovations (Bebbington 2012: 4). Movements for social and ecological justice can be a powerful force to generate knowledge for such transformations (Escobar 2008), as well as for the democratization and "bottom up" construction of new institutions for more power balanced and sustainable forms of resource management and use as the concept of constitutionality outlines (see Haller/Acciaioli/Rist 2015). To study the interactions of local and global dynamics in campaigns, their strengths, weaknesses, mutual misunderstandings, failures and successes is thus not only of analytical interest, but a way of actively participating in these changes (see Tsing 2005; Kirsch 2002, 2014).

2 For an impression of their scale, we suggest to search online for the satellite view of some of the mines covered in the case study chapters; see figures 1 to 7.

The present book is a result of four years of "digging" into the depths of the mining industry, its impacts and the strategies surrounding it. It started with a research exercise at the Institute of Social Anthropology at the University of Bern in 2011. Organized by the authors with the collaboration of Dr. Clare Oxby, who arranged a film-screening and a guest talk by Abdoulaye Attayoub of *Temoust Survie touarègue*.[3] During the exercise, about 20 students of social anthropology and geography worked in groups on case studies that they chose from a list of propositions. These essays, all organized along the same structure, are the base for most of the case study chapters in this book. The time-span of the investigation and writing process mostly ended around 2013, with some updates on later developments added subsequently. In some cases, the authors are the same students of the exercise who made an effort to improve their original essay to the present state. In others, new (co-)authors have improved or completely rewritten a case study, adding their own research and fieldwork experiences. Some contributions were written independent of the research exercise in order to explore additional aspects or cases of mining. One particularity of the book is that the authors include Bachelor, Master, and PhD-students, as well as senior researchers. Additionally, a number of people with experience in the concerned fields helped with their advice and critique. By combining this diversity of contributors and contributions to a sort of collective oeuvre, we hope to compensate to some extent for the lack of specific field research in several of the case studies, due to a lack of time and resources. With the process of circulating drafts, corrections and rewriting, this book has already created a sense of collective knowledge among the participants, and we hope that it can now provide a useful starting point for students, researchers, journalists, activists and other interested persons.

Further, the book is also a reference to the almost forgotten tradition of Action Anthropology. Though we are far away from the more radical propositions of prominent action anthropologists like Sol Tax and Karl Schlesier, some of us did participate in "action" during the writing of this book. Notably, *infoe CH* took part in extractive-industries-related campaigns through the Swiss MultiWatch coalition, which again forms part of the Swiss Coalition for Corporate Justice. Also, we had the opportunity to host several talks at the University of Bern; among them, Antonio Gustavo Gómez, the federal prosecutor of the Argentinean Supreme Court in Tucuman, who is investigating contamination and tax evasion of Xstrata's Alumbrera mine; José de Echave, former Vice-Minister of Environment of Peru, scholar and founder of an NGO specializing on mining; and Victoria Maria Veracierto, an anthropologist and activist from Argentina (see chapter on Uranio AG). Besides these "official" activities, some of the participants also turned up at marches, vigils and direct actions, such as several demonstrations in Zug, at the European Gold Forum in Zurich, and at the Global Commodities Summit in Lausanne (see Fig. 8).

All these diverse engagements influenced what we know about mining and corporations and provided an inside perspective into the mechanisms of NGO campaigns and their interaction with mainstream media, politics and corporate communication. With participation in these activities, as well as through contacts during research, exchanges were established with individuals and groups in places where mining is happening. This is a necessary first step for the possibility of carrying out actual Action Anthropology, for which *infoe CH* intends to provide a platform and institutional support. We think that transmitting some of the voices of directly affected people to places where they are normally not being heard is already a worthwhile form of action. The power disparity involved in this transaction makes it necessary to reflect on our position as researchers, as will be further elaborated in Part IV—Strategies for Action/Research—taking inspirations from Action Anthropology and more recent developments in engaged, collective research.

3 The film *Niger: la bataille de l'uranium*, directed by Nahan Siby et al. (2009), shows the conflict scenario around Areva's uranium mine in Arlit; Temoust is based in Lyon, France (see <www.temoust.org>).

1.2 Mining Impacts and Conflicts

Mining means the extraction of metals and minerals from the soil. Together with gas and oil, it is part of the so-called "extractive industries". Although there are common features, mining has important differences from oil and gas drilling, and there are only few companies which are active in both fields simultaneously. In this book, our focus is on large-scale mining carried out by transnational corporations (TNCs).

Contemporary large-scale mines are deep, extensive and long-lasting "open cuts" in the cultural and ecological landscapes of entire regions. They release huge amounts of toxic waste and use water and energy resources in quantities comparable to mid-size cities. The transformations which accompany the activities of such mines have an impact not only on the quality of livelihoods of local populations, but also on the regional and even national political economy and institutions. But the ways in which mining activities are perceived, and the discourses which legitimize or repudiate mining, differ widely. Among many other questions, debates are often about expectations of "development", different conceptions of what that should mean, but also on the relationship between human beings and "nature" as such. The resulting conflicts have complex feedback loops with expectations and perceptions of adverse and positive impacts at different levels and scales. Far from being local, they are as transnational or global as the mining corporations themselves are: key issues include global discourses about sustainability, corporate social responsibility, structural adjustment or climate change, and most of all, global market prices for raw materials and speculations about their future developments.

Fig. 2: Somaïr uranium mine in Arlit, Niger. Source: Google Maps, DigitalGlobe, CNES/Spot Image.

The boom of most metal prices and "commodities" accelerated in the 2000s and has brought a tremendous growth of investments in exploration and new projects. There are estimates that 50 to 80 percent of land targeted for future mining is inhabited by indigenous peoples (Whitmore 2012: 5). These "frontiers of extraction" (Bebbington 2012: 15-16, Ballard/Banks 2003: 288) come with a sharp increase of related conflicts. The (far from complete) "environmental justice atlas"[4] lists 283 out of 1265 environmental conflicts as mining conflicts. 216 of these are categorized as of medium or high intensity (out of 899), with a concentration along the Andean mountain range and in Central America, where massive, nationwide protest movements against mining projects frequently involve violence and killings. In Peru alone, a hot-spot of mining expansion, the *Defensoria del Pueblo* counted 95 out of a total of 133 socio-environmental conflicts as mining conflicts in the first half of 2014.[5]

In view of these numbers, it is no surprise that publications about mining conflicts have multiplied during the time we worked on this book. Apart from academics, NGOs and international finance organizations like the World Bank, publications increasingly come from mining industry bodies themselves. For instance, the annual survey of the Fraser Institute (Wilson/Cervantes 2013: 15), a Canadian consultancy, includes for the first time a section on "public opposition to mining": 36% of the responding companies reported important project delays due to resistance. The most prominent reasons cited were environmental impacts, often water contamination and water grabbing (59%), followed by "Indigenous or Aboriginal rights or title" (32%). Franks et al. (2014) carried out interviews with industry management from a sample of 50 conflict-ridden mining projects. Also here, pollution and access to natural resources (in particular to water) were the central issues in dispute in about 70% of the cases, while distribution of benefits was seen as a major underlying cause.

Setting up a modern large-scale open-pit mine can cost up to USD 5 billion and the monetary damages of delays and shut-downs can reach several hundred millions of USD. On the other side, many local communities experience the huge mines in their areas as a threat to their very existence and hence risk their lives to block them: half of the cases involved a physical blockade of the (projected) mine site, and in about a third of the cases a killing or damage to private property was reported (Franks et al. 2014: 2-3).

Mining conflicts are not only happening in "remote areas" of the Global South, but also in Europe: two of the most contested mega-projects worldwide are situated in Roşia Montana, Romania and Halkidiki, Greece.[6] Further, anti-mining groups are confronting mine projects in northern Spain and Sweden and are opposing expansion of open-pit coal mines in Germany.[7] And even Switzerland has seen some local controversies about gold exploration (see case study Val Medel in this book). In short, we are not talking about a few isolated cases where "something went wrong", but about a systemic problem of global proportions.

4 Ejatlas.org is a project of EJOLT ("a global research project bringing science and society together to catalogue and analyse ecological distribution conflicts and confront environmental injustice"). <www.ejolt.org> accessed 16.12.2014. See also Özkaynak et al. (2012).

5 Defensoria del Pueblo: "Estado de los conflictos sociales - Julio 2014." <www.defensoria.gob.pe> accessed 16.12.2014. For mining conflicts in Peru, see also: De Echave et al. (2009) and Bebbington (2012).

6 See Egresi (2011) and the campaign websites of "Save Rosia Montana" <www.rosiamontana.org> and "SOS Halkidiki" <www.soshalkidiki.wordpress.com>; accessed 17.12.2014.

7 For example, the gold mine project of Astur Gold in the Salave region of Asturias, Spain (Cambalache 2013); in Sweden: eco-activists and Sami indigenous people against the iron ore mine project in Kallak Gállok (<www.kolonierna.se>); in Germany: blockade camp at Hambach forest (<hambacherforst.blogsport.de>); all accessed 17.12.2014.

1.3 Introducing the Debates

Mining conflicts are at the very root of contemporary capitalism, in the sense that they are about the extraction of some of the most important raw materials at the beginning of different commodity chains. Mining is often discussed as one aspect of a global environmental and social justice crisis alongside other forms of natural resource extraction (i.e., oil and gas, wood, industrial agriculture etc.), land-grabbing and the appropriation of related common-pool resources (especially, water), loss of biological diversity and climate change. Thereby, mining is related to debates around (post-) colonialism, (neoliberal) development, globalization and unequal wealth distribution. In the 1990s, the "resource curse" hypothesis considered that abundant natural resources often turned out to be a curse because their extraction is typically accompanied by armed conflict, bad governance, corruption and extreme forms of exploitation (Auty 1993; Pegg 2006). Because of the sheer size of extractive industry operations and the "incredible wealth and destruction [they produce] at one and the same time" (Bebbington 2012: 5), the history and current condition of entire nations and regions cannot be understood in isolation from it; many authors therefore state an urgent need for improving regulation (e.g., Bebbington et al. 2008, and 2013: 7-8, 13; Ballard/Banks 2003: 294-296).

The increase in conflicts between mining corporations and local people is reflected in the academic debates. Mining is no longer the exclusive field of geologists, engineers, economists and lawyers. When Ballard and Banks in 2003 wrote the most recent review article on the anthropology of mining, they stated a lack of attention on large-scale mining. Since then, many anthropologists and other researchers have got involved and contributed to the ongoing debates.[8] There is a special issue in one of the leading anthropology journals *Current Anthropology* on the "Social Life of the Corporate Form" which contains several articles related to CSR and mining,[9] a special issue of *Journal of Cleaner Production* (2006, Vol. 14(3–4)), and numerous articles in *Resource Policy, Geoforum, Ecological Economics*, and others, treating the issue mostly from a resource geography or political economy/ecology angle.[10]

Social anthropology has a long tradition of research in mining, whereby the focus has been mostly on informally organized, artisanal and small-scale mining (see, e.g., Werthmann 2012 for Western Africa). The classic monographs are from June Nash (1979) and Michael Taussig (1986), interpreting Bolivian miners' rituals and bringing together concepts of sacrifice, commodity fetishism and the devil (see Godoy 1985: 208f.). Jonathan Parry (2008) takes this up in his discussion of the linkages between industrial development in eastern India and the sacrifice of humans and entire landscapes "for the good of the nation". In recent years, the interaction between informal miners and large-scale mining companies has received more attention, given that the latter are increasingly entering territories occupied by artisanal miners (see Bolay on Guinea, in this volume). Another recurrent theme in anthropological research is the demographic and social transformation in mining towns, often discussed together with rural-urban migration (Ballard/Banks 2003: 292). Mining tends to come with big expectations, followed by big deceptions. This happens for the individuals and groups who get involved in a "gold rush" frenzy (Cleary 1990) as well as for entire regions and nations which aspire to "reach modernity" through industrial-scale mining. Ferguson (1999) shows how mining first constituted the "expectations of modernity" in the towns in the Zambian Copperbelt and then led to the despair of the inhabitants following its break-down.

8 New publications include, e.g., O'Faircheallaigh/Ali (ed.) (2008), Li (2009), Rajak (2011), Sawyer/Gomez (eds) (2012), Thorp et al. (2012), Welker (2014), and Golub (2014)..

9 E.g., the one cited above by Catherine Coumans (2011), anthropologist and research coordinator of MiningWatch Canada.

10 E.g., Urkidi (2010), Hurley/Arı (2011), Hilson (2012), Haalboom (2012), Mutti et al. (2012), Ololade/Annegarn (2013), Prno (2013), Owen/Kemp (2013), Kronenberg (2014).

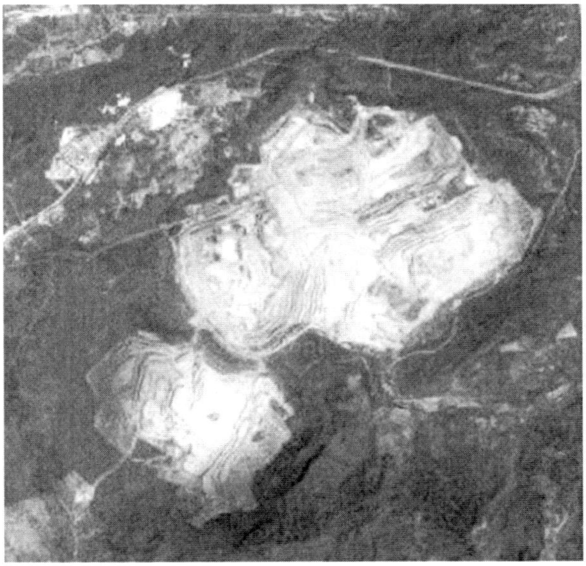

Fig. 3: El Cerrejón coal mine, Colombia. Source: Google Maps, DigitalGlobe.

With regard to the actors in mining conflicts, one can observe a shift of emphasis away from mine workers and unions (especially, in Marxist writings of the 1960s and '70s) to local and indigenous people's communities, rights and social movements. This reflects the weakening of unions and labour struggles due to the technological change in large-scale mining (needing fewer workers) and the extensive use of sub-contracting. The shift of emphasis corresponds to the increased attention on environmental problems and a resurgence of indigenous movements in the 1990s, too (Kirsch 2012: 203, 209; Martinez-Alier 2002). Further, it can be interpreted as expressing a shift in emphasis in capitalist forms from "accumulation by exploitation" to "accumulation by dispossession", i.e., from the exploitation of labour to the privatization and commodification of public and common goods (Harvey 2003, in Kirsch 2014: 7; see also Holden/Nadeau/Jacobson 2011).

The Anthropology of Large-Scale Mining

A recent major contribution to the debate on large-scale mining is Kirsch's "Mining Capitalism" (2014). Based on his long-standing involvement in the campaign to stop river pollution by the Ok Tedi mine in Papua New Guinea, Kirsch analyses corporate strategies which prevent, co-opt and de-legitimize critique, and states that the "dialectical relationship between corporations and their critics has become a permanent structural feature of neoliberal capitalism" (ibid.: 3). This approach corresponds to a "studying up", that is, a shift of focus away from specific local communities to corporate and state entities and other powerful actors, as well as to the rise of the so-called "audit culture"[11] and "corporate social responsibility" (CSR; see also Chapter 2) centred on concepts like "participation", "consultation" and "stakeholder dialogue". Many other relevant texts in

11 Kirsch (2014: 171) attributes the term to Michael Power (1994). See also Strathern (2002).

Fig. 4: Marlin gold-silver mine, Guatemala. Source: Google Maps, DigitalGlobe.

this field originated from research done in Papua New Guinea and Australia, where a number of anthropologists have been contracted by mining corporations to carry out social-impact assessments and community consultations since the 1980s. This involvement resulted from the need to better understand social impacts of mining after the armed conflict that led to the shut-down of the Panguna copper-gold mine in Bougainville and rising conflicts at the Porgera and Ok Tedi mines (Ballard/Banks 2003: 289, 306-7; Coumans 2011: 33-36). Some of the "corporate" anthropologists later "changed sides" and engaged in advocacy for local organizations (most notably Kirsch; see, for e.g., 2002), or they took a more critical position towards the industry (Hyndman 2001). Others defend an "honest broker" position between corporate interests and the communities, which they claim is not hampered by accepting to be paid directly or indirectly by the industry, as long as professional ethical standards are maintained (e.g., Filer 1999, 2012). Sometimes expressed in fierce personal polemics, this debate is crucial for the role and positioning of anthropology and social sciences with regard to mining corporations.

Critics say that CSR evolved as a reaction to civil society and activist critique. Further, CSR is described as a strategy to deal with the call for enforceable global standards by simulating reform, promoting voluntary standards and postulating that mining can be a driver for "sustainable development" (Kirsch 2014: 168-172; Moody 2007: 154ff.). This co-option of critical discourse by installing a notion of "sustainable mining" is described as a "corporate oxymoron" (Benson/Kirsch 2010b). CSR and the related socially responsible investment business have become entry points for numerous social scientists who occupy "space for dialogue and negotiation that was created by community struggle" (Coumans 2011: 30, 38). The term "greenwashing" is frequently used to imply that CSR is mostly a publicity stunt, which does not correspond to actual behaviour. But CSR has real impacts at the local level, too. Arguably, it improves leverage or bargaining power for local and civil society groups, however with side-effects: CSR can serve as an extension of corporate security and as moral self-justification to immunize management and staff against critiques (Welker 2009), and it may cause conflicts between different sectors of the population over the distribution of CSR-related benefits (Hilson 2012: 135). Or, to use Ferguson's term, CSR acts as an "anti-politics machine", similarly as development policies in dependent African states (see Ferguson 1990).

In a large part of the literature on CSR, a triad model of interactions between corporations, states and communities is used (Ballard/Banks 2003: 289). This view is well suited for efforts to "bring everyone at a table", which aims at mitigation and prevention of conflicts according to a "stakeholder dialogue" model. One risk of the triangle view is that it tends to conceal that local actors are typically in a very weak bargaining power position against allied corporate-state interests. Also, it may reinforce monolithical readings of "corporation", "state" and "community", all of which are internally differentiated and made up of sub-groups and individuals who have divergent interests (Ballard/Banks 2003: 290). Detailed ethnographic research points to unequal power relations and conflicts between and within communities, local elites and marginalization of women or youth, which might be intensified through the loss of land and subsistence options, salary jobs, changes in marriage and sexuality norms and an increase in sexually transmitted diseases (Ballard/Banks 2003: 302). Especially for understanding mining conflicts at the local level, it is thus indispensable to "[...] 'unpack' the term 'community' and shed light on the ways in which diverging local interests can lead to equally diverse responses—running from resistance through to negotiation and acquiescence" (Bebbington et al. 2013: 5), and how this changes over time. For example, Maconachie (in this volume) shows how groups of local youth in Sierra Leone have ambivalent attitudes towards mining companies, both welcoming the scarce opportunities for employment and economic independence, while challenging the distribution of benefits which is considered as unjust. An approach with focus on traditional decision-makers in the communities—often elderly men—falls short of recognizing such aspirations and their potential for conflict. Also gender perspectives on mining, often seen as an exclusively male field, are highly important, but relatively rare (Bebbington et al. 2013: 5; see also Lahiri-Dutt 2011, Jenkins 2014). Mastrángelo (2004) describes how the construction of the Alumbrera mine in Belén, Argentina, shifts power to local male elites and their system of patronage, to the detriment of the textile cottage industry, which used to be the traditional economic stronghold of women. Such transformations of local household and subsistence economies respond to larger market dynamics, which can be analyzed using theories of institutional change (e.g., Ensminger 1992; see section 1.4).

Unpacking the Corporate Black Box

While the study of internal differentiations within local communities is a stronghold of anthropology, to "unpack the corporation" proves to be considerably more difficult. Corporations are notoriously secretive about their internal affairs. Information is seen as an asset that grants them competitive advantage and corporate lawyers often "stifle attempts to reach out to and engage the public out of fear that any public statements might be used against the company in court" (Welker 2014: 46). There is little independent ethnographic research on how decisions are made within corporations, which is especially problematic in view of the far-reaching consequences that these decisions can have. Some of the best-informed researchers are "embedded anthropologists", who work for companies as consultants (as mentioned above for Papua New Guinea). However, this raises questions about their credibility: privileged access to information is coupled with loyalties and research results may become exclusive property of the company (Coumans 2011: 33). As the corporation remains a *terra incognita* to most anthropologists, it is no surprise that the mining company "lurks monolithically and often menacingly in the background of many anthropological accounts of communities affected by mining operations" (Ballard/Banks 2003: 290). A unique inside view is given by Welker's (2014) recently published ethnography, which covers both the headquarters of the major mining corporation Newmont in Denver, USA and the Batu Hijau min-

Fig. 5: Las Bambas mine under construction, Peru. Source: Google Maps, DigitalGlobe.

ing site in Sumbawa, Indonesia.[12] Based in the CSR department, she analyses the contradictions which arise in the attempt to adopt policies of corporate responsibility and ethics, showing how people "enact the corporation" in multiple ways, involving "[...] struggles over the boundaries, interests, and responsibilities of the corporation". Rather than with an abstract, profit-maximizing "*homo economicus* model of the corporation", people thereby identify themselves along the internal differentiation within the corporation. The perspectives, priorities and interests between the global headquarters and on-site mine management as well as with different disciplines such as geology, engineering, finance or security differ greatly (ibid.: 39 ff.). That Welker was granted access to a corporate headquarters was probably due to the fact that she got to the right person at the right time: the newly appointed head for CSR was an anthropologist who, as she believes, had an interest in an "experiment in enacting Newmont as a transparent corporation" (ibid.: 34). Even though her research was independently financed and not restricted by company censorship, there is thus a bias towards a CSR showcase mine, while other more problematic cases of Newmont, such as the controversial expansion of the Yanacocha mine in Peru, are side-lined.

Be it for a lack of access to such detailed inside views of corporate decision-making, or for methodological and political choices, most research is based on the simple assumption that the basic purpose of corporate actors is to maximize profits or "shareholder value", often in collusion with and under protection of powerful state and multinational actors, to the detriment of the environment and local people's well-being. Sawyer and Gomez (2012) observe this kind of corporate-state collusion in almost all the case studies covered (only Bolivia somewhat escapes the pattern). The role of International Finance Institutions (IFIs) such as the World Bank and International Monetary Fund in national law-making, the "revolving-door practice" of personnel between IFIs, corporations and state administrations, and the influence of "big business through the financing

12 See also Welker/Wood (2011) on shareholder activism and "responsible investment" in the USA.

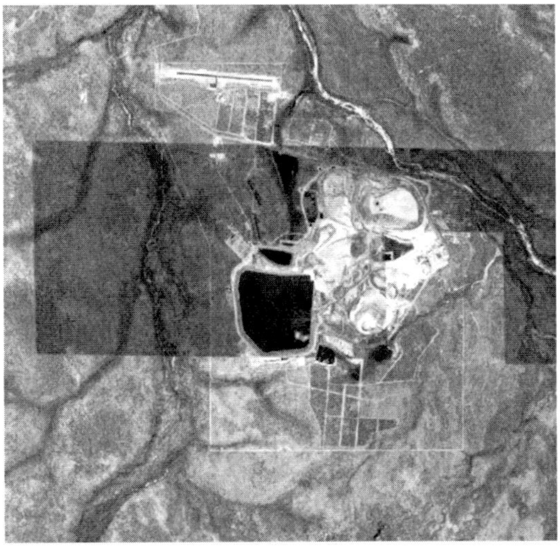

Fig. 6: Ranger uranium mine, Australia. Source: Google Maps, DigitalGlobe.

of political parties or politicians" lead them to speak of a state-IFI-MNC "nexus of power", which is pushing the underlying neoliberal agenda of "intensification and expansion of capitalist markets and trade" (2012: 38). In the African context, Ferguson shows how capitalist expansion is mystified by an "economistic language of international technocracy" which presents "structural adjustment" measures as "pragmatic" and "necessary" to improve "efficiency" and "economic growth"; a kind of "scientific capitalism" which treats politics not as an "arena for public participation and moral discussion", but simply as a technical factor "to be fixed at the 'correct' level" (2006: 78-79). This leads to the above mentioned notion of development as an "anti-politics machine", which we will take up again in Part III.

Corporation–State–Community Relations

With regard to the corporation-state-indigenous relations, Sawyer and Gomez state that "[a]ll the case studies suggest that MNCs, in patent alliance with the state, undermine indigenous interests, though tensions surface regularly between the two. MNCs and states often disagree over who should deliver basic services and infrastructure to rural areas" (2012: 40). In Australia, corporations opposed the government's move to cut spending in social services to impoverished Aboriginal communities, because they feared that this would fuel demands that they should step in to deliver such services. This observation by Altman (2012) points to a larger dilemma that comes with the increasing recognition by corporations that they need a "social licence to operate". In many mining regions, the presence of the state is weak and poverty rampant. Therefore, CSR claims are easily understood by the local population and by certain state bodies as an invitation to demand from the corporation the kind of development that the state does not or cannot deliver: infrastructure, schooling, healthcare, and social aid. The companies are in a dilemma: on the one hand, investments in state-like services may be a necessary condition to legitimize their operations; on the other hand, this opens the door for a potentially endless proliferation of demands. The sheer size of a large-scale mine and associated infrastructures, worker compounds, community relations offices etc., can make it look like a "state within the state", and the corporation thus

Fig. 7: Ambuja cement factory and limestone mines, Rawan, India. Source: Google Maps, CNES/Astrium.

faces the question of "how to act like a state without wanting to be the state?" Arellano-Yanguas (2012) describes this tension for Peru, whereby he discovers a paradox effect: the channelling of mining rents into local development led to an increase of social conflicts, exactly in those regions where the sums were highest (see also the case study of Peru in this volume). But many of these conflicts were not against mining as such, or against a company, but between different sectors of the society. Large sums become small when distributed over entire regions with diverse interest groups. Many local people realized that they first needed to protest to get a share of the benefits. Besides the outright opposition against the establishment of new mines, the author distinguishes two types of conflicts whereby contention is used as an entry point to negotiate the distribution of mining rents, by targeting either the company directly or government bodies through which mining taxes are transferred (2012: 91, 95f.). Similar observations are made by others (e.g., Welker 2009), but they also need to be taken with care, as "indigenous peoples might be stigmatized as feigning concern about environmental impacts in order to leverage additional economic benefits" (Kirsch 2012: 206). Individually and collectively, local people may be seriously worried about environmental damages and the well-being of future generations, but also ready to discount on this for immediate (financial) benefits if they are high enough and if the impression is that the development cannot be stopped anyway. This dilemma points to theories of collective action, which suggest that shared perceptions and ideological bases can contribute to articulating and enacting common interests (as further discussed in Part III; see also Haller et al. 2007).

A practical implication of CSR policies, coupled with the growing legal recognition of indigenous property rights in many countries (see Chapter 2), is a trend towards directly negotiated agreements between mining companies and indigenous communities. In the previously mentioned case of Papua New Guinea, this is indispensable for new projects, because indigenous landowners enjoy strong legal protection. The main conflict then seems to be in the identification of legitimate ownership claims in the context of fluid, extensive kinship networks (Ballard/Banks 2003: 298-301). Companies therefore employ anthropologists to map lineage ties, which are often linked to a sacred geography of burial and worship sites. Filer (1999), who has been involved in such research, argues that local land-owning communities do not necessarily have interests diametrically opposed to the mining companies and are in a very strong position to negotiate agreements that suit them (financially). But he also admits that the strong bargaining power of local populations in Papua New Guinea is quite unique (1999: 97). The fact that much of the reference research on mining impacts has been done in PNG might thus support the under-estimation of power asymmetries in the "stakeholder dialogue" approach (see also case study on oil exploration in PNG in Haller et al. 2007). An important but often neglected issue is the colonial process and the translation of local views, concepts and (common) property institutions into private and state property terms. Anthropologists are working on an already transformed notion of property that is then taken as being "customary" (see newer work on African contexts in Haller 2013; Peters 2013; Marfurt/Käser/Lustenberger 2015).

Important practical and theoretical contributions about negotiation processes between indigenous groups and corporations have been made by O'Faircheallaigh, who is following the development of so-called "Impact and Benefit Agreements" since many years, both as scholar and community advisor, with a focus on Australia and Canada. He shows that also in Australia, where the legal recognition of indigenous rights is relatively strong, communities need to increase their bargaining power to reach more or less satisfying agreements: it works best in the Northern Territory, where "Free, Prior and Informed Consent" (FPIC) is mandatory by law. In addition to the legal entitlements, the intervention of strong indigenous organizations in the negotiation helps, especially if they have a credible track record of organizing direct action and filing lawsuits to delay a project (2008; see also Gibson/O'Faircheallaigh 2010). Therefore, it has to be remembered that "[g]overnance as practised is thus a product of power relationships, and so the same set of formal rules is likely to result in quite different practices across space and time depending on the uneven geographies of power relations" (Bebbington et al. 2013: 9). Furthermore, there is a need to consider changes in the perception of the environment and of legal property entitlements that were introduced during colonial and postcolonial times, as well as power-driven and manipulative processes that suggest that even where FPIC is formally respected, it can still entail serious inequalities, as local elites might profit to the detriment of weaker sectors ("elite capture") (see also Haller et al. 2007, Haller 2013).

Adding to the complexities, not only the boundaries between corporate and state actors can become fuzzy in local settings, but also non-governmental and community actors may become involved in strategies that serve to facilitate and stabilize mining regimes. In the mentioned case in Indonesia, Newmont financed the construction of the police station, runs a community office and organizes the municipal trash disposal service. It also pays a police platoon stationed in the mining area as a backup for its own security guards. Welker (2009) traces the changes that lead local school teachers and youth to switch from blocking roads against the Newmont mine (in demand of benefits and jobs) to defending the mine against environmental activists, using the same method. The shift is influenced by a well-crafted CSR strategy that treats "the community" (or rather, allied sectors) as a partner for corporate security. CSR here serves to promote a biased and partial notion of "environmental protection" among corporate staff (downplaying mining impacts, exaggerating unsustainable resource use by locals). This evolves in interaction with NGO critiques and by using the services of consultancy firms which do not only sell CSR policies, but also strategy papers against anti-mining groups, and which do not refrain from spying upon and denigrating activists in covert operations. This last point leads to the question as to how far CSR can work as complementary to more militant forms of corporate defence. Ferguson (2005: 379) suggests that mining operations in the African context increasingly resemble the model of the offshore oil platform: as "extraction enclaves", they are almost completely detached from the local population, operated by a small number of skilled professionals who move back and forth from their walled-off compound, the extraction site and the airport without ever entering "real-life" Africa, protected by private armies strong enough to make sure that the population stays out.

Similarly, investments and profits do not "flow" to and from Africa, but "hop" between extraction enclaves and corporate headquarters, largely disconnected from weak states and a society which is left to "government by NGOs" to fix the worst disasters.

Human-Nature Relationship and the Good Life

The way mining is debated in the academic sphere not only reflects the "usual dominance" of the Anglophone countries, but this is even more amplified by the fact that Canada, Australia, the USA and UK have long mining industry traditions. However, there are macro-regional differences, related to different colonial pasts and corresponding discourses. Especially in South America, there have been urgent and productive debates in recent years.

On a macro-political level, the South American literature has been marked by the rise of so-called socialist governments in several countries (especially, Bolivia, Ecuador and Venezuela). A key concept in the debates is "extractivism", understood as the economic specialization on the extraction and export of raw material, and the need to transform this to a "post-extractivist" political economy (e.g., Svampa/Antonelli 2009, Moncloa/Gudynas 2012, Bebbington et al. 2013: 13). A special characteristic of the South American region seems to be the close exchange between academic and activist discourse, the latter often inspired or produced by indigenous peoples' movements (e.g., Alimonda ed. 2011, Arana Zegarra in De Echave et al. 2009). The notion of "living well" (*vivir bien, buen vivir* or *sumak kawsay*) has thereby taken the centre stage as a counter-discourse to mainstream development ideologies (see especially the works of Eduardo Gudynas, e.g., 2011), and is increasingly referred to in economic anthropology (e.g., Fischer 2014). In Bolivia and Ecuador, *buen vivir* has been enshrined in the new constitutions. Furthermore, Ecuador also officially recognized the "rights of nature", in what is described as a bio-centric approach rooted in indigenous cosmovisions (Gudynas/Acosta 2011), sometimes termed *pachamamismo*,

Fig. 8: Protest march against the Global Commodities Summit in Lausanne, Switzerland, with participants from Romania; 15 April 2013. Photo: Save Rosia Montana campaign.

after the Andean practice of worshipping *pachamama* (i.e., "Mother Earth"). At the same time, there are numerous contradictions arising from the fact that the governments of Bolivia and Ecuador, while adopting (and arguably co-opting) such discourses, continue to support and expand an extractivist economic model in which "living well" is underlined by a distributive state strategy—which is financed by negotiating better terms of trade for the export of raw materials (e.g., *Tierra Socialista* 2011; Radcliffe 2012).

Without going into further detail, we think that the questions arising from these debates touch on core issues of the human-nature relationship and can crucially contribute to an "anti-modern" anthropology that challenges the Eurocentric dichotomy between "culture" and "nature" (see Latour 1991; Escobar 2008; Galvin/Haller ed. 2008; Kohn 2013; Haller et al. 2013). Although we do not elaborate on this perspective in depth, the present book is an attempt to bring such considerations into the debate on mining, all too often dominated by "natural science". For the integration of South American and especially Andean indigenous knowledge systems into resource governance regimes, we especially point to the work of Stephan Rist (e.g., Rist/Dahdouh-Guebas 2006, Boillat et al. 2008, Delgado/Rist/Escobar 2010), whose advice and experience has contributed to this book.

State of the Debate in Switzerland

Switzerland has some history of mining, but by the 1950s it was largely abandoned. The issue became a niche field for a few geologists and NGO activists. Mining conflicts were largely ignored by the public as well as by academics, or they were treated merely as sub-themes in larger debates on development and North-South relations. This has changed in the last decade, due to the rapid growth of the commodities sector, concentrated in the cantons of Geneva and Zug. Commodity trading has a long history, linked to the Swiss banks that provide the necessary capital to broker large quantities of goods, which mostly never physically touch Swiss territory. But with the commodity boom and important trading companies moving their headquarters here, this extremely secretive business sector has seen a spectacular fifteen-fold growth in Switzerland between 1998 and 2009 (EvB/BD 2012: 41). In particular, the rise of Glencore has called the attention of solidarity groups and NGOs which are active in places where it operates. Through highly opportunistic trading, the infamous Marc Rich built up the base for this company in the tax haven canton of Zug, gradually expanding from trading into production. With the take-over of the major, Switzerland- and UK-based miner Xstrata in 2012/13, this has become "one of the world's largest diversified natural resource companies".[13] Another angle is the historical importance of Switzerland for the trade and refining of gold. Different from other commodities, gold is physically imported and exported, which makes questions about its origins even more pressing (see Part II – 2).

Although a number of smaller NGOs and solidarity groups had been following these developments and had issued some case studies and reports before, the large picture was brought to the fore thanks to a book published by the NGO Berne Declaration (EvB/BD) in 2012. Around that time, over 50 small and large NGOs came together to build the Swiss Coalition for Corporate Justice (SCCJ), for a campaign which demands that companies based in Switzerland should be legally responsible for the acts of their subsidiaries in other countries.[14]

13 Claim from <www.glencore.com>, accessed 09.01.2015. On the Forbes Magazine 2000 list of 2014, Glencore holds position 406 and is number 5 of the "diversified metals and mining" sector, after BHP Billiton, Rio Tinto, China Shenhua Energy and Freeport-McMoRan (see also Part III – B).

14 See: "Recht ohne Grenzen" campaign (i.e., "rights without borders"; the English name of the campaign is "Corporate Justice" <www.corporatejustice.ch>). SCCJ is framed along the European Coalition for Corporate Justice

At the time of writing, this coalition is launching the collection of signatures to bring its proposition to vote through the instrument of the popular initiative.[15] The demand so far has not found a majority support in the national parliament, but it does enjoy considerable sympathies beyond the "usual suspects" of green-left voters. A major concern for parts of the Swiss economic establishment is that the "Brand Switzerland" could suffer further damage due to scandals and conflicts fostered by the extractive sector, similarly to what happened with the facilitation of tax evasion by Swiss banks. According to this argument, mining corporations are free-riding on an image of "Swissness" associated with the democratic and humanitarian tradition, cheese and mountain beauty, apparently unspoiled by colonial legacy.

The Federal Government issued a report on the commodities sector in March 2013,[16] which was criticized by NGOs for its half-hearted recommendations. Generally, it states that the strengthening of multilateral voluntary agreements such as the Extractive Industries Transparency Initiative (EITI) (see Chapter 2) is seen as sufficient for the time being. Other official engagements of Switzerland concern the convention to ban the use of mercury, which is however only relevant to artisanal mining.[17]

Also academic research in Switzerland has started entering the mining field. The Commission for Research Partnerships with Developing Countries (KFPE) titled its 20th anniversary conference in August 2014 "Research Partnerships for Sustainable Solutions in Governing the Extractive Industries", with presentations of ongoing research by Stephan Rist (CDE), by the Swiss Tropical and Public Health Institute (on "health impact assessments in the extractive industry"), and the anthropo-technology project WANAMEI that aims to improve small-scale mining in the Madre de Dios region of Peru.[18] At the University of Basel, the Centre for African Studies hosted a two-day conference on "The Mining Industry in Southern Africa" in November 2014. For the social sciences, the present book project of *infoe CH* in collaboration with the Institute of Social Anthropology and the Centre for Development and Environment (CDE) at the University of Bern is the first comprehensive attempt to study large-scale mining with a comparative approach.

1.4 Approach and Structure of this Book

The complexities of the corporate–state–community relationships outlined above easily escape simple frameworks. However, to go beyond the description of individual cases we need a frame. The main question of the book for the comparison of the case studies in Part III refers to the strategies of corporations on the one hand, and the reactions and actions of local actors on the other—expressed in an interplay of imposition, negotiation and resistance. The analysis is informed by the New Institutionalist approach in social anthropology and political ecology, using a model of institutional change (see Ensminger 1992, Haller et al. 2007, modifications in Haller ed. 2010): we look at aspects of the ecosystem, population and market dynamics, technology and infrastructure

(<www.corporatejustice.org>). For a similar approach in Canada, see Canadian Network on Corporate Accountability (<cnca-rcrce.ca>), all accessed 17.12.2014. See Coumans (2011: 36-37) for a critique of the National Roundtable on CSR and Extractive Industry in Canada.

15 Konzernverantwortungs-Initiative ("Corporate Responsibility Initiative"): <konzern-initiative.ch> accessed 15.05.2015.

16 "Federal Council publishes background report on commodities" <www.news.admin.ch> accessed 15.05.2015.

17 "Mercury Convention To Be Finalised in Geneva", media release 09.01.2013 <www.eda.admin.ch> accessed 15.05.2015.

18 See: Commission for Research Partnerships with Developing Countries (KFPE) on <www.naturalsciences.ch>.

as well as other "external" variables that may influence decision-making by various actors—in particular, changes in prices of metals and minerals in relation to the necessary inputs for production are crucial to put a specific area on (or off) the map of the "extraction frontier". If an area promises high profits, we can expect corporate actors to seek access, in a situation of unequal power relations. This can be expressed in the ability of corporate actors to select the institutions—rules of the game—which give them more bargaining power to maximize their benefits to the detriment of other actors, often the people living in the area in question. Crucial institutions in this regard are property rights, whereby mining contributes to the transformation of collective or common property of land and related resources into state and private property. We suggest that in the context of CSR, the ideological reference to discourses of "modernity" and "development" serves to legitimize this transformation and to obscure the underlying power imbalances.

As "corporate actors", we broadly understand a field of allied actors that includes, but is not limited to, the Transnational Corporation; for mining, it may be framed as a set of strategic and tactical alliances formed around the common aim of accessing and exploiting the ore body in a profitable way, through technical means that require major inputs of capital (i.e., large-scale industrial mining, which is of a different scale and quality, and organized in different ways than so-called artisanal mining). The distribution of the derived profits is contested, and gives rise to tensions and competition within that field; also, its borders are not clearly defined—indeed, it may include local actors' groups that actively support mining.

The case studies in Part II follow the two-dimensional model already used for the *infoe* book on fossil fuels (Haller et al. 2007), comparing corporate strategies and local peoples' strategies in separate sub-chapters. The strategies of other actors such as IFIs, government bodies, NGOs, media and civil society organizations are not treated on their own, but appear in interaction with both companies and local people throughout the chapters. Many authors treat the role of the state in a separate subsection along with corporate strategies, and tend to emphasise the role of NGOs as allies for local people's strategies. This reflects a general notion of mining conflicts as a battlefield of "top down" (corporate-state) imposition versus "bottom up" (local people and their allies) resistance—a perspective that certainly has its merits, but needs to be critically evaluated according to the actual cases.

For the case studies covered in this book, the decision-making process within the different companies mostly remains a "black box", lacking detailed inside views. Our perspective is one of an informed outsider, holding on to a working hypothesis of a relatively homogeneous "corporate *homo economicus*" whose strategies aim at maximizing profits in one way or another. By comparing public statements on corporate strategy and some of the observable outcomes, we can guess which factors are at play when specific decisions are taken.

To avoid a simplistic reading, we suggest that mining conflicts be understood as a field of multiple contradictions and (in)compatible interests of many differently positioned actors, who apply strategies and build alliances to strengthen their position. These positions are not fixed, but can shift between outright resistance and negotiation or imposition and negotiation. Under the given power asymmetries, we expect the first to be prevalent for local people, and the second for corporations. If both sides tend towards negotiation, the conflict level is tempered, if they tend towards resistance and imposition, the conflict level is higher. A third possible position is acquiescence: local people apparently agree with the plans of the company, maybe as a result of successful negotiation, or they simply do not see any means to challenge them. There is still a latent potential for conflict, but it is not openly manifest as in the first two cases. Given the methodological difficulty to trace hidden or passive forms of resistance (the "weapons of the weak", as described by Scott 1985) without long-term ethnographic research, we treat most of all the collectively organized and explicit forms of resistance (i.e., as expressed in protest actions, public communications, lawsuits, direct action etc.).

Regarding the choice of institutional settings, we assume that local groups would participate in fora that emphasize an attitude of "negotiation", even if they are predominantly controlled by corporate actors (such as the informal "stakeholder dialogue" models presented in chapter 2) as long as they share the expectations of modernity and development. Contrarily, other fora may be chosen if promises are not kept and expectations belied, and at the same time local use of common property resources such as pasture land, water and other resources are undermined while externalized extraction costs are not adequately compensated; such could be legal regimes that emphasize local people's collective ownership rights, or neo-traditional or informal institutions that seem suitable to formulate demands, strengthen local autonomy and facilitate resistance activities, and thereby increasing local people's bargaining power.

Choice of Case Studies

The thematic focus we outlined when starting to work on this book was on large-scale, open-cut mining as promoted or controlled by Transnational Corporations (TNCs) and its impacts and interactions with local populations. Compared to the study on oil corporations and indigenous peoples (Haller et al. 2007), we made an important modification by opening up the focus from indigenous peoples and their organizations as local actors to the broader "local populations", or simply "local people". The main reason is that in some of the cases we wanted to include, no indigenous peoples are involved, or they do not identify themselves as such (even if they arguably could do so).[19] In the cases where indigenous peoples are affected, the shift brings a stronger emphasis on the interaction and alliance building between them and other sectors of the local population. Without doubt, it is essential to consider indigenous peoples' specific histories, ancestral relations to the territory, legal status and other differences to mainstream society.

The choice of case studies did not follow a very systematic pattern: in good part, it was driven by the urge to know more about a certain mine that had received controversial attention by media or NGO reports, or by the possibility to participate in ongoing campaigns, as described above. We also decided to focus on cases where reference persons and a sufficient base of documentation could provide a starting point for the students' research. Resulting from this approach, Switzerland-based corporations are over-represented. Geographically, we chose cases from different continents, with a focus on the Global South, but including one in a wealthy country (Australia). We realized only later that there would have been good examples also in Europe—an obvious gap that we addressed by including the Val Medel case in Switzerland. Nevertheless, we think that the case studies are diverse enough to show a wide spectrum for our qualitative analysis (see Table 1).

From the originally proposed cases, some simply did not materialize because no student chose to work on them, or because no one had the necessary language skills (e.g., for Russia, China). For others, a first essay was written by students but not finalized to be included in the book. Two case studies were dropped at a later stage (Alumbrera/Agua Rica in Argentina, and the Grasberg mine in West Papua, Indonesia). An attempt at an early stage of the project to invite a junior researcher from the Global South to carry out ethnographic research in the Swiss canton of Zug, a "hot spot" of the commodities trading sector, failed, because the proposal did apparently not correspond to the funding criteria of any institution in Switzerland.

19 Indigenous peoples' rights and identity are relevant for the case studies in Niger, Colombia, Guatemala, Peru, Australia and Philippines. A certain potential for indigenous rights claims could also be relevant for Zambia and India, although in the cases presented here, local actors hardly make reference to them.

Metals/ Minerals:	**Case Studies** *(italic: not included in comparison):*
Gold (Silver)	Switzerland (Val Medel; *Refineries*), Guatemala (Marlin), *Guinea (ASM)*
Copper (Gold)	Zambia (Mopani), Peru (Las Bambas), Philippines (Tampakan)
Uranium	Niger (Arlit), Australia (Ranger/Jabiluka), *CH/Argentina/CAR (Explorer)*
Diamonds	*Sierra Leone (Youth and Mining)*
Coal	Colombia (El Cerrejón)
Limestone	Chhattisgarh, India (Ambuja Rawan)

Table 1: Overview on Case Studies included in this book.

The case studies that started from students' essays and are now included in this book are: Mopani Zambia, Ranger/Jabiluka Australia, El Cerrejón Colombia, Ambuja Rawan Chhattisgarh (India), Marlin Guatemala, Tampakan Philippines, and Arlit Niger. The chapter on Las Bambas Peru is a summary of a previous MA thesis, and Val Medel Switzerland is partially based on a later BA thesis. Two students carried out field research for an MA thesis (Scherler 2013, Bucher 2014) about the Ranger/Jabiluka mines and completed the case study that had been written previously.

Additionally, a number of authors contributed chapters that focus on aspects not covered in the original choice of cases: Angela Lindt writes about the role of gold refineries in Switzerland, based on a research report she did for an NGO campaign demanding transparency of the Swiss gold import/export statistics as a necessary step to fill a crucial gap for the accountability of global gold trade flows. Thomas Niederberger follows the (online) traces and speculations around the uranium concessions in Argentina and the Central African Republic of the obscure, Switzerland-based explorer company Uranio AG. Roy Maconachie studies different and ambivalent expectations of youth involved with diamond mining in Sierra Leone and the spaces of participation they build "from below", and how these relate to CSR strategies and corporate-driven participation "from above". Matthieu Bolay presents findings from his PhD field study about the interactions of artisanal miners with a large-scale gold mine in Guinea, showing how CSR influences the identity formation of different types of informal mine workers. As the chapters on additional aspects do not cover all the elements needed for comparison, they are included only partly in the discussion in Part III.

Structure of the Book

In the continuation of this introductory Part I, *Chapter 2* discusses the legal framework for the protection of indigenous peoples' rights, the crucial concept of Free, Prior and Informed Consent (FPIC), and CSR-related mechanisms and initiatives, taking up several of the above-mentioned debates. *Chapter 3* deals with the environmental impacts of different methods of mining and concentration of the metals and minerals over the mine lifecycle. Regarding the metals and minerals, the focus is placed on gold-copper and uranium open-pit mines. Although this was not intended in the choice of case studies, it reflects their highly problematic nature. The cases cover the exploration, planning/construction, operation and expansion stages, with some approaching closure. *Chapter 4* gives a short introduction on the link between coal mining and climate change, with an example of a successful campaign to stop the planned construction of two coal power plants. Finally, *Chapter 5* provides an overview about the potential impacts of some of the most frequent mining related pollutants on human health, with a special focus on uranium mining in the second part.

Part II—the case study chapters—forms the core of this book. The case studies are grouped according to the geographic regions, starting with Europe, followed by Africa, South and Central America, Asia and Australia. Each block includes three to four chapters starting with the oldest active mine and ending with the most recent one. The structure of the case studies follows largely the model of Haller et al. (2007). The first part situates the case in the general context and in a long-term perspective, including the historic and colonial setting, geography and ecosystem, the composition of the local population, local economy, neo-traditional resource use and common property institutions, and up to the importance of mining for the national economy and the most relevant laws and regulations for the sector. The second part focuses on the actual case of the mine or mining project, divided into three chapters: "social and environmental impacts", "local people's strategies" and "corporate strategies and the state". The different character of the two parts is reflected in the sources: while the first relies on ethnographic and other background literature, the second depends more on newspapers and internet sources, NGO reports and corporate publications, because specific ethnographic material is often not available.

Part III presents the comparison of the case studies, divided into three chapters: after taking up the theoretical debates introduced above to sketch a framework for the comparison, we first discuss the context (roughly equivalent to the first part of the case studies) at the national and regional level and the specific problems arising with the type of mining in the cases; second, we characterize the involved companies and compare their "corporate strategies" at a general level and in the local setting; third, we identify and compare collective action of local people and their allies, or "local people's strategies", which are put in relation to the corporate strategies in the conclusion. By discussing and comparing the different case studies, we identify common problems and responses that point to a systemic level of mining-related conflicts. This part includes short summaries of the case studies regarding the discussed aspects, starting with the oldest mine (Mopani Zambia) and ending with the one in early exploration stage (Val Medel Switzerland).

Part IV is an essay on "strategies for action/research". Not directly related to the case studies, it reflects upon the process of writing this book, the production of collective knowledge, as well as about the tradition of action and advocacy anthropology and related experiences of engaged research. As an outlook on future possibilities and difficulties that similar endeavours may encounter, it addresses especially students who feel the need to get actively involved with ongoing struggles, while at the same time competing in the academic field.

References:

See Part III (page 503)

2. Legal Frameworks and Regulations for the Mining Industry

Helen Gambon and Madlen Kobi

2.1 Introduction

Violent clashes between mining workers and security police in Indonesia, polluted water, air, and soil around a copper mine in Zambia, or resettlement of communities for the expansion of a mine in Peru—such and other headlines reach us every day and are testimonies of the global activities of transnational mining companies. Their headquarters are often located in Europe, North America, and Australia, but they extract resources largely in countries where the monitoring of social and ecological standards is deficient. On the one hand, companies profit from loose legal frameworks to externalize the ecological and social costs of their activities. On the other hand, the countries of extraction are often dependent on the revenues from mineral extraction, which at the same time weakens their bargaining position vis-à-vis the mining companies. Although resource exploitation is regulated by a variety of international standards, non-compliance is a sad fact as Sawyer and Gomez note: "(...) despite the burgeoning number of international charters, state constitutions, and national laws across the world that assert and protect the rights of indigenous peoples, the majority of indigenous peoples find themselves increasingly subjected to discrimination, exploitation, dispossession, and racism" (Sawyer and Gomez 2012: 2).

Power relations between the main actors involved or affected by the mining industry are extremely unequal: we find marginalized indigenous local populations, government agencies at different levels, national and transnational mining companies, NGOs and unions, as well as research institutions. All these actors are internally heterogeneous and apply different strategies to achieve their goals. In order to show the diversity of strategies and references, the present chapter shall give an overview of the international legal framework for the mining industry. It will become clear that not only the actors are diverse, but also that there is a broad range of regulating instruments. Because many of the case studies presented in this book refer to indigenous peoples, section two will present the current international legal framework related to indigenous peoples' rights with a special focus on UN instruments as well as regional bodies and enforcement mechanisms. Indigenous peoples have been marginalized in many parts of the world, and their lands have been disproportionately affected by the extracting industries, as their ancestral territories often contain valuable natural resources. Therefore, in this chapter, we concentrate on indigenous peoples' rights. Non-indigenous local populations do not have the possibility to refer to these collective rights instruments; nevertheless, they are protected by general human rights. In section three, we will present the right to "Free, Prior and Informed Consent" (FPIC) that guarantees local communities' active participation in decisions concerning their territories, including mining. Section four is about the mining companies' reference to and implementation of international legal frameworks. Because the international laws presented earlier are not binding, we will include the international Corporate Social Responsibility (CSR) frameworks as well as the company internal sustainability reports in the analysis. After the presentation of the legal frameworks that the different actors refer to, we will in section six draw conclusions on the legal pluralism that characterizes activities in the mining industry.

2.2 International Frameworks Related to Indigenous Peoples' Rights

2.2.1 Indigenous Issues Within the International Community

General human rights instruments protect non-indigenous as well as indigenous people. In 1948, the United Nations adopted the Universal Declaration of Human Rights as a reaction to the human rights violations during World War II. As UN declarations are not legally binding for member states, two International Covenants were adopted in 1966: the first on Civil and Political Rights[1] and the second on Economic, Social, and Cultural Rights.[2] These treaties reflect the human rights discussions that were centred on the rights of individuals vis-à-vis states. These rights are considered to be universal, indivisible and egalitarian. The limitations of this individual-rights approach were later pointed out mainly by scholars indicating that certain rights can only be thought of collectively, such as the right to occupy a territory or to perpetuate a language (Bowen 2000: 12). Collective rights go beyond mere group rights (for example, LGBT[3] rights), as their members are not simply tied together through external discrimination, but by an internal cohesiveness. It is not equality and participation within the larger society that is striven for, but the protection and development of the collectives' particular cultural characteristics. Collective rights thus address ethnic, religious or linguistic minorities[4] (for example, French-Canadians, Muslims in India, or Afro-Colombians). Individual members of such collectives play subordinate roles; more important are representatives and institutions (Sanders 1991: 369f., 374).

The notion of "indigenous peoples' rights" emerged in the Americas, with the history of European colonization and invasion. The discussion found its way onto the international stage for the first time in the 1940s, after the International Labour Organization (ILO) studied problems of forced labour faced by "native populations" in South America. This eventually led to the ILO Convention No. 107 in 1957.[5] This treaty represents the first comprehensive international standard on indigenous rights. However, it was later criticized for its paternalistic and integrative approach and its underlying assumption that indigenous peoples were destined to disappear with "modernization."[6] The 1970s mark a turning point in the history of indigenous rights. With the expansion of the international indigenous peoples' movement and its focus on the United Nations system, indigenous issues received for the first time widespread attention from the international community. It was the time when the emphasis on assimilating indigenous peoples into dominant societies shifted towards the claim that indigenous peoples should have the right to maintain their cultural and economic difference (Bowen 2000: 12).

Indigenous peoples' rights are included both in the universal (individual) and minority (collective) rights standards. This could not prevent indigenous peoples being victims of marginalization, loss of land, discrimination, conflict and gross violations of human rights, including

1 Number of parties as of May 2015: 168 (ratified by all case study countries).

2 Number of parties as of May 2015: 164 (ratified by all case study countries).

3 Lesbian, Gay, Bisexual and Transgender/-sexual.

4 The most important international human rights standards that codify minority rights are the UN Convention on the Prevention and Punishment of the Crime of Genocide (1948), Article 27 of the International Covenant on Civil and Political Rights, and the UN Declaration on the Rights of Persons Belonging to National or Ethnic, Religious and Linguistic Minorities (1992).

5 The ILO Convention No. 107, entitled "Convention Concerning the Protection and Integration of Indigenous and other Tribal and Semi-Tribal Populations in Independent Countries", was ratified by 27 states and is still in force for 17 countries. Signing states are Angola, Bangladesh, Belgium, Cuba, Dominican Republic, Egypt, El Salvador, Ghana, Guinea-Bissau, Haiti, India, Iraq, Malawi, Pakistan, Panama, Syrian Arab Republic, Tunisia. 9 of 10 countries automatically renounced from ILO Convention 107 after ratifying ILO Convention 169 (see next section), while only one (Portugal) renounced completely. The treaty is no longer open for ratification.

6 ILO Convention No. 107 <www.ilo.org/indigenous/Conventions/no107> accessed on 16.12.2012.

massacres. The report from the Special Rapporteur of the Sub-Commission on Prevention of Discrimination and Protection of Minorities ("Martínez Cobo study" 1981-1984) marked a defining moment in the development of indigenous rights at the international level, as it recognized that indigenous peoples need the recognition of specific collective rights for their survival, including the right to their lands, territories and resources, to self-government and self-determination, and to be asked for Free, Prior and Informed Consent in decisions that may affect them (UNDG 2009: 10; UNPFII 2007).

2.2.2 UN Instruments, Mechanisms and Bodies

Since the arrival of indigenous issues on the international stage, a range of mechanisms has been developed to protect their collective rights. So far, the most important instrument is ILO Convention No. 169, adopted in 1989 and currently ratified by 22 states.[7] The "Convention concerning Indigenous and Tribal Peoples in Independent Countries" is the first and until today the only internationally binding legal instrument that recognizes individual and collective rights of indigenous peoples and establishes minimum standards for the implementation and compliance of these rights. It formally rejects the assimilationist orientation of the former standard, ILO Convention No. 107. Instead, it recognizes the aspirations of indigenous peoples to "exercise control over their own institutions, ways of life and economic development and to maintain and develop their identities, languages and religions, within the framework of the States in which they live" (Preamble). Self-identification as "indigenous" or "tribal" is fundamental for determining the groups to which the Convention applies, but it is made clear that "the use of the term peoples in this Convention shall not be construed as having any implications as regards the rights which may attach to the term under international law" (Art. 1). In the context of mining, several of the total 44 articles are of special interest: the Convention establishes that indigenous and tribal people have the right to traditionally occupied lands, as well as lands and other natural resources they need for their subsistence. While most states retain the ownership of sub-surface resources, the Convention declares that indigenous peoples must be consulted before any exploration or exploitation of such resources can take place. It also introduces the concept of free and informed consent, which applies to relocation.

Another very important instrument is the United Nations Declaration on the Rights of Indigenous Peoples (UNDRIP) that was adopted by the General Assembly on September 13, 2007 in a landslide affirmative vote of 144 in favour, 4 against and 11 abstentions.[8] It is the first universal, although not legally binding framework recognizing the rights of indigenous peoples on a wide range of issues, addressing both individual and collective rights. The Declaration is a product of over twenty years of negotiations, starting with the establishment of the UN Working Group on Indigenous Populations (WGIP) in 1982. The mandate of the Working Group included the development of standard-setting instruments regarding indigenous peoples' rights. In 1994, a final draft was adopted and forwarded to the UN Commission on Human Rights, which established

7 The majority of signatory countries are Latin American states. Last ratifications were made by Nicaragua and the Central African Republic (both 2010), the latter being the first African country to ratify the Convention. Switzerland has not signed the Convention yet, mainly because it is feared that minorities (such as the Yeniche) could identify themselves as tribal peoples and thus make legal claims. Signing states are Argentina, Bolivia, Brazil, Central African Republic, Chile, **Colombia,** Costa Rica, Denmark, Dominica, Ecuador, Fiji, **Guatemala,** Honduras, Mexico, Nepal, Netherlands, Nicaragua, Norway, Paraguay, **Peru,** Spain, Venezuela (case study countries in bold).

8 Opposing states were Canada, Australia, New Zealand and the US, known as CANZUS. They have since endorsed the UNDRIP, but expressed reservations regarding some of the key provisions of the Declaration, especially the right to Free, Prior and Informed Consent. Abstentions came from Azerbaijan, Bangladesh, Bhutan, Burundi, Colombia, Georgia, Kenya, Nigeria, Russian Federation, Samoa and Ukraine.

an intergovernmental Working Group to elaborate the draft Declaration (WGDD) in the following years. Leaving the paths of standard proceedings in the UN-system, indigenous peoples' delegates were allowed to participate in the development of the Declaration. It is the first and only UN-Declaration which was drafted with the right-holders themselves. This gives it very strong legitimacy. It has to be emphasized, however, that indigenous peoples did not see themselves as demanding "new rights", but rather the recognition and application of universal human rights to indigenous peoples[9] (Oldham and Frank 2008: 5-8, Morgan 2011: 10-17).

In that process, the debate about the definition of "indigenous peoples" re-emerged. Especially a group of African countries argued that a lack of definition would create ethnic tensions and instability within sovereign states. Opposing states[10] further exploited these frictions, leading to more than a decade of near deadlock over the Declaration. As still no consensus was reached in the final session of the WGDD in 2005, Chairperson Luis Enrique Chávez decided to produce a final "chair's text," later adopted by the first session of the Human Rights Council. Moreover, the main negotiators, Mexico, Peru and Guatemala, succeeded in reaching a consensus between indigenous representatives and the African Group, which consisted of nine amendments, including a reference to regional, historical and cultural peculiarities of indigenous peoples in different countries (Paragraph 23). Another provision that caused much debate was the right to "Free, Prior and Informed Consent (FPIC)", which has important implications for the mining context.

Within the UN system, three closely collaborating mechanisms are of special importance for indigenous peoples:

1) The United Nations Permanent Forum on Indigenous Issues (UNPFII) had its first session in May 2002 and annually meets for ten days. The UNPFII consists of 16 independent experts who serve for a three-year term and may be re-elected or re-appointed for an additional term. Eight of the members are nominated by governments and eight are nominated directly by indigenous organizations in their regions.[11] As an advisory body of the Economic and Social Council (ECOSOC), it has a mandate to discuss indigenous issues related to economic and social development, culture, the environment, education, health and human rights. The UNPFII's mandate also includes awareness-raising, the promotion and coordination of related activities within the UN system, as well as producing relevant information material about indigenous peoples' rights. In its sixth session in 2007, the Forum's special theme was "Territories, Lands and Natural Resources." The issue was brought up in coordination with other UN mechanisms, reflecting the importance the topic has gained on the international level.

9 The Declaration is mainly constructed from elements contained in "Human Rights: A Compilation of Instruments" (United Nations, New York, 1993), such as the Universal Declaration of Human Rights, the Human Rights Covenants, and conventions and declarations with respect to children (CRC), women (CEDAW), racial discrimination (CERD), genocide, minorities, and religious intolerance, as well as ILO Convention 169 (Technical review of the UN draft declaration on the rights of indigenous peoples E/CN.4/Sub.2/1994/2). Along these lines, indigenous peoples can refer to a range of other international standard-setting instruments, such as the UNESCO Convention Concerning the Protection of the World Cultural and Natural Heritage (1972); Convention on Biological Diversity (CBD, 1992), including the Akwé: Kon guidelines (2004); Agenda 21 (1992); Beijing Declaration and Platform of Action (1995); UNEP Malmoe Ministerial Declaration (2000); UNESCO Universal Declaration on Cultural Diversity (2001); UNESCO Convention for the Safeguarding of the Intangible Cultural Heritage (2003); UNESCO Convention on the Protection and Promotion of the Diversity of Cultural Expressions (2005).

10 The opposing states were Canada, Australia, New Zealand and the US, known as CANZUS.

11 The regions are Africa; Asia; Central and South America and the Caribbean; the Arctic; Central and Eastern Europe, Russian Federation, Central Asia and Transcaucasia; North America; and the Pacific—with one additional rotating seat among the three first listed above.

2) The Special Rapporteur on the Rights of Indigenous Peoples[12] was established in 2001 by the Commission on Human Rights (now the Human Rights Council) in order to examine the situation of indigenous peoples worldwide on the basis of country-specific visits and information submitted by governments, indigenous peoples and their organizations. He/she has the right to investigate specific cases of alleged violations and to formulate mediation recommendations. The mandate of the first Special Rapporteur, the Mexican Dr. Rodolfo Stavenhagen, ended in December 2007. He was succeeded by the Native American law professor James Anaya, who examined various cases in which mining, forestry, oil and natural gas extraction and hydroelectric projects have affected the lives of indigenous peoples (for example A/HRC/18/35), and focused his 2013 annual thematic report on the rights of indigenous peoples related to extractive industries (A/HRC/24/41). Since June 2014, Victoria Tauli Corpuz from the Philippines has assumed the mandate as Special Rapporteur.

3) The Expert Mechanism on the Rights of Indigenous Peoples (EMRIP) was created in December 2007 by the Human Rights Council. The Expert Mechanism is a subsidiary body of the Human Rights Council and provides it with thematic advice, studies and research on the rights of indigenous peoples. It is made up of five independent experts appointed by the Human Rights Council who are elected based on indigenous origin as well as gender balance and geographic representation. The Expert Mechanism holds an annual session in which representatives from states, indigenous peoples, indigenous peoples' organizations, civil society, inter-governmental organizations and academia take part. In September 2011, it completed its second study and advice on indigenous peoples and the right to participate in decision making. The Human Rights Council requested the EMRIP to further build on this study with a focus on extractive industries (A/HRC/EMRIP/2012/2).

Apart from these specific instruments, indigenous peoples and marginalized minorities also have the possibility to draw attention to human rights violations via the UN Universal Periodic Review (UPR), created by the Human Rights Council in 2007. All UN member states are examined on a regular basis on their human rights record and the fulfilment of states' obligations and commitments. Country reviews are based on three official documents: the National Report, a compilation of UN information including reports from UN mechanisms and special procedures relating to the human rights situation of the country under review, and a ten-page summary of stakeholders' information, the latter two being compiled by the Office of the High Commissioner for Human Rights (OHCHR). Although only states have the possibility of taking the floor during the review, there are some cases where indigenous organizations and NGOs were able to influence states to place queries and recommendations on their behalf. However, as the entire review of the human rights situation in a country is made during a single three-hour session, concerns of indigenous peoples can only be addressed superficially and this mechanism is not considered very effective (IWGIA 2011: 514-517).

This last example leads us to the main problems indigenous peoples have to overcome when addressing the international community to denounce the violation of their rights through mining activities. First, indigenous peoples still face difficulties in accessing the UN system. For a long time, they did not fit into the established categories of organizations and were thus completely excluded. Starting in the 1970s, special arrangements have been made to foster their participation. Among the 4045 NGOs which currently enjoy consultative status with ECOSOC, roughly 20 indigenous organizations (for example the Coordinating Body of the Indigenous Organizations of

12 The title of the Special Rapporteur was changed from "Special Rapporteur on the situation of the human rights and fundamental freedoms of indigenous people" to "Special Rapporteur on the Rights of Indigenous Peoples" in September 2010 in the course of renewing the mandate for another three years (Human Rights Council 2010: A/HRC/RES/15/14).

the Amazon Basin COICA, or the Indigenous People of Africa Coordinating Committee IPACC) are listed, which means that they can attend and contribute to a range of international and intergovernmental conferences.[13] The biggest obstacle nowadays to participate in the UN system is of an administrative nature: visa requirements of the United States, Switzerland or other countries hosting UN-meetings can often not be accomplished by indigenous persons. Obtaining a passport is difficult for economic or political reasons; evidence of adequate means of subsistence, such as credit cards, statements of account or certificate of employment are not easily provided by someone with a way of life primarily based on subsistence; and the need for personal application at consulates that are usually based at the administrative centres in the country's capital present an additional economic burden. There are numerous cases of denied visas, like the Ethiopian Delegation that wanted to participate at a conference in Sweden and was rejected for being "unable to demonstrate sufficient economic resources" (Resolution of the Terra Madre Conference 2011). In some cases, indigenous persons also risk being kidnapped or detained when travelling to international conferences.[14]

Second, the mechanisms and treaties established during the past 40 years regarding the participation of indigenous peoples by the international community only apply to the relationship between states and indigenous peoples. Even the ILO Convention No. 169, as the only legally binding instrument, is only directed at governments. Thus, indigenous peoples and other minorities can only denounce human rights violations committed by states and it remains an obligation of the state to incorporate human and indigenous rights into the national legal system and guarantee compliance. When it comes to mining, however, the willingness to do so is often limited. Mineral exploitation projects can generate large revenues for the state in the form of royalties and taxes. As the International Finance Corporation notes, the "risks of the private sector's influence over the government" to expedite project approval processes or to adopt laws that conflicts with a state's ability to comply with ILO Convention No. 169 "may be greater" (IFC 2007: 3).

2.2.3 Regional Mechanisms and Bodies

International organizations with the competence to set standards for human and indigenous rights are also found at the regional level. However, the efforts of these regional organizations to incorporate indigenous rights established at the UN level or to create their own instruments vary widely.

The most comprehensive set of human and indigenous rights bodies and mechanisms is found in the American Hemisphere. The world's oldest international institutional system dates back to 1889, when with the establishment of the International Union of American Republics the stage was set for what is now the inter-American system. The Organization of American States (OAS) was created by the 1948 Charter of the OAS that also acts as its Constitution. All 35 independent countries of the Americas have signed the OAS. In 1959, the OAS created the Inter-American Commission on Human Rights (IACHR)[15] as one of its principal and autonomous organs. The IACHR has its headquarters in Washington, D.C., and alongside with the Inter-American Court of Human Rights (IACourtHR)[16] is the main institution of the Inter-American Human Rights System.[17] The principal function of the IACHR is to promote the compliance with and protection

13 <csonet.org/?menu=100> accessed on 05.05.2015.

14 Personal communication to the author at EMRIP 2011; see also the statement of the Global Indigenous Youth Caucus at the 10th session of the UNPFII in 2011: 10th UNPFII: IYC Closing Statement <www.youtube.com> accessed on 1.10.2012.

15 Also referred to as "CIDH" (Spanish, Portuguese and French acronym).

16 Also referred to as "Corte IDH" (Spanish, Portuguese and French acronym).

17 <www.oas.org/en/iachr/mandate> accessed on 3.10.2012.

of human rights and to serve as a consultative organ of the OAS in these matters (Charter of the OAS: Art. 106). The Commission monitors the human rights situation in the member states (including visits to the countries), carries out initiatives on priority issues, and processes individual petitions. The latter may be submitted by individuals, groups or organizations that allege violations of the human rights guaranteed in the American Declaration of the Rights and Duties of Man (1948), the American Convention on Human Rights (Pact of San José, Costa Rica, 1969) and other Inter-American human rights treaties (IACHR 2010b: Section 1.4). In its annual report of 1972, the Commission affirmed that "for historical reasons, and based on moral and humanitarian principles, States have a sacred duty to provide special protection for indigenous peoples" (CIDH 1973: Section 5, own translation). As a consequence, the Office of the Rapporteur on the Rights of Indigenous Peoples was created in 1990 to strengthen, promote, and systematize the Commission's work in this area. The Proposed American Declaration on the Rights of Indigenous Peoples, initiated in 1989, is still not concluded as the goal to adopt each of the proposed provisions by consensus has not yet been achieved. In the Inter-American Human Rights System, besides indigenous peoples Afro-Descendants also enjoy special attention.[18]

The African Commission on Human and Peoples' Rights (ACHPR) was created by means of the African Charter and officially inaugurated in November 1987 in Addis Ababa, Ethiopia. A permanent Secretariat was established in Banjul, The Gambia, two years later. It is the main human rights body of the African Union (AU).[19] The ACHPR is composed of 11 commissioners and has the mandate to protect and promote human and peoples' rights and to interpret the provisions of the African Charter on Human and Peoples' Rights. In 2000, the ACHPR adopted a resolution establishing a Working Group with the mandate to examine the concept of indigenous populations/communities in Africa, to study the implications of the African Charter on the well-being of indigenous communities and to consider recommendations for the protection of their rights. The resulting "Report of the African Commission's Working Group on Indigenous Populations/Communities" was adopted by the African Commission in 2003 as its official conceptualization of the rights of indigenous peoples (www.achpr.org; IWGIA 2012: 564).

The Arctic Council was formally established as a high-level intergovernmental forum by the Ottawa Declaration in 1996. The idea behind its creation was to promote cooperation, coordination and interaction among the Arctic States, with the involvement of the Arctic Indigenous communities and other Arctic inhabitants on common issues, particularly sustainable development and environmental protection. Six indigenous peoples' organizations are Permanent Participants of the Council, made up of the eight member states.[20]

In Asia, the 2009 established ASEAN[21] Intergovernmental Commission on Human Rights (AICHR) on the contrary makes no reference to indigenous peoples in any of its documents. Very few states of the region recognize indigenous peoples on their territories, although they all voted in favour of the UNDRIP and an estimated number of 100 million people in Southeast Asia self-

18 One manifestation of this is the creation of the Rapporteurship on the Rights of Afro-Descendants and against Racial Discrimination by IACHR in 2005 <www.oas.org/en/iachr/afro-descendants> accessed on 3.10.2012.

19 The African Union was created in 2002 as the successor of the Organization of African Unity (OAU).

20 Member States are Canada, Denmark, Finland, Iceland, Norway, the Russian Federation, Sweden and the United States of America. Participating indigenous organizations are Arctic Athabaskan Council, Aleut International Association, Gwich'in Council International, Inuit Circumpolar Council, Russian Arctic Indigenous Peoples of the North and the Saami Council. <www.arctic-council.org> accessed on 3.10.2012.

21 The Association of Southeast Asian Nations (ASEAN) was established in August 1967 in Thailand with the signing of the ASEAN Declaration and counts today ten member states (Brunei Darussalam, Cambodia, Indonesia, Lao PDR, Malaysia, Myanmar, Philippines, Singapore, Thailand and Vietnam). The ASEAN Charter, providing a legal status and an institutional framework to the community, was adopted in 2007 and entered into force the following year (www.aseansec.org).

identify as indigenous. AICHR began in 2011 to draft an ASEAN Declaration on Human Rights. Nevertheless, the public has been excluded from this process, and although national consultations have started, the draft declaration has remained confidential (IWGIA 2012: 582f).

2.2.4 Interpretation and Enforcement of Indigenous People's Rights

Human and indigenous rights standards are written in a very general language, as it is a standard for documents that have to reflect and include all signatory states' positions. This means that these instruments have to be interpreted and their implementation monitored.

Within the UN system, the Human Rights Committee (HRC), the Committee on Economic, Social and Cultural Rights (CESCR), and the Committee on the Elimination of Racial Discrimination (CERD)[22] all examine the right of participation of indigenous peoples in relation to land and natural resources with regard to their respective treaties. These mechanisms conclude that the right to consultation (HRC) or consent (CESCR and CERD) on issues related to indigenous peoples' rights to lands and resources is guaranteed by provisions of these treaties.[23] However, all these general comments or recommendations of these UN supervisory bodies are not legally binding decisions (Ward 2011: 57).

Apart from UN bodies, other institutions are also involved in the implementation of international treaties. In the tripartite organization of the ILO, not only states, but also workers' and employers' organizations can submit information on the implementation of the Conventions. The Committee of Experts on the Application of Conventions and Recommendations (CEACR) is the body examining the application of ratified ILO Conventions and can make requests for more information about a given topic to governments in question. To deal with more serious situations, complaints can be submitted to the ILO through a workers' or employers' organization (ILO 2009: 188f). But also CEACR cannot enforce its recommendations (Ward 2011: 61). There are a range of examples, however, where domestic courts as well as the Inter-American Commission on Human Rights and the Inter-American Court of Human Rights used the ILO Convention 169 as a tool for interpretation or as a basis for decision, specifying the obligations of states under the regional human rights instruments (ILO 2009: 6, 10).[24]

The American Convention on Human Rights is a legally binding instrument, and so are the regulations of the Inter-American Court of Human Rights for member states of the OAS (Ward 2011: 61). Individual or collective lawsuits before the IACourtHR can only be filed via the IACHR. Over the years, numerous cases have been judged in favour of indigenous peoples. However, despite the binding nature of the decisions of the Court, many governments disrespect the judgements (GTZ 2008: 72).

The African Commission on Human and Peoples' Rights contributed to the interpretation of indigenous rights on the African continent by its Communication No. 155/96 of 2001 (Ward 2011: 66). The Commission concluded that fundamental rights contained in the African Charter, in particular the right to a satisfactory environment, to health, to shelter and to food, had been violated by the Nigerian National Petroleum Company in consortium with Shell. It appealed to

22 The three committees monitor the implementation of the International Covenant on Civil and Political Rights, the International Covenant on Economic, Social and Cultural Rights, and the Convention on the Elimination of All Forms of Racial Discrimination by their party states, respectively.

23 See: HRC General Comment No. 23 (1994), CESCR General Comment No. 21 (2009), and CERD General Recommendation No 23 (1997).

24 In case of the inter-American bodies, it has to be mentioned that they do not have jurisdiction to resolve disputes based on violations of ILO Convention No. 169. However, they have used the Convention to specify the obligations of states under the regional human rights instruments (ILO 2009: 6, 10). See, for example, the casebook "Application of ILO Convention No. 169 by Domestic and International Courts in Latin America" (ILO 2009).

the Nigerian government to stop all attacks on Ogoni communities by its military forces, to investigate human rights violations, and to provide "meaningful access to regulatory and decision-making bodies to communities likely to be affected by oil operations" (ACHPR Communication No. 155/96).[25]

As we have seen, human rights frameworks at the international and regional level do not apply directly to private companies, but work only through state enforcement. Their integration into domestic law is thus of crucial importance for determining whether people affected by mining projects can make legal claims against a company. Looking at the case studies, it becomes apparent that states adopt a wide range of legislative provisions for indigenous peoples, even those that have not ratified any of the ILO conventions.

It is interesting to observe, however, that some countries seem to have different agendas at the international and national levels. Guatemala, for example, does not stand out as a proponent of indigenous rights in the case of the Marlin Mine. The former Special Rapporteur on the Rights of Indigenous Peoples, James Anaya, examined the case and concluded in his report that "the Marlin Mine project has not been exemplary in meeting international human rights standards" (A/HRC/18/35/Add.3). At the same time, Guatemala was one of the main negotiators of the UNDRIP, contributing substantially through its diplomacy to the adoption of the Declaration in 2007 (Oldham and Frank 2008: 8). While some states may grant comprehensive rights regarding land and renewable natural resources to indigenous peoples, they are not willing to cede sovereignty over non-renewable resources. Especially, the so-called developing countries are often being compelled by international finance organizations like World Bank or IMF to create favourable conditions for foreign investment, such as mining projects (Sawyer and Gomez 2012: 34-35; Hilson 2012: 133). The resulting legal provisions are not always coherent with existing indigenous rights, leading to a situation of "legal pluralism." Moreover, many of the case study countries have rather weak governance structures with frequent power struggles between different institutions; for example, between mining ministries and environment ministries of the same state. This institutional pluralism contributes to legal pluralism and creates a complex net of responsibilities. Legal provisions on the same level have the same validity. This means that there is never one law above another law, or a constitutional provision that has more weight than the other. However, when it comes to the question, whether to respect indigenous land rights or to develop a mining project with promising revenues, the decision will likely be in favour of the latter. "Economic development" and "national interest" seem to be more important than the compliance of indigenous rights as many of the case studies from this book show.

2.3 Free, Prior and Informed Consent (FPIC)

At the international level, there is growing recognition of the right of indigenous people to participate in decision-making in relation to land and natural resources, reflected in the right to "Free, Prior and Informed Consent" (FPIC). While the ILO Convention No. 169 from 1989 speaks of "consultation" and excludes the competence of co-decision or the right to veto on extractive industries' projects, the UNDRIP from 2007 now calls for states to obtain "consent" from indigenous peoples for a range of issues. Among them are projects regarding the exploitation of mineral or other resources. FPIC is also required for displacement and relocation, the storage or disposal of hazardous materials in their lands or territories, and generally administrative and legislative

25 For a more detailed discussion of the case, see the Nigeria Case Study in the Infoe publication by Haller et al. (2007).

measures that affect indigenous peoples.[26]

The IACHR published a report based on the legal instruments of the Inter-American system titled "Indigenous and Tribal Peoples' Rights over their Ancestral Lands and Natural Resources" (2010a). In this report, the Commission concludes that indigenous peoples' consent is mandatory in a series of circumstances such as "development or investment plans or projects that imply a displacement of indigenous peoples or communities from their traditional lands" or in cases where the execution of such plans "would deprive indigenous peoples of the capacity to use and enjoy their lands and other natural resources necessary for their subsistence," as well as for storage or disposal of hazardous materials (IACHR 2010a: 120).

FPIC has become a key issue among multilateral development agencies and NGOs, and even the finance sector increasingly relates to the concept. The mining industry, however, has been largely reluctant to commit to FPIC. This is mainly due to the fear that FPIC can be interpreted as the right to veto mining operations by indigenous peoples (Sosa 2011: 2, 7). While mining companies generally agree on the importance of the principles of "free", "prior" and "informed" for achieving the social license to operate, the issue of "consent" remains highly controversial. In 2008, the International Council on Mining and Minerals (ICMM) adopted a position statement regarding Mining and Indigenous Peoples' Issues, through which ICMM member companies commit to "engage and consult with Indigenous Peoples in a fair, timely and culturally appropriate way throughout the project cycle" (Commitment 3). They further commit to "seek broad community support" for their activities and recognize that "following consultation with local people and relevant authorities, a decision may sometimes be made not to proceed with developments or exploration, even if this is legally permitted" (Commitment 9). Although the ICMM declares to comply with national laws requiring FPIC, it makes clear that its members are not willing to give a "blanket endorsement" for FPIC at the current stage (ICMM 2010: 24).

Similar concerns are brought up by states, which see their sovereignty over the development of natural resources contested by the "consent" part of the concept. In fact, only three states have incorporated the principle in their national legislation: the Philippines (Indigenous Peoples Rights Act, 1997), Canada (Yukon Oil and Gas Act, 2002), and Australia [Australian Aboriginal Land Rights (Northern Territory) Act, 1976]. Some Latin American countries, such as Peru and Bolivia have national laws granting the right of *consulta previa* (previous consultation), while Ecuador, Colombia and Venezuela even codified the right to consultation in relation to the exploration and exploitation of natural resources on indigenous lands in their constitutions.

There is no universally accepted definition of FPIC, and several key questions remain subjects of discussion. It is not clear what "consent" means and who should have the right to provide consent. For mining projects, the question of "prior to what?" is crucial, as indigenous communities often understand "prior" to mean before the beginning of exploration, while for mining companies this means rather before developing the actual mine (Sosa 2011: 9ff). According to the Special Rapporteur on the Rights of Indigenous Peoples, it is misguiding to focus the debate on FPIC alone: "neither consultation nor consent is an end in itself, nor do they constitute stand-alone rights. (...) [P]rinciples of consultation and consent together constitute a special standard that safeguards and functions as a means for the exercise of indigenous peoples' substantive rights" (A/HRC/21/47).

As we have seen, the demand for FPIC is growing, so states as well as companies have to deal with the implementation of such consultations. However, the way such consultations shall take

26 Two articles also mention the duty of states to provide effective mechanisms to redress taking of property without FPIC of indigenous peoples: Article 11 refers to cultural, intellectual, religious and spiritual property. Article 28 to lands, territories and resources which indigenous peoples have traditionally owned or otherwise occupied or used, and which have been confiscated, taken, occupied, used or damaged.

place and how affected indigenous and local communities are being asked about their preferences is still under debate. Four major companies initially considered for this study (Glencore, Xstrata, Anglo American, and Freeport McMoRan) all mention in their sustainability reports achievements in consulting affected communities before starting exploitation of mines. Xstrata, for example, claims that: "community consultation is a key part of our sustainability approach and risk management and business planning process and is carried out in a transparent, culturally appropriate manner. We listen to community members to understand their needs, concerns and perspectives" (Xstrata 2010: 85). Whether these needs and perspectives are respected in case they are opposing the company's idea of development, remains unclear. Only long-term participant observation at the mining site would reveal how companies actually deal with "consultation" and "consent," and in particular, with the lack of consent. There are numerous allegations of "fake consultations," and "manufactured consent" (Whitmore 2006: 312). However, FPIC is only one aspect concerning the ecological and social responsibility of companies that will be addressed in the following section.

2.4 Corporate Social and Environmental Responsibility

In the negotiation and legitimization of their claims for exploitation or refusal of a mining activity, affected local people as well as companies refer to a wide range of different legal instruments. Some of them are not part of constitutional law and are categorized as "soft law," "project law," or "standard-setting" (Benda-Beckmann et al. 2009: 4). The focus of the present section is on such kinds of emerging standards and their significance for companies' social and ecological responsibility in the mining sector. "Soft law" includes treaties, principles, guidelines, frameworks, as well as UN declarations. It differs from "hard law" in its legally non-binding character. Even if a state or a company subscribes to a "soft law"-regulation, it can almost not be denounced for non-compliance. Often, there are no responsible institutions (like courts) where complaints could be lodged. Moreover, companies are internally divided, which complicates decision-making processes and the allocation of "responsibility." They are entangled in a dense net of sub-contractors, marketing specialists, managers, service employees, and workers contracted at the mines. In fact, "(...) the field is saturated with experts of various sorts and with an array of nonprofit organizations that benefit from, are drawn into, or advocate the idea of CSR" (Shamir 2004: 645). Anthropologists are involved on the mining extraction sites as community relaters or external advisers, too (Coumans 2011). The general discourse about the "responsible" corporation has become pervasive in the last decade.

The term "Corporate Social Responsibility" (CSR) refers to methods and strategies of companies to make their business activities look environmentally and socially conforming (cf. Henderson 2001; Prieto-Carrón et al. 2006: 978; Multiwatch 2012; Sawyer and Gomez 2012: 41).[27] For this paper, the term CSR is mainly used as the articulated interest of companies to include local populations' participation and environmental concerns into their business performance. This includes the application of human and labour rights when dealing with workers and locally affected peoples, but it has been expanded into other activities like the voluntary establishment

27 The separation between Corporate Social Responsibility (CSR) and Corporate Responsibility (CR) is not always clearly drawn, although the term Corporate Responsibility is broader and includes performance in corporate governance and corporate citizenship while CSR is more focused on social and environmental issues. Some actors like the Swiss NGO Berne Declaration see CSR as a voluntary contribution to society while Corporate Accountability (CA) is more compulsory and would include, for example, the compliance with OECD guidelines. <www.kampagnenforum.ch/blog> accessed on 1.7.2014.

of foundations, local development projects and the presentation of the company's performance in sustainability reports.

It is commonly assumed that CSR initiatives emerged as a response to the changing business conditions in today's globalized world and to the popularity of new media. Public pressure through civil society's awareness creates an increased concern about the public image of companies and as a reaction, companies adapt a wide range of CSR strategies (Henderson 2001: 23; Jenkins 2005: 528). Additionally, Frynas assumes that not only external pressure, but also the creation of stable company internal working conditions has an influence on CSR performance (2005, 583). Voluntary standards, in the extractive industry especially focused on labour conditions, have expanded rapidly in recent years. Community rights and the security of the person are particularly emphasized, too (Ruggie 2007: 836). Companies are not so altruistic as to establish projects just for the well being of the persons and communities they work with. Instead, self-representation as a generous and prospective company is among the most important objectives of CSR activities. Nurturing this image promises higher revenues in today's society where green and social values are increasingly esteemed by customers, employees, and investors (Blowfield 2005: 518; Moon 2007: 299; Hilson 2012: 132).

Corporation-State Relationship

A particular feature of transnational mining is that both corporations and states are responsible to comply with international standards. States are the legitimate actors in the international legal context while companies are embedded in their home states' regulations. In recent years, the role of the state as actual protector of human rights in mining operations is more emphasized due to the limitations of company-elaborated voluntary CSR initiatives.[28] If states would take a more active role in promoting human rights compliance among companies registered on their territories, then voluntary CSR standards would lose the importance they have gained in the last decade.

The state has multiple responsibilities: when the binding obligations derive from the membership of the companies' home state to an international treaty, states become important actors in the formulation and safeguarding of rights and duties (cf. Ruggie 2007; McCorquodale and Simons 2007). But also the relationship between companies and the states from where resources are extracted—often shaped by mutual dependence—calls for special attention: poor states need the revenues from mining activities and companies rely on the state for legal admission to exploit natural resources and to deal with local communities. However, the position of countries with a low GDP is often weak. A report from OECD observes that for the period of the 1990s "(...) the more that states rely on exporting minerals, the worse their standard of living is likely to be" (OECD 2002: 5). This phenomenon is often named the "Dutch disease" or "resource curse" (Whitmore 2006: 311). Many mineral-dependant states, especially in sub-Saharan Africa, are not only poor, but highly indebted, which makes them dependent on international financial institutions (IFI) like the International Monetary Fund or the World Bank. When giving loans, IFIs often prescribe that foreign companies get access to exploit natural resources, without setting binding social and environmental standards (Sawyer and Gomez 2012: 35). Moreover, weak states are often unable to provide sufficient infrastructure for their citizens. Foreign investment for road construction or even the establishment of education or health centres is thus welcomed.[29]

28 In Switzerland, parliamentarians are demanding a change of the political frame from CSR to "Economy and Human Rights", where the state will have to exert a more active role in the implementation of human rights by companies under its sovereignty. Nationalrat 2012. <www.parlament.ch>, accessed on 4.10.2012. Further, the campaign "Corporate Justice" seeks to ensure that Swiss companies are compelled to integrate the protection of human rights and the environment in their business practices. <www.corporatejustice.ch> accessed on 8.5.2015.

29 As the case studies of Zambia, Guatemala, Peru and the Philippines show; see Part II.

2.4.1 International CSR framework

In the following paragraphs, we give an overview of the most important international guidelines relevant in the mining sector. Their use and impact will be analyzed in relation to the case studies in Part 3 of this book. When establishing their CSR strategy, companies choose from a wide range of regulations from the international to the grassroots level. Although there are some mining-specific guidelines, mining companies often refer to general international guidelines for transnational companies when they publish their annual sustainability reports. The guidelines presented here can roughly be divided into the following categories:

1. International guidelines, emphasizing the responsibility of the home states of transnational companies; for example the OECD Guidelines, the OECD Guidance, and the Guiding Principles on Business and Human Rights.
2. Multi-stakeholder initiatives where governments, companies, and NGOs jointly work on environmental and social principles, like the Global Reporting Initiative, the Voluntary Principles on Security and Human Rights, and the EITI initiative.
3. International standards, emphasizing the responsibility of transnational companies, like the UN Global Compact and the International Standard Organization ISO.
4. Industry-specific guidelines elaborated, for example, by the International Council on Mining and Metals ICMM.

International Guidelines emphasizing the responsibility of the home states of TNCs

The OECD Guidelines for Multinational Enterprises[30] are among the most widely known principles for regulating activities of transnational companies (cf. Heydenreich and Milke 2012). The OECD guidelines are not especially designed for the mining sector, but are relevant for multinational companies of different industries. They were established in 1976, the last of five revisions taking place in 2011.[31] States have the responsibility of promoting the observance of these guidelines by transnational companies operating in or from adhering countries.[32] In case of non-compliance, NGOs or other institutions from a signing state can hand in a complaint against activities of a company of a member state at the OECD contact points established since 2000 and located in the 42 adhering countries (Ruggie in *Business Ethics* 2011: 5). This complaint mechanism differentiates the OECD Guidelines from other CSR standards which do not envisage a defined procedure in case of guideline violation. However, the OECD contact points are often located in the foreign economy departments where they clash with the economic interests of the companies and the states. Critics suggest restructuring the current national contact points to guarantee more independence, for example in Germany (Heydenreich 2004; Heydenreich and Milke 2012) or in Switzerland (Arbeitsgruppe Schweiz-Kolumbien 2009). Even if the OECD guidelines are promoted by OECD member states, companies are rarely accused for damaging activities, because the procedure at the OECD contact point is complicated and barely accessible for affected populations as the case study of Colombia shows, for example.

Similarly, in the UN Guiding Principles on Business and Human Rights (A/HRC/17/31), the responsibility for social and environmental consequences of companies' activities is also placed on the home state of the companies. The Guiding Principles derive from a process that started in 2005 with the establishment of a Special Representative of the Secretary-General on the issue of human rights, transnational corporations, and other business enterprises. Named after this

30 The OECD Guidelines can be accessed on the OECD's website: <www.oecd.org> accessed on 24.5.2012.

31 <www.oecd.org/about> accessed on 21.5.2012.

32 The 34 OECD countries and 9 non-OECD countries (Argentina, Brazil, Colombia, Egypt, Latvia, Lithuania, Morocco, Peru and Romania): <www.oecd.org/dataoecd/17/44/1900962.pdf> accessed on 21.5.2012.

Special Representative, John Ruggie, the process has elaborated the framework "Protect, Respect, Remedy," addressing companies' responsibility to protect and respect human rights and to remedy in case of having violated basic rights. As with other UN-initiated instruments, the responsible institution to comply with these principles is the state. Companies as private actors can thus neglect their responsibilities by shifting them to the respective states. Moreover, no institutional complaint mechanisms like with the OECD Guidelines exist yet.

As a consequence of the Ruggie-process, the practice "human rights due diligence" spread among transnational companies. States and companies are encouraged to actively invest in the implementation of human rights in their activities. De Schutter et al. (2012) found little reference to human rights in their analysis of states' "due diligence" performance compared to consumer protection, environment protection, money laundering, or human trafficking. Every sector has its own standards. One example of "due diligence" in the mining sector is a new instrument that the OECD elaborated in a multi-stakeholder dialogue, the "OECD Guidance for Responsible Mineral Supply Chains in Conflict and High Risk Areas."[33] It states that "[w]hile implementation of due diligence should be tailored to particular company activities and relationships, such as their position in the supply chain, all companies should conduct due diligence aimed at ensuring that they do not contribute to human rights abuses or conflict" (OECD 2011: 15). This new guidance is particularly directed at the conditions in the extraction and trade of the minerals tin, tantalum, and tungsten as well as gold. However, "[o]bservance of this Guidance is voluntary and not legally enforceable" (OECD 2011: 16).

Multi-Stakeholder Initiatives

A number of CSR standards emerged through a collaboration of states, companies and NGOs. One of the most important CSR instruments in the extractive industries is the Voluntary Principles on Security and Human Rights.[34] The Principles were initiated in 2000 by the US and the UK governments, by companies from the extractive and energy sectors and by NGOs. The objective is to engage in a "multi-stakeholder dialogue" about security and human rights issues, particularly concerning risk assessment and public and private security relations. According to Welker (2009: 147), these principles are a centrepiece of CSR in the extractive industry due to the notorious human rights abuse through state and private security forces involved in conflict prevention, mitigation, or even repression.

Another "multi-stakeholder" approach for the extractive industries is the EITI initiative[35] that emerged shortly after the above-mentioned Voluntary Principles. The participants are governments, companies, NGOs, and institutional investors who pledge to improve governance in resource-rich countries by securing verification and publication of company payments and revenues from oil, gas, and mining. While several international NGOs like Oxfam, Transparency International, Global Witness and the "Publish What you Pay" coalition are involved in this campaign,[36] the local communities actually affected are mostly absent from the supposedly "multi-stakeholder" dialogue.

The Global Reporting Initiative GRI, an outcome of UN-based initiatives like the UN Global Compact, provides a sustainability reporting framework adaptable by transnational companies in

33 "This Guidance builds on and is consistent with the principles and standards contained in the OECD Guidelines for Multinational Enterprises and the OECD Risk Awareness Tool for Multinational Enterprises in Weak Governance Zones" (OECD 2011: 16). For the full text of the guidance, see <mneguidelines.oecd.org/mining> accessed on 8.5.2015.

34 The Voluntary Principles on Security and Human Rights were publicly announced in the year 2000: <voluntaryprinciples.org/timeline> accessed on 21.5.2012.

35 Eads et al. 2009; <eiti.org/eiti/history> accessed on 21.5.2012.

36 <www.publishwhatyoupay.org/about/stock-listings/cardin-lugar-amendment-dodd-frank-1504> accessed on 25.1.2013.

addition to the obligatory financial reporting. Companies like Glencore and Xstrata already apply the GRI-framework in their sustainability reports to present their performance and also other companies use the GRI framework informally or partially for their own reports (Ruggie 2007: 836-837). Similar to the UN Global Compact, GRI also aims at providing information concerning a company's performance regarding human, environmental, as well as labour rights.[37]

International Standards emphasizing the Responsibility of TNCs

The most famous UN-initiated CSR-process is the UN Global Compact established in the year 2000 to encourage businesses and non-business members to adopt universal values in business operations. At the core of the UN Global Compact are ten principles on human rights, labour, environment and anti-corruption.[38] The principles of the Global Compact are formulated in a very general way. In its own words, the Global Compact helps companies to "(...) ensure that markets, commerce, technology and finance advance in ways that benefit economies and societies everywhere."[39] Even if the UN Global Compact may contribute to companies' awareness of the negative impacts of their activities, the agreement is highly questioned by leading NGOs like Action Aid: "As long as resource extraction giants like the British Anglo American, highly criticized by civil society protest, are members of the Global Compact, the agreement is not worth the paper it is written on."[40]

A rather new player in the definition of CSR principles is the International Standard Organization ISO.[41] ISO 26000:2010 is a "best practice" norm that contains voluntary principles for companies, uniting existing private and public CSR initiatives. It was developed in cooperation with the ILO, the UN Global Compact and the OECD. Core subjects of the ISO 26000 norm include human rights, labour practices (for example, social dialogue), the environment (for example, sustainable resource use, protection of the environment), fair operating practices (for example, respect for property rights), consumer issues, community involvement and development. Some mining companies have started referring to this ISO norm to underline their efforts. More popular among mining companies is ISO 14000, which is a "framework for a holistic, strategic approach to the organization's environmental policy, plans and actions."[42] Because of different requirements for ecological performance in different industries, the standards are not further detailed. A company-internal Environmental Management System (EMS) is required so that a company can get the certification, because EMS supposedly enables a negotiation about the company's internal ecological standards (Kollmann 2008: 403). If a company shows that it has an EMS and is eager to improve environmental standards, it can get the ISO 14000 certification.

Industry-Specific Guidelines

An industry association that is itself member of the above-mentioned EITI initiative is the International Council on Mining and Metals ICMM,[43] (former Global Mining Initiative GMI). It was launched in 1999 by several CEOs from mining companies to address the low esteem of mining operations in public opinion. In 2001, the GMI was reorganized into the ICMM, composed

37 <www.globalreporting.org> accessed on 22.5.2012.

38 The UN Global Compact is based on the Universal Declaration of Human Rights, the ILO's Declaration on Fundamental Principles and Rights at Work, the Rio Declaration on Environment and Development, and the UN Convention Against Corruption <www.unglobalcompact.org/AboutTheGC/TheTenPrinciples> accessed 22.5.2012.

39 <www.unglobalcompact.org/AboutTheGC> accessed on 22.5.2012.

40 Action Aid, cited in: <www.evb.ch/p25012945.html> accessed 22.5.2012, own translation.

41 <www.iso.org/iso/home> accessed on 24.5.2012.

42 <www.iso.org/iso/iso_14000_essentials> accessed on 19.6.2012.

43 <www.icmm.com> accessed on 22.5.2012.

of 22 of the leading companies from the mining and metals industry. The reporting of companies works similar to the UN Global Compact presented above and is based on a set of ten principles deriving from different international frameworks.[44] These principles are based on a respectful performance concerning environment and local populations, but again, the main problem is that companies are not punished if they do not comply with these principles. It is one of the few initiatives entirely managed by the mining industry and indigenous peoples' organizations and NGOs criticize the overwhelming power of the corporate sector even if the initiative was meant to be "inclusive" (Whitmore 2006, 310).

A number of other CSR guidelines appeared in the last decade but will not be analyzed in detail here; for example, the framework of the Commission and the Parliament of the European Union (Henderson 2001: 149-152);[45] the Kimberley process targeting "blood diamond" trade;[46] the guidelines of the International Finance Corporation IFC,[47] which provide loans to the mining sector through the World Bank;[48] and the Intergovernmental Forum on Mining, Minerals, Metals and Sustainable Development, a platform for dialogue between states.[49]

Additionally, the entry to the stock market of certain companies (for example, Glencore) requires a public reporting on measures to control sustainability risks and thus increases corporate transparency. Even if the Swiss government emphasizes its commitment to engage in the open publication of companies' payments through supporting EITI and the OECD guidelines (Schweizerische Eidgenossenschaft 2013), other states are ahead in regulating transnational financial flows. In the US for example, the 2012-established Dodd-Frank-Act 1502 obliges all companies listed on the stock-exchange to openly present their payments to foreign governments in order to impede corruption, especially in the mining industry (Vonplon 2012; Publish What You Pay 2013).

2.4.2. Corporate Sustainability Reports—Greenwashing and Philanthropy

As shown in the above section, companies can refer to many guidelines promising "ecological and social sustainability" elaborated in so-called "stakeholder dialogue initiatives" between national governments, international organizations, NGOs and the extractive industry. Besides these activities, transnational companies often publish their own sustainability reports to highlight their performance with regard to their employees, environment and affected populations. For the compilation of data and content of the sustainability reports, companies often mandate external advisory agencies. This creates dependencies of such agencies, which are obliged to write satisfying reports in order to keep their assignment. The independence of such auditing institutions and the published reports is thus questioned (Newell 2005: 554).

Sustainability reports emerged as a reaction to the constant critique of companies' "unsustainable" performance concerning ecology and local populations. The industry started to provide its own definitions of sustainability, which today is rather rooted in economic than social sustainability as Kirsch elaborates in his paper on "Sustainable Mining" (2010). He states that

44 Including the Rio Declaration, the Global Reporting Initiative, the UN Global Compact, OECD Guidelines on Multinational Enterprises, World Bank Operational Guidelines, OECD Convention on Combating Bribery, ILO Conventions 98, 169, 176, and the Voluntary Principles on Security and Human Rights. <www.icmm.com> accessed on 22.5.2012.

45 <reportingcsr.org/european_union> accessed on 22.5.2012.

46 <www.kimberleyprocess.com> accessed on 22.5.2012.

47 The IFC has recently adopted a new Performance Standard for Indigenous Peoples, particularly focusing on the implementation of Free, Prior and Informed Consent standards. Performance Standard No.7 'Indigenous Peoples' was released in August 2011 and came into effect in January 2012 (Sosa 2011: 7).

48 <www1.ifc.org> accessed on 22.5.2012.

49 <www.globaldialogue.info/intro> accessed on 24.5.2012.

"[a]lthough the concept of sustainability may previously have been used to critique the environmental impacts of the mining industry, it has now become a means to promote mining" (Kirsch 2010: 91). The use of terms has become increasingly important for companies to show their interest in environmental protection and compliance with the rights of local populations. However, companies have instrumentalized the term "sustainability" for their own purposes (Kirsch 2010; Benson and Kirsch 2010).[50] Shamir (2004: 646-647) notes that "[t]hrough such initiatives, the notion of social responsibility undergoes a process of bureaucratization and standardization that transforms the heretofore politically loaded and morally debated notion of corporate responsibility into a measurable set of indicators that can then be exchanged and traded among shareholders and investors as any other commodity that adds 'value' to the firm." Both the notion of "sustainability" and the frequent reference by companies to a supposed "multi-stakeholder-dialogue" are instruments used to get approval for the establishment of mining extraction sites and we have to be aware of this double meaning when reading information published by companies.

Sustainability reports preferably show the positive impact of the company on the local population where they are operating. The reports are a way of promotion as the foreword of the Glencore Sustainability Report 2010, signed by CEO Ivan Glasenberg, exemplifies: "We also place great importance on the economic benefits we bring to the communities surrounding our operations. We provide vital employment and income to some areas where economic opportunity can be scarce. We also contribute to local development, in close consultation with community leaders and stakeholders. We report on our community development initiatives and programs extensively in this report" (Glencore 2010: 8). By frequently using the term "we", the company underlines its achievements and creates a common identity for both employees and investors. In a similar way, the Sustainability Report 2010 of Xstrata presents the company's Statement of Business Principles that "(...) stipulates our behaviour in protecting human rights in the day-to-day management of our business. We expect our employees, contractors, suppliers, joint venture partners and other business partners to be familiar with and to comply with these principles" (Xstrata 2010: 33).[51] Apart from offering insights into the financial performance of a company, these reports highlight the compliance with labour rights, human rights and environmental protection. However, the publishing of such reports is a recent phenomenon: while Xstrata and Rio Tinto started to publish sustainability reports in 2002 and Anglo American in 2003, the first sustainability report of Freeport McMoRan was only published in 2008 and Glencore's in 2010. Sustainability reports usually include sections about development and community projects initiated by the companies in the rather poor countries where extraction sites are located.[52]

In some cases, companies establish their own foundations that take care of the implementation of such development projects. Goldcorp Guatemala tackles local poverty with its Sierra Madre Foundation and Ambuja Cements runs its own foundation in the Indian case study. Partnerships with NGOs have become increasingly popular, too (Prieto-Carrón et al. 2006: 981). Besides, these foundations can apply to international institutions such as the International Finance Corporation to get project funding. With this additional money, many companies facilitate the construction of roads, schools or health care centres in order to support the local population. These investments are effectuated in the name of the company despite being tasks of the state.

50 Moon (2007) analyzes the impact of CSR strategies on sustainable development and especially points to the ambiguous meaning of both concepts.

51 This analysis was done prior to the company merger of Glencore and Xstrata in December 2012. However, the analysis and the relevance of these statements remains valid as such and similar sentences can be found in any CSR publication.

52 Prieto-Carrón et al. point to the necessity of defining poverty when evaluating CSR activities. Different kinds of businesses may have impacts on different forms of poverty (2006: 980-981).

However, such development projects are criticized, because on the one hand, through their main industry activities, the companies destroy livelihoods, for example by contaminating rivers or resettling communities. On the other hand, companies pretend to help the local population by contributing to infrastructure development. The involvement of companies in local policies can cause conflicts between different communities as the case study of Peru shows. Irrespective of what the actual implementation looks like, the trend to publish sustainability reports shows the rising importance of human and environmental issues for the companies' positive branding. CSR investments can thus help to get approval for the extraction activities through their effective community relations.

2.5 Conclusion: Legal Framework and Actors' Strategies

As we have seen, despite the growing recognition of indigenous peoples' issues at the international and national level, violations of human rights, especially of indigenous rights, still occur. Also the growing number of guidelines and voluntary principles regulating companies' CSR performance cannot ensure an ecologically and socially viable resource extraction. There are several explanations for the still existing discrepancy between regulated discourse and failing implementation.

The extractive industry is embedded in a legal, institutional and ideological pluralism that mainly emerges through the complex net of responsibilities between states and companies as well as within companies. Discussions between states, transnational mining companies and affected local populations are shaped by the extensive range of hard and soft laws discussed above and the underlying power relations. It is thus still extremely difficult for indigenous people and local populations to assert their rights, and even more in the particular context of exploitation of non-renewable resources. Affected populations face very powerful actors with enormous financial, technical and human capital on the one side (transnational mining companies), and with large economic interests at stake on the other (states, IFIs).

Frameworks and guidelines of international bodies (such as the UN or OECD) related to local populations' rights vis-à-vis activities of transnational companies are actually being elaborated by states. However, states' strategies are often two-sided: at the international level most of the case study countries support the development of instruments to protect indigenous peoples' rights. At the domestic level, however, regulatory frameworks are often deficient and "major legislative and administrative reforms are needed in virtually all countries in which indigenous peoples live to adequately define and protect their rights over lands and resources and other rights that may be affected by extractive industries" (A/HRC/21/47). The Special Rapporteur on the Rights of Indigenous Peoples states further that in the same countries, extractive industries are permitted to freely operate on indigenous lands, a situation he considers "alarming and in need of urgent attention" (ibid.). The vagueness of the existing legal regulations—be it conventions, constitutions or laws—leaves room for interpretation of the rights of indigenous peoples and the responsibilities of transnational companies and states, both of which take advantage of this lack of conceptual clarity to interpret the provisions in favour of their interests.

Even though the power of the state, the possibilities of implementing human rights and CSR standards and the freedom of speech enjoyed by civil society groups differ in every country, companies are nevertheless urged to deal with these difficulties and respect human and environmental standards. In this sense, the UN Guiding Principles on Business and Human Rights state that the responsibility of companies to respect human rights, including indigenous rights, exists "independently of States' abilities and/or willingness to fulfil their own human rights obligations, and does not diminish those obligations. And it exists over and above compliance with national law and regulations protecting human rights" (A/HRC/17/31).

However, despite the variety of voluntary international principles and multi-stakeholder dialogues presented here, no overall binding standards for socially and environmentally less destructive mining have been established yet (Ruggie 2007: 836). One important aspect why many guidelines and principles are not followed is the internal multi-layered system in transnational companies: these companies employ an enormous amount of people and responsibilities are divided into different sections in the headquarters and on the extraction sites. Even if the general guidelines may reflect the companies' willingness to improve standards, it seems to turn out to be difficult to implement them. Another common way of how to partly be ignorant of the local conditions is through delegating responsibilities to subsidiary companies, as well as to sub-contractors operating in the extraction sites under other names than the main company. The interaction between the different levels inside a company (headquarter vs. extraction sites; CSR office vs. investment board; mother company vs. subsidiary institutions or sub-contractors) create a situation of confusion and diffusion of responsibilities.

Besides a company's internal diffusion of responsibility, one of the biggest challenges in the compliance with human rights in resource extraction is the intertwined responsibilities of the states, the companies and the local populations. The consequential legal pluralism brings advantages to the companies while at the same time diminishes local populations' possibilities to sue for social and environmental damages caused by mining activities. Not only do indigenous peoples and other local populations often lack the financial and technical means to file a complaint if they feel that their rights have not been respected, but additionally, it is very challenging to get an overview of all the bodies, mechanisms and instruments involved, how they work, what is their jurisdiction, what are their limits, etc. Such political and economic conflicts in that case often transform into conflicts between different legal systems (Benda-Beckmann 2006: 180).

Hence, a transparent communication by companies about their activities vis-à-vis the national and international community is required. However, as Sawyer and Gomez note, "[t]he primary concern about the viability of CSR is that, given the overwhelming influence that transnational capital can have over the state and international agencies, self-regulatory measures are unlikely to serve as an effective monitoring mechanism" (Sawyer and Gomez 2012: 43). The existing legally binding indigenous rights instruments apply to the relationship between states and indigenous peoples, but do not have any regulative force with regard to the relationship between the indigenous peoples and the transnational mining companies. At the same time, companies are increasingly assuming the responsibilities of the state, by providing labour, education, and health services that the state is not able or willing to supply.[53] This vests even greater power in mining corporations, leading to a dangerous void of action possibilities for local populations that can result in violent conflicts.

Stakeholder dialogues are an often proposed measure to overcome potential or actual conflicts between local populations and companies. They are considered as being neutral and inclusive forums for all the different actors involved in a mining project. However, "(...) some stakeholders are often missing from lists of stakeholders, or physically absent from stakeholder's meetings and forums. They are precisely those in developing countries who do not normally have a voice in society" (Prieto-Carrón et al. 2006: 984). Thus, multi-stakeholder initiatives like EITI, the Global Reporting Initiative, and the Voluntary Principles on Security and Human Rights rather represent interests of the companies and some international NGOs that have adapted to the international way of discussions, dialogues and consensus-finding. Indigenous peoples and local communities, however, are often excluded from conferences of parties, stakeholder meetings and forums due to administrative, institutional, financial and educational reasons. Local populations are thus under-

53 See also Hilson's critique on such a "de facto government" (2012: 132, 134).

represented in the international arena where the actual guidelines for the extractive industry are discussed.

If Free, Prior and Informed Consent (FPIC) becomes compulsory for mining projects on indigenous territories, corporations will have to disclose more profound information about the social and environmental impacts of their activities and enter into a meaningful dialogue with the local communities. Although no universal definition of FPIC exists and many questions regarding its implementation persist, the incorporation of the affected population in decision-making processes throughout the project cycle presumably leads to more stable production, and thus profit, in the long run. The admittedly time-intensive approach with unclear results—indigenous peoples might withhold their consent—is expected, if applied comprehensively, to reduce tensions and conflicts substantially, to establish a good relationship between mining companies, states and local populations, and thus to increase the potential for further mining projects. However, neither consultation nor consent should be considered as an end in itself, but as mechanisms that safeguard the rights of affected local populations.

We have seen that the range of standards and guidelines for the performance of mining companies is broad. This reflects the growing demand by many institutions for regulations to mitigate past and future environmental and social impacts of extraction activities. Nevertheless, a visit to the actual mining extraction sites often reveals a non-compliance with most of the above described frameworks, as shown in the case studies covered in this book.

References

Benda-Beckmann, Franz von 2006: Rechtsethnologie. In: Beer, Bettina und Hans Fischer (eds.): *Ethnologie. Einführung und Überblick.* Berlin: Dietrich Reimer. 179-196.

Benda-Beckmann, Franz von, Benda-Beckmann, Keebet von, and Julia Eckert 2009: Rules of Law and Laws of Ruling: Law and Governance Between Past and Future. In: Benda-Beckmann, Franz von, Benda-Beckmann, Keebet von, and Julia Eckert (eds.): *Rules of Law and Laws of Ruling. On the Governance of Law.* Farnham: Ashgate. 1-30.

Benson, Peter and Stuart Kirsch 2010: Corporate Oxymorons. *Dialectical Anthropology* 34:45–48.

Blowfield, Michael 2005: Corporate Social Responsibility: Reinventing the Meaning of Development? *International Affairs* (Royal Institute of International Affairs 1944) 81(3): 515-524.

Bowen, John R. 2000: Should We Have a Universal Concept of "Indigenous Peoples' Rights"? Ethnicity and Essentialism in the Twenty-First Century. *Anthropology Today* 16 (4): 12-16.

Coumans, Catherine 2011: Occupying Spaces Created by Conflict: Anthropologists, Development NGOs, Responsible Investment, and Mining: With CA comment by Stuart Kirsch. *Current Anthropology* 52 (S3):S29-S43.

Frynas, Jedrzej George 2005: The False Developmental Promise of Corporate Social Responsibility: Evidence from Multinational Oil Companies. *International Affairs* (Royal Institute of International Affairs 1944). Special Issue: Critical Perspectives on Corporate Social Responsibility 81 (3): 581-598.

Gomez, Edmund Terence and Suzana Sawyer 2012: State, Capital, Multinational Institutions and Indigenous Peoples. In: Sawyer, Suzana and Edmund Terence Gomez (eds.): *The Politics of Resource Extraction. Indigenous Peoples, Multinational Corporations, and the State.* London: Palgrave MacMillan. 33-45.

GTZ 2008: Indigene Völker und Konflikt. Stärkung indigener Organizationen in Lateinamerika. *PROINDIGENA.* Eschborn: Gesellschaft für Technische Zusammenarbeit (GTZ) GmbH.

Haller et al. 2007: Fossil Fuels, Oil Companies and Indigenous Peoples. Strategies of Multinational Oil Companies, States, and Ethnic Minorities. Action Anthropology/Aktionsethnologie Series, Vol. 1. Münster: LIT Verlag.

Henderson, David 2001: Misguided Virtue. False Notions of Corporate Social Responsibility. London: The Institute of Economic Affairs.

Heydenreich, Cornelia 2004: Die OECD-Leitsätze für multinationale Unternehmen - Erfahrungen und Bewertungen. In: Fonari, Alexander (ed.): *Menschenrechts-, Arbeits- und Umweltstandards bei Multinationalen Unternehmen.* München: Germanwatch Regionalgruppe Münchner Raum und Europäische Akademie Bayern. 41-48.

Hilson, Gavin 2012: Corporate Social Responsibility in the Extractive Industries: Experiences From Developing Countries. *Resources Policy* 37: 131-137.

IACHR 2010a: Indigenous and Tribal Peoples' Rights over their Ancestral Lands and Natural Resources. Norms and Jurisprudence of the Inter-American Human Rights System. Doc 56/09. IACHR.

IACHR 2010b: Petition and Case System. Informational Brochure. Washington, D. C.: IACHR.

ICMM 2010: Indigenous Peoples and Mining. Good Practice Guide. London: ICMM.

IFC 2007: ILO Convention 169 and the Private Sector. Questions and Answers for IFC Clients. IFC Quick Note.

ILO 2009: Application of ILO Convention No. 169 by Domestic and International Courts in Latin America. A Casebook. Geneva: International Labour Office.

IWGIA 2011: The Indigenous World 2011. Copenhagen.

IWGIA 2012: The Indigenous World 2012. Copenhagen.

Jenkins, Rhys 2005: Globalization, Corporate Social Responsibility and Poverty. *International Affairs* (Royal Institute of International Affairs 1944). Special Issue: Critical Perspectives on Corporate Social Responsibility 81 (3): 525-540.

Kirsch, Stuart 2010: Sustainable Mining. *Dialectical Anthropology* 34: 87-93.

Kollmann, Kelly 2008: The Regulatory Power of Business Norms: A Call for a New Research Agenda. *International Studies Review* 10: 397-419.

McCorquodale, Robert and Penelope Simons 2007: Responsibility beyond Borders: State Responsibility for Extraterritorial Violations by Corporations of International Human Rights Law. *Modern Law Review* 70 (4): 598-625.

Moon, Jeremy 2007: The Contribution of Corporate Social Responsibility to Sustainable Development. *Sustainable Development* 15: 296-306.

Morgan, Rhiannon 2011: Transforming Law and Institution. Indigenous Peoples, the United Nations and Human Rights. Farnham, Surrey: Ashgate Publishing Company.

Newell, Peter 2005: Citizenship, Accountability and Community: The Limits of the CSR Agenda. *International Affairs* (Royal Institute of International Affairs 1944) 81(3): 541-557.

Oldham, Paul and Miriam Anne Frank 2008: "We the Peoples…". The United Nations Declarations on the Rights of Indigenous Peoples. *Anthropology Today* 24 (2): 5-9.

Prieto-Carrón, Marina, Peter Lund-Thomsen, Anita Chan, Ana Muro, Chandra Bhushan 2006: Critical Perspectives on CSR and Development What We Know, What We Don't Know, and What We Need to Know. *International Affairs* (Royal Institute of International Affairs 1944) 82(5): 977-987.

Ruggie John Gerard 2007: Business and Human Rights. The Evolving International Agenda. *The American Journal of International Law* 101 (4): 819-840.

Sanders, Douglas 1991: Collective Rights. *Human Rights Quarterly* 13 (3): 368-386.

Sawyer, Suzana and Edmund Terence Gomez 2012: Transnational Governmentality in the Context of Resource Extraction. In: Sawyer, Suzana and Edmund Terence Gomez (eds.): *The Politics of Resource Extraction: Indigenous Peoples, Multinational Companies and the State.* Basingstoke: Palgrave McMillan. 1-8.

Shamir, Ronen. 2004: Between Self-Regulation and the Alien Tort Claims Act: On the Contested Concept of Corporate Social Responsibility. *Law and Society Review* 38 (4):635-664.

UNDG 2009: Guidelines on Indigenous Peoples' Issues. UN Document HR/P/PT/16. New York: United Nations.

UNPFII 2007: Indigenous Peoples, Indigenous Voices. New York: UN Department of Public Information.

Vonplon, David 2012: Freiwilliges Risiko. *Handelszeitung* (4. October):. 26.

Ward, Tara 2011: The Right to Free, Prior, and Informed Consent. Indigenous Peoples' Participation Rights within International Law. *Northwestern Journal of International Human Rights* 10 (2): 54-84.

Welker Marina 2009: Corporate Security Begins in the Community: Mining, the Corporate Social Responsibility Industry, and Environmental Advocacy in Indonesia. *Cultural Anthropology* 24(1): 142-179.

Whitmore, Andy 2006: The Emperors New Clothes: Sustainable Mining? *Journal of Cleaner Production* 14: 309 – 314.

Conventions, Charters and Declarations

ILO 1989: Convention No. 169 concerning Indigenous and Tribal Peoples in Independent Countries.

ILO 1957: Convention No. 107 concerning the Protection and Integration of Indigenous and other Tribal and Semi-Tribal Populations in Independent Countries.

OAS 1948: Charter of the Organization of American States (A-41).

UN Doc. A/52/18, annex V

CERD 1997: General Recommendation No. 23. Indigenous Peoples.

UN Doc. A/RES/48/163

United Nations General Assembly 1993: International Decade of the World's Indigenous People.

UN Doc. A/RES/59/174

United Nations General Assembly 2004: Second International decade of the World's Indigenous People.

UN Doc. A/RES/61/295.

United Nations General Assembly 2007: Declaration on the Rights of Indigenous Peoples.

UN Doc. A/HRC/RES/15/14

Human Rights Council 2010: Resolution adopted by the Human Rights Council. Human Rights and Indigenous Peoples: Mandate of the Special Rapporteur on the Rights of Indigenous Peoples.

UN Doc. A/HRC/17/31

Human Rights Council 2011: Report of the Special Representative of the Secretary-General on the issue of human rights and transnational corporations and other business enterprises, John Ruggie. Guiding Principles on Business and Human Rights: Implementing the United Nations "Protect, Respect and Remedy" Framework.

UN Doc. A/HRC/18/35

Human Rights Council 2011: Report of the Special Rapporteur on the Rights of Indigenous Peoples, James Anaya. Extractive Industries Operating within or near Indigenous Territories.

UN Doc. A/HRC/18/35/Add.3
Human Rights Council 2011: Observations on the Situation of the Rights of the Indigenous People of Guatemala with Relation to the Extraction Projects, and other Types of Project, in their Traditional Territories.
UN Doc A/HRC/21/47
Human Rights Council 2012: Report of the Special Rapporteur on the Rights of Indigenous Peoples, James Anaya.
UN Doc A/HRC/24/41
Human Rights Council 2013: Report of the Special Rapporteur on the Rights of Indigenous Peoples, James Anaya. Extractive Industries and Indigenous Peoples.
UN Doc. A/HRC/RES/6/36
Human Rights Council 2007: Expert Mechanism on the Rights of Indigenous Peoples.
UN Doc. A/HRC/EMRIP/2012/2
Expert Mechanism on the Rights of Indigenous Peoples 2012 : Follow-up Report on Indigenous Peoples and the Right to Participate in Decision-making, with a Focus on Extractive Industries.
UN Doc. CCPR/C/21/Rev.1/Add.5
Human Rights Committee 1994: General Comment No. 23. The Rights of Minorities (Art. 29).
UN Doc. E/C.12/GC/21
Committee on Economic, Social and Cultural Rights 2009: General Comment No. 21. Right of Everyone to Take Part in Cultural Life.
UN Doc. E/CN.4/Sub.2/1994/2
United Nations Economic and Social Council 1994: Discrimination against Indigenous Peoples. Technical Review of the United Nations Draft Declaration on the rights of Indigenous Peoples.

Internet Sources

African Commission on Human and Peoples' Rights 2001: Communication No. 155/96. The Social and Economic Rights Action center and the Center for Economic and Social Rights v. Nigeria. <www1.umn.edu> accessed on 3.10.2012.

Arbeitsgruppe Schweiz-Kolumbien 2009: *Ernüchternde Erfahrung mit dem Schweizer Nationalen Kontaktpunkt. OECD Eingabe Gegen Xstrata, Umsiedlungen bei der Kohlenmine El Cerrejón.* <www.askonline.ch> accessed on 27.9.2012.

Arctic Council: About the Arctic Council. <www.arctic-council.org> accessed on 3.10.2012.

Berne Declaration 2012: *NGO kritisieren 'Blue Washing' durch Global Compact.* <www.evb.ch> accessed 22.5.2012.

Nationalrat 2012: *Curia Vista. Geschäftsdatenbank. Eine Ruggie-Strategie für die Schweiz.* <www.parlament.ch> accessed on 4.10.2012.

Business Ethics 2011: Business and Human Rights: Interview with John Ruggie. <business-ethics.com> accessed on 21.5.2012.

CIDH 1973: *Informe Anual de la Comisión Interamericana de Derechos Humanos* 1972. Doc. 41 rev. 2 OEA/Ser.L/V/II/.29. <www.cidh.org> accessed on 3.10.2012.

Eads, Christopher, Paul Mitchell, and Francisco Paris (eds.) 2009: Advancing the EITI in the Mining Sector. A Consultation with Stakeholders. <eiti.org> accessed on 21.5.2012.

EITI 2012: Extractive Industries Transparency Initiative. <eiti.org/eiti/history> accessed on 21.5.2012.

European Union 2012: The Website About CSR Reporting of Companies. <reportingcsr.org> accessed on 22.5.2012.

Global Reporting Initiative. 2012. <www.globalreporting.org>accessed on 22.5.2012.

Heydenreich, Cornelia and Klaus Milke 2012: Green Economy und CSR-Debatte: Grün und Fair. *Informationsbrief Weltwirtschaft & Entwicklung* (W&E), Luxemburg. <www.germanwatch.org> accessed on 19.6.2012.

Intergovernmental Forum on Mining, Minerals, Metals, and Sustainable Development 2012. <www.globaldialogue.info> accessed on 24.5.2012.

International Council on Mining and Metals 2012. <www.icmm.com> accessed on 22.5.2012.

International Finance Corporation 2012: Oil, Gas, and Mining. <www1.ifc.org> accessed on 22.5.2012

ISO 2012. International Organization for Standardization. <www.iso.org/iso/home.htm> accessed on 24.5.2012.

Kampagnenforum. 2012: *'CSR, die weh tut'. Ein Interview mit Andreas Missbach von der Erklärung von Bern.* <www.kampagnenforum.ch> accessed on 19.6.2012.

Kimberley Process 2012. <www.kimberleyprocess.com> accessed on 22.5.2012.

Multiwatch 2012: *Unternehmensverantwortung.* <www.multiwatch.ch> accessed on 24.5.2012.

OAS: Afro-Descendants. Rights of Afro-Descendants and Against Racial Discrimination. <www.oas.org> accessed on 3.10.2012.

OAS: IACHR. What is the IACHR? <www.oas.org> accessed on 3.10.2012.

OECD/OCDE 2002: The Relevance of the OECD Guidelines for Multinational Enterprises to the Mining Sector and the Promotion of Sustainable Development. OECD Global Forum on International Investment. Conference on Foreign Direct Investment and the Environment. Lessons to be Learned from the Mining Sector. Paris: OECD headquarters. <www.oecd.org> accessed on 21.5.2012.

OECD 2011: OECD Due Diligence Guidance for Responsible Supply Chains of Minerals from Conflict-Affected and High-Risk Areas. OECD Publishing. <www.oecd.org> accessed on 19.12.2012.

OECD 2012. OECD Guidelines for Multinational Enterprises. <www.oecd.org> accessed on 24.5.2012.

Publish What You Pay 2013: The Cardin-Lugar Amendment (Dodd Frank 1504). <www.publishwhatyoupay.org> accessed on 25.1.2013.

Resolution of the Terra Madre Conference 2011: Resolution and Statement of Profound concern Regarding Sweden's Failure to Implement Article 36 of the UN Declaration on the Rights of Indigenous Peoples for the Participation of Delegates to the Indigenous Terra Madre Conference Sapmi Territory Jokkmokk Sweden. <www.treatycouncil.org> accessed on 3.10.2012.

Schweizerische Eidgenossenschaft 2013. *Grundlagenbericht Rohstoffe. Bericht der interdepartementalen Plattform Rohstoffe an den Bundesrat.* Eidgenössisches Department für Auswärtige Angelegenheiten EDA, Eidgenössisches Finanzdepartment EFD, Eidgenössisches Department für Wirtschaft, Bildung und Forschung WBF. <www.eda.admin.ch> accessed on 2.4.2013.

Schutter de, Olivier, Ramasastry, Anita, Taylor, Mark B. and Robert C. Thompson 2012: Human Rights Due Diligence: The Role of the State. <accountabilityroundtable.org> accessed on 19.8.2013.

Sosa, Irene 2011: License to Operate. Indigenous Relations and Free Prior and Informed Consent in the Mining Industry. Sustainalytics. <www.sustainalytics.com> accessed on 13.11.2012.

United Nations 2012: About the UN Global Compact. <www.unglobalcompact.org/AboutTheGC/TheTenPrinciples> accessed on 22.5.2012.

Voluntary Principles on Security and Human Rights 2012: Timeline. <voluntaryprinciples.org> accessed on 21.5.2012.

10th UNPFII: IYC Closing Statement. <www.youtube.com/watch?v=0GZno7iGWL0> accessed on 1.10.2012.

Sustainability Reports

Glencore 2010: Sustainability Report 2010. Glencore Corporate Practice. Baar: Glencore. <www.glencore.com/documents/Sustainability_Report.pdf> accessed on 22.5.2012.

Xstrata 2010: Sustainability Report 2010. Zug: Xstrata. <www.xstrata.com> accessed on 22.5.2012.

Anglo American 2011: Sustainable Development Report 2011. What it Takes: Partnership and Innovation. London: Anglo American. <www.angloamerican.com> accessed on 22.5.2012.

Freeport-McMoRan 2010: Working Towards Sustainable Development Report. Arizona: Freeport-McMoRan. <www.fcx.com> accessed on 22.5.2012.

Rio Tinto 2009: The Way We Work. Our Global Code of Business Conduct. London: Rio Tinto. <www.riotinto.com/documents/The_way_we_work.pdf> accessed 22.5.2012.

Personal Communication

Information from an informant working in the mining sector in southern Africa. May 2012. (For safety reasons, the identity of the informant cannot be disclosed.)

Information from an indigenous representative at EMRIP 2010, Geneva. (For safety reasons, the identity cannot be disclosed).

3. The Mining Process and its Environmental Impacts

Isabella Brom [1]

3.1 Introduction

The mining process has manifold impacts on the environment. These include effects of scales from short- to long-term, as well as from local to global, which interact in complex ways. The present chapter provides a general description of the mining processes relevant to this book. It supports a better understanding of the technical aspects discussed in the case studies and gives an overview of their most important environmental impacts. In accordance with the case studies, the emphasis is on large-scale open-pit mines for gold, copper and uranium, with some references to coal.

3.1.1 The Lifecycle of Mining

The lifecycle of a mining project can be divided into four different phases, each of which leads to specific impacts on the environment (ELAW 2010: 3f., 7; ABS 2003).

- *Exploration phase:* The exploration of possible mineral ore deposits usually marks the first stage of a mining project. Their location, extent and value are evaluated through surveys, field studies and geological drilling or other exploratory excavations. This phase involves contributions from different disciplines such as basin modelling, petrology, geochronology, structural geology, geophysics and geochemistry.
- *Development/expansion phase:* The development or expansion phase involves site preparation and clearing. With heavy support equipment such as earthmovers, the area dedicated to mining is cleared of vegetation and topsoil. New infrastructure such as roads, mining towns and transect lines cut through the living environments of local communities. Similar preparations are needed for the expansion of a mine in operation.
- *Active mining/exploitation phase:* The actual mining stage includes ore extraction or excavation, the processing of a mineral (beneficiation) and, if needed, metallurgical extraction of the metals. At this stage the most severe impacts occur, often including an alteration of the landscape, a loss of wildlife and chemical and physical contamination of air, water and soil. The two key processes—excavation and beneficiation—of a material are shared by all mines, regardless of what metal or mineral is quarried. They can however differ in the techniques applied (ELAW 2010: 3f., 7; ABS 2003).

1 This chapter was composed by Isabella Brom and supported by Thomas Niederberger. A number of students and researchers of the ETH Department of Environmental Systems Science (Zurich) have contributed comments and suggestions: Sandra Rüd-Probst, Annika Linkhorst, Jasmin Mertens and Nina Küng. The BA thesis of Severin Hellmüller (2012) was useful for comparison.

- *Mine closure and rehabilitation:* Open mining pits, tailing dams or waste rock piles can, if left untended, release toxic substances for decades or centuries after the closure of a mine. There are different levels of restoration efforts after mine cessation. Complete rehabilitation of a mine includes the respective area to be recovered as close as possible to its original condition.

3.1.2 Open-pit and Underground Mining

Historically, the image of large-scale mining has been dominated by underground, i.e. close-pit mining, whereby workers access layers of highly concentrated, high-grade metal ores through shafts. This type of mining continues to exist, but is increasingly superseded by the open-pit, i.e. open-cast or open-cut mining technique. The choice between the two methods depends on shape, depth, size, stability and grade of an ore body as well as on economic aspects. However, in certain cases the two techniques are also applied in combination at the same site.

Open-pit mining is a form of strip-mining, where heavy machinery such as bulldozers, excavators and earthmovers are used to remove layer after layer of topsoil and rock (overburden) to access the ore. Depending on the geological depth of the ore, pits of several hundred meters depth and radius are created. Commonly these pits extend below groundwater level and water needs to be pumped out continuously and tends to form a pit lake after mine closure.

Close-pit mining typically has less visible impacts on the environment than open-pit mining, since it requires smaller areas of land to be cleared and produces smaller amounts of waste rock. It is however more expensive, requires a bigger workforce and causes complex safety issues like ground support and ventilation (ELAW 2010: 5). The open-pit method is most often used to quarry bauxite, iron ore and coal deposits. Also gold and copper mines are increasingly applying the open-pit method, as this allows the exploitation of low-grade ore in a more cost-effective way.

In 1990, 88% of copper mines in the United States were operated with the open-pit excavation method. Typically, these mines produced a variety of by-products such as cobalt, nickel, molybdenum, silver or gold (Ayres et al. 2002: 17, 30). In 2011, no underground copper mines were in operation in the USA, and 97% of all non-fuel minerals were mined with surface methods (Ewell 2013: 1). In the US copper mines, the ore content of the excavated rock has decreased from 10 to 20 percent in 1850 to an average of 0.5 percent in 1970. Globally, the average in the operating copper mines is estimated at 0.6 percent (Namorado Rosa in Bardi 2013: 136-138). Similar conditions apply for uranium: as reserves have become scarce and most of the recently located deposits are of a much lower grade than the discoveries in the 1970s in Australia and Canada, more and more deep-lying, low-grade deposits are excavated (Mudd 2006: 3). Today's mines exploit concentrations of natural uranium of 0.05 to 0.2% of the rock, which means that up to 200 tons of rock need to be extracted and processed to produce a single kilogram of uranium ore (IPPNW 2010). For gold, the ore grade exploited in Australia, Brazil and South Africa has decreased from 15 to 25 gram per ton around 1900 to current levels of 1 to 5 grams (Mudd 2007: 47).

An illustration of a simplified mining process is given in figure 1. In the first step, mining produces ore and waste. Ore is distinguished from waste because it contains "economically recoverable levels of metals or minerals". The latter need to be separated from unwanted (*gangue*) minerals by processing, which results in "ore minerals", which are refined to pure metal by metallurgical extraction (Lottermoser 2007: 4). Of course, there are many differences here depending on the mined mineral or metal and the process for its extraction. Some minerals (e.g., coal) need only minimal processing, while some metals like copper and gold need very elaborate metallurgical treatment. It is important to note that not all the production steps must happen in the same location—sometimes long conveyor belts transport crushed ore from the mine to a factory for further refining, or the mine produces a slurry-like concentrate which can be pumped through pipelines over long distances for further processing. Also, the final product sold on the market can be a high-grade concentrate, which will be further refined in a completely different place.

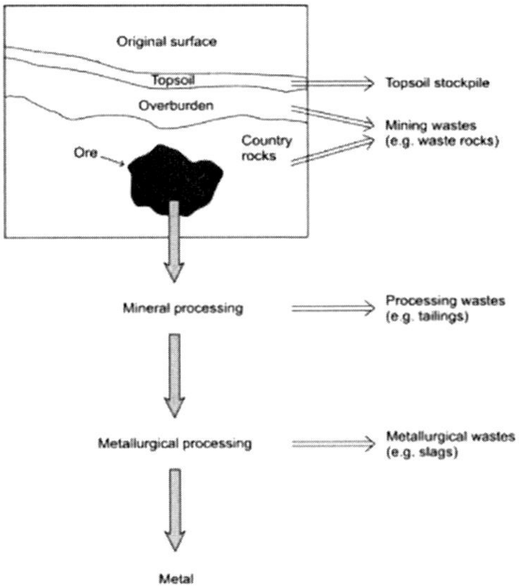

Fig. 1: Schematic mining process (Lottermoser 2007: 4).

3.1.3 Beneficiation and Metallurgical Extraction

The mineral processing, concentration or beneficiation physically and/or chemically separates the desired minerals or metals from unwanted ones (see Fig. 2, 3). In order to do so, the retrieved ore needs to be crushed and grinded first. Processing techniques include gravity concentration, magnetic separation, electrostatic separation and flotation. The concentrated ore mineral can then be further refined through pyrometallurgical processes. Further procedures used in benficiation are the process of solvent extraction or leaching, called hydrometallurgy, and electrometallurgy. What combination of methods is applied depends on the geology, the target minerals as well as commercial viability. Below the frequently used beneficiation techniques which apply to the case studies covered in this book are briefly introduced:

Leaching: In the leaching process, finely ground ore is stocked in a leach pile and treated with solvents to dissolve the target metal. The mixture is then collected for further processing (ELAW 2010: 6). Copper is commonly retrieved through hydrometallurgy; i.e., dump, heap, vat or in-situ leaching. In dump leaching, ore is alternately treated with dilute sulphuric acid and exposed to air to oxidize the sulphides. The acid reacts with pyrite, which creates further sulphuric acid and ferric sulphate, as well as copper in concentrated form. Heap leaching uses a similar technique, except that it is applied to previously crushed and heaped oxide ores. Vat leaching is a rapid process where crushed ore is flooded with sulphuric acid in concrete containers. It requires large quantities of highly concentrated acids to recover approximately 75% of the copper (Ayres et al. 2002: 17). A very cost-effective leaching method is the in-situ leaching (ISL): the sulphuric or hydrochloric acid is directly injected into the ground to dissolve the minerals and is then pumped back to the surface for processing (see case study Zambia; Simpere et al. 2010). Similar leaching techniques are also used for uranium (Frost 1998).

For gold/silver extraction, cyanide leaching is widely applied. The highly leaching-efficient but toxic cyanide solution separates the gold particles from the rock by forming soluble gold cyanide complexes. In a tank with activated carbon, the gold-cyanide complexes adsorb to the carbon and

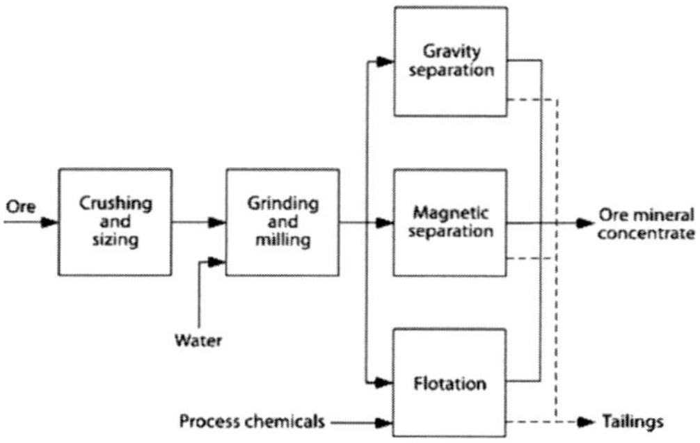

Fig. 2: Simplified chart of mineral processing (Lottermoser 2007).

can then be collected in solid state. Other gold extraction techniques include gravity concentration, flotation, panning and pyrometallurgy. Due to its high toxic potential, the amalgamation process with mercury is out of use in modern large-scale mines, but it remains the preferred method of small-scale, artisanal miners due to its wide accessible, easy application and low costs (Lacerda 1997).

3.2 Environmental Impacts of Mining

The following section entails a broad overview of different impacts on the environment during the lifecycle of a mine. Its focus rests on water quality, the problem of acid mine drainage and tailings and a number of specific impacts with regard to the main metals and minerals covered in this book.

The effects of a mine on the environment depend on the ore mineralogy, mineral concentration, waste disposal methods and climate. In the process of setting up large-scale open-pit mines, great quantities of rock are removed, considerably altering the local environment, landscape and waterways. In addition, a large number of inputs are needed for mining and refining: water, electricity, fuel, explosives and different chemicals. Less visible, but as important, is the release of toxic emission into the air, water and soil. Climatic conditions and hydrology strongly influence the release of toxic substances into the different elements. The substances released can be natural components of the rock, set free through mining and refining, or are chemical additives stemming from the processing procedures (Higueras et al. 2004: 1858; Hoskin et al. 2000).

3.2.1 Impacts during the Lifecycle of Mining

Damages to the natural environment start with the exploration phase, due to test drillings, forest clearing and the creation of access tracks for vehicles (ELAW 2010: 3f.). In the development phase, large infrastructure is built including the mining pit, concentrators, smelters and other installations required for beneficiation, as well as tailing dams, power plants, high voltage lines, the workers' compound and fortified access roads. It can also be necessary to change the course of waterways, to clear forest areas or glacier cover.

The most severe impacts occur during the active mining or exploitation phase, due to the release of potentially harmful substances into the air, water and soil, and the large quantities of waste rock and tailings that have to be handled. Figures 1, 2 and 3 give an illustration of the major sources of emissions. Appropriate waste management requires the solid or liquid material to be disposed in a way that the mobilization and release of toxic particles is reduced on a long-term basis (ELAW 2010: 6). Taking into account the large amounts, this is a major challenge. For instance, the global average of waste production of the copper mining industry lies at of 37 tons of waste per ton of concentrate (Ayres et al. 2002: 26). For gold, up to 99.999% of the extracted ore ends up as waste, i.e., tailings (Lottermoser 2007: 156).

Once a mine is closed, there are various ways in which contaminants can be dispersed, even if the tailings are covered or stored underground. It is usually impossible to entirely prevent any contact with groundwater flows or weathering on the surface. Acid mine drainage can further continue for many years, which can cause continuous contamination of large regions (Frost 1998). Examples from Australia (see box) show that even for a first-world country the challenges to prevent contamination after mine-closure are considerable.

After mine closure

Large-scale open-pit mining operations often leave behind wastelands with featureless waste dumps and poorly covered pit holes. The Migori Gold Belt in Western Kenya serves as an example, where numerous gold mines have caused littered landscapes, open drains, contaminated streams and deep open pits filled with water. The removal of trees for timber has caused severe erosion. Today, vegetation hardly grows in the district due to Acid Mine Drainage (AMD) that has rendered the soil infertile (Ogola et al. 2002: 145f.). The abandoned Rum Jungle, Radium Hill and Mary Kathleen uranium mines in Australia have all been proven to still have extensive impacts on the environment, even decades after their closure. At the Rum Jungle site, AMD continues to seep into the Finniss River due to poor remediation works. At Radium Hill, erosion and tailings require ongoing monitoring and maintenance, and at Mary Kathleen, the seepage of saline, metal and radionuclide-rich water from tailing dams has caused damage to local waterways since its closure in mid-1980 (Mudd 2006: 5).

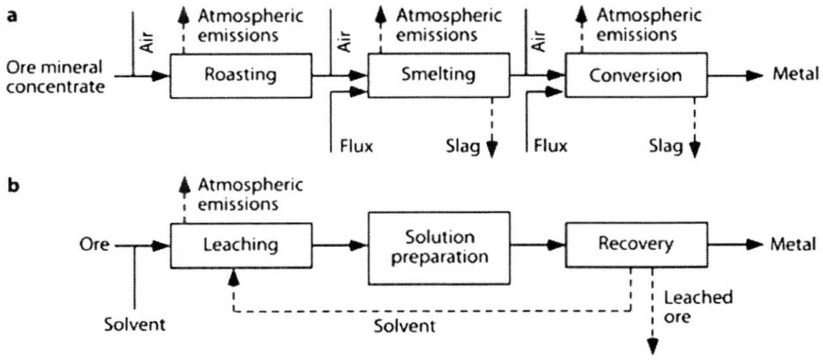

Fig. 3: Pyrometallurgical (a) and hydrometallurgical (b) flow-charts from ore to metal (Lottermoser 2007: 6).

3.2.2 Energy Use and other Inputs

The amount of energy used by the mining sector remains uncertain. Rábago et al. give a cautious estimation of 4 to 7 percent of total global energy use (2001: 17). Others think it is close to 10 percent (Bardi 2013: 155). The most energy intensive industries are by far steel and aluminium. On average, the energy used for mining and quarrying itself, mostly fuel for the machinery, is just 3 to 7 percent, but this makes around 17 percent of the total cost of supplies. A larger part is used for processing the ore. The sources are about one-third each for fuel/diesel oil, electricity and coal/gas/gasoline. Electricity is used mostly for pumping, grinding and refining (Rábago et al. 2001: 22). As mining companies are typically among the largest consumers of electricity in the regions where they operate, it is no surprise that they often choose to invest in power plants, or to negotiate favourable contracts with electricity providers, in order to ensure constant electricity supply at low prices (ibid.: 30). Mining further requires large amounts of water, which is again linked to energy use due to the water often being pumped from far or deep lying aquifers, or it has to be pumped out of rain- or ground water filled open pits and shafts.

As mentioned above, there is a general trend of decreasing ore grade. Consequently the energy input needed for the same amount of recovered metals and minerals is increasing for various reasons: More overburden are moved, deeper pits and shafts are built and the processing difficulty is increasing. Therefore, the grinding needs to be finer to recover the metals which are disseminated as small grains in the rock. For example, the energy needs of Chile's copper industry increased by 50 percent from 2001 to 2010, while production increased by just 14 percent in the same period. To produce one ton of copper, 24 gigajoules were used. For the mines in the USA, it was 112 gigajoules or four times higher (Namorado Rosa in Bardi 2013: 139). The production of 1kg of gold consumes an average of 143 gigajoules of energy, 690 kilolitres of water and releases 11.5 tons of CO_2. The relationship between decreasing ore grade and higher input is well illustrated by cyanide use in gold mining: While high-grade mines (more than 6 g/t) needed 100 kg of cyanide to produce 1 kg of gold, low-grade mines (less than 2 g/t) needed up to 1000 kg, i.e., 10 times more (Mudd 2007: 52-53).

3.2.3 Mine Waste

Waste rock: Open-pit mines generate large volumes of waste rock or overburden, which is the excavated rock that does not contain enough minerals or metals to be considered for further processing. For gold, the quantity typically lies between 2 and 10 times the amount of extracted ore (Mudd 2007: 54). Overburden is frequently deposited in on-site piles and used to backfill the pits after closure. In a well-executed remediation, it is covered with soil and vegetation. A major risk associated with the disposal of waste rock is the potential of toxic leachate seeping into the ground. In some countries (e.g., Australia) mining companies are required to develop a waste rock management plan based on tests and drilling programs in order to assess the potential of acid generation from the waste rock (Frost 1998).

Tailings: In open-pit mines with little metal content, up to 99% of the extracted ore is disposed as a slurry-like waste called tailings. After the beneficiation process, in which the ore is ground to a fine powder and treated with chemicals, these tailings contain large amounts of toxic substances (ELAW 2010: 5; ABS 2003). Depending on the rock geology, these can include arsenic, lead, mercury, cadmium and other heavy metals occurring as stable minerals in most metal deposits (Ogola et al. 2002: 149). In uranium mines, the rock contains radioactive elements that also end up in the tailings.

Storage and disposal of tailings are challenging for waste management. In dry climate, it can be possible to let them dry out to use as backfill later. More commonly, they are pumped into "tailing ponds". Sometimes natural basins can be used as tailing ponds, which significantly reduces the costs, but usually artificial reservoirs have to be constructed with one or several dams that hold the tailings back. To reduce filtration into the groundwater, the floor of the tailing pond should be isolated with a geo-membrane, usually a polyethylene sheet of 1 to 1.5 mm. However, this cannot prevent filtrations completely, as the membrane may have holes or decompose slowly over the years (Erzinger et al. 2008: 29). In some cases, tailings are discharged on the seafloor or in rivers, which is a very cost-saving but destructive method (McKinnon 2002: 43f., see box STD/RTD).

The management of tailings is heavily influenced by the local climate. In arid climates, tailing ponds may dry out and toxic substances can be spread by the wind, contaminating the surrounding area. In wet, tropical regions with small evaporative losses, water management is more complicated, and in periods of heavy rain, tailing dams can overflow. Tailing dams also require continuous monitoring and rehabilitation after mine-closure. Unstable dams pose an imminent risk by leaking or breaking (McKinnon 2002: 44) and earthquakes can damage dam structures (Johnston/Prendergast 1999: 48f.). In the past, numerous cases of dam failure have caused detrimental environmental disasters, killing aquatic life and poisoning drinking water supplies of communities downstream of the tailing pond (ELAW 2010: 11).

Submarine (STD) and Riverine Tailings Discharge (RTD)

Submarine Tailings Discharge (STD) is the process of disposing slurries in deep waters where they are intended to flow to a deposition site on the ocean floor and to be trapped by stratified surface layers. However, substantial parts of the tailings get carried away by deep-water currents. Consequently, large areas of seafloor are covered by tailings, containing chemicals used in production and heavy metals such as arsenic, cadmium, mercury and lead (McKinnon 2002: 44f.). STD is practiced at the Lihir gold mine in Papua New Guinea, where approximately 341 to 432 tons of processing wastes will be disposed into the sea over the life of the mine (ibid.: 41, 45). According to the ELAW (2010: 6), STD has poor environmental records, and is thus illegal in the USA, Australia and Canada.

River Tailing Discharge (RTD) is a method referring to the disposal of tailings into river valleys. Tailings contaminate the water downstream as well as the surrounding vegetation (Cardiff et al. 2012: 9). A drastic example of RTD is the Freeport McMoRan's Grasberg gold-silver mine in West Papua. According to some estimates, the mine produces an astonishing quantity of up to 200,000 tonnes of tailings per day, which are dumped into the Otomina and Ajkwa rivers (Cardiff et al. 2012: 16).

3.2.4 Impacts on Water Systems

The effects on water quality and availability are considered the most significant impacts of mining operations (ELAW 2010: 7). Impacts on ground and surface waters are especially problematic in water-scarce regions. Ecosystems and livelihoods of local inhabitants are affected by the extraction of large quantities of water from rivers and ground water reserves on a long-term scale (Barber/Jackson 2011: 48). The dewatering of open-pits, the water consumption for processing ore (e.g. flotation; hydrometallurgy) and other activities (e.g. dust prevention) can lead to lower water tables, drying up of springs and waterholes and the development of sinkholes. Poor water management and extensive clearing of forests causes erosion, which again can result in siltation of nearby floodplains or waterways through sediment-loaded run-off. The occurrence of run-off, its volume, and the associated erosion are determined by a variety of factors such as geology, available material, climate and the extent of rain events (ELAW 2010: 9f.).

Water Contamination and Acid Mine Drainage

The most damaging types of mine wastewater are Acid Mine Drainage (AMD) and leachate from mine tailings. AMD is based on the oxidation of pyrite and other sulphides, which release highly toxic free metal ions in acidic solution. This oxidation reaction can happen in any part of a mine where sulphides are uncovered, such as disposed tailings, waste rock, open pits and mine tunnels. Sulphide minerals are mainly associated with coal and certain metals like gold, copper and silver. Unchecked AMD frequently causes a decrease in pH of groundwater, waterways and soil, as well as heavy metal contamination of biota, ecosystems, drinking water and irrigation systems. AMD can persist for decades after closure of a mine and the disposal of wastes (ELWA 2010: 6, 8; Frost 1998).

In their report, "Comparison of Predicted and Actual Water Quality", Kuipers et al. (2006) found that 76 percent of the 25 metal mines in the USA they analysed did not match up to water quality standards and were polluting rivers and groundwater with lead, mercury, cadmium, arsenic and cyanide. In mines with high risk of AMD and those close to ground or surface water, this rate was around 90 percent. Predictions made in Environmental Impact Assessments were not adequately checked with the results of past predictions and measures to prevent water contaminations failed in 65 percent of the cases.

3.2.5 Metal-specific Environmental Risks of Mining

Certain environmental risks are directly related to the characteristics of the mined mineral, or to the specific process used for beneficiation of an ore. Below the impacts of the majority of the minerals relevant to the case studies are outlined:

- *Copper:* The roasting and smelting processes produce large amounts of sulphur dioxide, the principal airborne pollutant related to copper mining (World Bank 1998: 291), together with other toxic gases. The sulphur used in the heap leaching process is highly likely to leach into nearby waterways and ground water storage (Ayres et al. 2002: 27ff.).
- *Gold:* The most controversial aspect specific to gold mining is the use of sodium cyanide for beneficiation. This is the most widespread method applied and currently has no commercially viable alternatives (Mudd 2007: 46). Cyanide is a well-known, highly toxic agent, causing strong controversies in regard to its use in large quantities. Studies close to the industry argue that cyanide is not problematic if safety standards are upheld, since it decomposes rapidly and does not accumulate in the food chain (Logsdon et al. 1999). Other authors claim however that even low concentrations and a variety of cyanide break-down compounds pose serious risks, especially for fish and birds. Also, numerous examples of accidental spills and break-downs of tailing dams, such as the Baia Mare disaster in Romania support their case (e.g., Eisler/Wiemeyer 2004; Moran 2002).
- *Uranium:* A main concern for uranium mining is the release of radionuclides into the environment, increasing the natural background radiation levels, endangering the health of people and the ecosystem. Waste handling in uranium mines is hence a complex and expensive process, in particular the decontamination of radioactive water (Mudd 2003: 182; 2006: 3; ELAW 2010: 9). In reality, if uranium-contaminated water is discharged into the environment, decontaminating measures are indispensable but are often insufficiently done (Frost 1998). The tailings of uranium mines are an additional risk to the environment, retaining approximately 85% of the ore's original radioactivity. However, the application of common solutions such as tailing dams entail the risk for the dams themselves to become radioactive and consequently require additional isolation (Mudd 2006: 2).
- *Coal:* Differently from copper, gold or uranium, coal does not require toxic chemicals for its beneficiation. Physical environmental degradation through soil erosion, waste rock, AMD and increased sediment load still occurs and usually covers much larger areas than in other types of mining. As mining companies increasingly exploit low-grade ore and ex-

cavate to increasingly deep levels, waste volumes also increase (Sahu/Dash 2011: 132ff.). Agricultural areas and forests are removed or degraded completely, often leading to a loss of biodiversity. Rain or groundwater that enters the mine gets contaminated with nitrate, sulphate and high levels of suspended solids, including heavy metals, and seeps as contaminated water into the soil and waterways. Also air quality is affected by windblown dust from coal transportation to washing facilities and export ports (Chitade/Katyar 2010: 7172-75; Tiwary 2000: 186-188; Castleden et al. 2011: 333).

3.2.6 Impacts on Flora and Fauna

Mining can cause a severe loss and fragmentation of habitats through roads and cleared areas, which renders the migration of animals difficult to impossible. The isolation of species due to habitat dissection can lead to a decline in environmental quality and a consequent reduction in gene pools and in extreme cases can eliminate species that rely on large habitat patches (ELAW 2010: 13, 14). Strong interference with the balance of ecosystems is also caused by the removal of topsoil and vegetation, the extraction of large amounts of waste rock, and heavy metals and other contaminants released into the air, water and soil. These are taken up by algae, bacteria, plants and small animals and accumulate in the food chain. The pollution and sediment run-off in streams, lakes and other waterways cause a degradation of aquatic life. In combination with land clearing, this results in a deterioration of food sources for all kinds of wildlife, especially predators. The ability of fauna and flora species to adapt to a changed environment is limited, and if a critical habitat is altered or destroyed, the impact of a mining project can include the complete elimination of certain species in the respective area (ibid.). This is becoming increasingly concerning, as more mines are built in sensitive tropical regions with high biodiversity, often close or even overlapping with nature reserves (Miranda et al. 2003: 16).

3.3 Outlook

This chapter outlines the mining processes relevant to the case studies in more details and gives an overview of the associated impacts. Modern mines are often very large and resource-intensive projects, which cannot run without using large amounts of water, electricity, fuel, land and often chemical agents. The contamination of air, water and soil, and the resulting accumulation of toxic substances in the food chain are universal, but the extent and gravity depend on multiple factors. Open-pit and closed-pit mines are associated to different sets of problems and management challenges. These however vary also depending on the metal or mineral mined, the ore concentration and the geological and climatic conditions. There are technical means to prevent and mitigate some of the worst effects, they are however associated with considerable costs. Mining companies are hence often reluctant to implement them, if they are not legally obliged to do so. Overall, it can be concluded that the impacts of mining on the environment are extensive, very complex and only marginally controllable. Because the availability of highly concentrated, easily accessible metal and mineral ores is decreasing steadily and exploitation moves to low-grade, not easily accessible ores, the impact of mining is bound to become ever more destructive, especially considering the destruction of natural habitats and landscapes, the quantity of waste produced and the inputs needed. Despite the recent trend towards a more "sustainable" mining, the tendency are increasingly deteriorating impacts.

References

ABS (Australian Bureau of Statistics) 2003: Mining and the Environment. <www.abs.gov.au> accessed on 02.04.2014.

Ayres, Robert U., Ayres Leslie W. and Ingrid Rade 2002: The Life Cycle of Copper, its Co-Products and By-Products. *Mining Minerals and Sustainable Development* Report No. 24. IIED/WBCSD.

Bardi, Ugo 2013: *Der geplünderte Planet – Die Zukunft des Menschen im Zeitalter schwindender Ressourcen.* München: Oekom Verlag.

Barber, Marcus and Sue Jackson 2011: Indigenous Engagement in Australian Mine Water Management: The Alignment of Corporate Strategies with National Water Reform Objectives. *Resources Policy* (37): 48-58.

Bridge, G. 2004: Contested Terrain: Mining and the Environment. *Annual Reviews Environmental Resources* (29): 205-59.

Bosman, Carin 2009: The Hidden Dragon: Nitrate Pollution from Open-pit Mines — A Case Study from the Limpopo Province, South Africa. Abstracts of the International Mine Water Conference.

Cardiff, Scott, Coumans, Catherine, Hart, Ramsey, Sampat, Payal and Bill Walker 2012: Troubled Waters. How mine waste dumping is poisoning our oceans, rivers and lakes. Washington DC/Ottawa: Earthworks & Mining Watch Canada.

Castleden, William M., David Shearman, George Crisp and Philip Finch 2011: The Mining and Burning of Coal: effects on health and the environment. *Medical Journal of Australia* 195(6): 333-335.

Chitade, Anil Z. and S.K. Katyar 2010: Impact Analysis of Open Cast Coal Mines on Land Use / land cover using remote sensing and GIS technique. *International Journal of Engineering Science and Technology* 2(12): 7171-7176.

Eisler, Ronald and Stanley N. Wiemeyer 2004: Cyanide Hazards to Plants and Animals from Gold Mining and Related Water Issues. *Reviews of Environmental Contamination and Toxicology* (183): 21–54.

ELAW, Environmental Law Association Worldwide 2010: Guidebook for Evaluating Mining Project EIAs <www.elaw.org> accessed on 17.08.12.

Erzinger, Florian, Louis González and Angel M. Ibarra 2008: *El lado oscuro del oro – Impactos de la minería en El Salvador.* Cáritas El Salvador: UNES.

Ewell, Mary E. 2013: Mining and Quarrying Trends—Minerals Yearbook. Washington, DC: U.S. Geological Survey (USGS). <minerals.usgs.gov/minerals> accessed on 02.04.2014.

Frost, Stan E. 1998: Waste Management in the Uranium Mining Industry. The Uranium Institute. Twenty Third International Symposium 1998. <www.world-nuclear.org> accessed on 01.07.12.

Hellmüller, Severin 2012: *Metallische Rohstoffe – ökologische Probleme und Engagement Schweizer Rohstofffirmen.* Bachelorarbeit im Departement Umwelt– und Systemwissenschaften, ETH Zürich.

Higueras, Pablo, Roberto Oyarzun, Jorge Oyarzun, Hugo Maturana, Javier Lillo and Diego Morata 2004: Environmental assessment of copper–gold–mercury mining in the Andacollo and Punitaqui districts, northern Chile. *Applied Geochemistry* (19): 1855-1864.

Hoskin, W., G. Bird and T. Stanley 2000: Mining – Facts, Figures and Environment. *Industry and Environment* (23): 4-8.

Johnston, Arthur and Bernard Prendergast 1999: Assessment of the Jabiluka Project. Report of the Supervising Scientist to the World Heritage Committee. Supervising Scientist Report 138. Canberra: Supervising Scientist.

IPPNW 2010: How is uranium mined? Uranium Fact Sheet 3. <www.ippnw.org> accessed on 29.06.12.

Kuipers, J.R., Maest, A.S., MacHardy, K.A., and Lawson, G. 2006: Comparison of Predicted and Actual Water Quality at Hardrock Mines: The reliability of predictions in Environmental Impact Statements. Washington DC: Earthworks.

Lacerda, L.D. 1997: Global Mercury Emissions from Gold and Silver Mining. *Water, Air and Soil Pollution* 97 (3-4).

Ledin, M. and K. Pedersen 1996: The environmental impact of mine wastes – Roles of microorganisms and their significance in treatment of mine wastes. *Earth-Science Reviews* (41): 67-108.

Logsdon, M.J., Hagelstein, K., Mudder, T.I. 1999: The management of cyanide in gold extraction. Canada, Ottawa: International Council on Metals & the Environment (ICME).

Lottermoser, Bernd 2007: Mine Wastes – Characterization, Treatment and Environmental Impacts (2nd edition). Berlin/ Heidelberg: Springer.

McKinnon, Elizabeth 2002: The environmental effects of mining waste disposal at Lihir Gold Mine, Papua New Guinea. *Journal of Rural and Remote Environmental Health* 1(2): 40-50.

Miranda et al. 2003: Mining and Critical Ecosystems: Mapping the Risks. Washington D.C.: World Resources Institute. <pdf.wri.org/mining_critical_ecosystems_full.pdf> accessed on 02.04.2014.

Moran, Robert E. 2002: De-coding Cyanide – An assessment of gaps in cyanide regulation at mines. <www.earthworksaction.org> accessed on 20.03.2014.

Mudd, Gavin M. 2003: Uranium Mining in Australia: Environmental impact, radiation releases and rehabilitation. Darwin NT: Environment Centre of the Northern Territory (ECNT).

Mudd, Gavin M. 2006: Uranium Mining: Australia and Globally. Fact sheet 6. <www.energyscience.org.au> accessed on 28.10.11.

Mudd, Gavin M. 2007: Global Trends in Gold Mining: Towards Quantifying Environmental and Resource Sustainability? *Resources Policy* (32): 42-56.

Mudd, Gavin M. 2009: The Environmental Sustainability of Mining in Australia: Key Mega-trends and Looming Constraints. *Resources Policy* (35): 98–115.

Ogola, Jason, Winnie V. Mitullah and Monica Omulo 2002: Impact of Gold Mining on the Environment and Human Health: A case study in the Migori Gold Belt, Kenya. *Environmental Geochemistry and Health* (24): 141-158.

Rábago, K. R.; Lovins A. B.; Feiler T. E. 2001: Energy and Sustainable Development in the Mining and Minerals Industries. IIED report. <pubs.iied.org> accessed on 20.03.2014.

Sahu, H.B and Er. S. Dash 2011: Land Degradation due to Mining in India and its Mitigation Measures. 2nd International Conference on Science and Technology. IPCBEE (6): 132-136.

Tiwary, Ramakant K. 2000: Environmental Impact of Coal Mining on Water Regime and its Environment. *Water, Air and Soil Pollution* (132): 185-199.

World Bank Group 1998: Copper Smelting. Pollution Prevention and Abatement Handbook. <www1.ifc.org> accessed on 20.08.12.

4. Coal Mining and Climate Change

Kathrin von Allmen

Between 1970 and 2004, the global surface temperature increased by 0.2 to 2 degrees for most regions on Earth (IPCC 2007: 32). This dramatic increase is primarily due to industrialization and the increasing CO_2 emissions[1] that accompany this process. Among the total use of fossil fuels, coal accounts for the biggest share. Compared to gas, the burning of coal generates twice the amount of CO_2 emissions, although the share of gas and coal in the world energy supply is comparable. Of the total CO_2 emissions generated by the world's primary energy suppliers, coal alone is responsible for 40% of the output (Peterson/Quadrelli 2007: 5940). In addition to the burning of coal for energy generation, the mining of coal itself contributes to climate change because extraction, processing and transport require a remarkable amount of fossil energy. Some impacts like deforestation or glacier blasting directly result from the need for making land available for mines. Other environmental impacts include pollution of water, soil and air through coal particles and acid rain (Tiwary 2001; Tsosie 2007).

Ironically, the mining industry is itself vulnerable to climate change related problems, such as water and energy shortages or risks related to flooding and worker health. They are thus taking measures to adapt to climate change (see, e.g., ICMM 2013; Mining Association of Canada 2014). However, a reduction of coal production seems not viable economically. Coal has considerable advantages over other energy resources. The mining techniques are well known, and the stocks are immense. The age of coal does not seem to be over: its global consumption has increased from 5.3 billion tons in 1990 to 8.2 billion tons in 2012, and is projected to reach 9 billion tons by 2019.[2]

Global climate change leads to more extreme weather events like storms, heat-waves, floods, fires and droughts. This contributes to a higher number of diseases, injuries, deaths and a deterioration of agriculture and infrastructure (Bell 2011: 99). Climate change affects everyone, but it does not affect everyone *equally*. People living in areas that are more vulnerable to extreme weather events are particularly affected; for example, in coastline-areas, around the Arctic, or in regions that have long dry and wet seasons. Also indigenous people, foragers and small-scale farmers who depend directly on land, forests and fishing grounds for survival are more exposed to changing climate patterns (Tsosie 2007: 1628). As many of them live in areas where mining and extraction of fossil fuels takes place, they not only suffer disproportionately from global climate change, but they are confronted directly with pollution and other negative impacts. Therefore, climate change effects can be characterized as violating fundamental human rights, such as the right for life and the right for physical security (Bell 2011: 100). In terms of "climate justice", there is a disparity between industrialized and developing nations: The main driver of climate change is the energy consumption of the industrial nations, which themselves do not suffer to the same

1 The parts per million (ppm) of CO_2 have increased from a constant pre-industrial level of 280 ppm to a level of 379 ppm in 2005, accounting for an increase of more than 30 percent. While pre-industrial CO_2 emissions from fossil fuel combustion were almost zero, they increased to 26.6 Gt in 2004 (Peterson/Quadrelli: 5940).

2 U.S. Energy Information Administration: "International Energy Statistics — Coal" <www.eia.gov>, and International Energy Agency: "Global coal demand to reach 9 billion tonnes per year by 2019" <www.iea.org/newsroomandevents/pressreleases> accessed on 24.08.2015.

Fig. 1: Blockade of the railways at the El Cerrejón coal mine, Colombia. Photo: <notiwayuu.blogspot.ch>.

extent from climate change effects as people in less industrialized regions (Moore/Russell 2011: 19-21).

A concrete example of how coal mining, climate change and global justice can play together in building social movements is the case of Repower. This semi-public energy company, under majority-control by the Swiss Canton of Graubünden, was planning the construction of two new coal power plants, one in Brunsbüttel, Germany and one in Saline Joniche, Italy. The projects were confronted with local resistance in both places. On the 27th of August 2011, 36 associations from Switzerland, Germany and Italy joined efforts in a rally against Repower's plans in Chur, the capital city of Graubünden.[3] This also opened up the opportunity to denounce the social and environmental impacts caused by coal extraction in the El Cerrejón mine in Colombia and how the Wayuu indigenous peoples were forced to leave their land in the name of development (see case study Colombia).[4] Repower had the intention to fuel its power plants with imported coal, purchased at the open market and transported by ship to the plants. As the El Cerrejón mine is a major exporter of low-cost thermal coal, it is probable that Repower would have bought coal from there. After the Chur rally, the opposition in Graubünden went the usual "Swiss way" of showing discontent, by launching a popular initiative (on the level of the canton) which demanded a ban on majority public owned companies investing in coal power plants. For once, they had success: In September 2013, a slight majority of just over 50 per-cent voted in favour of the initiative and forced Repower to drop its plans.[5]

3 Including *Zukunft statt Kohle* and *WWF Graubünden*. See <www.zukunftstattkohle.ch> and <www.stromohnekohle.ch>, accessed on 27.09.2013.

4 Karmen Ramirez, co-author of the Colombian case study (this volume), gave a speech at the rally and a newspaper interview. See: *Die Südostschweiz* (29.8.2011): „Kohleabbau zerstört Lebensraum". <www.zukunftstattkohle.ch> accessed on 6.02.2013.

5 Tagesanzeiger (22.9.2013): „124 Stimmen zwingen Repower zur Aufgabe des Kohleprojekts." <www.tagesanzeiger.ch> accessed on 27.09.2013.

The example shows ways to address the complex issue of climate change through creating awareness, mobilizing people and linking specific settings. Among the many factors that contributed to the successful suspension of the two power plants, we would like to highlight two: First, the action strategy was characterized by the involvement and propositions of "frontline communities"[6], i.e., directly affected groups from Brunsbüttel and Saline Joniche, and to a lesser extent from Colombia, which made the abstract issue become concrete for the people in Graubünden. Second, the coalition was able to reach out to a large number of individual supporters thanks to the variety of involved organizations, some of which, such as the WWF, have an important membership base (Moore/Russell 2011: 49). The example further shows the importance of direct democratic instruments, especially where they guarantee a co-determination in issues that touch the "economic sphere", which was given in the case of Repower thanks to its semi-public ownership structure.[7]

Finally, in the whole debate many people in Graubünden felt that it was unfair that the people being directly affected in the (not so far-away) places where the plants were to be built did not have the right to decide on them. It would need more research to evaluate how importantly this ethical argument weighted in the decision. If this feeling could be extended to include the (far-away) places where the actual mining is carried out, it may provide a strong base to address the basic questions of social, environmental and climate justice.

References

Bell, D. 2011: Does Anthropogenic Climate Change Violate Human Rights? *Critical Review of International Social and Political Philosophy* 14(2): 99-124.

Erklärung von Bern / Berne Declaration (EvB/BD) (ed.) 2012: *Rohstoff: Das gefährlichste Geschäft der Schweiz*. Zürich: Salis.

ICMM 2013: Adapting to a changing climate: implications for the mining and metals industry. London: International Council on Mining and Metals (ICMM). Available from <www.icmm.com/document/5173>.

IPCC 2007: Climate Change 2007: Synthesis Report. <www.ipcc.ch/publications_and_data> accessed on 20.3.2013.

Mining Association of Canada 2014: Climate Adaptation in the Canadian Mining Sector. Ottawa: Delphi Group, July 7. Available from <mining.ca>.

Moore, H. and J.K. Russell 2011: Organizing Cools the Planet: Tools and Reflections to Navigate the Climate Crisis. Oakland: PM Press.

Peterson, S. and Quadrelli, R. 2007: The energy-climate challenge: Recent trends in CO2 emissions from fuel combustion. *Energy Policy* 35: 5938-5952.

Tiwary, R.K. 2001: Environmental Impact of Coal Mining on Water Regime and Its Management. *Water, Air and Soil Pollution* 132(1-2): 185-199.

Tsosie, R. 2007: Indigenous People and Environmental Justice: The Impact of Climate Change. *University of Colorado Law Review* 78: 1625-1677.

6 This term is taken from the useful manual *Organizing Cools the Planet* (Moore/Russell 2011), which sums up strategies and tools developed by North American climate action groups and campaigns.

7 See also the case study Val Medel, Switzerland, for an example of a popular vote about gold mining concessions on the communal level (this volume).

5. Impacts of Mining-related Pollutants on Human Health

Jacques Diezi and Günter Baitsch[1]

The environmental impacts of mining processes, introduced in chapter 3, have manifold implications on the health of local populations and mine workers. The contamination of drinking water supplies (surface and groundwater) in residential areas and the exposure of people to high concentrations of sulphur dioxide and heavy metals in the air are common public health risks associated with mining. The deposition of airborne emissions contaminating soils and agricultural products is a further concern, with increased risk of tuberculosis, asthma, chronic bronchitis, gastrointestinal diseases and various forms of cancer (ELAW 2010: 17). Eventually, mining contributes to climate change, which is a global concern for human livelihoods and health (see chapter 4). Nevertheless, the potential health risks of mining operations are often underestimated and under-researched (ELAW 2010; Castleden et al. 2011). For a full picture, not only the toxicity of mining related pollutants need to be considered, but also the social impacts of mining, which can affect physical and mental health. For instance, a study on coal mining affected communities in the USA shows an elevated risk of mental illness such as depression and anxiety, in addition to increased rates of lung-, kidney- and heart-related diseases (Castleden et al. 2011: 333).

The complex interactions and accumulative effects of various sources of illness cannot be satisfyingly addressed here. Even more, a complete health impact assessment would need to be sensitive to socio-cultural issues and non-Western perceptions of, and knowledge about, health and illness. In this chapter, the scope is limited to an introduction of the current knowledge regarding the toxicity of four major pollutants associated with most forms of metal mining: Arsenic, cadmium, lead and mercury. Second, the specific health impacts of uranium mining are looked at. The perspective adopted is one of Western medicine and science. For practical uses, we recommend readers to refer also to the extensive "Mining & Health" assessment toolkit[2] and Conant and Fadem (2008: chapter 21).

1 Jacques Diezi, professor (hon.) at the Department of pharmacology and toxicology of the University of Lausanne, provided the sections on the toxicity of arsenic, cadmium, lead and mercury. Günter Baitsch is a practising physician and contributed the section on uranium mining impacts. Both are board members of the Swiss section of International Physicians for the Prevention of Nuclear War (see <www.ippnw.ch>). The review was done by Peter Diehl (WISE) and Urs Ruegg. Editing and introductory section by Thomas Niederberger with material of Isabella Brom.

2 Coumans/Moodie/Sumi: "Mining & Health: A Community-Centred Health Assessment Toolkit." Nov. 6, 2009. <www.miningwatch.ca/es/node/6394> accessed in June 2015.

5.1 Arsenic

Arsenic (As) toxicity was known by the Romans and other contemporary civilizations in China and India. It has been used extensively across history as an agent of murder ("the poison of kings and the king of poisons"), but its characterization as "a major health concern" for world populations by the World Health Organization (WHO) is relatively recent. Arsenicals have been used in the past as drugs, one compound (As trioxide) is still used in some types of leukaemia.

Sources and exposure: Arsenic (As) is a metalloid widely distributed in rock and soil, leaching naturally into well and spring water. As exists in its elemental form, and in inorganic trivalent (As^{3+}, arsenite) or pentavalent (As^{5+}, arsenate) states. Arsine (AsH_3) is a gaseous hydride of trivalent As with distinct toxicity. Organic As compounds, either tri- or pentavalent, as are formed in animals, including humans, show much lower toxicity. In general, toxicity increases in the following sequence: organic arsenicals < As^{+5} < As^{+3} < Arsine. Mechanisms of As toxicity include enzyme inhibition by covalent binding to sulfhydryl groups (As^{3+}) and interference with cell metabolism (As^{5+}).

Occupational exposure occurs in many industrial processes (alloying agent, processing of glass, pigments, textiles, paper, wood preservatives, ammunition), and high concentrations of As in air can be found in the working and general environment around non-ferrous metal smelters and power plants fired with low-grade coal.

Drinking water and food are the main sources of human exposure to As. Drinking water constitutes the most important threat to public health in countries where ground water contains naturally high levels of As (e.g., several South-American countries, China, India's West Bengal). The situation in Bangladesh has received much attention: there, 45 million people are at risk of being exposed to As concentration in drinking water higher than the WHO recommended limit of 10 µg/L, and 20 million at risk of being exposed to >50 µg/L, the national standard. However, significant progress has been made in reducing As levels in drinking water, and consecutive improvement in skin lesions have been reported recently. A recent analysis of the European Food and Safety Authority (EFSA) of the occurrence of As in food in Europe indicated that 98% of drinking water samples show As levels lower than the 10 µg/L WHO guideline. Inorganic As dietary daily exposure ranged between 0.2 and 2.1 µg/kg body weight (b.w.), resulting mostly from rice, grain-based processed products, dairy products, and drinking water consumption, and was relatively higher in infants.

There are currently no recommended maximum levels of As in food at the European Union level, though some States have guidelines. The Joint Expert Committee on Food Additives (JECFA) has recently withdrawn its former Provisional Tolerable Weekly Intake of 15 µg/kg b.w., considering the very small margin between this limit and the occurrence of the first toxic effects.

Toxicokinetics: Water soluble forms of inorganic As are well absorbed (ca. 90%) by the gastro-intestinal tract. As is then distributed in the whole organism (crossing the placenta and the blood-brain barrier, accumulating in nails and hair), and metabolised to a large extent by methylation and excreted in urine. Methylation of As^{3+} increases the toxicity of the compound. Inhaled As (mostly inorganic As trioxide) is also absorbed in the lung, and As compounds can be absorbed through the skin. Excretion half-life ranges from one to three days.

Human toxicology: All forms of inorganic As exhibit similar toxic properties, except arsine gas. A large range of toxic effects are elicited by inorganic As, particularly in humans. Acute exposure to high doses of As (70-180 mg) may be fatal, by a toxic effect on the heart and the gastrointestinal tract. Later manifestations of exposure to high doses show toxicity to the nervous system and the blood-forming organs. Inhalation of arsine gas in industrial surroundings causes haemolysis, and has been fatal in 25% of reported human cases. *Chronic toxicity* concerns several organs and systems, mostly the skin, the cardio-vascular system, the gastro-intestinal tract and the central and peripheral nervous systems. The skin lesions occur early and are often diagnostic of As poisoning. Usually, abnormal skin pigmentation appears first (6 months to 3 years after beginning of the exposure, typically resulting from high doses in drinking water), followed by palmar and plantar hyperkeratosis after several years. Toxicity of inorganic As to the cardiovascular system include alterations in the cardiac rhythm (arrythmias) and ischemic heart disease. Peripheral vascular disease has been observed in populations exposed to large doses of As in drinking water, particularly in Taiwan, where a high incidence of severe feet lesions leading to gangrene ("blackfoot disease") has been reported. Gastrointestinal dysfunctions follow acute or sub-acute exposure to high doses of As, and may progress to intestinal haemorrhages and death. Peripheral neuropathy, involving both motor and sensitive nerves, can be a consequence of repeated exposure to low levels of As. Loss of sensitivity in hands and feet is often followed by muscular weakness. Many additional toxicities of As have been described, such as anaemia and decrease of blood leucocytes, inhibition of haemoglobin synthesis, immunotoxicity, lesions of renal blood vessels, developmental effects, diabetes.

Carcinogenicity: The International Agency for Research on Cancer (IARC, WHO) concludes that there is sufficient evidence in humans of the carcinogenicity of mixed exposure to inorganic arsenic compounds, which cause cancer of the lung, the urinary bladder and the skin. A positive correlation has also been found between As and inorganic As compounds and cancer of the kidney, liver and prostate. Arsenic and inorganic arsenic compounds are thus carcinogenic to humans (group I).

Treatment: Besides monitoring of fluid levels to prevent fatal hypovolemia, administration of chelating agents is effective in the therapy of acute As poisoning, but this shows little or no benefit in chronic poisoning.

Reference value: 0.01 mg/L in drinking water (WHO, EU, Switzerland).

5.2 Cadmium

Cadmium was discovered only in the early 19th century, and is used on an industrial scale since the 1930s. The knowledge about its toxicity is thus more recent than for arsenic, lead and mercury.

Sources and exposure: Cadmium (Cd) is a transition metal, associated with zinc, lead and copper ores and produced as a by-product of metal smelting. It is mostly used for production of batteries and, in smaller amounts, as a pigment for paints and plastics and in electroplating and galvanizing alloys. The oxidation state of all Cd compounds is +2; Cd^{2+} does not undergo oxido-reduction reactions, and does not form organo-metallic complexes of toxicological significance. Cd^{2+} contamination in soil, water and air originates from mining and refining, production and application of phosphate fertilizers, industrial emissions, fossil fuel combustion, and waste incineration. It disperses in the environment through air, soil and water. The mobility in soil depends on its composition, since Cd^{2+} generally binds strongly to organic matter, but can be taken up by plants and thus enter the food chain. Soluble Cd^{2+} salts (e.g., chloride, sulphate) migrate in water, whereas insoluble forms will absorb to sediments. Environmental contamination results in slow but steady accumulation in vegetables. High levels of Cd^{2+} are also found in shellfish and in liver and kidneys of animals. Tobacco smoking is an important non-occupational source of Cd^{2+} intake. It has been estimated that daily exposure to Cd^{2+} from all sources in Europe and the USA amounts to 10-30 μg per day. Occupational exposure to Cd^{2+} results mostly from inhalation (dust and fumes), and is found in relation with lead, zinc and iron ore processing, cement, fossil fuel and paint pigment production, as well as in the Ni-Cd battery and electroplating industries.

Toxicokinetics: Cd^{2+} does hardly cross the skin. Through inhalation, about 5 to 60% of inhaled Cd^{2+} enters into blood. By gastrointestinal absorption, about 5 to 10% of a given dose is absorbed in the gastrointestinal tract. Absorption is enhanced by diets deficient in calcium or iron, or low in protein. Women of childbearing age show higher Cd^{2+} blood levels, as a likely consequence of lower iron stores and enhanced intestinal Cd^{2+} absorption. Cd^{2+} accumulates in the body and is very poorly excreted. After absorption, it binds to high molecular weight proteins in blood, and is taken up by various tissues, mostly liver and kidneys (about 50-75% of retained Cd^{2+} is found in these organs). Blood levels of Cd^{2+} in non-occupationally exposed, non-smoker populations are usually <1 µg/L. Cd^{2+} accumulates in the placenta and, at high levels of exposure, can lead to foetal alterations, as has been shown in animal experiments. Cd^{2+} concentrations in breast milk average 5-10% of the mother's blood levels. The biological half-life of Cd^{2+} in the human body has been estimated to range between 10 and 30 years.

Human toxicology: Acute toxicity from Cd^{2+} is rare. High doses by the oral route cause vomiting and other gastrointestinal symptoms. Acute inhalation of Cd^{2+} fumes may produce pneumonitis and pulmonary oedema, and may be lethal at very high doses. Toxicity consecutive to long-term exposure to low doses of Cd^{2+} by oral or inhalation routes include 1) nephrotoxicity, 2) skeletal alterations (osteoporosis), 3) cardiovascular diseases, 4) chronic pulmonary diseases following respiratory exposure.

1) *Nephrotoxicity* results from toxic effects of Cd^{2+} on the proximal tubular cells and on the glomeruli. Cell necrosis, consecutive interstitial inflammation and subsequent fibrosis lead to chronic nephropathy and functional impairment. Nephropathy is related to Cd^{2+} accumulation in the renal tissue. Proximal tubule toxicity of Cd^{2+} is associated with reduced tubular reabsorption and increased urinary excretion of several ultrafiltered solutes, which are used as biomarkers of initial renal toxic effects of Cd^{2+}. Increased excretion of low molecular weight proteins has been reported in workers who were occupationally exposed to Cd^{2+} and populations living in highly polluted environments. At higher exposure levels, development of renal insufficiency has been observed, requiring dialysis or renal transplantation, and is associated with increased mortality. A higher frequency of renal stone formation has also been reported in workers exposed to Cd^{2+}.

2) *Skeletal effects*: Major skeletal alterations can appear following chronic Cd^{2+} exposure, mainly as a consequence of renal dysfunction. Painful and debilitating bone disease ("Itai-Itai disease") was observed in the late 1940s in elderly Japanese women exposed to high intake of Cd^{2+} from polluted water and food (rice), and with poor nutritional conditions. Even at lower exposure levels from environmental Cd^{2+} sources, increased bone alterations (osteomalacia, osteoporosis), height reduction and bone fractures have been reported in human populations. Several mechanisms have been proposed to explain the bone lesions, in particular renal calcium loss, alterations of bone formation by direct Cd^{2+} toxicity, and reduced activation of vitamin D by renal cells.

3) *Cardiovascular disease*: Some epidemiological studies have suggested that Cd^{2+} may be an etiologic agent for cardiovascular diseases, but firm evidence is still lacking.

4) *Chronic pulmonary disease*: Inhalation of Cd^{2+} via fumes or dust entails pulmonary toxicity, the severity of which is positively correlated with doses and duration of exposure. Initial chronic bronchitis may progress to pulmonary fibrosis and related functional impairments (dyspnea, emphysema). Acute inhalation of high doses Cd^{2+} and chronic respiratory exposure at lower levels have both been reported to increase mortality in occupationally exposed workers.

Carcinogenicity: Cd^{2+} and Cd^{2+} containing compounds are classified as human carcinogens (group I) by the IARC (WHO) and the National Toxicology Program (NTP, US Government). Conclusive evidence is based on epidemiological and mechanistic studies. Cd^{2+} and Cd^{2+} containing compounds cause cancer of the lung, and positive associations have been found with cancers of the kidney and of the prostate. Potential mechanisms of carcinogenicity include disturbances of DNA repair and tumour suppressor proteins, leading to chromosomal damage and genomic instability. Cd^{2+} is only weakly mutagenic.

Treatment of Cd^{2+} poisoning is symptomatic. Chelation therapy is ineffective, and may result in adverse effects.

Reference values: Chronic exposure to Cd^{2+} by inhalation: 0.01 µg/m^3 (Minimal Risk Level, ATSDR, USA) to 0.3 µg/m3 (*Valeur Toxicologique de Référence*, ANSES, France);

Chronic exposure to Cd^{2+} by food: 0.1 μg/kg b.w. per day (Minimal Risk Level, ATSDR, USA) to 2.5 μg/kg b.w. per week (Tolerable Weekly Intake, EFSA, EU); Tap water, max. Cd^{2+} concentration: 3 μg/L water (WHO) to 5 μg/L water (EFSA, EU, Switzerland).

5.3 Lead

Lead (Pb) has been used by humans for several millenaries and its toxicity had been already identified in Antiquity. It remains an important threat to human and environmental health, and is one of the ten chemicals of major public health concern (WHO 2010).

Sources and exposure: Besides its metallic form, Pb exists as divalent, inorganic tetravalent and organic tetravalent forms. The divalent cation is the primary environmental form, whereas the tetravalent organic compounds have been used as gasoline additive (tetraethyl lead). Commercially traded lead is obtained either from mined lead ore or from recycled scrap lead products. Sources of exposure for populations and the environment are numerous. Large quantities of Pb are used, e.g. in lead-acid batteries, cable sheeting, ammunition, lead pipes and solders. Various inorganic Pb compounds are used as pigments in paints, as additives in ceramic glazes and glass as well as plastic stabilizers. Though general exposure of populations to Pb has decreased following removal of leaded gasoline additive, regulatory limitation to Pb use in paints and replacement of water pipes, total Pb industrial use has increased by nearly 50% between 1960 and 1990 in six major consuming countries. Occupational exposure to Pb, occurring mostly by inhalation of dust and fumes, has decreased following more protective regulations. The general population is exposed to Pb mostly through food and drinking water, though many other sources have been identified, such as old paints, contaminated household dust and soil around old houses.

Toxicokinetics: Gastro-intestinal Pb absorption averages 5-15% in adults and 40-45% in children. More than 90% of inhaled Pb (dust particles) may be absorbed by the lung. Intestinal absorption is markedly enhanced by fasting and by diets deficient in calcium, iron, zinc and manganese, particularly in children. In blood, Pb is primarily (99%) bound in red blood cells. Initially, Pb distributes to soft tissues, mostly the kidneys and the liver. Over time, it is redistributed to bone, teeth and hair. Excretory half-life from blood approximates 30 days. Bone Pb half-life in humans ranges between 20 to 30 years and, therefore, bone is the major deposit of Pb. In fact, about 94% of the total body burden of Pb in adults is found in the bones, a larger value than in children (73%). Pb released from bones contributes to maintain Pb blood levels after the end of exposure, and constitutes an endogenous source of exposure. Pb bone release is increased in women with bone resorption from pregnancy, lactation, menopause and osteoporosis. Pb readily crosses the placenta, and Pb levels in the cord blood are close and proportional to maternal Pb blood levels. Pb is found in breast milk, but breast milk/maternal blood concentration ratios are generally < 0.1. Plasma Pb is excreted mostly in urine by glomerular ultrafiltration, but only the very small, unbound fraction of circulating Pb is ultrafiltered by renal glomeruli. A small proportion of Pb is excreted by the bile.

Human toxicology: Pb induces a wide range of toxic effects, depending on the level and duration of exposure. The main targets of toxicity include 1) the central nervous system (children are particularly sensitive), 2) the cardiovascular and renal systems, 3) the reproductive organs and developmental patterns, and 4) there are haematological effects. Several additional Pb induced toxic effects have been reported, notably with regard to endocrine organs (reduced thyroid and testicular hormones), the gastro-intestinal tract (constipation, lead colic at high Pb levels) and, possibly, the immune system. The toxic effects of very low levels of exposure on human populations have been better recognized and delineated in the last two decades.

1) *Neurotoxicity:* The main target of Pb is the nervous system, both the central and the peripheral one. Severe poisoning in children (blood Pb levels > 70 μg/dl) results in lead encephalopathy, characterized by a number of functional alterations of the central nervous system. Mortality of

Pb encephalopathy is around 25%, and most survivors suffer from long-term sequelae (seizures, cognitive deficits). Pb encephalopathy occurs also in adults, but at higher blood levels. Neurotoxic effects can occur, most notably in developing brains of children, at much lower levels of exposure to Pb than those in adults, inducing global encephalopathy, as shown by a number of recent studies. Pb related alterations in brain development in children result in deficits in cognitive skills, decreased IQ and poor schooling performances. Neurotoxic effects of Pb at low doses in adults include the development of essential tremor (5-10 μg/dl), decreased cognitive function, psychiatric effects and hearing loss (limited evidence at 10 μg/dl). "Lead palsy", a neuromuscular alteration, has been described in conditions of occupational exposure to Pb.

The NTP (USA) concluded that there is sufficient evidence that blood Pb levels < 5 μg/dl are associated with adverse neurological effects in children. Both the EFSA (EU) and the JECFA concluded that a threshold dose for the key effects of Pb could not be established, while the French *Agence Nationale de Sécurité Sanitaire* (ANSES) proposed a "critical Pb level" at 1.5 μg/dl, which would be the limit for all toxic Pb effects.

2) *Renal and cardiovascular toxicity:* Cardiovascular and renal alterations are known to result from Pb exposure. Their occurrence has been recently correlated with lower Pb blood (or bone) levels than previously considered. Thus, the NTP concludes that, in adults, a risk of arterial hypertension and of decreased renal glomerular ultrafiltration is established for Pb blood levels < 10 μg/l and < 5 μg/l, respectively. Cardiac mortality and ECG changes might also be increased at Pb blood levels < 10 μg/dl.

3) *Reproductive and developmental toxicity:* According to the NTP, there is sufficient evidence to conclude that delayed puberty and reduced post-natal growth due to Pb exposure may occur in children at blood levels < 10 μg/dl. Reduced foetal growth has been convincingly observed at maternal blood levels < 5 μg/dl, while at higher concentrations (> 15-20 μg/dl) in adult men, adverse changes in sperm parameters and delayed time of conception have been reported.

4) *Haematological effects:* Pb has long been known to alter the haematological system, inducing hypochromic anaemia—by inhibition of haemoglobin synthesis and reduction of red blood cell lifespan. Inhibition of one of the haeme synthetizing enzymes (delta-aminolevulininic acid dehydratase, ALAD) occurs already at very low Pb levels in children and adults, with no apparent threshold, and appears the most sensitive indicator of Pb exposure. Threshold blood Pb levels associated with decreased haemoglobin levels (anaemia) have been estimated at 50 and 40 μg/dl in adults and children, respectively.

Carcinogenicity: The IARC concluded that there is limited evidence in humans for the carcinogenicity of inorganic Pb compounds, but sufficient evidence in experimental animals for the carcinogenicity of these compounds. Inorganic Pb compounds are therefore considered as probably carcinogenic to humans (group 2A).

Treatment: Administration of chelating agents is warranted for adults and children with high blood Pb levels and/or acute symptoms of Pb poisoning. However, chelators do not reduce the chronic effects of Pb beyond the benefits of blood Pb decrease alone, and, in children, have no benefit on Pb induced learning and behaviour alterations.

Primary prevention: Further efforts aiming at a primary prevention of exposure to Pb, particularly for children and women of childbearing age, are required. Interventions include eliminating non-essential uses of Pb (e.g., Pb in paint), safe recycling of Pb containing waste, educating the public about the importance of safe disposal of acid-Pb batteries and computers, and appropriately monitoring blood Pb levels in children, women of childbearing age and workers (WHO 2010).

Reference values: The former "Provisional Tolerable Weekly Intake" (PTWI) of 25 μg/kg b.w. set by the Joint Expert Committee on Food Additives is no longer appropriate, considering the recent evidence of an absence of a threshold for a number of critical toxic end points. The JECFA and EFSA recently concluded that it is currently not possible to establish a new PTWI that would not affect health. For drinking water, a common lead limit in use is 0.01 mg/L (WHO, EU, CH).

5.4 Mercury

Mercury (Hg) has been used industrially since many centuries, and its toxicity already recognized by Hippocrates (5th century BC). It has been also extensively used as a therapeutic agent. Its deleterious effects on health and environment have been well recognized nowadays, and exposure to Hg is currently considered by WHO as a major public health concern.

Sources and exposure: Mercury exists in different forms: elemental (Hg^0), inorganic (mercurous Hg_2^{2+} and mercuric Hg^{2+} cations) and organic (e.g., methylmercury: $MeHg^+$). Hg^0 vapours in the atmosphere originate in part from natural degassing of the earth's crust and volcanic activities, but a large proportion results from industrial activities (fossil fuel combustion, mining, smelting, solid waste incineration). Hg^0 may then be oxidized to inorganic mercury (Hg^{2+}) and returned to the earth's surface in rainwater. Hg^{2+} may then be methylated by micro-organisms in aquatic environments, and lipid-soluble organic Hg enters into the aquatic food chain, from plankton to fish, with high bio-accumulation factors. Human populations may be exposed to Hg through air, water and food. Atmospheric air and drinking water levels are usually far below toxic limits. Human exposure to organic Hg originates mostly from diets rich in fish, shellfish or marine mammals. The main source of exposure to inorganic Hg is dental amalgam, releasing Hg^0 which is inhaled and converted to inorganic Hg after lung absorption.

Safety limits for dietary exposure to Hg have been established by the Joint Expert Committee on Food Additives. Provisional Tolerable Weekly Intake (PTWI) for $MeHg^+$ has been set at 1.6 µg/kg b.w. and for inorganic Hg at 4 µg/kg b.w. A study carried out by the EFSA on dietary Hg exposure in European countries has concluded that exposure of toddlers, children in some surveys and high consumers of fish may notably (up to six-fold) exceed the PTWI for $MeHg^+$, while dietary exposure to inorganic Hg did not exceed the PTWI (though contribution of dental amalgam may add to the total body burden of inorganic Hg beyond the PTWI). Occupational exposure to Hg occurs mainly by inhalation of elemental Hg, as in chloralkali industry, in the manufacture of scientific and medical devices, in dentistry (Hg amalgams for tooth restoration), in mercury mining and in the extraction of gold from ore concentrates.

Toxicokinetics: Hg^0 vapour is largely (ca. 80%) absorbed in the lungs, and oxidized to inorganic Hg in tissues. Hg^0 crosses the blood brain barrier and the placenta, in contrast with inorganic Hg. Inorganic Hg is excreted in urine and faeces with a half-life of about one to two months. Inorganic Hg is poorly (5-15%) absorbed in the intestine. Accumulation of absorbed Hg^{2+} occurs mainly in the kidney, a major target of toxicity. Orally ingested $MeHg^+$ is almost completely absorbed in the intestine. It distributes to all tissues, including the brain and the placenta. It accumulates in hair, and is eliminated mainly by the faeces with a half-life of approximately two months.

Human toxicology: The main toxicity targets are the kidney for inorganic Hg, and the nervous system for $MeHg^+$. Inhaled Hg^0 distributes to the brain before oxidation to inorganic Hg, and may elicit neuro- and nephrotoxicity. The mechanisms of Hg cellular toxicity remain in part unknown, though Hg binding to sulfhydryl groups of proteins, cell oxidative stress, impaired calcium homeostasis, effects on neurotransmitters are among the mechanisms which may explain Hg toxicity.

- *Metallic Hg*: inhaled Hg^0 vapour in high doses is acutely toxic to the lung (pneumonitis, decreased respiratory function) and to the brain. Chronic exposure to Hg^0 causes central nervous system toxicity (hand tremor, emotional and cognitive disturbances, memory loss), with symptoms becoming irreversible after prolonged exposure. Gingivitis, cardiovascular and renal alterations are also consecutive to Hg^0 inhalation, particularly after occupational exposure.
- *Inorganic salts of Hg*: after oral intake, they are extremely irritant to the gastrointestinal tract (diarrhoea, intestinal bleeding and necrosis). The main target of toxicity is the kidney, where inorganic Hg salts induce proximal tubule necrosis and acute renal failure. Chronic exposure may lead to immunological lesions of the glomeruli.

- *Organic Hg:* the toxicity target of methylmercury is the central nervous system. As shown by two episodes of severe poisoning by $MeHg^+$ in humans in Japan ("Minamata disease", due to contaminated fish consumption) and in Iraq (accidental consumption of grains treated with $MeHg^+$), symptoms of human $MeHg^+$ poisoning include visual disturbances, trouble in motor and sensory functions, blurring of speech, cognitive deficits and, following severe exposure, paralysis and death. The developing brain (foetus, children) shows increased sensitivity to the toxic effects of $MeHg^+$, and severe foetal neurological alterations can occur in the uterus even in the absence of symptoms in the mother.

Carcinogenicity: The IARC (1993) concluded that methylmercury compounds are possibly carcinogenic to humans (group 2B), while metallic and inorganic Hg compounds are not classifiable as to their carcinogenicity to humans (group 3).

Treatment of acute poisoning with inorganic and elemental Hg aims at reducing the local and general exposure to Hg by administration of chelating agents, and require specific interventions in severe cases of intestinal necrosis or acute renal failure. Chelation therapy does not provide clinical benefits in case of organic Hg poisoning.

Prevention: The WHO underscores the need to reduce or eliminate release of Hg and its compounds into the environment. The use of Hg-free alternatives for various scientific and medical instruments, safe waste-handling and clean-up procedures, development of policies and regulations on mercury are some of the needed undertakings. Recommendations from various health authorities have been directed at women who may become pregnant, pregnant women, nursing mothers and young children to avoid eating fish with high $MeHg^+$ content.

Reference values: Total Hg in drinking water: 1 µg/L (WHO); Provisional tolerable weekly intake (food): methylmercury: 1.3 µg/kg b.w. (EFSA), to 1.6 µg/kg b.w. (JECFA); inorganic mercury: 4 µg/kg body weight (JECFA, EFSA); Hg in air: 1 µg/m^3 (annual average) (WHO), 0.2 µg/m^3 for long-term inhalation exposure to Hg^0 vapours (WHO).

5.5 Impacts of Uranium Mining on Human Health

Uranium mining has complex consequences on the health of local populations. Carried out on all five continents, the main victims are often indigenous people. The contamination of the environment affects the population and, particularly, the children via water, airborne particles and food, be it due to ionizing radiation, chemicals or heavy metals. Mine workers are frequently affected by lung and other carcinomas, pneumoconiosis, and various other diseases. In this chapter, the major impacts of uranium mining on human health are summarized. Additionally, impacts on the non-human environment and the social well-being of the affected populations need to be considered, which is done in more detail in the respective case studies.[3]

Uranium[4] is at the beginning of the nuclear chain that starts with exploration and mining and ends with the use of uranium in nuclear power plants or weapons. At every step of this nuclear chain, there are significant health and social consequences, resulting both from the ordinary operation as well as from accidents of different scales. Furthermore, there are associated prob-

3 See chapters Australia and Niger. Note that uranium contamination can also occur in other types of mining, if the mineral ore contains uranium that is liberated as a by-product or waste. See, for e.g., South Africa's Witwatersrand gold fields: Oliver Balch: "Radioactive city: how Johannesburg's townships are paying for its mining past", *The Guardian* 6 July 2015; and the upcoming documentary film "Legacy Warnings!" <www.ujuzi.de/legacywarnings>.

4 Natural uranium ore contains 0.7% of the fissionable 235U isotope and 99.3% of the less radioactive 238U isotope. 235U has to be enriched to 2-3% for nuclear power plants and to 90% for nuclear weapons. What is left behind after this refinement is depleted 238U.

Fig. 1: The Rössing uranium mine near Swakopmund, Namibia, in 2009. Photo: "Ikiwaner", *Wikimedia Commons*.

lems such as the use of depleted ^{238}U in anti-tank weapons, the intermediate storage as well as the unresolved decommissioning and disposal of nuclear waste and power plants, and finally, the danger of nuclear war.

Although uranium has been mined for more than 45 years, there is a lack of reliable long-term studies regarding health; i.e., studies that show a strong cause-effect relationship between exposure type and frequency of diseases, and that have been reproduced at least qualitatively. A major reason for this lack of information is the fact that mining companies are not interested in and even refuse participation in epidemiological studies. Furthermore, about 70-80% of the mining areas are located in homelands of indigenous people, who are politically marginalized and whose traditional health knowledge is ignored by Western medical research. At the same time, hospitals and other Western health care services that could produce relevant data are often absent or lack funding. But even though not for all impacts described in the following there is sufficient proof available according to Western scientific criteria, there is nevertheless a high probability for cause-effect relationships because of the observed accumulation of similar changes in all mining areas.

Three main methods for uranium mining can be distinguished[5] 1) *underground mining*—this method is becoming rare, but there are still several examples in Africa, mostly in the DR of Congo. The underground Wismut mine in Eastern Germany (especially relevant for medical studies) was closed in 1990 after the decay of the Soviet Union; 2) *open-pit mining*—this is the most common procedure. Famous examples are the Rössing mine in Namibia and the Olympic Dam mine in Australia; 3) *in-situ leaching*—hereby, a chemical solution is pumped into the ore body to extract the uranium.

5 See chapter 3 for more details on mining processes.

How to Assess Health Impacts?

The following questions sum up relevant issues for an assessment of health impacts:

- How is the mining performed? Are there adequate measures to prevent exposure to dust; for example, by watering the surroundings during drilling and blasting?
- What happens with the contaminated water? Is it adequately contained in tailing ponds sealed off from the surroundings, or does it seep into water bodies used for irrigation, as fishing pond or as drinking water sources for humans and livestock?
- What happens with the tailings? Do they dry out or form mud lakes? For example, in the Sahara desert, tailings dry out and dust is carried away by sand storms over large distances, while in humid tropic regions tailing ponds may overflow in times of heavy rain.
- Which lubricants and other chemicals are used in drilling and further processing of the ore, and how are they disposed? Is contaminated material reused, e.g. for building and road construction? In Niger for instance, scrap metal from the mine is sold on the market and even used to fabricate cooking equipment.
- Who is employed as workers, and how are the workplace conditions? In India, for example, migrant workers are hired, whereas in China prisoners work for short periods under unknown exposure conditions. Thereby, consequences on workers' health are impossible to assess, as opposed to those on people hired on a long-term basis.

Of particular interest from a medical point of view are: 1) the different effects of uranium as heavy metal ion, on the one hand, and of its radioactive decay products on the other hand (Brugge 2013); 2) combination effects of uranium and ionizing radiation with other heavy metals and arsenic contained in the rock and released through mining, as well as with chemicals used in processing or with other health hazards such as tobacco, alcohol or drug abuse.

Endpoints that must be considered for medical studies are: 1) morbidity—thereby, illnesses caused by previous uranium exposure are often not recognized as such, because of the long time span between exposure and the manifestation of the disease; 2) mortality—here, a cause-effect relationship is relatively easy to establish; 3) general life expectancy, child mortality and number of children, in comparison to a non-exposed population under similar living conditions.[6]

A critical question for the assessment of impacts is if regular and systematic measurements of exposure of local people to ionizing radiation are made. Of note is that this is often done only initially, but later-on there are only estimations available based on occasional measurements at the working place or in the environment.

Established Impacts on Human Health

Local residents living around uranium mines and processing plants risk illness due to radioactive dust, contaminated drinking water, food, construction material and others, as well as transport of uranium-containing material through residential areas. Observational studies describe defects and miscarriages, disabled children, increased child mortality, cerebrovascular and cardiovascular disease, type 1 diabetes, depression, suicide, leukaemia and other carcinomas as common problems. For mine workers, lung silicosis due to dust exposure is the most common disease, except when leaching is used. Although many observational studies are not adequately supported by validated, reproduced studies, a number of well-established health effects have been identified, both for uranium as a heavy metal and as a source for radiation:[7]

6 Also, the life expectancy of uranium mine workers was compared to that of the general population nearby the mine—whereby the first was shown to be much lower. For the Wismut mine in Germany, there are solid studies on that.

7 Where not indicated otherwise, this section refers to the studies summarized by WISE Uranium Project: "Health Impacts for Uranium Mine and Mill Residents – Science Issues", <www.wise-uranium.org/uhm.html> and "Uranium Mine and Mill Workers – Current Issues" <www.wise-uranium.org/uim.html>, accessed in July 2015.

Uranium as a heavy metal: Encephalopathy; renal failure; high blood pressure; nervous disorders; memory and concentration problems; muscle and joint pain; skin rashes and dermatitis; systemic lupus erythematous (Lu-Fritts 2014); fertility problems in men and women; damage of the genome of male miners; increased risk of illness during pregnancy; harm of the foetus including brain damage or death. Specifically for children, well-established problems are: nervous system and kidney damage; learning disabilities and decreased intelligence; speech, language and behavioural problems; poor muscle coordination; decreased muscle and bone growth; hearing damage.

Uranium as a source of ionizing radiation: Lung cancer in mining workers, who are exposed to radon, is undoubtedly an accepted disease that might lead to premature death (Brugge 2013). Radium causes bone cancer, cancer of the sinuses and mastoid-air cells and leukaemia (Brugge 2013). Genetic damage that can lead to many types of cancer have been described; quite common is nasopharyngeal carcinoma. As a leading cause of death, however, they are not significant. For the well documented Wismut mine, which employed more than 80,000 workers at its peak in the 1970s, 7200 cases have been recognized as occupational disease due to uranium until 1999. Until 2003, 3000 people died of lung cancer. More than 25 years after the closure of the mine in 1990, more than 300 deaths occur per year, mostly from lung cancer or leukaemia (Möhner 2006).

Conclusion: Leave it in the Ground

Uranium mining is at the beginning of a nuclear chain that can lead to serious consequences on health at all steps. Currently, the measures taken to address contamination risks in the mine operation concern "secondary prevention": in particular, the conditions of mining should be ameliorated, e.g., by aeration of sites—to remove radon gas—and by using water when blasting or drilling—to reduce emerging dust containing radioactive isotopes and heavy metal ions, called "hot particles". However, given the limitations of such measures, and the risks involved in further steps of the nuclear chain, from a medical point of view the most adequate and efficient "primary prevention" is clearly to leave the uranium in the ground.

References

General

Castleden, William M., David Shearman, George Crisp and Philip Finch 2011: The mining and burning of coal: Effects on health and the environment. *Medical Journal of Australia* 195(6): 333-335.

Conant, Jeff and Pam Fadem 2008: A Community Guide to Environmental Health. Berkeley, CA: Hesperian Health Guides. Available from <www.hesperian.org/publications_download_EHB.php>

ELAW, Environmental Law Association Worldwide 2010: Guidebook for Evaluating Mining Project EIAs <www.elaw.org> accessed on 17.08.12.

Arsenic

Agency for Toxic Substances and Disease Registry (ATSDR) 2007 (updated): Toxicological Profile for Arsenic. US Department of Health and Human Services.

Casarett Louis J. and John Doull 2013: Toxicology: The Basic Science of Poisons, 8th edition. New York: McGraw-Hill, 986-989.

European Food Safety Authority (EFSA) 2014: Dietary exposure to inorganic arsenic in the European population. *EFSA Journal* (12): 3597.

International Agency for Research on Cancer (IARC) 2012: Arsenic, metals, fibres and dusts. *IARC Monographs* (100C): 41-93.

Joint FAO/WHO Expert Committee on Food Additives (JECFA) 2011: Arsenic (addendum). *WHO Food Additives Series* (63): 153-316.

World Health Organization (WHO) 2012: Arsenic. *WHO Factsheet* 372.

Cadmium

Agence Nationale de Sécurité Sanitaire (ANSES) 2012: *Valeur toxicologique de référence pour le cadmium et ses composés.*

Agency for Toxic Substances and Disease Registry 2008: Toxicological profile for cadmium. U.S. Department of Health and Human Services.

Casarett Louis J. and John Doull 2013: Toxicology: The Basic Science of Poisons, 8th edition. New York: McGraw-Hill, 990-993.

European Food Safety Authority (EFSA) 2011: Statement on tolerable weekly intake for cadmium. *EFSA Journal* (9): 1975.

International Agency for Research on Cancer (IARC) 2012: Arsenic, metals, fibres and dusts. *IACR Monographs* (100C): 121-145.

National Toxicology Program (NTP) 2014: Cadmium and Cadmium Compounds. Report on Carcinogens, 13th edition. U.S. Department of Health and Human Services, Public Health Service.

Lead

Agence Nationale de Sécurité Sanitaire (ANSES) 2013: *Expositions au plomb: effets sur la santé associés à des plombémies inférieures à 100 µg/L.*

Agency for Toxic Substances and Disease Registry (ATSDR) 2007: Toxicity Profile for Lead. US Department of Health and Human Services.

Casarett Louis J. and John Doull 2013: Toxicology: The Basic Science of Poisons, 8th edition. New York: McGraw-Hill, 993-996.

European Food Safety Authority (EFSA) 2010: Scientific Opinion on Lead in Food. *EFSA Journal* (8): 1570.

International Agency for Research on Cancer (IARC) 2006: Inorganic and Organic Lead Compounds. *IARC Monographs* Vol. 87.

Joint FAO/WHO Expert Committee on Food Additives (JECFA) 2011: Lead (addendum). *WHO Food Additives Series* (64): 381-497.

National Toxicology Program (NTP) 2012: NTP Monograph on Health Effects of Low-Level Lead. U.S. Department of Health and Human Services, Public Health Service.

World Health Organization (WHO) 2010: Exposure to Lead: a major public health problem.

Mercury

Agency for Toxic Substances and Disease Registry (ATSDR) 1999: Toxicological profile for mercury. US Department of Health and Human Services.

______2013: Addendum to the toxicological profile for mercury (alkyl and dialkyl compounds).

Casarett Louis J. and John Doull 2013: Toxicology: The Basic Science of Poisons, 8th edition. New York: McGraw-Hill, 996-999.

European Food Safety Authority (EFSA) 2012: Scientific opinion on the risk for public health related to the presence of mercury and methylmercury in food. *EFSA Journal* (10): 2985.

International Agency for Research on Cancer (IARC) 1993: Beryllium, cadmium, mercury and exposure in the glass industry. *IARC Monographs on the Evaluation of Carcinogenic Risks to Humans* Vol. 58.

Joint FAO/WHO Expert Committee on Food Additives (JECFA) 2011: Mercury (addendum). *WHO Food Additives Series* (63): 605-684.

World Health Organization (WHO) 2007: Exposure to mercury: a major public health concern.

Uranium

Brugge, D. 2013: Uranium and its decay products radon, radium and thorium. Heavy metal toxicity. Speech delivered on the occasion of the Congress on uranium mining, Dar-es-Salaam, Tanzania, Oct. 2013.

Indian Doctors for Peace and Development (IDPD) 2004: Study on health status of indigenous people around JADUGODA uranium mines in India. Available from <www.idpd.org>.

Lu-Fritts PYet al. 2014: Association of systemic lupus erythematous with uranium exposure in a community living near a uranium-processing plant: A nested case-control study. *Arthritis & Rheumatology* (66): 3105-12.

Möhner et al. 2006: Leukemia and exposure to ionizing radiation among German uranium miners. *American Journal of Industrial Medicine* (49): 238-48.

WISE Uranium Project: "Health Impacts for Uranium Mine and Mill Residents – Science Issues", <www.wise-uranium.org> accessed in July 2015.

PART II
Case Studies

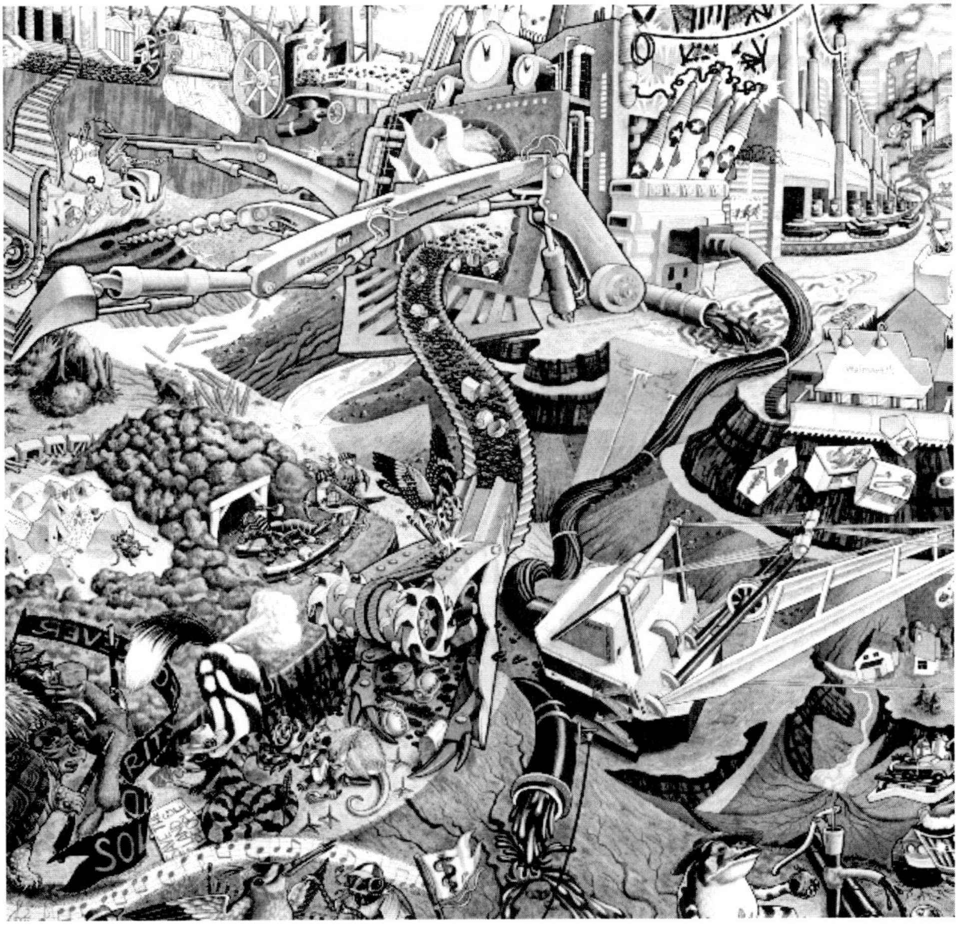

Gold Exploration in Val Medel, Switzerland

Gregor Kaufeisen and Annatina Foppa[1]

Switzerland is generally not considered to be rich in natural resources. Nevertheless, the mining industry enjoys a long history here, and the plans for commercial gold extraction in eastern Switzerland come as a surprise only at first sight. This case study focuses mainly on the discussions among the local population in Val Medel about the impact of such a project on their community, leading to a rejection of the gold mining proposal at first and to the approval of an exploration concession in a second step.

In 2011, SwissGold Exploration Ltd., a subsidiary of the Canadian NV Gold Corporation, requested to renew a previous concession for gold exploration in Val Medel, a valley in the Surselva region of the canton of Grisons. In April 2012, the municipality voted in a referendum against the granting of an exploration and extraction concession for gold (NZZ 2012). The decision was more pronounced than expected: not only that two-thirds of Medel's women and men spoke out in principle against gold mining in the municipality, but also about 8 out of 10 eligible voters participated—twice as many as usually in popular votes.[2] The exact question had been, whether the Medel population wants to give the municipal council the authority to grant companies concessions to commercially mine for gold in the municipality. The answer was a resounding "No", rejecting the "gold fever" or "gold rush" that the media had claimed prior to the vote. However, in a second vote in 2014, a change in the municipal mining law that allows the municipal council to grant exploration concessions without including exploitation found a majority.

The debates about gold mining in Val Medel show concerns about the future development of a valley confronted with structural problems of depopulation and lack of employment, as well as the importance of decision-making procedures.

Politics in Switzerland: "And then the people say Yes or No"[3]

In contrast to representative democratic systems, the system of "semi-direct democracy" in Switzerland is based on a relatively high participation of the population. Important political issues (i.e., changes on the constitutional level) have to be submitted to a popular vote or "referendum" (Linder and Lutz 2006: 109-110). Switzerland is a federal state of 26 cantons which enjoy extensive autonomy and participation in decision-making at the federal level (Vatter 2006: 81). The form of organization, the powers and hence also the autonomy of the municipalities or com-

1 This chapter is based on interviews and participatory observation in Val Medel for a BA thesis, conducted by Gregor Kaufeisen during several visits between 29 August 2012 and 27 February 2013. Annatina Foppa completed the results with additional sources in Romansh. The authors thank Helen Gambon for her support in the completion of the chapter. Translation from German by Carmen Hidalgo Staub. Editing and update by Thomas Niederberger.

2 Canton Grisons. Poll Results. <www.gr.ch/DE/publikationen/abstimmungenwahlen/resultate> accessed on 23.1.2014.

3 If not specified differently, all quotes are from documented conversations with individuals of the Medel population who wish to remain anonymous.

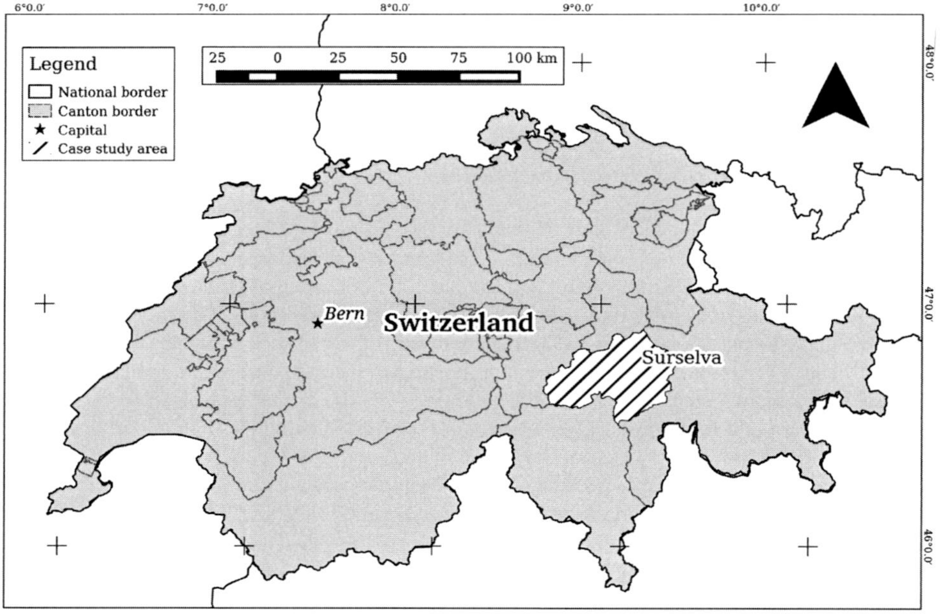

Map 1: Location of the Surselva District in the West of Canton Grisons, Switzerland.

munes ("Gemeinde") are regulated at the level of the cantons (Horber-Papazian 2006: 234). In the canton Grisons (as elsewhere), the municipal assembly is the supreme organ of a municipality and has the authority to determine the organization and distribution of tasks. In part, the municipal assembly meetings have been replaced by the ballot box (Vital 1988: 16-22). This is also true in the municipality Medel: although the municipal assembly, i.e. the congregation of all interested voters, still exists, changes to the municipal constitution have to be approved by a popular vote trough the ballot box (which is anonymous and can be done by letter without attending personally, increasing the participation level considerably). The daily affairs of the administrative and police authority are managed by the municipal council, which consists of the elected mayor plus four individuals (Vischnaunca da Medel /Lucmagn 2012b: 10-12). This political structure of so-called "municipal autonomy" allows for a relatively high codetermination of the population on local issues. Nonetheless, decisions at the municipal level are valid only insofar as they do not break higher (cantonal and federal) laws. There is no mining law at the federal nor at the cantonal level in Grisons, therefore the relevant superior laws to be considered in this case are the Environmental Protection Act, the Nature and Cultural Heritage Protection Act, as well as land use planning regulations (Tiess 2009: 129).

Apart from the formal decision-making, the Medel municipal council has put forth efforts to increase the participation of the population through so-called "future workshops". However, these events were attended rather sparsely. Regarding the gold mining proposal, the process of opinion formation appears to have happened mostly in an informal setting and face-to-face discussions.

The Initial Situation: "It Must Go Forward"

The municipality Medel/Lucmagn is located in the middle of the Surselva, in Val Medel in eastern Switzerland. Val Medel is a tributary valley of the anterior Rhine valley. The municipality covers an area of 135 square kilometres and includes several smaller villages and hamlets. The people speak Romansh, the fourth Swiss national language and mother tongue of less than one percent of the Swiss population (BFS 2013: 8). Over half of the employed population in Val Medel works in agriculture and forestry and only one-third in services.[4] This stands in stark contrast to the overall situation in Switzerland where 7 out of 10 people work in services, and less than one percent in the agricultural sector (BFS 2013: 11). From a geographical, cultural and economic point of view, Val Medel is situated at the margins. The mining issue thereby falls into a wider debate about the development of the "rural periphery" (Pütz et al. 2008) or *strukturschwache Gemeinden* (i.e. "structurally weak communes"). As the maintenance of infrastructure and public services is especially expensive in sparsely populated mountainous regions, this debate is crucial in the Canton of Grisons, which consists exclusively of mountain valleys. Economic liberalization and the pressure to cut public spending have led to a steady decline in public services and associated employment since the mid-1990s. This development is both a result of and further amplifies the depopulation of those regions which do not benefit from a strong tourism sector.

In 2011, the year before the vote on gold mining, the municipality of Medel did not record a single birth or wedding, and the population had declined further from 439 to 426 inhabitants (Vischnaunca da Medel/Lucmagn 2012a: 41). Over the last few years, the postal office, the kindergarden, the secondary school and the home for the elderly were all closed down, and the

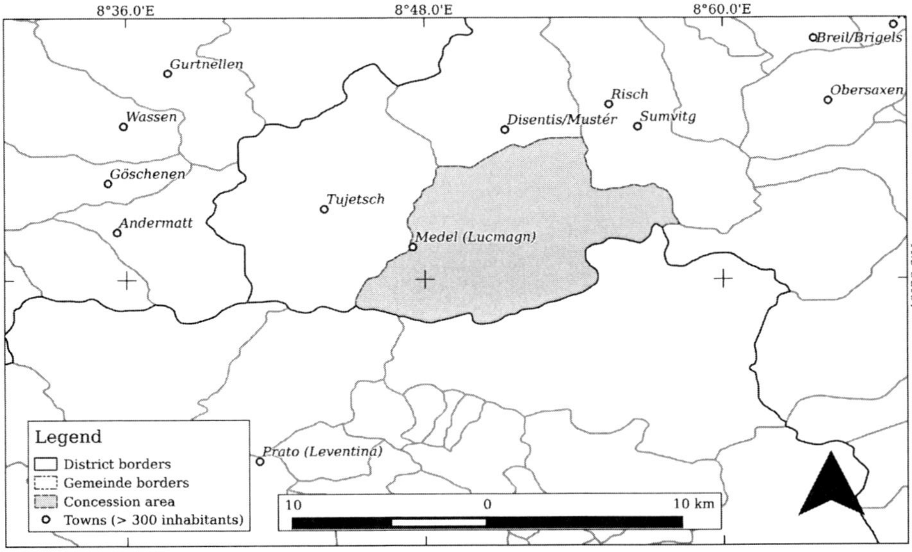

Map 2: The concession area covering the municipality of Medel in the Surselva District.

4 Canton Grisons. Description of the Territory of Medel (Lucmagn). <www.statistik.gr.ch> accessed on 29.9.2013.

Fig. 1: View from Alp Solivas to Medel Valley. Source: Community Administration Medel/Lucmagn.

primary school was expected to follow soon (Schmidt 2014: 78). At the future workshop organized in 2011, the Medel population was asked to reflect on the future of the valley on this background. The participants formulated three objectives: to stop the rural exodus, to create new jobs, and to make the valley more attractive for young families (Val Medel, 2011: 4). Based precisely on the results of this workshop, the municipal council later declared its support for an exploration and possible subsequent mining of gold, as this would create new jobs and promote immigration (Suprastonza da Medel/Lucmagn 2012: 2). However, the main prospects for economic development, personally put forward by the mayor, are considered to be the construction of holiday houses, eco-tourism and related offers for urban clients such as *anti-burn-out* retreats (Schmidt 2014: 79).

The Gold Deposit: "There Have Always Been Investigations"

The search for precious metals has been going in the region on for a long time, as the name Val Medel indicates. "Medel" is based on the Latin word "metallum". In the mid-14th century, the mining of silver was documented in the village of Bleiglanz, which belonged to the manorial estate of the monastery Disentis. Starting in 1366, the Val Medel silver mines were leased to entrepreneurs from Lucerne, Uri, Schwyz and Unterwalden (Bächtiger 1982: 7, Krähenbühl 1986: 14). However, the old galleries are lost since long.

The metal deposits only became a focus of interest again in the 1980s, when the Canadian company Narex International Exploration Inc. first collected samples to examine the rock's gold content (Trueb 1992: 28). In 1986, Narex had acquired an exploration permit covering the area of five municipalities, including Medel. In the following years, the license was divided into two and changed hands several times.[5] The exploration was largely abandoned until 2011, when the SwissGold Exploration Ltd., a subsidiary of the Canadian NV Gold Corporation, took it up again. Though the ownership had changed, the individuals behind the scene were partly the same as those already trying to locate gold-bearing rock in the Surselva since the 1980s. The exploration permits were to expire in 2012, but had until then been renewed by the municipal councils without much notice. Indeed, the neighboring municipalities of Medel granted the company a new five-year exploration permit within their territory at the end of 2012 without the need for a referendum. The situation in Medel was different because the municipality had adopted a new mining law in 2008, which stated that the municipal council may grant an exploration license only if the population also approves the commercial extraction in a popular vote (Vischnaunca da Medel/Lucmagn 2008: 2, Article 4, Section 2). This local particularity was later removed again with a partial revision of the mining law by the municipal assembly in 2014 (see last section).

The differences between the municipalities can be explained by the mentioned lack of higher-level regulations. Neither the federal state nor the canton of Grisons have adopted any specific laws on mining (Schweizer Geotechnische Kommission 1997: 443). Therefore, the mining sovereignty remains with the municipalities which have to formulate their own mining laws. Grisons is the only case where mining is not regulated at the cantonal level, which makes the exceptional story of the Val Medel debate possible.

The Company: "Mining is Like Gambling"

The owner of the Val Medel permit is the NV Gold Corporation, based in Vancouver, Canada and traded at the TSX Venture Exchange, which is popular for junior exploration companies. NV Gold owns another permit in Nevada, USA. According to its website[6], the purpose of the company is the maximization of shareholder value through the exploration of gold and copper deposits. Some people on the board of directors have longtime experience in other exploration companies. SwissGold Exploration was founded in 2011 as a fully owned subsidiary of NV Gold Corporation and consists of only two individuals. Its legal seat is a letter box in Curaglia, the main town of the municipality Medel. According to an excerpt from the commercial register, the purpose of the company is to control the entire production of gold—from exploration to exploitation, up to the sale.

Nevertheless, such exploration companies often hope to be taken over by a big player in the industry as soon as they can demonstrate promising gold deposits. A representative of SwissGold Exploration says in an interview: "mining is like gambling." The suspicion that the interests behind the exploration in Medel are mostly speculative is nurtured by experience with similar "penny stock" explorer companies which run on "[...] the so-called pump-and-dump scam [...] whereby the shares are pumped high with a success story only to be sold off [dumped] profitably" (Stefano in *Handelszeitung* 29.9.2011; own translation).

5 Investigations revealed the following chronology: in 1986, Narex fused with the Canadian Micham Exploration Inc. Their corporate entity was the Société d'Exploration d'Or Helvetique Ltée, with its subsidiary Miniera Val d'Aur SA. Narex and Micham later split the exploration licenses, with Micham taking over the southern part of Val Medel and renaming itself as International Micham Resources Inc. In 1996, after this company went bankrupt, a private individual, who now also sits on the Board of Directors of NV Gold Corporation, took over the Val Medel licenses. As to the other part of the original exploration licenses, the Geneva-registered MinAlp SA received a renewed permit in 2006. In 2009, MinAlp SA was sold to the Canadian Murray Brook Mineral Inc., which later on was disallowed the license.

6 <www.nvgoldcorp.com> accessed 5.12.2013.

The Referendum Campaign: "Something Needs to Be Done"

While the exploration had thus been going on for several years, it became an issue only when the population had to express its position in 2012. With a vote in favour of the exploration it would at the same time agree to an eventual later exploitation, i.e., to the actual gold mining. It is this point that triggered most people to say "No". Only few were against an exploration as such. However, the reservations were primarily directed against possible exploitation and against the regulation that the municipal council, and not the people, would be in the position to grant the final permits. This issue was also the main topic of conversation at an information session in early March 2012, less than one month before the vote in April.[7] This evening session—although "poorly attended" as a representative of the municipal council said—is often the starting point for Val Medel inhabitants recounting when they first learned about the popular vote.

During the event information was provided by the municipal council and a representative of SwissGold Exploration Ltd. The latter held his presentation in English, but this was not the main point of criticism that evening. According to local residents, "the ones of the gold" were unable to tell much. In response to the questions raised on where the tunnels would be dug, how deep they would be, and where the landfill would go, the answer was always: in order to know that, more explorations are needed. In fact, according to the municipal council, it was assumed that the company would invest up to 20 million Swiss francs in the exploration process. Only after five to seven years it would then be possible to decide whether and where gold mining is feasible (Suprastonza da Medel 2012: 4). The population learned at the information session that during investigations in the previous year, 7 out of 86 rock samples had a gold content of more than 10 grams per ton of ore,[8] which surpasses the commercial minimum threshold of 5 grams;[9] "several hundred people" would be employed by a prospective gold mine; the method would combine underground and open-pit mining, but precisely what processes would be used was still undecided.[10]

"I am not completely against it—but with this scarce information..." said an inhabitant who belongs to a group of critical people who joined together to confront the project under the name "Pro Val Medel". The meetings of this loosely organized group were usually held in an informal setting and served to exchange opinions and information, and to discuss possible actions. Besides acquiring knowledge about mining in order to close the gaps that were left open in the information event, Pro Val Medel contacted environmental organizations and distributed a leaflet with twelve questions about the project (for example: "How is our quality of life?") to all households in the municipality. In response, a counter leaflet was distributed by 14 proponents of the project, taking a stance on the concerns of Pro Val Medel, asking i.e.: "What are our chances to slow down the population decline?" The "lively discussion" (Vischnaunca da Medel 2013: 1) surrounding the vote was manifested in these leaflets. A striking feature of both leaflets is their sole focus on Val Medel. Concerns or advantages are expressed only in regard to the own community. At most, a

7 Radiotelevisiun Svizra Rumantscha RTR 2012: Curaglia: In lieu d'aventuriers d'aur? <www.rtr.ch> accessed on 29.9.2013.

8 NV Gold Corporation. Sampling Results from Medel Property In Switzerland Very Encouraging - Include Samples of over 10 G/T Gold. <www.nvgoldcorp.com> accessed on 10.4.2013.

9 This value refers to underground mining. With open-pit mining, a gold content of 1 gram per ton of rock is already feasible today (see NZZ 2012).

10 According to a member of SwissGold Exploration, the flotation method shall be used. This mainly mechanical method to separate gold from ore needs only few chemicals. Considering the many years that would pass until starting the extraction, the progress in mining techniques, as well as uncertainties regarding who would actually undertake it, it is obvious that no certain informations can be given at that point.

Fig. 2: Alp Muschaneras, Medel Valley. Source: Community Administration Medel/Lucmagn.

comparison with a tunnel construction project in a neighboring municipality is drawn, but there are no references to gold mining at the global level. Such comparative thoughts were only articulated in some of the personal interviews.

None of the 11 associations in Medel[11] exposed itself with an opinion on the gold mining project. On the contrary—a proponent said that he had wanted to initiate the discussion in his association, but "nothing came back". Information and opinions seem to circulate in the population primarily through discussions in an informal setting. Internet was consulted as a source of information too. With indignation, a resident discovered a news report from fall 2011 on the Val Medel project, where the mayor of Medel described the process of granting exploration permits as "merely a formal act" (Chapman in *TagesAnzeiger* 6.9.2011). The mayor seemed to be pleased with the progress. The former manager of the international consulting firm PricewaterhouseCoopers is a central figure of the whole story and one of the few who stated his conviction so clearly. However, other supporters of the project retrospectively formulated their approval with more reluctance. One supporter said that he was not "on fire" either, but that he could not reject everything, especially now that finally someone wanted to invest in the valley.

11 Most Medel residents are member of one or more associations (e.g., for hunting, music, farming), which form a central part of social life in rural Switzerland.

The Arguments: "Our Gold is the Beautiful Landscape"

The future of the Medel valley was the most significant concern of both proponents and opponents of the gold mining proposal, but with different tendencies: one side emphasized the population decline, while the other side referred to the protection of nature and landscape. The first saw the 200 to 500 workplaces[12] as an opportunity for the preservation of the community, the others raised the question of how so many immigrant workers could be integrated into the small Romansh village community. Both sides drew a comparison between the project and the large construction site of a new tunnel through the Alps (NEAT) in a neighboring municipality, but interpreted it differently. Often, a comparison was also drawn with the construction of a hydroelectric power plant in the upper part of the Medel valley in the 1960s. Due to the construction, the population rose sharply for nearly a decade but levelled off thereafter at the same level as before (Vischnaunca da Medel/Lucmagn 2013: 42). The tax revenue of the plant today accounts for over a third of the total revenue of the municipality and allows a relatively low communal tax rate. That the whole municipality benefits from the hydro plant was used as a pro-argument for the mining project. The opponents countered this by saying that the dimensions of the hydropower project were clear from the beginning, but with the gold project these were more difficult to assess, also because the price of gold is subject to strong fluctuations.

Overall, in the concerns about the future development of the valley the potential financial benefits seem to have played a minor role compared to arguments for "protecting the homeland". This is also reflected in a general skepticism about the role of "outsiders" or "foreigners": referring to everyone grown up outside the valley, this regarding not only the company with Canadian background, but also the mayor and one of the prominent opponents who are both originally from the Swiss "lowlands". Their commitment was put into question as they also have secondary domiciles outside of Medel and could always move back there. "But we, we live here" said one of the villagers. According to him, the original Medel inhabitants know best what is realistic in their valley and what is not. Skepticism and even dismissal of "outside projects" can be found also with other proposals, like the new national park (see below), or the administrative fusion with the neighboring municipality. In a leaflet, the latter was opposed because "they have little or no relation to our valley". The loss of autonomy creates tensions, and for a part of the population the resentments against "outsiders" or newcomers seem to have grown as a result of the controversy. In one case, the owner of a vacation home declares that he had received verbal threats in response to his commitment against the mining project. At the same time, new relationships and stronger links between original and recent inhabitants emerged from the joint resistance against the project, too.

Three supraregional environmental organizations reacted on the impetus of the group Pro Val Medel with a joint press release shortly before the vote in 2012, in which they expressed a critical position about gold mining in Val Medel (WWF Graubünden 2012). Apart from considerable national media interest stirred up by this, the direct influence of environmental organizations within the community seems to have been low. However, the general concern for landscape and nature as well as the insistence on sustainable and eco-friendly tourism weighed heavily in the arguments. Many people saw a conflict between mining and the maintenance of the natural environment; a tradeoff that a resident summarized as follows: "our gold is the beautiful landscape." Another project, which seems to be diametrically opposed to gold mining, is the new national park "Parc Adula". According to the current proposal, almost the entire territory of the municipality Medel would lay within the so-called buffer zone of Parc Adula, but only a small part within the core

12 After having visited a wolframite mine in Austria, the municipal council later corrected this number down to approximately 70 workers.

Fig. 3: Muraun, Cazirauns and Caschleglia, Medel Valley. Source: Community Administration Medel/Lucmagn.

zone. While the buffer zone is supposed to mediate the influences of man and nature, there are clear restrictions in the core zone to protect flora and fauna.[13] However, for proponents of the mining project there seems to be no contradiction with also having a national park close-by; for instance, the "pro-exploration" mayor of Medel also sits in the executive committee of project Parc Adula. Legally, a gold mine in the buffer zone would probably be possible, as a member of the environmental organization Pro Natura explains. Whether the park will be realized also depends on the popular vote of the affected municipalities, which is expected in 2016. Given the skepticism mentioned above and little enthusiasm among local residents, the outcome of the vote was quite uncertain at the time of writing.

The Consequences: "I Still Greet Everyone"

On April 1, 2012 the residents of Medel/Lucmagn entitled to vote decided at the ballot box against the exploration permit for SwissGold Exploration with 181 No to 90 Yes votes. This marked a tentative end to what had started with the information event less than a month earlier. A short time, during which the draft proposal and the two leaflets, various letters to the editor of local newspapers, the press release by environmental organizations, and a politically motivated story[14] were published. Moreover, it was a time in which numerous reports on Medel could be

13 Parc Adula. Nationalparkprojekt. Eine kleine Schweiz im Herzen der Alpen. <www.parcadula.ch> accessed 03.01.2014.
14 In "Curaglia anno 2025" a tourist with roots in the Val Medel describes a scenario in which the population has to be

found in local, national and even some international media. "Not normal" is how a resident describes the media attention surrounding the gold mining project, which subsided shortly after the vote. And so the community returned to daily life. After this public attention and the controversial discussions, earlier disagreements amongst the residents are commented on with: "everyone is entitled to have an opinion" or "I still greet everyone." Most disputes seem to have been settled.

The mayor was re-elected a few weeks after the vote without counter candidacy, but lost support at least with part of the population in Medel due to his position during the gold mining debate. Furthermore, a member of Pro Val Medel was newly elected into the municipal council. The group itself dispersed after winning the poll but said it could step back into action if the subject would come up again. The president of the Canadian NV Gold Corporation wrote a letter to the "residents of Val Medel", expressing full respect for the decision but with the hope that "maybe someday you will allow us to conduct explorations in your valley." During the debate and after the vote, the shares of NV Gold had dropped sharply from around 0.30 to a bottom of 0.02 to 0.10 Canadian dollars, where they remained until rising again when a new agreement on the Medel permit came in sight in the firsts months of 2014.[15]

The Future: "What Do We Want?"

Unlike its neighbouring municipalities, the Medel population could vote over a mining permit —but at a time when the dimensions of the potential extraction were mostly speculative. Besides general concerns for nature and landscape, the lack of information on the project and the fact that a future exploitation permit could be granted by the municipal council without a popular vote were the major reasons for the "No" at the ballot box. Following this decision, the municipal council concluded that the existing law was "not acceptable for the majority" (Vischnaunca da Medel/Lucmagn 2013: 1) and should therefore be revised. In 2014, the municipal council proposed a partial revision of the mining law, according to which the population would vote before the exploitation phase of a mining project rather than before the exploration. Thereby, more information on the actual project would be available. The proposed modifications would also lead to a harmonization with the mining laws of the neighbouring municipalities. On the 24th of April 2014, the municipal assembly of Medel approved the revision of the mining law with 15 Yes to 13 No votes. Though the outcome was tight and the discussion "intense" (*Suedostschweiz* April 26, 2014), this vote was completely unlike the one two years ago in terms of media coverage and in terms of people's attention—almost ten times less. On the one hand, this reflects the generally lower turn-out in assemblies compared to ballot box votes. On the other hand, the issue was less disputed, as with the revised mining law the people could vote again before an eventual mining permit.

The first version of the mining law had even been adopted by the municipal assembly in 2008 without much debate. Its practical relevance did not become clear until four years later. The same can become true for the 2014 revision. If there is again a popular vote on a specific mining project, it will surely get more attention, as this would be one of the most important decisions to take for the future of Val Medel. For the time being, NV Gold has announced that it has received a new exploration permit in October 2014 with a remark that "[t]he votes by the residents [reflect] the acceptance of mining as a possible industry in their communities". The permit is valid for 5 years and the company anticipates resuming its exploration program in 2015.[16]

evacuated due to an incident at the gold mine. The story was printed and distributed to all households in the municipality.

15 TMX Money: NV Gold Corporation <web.tmxmoney.com>, accessed 15.04.2015.

16 NV GOLD Reissued Exploration Permit For Swiss Property. <www.nvgoldcorp.com>, accessed 15.04.2015.

Conclusion

The gold mining proposal in Val Medel is a good example of the local political processes at work when it comes to large-scale projects in Switzerland. The direct democratic structures and communal autonomy recognize the local population as supreme authority whose consent is necessary and binding—in the special case of Medel, this even applied to an exploration concession at the earliest stage of a potential gold mine. Consent here includes the "right to veto" as a matter of course, which neither the state nor mining companies can overrule in principle. It is noteworthy that resistance against the project came mainly from the local population itself, while environmental organizations played a marginal role; one could argue that they were simply not needed, as the balance of power is favourable for local ad-hoc campaign groups. However, the case also illustrates that decision-making at the local level can increase the possibilities for (even small) interest groups to influence the law making process. In the debates, it was revealed that representatives of SwissGold Exploration "made proposals to adapt the mining laws in certain aspects" (Suprastonza da Medel 2012: 1). After the negative vote, these proposals were not further discussed. The mayor promised to solicit advice from independent experts rather than from the company in the future, as he had come under suspicion of having personal economic interests in the project (Schmidt 2014: 85). Nevertheless, the 2014 revision concerned mainly the same points that the exploration company identified as problematic. Given the large financial power of transnational mining companies, the question arises to what extent critical local groups would be able to withstand a stronger opponent. The example thus seems to confirm the increasing demands for national regulations on the usage of the subsoil (also with regard to new extractive technologies such as hydraulic fracturing for natural gas).[17]

One can draw interesting comparisons from the case with regard to "Free, Prior and Informed Consent" (FPIC; see Part I – 2), an instrument to protect the rights of indigenous peoples. Though the people in Medel do not self-identify as "indigenous", and the term FPIC was never used in the debates, the Swiss structures of political decision-making are similar to the principles of FPIC. Thereby, a crucial challenge lies in the provision of transparent information at an early stage of project development. On the background of scarce information, the main opposition in 2012 did not arise against the project as such, but against the fact that with a "Yes" to the exploration permit, the Medel population would have ceded the power to grant an exploitation permit to the municipal council. As most features of a potential mine can not be known previous to completion of the exploration and planning, it can be argued that it was a mistake to vote on the issue in Medel so early. And indeed, this special feature of the local mining law was abolished in 2014 by the municipal assembly—indicating that the previous resistance was directed less against gold mining as such, but rather against the decision-making procedure. However, the early vote had the advantage to launch the debate and raise awareness on the issue. The local people are now certainly better prepared to take an informed decision should they really be confronted with a fully developed mining project one day. Also, the case shows that a decision at the local level can turn (or be turned) quickly to the contrary. For the FPIC concept, this indicates that it is not sufficient to obtain consent from the local population only at a certain point in time. Rather, for an informed decision it should be at least once before the exploration phase and again at a later stage of planning, before the actual construction begins. Though the controversy has created some divisions within the community, these seem to be mostly temporary. The case therefore shows how crucial questions of due procedure and recognition of local decision-making about large-scale projects are to prevent and conflicts from early on.

17 See for example the recommendations on the use of subsoil by the Federal Geological Commission (EGK 2014).

References

Bächtiger, Kurt 1982: Die Erzvorkommen und Lagerstätten Graubündens und der ehemalige Bergbau. *Bergknappe* 21 (3): 2-12.

Bundesamt für Statistik BFS 2013: Taschenstatistik der Schweiz. Neuchâtel: Bundesamt für Statistik.

Chapman, Matthias 2011: Goldabbau in der Schweiz – es geht vorwärts. *Tagesanzeiger,* 6. September.

Eidgenössische Geologische Fachkommission (EGK) 2014: Februar 2014. Empfehlungen zur Nutzung des tiefen Untergrundes. <www.vbs.admin.ch> accessed on 12.02.2014.

Horber-Papazian, Katja 2006 (1999): Les communes. In: Klöti, Ullrich, Peter Knoepfel, Hanspeter Kriesi, Wolf Linder, Yannis Papadopoulos und Pascal Sciarini (Hrsg.): *Handbuch der Schweizer Politik*. Zürich: Neue Zürcher Zeitung. 233-258.

Krähenbühl, Hans 1986: Der Bergbau im Bündner Oberland. Verschiedene Abbaustellen. *Bergknappe* 37 (3): 12-14.

Linder, Wolf und Georg Lutz 2006 (1999): Direkte Demokratie. In: Klöti, Ullrich, Peter Knoepfel, Hanspeter Kriesi, Wolf Linder, Yannis Papadopoulos und Pascal Sciarini (Hrsg.): *Handbuch der Schweizer Politik*. Zürich: Neue Zürcher Zeitung. 103-124.

NZZ 2012: Vorerst kein Goldrausch im Bündner Oberland. *Neue Zürcher Zeitung*, 1. April.

Pütz, Marco, Daniel Kübler, Jochen Monstadt, Karin Enzler, Felix Kienast, Irmi Seidl, Fritz Zollinger 2008: Decentralized Settlement as a Challenge for Spatial Development: A case study of the Surselva region in Graubünden. *disP - The Planning Review* 44(173).

Schmidt, Christian 2014: Angst vor dem Gold. *Reportagen* 18: 74-91.

Schweizer Geotechnische Kommission (Hrsg.) 1997: Die mineralischen Rohstoffe der Schweiz. Zürich: Schweizerische Geotechnische Kommission.

Stefano, Roberto 2011: Im Goldrausch. *Handelszeitung* 39, 29. September: 30.

Suedostschweiz 2014: Medel: "Goldgesetz" knapp angenommen, 26. April: 7.

Suprastonza da Medel/Lucmagn 2012: Botschaft zur Volksabstimmung vom 1. April 2012.

Tiess, Günter 2009: Rechtsgrundlagen der Rohstoffpolitik. Ausgewählte Länder Europas. Wien: Springer.

Trueb, Lucien 1992: Gold : Bergbau, Verhüttung, Raffination und Verwendung. Zürich: Neue Zürcher Zeitung.

Val Medel 2011: Luvratori pigl avegnir—Ergebnisbericht Zukunftswerkstatt im Val Medel.

Vatter, Adrian 2006 (1999): Föderalismus. In: Klöti, Ullrich, Peter Knoepfel, Hanspeter Kriesi, Wolf Linder, Yannis Papadopoulos und Pascal Sciarini (Hrsg.): *Handbuch der Schweizer Politik.* Zürich: Neue Zürcher Zeitung. 79-102.

Vischnaunca da Medel/Lucmagn 2008: Lescha d'explotaziun per la vischnaunca da Medel/Lucmagn.

Vischnaunca da Medel/Lucmagn 2012a: 57. Rapport da gestiun e quen per 2011.

Vischnaunca da Medel/Lucmagn 2012b: Constituziun communala per la vischnaunca da Medel/Lucmagn.

Vischnaunca da Medel/Lucmagn 2013: 58. Rapport da gestiun e quen per 2012.

Vital, Peter Andri 1988: Das Verfahren in der bündnerischen Gemeindeversammlung. Zürich: Juris.

WWF Graubünden 2012: Abstimmungen im Val Medel erfolgen ohne genauere Projektkenntnisse zum Goldabbau. <wwf-gr.webofsections.ch> accessed on 1.5.2013.

‘Speech is Silver, Silence is Golden’: Gold Refining in Switzerland

Angela Lindt[1]

Switzerland became one of the most important sites for the international commodity trade, when multinational companies relocated their headquarters to low-tax communes in Switzerland, and cities like Geneva or Zug were turned into leading hubs for oil or other raw material businesses (Flütsch 2012). While most of the commodities traded in Switzerland never enter the country, there is one raw material that is imported in enormous amounts: gold. In 2011, more than 2,600 tonnes of unwrought gold, adding up to the value of more than 96 billion Swiss francs, were imported into Switzerland (FCA 2012). To the public, however, these huge sums are largely unknown, which is related to the fact that the Swiss policy supports keeping information about the gold sector secret.

Switzerland, particularly through its banks, has a special relationship with precious metals. The Swiss National Bank (SNB) holds the seventh largest official stock of gold with more than 1000 tonnes (WGC 2012b). Moreover, calculated per capita, Switzerland tops the list with 135 grams of gold per inhabitant (NZZ Folio 04.2009). Apart from that, there are a lot of Swiss private banks holding gold stocks for their clients. And last but not least, there is a huge gold processing industry, which goes beyond the traditional luxury watchmaking sector: four of the nine largest global gold refineries are located in Switzerland, processing gold mined from all over the world, without Swiss citizens taking notice of it. For the insider, however, it is evident why the global gold industry’s spearhead organization for “peer networking, business development and deal-making”, the Denver Gold Group, holds its annual “European Gold Forum”[2] in Zurich.

But let us start at the beginning: the origins of gold. Gold mining is one of the dirtiest resource extraction businesses, often accompanied by enormous environmental pollution and serious human rights violations—whether produced in big industrial gold mines with the use of highly toxic cyanide, often leading to land conflicts with local communities, or produced by artisanal and small-scale miners, who use mercury to extract the metal. Apart from hazardous mining methods, some of the gold mines are located in regions characterized by civil wars or the absence of any rule of law. Examples of this include artisanal and small-scale gold mines in Democratic Republic of Congo (DRC), which significantly contributed to the financing of the armed conflict in the country, or informal gold mines located in the Peruvian jungles which are associated with increasing corruption, money laundering and tax fraud. In this context, Switzerland remains the ideal trading centre for gold: since 1981, the Swiss customs authority, based on a decision of the Federal Council, does not list the origin of the annually imported gold, but just publishes the total amount. As a result, the public has no idea as to where the gold manufactured and traded in Switzerland comes from.

1 This chapter is based on research by a group of Swiss NGOs, as part of the “Corporate Justice” campaign. A review was provided by Eva Schmassmann. For the original report (in German) see: Lindt/Schmassmann 2012.

2 Denver Gold Group. www.europeangoldforum.org accessed on 02.02.2013.

Overview of the Gold Mining Industry and Gold Trade

In 2011, around 2700 tonnes of gold were mined worldwide (USGS 2012: 66). The leading gold producing countries in 2011 were China (355 tonnes), Australia (270 t) and the United States (237 t), followed by Russia (200 t), South Africa (190 t), Peru (150 t) and Canada (110 t) (ibid: 67). It is estimated that industrial companies today mine more than three-quarters of all the gold extracted worldwide. The biggest of them are transnational corporations with headquarters in the traditional gold-producing countries—South Africa, United States, Canada or Australia (Hosp 2012). Apart from mining gold, recycled sources play an important role. In 2011, for example, about 40 per cent of the yearly gold demand or, in absolute terms, 1,612 tonnes, were supplied by gold scrap (WGC 2012a: 23).

The global mining capacity of gold today is about one hundred times bigger than it was 120 years ago (Eibl 2008: 23). About 80 per cent of all the gold ever mined in history was extracted in the 20th and the 21st centuries (Südwind 2010: 6). These facts illustrate the enormous gold rush in the past decades. At the same time, the extracted amount of gold seems relatively small: all the gold ever refined throughout the history of mankind would result in a cube only measuring 20-meter in width (Reisenberger/Seifert 2011: 8). Due to the intensive extraction of gold in the past century, deposits of high gold concentration decreased enormously: today, in an average industrial gold mine there are about 0.8 to 2 grams of gold per tonne of rock. Between 1830 and 1920, one could find about 22 grams per tonne (ibid: 14). But because of advancement in processing techniques, especially the invention of the heap-leaching process using cyanide to extract the metal in huge open-pit mines, it is worth to run mines in areas with very low gold concentration (Roth 2012b: 18f.). But most importantly, it is the enormous rise in gold prices in the world market within the last ten years, which still makes it profitable to mine for gold notwithstanding declining gold concentration. As Figure 1 below illustrates, the price of gold has almost sextupled from almost USD 280, at the beginning of 2002, to approximately USD 1,700, at the end of 2012 (WGC 2013). Triggered by the international financial crisis, the demand for gold as a safe investment grew enormously. One effect of this price increase was an even bigger run on remaining gold deposits worldwide (Lindt/Schmassmann 2012: 8).

Fig. 1: Gold Price 2002-2012; Source: World Gold Council.

Profits for the mining industry are secured by the enormous rise in the price of gold. Technical progress led to the increased extraction of gold in areas where it was not possible before. At the same time, negative effects for the environment and for human beings grow due to the increasing use of toxins to extract the last remaining gold particles. To extract one single kilo of gold, 141 kilos of highly toxic cyanide is required, on an average (Südwind 2010: 9). It is estimated, that per year, 182,000 tonnes of cyanide is used worldwide in gold mines (Jensen/McBay 2009: 122). The use of highly toxic substances is just one of the many problems linked to open-pit gold mines. In heap-leaching mines, enormous amounts of rock have to be removed, crushed, piled into heaps and irrigated with a cyanide solution, which trickles down through the ore and bonds the tiny gold particles. To extract gold for one single wedding ring, 20 tonnes of rocks have to be processed (BD 2011: 183). As a result, huge areas of land disappear due to mining activities. This often leads to land disputes with local communities.

Toxic substances like uranium, mercury, lead or arsenic are dissolved by the cyanide solution together with precious metals like gold or copper. The contaminated rocks are stored as waste in large open-air stockpiles and detention reservoirs. In the past, bursting and leaking dams of these deposits repeatedly resulted in severe pollution of rivers and groundwater near the mines. The Rumanian Baia Mare goldmine dam failure of January 2000 is just one well-known case of severe environmental damage due to heap-leaching processes in gold mines (Verseck 2010; Luyken 2000). In addition to the contamination of rivers and drinking water supplies, open-pit gold mines use enormous amounts of water (see Part I – 3 and case study Marlin, Guatemala). Apart from industrial gold mines, there are artisanal and small-scale miners in different parts of the world. According to the United Nations Industrial Development Organization (UNIDO), this sector employed between 10 and 15 million people in 2004, of whom 300,000 were children (Global Mercury Project 2004: 13). According to recent studies, the total number of artisanal and small-scale miners rose in the last eight years up to 20 million people due to the dramatic rise in the price of gold. UNIDO estimates that between 650 and 1,000 tonnes of mercury are released by these artisanal and small-scale mines into the environment annually (ibid: 1).

Whether gold is mined industrially or by artisanal and small-scale miners, the extracted gold has to pass through a gold refinery in order to attain a higher degree of purity before it is released for sale or for further processing. After the refining process, the gold bars are sold in the international gold markets to bank clients, to jewellery and watch manufacturers, or to the electronic and dental industry. In 2011, 48 per cent of the global gold demand was generated by the jewellery industry, 40 per cent was sold in the investment sector, which includes gold bars, coins and Exchange Traded Funds (ETF) and 11 per cent was bought by the electronic and the dental industries (WGC 2012a: 22).

International Guidelines, Due Diligence Standards and Regulations

In recent years, various international organizations recognized the ecologically disastrous nature of the gold mining and trading sectors and started developing initiatives and guidelines to minimise its negative impacts. An important starting point was the report of the UN Group of Experts on the Democratic Republic of the Congo, which stated that the trade in minerals and other natural resources, including gold, exacerbates the armed conflict in the east of DRC by supporting, directly or indirectly, the rebel groups involved.[3] In its final report of November 2010, the

3 UN Security Council Committee Established Pursuant to Resolution 1533 (2004) Concerning the Democratic Republic of the Congo. Note by the Chair. Due Diligence Guidelines of the Group of Experts on the DRC. <www.un.org/sc/committees/1533/egroupguidelines.shtml> accessed on 02.10.2012.

Fig. 2: Informal artisanal gold mining, Madre de Dios region, Peru. Photo: Terre des Hommes Suisse.

Group of Experts suggested due diligence guidelines to the UN Security Council, including recommendations for individuals and entities who are trading, processing and consuming minerals originating in DRC or neighbouring countries.[4] Shortly after, these due diligence guidelines were taken over by the Organization for Economic Co-operation and Development (OECD), resulting in the OECD Due Diligence Guidance for Responsible Supply Chains of Minerals from Conflict-Affected and High-Risk Areas and its Supplement on Gold (OECD 2012).

The OECD's guidance and supplement documents on gold, as well as the Group of Experts' recommendations, focus on the direct responsibility of companies and invite them to check their supply chains. By strengthening company management systems, identifying and assessing risks in the supply chain and by implementing a strategy to respond to those risks, companies should be able to avoid conflict-financing businesses (OECD 2012; UN Security Council 2010: 86ff.). The OECD guidelines have no legally binding character. Due to the fact that it has been framed relatively recently, its effectiveness is not yet clear. It remains to be seen to what extent the efforts of the OECD and the UN Group of Experts become successful in restricting the trade with minerals from conflict areas in the Great Lakes Region. But, currently, it is already evident that

4 UN News Center. Group of Experts on the Democratic Republic of the Congo Delivers Due Diligence Guidelines to Security Council. <www.un.org/News/dh/infocus/drc/GOE_press_backgrounder.html> accessed on 28.12.2012.

the due diligence guidelines and those in the Supplement on Gold are insufficient to improve the situation in gold mining areas other than DRC, because they are often interpreted as being limited to the Great Lakes Region. "Conflict-affected areas" are defined in the OECD Guidance document as "areas identified by the presence of armed conflict, widespread violence, including violence generated by criminal networks, or other risks of serious and widespread harm to people". "High-risk areas" are "characterised by political instability or repression, institutional weakness, insecurity, collapse of civil infrastructure, widespread violence and violations of national or international law" (OECD 2011: 66). However, the OECD did not provide any further guidance on how to define these regions. Consequently, it is up to the company to categorise the region and decide whether increased due diligence measures are necessary. Because of this interpretation and misreading of the guidance, as mentioned above, gold mines located outside this region, but inside other conflict areas, for example in Colombia, are implicitly excluded from the guidance's purview in the understanding of gold trading firms. Therefore, gold processing companies, including Swiss gold refiners, claim to process no "conflict gold", because they did not process gold from the DRC. According to this view, gold is "conflict-free", as long as it was not mined in the Great Lakes Region. Another big shortcoming of the OECD guidelines is the fact that they do not incorporate requirements concerning environmental and human rights standards but merely focus on the direct violence resulting from the business of gold mining. Given the vagueness regarding the definition of what conflict areas or high-risk areas are, together with the lack of a powerful supervisory authority and a legally binding character, the effect of the OECD guidance will probably remain largely unsatisfactory.

In addition to the guidelines of multilateral organizations, there also exist some initiatives of industry associations. It is remarkable that almost all of these initiatives originated in the jewellery or electronic industry and not in the finance sector. Especially, consumers of jewellery seem to be more sensitized to questions concerning where the purchased gold comes from and under what conditions it is mined. The most important among these guidelines are the Principles and Code of Practices launched in 2006 by the Responsible Jewellery Council (RJC). This code includes principles concerning business ethics, human rights and social performance, environmental performance and management systems. In 2012, the RJC complemented it with a Chain-of-Custody (CoC) Standard. The RJC, an association of mining firms, as well as diamond and gold jewellery companies, set itself the target "to advance responsible ethical, social and environmental practices, which respect human rights throughout the diamond, gold and platinum group metals jewellery supply chain, from mine to retail".[5] To become a certified member of the RJC, companies have to demonstrate independently verified conformity to the RJC Code of Practices. Third party auditors verify compliance with the principles (RJC 2012a). The large majority of certified members are jewellery manufacturers and retailers as well as gold and diamond traders and some refiners. Because the Code of Practices says nothing about the conditions under which the manufactured gold has to be mined but only refers to the proper processing steps of the companies, i.e. the refining process or the manufacturing method of the jeweller, it is easy for retailers or refiners to achieve RJC-certification. With its CoC Standard, RJC tries to go one step further to reach its goal of a completely environmentally and socially responsible supply chain. To achieve this standard, companies have to ensure that their processed gold comes only from legitimate and conflict-free sources (RJC 2012b: 8). The CoC Standard is still at an early stage and until the date of writing no complete supply chain from the mine to the retailer has been certified. The first refiners certified with the CoC certificate have been so only for gold from scrap. Nevertheless, this could be a step towards more transparent origins of gold, at least in the jewellery sector.

5 Responsible Jewellery Council. About. Mission. <www.responsiblejewellery.com/about-us> accessed on 03.01.2013.

A much more rigorous approach is the ban on use of cyanide in mining. With the current technology, this is in effect equal to a ban on industrial gold mining. Accordingly, various local and international environmental campaigns could put pressure for such a move, with some success.[6] An intermediate step is strict legal regulation, including requirements for assurance and payment of compensations in case of disasters, which can disincentivise industrial gold mining. The industry generally reacts to such bans and regulations by trying to overturn them at other legal levels and through court appeals. In reaction to the increasing criticism, particularly after the catastrophic Baia Mare cyanide spill of 2000 in Romania, the industry has also set up a voluntary initiative called International Cyanide Management Code, in collaboration with the United Nations Environmental Program (UNEP) and the International Council on Metals and the Environment (ICME).[7] However, critics point out that the code's effectiveness is doubtful. For example, a cyanide spill at Newmont Mining's Ahafo mine in Ghana, in October 2009, happened shortly after the mine was certified as being in full compliance with the requirements of the code.[8]

There are also moves to regulate or ban the use of mercury at the international level, which would have a major impact on artisanal gold mining. An important step in this direction is the Global Mercury Convention, adopted by 140 states in January 2013 (Britt 2013). However, to restrict mercury use without simultaneously restricting the use of cyanide, and without providing alternative sources of livelihood to artisanal miners, would just be to the advantage of large-scale industrial mining over artisanal mining. This would risk exacerbating the displacement of the latter to serve the interests of transnational corporations and could trigger more conflicts, as millions of people depend on artisanal mining for their livelihood.

Summing up, there are some initial approaches towards both voluntary and legally binding regulations, but much remains to be done to increase pressure on gold mining companies to reduce negative impacts of their operations on local population and the environment.

Gold Refiners in Switzerland

Gold refiners are at the interface between the gold mining and the gold manufacturing industry. In industrial gold mines, the freshly mined gold, called gold doré, is cast into bars with a purity of approximately 90 per cent. These bars contain silver, copper and other metals apart from gold. Gold refiners process this gold doré to pure gold. They produce gold bars and coins for the finance sector, blanks for the jewellery and watch industry and other semi-finished industry materials. Switzerland hosts four of the nine leading gold refineries worldwide (Lindt/Schmassmann 2012: 12). Further important gold refiners are located in the United States, South Africa, China and Australia (Olden 2010: 7). Refiners located in gold mining countries benefit from short transport distances. The absence of gold mines located near the four large Swiss refiners is compensated for by the proximity to the financial market and the major Swiss banks. Further advantages include political stability and security as well as efficient logistics in Switzerland (Roth 2012a: 19).

Three of the Swiss refiners had close connections with leading banks in their history. Metalor,

6 Campaigns to ban cyanide for mining include those launched by Observatorio de Conflictos Mineros de América Latina (OCLA) for South America see: <www.conflictosmineros.net/component/cianuro>. For the campaign by "Save Rosia Montana" and others for Romania and the European Union see: <www.rosiamontana.org>; <www.bancyanide.ro>. Bans or moratoria on cyanide use were decided at some level in the US state of Montana, a number of Argentinean provinces, Costa Rica, Germany, Czech Republic, Hungary, Turkey and others. An overview centred on the US, see: Laitos (w.y.).

7 For more information see: <www.cyanidecode.org> accessed on 05.02.2013.

8 EARTHWORKS/WACAM: Denver-based Newmont Mining Co. fined millions for cyanide spill at Ghanaian mine. (21.01.2010). <www.earthworksaction.org> accessed on 05.02.2013.

located in Neuchâtel and estimated to be the biggest of the Swiss gold refineries, was until 1998 owned by the Swiss Bank Corporation, today known as UBS (BD 2011: 184). Valcambi, situated in Balerna near the Italian border in the canton of Ticino, was until 2003 in the possession of Credit Suisse. Today, Newmont Mining, one of the world's largest producers of gold, has a 60 per

Fig. 3: The Valcambi gold refinery in Balerna, Switzerland at night. Photo: Jacob Balzani Lööv.

cent holding in Valcambi.[9] The third Swiss refinery, Argor-Heraeus, belonged to the Union Bank of Switzerland and, after merger with Swiss Bank Corporation, to the UBS. As the bank decided to retreat from the refining sector in 1999, the refinery was taken over by the Argor-Heraeus management.[10] Close relationships between refiners and banks persisted even after the sale of the bank interests: Credit Suisse, for instance, announced in 2011 the purchase of all its traded gold bars from Valcambi (Credit Suisse 2011: 19). In the mid-1970s, all these three refineries, faced competition from PAMP (Produits Artistiques de Métaux Précieux), the fourth of today's leading refineries in Switzerland, which was founded by the Lebanese gold trading family Shakarchi, with a production site in Castel San Pietro, canton of Ticino (BD 2011: 185). PAMP is part of the company MKS Finance SA (MKS), headquartered in Geneva. MKS is said to be the leading bank-independent, private company involved in the physical gold trade (Eibl 2008: 100).

Precise quantities of gold refined by different companies in Switzerland are not known, neither the proportion of globally mined gold refined in Switzerland. The estimated proportion ranges from at least one-third of all the gold mined globally (Lindt/Schmassmann 2012: 12) to 40 per

9 Valcambi Suisse. History. <www.valcambi.com/about-valcambi/history> accessed on 04.01.2013.

10 Argor-Heraus. About us. History. <www.argor.com> accessed on 02.10.2012.

cent of the total gold refined worldwide (BD 2011: 184). None of the four gold refiners publish information about how much gold they process every year. It is estimated that Metalor has a gold refining capacity of 650 tonnes per year, followed by Valcambi with 600 tonnes, PAMP with more than 450 tonnes and Argor-Heraeus with 400 tonnes per year (ibid: 13). It is not possible to draw accurate conclusions with respect to the proportion of gold doré from these data, because the refining capacity also includes recycled gold. It is however not unreasonable to assume that a large proportion of the manufactured gold is gold doré, i.e., mined gold. This presumption is confirmed, for example, by the fact that Valcambi refined 435 tonnes of gold doré in 2009 (Newmont Mining 2010).

All four Swiss gold refiners seem to have recognized the risk to the reputation of their companies from the origins of the gold they process. All of them participate actively in international initiatives of the World Gold Council and the London Bullion Market Association, also concerned with voluntary due diligence standards. All four refineries use this engagement to show publicly their efforts to develop corporate social responsibility and sustainability programs. Metalor, PAMP and Argor-Heraeus are certified Responsible Jewellery Council members, and thus fulfil the RJC Code of Practices. Valcambi is not a certified member of RJC until now.[11] Metalor achieved in May 2012 the RJC Chain-of-Custody Standard, becoming the first refinery worldwide to do so, but just for its recycling and so-called "grandfathered"[12] gold section (RJC 2012c). Metalor's gold doré, as well as the gold doré processed by the three other Swiss refineries, until now, do not fulfil the RJC Chain-of-Custody Standard. However, Metalor seems to be the most advanced Swiss refinery with regard to the determination of a company internal policy concerning the origin of gold. Currently, for example, the refiner claims to be making efforts to manufacture only Peruvian gold from artisanal and small-scale mines complying with Peruvian laws.[13] Characteristically, Metalor made these efforts public two days after being criticized in a newspaper for buying Peruvian gold extracted in informal and partly illegal small-scale mines (Brocard/Maurisse 2012).

According to the refineries, Gold doré manufactured in their plants should be in line with the recommendations of the OECD Due Diligence Guidance (Valcambi Suisse 2012: 2; Argor-Heraeus 2011: 24, 31). But because most of the Swiss refineries, like many other companies, interpret the guidance in the sense that the gold should not come solely from DRC or neighbouring countries, there is no guarantee that gold from other conflict regions or high-risk areas is not refined in Switzerland. Furthermore, as noticed above, the OECD guidelines do not specify that an origin of the gold be reconcilable with environmental and social standards. Gold coming from mines with negative impacts on nature and human beings due to its use of cyanide or mercury or gold from mines where violent conflicts are triggered by local resistance against mining projects, do not contradict the OECD Due Diligence Guidance. As a consequence, it is questionable when Swiss gold refiners praise themselves for their conformity to the OECD guidance or to the RJC Code of Practices as it is obvious that these initiatives do not guarantee responsible origins of gold.

It is not particularly surprising that Swiss gold refiners today emphasize their commitment to international initiatives for "conflict-free" gold. All of them were in the last ten years involved in different scandals related to gold originating from dubious and partly illegal sources. Damage

11 Responsible Jewellery Council (RJC). Members. All Gold and/or Platinum Group Metals Trader, Refiner Hedger Members. <www.responsiblejewellery.com> accessed on 08.01.2013.

12 "Grandfathered" gold is defined as "Gold investment products (ingots, bars, coins, and grain in sealed containers) held in bullion bank vaults, central bank vaults, exchanges and refineries with a verifiable date prior to 1 January 2012, which will not require a determination of origin. This includes stocks held by a third party on behalf of the listed entities." (OECD 2012: 7).

13 Metalor. Metalor – The Leading Gold Refiner Worldwide for its Efforts on Conflict Gold and Improvement of Artisanal Mining, 25.09.2012. <www.metalor.com> accessed on 25.01.2013.

to their public image was considerable. Argor-Heraeus and Metalor were directly mentioned by the UN Group of Experts for refining gold originating in the DRC; Metalor was said to be one of the principal buyers of gold traded via Uganda (UN Security Council 2005: 32). In 2005, Metalor suspended further deliveries from Uganda (BD 2011: 184). In 2006, the UN Security Council accused Argor-Heraeus of violating the mineral trade embargo imposed by the UN Security Council resolution 1596 (UN Security Council 2006: 26).

The other Swiss refiners did no better: two journalists disclosed in 2008 that Valcambi refined gold originating from bush mines in Mali, where thousands of children are working under miserable conditions (Callimachi/Klapper 2008). According to research by Swiss NGOs, Valcambi currently refines a huge part of the gold extracted from the Yanacocha mine in Cajamarca, Peru. No surprise, as both the mine and the refinery are under majority control of the US-based Newmont Mining Corporation (Lindt/Schmassmann 2012: 22f.). An expansion project of this gold mine, the project Conga, is strongly opposed by the local population. Between November 2011 and July 2012, the Peruvian Government several times declared a state of emergency in the province of Cajamarca. Serious violations of human rights took place and three protesters were killed in July 2012 alone (BBC News 2012). In 2011, the Yanacocha mine exported more than 60 tonnes of gold amounting to nearly USD 1.5 billion into Switzerland (SUNAT 2012). The fourth Swiss refiner PAMP seems to choose its gold suppliers in a similar uncritical manner: towards the end of September 2012, Swiss newspapers made public that PAMP had recently bought gold from a dubious Peruvian gold trader accused of money laundering and illegal gold mining in the south-eastern jungle of Peru (Brocard/Maurisse 2012; Weiss 2012). Due to the fact that none of the four gold refineries disclose the origin of the processed gold, it is unknown if these reported cases of problematic sourcing of gold are isolated instances. But when recollecting the widespread negative impacts of gold mining, it must be assumed that a substantial part of gold refined in Switzerland was mined under unacceptable conditions.

To summarise, gold refined in Switzerland is not that conflict-free, as the four refineries would like us to believe. At least, they seem to have learnt that conflict minerals from DRC are bad for business, especially after the criticism of the UN Security Council. Their commitment to the development of international due diligence guidelines like the RJC-CoC Standard may be laudable, but as long as it is not clear from which countries and from which specific mine the gold comes, it cannot be ruled out that there is a lot of dirty gold manufactured in Switzerland.

Swiss Foreign Trade Statistics

Apart from secrecy about the origin of gold processed in Switzerland, there is also a general lack of information about gold traded in Switzerland. The Swiss foreign trade statistics apply special rules for listing the gold trade. Whereas every other imported or exported product is listed yearly as per country of origin or country of destination respectively, gold and silver in unwrought form or as coins are not listed separately as per trade partner, but just in total amounts. The imported amounts of unwrought gold are enormous, as Table 1 illustrates.

Apart from gold doré, there is obviously a lot of bullion gold imported into Switzerland too, which, in parts, is also included in the listed unwrought gold. Even if Zurich became, as a result of the globalized markets, less important as a trading centre of bullion gold since the turn of the millennium, Switzerland is (not least because of the important gold refineries) still assumed to be one of the major importers of gold worldwide (BD 2011: 180). However, the confederate customs administration does not separate gold doré from bullion gold in its statistics. Therefore, it remains unclear how much of the imported unwrought gold is in fact gold doré, i.e., coming directly from a gold mine, and how much is bullion gold intended for direct sale or trade.

Year	In kilograms	In Swiss franks
2011	2.671.540	96.049.821.402
2010	2.326.899	70.788.774.622
2009	2.185.808	55.193.421.861
2008	2.084.813	50.038.121.564
2007	1.614.342	32.062.832.506

Table 1: Import of Unwrought Gold into Switzerland;
Source: Federal Customs Administration (FCA)[14]

The decree to summarize all the imported gold and not to list it as per country of origin was implemented by the Federal Council in 1981. The Swiss government officially based its decision on the regulation of the foreign trade statistic, where it is stipulated that the Directorate General of Customs can give only summaries of certain statistical data "if the detailed publication significantly harms Swiss interests".[15] In the Federal Council's opinion, the publication of the gold imports would have significant negative impacts on Zurich as a trading centre for gold, because Great Britain, with its competing gold trading spot at London, did not publish details of its gold imports either (Curia Vista 1989). In retrospect, however, it is clear that the main reason for the Federal Council's decision was that gold trade with sensitive partners, above all with South Africa, had to be covered up. Recent studies come to the conclusion that an estimated 80 per cent of all the gold mined in South Africa in the 1970s and 1980s were traded by Swiss banks and refined in Switzerland (Bott et al. 2005: 297). During this time, South Africa was by far the largest gold producing country worldwide. According to this report, Swiss gold dealers played an important role in supporting the South African Apartheid government and crucially helped the regime in surviving despite an international embargo on trade (ibid: 287, 297). Apart from South Africa, another important gold trading partner of Switzerland in the 1970s and the following years was the Soviet Union (Eibl 2008: 99). The possibility of maintaining this business relationship secret during the Cold War was surely attractive.

Until today, the practice of keeping the origin of the imported and exported gold a secret remains in force. In 1989, a Member of Parliament made an attempt to reintroduce the publication of detailed statistics of the gold trade (Curia Vista 1989). The Federal Council rejected the proposal with the justification that Swiss financial markets needed to be protected. In the Swiss parliament, the proposal was not even discussed (ibid.). In 2012, Member of Parliament Cédric Wermuth asked the Federal Council how the government could continue to justify the special listing of gold in the foreign trade statistics and how they assessed the reputation risk for Switzerland arising from the gold trade (Curia Vista 2012). The Federal Council recognized at least the change in the political, economical and social context since the proposal of 1989, and promised in its answer to re-examine the question of the gold trade data's publication (ibid.). At the time of writing, a

14 The tariff number of unwrought gold is 7108.1200. One can find only the list of the past year's data on the Internet. Figures of the preceding years can be requested from the Federal Customers Administration (FCA).

15 Verordnung SR 632.14: Schweizerischer Zolltarif, Verordnung über die Statistik des Aussenhandels vom 12. Oktober 2011, Art. 16, Abs. 2. <www.admin.ch/ch/d/sr/6/632.14.de.pdf> accessed on 19.06.2012.

inter-departmental working group was preparing a proposal for the Federal Council, to be decided by the end of 2013, about the possibility to publish a more detailed import/export statistic for the gold trade.[16]

To a certain extent, it is possible to find information about trade in gold in the foreign trade statistics of the most important gold mining and exporting countries. To gain information about the Chinese or the Russian foreign trade statistics is difficult. Statistics from the United States give detailed information, and even distinguish between gold doré and bullion gold.[17] Peru publishes detailed information, and so the amount of gold exported to Switzerland is known. Even the exports of the ten leading exporting firms by the country of destination can be obtained publicly (SUNAT 2012). Nevertheless, without official Swiss information, it is difficult to get a rough idea about Switzerland's most important gold suppliers. The publication of the data regarding where the gold refined and traded in Switzerland comes from would be a significant step towards more transparency. Today, there is no transparency in the Swiss gold business.

Conclusion

The secrecy of the Swiss administration is almost legendary and this does not apply only to the finance and banking sectors. Discretion has led to the concealment of an entire industry sector: Swiss companies are among the leading gold refineries worldwide but at the same time they are hardly known. Enormous amounts of gold are imported into Switzerland every year, but almost nobody has an idea about where this gold comes from and where it goes. In this context, the motto seems to be speech may be "silver" but silence is certainly "golden". This situation would not pose a problem if gold were an ordinary product. But due to the hazardous circumstances under which gold is mined, critical observers, NGOs as well as some Swiss government representatives themselves point out that trade and import of the precious metal bears a considerable reputation risk for Switzerland—on top of the growing international pressure due to Switzerland's policy of facilitating tax evasion. Martin Dahinden, Director of the Swiss Agency for Development and Cooperation, called the commodities sector a "time bomb" that could undermine all the efforts of the Swiss Government for Human Rights and sustainability.[18]

The Swiss gold refineries constitute a major transit point, linking gold mines to traders and consumers. Therefore, the Swiss section of the NGO Society for Threatened Peoples (STP) launched a campaign that included a petition to the Swiss authorities to ensure transparency and legally binding standards for gold processing and trading. The publication of detailed information regarding the imported and exported amounts of gold listed per trading partner would be a reasonable and practically enforceable measure, as a first step towards more transparency. Furthermore, the campaign calls for changes in the Swiss law to ensure that refineries and traders only accept gold produced in compliance with effective human rights and environmental standards, because the success of voluntary guidelines has been very limited in this respect. Without taking these steps, Switzerland risks providing a legal loophole and becoming a hub for gold mined under dubious and miserable conditions, in a context of increasingly strict international regulations. At the international level, the use of cyanide and mercury in gold mines should be strictly controlled. Where local communities, countries or states prohibit its use altogether, companies should respect

16 Matthias Pfammatter, Eidgenössische Zollverwaltung, Sektion Statistik, by E-Mail, 13. August 2013

17 U.S. Geological Survey. Minerals Yearbook. <minerals.usgs.gov> accessed on 12.08.2013.

18 Martin Dahinden at a panel discussion at ETH Zurich, 02.11.2011, cited in Swiss Television SRF newscast 10vor10 of 24.11.2011, "Das schmutzige Rohstoff-Geschäft": <www.srf.ch> accessed on 02.05.2013. For the NGOs position, see the Swiss "Corporate Justice" campaign, <www.corporatejustice.ch> accessed on 06.02.2013.

this decision. To minimise conflicts with local communities, conventions like the ILO convention 169 (see Part I – 2) should be ratified and implemented by the nations concerned. Additionally, a lot of responsibility lies with the refineries, as they operate not only with manufactured gold scrap but also large amounts of gold doré coming directly from mines worldwide. If they demand compliance with effective human rights and environmental standards from their suppliers, they could contribute a lot to improve the situation in the gold mining sector. To bear the additional costs, strong disincentives would be needed for the refineries.

The general public should be made aware of the problematic situation in the gold mines and the important role of Switzerland in the entire refining and trading process. One possibility to do so is to target jewellery retailers and their consumers, as does the "No dirty Gold" campaign in the US.[19] As the experience of past years have shown, refining companies do react to public criticism. If their image is under threat, they start talking about where their gold comes from and stop processing gold from publicly exposed illegal sources. Private companies are understandably not very motivated in disclosing all their trading partners, but as long as the situation in the gold mining sector has not improved, it is hard to believe that the refineries have exclusively reliable sources. Thus, the timid regulatory approaches towards "conflict-free" and "sustainable sources" have to be more than pure lip service and public marketing strategies.

Update

In February 2014, the Swiss Federal Customs Administration (FCA) published for the first time since more than thirty years the amounts of gold and silver exported and imported in the preceding month broken down by country (FCA 2014a). This can be read as a follow-up to the "Background Report on Commodities" already published in March 2013, where the Federal Council considered including the gold trade in the official Swiss foreign trade statistics (Swiss Federal Council 2013a: 44). In December 2013, the government finally approved a change of practice in order to "contribute to transparency in precious metals trading" (Swiss Federal Council 2013b).

However, the published statistics show serious limitations: contrary to the current practice of other gold trading countries such as the USA, the FCA does not differentiate between bullion gold and gold doré. Therefore, it is not a surprise that the United Kingdom with its important bullion gold trading centre London dominated the imports in the first three months of 2014 with 41 per cent of the total amount measured in weight (FCA 2014b). Nevertheless, on the top positions of the import list, we find gold producing countries such as the USA, Chile, Russia, Argentina and Peru (ibid.). Another weakness of the statistics is that it does not contain information about the country where the imported gold has been mined before being traded at the bullion market nor from which mines the directly imported gold doré comes from. In summary, the newly published foreign gold trade statistics are just one first step toward more transparency, but leave many open questions.

19 This campaign was launched by the Washington DC based NGO Earthworks. See: <www.nodirtygold.org> and <www.earthworksaction.org> accessed on 02.02.2013.

References

Berne Declaration (BD) 2011: *Rohstoff. Das gefährlichste Geschäft der Schweiz*. Zürich: Salis.

Bott, Sandra, Sébastien Guex and Bouda Etemad 2005: *Les relations économiques entre la Suisse et l'Afrique du Sud durant l'Apartheid (1945-1990)*. Lausanne: Éditions Antipodes.

Brocard, Martina and Marie Maurisse 2012: Die Schweiz profitiert von blutigem Gold aus Peru. *SonntagsZeitung* 39 (23 September): 54.

Eibl, Christoph 2008: *Alles, was Sie über Gold wissen müssen. Hintergründe, Fakten und Anlagetipps für den wichtigsten Rohstoff der Zukunft*. München: Finanzbuchverlag.

Flütsch, Andreas 2012: Lausanne statt London. Das «WEF der Rohstoffbranche» am Genfersee. *Der Bund* 163 (24 April).

Global Mercury Project 2004: Protocols for Environmental and Health Assessment of Mercury Released by Artisanal and Small-Scale Gold Miners. Wien: GEF/UNDP/UNIDO.

Hosp, Gerald 2012: Kaltes Gold aus Sibirien. *Neue Zürcher Zeitung* 47 (25 February): 33.

Jensen, Derrick and Aric McBay 2009: What We Leave Behind. New York et al: Seven Stories.

Reisenberger, Brigitte and Thomas Seifert 2011: *Schwarzbuch Gold. Gewinner und Verlierer im neuen Goldrausch*. Wien: Deuticke.

Roth, Pascal 2012a: Schweizer Scheideanstalten. Goldraffinerien von Weltruf. *Stocks, Das Schweizer Anleger-Magazin* 20 (November 2012): 18–19.

Roth, Pascal 2012b: "Zentralbanken kaufen 500 Tonnen". Interview mit Marcus Grubb, Direktor beim World Gold Council. *Stocks, Das Schweizer Anleger-Magazin* 20 (November 2012): 16–20.

SÜDWIND. Institut für Ökonomie und Ökumene (ed.) 2010: *Schmuck – Liebesbeweis, Broterwerb und Ausbeutung. Vom Rohstoff zur Ladenkette*. Bonn: SÜDWIND.

Verseck, Keno 2010: "Dieser Unfall hat unser Leben vergiftet". *WochenZeitung* 6 (11 February): 27.

Weiss, Sandra 2012: Schmutziges Gold aus Peru in Schweizer Banken. *Der Bund* 163 (17 September): 2.

Internet Sources

Argor-Heraeus 2011: What is the Value of Growth? Sustainability Report. <www.argor.com/fileadmin/Sustainability/Report_sostenibilita_2011_ENG.pdf> accessed on 25.01.2013.

BBC News 2012: Peru: Three Die in Clashes over Conga Gold Mine Project. *BBC News Latin America & Carribean* (04.07.2012). <www.bbc.co.uk/news/world-latin-america-18700522> accessed on 09.01.2013.

Britt, Chantal 2013: Global Mercury Convention Adopted. *Swissinfo.ch* (19.01.2013). <www.swissinfo.ch> accessed on 21.01.2013.

Callimachi, Rukmini and Bradley Klapper 2008: Tausende von Kindern schuften in afrikanischen Goldminen. *The EpochTimes Deutschland* (11.08.2008). <www.epochtimes.de> accessed on 04.05.2012.

Credit Suisse 2011: Goldanlagen einfach gemacht. Credit Suisse ETFs on Gold. <www.credit-suisse.com/asset_management/downloads/marketing/pb_csetf_gold_ger.pdf> accessed on 19.06.2012.

Curia Vista, Geschäftsdatenbank der Bundesversammlung 1989: *Postulat 89.788, Goldhandelsstatistik*. Veröffentlichung. Eingereicht von Rolf Seiler am 14.12.1989. <www.parlament.ch/afs/data/d/gesch/1989/d_gesch_19890788_002.htm> accessed on 13.05.2012.

Curia Vista, Geschäftsdatenbank der Bundesversammlung 2012: *Interpellation 12.3442, Ein- und Ausfuhr von Gold. Offenlegung der Statistik*. <www.parlament.ch> accessed on 03.01.2013.

Laitos, Jan G. (w.y.): The Current Status of Cyanide Regulations. *Engineering & Mining Journal*. <www.e-mj.com> accessed on 05.02.2013.

Lindt, Angela and Eva Schmassmann 2012: *Gold. Die Rolle der Schweiz in einem schmutzigen Geschäft*. Gesellschaft für bedrohte Völker (GfbV). <assets.gfbv.ch/downloads/goldbericht.pdf> accessed on 12.12.2012.

Luyken, Reiner 2000: Gold ist ein schmutziges Geschäft. Wie ein australisches Unternehmen in Rumänien eine Umweltkatastrophe auslöste. *Die Zeit Online* (24.02.2000). <www.zeit.de> accessed on 12.12.2012.

Newmont Mining 2010: Valcambi Refinery Driving Downstream Business. <www.newmont.com> accessed on 07.01.2013.

NZZ Folio 2009: *Was Sie über Gold wissen sollten*. 4 (April 2009). <www.nzzfolio.ch/images/upload/0409_Grafik_Karte_internet.pdf> accessed on 05.09.2012.

OECD 2011: OECD Due Diligence Guidance for Responsible Supply Chains of Minerals from Conflict-Affected and High-Risk Areas. Second Edition. *OECD Publishing*. <www.oecd.org/daf/internationalinvestment/guidelinesformultinationalenterprises> accessed on 23.12.2012.

OECD 2012: OECD Due Diligence Guidance for Responsible Supply Chains of Minerals from Conflict-Affected and High-Risk Areas. Supplement on Gold. <www.oecd.org> accessed on 23.12.2012.

Olden, Philip 2010: OECD Due Diligence Guidance for Responsible Supply Chain Management of Minerals from Conflict-Affected and High Risk Areas. Implications for the Supply Chain of Gold and other Precious Metals. Review produced for the OECD. <www.oecd.org> on 10.12.2012.

Responsible Jewellery Council (RJC) 2012a: Building a Responsible Diamond, Gold and Platinum Metals Supply Chain from Mine to Customer. <www.responsiblejewellery.com/files/RJC_Brochure_ENG3.pdf> accessed on 18.06.2012.

Responsible Jewellery Council (RJC) 2012b: Chain-of-Custody (CoC) Standard. <www.responsiblejewellery.com/files/S002_2012_RJC_CoC_Standard_PM.pdf> accessed on 19.12.2012.

Responsible Jewellery Council (RJC) 2012c: RJC CoC Certification Information – Metalor Technologies SA (15.05.2012). <www.responsiblejewellery.com/files/RJC-CoC-Certification-Information-Metalor-Technologies-SA.pdf> accessed on 04.01.2013.

Superintendencia Nacional de Aduanas y de Administración Tributaria (SUNAT) 2012: Regimenes Definitivos. Exportación definitiva. Principales Exportadores según País de Destino. Periodo Enero - Diciembre 2011. <www.aduanet.gob.pe/aduanas/informae/XPaisEmpMensual_01122011.htm> accessed on 21.06.2012.

Swiss Federal Customs Administration (FCA) 2012: Ein- und Ausfuhr von Gold, Silber und Münzen (14.05.2013) <www.ezv.admin.ch> accessed on 12.08.2013.

Swiss Federal Customs Administration (FCA) 2014a: Foreign Trade of Gold: First Publication Broken Down by Country Since 1980. Press release (20.02.2014). <www.news.admin.ch> accessed on 26.04.2014.

Swiss Federal Customs Administration (FCA) 2014b: Schweizerischer Aussenhandel. Ein- und Ausfuhr von Gold, Silber und Münzen. Periode Januar – März. (24.04.2014). <www.ezv.admin.ch> accessed on 26.04.2014.

Swiss Federal Council 2013a: Background Report. Commodities Report of the Interdepartemental Platform on Commodities to the Federal Council (27.03.2013). <www.news.admin.ch> accessed on 26.04.2014.

Swiss Federal Council 2013b: Gold Imports and Exports. Statistics Broken Down by Country From 2014. Press release (13.12.2013). <www.news.admin.ch> accessed on 26.04.2014.

UN Security Council 2005: Letter Dated 25 January 2005 from the Chairman of the Security Council Committee Established Pursuant to Resolution 1533 (2004) Concerning the Democratic Republic of Congo Addressed to the President of the Security Council. Document S/2005/30. <daccess-dds-ny.un.org> accessed on 03.01.2013.

UN Security Council 2006: Letter Dated 26 January 2006 from the Chairman of the Security Council Committee Established Pursuant to Resolution 1533 (2004) Concerning the Democratic Republic of the Congo Addressed to the President of the Security Council. Document S/2006/53. <www.securitycouncilreport.org> accessed on 03.01.2013.

UN Security Council 2010: Final Report of the Group of Experts on the Democratic Republic of the Congo. Document S/2010/596. <www.un.org> accessed on 03.01.2013.

U.S. Geological Survey (USGS) 2013: Minerals Yearbook. <minerals.usgs.gov> accessed on 12.08.2013.

U.S. Geological Survey (USGS) 2012: Mineral Commodity Summaries 2012. <minerals.usgs.gov/minerals/pubs/mcs/2012/mcs2012.pdf> accessed on 24.09.2012.

Valcambi Suisse 2012: Responsible Gold Policy Statement. <www.valcambi.com> accessed on 25.01.2013.

World Gold Council (WGC) 2012a: Gold Demand Trends. Full Year 2010 (February 2012). <www.gold.org/download/pub_archive/pdf/GDT_Q4_2011.pdf> accessed on 08.06.2012.

World Gold Council (WGC) 2012b: World Official Gold Holdings. International Financial Statistics (June 2012). <www.gold.org> accessed on 23.10.2012.

World Gold Council (WGC) 2013: Interactive Gold Price Chart and Downloads. Spot Gold Price in USD in Oz. <www.gold.org/investment/statistics/gold_price_chart> accessed on 04.01.2013.

The Rise and Fall of the Explorer Company Uranio AG

Thomas Niederberger

This case study is about a small company with domicile in Switzerland, named Uranio AG, which claimed to own large concessions for uranium exploration in Argentina and the Central African Republic (CAR) between 2006 and 2011. The company profited from the "gold rush" sentiment around the supposed revival of nuclear energy (pre-Fukushima). Its business model was however challenged by a constellation of local resistance in Argentina, power struggles in the CAR, and small shareholders using the Internet to "mine" and share all kind of unfavourable information about the company's claims.

The attention on large, transnational mining corporations should not make us forget the importance of a multitude of smaller actors in the mining industry. Many of the thousands of small companies active in consulting, subcontracting and exploration are as transnational as the large corporations. Dougherty (2011) points to the structure of the mining sector itself to understand why local resistance is building up against specific projects. He shows that in the case of gold mining in Guatemala the proliferation of "junior" mining firms is crucial. These juniors work mostly at the exploration stage, are financed by venture capital and operate under heavy competition. They typically "under-perform in terms of environmental and social standards". When "senior" companies buy a "junior" or its projects to fill their project pipeline, they also inherit the problems and resistance the junior has provoked so far (2011: 12).

From another angle, Tsing (2000) discusses the case of the Canadian explorer company Bre-X as an example of an "economy of appearances", whereby dramatic performance and spectacle are used to animate expectations about an imaginary "gold discovery" in a frontier region in Indonesia. The Bre-X "scandal" is now known as one of the biggest frauds in the history of mining; as Tsing shows, it is rather an articulation of international (mostly Canadian) finance capital with "greedy elite dreams" and a "franchise cronyism" in Indonesia and "migrant dreams of a regional frontier culture in which the rights of previous rural residents could be wiped out entirely to create a Wild West scene of rapid and lawless resource extraction" (ibid.: 121). In that sense, the Bre-X case builds on the old narratives of "discovery" and "gold rush", which resonates well with the story of Uranio AG.

From Argentina via "Wallstreet Online" to Bakouma

"Have you heard about that Swiss company, Uranio AG?" I was asked by a friend while travelling in the Argentinian province of La Rioja; "they own thousands of hectares of concessions for uranium in our province". He told me that this company, of which I had never heard before, had stirred the local movement against mining, and, in particular, mining of uranium. My friend took me to a 2000-people strong anti-mining march in the provincial capital of La Rioja on 20 August 2010. In several small towns like Chamical, Olta and Chepes, local people had formed assemblies

Fig. 1: "Uranium kills and contaminates—Governments and Capital too". Banner at an anti-mining march in La Rioja, 20 August 2010. Photo: Thomas Niederberger.

to confront the authorities and to demand explanations. The latter had never admitted that the concessions actually existed, but the locals referred to the information from Uranio's website, as well as to observing engineers doing exploration on site.

Back in Switzerland, I soon found out that Uranio AG in its legal domicile did not even have a proper office. Its address in Basel was a shiny law firm called "Neovius Advokaten", where Uranio's president, Peter Goeggel, offered his services to a whole bunch of other companies. Uranio was thus one among many "letterbox companies", or a "shell", as it is often used to hide questionable business practices and avoid legal responsibilities or taxes. Looking for information online, I stumbled upon two shareholder forums called "Wallstreet Online" and "Stocks", where the threads about Uranio AG had hundreds of comments, with links to all kinds of related news and documents. The discussion started in September 2007, after a very promising press release announced Uranio AG as the "first uranium value of Switzerland", to be traded at the "open market" of the Frankfurt Stock Exchange, Germany. The company claimed to own 29 concessions for exploration, with option for exploitation of uranium in the Argentinian provinces of Jujuy, Salta, La Rioja and San Luis on a total of 224,000 hectares (later reduced to 179,000 hectares), through its 100 per cent ownership of the Argentinian Uranio del Sur S.A. A report from the consultancy

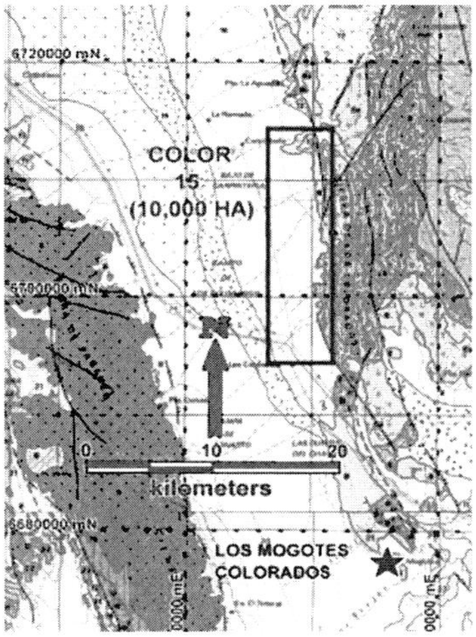

Fig. 2: Geological map of Uranio AG's concession in Los Colorados, La Rioja, taken from the company's website.

firm Yalta AG (2007) presented the company as a spinoff from the Canadian "Rome Resources Ltd.", with domicile in British Columbia, which remained the main shareholder. Especially mentioned was the fact that two "internationally renowned" geologists formed part of the board of directors of Uranio.

From the very beginning, many comments on "Wallstreet Online" and other online forums were sceptical about the seriousness of Uranio AG,[1] pointing out that the pattern looked similar to a "penny stock" scheme, using the less regulated "open market".[2] Wikipedia explains the "penny stock" or "pump and dump" scheme like this: "manipulators first purchase large quantities of stock, then artificially inflate the share price through false and misleading positive statements. [...] When buying pressure pushes the share price up, the rise in price entices more people to believe the hype and to buy shares as well. Eventually, the manipulators doing the 'pumping' end up 'dumping' when they sell their holdings."[3] However, other comments on the forum were enthusiastically pointing to the increasing demand in uranium, especially from China. The debate was thus between "bulls and bears", with both sides trying to convince the others, using information from the Internet and with reference to the up-and-down movements of the share value. The discussion was mostly in German, sometimes English, and citing articles in Spanish and French, often through Google translator (provoking more comments on how to interpret the confusing results).

The price of Uranio shares rapidly soared up to over 1.5 Euro in the second half of 2008, after the shareholders expectations suddenly shifted away from Argentina to the Central African

1 See web forums (all accessed 26 December 2012): Wallstreet Online; fondscheck.de; stockhouse.com.

2 Börse Frankfurt: "Open Market is not an organized market in the sense of the German Securities Trading Act (WpHG – Wertpapierhandelsgesetz). For issuing companies, the Open Market admission criteria are less strict than those for the two EU-regulated segments; Official Market and Regulated Market. Moreover, there are no follow-up obligations." <www.boerse-frankfurt.de/en/basics+overview/market+segments/open+market>, accessed 26 December 2012.

3 Wikipedia.org: "Penny Stock", accessed on 24 July 2013.

Fig. 3: Chart of Uranio AG's share value. Source: "Wallstreet Online".

savannah. In September 2008, Uranio had announced a reverse takeover of "Xenon S.A.", with domicile in the CAR, from "Lana International Ltd.", with domicile in the Seychelles. At the same time, the company increased the number of shares from 18 million to 225 million to finance the takeover. The acquisition was presented as a real coup. Uranio now owned, through Xenon, 100 per cent of "Les Mines de Centrafrique S.A.", holder of two uranium mining licenses over 200,000 hectares in the Bakouma region of the CAR. The concessions were directly adjacent to those that the French major Areva had bought through "UraMin" in 2007 at a staggering 2.5 billion dollars. A new consultancy report mentioned that the deal was made possible "through the assistance of Richard Ondoko, who was a consultant for UraMin" (Equity Development 2009). The move provided plenty of material to capture the small shareholders' imagination on "Wallstreet Online": Would Areva buy Uranio's newly acquired concessions too? Or, rather, "the Chinese", who reportedly had cast an eye on the region? However, the frenzy of the "bulls" did not last for long. In 2009, the shares fell to 0.04 Euro, where they remained for some time, to finish at 0.001 in 2012. So much for the market.

In the first months of 2010, the verbal fight on the Internet fora produced dozens of comments in only a few days, centring around the question: do Uranio's concessions actually exist, and if so, what interests are behind it? A second forum was opened on "Wallstreet Online" in February 2010, titled "Uranio AG—massively overvalued—strong sell". The commentators now assumed that Richard Ondoko was the main shareholder of Uranio. They dug out articles on African news-blogs, which again quote largely from "Africa Intelligence" (a pay-wall protected news-agency). Various articles link Ondoko, a businessman from the Democratic Republic of Congo, to Saifee Durbar, a businessman of Indo-Pakistani origin with domicile in London. The latter is described as maintaining close relations with the then minister of mines, Sylvain Ndoutingaï, and the President of CAR, François Bozizé.[4] The discussion became so heated that Uranio in February 2010

4 See various articles (all accessed last on 26 December 2012): *Le Confident*: "Les dessous de la mise en demeure de Richard Ondoko" <www.leconfident.net>; *Centrafrique-Presse*: "Feuilleton de l'affaire DURBAR" and "Bozizé aurait

felt obliged to deny the "rumours" by publishing a letter from Minister Ndoutingaï to Ondoko, confirming the ownership of the Bakouma concessions. However, in July 2010, Uranio announced the need for more capital to negotiate for the prolongation of the Bakouma concessions, and in March 2011, that the negotiations were still without results and therefore the emphasis of the explorations would shift back to Argentina.[5] Meanwhile, the tone of comments had calmed down and their frequency became less. An article in *Afrique Magazine* from February 2011 went unnoticed. There, Ondoko presented himself as majority shareholder of Uranio AG and mentioned that Areva had tried to take over all the Bakouma concessions in 2009, without success. He further comments about his mighty competitor: "Two currents confront each other within Areva. The reformists [...] who 'think business'. And the conservatives who represent that old Françafrique and continue to belief that one can deal by pushing political buttons, multiplying the pressures..." (Bourgeois 2010, *own translation*). A short notice by *The Economist* completes the picture: Areva in September 2010 declared that it will start work in Bakouma, after having "acquired several blocks adjacent to its concession from Uranio AG [...] owned by Richard Ondoko".[6] It seems that Ondoko had lost the power struggle, or struck a back-door deal, about which the small shareholders were kept in the dark. Uranio AG and Ondoko were last mentioned in an article about the huge scramble for nuclear power plant construction contracts in South Africa (Orange 2014).

Driven Out by Angry Farmers in Jujuy

The online commentators also dug out information about the concessions in the Argentinian provinces of Jujuy and La Rioja. Many of the links are from anti-mining sites, including the influential news-service *No A La Mina*. Apparently, in Argentina at that time everyone still believed that the Canadian "ROME Resources" owned Uranio. Different commentators interpret the information either as proof of a fraud or as proof of the actual existence of the concessions. Someone discovered that the High Court of Jujuy had apparently cancelled the concessions in that province.[7] An article from June 2010 reports that the mayor of Olta in La Rioja, after being confronted by local youth, had denied that uranium concession existed in his town—only to find himself contradicted by the governor of the province shortly afterwards.[8] These local youth, who took the initiative to distribute leaflets to inform their neighbours about what was planned on their land, were among those I had met at the anti-mining demonstration of 20 August 2010 in La Rioja. At the march, a member of the Environmental Assembly had confirmed that the Ministry of Mines of La Rioja completely refused to answer questions from citizens with respect to uranium concessions. Their only source of information was the company's website and direct observation of ongoing exploration in the concerned territories.[9]

commandé un Falcon 900" <centrafrique-presse.over-blog.com>; *Africa Times News*: "L'ex ministre centrafricain Saifee Durbar arreté en France" <www.africa-times-news.com>.

5 Uranio AG: "Press Release" <www.uranio.ch>, accessed 10 February 2012.

6 The Economist - Africa Business Briefing, 1 September 2010 <store.eiu.com/article.aspx?productid=1760000176&articleid=1287433913>, accessed 24 July 2013.

7 Referring to a note in the *Latin American Herald Tribune*: "Court Halts Open-Pit Mining in Northern Argentina" <www.laht.com> accessed 26 December 2012.

8 La RiojaVirtual: "Una minera canadiense explotará uranio en varias zonas de la provincia", 25 July 2010. <riojavirtual.com.ar>, accessed 10 February 2012.

9 Interview with Ramon Navarro, 20 August 2010.

Almost two years later, we had the opportunity to hear about what had happened on the ground in the province of Jujuy from Victoria Maria Veracierto,[10] who at that time was doing research on water management in the Puna and Quebrada region of that province. Actually, among the very first ones to know about the exploration were the indigenous peasants in the village of Juella, where she was staying. In 2008, they had observed pick-up trucks driving to the mountains, followed by the noise of explosions on various days. On the way back, the cars were forcefully stopped. It turned out that the occupants were engineers carrying out exploration works for Uranio del Sur. The people of Juella reacted by organizing a series of protests, a roadblock and a march to Tilcara, a major tourist centre in the region. There, an assembly was formed and representatives of the provincial government, including the "mining judge" (*Jueza de Minas*) who had issued the concessions, were called to a meeting. During the meeting, infuriated citizens turned up, chased the officials out of the municipality's building and forced them to escape from Tilcara (see also MU 2008).

At the judicial level, eight individuals demanded a temporary restraining order from the provincial court, to suspend all exploration concessions in the Quebrada de Humahuaca UNESCO World Heritage patrimony. After two negatives, the Jujuy High Court gave merit to the cause. Though it did not cancel the concession, as was falsely reported, it rebuffed the provincial court's decision on the ground that it had not taken into account the precautionary principle of environmental law. Also, judge Héctor Tizón considered that it was contradictory to allow environmentally damaging activities like uranium mining in a UNESCO protected area, that the UNESCO label was revocable and that this could put in danger the tourism sector. Thus, while formally the court simply sent the hot potato back to where it came from, the justification was considered historic, as Veracierto explains: for the first time in Argentina, the precautionary principle of environmental law—to prevent damage before it happens—was given priority above the mining law. It has thereby become a precedent for the environmental movement in Argentina, which since then is trying to stop mining projects in an early stage using the legal instrument of the temporary restraining order (Aranda 2010).

After the Rush

After Uranio's shares had lost their value, the company's domicile was transferred from Basel to Zug. Bankruptcy procedures were opened directly after that, on 20 June 2011, and the company was dissolved on 29 May 2012.[11] It remains unclear whether the main interest behind Uranio AG actually was the exploration of uranium, though it was certainly part of the scheme, especially in the first two years. The second stage from 2009 onwards looks more like a "penny stock" sale-out, which used the concessions to animate small shareholder speculation about fabulous profit opportunities. These concessions were apparently obtained through dubious channels and challenged in different ways. It looks as if a "penny stock" scheme was then used to transfer these "dirty" concessions into unregulated offshore cash to reimburse the partners who had arranged the deals. To further illuminate what happened would need experts in law and finances, rather than an anthropologist. The business model of Uranio AG could well be technically legal. At least, it was never challenged for fraud before a Swiss or German court. The damage is limited to losses for shareholders, each too small to justify the costs of legal action in a very unpromising setting

10 Veracierto is an anthropologist and lecturer at the University of Buenos Aires. Guest lecture organized by *infoe CH*, 19 March 2012, University of Bern.

11 Handelsregister Kanton Zug: "URANIO AG" <www.hrazg.ch> accessed 26 December 2012.

Fig. 4: Meeting after the march against large-scale mining, La Rioja, 20 August 2010. Photo: Thomas Niederberger.

of diffuse responsibilities spread across various jurisdictions. The scheme used the "serious" images of Switzerland and Germany (for the stock market), while profiting from weak regulations. The case went unnoticed by media or watchdog NGOs that focus on worst-case examples of large corporations.

The former CEO and president of Uranio, Peter Goeggel, reappeared on the board of Auris AG "[...] a holding company in the mineral exploration industry, which [...] wants to explore and evaluate the natural wealth of Africa, in particular precious metals on its properties in the Democratic Republic of Congo."[12] Auris claims ownership of exploration concessions on 325,000 hectares in the Nord Kivu province in the north east of the DRC. The region is known for artisanal gold mining (Auris' website shows photos of local artisanal miners' discoveries) and has attracted the interest of major mining corporations in recent years.[13] Also, it is among the worst hit by "resource

12 Auris AG: "Company Profile" <www.aurisag.ch> accessed 27 December 2012.

13 Interview: Maurits van den Berg, Auris AG, in "financial.de", 21 October 2011.

wars". On the "Wallstreet Online" forum, Auris does not find many friends because commentators remember the bad example of Uranio AG.[14] In a newspaper article of 2012, the businesses of advocate Goeggel are described with yet another example of a shady explorer company called Moku Goldmines, pursuing a similar strategy in the same region (Heim 2012).

In Argentina, the temporary restraining order demanded by the local neighbourhood assembly against the exploration permit was sent back and forth between the local court, the Jujuy High Court and the Supreme Court. The question remains unresolved, and the only judge supportive of the measure in the Jujuy High Court, Héctor Tizón, has died in the meantime.[15] From the Central African Republic news trickled in that a rebel army called *Seleka* was about to take the capital city Bangui, which they finally did in March 2013. President Bozizé was forced to leave the country to Cameroon.[16] Richard Ondoko was again mentioned in the context of a debate about sending troops to the CAR to defend South African mining interests.[17] Interestingly, the French had not done much to defend Bangui against the rebels, except protecting the airport—contrary to Mali, where they started a major military intervention at about the same time (amongst other reasons, to protect Areva's uranium mines in neighbouring Niger). In the CAR, according to Areva's website, the company had decided to suspend mining operations in September 2012 "due to low uranium prices since the Fukushima accident and the current security issues".[18] The CAR was meanwhile going through a humanitarian crisis. Various factions within the camp of, and opposed to, Seleka were fighting over the control of the capital city and the hinterlands. By the beginning of 2014, the country was in full civil war, with ethnic violence between Christian and Muslim population groups killing thousands of people and leaving many more homeless.

Conclusion

The story of Uranio shows how speculation in future gains, underlined with mineral concessions of a rather doubtful kind, can itself become a commodity—long before major companies come in and actual mining starts. The Argentinean part represents an example of how resistance against mining projects at the early exploration stage can be quite successful. In that stage, the company is weak, without much capacity to buy local support. In the case of CAR, we do not know about local peoples' reaction, as their capacity in making their voice heard seems less strong. Here, the role of online blogs and news-agencies in providing information about the officials of African governments and ministries and their ties to business networks was more important.

Further, the case shows a strange collusion of interests between the locals and the small shareholders. While the company was forced to publish enough information to please the shareholders, the Argentinean authorities blocked all requests, making the deal look even more dubious. The small shareholders self-organized through online forums to access more accurate information, and ended up reading about local peoples' protests. As a possibility for the future, we can imagine that local people enter into direct contact with self-organized shareholders searching for "ground knowledge", to make them understand that their investment is not welcome, and that the concessions they are speculating upon lack legitimacy.

14 Wallstreet-online.de: "Auris AG - die neue Moto Gold Geschichte", accessed 27 December 2012.

15 Personal communication Maria Victoria Veracierto, by email, 8 February 2013.

16 Reuters: "Rebels capture Central African Republic capital, Bozize flees", 24 March 2013. <uk.reuters.com>.

17 Mail & Guardian: "Who wants SA back in Bangui?", 24 April 2013. <mg.co.za>.

18 Areva: "resources centreafrique" <www.Areva.com>, accessed 24 July 2013.

References

Aranda, Dario 2010: Nace un megaproblema para la megaminería. *Pagina/12*, 24 April, Buenos Aires. <www.pagina12.com.ar>.

Bourgeois, Jérôme 2010: Uranium: un enjeu africain. *Cahier Business - Afrique magazine*, February 2010(293).

Dougherty, Michael L. 2011: The Global Gold Mining Industry, Junior Firms, and Civil Society Resistance in Guatemala. *Bulletin of Latin American Research* 30 (4): 403-418.

Equity Development 2009: Uranio AG. London, 21 October. <www.equitydevelopment.co.uk>.

Heim, Michael 2012: Basler Rohstofffirma wollte im Kongo Gold finden. *Schweiz am Sonntag*, 12 August 2012.

MU 2008: Apunando a las mineras: Tilcara le ganó a la minería a cielo abierto. *el periódico de lavaca,* August, Vol. 17. <media.lavaca.org/pdf/mu/mu17.pdf>.

Orange, Martine 2014: Uramin: la diplomatie parallèle d'Areva en Afrique. *Mediapart.fr,* 13 December.

Tsing, Anna 2000: Inside the Economy of Appearances. *Public Culture* 12(1): 115 - 144.

Yalta AG 2007: Research Report. <www.yaltaadvisory.de/fileadmin/user/.../yalta_research_uranio.pdf> accessed 27.12.2012.

Online Fora

(all checked last on 26. December 2012)

Fondscheck.de: "Uranio AG. Trendwende am Uranmarkt eingeläutet" <www.fondscheck.de>.

Stocks.ch: "Uranio AG – Top oder Flop?" <forum.stocks.ch>.

Stockhouse.com: "Is this a real company?" <www.stockhouse.com>.

Wallstreet Online:
"Uranio AG – der erste Schweizer Uranwert".
<www.wallstreet-online.de/diskussion/1129754-1-10/uranio-ag-inh-der-erste-schweizer-uranwert>
"Uranio AG—massively overvalued—strong sell" .
<www.wallstreet-online.de/diskussion/1155972-1-10/uranio-ag-massive-ueberbewertung-strong-sell>

The Mopani Copper Mines in the Copperbelt District, Zambia

Eva Schober, Amadea Tschannen, Thomas Niederberger[1]

1. Situating the Case Study

The social, political and economic reality of the Zambian state and its people is linked to a high degree with one single mineral, copper. Zambia holds the largest reserves of copper in Africa, 6% of known copper reserves in the world, and currently is the seventh largest producer of recovered copper worldwide (Cunningham 1981: 16; ICSG 2012: 10). Its economy has been dominated by the export of copper since the 1920s, when large deposits were discovered in the Copperbelt district. The Lamba, who are the original inhabitants of the region, were largely displaced with subsequent urbanization and represent a highly marginalized minority today. Instabilities of world commodity prices have had profound impacts on the country. In the 1990s, after a period of low copper prices and a debt crisis, large parts of the national mining company Zambia Consolidated Copper Mines Ltd. (ZCCM) were sold to private investors such as Glencore International, the current majority owner of Mopani Copper Mines (MCM). MCM runs four underground mines, a smelter and a cobalt plant in the Copperbelt district's mining towns Kitwe and Mufulira. Glencore's Mopani mines supposably cause major social and environmental problems. Furthermore, the case exemplifies tax avoidance strategies, on the one hand, and international NGOs' counter-strategies on the other.

1.1 History and Current Context

Zambia's history of the copper industry starts with the colonial era. In the 1850s, the missionary and explorer David Livingstone visited the region, which is today known as the Copperbelt, and he published the book Missionary Travels and Researches in South Africa (Livingstone 1858). The promising commercial possibilities he described convinced the British government to finance a new expedition and to support Livingstone's research. He thus became not only a missionary for Christianity, but also for industrial revolution (Roberts 1981: 151f.). In the late 19th century, the British South African Company (BSAC) took control of the region. Cecil Rhodes, a British millionaire and founder of the company, convinced the British government to allow his company to make exclusive claims in the territory, which was named "Rhodesia" in 1895 (Roberts 1981:

1 Review by Stephy-Mathew Moozhiyil.

157, 162). During the so-called "Company rule", the northern part of Rhodesia, which was to become Zambia, was important mainly for recruiting cheap labour for the mines in Katanga under Belgian control (in today's Democratic Republic of the Congo, DRC) and as a link to other regions in southern Africa. In 1897, a railway passing through Northern Rhodesia was built to transport copper from Katanga to Beira (in today's Mozambique). In the following years, white immigrants settled and built large farms along this route. With the railway, cheap products from overseas were brought to the local market now, which led to growing competitive pressure for the local small-scale farmers. The locals were thereby pressured into accepting wage labour to pay taxes (ibid.: 175ff.).

The Company rule ended in 1923 and the administration of Northern Rhodesia was formally assigned to the Colonial Office in London. The BSAC retained its mineral rights in the north-western part. In order to attract European immigrants, exclusive areas for white farmers were created along the railway, the best location for large-scale commercial farming. For this purpose, the colonial government relocated about 60,000 indigenous people into native reserves. In the 1920s, vast copper deposits were discovered in the so-called "Copperbelt", near the Kafue River. New technical developments allowed more profitable mining (Fraser 2010: 4). By 1930, the major mines in Lusanshya, Nkana, Mufulira and the other mining towns in the region employed more than 30,000 workers and by 1945, Northern Rhodesia was one of the major copper producers in the world (Ferguson 1999: 49). Within a few years, areas formerly viewed as unused "bush land" had transformed into industrial and urban zones (Fraser 2010: 4). The increasing demand for unskilled mine labourers led to vast labour migration and urbanization (Roberts 1981: 182ff.). At first, the labour force was strictly segregated between unskilled contract workers (the "Blacks") and the skilled and better paid positions of the "Whites". The biracial politics of the European Mine Workers' Union prevented black Africans from entering a higher position (Powdermaker 1962: 90). It was not until the 1950s when black unions were formed. They were united in 1949 in the Northern Rhodesia African Mineworkers' Union and their status was protected by a government ordinance, which assured that there would be no distinction between European and African unions (Roberts 1981: 205).

In 1953, Northern and Southern Rhodesia, together with Nyasaland (today's Malawi), were grouped in the Central African Federation (Taylor 2006: 12). The colonial government intended thereby to connect the modern industry of the south with the mineral resources and labour force of the north (Roberts 1981: 195). Each state retained its own government and kept power also regarding the so-called "Native Policy Issues", while the federal government was responsible for trade, defence, industry and finance (Taylor 2006: 13). The formation of the Central African Federation was the beginning of the independence movement in today's Zambia. The Native Policy Issue was the crucial point because the people of Northern Rhodesia feared that the extensive land expropriation policy of Southern Rhodesia would also be applied to the North. In 1951, the Northern Rhodesian African National Congress (ANC) was formed on the plank of a welfare society, which defended the interests of the mining communities (Roberts 1981: 204ff.). In 1958, the new constitution of Northern Rhodesia gave voting rights to 25,000 black Africans. Within the Congress, a group of young politicians formed around Kenneth Kaunda to build a new opposition party, the United National Independence Party (UNIP). In the elections of 1962, UNIP and ANC won the majority in the parliament and took over several departments. Two years later, after the first free elections, Kenneth David Kaunda became Prime Minister and the independent state of Zambia came into being on 24 October 1964 (Roberts 1981: 220ff.).

After Independence

After some years of a coalition between the ANC and the UNIP, Kenneth Kaunda introduced a single-party system, based on which he ruled the country in an autocratic manner for the following thirty years. His political style, which he called "Zambian Humanism", was inspired by socialist philosophy. Core features were central planning of the economy and nationalization of key sectors, including the copper mines, in order to develop and subsidize the non-core sectors and reinforce the social security system. Hence, Zambia's economic success remained highly dependent on the mining sector (Fraser/Lungu 2007: 7f.). By the end of the 1980s, the UNIP faced growing opposition, mainly from the mine workers' coalitions.

The interference of the party in industrial activities was said to be one of the main causes for diminishing productivity (Lungu 2008: 405). Together with a sharp drop in copper prices, starting in 1975, this triggered an economic and political crisis, followed by urban food riots in the Copperbelt in 1986. The downturn of the economy created momentum for a new party, the Movement for Multiparty Democracy (MMD), founded around the labour union leader Frederick Chiluba, and ultimately led to the fall of Kaunda's government in 1991 (Ferguson 2006: 81). The MMD won the elections and introduced important constitutional changes. The liberalization of the Zambian economy and the privatization of the national enterprises were declared as the party's main political program (Lungu 2008: 405). Foreign donor institutions encouraged the country's transition to multi-party democracy and liberal capitalism. Despite some initial success, the expectations with regard to the reforms remained unfulfilled. Employment and national wealth further diminished, and allegations of rampant corruption came up (Fraser 2007b: 7). In 2007, former president Chiluba was found guilty by a London court of stealing USD 23 million of public money.[2]

In the past couple of years, Zambia saw GDP growth rates of 5 to 6%, due to increased copper prices and foreign investments.[3] But poverty and corruption are still major problems. Michael Sata, who was elected as president in 2011, promised to bring back prosperity to Zambia by investing the profits from the mining industry in the national economy and public sector. However, most of his promises seem to have vanished shortly after the elections.

1.2 Geography and Ecosystem

Zambia is situated in the southern part of Africa, landlocked between eight countries: Angola, Namibia, Botswana, Zimbabwe, Mozambique, Malawi, Tanzania and the Democratic Republic of the Congo. Its surface covers 743,390 km2, with 14 million inhabitants, which makes a relatively low population density of 18 people per square kilometre (Taylor 2006: 2f). The country lies on the great central African plateau, in the southern tropics, mostly higher than 900 meters above sea level. Its tropical climate is moderated in most parts by the high altitudes and long rainy season. Zambia's rainy season lasts from November until March/April. The temperature ranges from 19-25° C in this period. Annual rainfall amount is highest in the north and north-west (from 1000–1500 mm) and decreases towards south, near the capital Lusaka (640–1000 mm) (Taylor 2006: 5f.). From May to August, the climate turns cool and dry. From August to November, the temperature rises and it gets generally hot. Most of the country's soils are composed of old crystalline rocks, such as granite. These soils are acidic, sandy and poor in nutrition. In the Southern Province, the soils are of higher quality for agriculture than in the north. These areas are suitable for cattle rearing and also maize and other cash crops grow easily. On the north-eastern and the north-western plateaux, the soils are profoundly washed out because of the short but heavy rains in the wet season. These areas are partly covered by smallish trees, bush- and grassland. Depressions of the plateau, which are exposed to annual flooding, such as the Kafue Flats and the upper

2 Matheson, Ishbel 2007: "Chiluba's legacy to Zambia." *BBC News* 04.05.2007. <news.bbc.co.uk> accessed on 26.01.2013.
3 World Bank 2013: "Zambia overview." <www.worldbank.org> accessed on 15.07.2013.

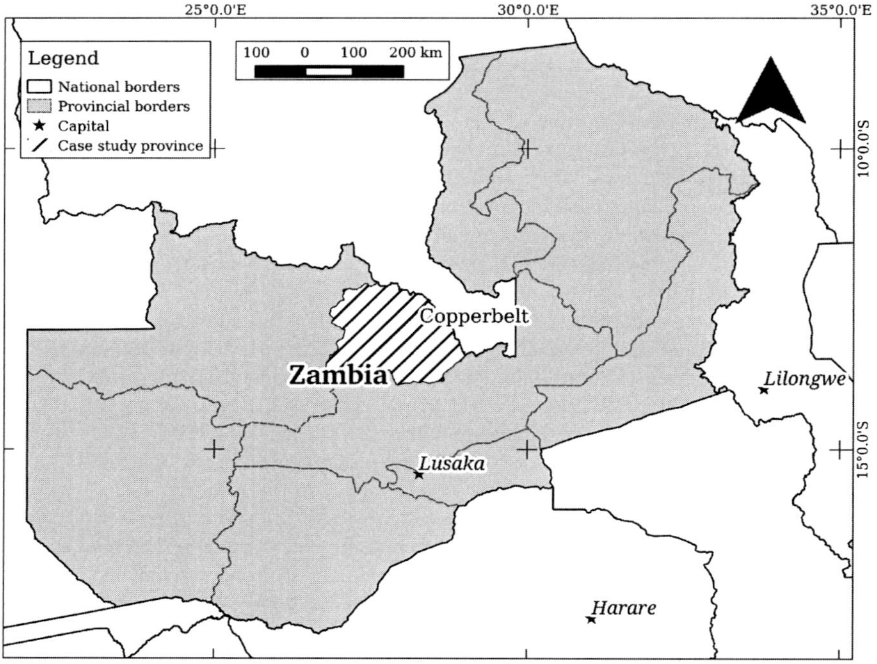

Map 1: The Copperbelt Region within Zambia.

Zambezi floodplains, also provide excellent grazing areas (Roberts 1981: 8f.).

The focal region of this chapter, the Copperbelt lies in the centre of Zambia, at the border of the DRC. It covers roughly 800 km^2, from Luanshia in Zambia in the southeast to Kol Ouezl in the DRC in the north-west, and is situated at an altitude of 1200–1300 metres on the Zambian plateau. The region receives a relatively high annual amount of rain (over 1000 mm) and the soils are poor in nutrition and therefore unfavourable for intensive cultivation. Prior to the discovery of extensive mineral deposits, the region was only sparsely populated. Most of the Copperbelt's numerous streams lead into the Kafue River, which flows into the Zambezi River further south (Gabel 1976: 3; Ferguson 1999: 148).

1.3 Local Population and Resource Use

Before the Company rule (starting in 1890), the region of todays Copperbelt district was mainly populated by different tribes related to the Bemba group, including the Lamba (Powdermaker 1962: 24; Roberts 1981: 128f.). Roberts (1981) reports that in 1890, Joseph Thomson, a representative of the British South African Company (BSAC), negotiated a treaty with chief Mushili I., securing exclusive mineral rights in the Lamba territory (Roberts 1981: 159, 192). In return, Thomson promised to the Lamba protection against other groups like the Ngoi warrior herdsmen from the South and the Chikunda raiders, who were former soldiers of the Portuguese (ibid.: 117f., 125). Thereby, the Lamba were displaced to native reserves in areas far from the mines and railway lines, which reduced their territory by 80% (Siegel 1989: 354f.).

Social Organization of the Lamba

A classical ethnography of the Lamba was published in 1931 by Clement Martyn Doke, a linguist with a missionary background. He describes a tribe organized in independent chiefdoms, with village headmen, several group-chiefs and a paramount chief (Doke 1931: 50f.). The power of the paramount chief was directly related to the belief in supernatural beings, mainly the guardian spirits of the deceased former chiefs (Powdermaker 1962: 30; Doke 1931: 238). These spirits gave the chief the power to influence the prosperity of the land and the welfare of the people. Chieftainship was succeeded through matrilineal descent through the mother's clan (Doke 1931: 35f.). Generally, the Lamba followed matrilineal descent rules and mothers played a major role in their children's lives, especially regarding marriage (Powdermaker 1962: 31; Doke 1931: 160). Marriage within the same clan was prohibited and residency was matrilocal. Divorce and remarriage were quite common, whereby loss of property through divorce was not an issue as there was no private property in land (ibid.: 170ff.). Rather, access to land was regulated through membership in the matrilineage. The Lamba, like most of the ethnic groups in the area, believed in witchcraft as well as in positive and destructive magical powers (Powdermaker 1962: 41; Doke 1931: 302f.).

Local Resource Use and Institutions

In 1989, anthropologist Brian Siegel described the Lamba people practising subsistence economy, using slash-and-burn cultivation (Siegel 1989: 351). Citing Siegel, it seems that some of Doke's observations were still accurate at that time. According to him, the Lamba cultivated maize, sorghum, different vegetables and fruits. They held domestic animals such as goats, chicken and wild pigs, but no cattle—an important feature distinguishing them from other tribes (Doke 1931: 27, 95-97). The men also practiced seasonal hunting. Doke observed that the Lamba practised a rather mobile village settlement pattern, moving elsewhere when soils became less fertile after some years of cultivation. Arriving at a new place, woods were cleared and dried for burning. The resulting cultivated land was partly ash-covered and roughly hoed soil (ibid.: 85–87). The land of the Lamba was group-based common property and managed by the paramount chief in collaboration with the group chiefs. Land use and access was governed according to strict rules. Hunting or fishing outside the paramount chief's territory was not allowed and could even cause war between different groups (ibid.: 49, 75). Although there are only a few ethnographic accounts, we can assume that the traditional resource use and institutions of the Lamba must have changed dramatically with the European settlements and industrialization. Especially because of the institutional change from common to state property of land and related resources, which meant the loss of territory and mobility (see Haller 2013).

Urban Population

The construction boom of new large-scale mines at the end of the 1920s was accompanied by migration from the rural areas to the emerging urban centres in the Copperbelt. Employment in the copper mines more than tripled between 1927 and 1930. Mining companies contracted villagers from the rural areas or hired men migrating from other mining towns in Southern Rhodesia or the Belgian Congo. Some companies preferred to hire single men, migrating from town to town or between the rural and the urban areas, on short-term contracts. Nevertheless, Ferguson highlights that many companies, wishing for permanent urban settlement of their labourers, were encouraging the presence of women and families as a means of gaining a competitive advantage over other labour markets in the south (1999: 48f.). This led to heavy growth of the urban population, based on prosperity of the mining industry. However, many workers maintained connections with their villages by marrying a woman from their home towns or by moving back to rural areas right after retirement (ibid.: 77, 72). "Going back to the land" was strongly encouraged by the mining companies; especially, at the time of economic decline during the 1980s and 1990s (ibid.: 124ff.). The miners who went back to their home villages often faced jealousy and mistrust (ibid.: 141ff.). In order to avoid such problems, a minority of the retired miners were also settling in the nearby rural areas among the Lamba (Siegel 1983, cited in ibid.: 133) Census data show that the phenomenon of urban-rural migration persisted: "By 1990 [...] working-age men continued to be strongly overrepresented in the population, and older people (especially, women) strongly underrepresented" (ibid.: 73).

With large-scale urbanization of the Copperbelt and massive immigration, the original local population of the region, including the Lamba, has become an almost forgotten minority (Roberts 1981: 108, 128). Today, most of the people living in the area of the mines are descendants of workers who immigrated during the last 70–80 years from other parts of former Southern Rhodesia and neighbouring countries. Today, with 62.5 persons per square kilometre the Copperbelt district has the highest population density after Lusaka, compared with the national average of 17.5 persons per km^2. Most people live in the area of Kitwe, one of the major mining towns, with 362,423, and in Mufulira with 137,272 inhabitants in 2000. Lamba-speaking people represented 1.9% of the total population of the country (Central Statistical Office of Zambia 2000: 14, 42; 2011: 4f.). In 1989, around 125,000 Lamba were living in the Copperbelt district, many of them as subsistence farmers in the rural areas (Siegel 1989: 351). There are about fifteen Lamba groups living across the sparsely populated rural region of the Copperbelt district today. According to Tuesday Musaka, an indigenous Lamba, the integration of the tribes into the urban context remains minimal. Those groups who continue to live in the rural area are very marginalized, and labour connections to the mining towns hardly exist. Illiteracy and insufficient education represent high obstacles for Lamba people, making it almost impossible to find employment in the urban centres and in the mining industry.[4]

1.4 National Economy and Legal Framework

The agricultural sector absorbs about 85% of Zambia's labour force, most of them as subsistence farmers. Nevertheless, agriculture only accounts for 1/5 of the national GDP today (Hart Group 2013: 19). With the establishment of large-scale copper mines in the 1920s, Northern Rhodesia developed into a mono-sectorial economy. In and around the prosperous copper-towns agriculture was largely replaced by wage labour (Davis 1976: 178ff.). Until now, urban centres in the mining regions have the highest economic growth rates and the lowest poverty rates. Meanwhile, these regions are very vulnerable to external factors like alterations of copper prices and changes in economic policies. Due to repeated failures to make the copper industry support poverty reduction and development in other sectors, today 70–90% of the people in rural regions live below the poverty line of one US-dollar per day.[5]

For 2013, the Zambia Extractive Industry Transparency Initiative (ZEITI) reports that the mining sector is directly accountable for 11% of the GDP; indirectly however, for up to 50% (Hart Group 2013: 19, 38). In 2010, 78% of Zambia's export products were primary or secondary products of copper (at an estimated value of USD 5.8 billion) and 4% were cobalt products (USD 304 million), an associated mineral in the production process of copper (ibid.). Astonishingly however, the mining sector contributed only 8% to the state's total tax revenue, according to estimations for 2008 (World Bank 2011: 15). Interrupted by the world economic crisis in 2008–09, copper prices and foreign direct investment have been rising steadily after the privatization program of the late 1990s. According to the Ministry of Mines, Energy and Water Development (MMEWD),

4 Email correspondence with Tuesday Musaka, an informant from a Lamba village.

5 World Bank 2013: "Zambia overview" <www.worldbank.org/en/country/zambia/overview> accessed on 27.03.2013.

Fig. 1: Workers of MCM in Kitwe in the elevator that brings them to their workplace 1.5 km under the earth's surface. Photo: Meinrad Schade.

the total output of copper reached 881,108 tonnes in 2011, with a projected output of 1.3 million tonnes by 2015. Today, Zambia is the world's seventh biggest exporter of copper and the second biggest exporter of cobalt worldwide and is likely to figure among the top five copper producers in a few years (Hart Group 2013: 19ff.). According to the Central Statistical Office of Zambia (CSOZ), Switzerland is the biggest export partner, with more than 50% of the exported goods, directly followed by China.[6] This reflects Switzerland's importance for copper trading.

Nationalization and Re-Privatization of the Copper Industry

After independence in the 1960s and 1970s, a rapid economic growth started, driven by high copper prices. Zambia developed into one of the wealthiest states of Africa and was described enthusiastically as a role model for growth, built on political and economic independence. In the 1970s, Zambia's GDP was one of the highest on the continent. A quarter of the population was living in urban areas and almost the same number was formally employed. During that time, the mining sector contributed up to 50% of Zambia's foreign trade and of the government's revenues (Cunningham 1981: 16; Fraser/Lungu 2007: 1ff.). Following the "Zambian Humanist" ideology

6 Mfula, Chris: "Zambia plans to double mining contribution to GDP." *Reuters*, 10.01.2012.

under the presidency of Kenneth Kaunda, the copper mines were nationalized: mining companies were forced to transfer 51% of their shares to the state and important parts of the profits from the copper export were invested in healthcare, education and the social security system (Fraser/Lungu 2007: 7f.). The copper boom fostered attempts to liberate the country of the inherited political and economic dependency on European countries. Zambia was one of the founding nations of a union of copper producing countries (together with Chile, Peru, and DRC/Zaire). However, the union was not able to gain enough influence on the world market's copper prices, as "[...] the states involved were simply too poor to force concession from rich countries unwilling to support a project designed to reduce dependence" (Fraser 2010: 7).

After some years of economic growth, the euphoria about the nationalized mining sector vanished. The close relationship of the mining companies to the state seemed to encourage ineffective management and declining productivity. The government failed to invest the profits into diversified, sustainable development and therefore, Zambia was severely struck by the oil crisis and the subsequent collapse of copper prices in 1974–75. Many national mines had to close down, others were grossly neglected and no investments in maintenance made (Soest 2009: 89). Nevertheless, the salaries of the urban population and the miners remained relatively high, not least because of the efforts made by the labour unions. To guarantee the high living standards of its people, Zambia was forced to take up loans from the International Monetary Fund (IMF), coupled with the implementation of a first World Bank structural adjustment program in 1983. This program included the devaluation of the currency in order to reduce import demand and make exports cheaper, the removal of trade barriers and the implementation of an overall liberalization policy (Esomba 1996: 97). Ferguson shows in his analysis of the World Bank's and the IMF's structural adjustment programs that they had serious social consequences: sky-rocketing food prices, abandoned health care, decline of employment and incomes, led to the impoverishment of Zambia's population. He argues that the neoliberal ideology, based on a cold, technocratic thinking, led to a "de-moralizing" of the policies on the whole continent (2006: 70f.).

Zambia's external debt grew from USD 627 million in 1970 to USD 7.2 billion in 1990. The inflation rate was at 107% in 1990 (Soest 2009: 88). The state reduced expenditures and abandoned public as well as social infrastructure. In three decades, Zambia's per capita income had shrunken by 50%. Under the pressure of the international donor agencies, the neoliberal economic strategy was enforced during Frederic Chiluba's presidency (1991–2001). Following a second structural adjustment program, the national mining company ZCCM was ultimately sold to private investors, as copper prices were not expected to rise again (Fraser/Lungu 2007: 9f Simpere 2010: 4; Craig 2001: 397). The main goal of the privatization process was to attract external investment in order to make the mines profitable again. Therefore, the government introduced policies designed to create a highly investment-friendly environment. The ZCCM was unbundled into seven units and sold to six foreign companies between 1997 and 2000 (with a second selling round in 2006). ZCCM was transformed into an investment Holding company, which holds minority interests in most of the privatized mines (Fraser/Lungu 2007: 11–14; Hart Group 2013: 17). In 2000, at a historic low point of copper prices, MCM, one of the two most significant bundles of ZCCM, was sold to Glencore (73.1%) and First Quantum (16.9%). The state-owned ZCCM retained a minority share of 10% (Fraser/Lungu 2007: 14).

During the selling process, world copper prices kept declining and Zambia's negotiating position weakened. Figure 2 shows the development of copper prices at the London Metal Exchange from 1985 to 2013. Copper prices began to rise very rapidly in 2005, just after the last bundle of ZCCM had been sold to foreign corporations. Interrupted by a slump during the 2008–2009 financial crisis and a peak at USD 10,000 per tonne shortly afterwards, the copper price is hovering between 7,000 and 9,000 USD per tonne since 2006, which is about four times higher than it had been in 2000. Yet, besides the market pressure, the privatization process was also marked by high degrees of corruption. Experts estimated that the selling price of ZCCM should have been more

Fig 2: Evolution of copper prices from 1985–2013 in USD/tonne (LME).

than five times higher than the final USD 627 million, so that there has been undoubtedly some trade-off (Simpere 2010: 5).

Legal Framework and Taxation

During the Company rule, the British South African Company (BSAC) owned all the mineral concessions. As they were based on contracts with local chiefs, whose territories were not clearly demarcated, their legal status was very questionable. In the negotiations with some of the larger chiefdoms like the Lozi and the Lamba, the company used false pretences and dubious methods to gain access to the territory (Roberts 1981: 159; Taylor 2006: 11). The company remained in possession of the mineral extraction rights after the country had become a British Crown Colony. In the 1930s, copper mining investment companies like the Anglo American Corporation and Rhodesian Selection Trust had to pay royalties to the BSAC, as owner of the concessions. Therefore, a large part of the copper revenue went to the BSAC headquarters in London, which paid its taxes to the British government (Roberts 1981: 186, 192f.). After Independence, the new Zambian government challenged BSAC's legality and subsequently nationalized the concessions against a symbolic payment (Siegel 1989: 192f.; 222). Moreover, according to the socialist idea, land possession was nationalized in order to benefit the whole population.

Twenty years after the forceful displacement of most of the Copperbelt's local population in the 1920s, the government designated some additional 40,000 km2 as "Native Trust Land" for the exclusive use of the Northern Rhodesian native people. On the other hand, the so-called "Crown Land", which had been identified as suitable for European agriculture and mineral quarrying, remained mostly controlled by white settlers. This changed after independence, during Kaunda's presidency. In practice, many inequalities remained: "Racial segregation ended with colonialism and subsequent divisions have been based on income (or class) rather than race. But the impact of colonial thinking and town planning has left an indelible mark on the Copperbelt" (Hansungule et al. 1998: 10). As an important part of the neoliberal strategy of the 1990s, national property like land and minerals was privatized again.

With the Land Act of 1995, an attempt was made to formalize land rights. It holds that the customary land tenure should be transformed into legal land titles by consultation with the local chief. Zambia's administrative authorities apparently are confronted with many institutional obstacles and lack the capacity to regularize customary tenure. Instances have been reported where people without legal land titles are evicted, as they are seen as "squatters" on their own land.[7] While the mining companies used to provide land and housing for their employees in earlier times, today's labourers are struggling not only for employment, but with the sale and expansion of mining operations, they also have to compete over land with multinationals.

With regard to environmental legislation, the revised Mines and Minerals Act provides that companies applying for large-scale mining should submit an environmental management plan, which includes measures to prevent pollution and overexploitation of water, land and other resources. Furthermore, they have to prepare a proposal on how to promote local business as well as employment and training of local citizens (Hansungule et al. 1998: 27). However, this law is considered vague, and it is followed only reluctantly by some companies. Moreover, the corporations have to submit annual reports to the Mines Safety Department in order to inform about their performance in safety and environmental terms. As the authorities rarely pursue preventive audits but only react to incidents, the effectivity of this legislation is questionable (Fraser/Lungu 2007: 26).[8] Thus, despite the existing legal framework, precarious working conditions and environmental pollution are frequent.

During the selling process of the ZCCM, personalized contracts were negotiated between the state and the mining corporations, which defined the terms, rights and responsibilities of both parties. These so-called "Development Agreements" (DA) were supposed to ensure beneficial conditions for both parties. In practice, they paved the way for environmental destruction and minimal benefits for Zambia, as they include exemptions and privileges for the companies in fiscal, environmental and social terms (Simpere 2010: 5; Fraser/Lungu 2007: 15). The DAs stipulate periods of 15–20 years during which the corporations are exempted from fines for environmental pollution, which does not exceed the level established by ZCCM in the preceding era (Fraser/Lungu 2007: 34).

Tax revenue is the main benefit the mining industry brings to the Zambian state, as there exist very few linkages to other industries, and employment generation is small (Gregow/Hermele 2011: 15, 17). In addition, foreign companies own most quarrying licences. Due to those circumstances, fair and accurate tax rates are crucial.[9] For Mopani, as for most other companies, mineral royalties were set in the DAs at only 0.6% of gross revenue. Taxation of import of material and machinery was suspended completely (Fraser/Lungu 2007: 70). As a result, the mining corporations were making huge profits during the last decade. Most probably, this is a result of the very weak negotiating position of the Zambian government during the selling process, caused by indebtedness and low copper prices (ibid.: 17). The revised Mines and Minerals Act of 2008 repealed the Development Agreements for all corporations, in reaction to the imbalance in the distribution of gains (Mwambwa et al. 2010: 7). The corporate income tax was raised from 25% to 30% and royalty fees from 0.6% to 3%, which is still low compared to other countries like Chile, with 5 to 10% (Soest 2009: 77ff.; BD 2011b: 2). When copper prices were slumping in 2009 as a result of the international financial crisis, the state again made concessions to the companies, fearing they would otherwise withdraw from the country.

7 For a detailed study on the "squatter" question, see Hansungule et al. (1998). For an example from the Copperbelt Province Luapula, where more than 500 people were evicted to make place for mineral prospecting, see: IRIN, 30.08.2010: "Zambia: Land and mineral rights in conflict". <www.irinnews.org> accessed on 23.07.2013.

8 Human Rights Watch 2013: "Zambia: Safety Gaps Threaten Copper Miners" <www.hrw.org> accessed on 30.12.2013..

9 For an overview on taxes in the mining industry, see Mwambwa et al. (2010: 4).

Mwambwa et al. show that the taxes from the mining industry contribute only marginally to Zambia's overall budget (2010: 8). In fact, PAYE ("Pay as you earn", a tax on employees' income) was the most significant contributor of direct taxes in 2010. Other taxes are variable company taxes (which brought high gains in 2006 but became less important again in 2008–9), mineral royalties and a windfall tax, which was only collected in 2008.[10] Most of the companies (among them, the MCM) resist paying any variable profit taxes, as they claim not to make profits. Only two companies paid windfall tax in 2008–9 (ibid.: 7f.). Zambia's company taxation seems inefficient and obscure: The national revenue authority (ZRA) audited some companies' budgets for the first time in 2010. Apparently, there exist high obstacles for the ZRA, which lacks the necessary financial resources and experts to examine the corporations' opaque accounting strategies. Only recently, Zambia's participation in the Extractive Industry Transparency Initiative (EITI) has brought some improvements in transparency.[11] The EITI reports should disclose discrepancies in the revenue flow between the corporations and the state. Yet, the effectiveness of the EITI principles is limited, as they are not binding, and violations are not sanctioned (Gregow/Hermele 2011: 22).

Current Situation

In 2011, Michael Sata was elected president, initially with strong support in the mining regions. In the election period, he had promised to keep a bigger share of the mining profits in Zambia through the introduction of a windfall tax of 25%, the improvement of working conditions in the mines, anti-corruption policies as well as better regulation for the growing number of Chinese firms in the mining industry.[12] Zambia is facing major investments by Chinese corporations—a logical development, as China is the world's second largest importer of copper. After the elections however, many of the promises stay unfulfilled. In consideration of the growing importance of the Chinese business community, Sata had moderated his anti-Chinese rhetoric. Furthermore, concerns were expressed about the autocratic style of Sata, and he was accused of nepotism, occupying key positions in the state apparatus with close kin and loyal followers.[13]

10 The revenue-based windfall tax was replaced by a variable profit tax in 2010. This generates less income for the state, as it applies only to profits and thus is often subject to sophisticated tax evasion strategies as described in part 2 (Mwambwa et al. 2010: 4, 10).

11 Zambia is an "EITI compliant country". <eiti.org/Zambia> accessed on 18.10.2013.

12 Wülser, Patrick 2013: "Michael Sata will neuer Präsident von Sambia werden." *Neue Zürcher Zeitung*, 23.09.2011.

13 Mfula, Chris 2012: "Zambia plans to double mining contribution to GDP." *Reuters*, 10.01.2012; Rotberg, Robert: "Zambia, King Cobra disappoints and confounds." *Think Africa Press*, 16.06.2012. <thinkafricapress.com> accessed on 26.03.2013 and Scheen, Thomas 2012: "Die Geburt einer Diktatur." *Frankfurter Allgemeine*, 26.06.2012.

2. The Mopani Copper Mines

Mopani Copper Mines plc (MCM) is the largest mining company operating in Zambia. The mining sites cover about 190 km^2 and include various mines and industrial installations in Mufulira and Kitwe.[14] The Mufulira Mine comprise a copper extraction site, a refinery and a smelter; the Nkana Mine in Kitwe four underground mines, a concentrator and a cobalt extraction plant. The capacity of the Mufulira smelter was increased from 650,000 tonnes of concentrate to reach 870,000 tonnes by the end of 2010. According to Glencore's website, metal production capacity was at 250,000 tonnes of copper and 2400 tonnes of cobalt yearly. This information is provided without further details.[15] The ZEITI reports the production of Mopani being 95,000 metric tonnes of copper and 790 metric tonnes of cobalt in 2010 (Hart Group 2013: 60).

Mufulira and Kitwe are situated in the Copperbelt, along the Kafue River, where large-scale mining had started in the 1920s. The Mufulira Copper Mine was opened in 1930, by the Rhodesian Selection Trust, an American-owned company. The Nkana Mine opened a few years later (Roberts 1981: 186). MCM was sold by ZCCM for USD 43 million to Glencore and First Quantum in 2000, when copper prices were at a bottom point. The Switzerland-based Glencore, holds a share of 73.1%. With a turnover of USD 186 billion in 2011, Glencore was the company with highest turnover in Switzerland and is one of the top resource traders worldwide (Mercier/Peyer 2012: 1). Besides, First Quantum Minerals Ltd., a Canadian company, holds 16.9% and the state-owned Zambia Consolidated Copper Mines Ltd. (ZCCM) holds 10% of the shares (Fraser/Lungu 2007: 14).

In 2011, the previously privately held Glencore entered the stock exchange (listed in London and Hong Kong), and therefore was exposed to more public scrutiny. In the same year, an alliance of non-governmental organizations accused Glencore and First Quantum of violations of the OECD guidelines for multinational enterprises (BD 2011b). In May 2013, Glencore merged with its long-time partner company Xstrata to form one of the largest mining and commodity trading corporations worldwide.

2.1 Environmental and Social Impact

"[...] Mopani Copper Mines plc wishes to ensure that the continued development and exploitation of commercial deposits of copper and cobalt [...] will secure the maximum benefit for, and adequately contribute to the advancement and the social and economic welfare of the people in Zambia, including the people in the vicinity of the contracted area in a manner consistent with their needs and the protection of the environment. [...]"[16]

In contradiction to the statement above, different NGOs and media reports accused Mopani Copper Mines of causing negative social and environmental impacts—including dangerous working conditions, precarious employment and deficient health and welfare systems. Waste management was considered to be far from effective and people living in the vicinity of the mines claim that they are suffering from respiratory problems caused by *centa*, the local name for the toxic emissions (BD 2011a: 103).

14 This area is equivalent of 3/4 of the canton of Zug in Switzerland, where Glencore has its headquarters (BD 2011a: 97).

15 Glencore International plc: "Mopani". <www.glencore.com/mopani> accessed on 25.10.2012.

16 A statement from the Development Agreements between Mopani and Zambia (cited in Fraser/Lungu 2007: 16).

2.1.1 Environmental Impact

The decades of metal mining resulted in severe degradation of the environment in the Copperbelt. During the ZCCM era, environmental concerns were of minimal concern. When the private corporations took over, they refused the liability for inactive dams and dumps or any environmental pollution from previous operations. The unused dams and dumps continue to discharge acid run-off into the water. Their rehabilitation would mean huge costs for the Zambian state (Fraser/Lungu 2007: 34). The current impacts of mining need to be evaluated against this background. The information in this section largely comes from a report of Friends of the Earth prepared by A. S. Simpere, in collaboration with the Zambian NGO Centre for Trade Policy and Development (CTPD).

Environmental impacts of mining operations start with the input of resources. Copper mining and refining consumes immense amounts of water—in average a copper mine uses 13–28 m^3 of water per second for its operations. This is more than the average monthly water consumption of a Zambian citizen. According to some estimates, about half of the national electricity consumption is due to the mining industry. The extraction of one tonne of copper produces 110 tonnes of waste and 200 tonnes of displaced material (Simpere 2010: 7ff.). Large parts of the waste from the Mopani mines are deposited in a huge tailing pond next to the town of Mufulira.

In 2003, MCM adopted the "in-situ leaching" procedure at their production site of Mufulira. In order to reduce the extraction costs, manpower gets partly replaced by several physical, chemical and electrochemical processes. To dissolve the minerals, highly caustic sulphuric acid is pumped into the underground ore body through a borehole. Together with the dissolved copper, the solution is then pumped out again for further processing (Knippertz 2005: 13f). MCM was the first operator to adopt this technique in the region.[17] According to a World Bank report, in-situ leaching is a very contestable practice from an environmental point of view, as it produces a lot of

Fig. 3: Slag dump of Mopani Copper Mines. Photo: Meinrad Schade.

17 See also the documentary film by Odiot and Gallet 2012: *Zambia: Good Copper, Bad Copper.*

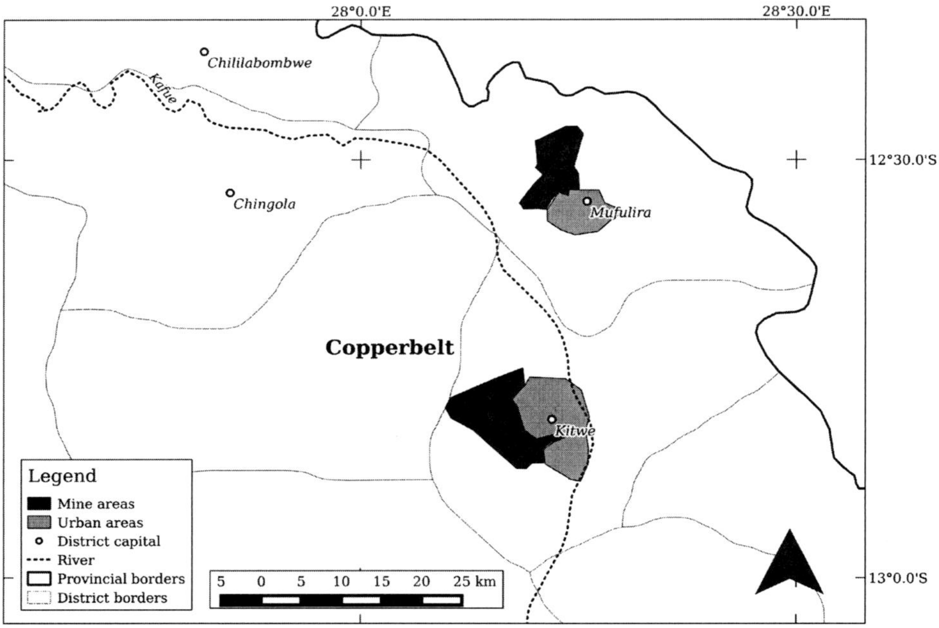

Map 2: Mufulira and Kitwe in the Copperbelt Region.

by-products such as heavy metal contaminated waste-water and sulphur dioxide vapours (World Bank 1998: 291f.).

After the extraction of metals from the ore, the slurry-like wastes are transported with pipelines to the final deposit, the "tailing ponds". Without any special protection measures, this pipeline runs directly through the inhabited areas of Mufulira. Especially during the rainy season, the pond overflows and toxic fluids run into the river system. Although a water filter has been installed some years ago, purification seams not to work effectively. In January 2008, local newspapers reported about an accidental discharge of water containing high concentrations of sulphuric waste through a leaking pipe.[18] Direct contact with the contaminated water caused intoxication of about 800 people in Mufulira. Many people were treated in local hospitals subsequently because of gastrointestinal problems, such as vomiting and abdominal pains. Although Mopani replaced the pipeline, they omitted the cleaning of the damage (Simpere 2010: 18). All these mentioned aspects contribute to the degradation of the aquatic system as well as the contamination of surface and groundwater (Simpere 2010: 18–19). The investigations of Knippertz proved a high contamination of the Kafue River with copper, cobalt and heavy metals. Vegetation as well as the number and variety of animals have diminished significantly (2005: 32, 78). The water pollution entails effects in the whole area downstream of the plant, which may endanger even the Kafue National Park, one of the most significant protected areas worldwide (Simpere 2010: 18f.). It can

18 Lusakatimes 2008: "MCM acid spillage pollutes domestic water." 02.01.2008 <www.lusakatimes.com> accessed on 30.01.2013.

Fig. 4: View into an open pit section of a copper mine of MCM in Kitwe. Photo: Meinrad Schade.

be assumed that the same is true for ecosystems further south. Haller shows that due to very low precipitation, the semi-dry region of the Kafue flats south-east of Lusaka is highly dependent on water from the Kafue River, which originates from sources in the Copperbelt province. The Kafue flats, a sensitive and rich ecosystem in terms of aquatic species and migratory birds, are nurtured up to 80% from inflows of the northern regions (Haller 2013: 40ff.).

According to Simpere, the mines and the smelter of Mopani produced sulphur dioxide (SO_2) emissions that were 72 times higher, and lead emissions 90 times higher than the permitted amount (2010: 15). The amount of SO_2 emitted at Nkana in Kitwe ranged from 6,000 to 1,200 tonnes per month, which by far exceeded the amount permitted by the Zambian State as well as by the World Bank (Knippertz 2005: 67, Flütsch 2012).[19] These emissions allegedly caused severe health problems for the people living in the vicinity of the mines. Inhabitants reported respiratory problems and indicated that the air feels heavy and leaves a metallic taste in the mouth (BD 2011a: 103f.). In the regional hospitals, the number of asthmatic cases and lung infections has increased substantially (Simpere 2010: 16). Furthermore, the pH levels of soils is decreased substantially, due to the acid rains, resulting from the high SO_2 emissions, which make agricultural use nearly impossible. According to a local environmentalist, the only crops growing today are mangoes, avocados and cactuses (Fraser/Lungu 2007: 33). In March 2012, Zambia's environmental agency closed a copper concentrator of Mopani at Mufulira West on the grounds of excessive sulphur emissions that affected the health of the 3,000 inhabitants living nearby (Flütsch 2012).

19 This incident was later used as proof in the OECD complaint against Glencore by a number of NGOs. See also the documentary film: "Reporter – Glencores Hinterhof", *Schweizer Radio und Fernsehen SRF*, 25.10.2012.

2.1.2 Social Impact

Historically, the copper industry created the wealth of a new urban population in the Copperbelt, at the expense of the traditional local population like the Lamba tribe. The colonial rule caused a breakdown of the common property resource use systems of the Lamba, after they lost the large majority of their territory (Siegel 1989: 355). The consequences were existential, as the traditional cultivation method depended on frequent shifting of the fields to allow the soil to recover and the trees to regrow (Roberts 1981: 183). Brian Siegel states that "[...] no compensation was ever paid for [the loss of their territory], apart from some small celebration hosted by Chirupula Stephenson around or before 1930 [...]", and that the Lamba felt that "[...] their wealth was stolen from them by the mining companies, backed by the Northern Rhodesian government."[20]

The colonial rulers tried to constrain the Lamba in such a manner that they were forced to accept wage labour in the mines. Nevertheless, the "integration" of the Lamba into urban life and work meant an alienation from the traditional lifestyle. The population in the towns as well as the British Colonial Office stereotyped them as being lazy, weak and backward (Siegel 1989: 363). During the Great Depression in the 1930s, some Lamba men started to brew beer as an alternative income source and sold it to the mine workers. For the women, prostitution became a source of income. To curb these unwelcome activities, the colonial administration degraded a Lamba chief in 1938. This intervention in internal group affairs led to increasing mistrust towards the administration. Only in the 1950s, mainly the young Lamba men started to seek wage labour in the mining towns in larger numbers (accounting for around 50 to 60% of the total Lamba population), because of famines and demographic as well as institutional changes in their villages. At that time, they represented fewer than 3.4% of the mine workers in Mufulira, and found themselves in a highly marginalized position (ibid.: 362).

Impact on the Urban Population

With its long history of mining, the livelihoods of many Zambians are directly linked with the extraction industry, and the majority of families living in the mining towns depend on wage labour. When copper prices were at their height, at the beginning of the 1970s, the industry's share in employment was at 50% (Fraser/Lungu 2007: 7f.).

The drop in copper prices subsequent to the oil crisis and the following privatization of the mines left a high number of labourers unemployed. Between 1991 and 2004, employment in the mines was reduced by two-thirds from 56,582 to 19,900 (Fraser/Lungu 2007: 21). The new investors did not feel responsible to assist the redundant employees, previously employed by ZCCM, and the state's priority was to repay external debts. Redundant labourers, once belonging to an economic elite, were left with minimal pensions, often only paid for about 5 years on an irregular basis (Fraser/Lungu 2007: 21f.). In 2008, Mopani again dismissed about 3,000 workers without advance notice, justifying it with low copper prices (BD 2011a: 107; Simpere 2010: 13).

Labour Conditions and Sub-Contracting

According to the media spokesman of Glencore, Simon Buerk, in total about 17,000 workers are employed by MCM, including staff of contractor companies.[21] Simpere reports that in 2006 more than half of MCM's mine workers were sub-contracted through a subsidiary company. This has severe impacts on the social situation: While permanent employees and their families benefit from good social security services, higher salaries and entitlement to pensions, this is not the case for sub-contracted workers. The access to company services such as hospitals, malaria and HIV

20 Email correspondence with Brian Siegel, 15.05.2013.

21 Email correspondence with S. Buerk, 04.01.2012.

prevention programs as well as free education for children is denied to them (Simpere 2010: 13). Additionally, sub-contracted workers are often committed to do the most dangerous tasks and run higher risks of accidents if payment is done per unit of work (Fraser/Lungu 2007: 24). A newspaper report from 2008 mentions salaries of about USD 525 per month for a permanent employee, whereas a sub-contracted worker earned only a third of the amount—hardly enough for a family to cover daily expenses.[22]

Before privatization, the companies offered their employees retirement preparation programs in agricultural work and fishery, as they were expected to go back to their home-villages or to a place in the countryside where they had relatives (Ferguson 1999: 123ff.). The family and the village community thus represented a relief to the welfare-system (Roberts 1981: 188). Although no comprehensive retirement program exists anymore, Mopani recently introduced similar projects for its employees. Nevertheless, the companies generally pay minimal pensions and sub-contracted labourers are often left without any benefits at all (Fraser/Lungu 2007: 21ff.).

Summing up, we can say that employment in the Copperbelt is highly dependent on the world market commodity prices and alternative labour opportunities hardly exist. The general social problems, which arise in this "mining town situation" are intrinsically linked to the mining activity, but despite everything, the company provides at least the few employment opportunities and services existing in the region.

2.2 Local People's Strategies

In the early colonial period, the Lamba sought to avoid tax and labour demands with passive resistance. When the towns were growing around the mines, they remained in their villages and started only slowly to adapt to the urban reality. According to Siegel, some of the Lamba people were seeking occasional wage-labour in the mining towns or on white settlers' farms, but most of them avoided working directly for mining companies. Others tried to sell agricultural products, but that was not welcomed by the colonial authorities as it competed with the commercial farms of the "Whites", who supplied the mining towns (1989: 355f.). There are no incidents of open resistance or rebellion on record, and till today, there seems to be no organized resistance. The Lamba could claim, with much reason, the status of Indigenous Peoples, and demand respect for the corresponding rights. We did not find any records on that either, but this is also not surprising, seeing the fact that the International Labour Organization (ILO) Convention 169, one of the few legal international frameworks supporting indigenous rights, is not ratified anywhere in Africa. Apparently, the autochthonous peoples' interactions with the mining industry are minimal. Musaka states that, "[...] most people working in the mines have no clue about the Lamba, as they are considered to be a small, weak, uneducated [minority]."[23] Ferguson describes an urban subculture in the late 1990 called *lambawaza*, which used hanging-on and hawking as strategies to adapt to the miserable economic and social realities. The term applied to town "loafers" of any ethnic group, but is thought by some to originate from a reference to the Lamba (1999: 250f.). This shows that the stereotypes attributed to the Lamba, as described by Siegel (1989), still persisted.

Many locals in urban areas live under precarious economic conditions. The dump sites of mines, which are not commercially used anymore, still contain ore with small amounts of copper. Some of the unemployed men collect and sell this copper residue. At the entrance to the dump sites security guards, employed by official mining companies, collect entrance fees and ensure

22 Puntas Bernet, Daniel: "Wem gehört das Kupfer?" *Neue Zürcher Zeitung*,11.04.2008.

23 Email correspondence with Tuesday Musaka, an informant from a Lamba village, on different occasions.

that no unauthorized person enters the dangerous territory. With their own hands and shovels, the men gather the shimmering stones and later carry them in 50 kg bags to the informal merchant. On a good day, about USD 50 can be earned in this way.[24] Women and children are also expected to contribute to the family income. Some work on the dump sites alongside the men. Yet, the selling process and prices are controlled by men. Other women adopt historical survival strategies associated with the Lamba, like beer brewing or prostitution—the latter under dangerous conditions, as HIV/AIDS is a major public health concern in the area (Mususa 2010: 197).

Strikes and Electoral Politics

In the mining towns of the Copperbelt, local people's demands were largely expressed through labour unions in the past. The unions played an important role in the democratization process of the country and in the negotiation of the workers' rights. With the neoliberal turn of the 1990s and the privatization of state-owned companies, the position of labour unions has weakened. Mulenga (2008: 248) indicates that the "[...] neoliberal policies of the IMF and World Bank have created a labour-unfriendly environment that has eroded their chances of influencing policy decisions". At present, it is difficult for mine workers to organize themselves and the unions are in a vulnerable position. Labour laws are weak or not implemented, and formal regulatory bodies like the Mines

Fig. 5: Workers of MCM in Kitwe cover their ears, waiting for a detonation. Photo: Meinrad Schade.

Safety Department are severely under-funded. Some companies simply discharge employees who are members of a union. Moreover, the high number of sub-contracted workers makes it especially difficult for unions to strengthen the position of the labourers (Fraser/Lungu 2007: 24f).

Nevertheless, there are successful strikes, also at Mopani. Alastair Fraser reports in a blog entry about a strike wave in 2007 that "[...] [i]n all cases, workers are downing tools and walking out in 'wildcat' strikes, rather than waiting for the official union and legal processes that might secure

24 Puntas Bernet, Daniel: "Wem gehört das Kupfer?" *Neue Zürcher Zeitung* (11.04.2008).

a 'legal' strike".[25] In January 2012, a strike of 3000 workers at Mopani's Nkana mine disrupted production for more than 12 hours. The workers demanded a 20% rise of their salaries because of increasing living costs.[26] The 2012 wave of strikes seems to be a reaction to the new president's unfulfilled promises. As a presidential candidate, Michael Sata was strongly supported in the Copperbelt Province, and especially by mine workers, following his promises to improve working conditions in the mining industry and raise tax incomes (BD 2011a: 100).

In spite of insufficient documentation, this could indicate that electoral politics might play an important role in absorbing discontent and protest potential in the Copperbelt. At the same time, the Mineworkers' Union of Zambia (MUZ) is trying to rebuild its former strength against a multitude of newer unions that have come up after privatization. The MUZ still remains the main labour union, and it has some international leverage through its affiliation with IndustriALL, a global federation of unions representing 50 million workers in 140 countries. Both MUZ and IndustriALL have pronounced a critical position towards the merger of Glencore with Xstrata.[27]

NGO Coalitions

The Swiss NGO Berne Declaration (BD) reports that the resistance against the mining giants is evolving on a local as well as an international level. Organized opposition in the Copperbelt has become more frequent. Recently, local grassroots organizations are forming, and different organizations are joining forces against Mopani from Lusaka to Bern and Brussels to London.

In 2010, inhabitants of Kitwe founded the organization "Green and Justice", with the objective to promote local resistance and enhance awareness of the severe environmental and social impacts of Mopani's operations. The organization collaborates with Zambian environmentalists, European activists and external advisers such as Saviour Mwambwa (BD 2011a: 119ff.). Mwambwa, a Lusaka-based economist, is the founder of the well-established and internationally connected organization CTPD. The organization's target is to encourage fairer trade policies and practices in Zambia by analysing the flow of mining profits.[28] Mwambwa has become a crucial actor in the promotion of revenue transparency and accountability of mining firms in Zambia. In 2007, he uncovered the Development Agreements (see section 1.5), which had been carefully kept secret before, by presenting them to the Zambian parliament.[29] It can be assumed that this move influenced important policy changes such as the revision of the Mines and Minerals Act, the introduction of a windfall tax in 2008, and the promotion of transparency and auditing of the extractive industry through initiatives like the ZEITI.

A documentary film of the Swiss reporters Alice Odiot and Audrey Gallet from 2012 shows how European lawyers suggest establishing a local NGO in order to group victims and get relevant proof of the 2008 pipeline disaster, with the goal of filing a case at a court in London. Mwambwa was functioning as intermediary between the people from Mufulira and lawyers as well as politicians such as Eva Joly, a member of the European parliament. As the foundation of Green and Justice has been partially encouraged by European activists, it cannot simply be called a "local NGO". Rather, it is an example for Ferguson's argument that grassroots organizations built by "the civil society" in the African context are seldom just local movements, but are inter-

25 Alastair Fraser / Minewatch Zambia: "Copperbelt Strikes Spread." March 2007. <minewatchzambia.blogspot.com> accessed on 30.01.2012.

26 The Financial 2012: "Zambia Mopani Copper Miners Reject 12% Pay Rise." January <insurance.finchannel.com> accessed on 19.10.2012.

27 IndustriAll 2012: "MUZ leaders commit to union rebuilding"; and: "Glencore and Xstrata epitomize the resource curse: Unions oppose their merger". <www.industriall-union.org> accessed on 15.05.2013.

28 Centre for Trade Policy and Development (CTPD): <www.ctpd.org.zm> accessed 30.08.2013.

29 See documentary film of Odiot / Gallet 2012: *Good Copper, Bad Copper.*

nationally connected to a great degree. He highlights that "[the] role played by NGOs in helping Western development agencies to 'get around' uncooperative national governments [...]" should not be missed in the discussion about those forms of "resistance" (2006: 101f.).

Also in Switzerland, some opposition against Glencore has started. As a result of Glencore's stock market launch in 2011, its major shareholder and CEO Ivan Glasenberg became a multi-billionaire overnight. For his residential community Rüschlikon this meant a sudden increase of income tax of CHF 360 Million (about USD 320 Million). As a consequence, the municipal authority decided to lower the already low tax level by 7% (in Switzerland, a major part of tax collection is regulated at the local municipal level). In reaction to a series of media reports about Glencore's business practices, wherein Mopani Zambia played a prominent role, some locals founded the organization "Solidarität Rüschlikon" and launched a community initiative to reduce personal taxes by only 5% instead of 7%. The initiative demanded that the remaining 2% (CHF 1 Million) should be returned to countries such as Zambia, where Glencore operates its mines. In the communal assembly, the initiative faced stiff opposition and was finally dismissed.[30] Later, similar initiatives were launched in several other communes, which profited indirectly from the tax influx, through a mechanism that equilibrates the budgets between richer and poorer municipalities. The municipality of Hedingen voted in favour of the initiative, and thereby decided to donate 110,000 CHF (about 10% of the additional tax revenue) in to aid projects for mining-affected communities.[31] The activities of solidarity groups and NGOs thus fostered an increasingly critical media coverage of Glencore, which had been an unknown giant for most people in Switzerland only a few years ago, and even motivated some to organize and act on the local level.

Two OECD Complaints against Mopani

The owners of Mopani Copper Mines, Glencore and First Quantum Minerals, were accused of violations of the OECD guidelines twice. In 2001, several NGOs submitted a complaint to the Canadian OECD Contact Point because MCM planned to evict subsistence farmers from their territory in order to expand its mine shafts. This initially was successful—MCM stopped the eviction and started a consultation process with the community. The example was later used to illustrate the effectiveness of the OECD mechanisms. Still, after 2006 the evictions continued, with an estimated number of 600 affected individuals (Wanless et al. 2008).

In April 2011, a coalition of several NGOs, including Berne Declaration, again accused the company of violations of the OECD guidelines. The complaint illustrates how Glencore transferred profits through offshore subsidiaries to its headquarters in the "tax haven" canton of Zug in Switzerland and thereby avoided paying profit tax in Zambia. After a series of meetings with both parties, the OECD National Contact Point (NCP) of Switzerland published a vague final statement in November 2012, mentioning that Glencore had denied the allegations and the parties "[...] agreed to disagree", without any further details.[32] It has to be noted that the NCP is part of the State Secretariat for Economic Affairs (SECO), which has the mission of pursuing national economic interests.

30 Christian Rensch 2013: "Geld aus Afrika—Schweizer Profiteure der Glencore Millionen." *SRF.*

31 Hotz, Stefan 2013: "Hedingen spendet für Opfer des Rohstoffabbaus." *Neue Zürcher Zeitung*, 22.9.2013.

32 NCP 2012: "Final Statement Mopani." <www.seco.admin.ch> accessed on 16.08.2013; and OECD Watch: "Conclusion of the second OECD complaint against Mopani" <oecdwatch.org/cases/Case_208> accessed 30.08.2013.

2.3 Corporate Strategies and the State

Glencore, the majority owner of MCM, is a very inconspicuous company based in Zug, Switzerland. Marc Rich, the founder of today's Glencore, was a pioneer in the oil and later in the mineral trading sector and experienced a high-speed rise in the 1970s due to his extremely opportunistic business strategies. Accused of disregarding the oil embargo on Apartheid-ruled South Africa, resource delivery from Iran while it faced sanctions, as well as tax evasion, he was seeking shelter in Switzerland from a 325-year prison sentence in the United States (BD 2011a: 76f.). In 1993, when Rich lost an internal power struggle against rival managers and left the company, it had a turnover of USD 25 billion. By 2010, this had nearly increased sixfold to USD 145 billion. Glencore was mainly a trader then, but had also acquired mines, production and transport facilities. Even though most of the traded resources never physically arrive in the country from where they are bought and sold, Glencore (and similar trading companies) make Switzerland figure as top consumer for a wide range of raw materials (including copper) in international trade statistics. In 2010, Glencore controlled 50% of the freely traded copper, 60% of zinc, 45% of lead, 28% of coal and 9% of wheat, amongst others (ibid.: 129f.).

With the takeover of Xstrata by Glencore in May 2013, the new corporation has become "[...] one of the world's largest global diversified natural resource companies [...] supported by a global network of more than 90 offices located in over 50 countries. [Their] [. . .] diversified operations comprise over 150 mining and metallurgical sites, offshore oil production assets, farms and agricultural facilities."[33] Glencore now has about 190,000 employees, including contractors, and is projected to earn a net income of USD 10 billion per year. As its distinctive advantage, it presents a broad diversification in produced and traded commodities, from oil to agricultural products to different metals (BD 2011a: 136; 2011b).[34] After the merger, CEO Ivan Glasenberg announced that the company would now focus on less expensive "brownfield" expansion projects (expansion of formerly used mine sites) instead of investing in new "greenfield" mines.[35] This corresponds to Glencore's earlier strategy, of which the takeover of Zambias' Mopani copper mines is an excellent example.

Corporate Social Responsibility and Media Strategy

During the last two decades companies started to counteract the mounting public criticism by developing Corporate Social Responsibility (CSR) programs. Also Glencore is under growing public observation since its stock market launch in April 2011. From near-total secrecy, it has just started to care about its public corporate image. The Sustainability Report of Glencore from 2010 was the first of its kind published by the company (Glencore 2010). The company describes on more than 100 pages their efforts around the world, including several social and environmental projects in Zambia. Mopani (MCM) itself does not publish a separate Sustainability Report, nor does it have a website. There is no independent assessment or review on-the-ground of Glencore's claims, therefore it remains unclear to what extent the community really benefits from the mentioned programs and investments. According to the report, Mopani, together with other stakeholders including the Zambian government, has been supporting an anti-HIV program for more than ten years. MCM also provides medical facilities. However, it seems that they are basically reserved for its regular employees, while contract workers pay higher fees than in public hospitals

33 Glencorexstrata.com: "At a glance" <www.glencorexstrata.com > accessed 14.11.2013.

34 Blas, Javier: "Glencore and Xstrata deal could reshape industry." *The Financial Times* 07.02.2012. <www.ft.com> accessed on 12.12.2013.

35 MacDonald, Alex: "Glencore to avoid Greenfield projects; focus on Brownfield projects." *The Wall Street Journal* 05.03.2013. <www.online.wsj.com> accessed on 12.12.2013.

(BD 2011a: 106). Furthermore, an anti-malaria program is listed, which includes environmental management to eradicate mosquito-breeding areas, the use of larvicide and indoor residual spraying (Glencore 2010: 39).[36] Mopani also runs an agricultural training program for its retired employees who were under regular contracts. A number of former employees and their families were provided with farming equipment and advice to run a cooperative farm (ibid: 55). This relates to the practice prior to privatization, when employees were trained in agricultural work as a preparation to go back to their villages of origin after retirement.

A comparative study of mining companies' CSR strategies, published by the French institute Novethic, placed Glencore on the second to last position (Peyer/Mercier 2012: 57). Glencore's only commitment to international CSR standards was a timid participation in the Extractive Industries Transparency Initiative (EITI) process.[37] Peyer and Mercier state that Glencore referred to vague definitions of general normative concepts like the universal human rights, and created its own standards, which included occasional donations and charity rather than actual social or environmental programs (ibid.: 59).

Similar to CSR, the media strategy of Glencore is also in an early stage of development. Prior to its stock market launch in 2011, the company appeared very rarely in the media (BD 2011a: 150). Its strategy seemed to be to keep quiet as long as possible. Only recently, Glencore started to communicate proactively and to respond to criticism from the media and the NGOs. On several occasions, the CEO Ivan Glasenberg gave personal interviews to different Swiss media; for example, an interview where he countered the accusations that Glencore profited from child labour in artisan mining in the Democratic Republic of the Congo (Mercier/Peyer 2012: 2f.).[38] To counter such criticisms, Glasenberg uses a development discourse: in several interviews, he argues that the poorest countries in the world need the investment of international companies like Glencore in order to provide employment, education and social security. He highlights that his company is advancing development and invests in schools, hospitals and sanitation infrastructure; for example, in Zambia.[39]

Alliance with the European Investment Bank

Considering the environmental impact of copper extraction, Mopani Copper Mines plc started a sulphur emission capture project at the smelter in Mufulira in 2005. For the installation, described as a "development project", Mopani was granted a loan of USD 50 million from the European Investment Bank (EIB). The total costs of the project were USD 130 million. The project was finished in 2007, with more than half of the sulphur emissions captured and turned into sulphuric acid.[40] However, some investigations show that the total emissions have not been reduced, because the production amount has been doubled simultaneously (Simpere 2010: 14f.). The EIB investment was criticized by a number of NGOs, European politicians and lawyers. In the updated project description regarding Mopani, the EIB stated that there would be no further loans for Glencore or its subsidiary companies (BD 2011b). According to a press release of the company, concerning the

36 Malaria is a concurrent problem since the beginning of mining in the Copperbelt in the 1930s, as the mines were built nearby malarial swamps (Siegel 1989: 355).

37 Eiti.org: "EITI Supporters: Glencore" <eiti.org/supporters/companies/glencore> accessed 25.10.2013.

38 A French documentary (RTS TV5 Monde, 31.05.2012) accused Mopani of buying copper from artisanal mines quarried by children in the south of the DRC across the border of Zambia. Glasenberg denied this in his first TV interview, in the Swiss TV Newscast "10vor10" on 16.04.2012 ("Kinderarbeit und verschmutzte Gewässer" <www.videoportal.sf.tv> accessed on 05.07.2012).

39 E.g. Spieler, Martin: "Letztlich macht es keine Differenz, ob ihr Vermögen eine Milliarde beträgt oder sechs" *SonntagsZeitung*, 05.05.2013.

40 European Investment Bank (EIB): "Mopani Copper Project, May 2011", <www.eib.org> accessed on 28.01.2012.

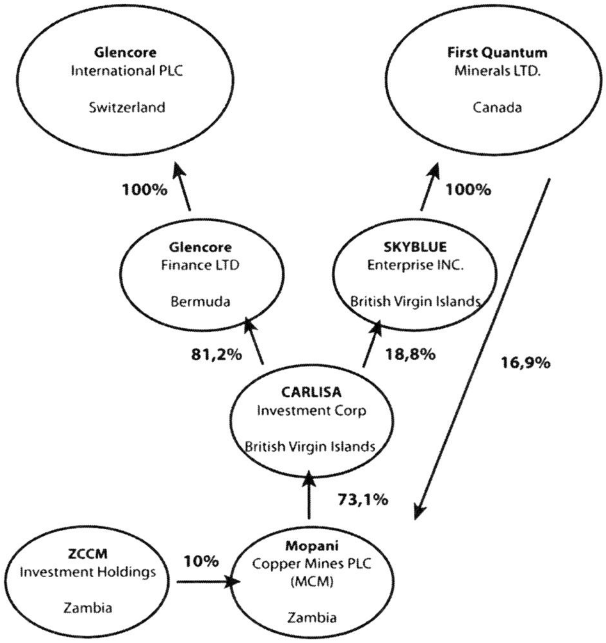

Fig. 6: Mopani-Glencore ownership structure; Source: BD 2012: 285, modified by E. Schober.

sulphur dioxide emission, the construction of a second acid plant should be finished in 2013. By then, over 97% of the sulphur dioxide emission should be captured.[41]

Tax Evasion

In 2010, Mopani paid about USD 73 million (ZMK 364,500,386 at USD 1= ZMK 5000) in taxes to the Zambian government, making it the forth-largest contributor of the mining sector to the Zambian tax income (Hard Group 2010: 53). Nevertheless, for Mopani, like for many other multinational mining corporations, the amount paid in taxes is far lower than the amount that should have been paid, due to opaque business strategies and insufficient auditing by state authorities (Mwambwa et al. 2010: 4, 10).

According to a spokesman of Glencore, Mopani has spent over USD 500 million on contractors regarding staff and equipment in the first nine months of 2011.[42] Critics highlight that companies like Mopani did not pay any taxes for equipment and machinery for a long time, due to the Development Agreements and other concessions (Mwambwa et al. 2010: 14). These contradictory statements indicate that the spending could be strategic: It is a common tax avoidance strategy to reduce the declared profits (and thereby higher profit tax rates) by overstating costs (Fraser/

41 Email correspondence with the spokesperson of Glencore S. Buerk, 04.01.2012. (Attachment: "Press Release Mopani to complete Sulphur Dioxide Emissions").

42 Email correspondence with S. Buerk, 04.01.2012.

Lungu 2007: 11). Profits are transferred through complicated offshore constructions (see Fig. 6) to "tax havens" such as Zug, Switzerland, where so-called "holding companies" pay minimal taxes. Even though Switzerland is Zambia's biggest export partner for copper, Zambia does not figure as such on the import statistics of Switzerland. Export products are sold internally to subsidiaries in the British Virgin Islands and Bermuda at higher prices than the usual London Metal Exchange (LME) rates ("transfer pricing"). Meanwhile, the physical products are shipped directly to customer countries like China (BD 2011a: 285f.). Mopani's internal sales are apparently always conducted when prices are at a bottom point, to report smaller profits. Furthermore, the production costs are overestimated and the produced quantity underestimated. According to a tax inspector, Zambia loses USD 124 million annually through these manipulations (ibid.: 288).

Interestingly, Glencore also has not paid any income tax in Switzerland, at least for the period from 2010 to 2013. After the merger, Glencore did not pay tax to the federal government because it made a book loss after value adjustments on some of Xstrata's mining projects. Moreover, the shareholders benefit from tax-free dividends. But the canton of Zug and the Swiss state profit from considerable income taxes paid by management and employees—in 2012, they amounted to CHF 121.5 million for the total commodity sector companies registered in Zug. The share of Glencore remains unknown.[43]

3. Conclusion

Since the early 20th century, the development and the living conditions of the population of Zambia, and in particular, the Copperbelt district, have been entangled to a high extent with large-scale copper mining. The profit from the exploitation of mineral resources was always distributed very unequally, but for some time it seemed that Zambia, once one of the most prosperous economies in Africa, could escape from the notorious "resource curse". This implies that countries of the "South" with abundant resources remain dependent on exporting their wealth to the richer countries of the "North" (Gregow/Hermele 2011: 37). With the decline of the "socialist" experiment, falling copper prices and increasing public debt, Zambia committed itself to the neoliberal ideology and introduced policies which fostered the absorption of profits by multinational companies and elites at the expense of local people and the environment. The corporations were granted many advantages and low taxation in the Development Agreements. But the promised benefits for the country did not materialize: job creation and the overall linkages between the mining industry and the local economy are low. The taxation system relies on income generated through individual workers and is unable to prevent capital flight. Neither the government nor the legal institutions are able to protect the workers and the environment sufficiently, while there is no mechanism at all to compensate the Lamba as the former autochthonous landowners. Zambia thus failed to translate its mineral wealth into the development of a diversified, stable economy and an institutional setting that would allow its people to live in relative prosperity.

Given this situation, one could expect a good base for social and environmental justice movements. Yet, local people's strategies are difficult to recognize for this case, with the autochthonous population extremely marginalized and stigmatized, mine workers' unions weakened, local NGOs just emerging, and discontent apparently largely channelled into party politics, which is seemingly less effective. On the side of Glencore, the basic strategies show quite a clear-cut example of profit maximization, evasion from tax-paying and environmental as well as social responsi-

43 Soukup, Michael: "Die Rohstoffbranche zahlt in Zug nur 36 Millionen Franken an Steuern." *TagesAnzeiger* 20.09.2013; and Flütsch, Andreas: "Glencore zahlt in der Schweiz seit dem Börsengang «null» Steuern." *TagesAnzeiger* 14.03.2013.

bilities. Only recently, this is complemented with some CSR programs in reaction to criticism by international NGOs. In relation to Mopani, Glencore and its partner First Quantum Minerals have been accused of violations of the OECD guidelines twice since 2001, without much result. But the case also confirms that international networks of NGOs and solidarity groups can be successful in highlighting the injustice linked with the exploitation of natural resources, to the effect that people even on the profiting side begin to ask questions.

References

Berne Declaration (BD) (ed.) 2011a: Rohstoff, das gefährlichste Geschäft der Welt. Zürich: Salis. (Available online in English as: "Commodities—Switzerland's most Dangerous Business", <www.evb.ch>).

Craig, John 2001: Putting Privatisation into Practice. The Case of Zambia Consolidated Copper Mines Limited. *Journal of Modern African Studies* 39 (3): 389–408.

Cunningham, Simon 1981: The Copper Industry in Zambia. Foreign Mining Companies in a Developing Country. New York: Praeger Publishers.

Davis, Merle J. 1976: Modern Industry and the African. An Enquiry Onto the Effect of the Copper Mines of Central Africa Upon Native Society and the Work of the Christian Missions. London: Frank Cass Ltd.

Doke, Clement Martyn 1931: The Lambas of Northern Rhodesia. A Study of Their Customs and Beliefs. London: Harrap.

Esomba, Stephen Nangoh 1996: Zambia Under Kaunda's Presidency. The Conditions, Experiment with Socialism and the Final Lap to Democracy. Hamburg: LIT Verlag.

Ferguson, James 1999: Expectations of Modernity, Myth and Meanings of Urban Life on the Zambian Copperbelt. Los Angeles: University of California Press.

Ferguson, James 2006: Global Shadows: Africa in the Neoliberal World Order. Durham/London: Duke University Press.

Fraser, Alastair 2010: Introduction. Boom and Bust on the Zambian Copperbelt. In: Fraser, Alastair and Miles Larmer (eds.): *Zambia, Mining and Neoliberalism.* New York: Palgrave Macmillan, 1–30.

Gabel, Creighton 1976: Archaeology in the Western Copperbelt. *South African Archaeological Bulletin* 22 (85): 3–14.

Haller, Tobias 2013: The Contested Floodplain. Institutional Change of the Commons in the Kafue Flats, Zambia. Plymouth: Lexington Books.

Heyden, Constantin von der 2004: Groundwater Pollution on the Zambian Copperbelt: Deciphering the Source and the Risk. *Science of the Total Environment* 327 (1–3): 17–30.

Knippertz, Martin 2005: Analyse von Rehabilitationspotenzialen kupferbergbaulich beanspruchter Flächen in Sambia und der Mongolei. Aachen: Geographisches Institut der RWTH.

Livingstone, David 1858: Missionary Travels and Researches in South Africa. New York: Harper & Brothers.

Lungu, John 2008: Copper Mining Agreements in Zambia. Renegotiation or Law Reform? *Review of African Political Economy* 117: 403–415.

Mulenga, Friday E. 2008: Fighting for Democracy of the Pocket. The Labour Movement in the Third Republic. In: Gewald, Jan-Bart, Marja Hinfelaar and Giacomo Macola (eds.): *One Zambia, Many Histories. Towards a History of Post-colonial Zambia.* Leiden: Brill, 243–258.

Mususa, Patience 2010: Contesting Illegality. Women in the Informal Copper Business. In: Fraser, Alastair and Miles Larmer (eds.): *Zambia, Mining and Neoliberalism.* New York: Palgrave Macmillian, 185–204.

Powdermaker, Hortense 1962: Copper Town. Changing Africa. The Human Situation on the Rhodesian Copperbelt. New York: Harper & Row.

Roberts, Andrew 1981: A History of Zambia. London: Heinemann Educational Books Ltd.

Siegel, Brian 1989: The "Wild" and "Lazy". Lamba: Ethnic Stereotypes on the Central African Copperbelt. In: Vail, Leroy (ed.): *The Creation of Tribalism in Southern Africa.* London: James Curry Ltd., 350-371.

Soest, Christian von 2009: The African State and its Revenues. How Politics Influences Tax Collection in Zambia and Botswana. Baden-Baden: Nomos Verlagsgesellschaft.

Taylor, Scott D. 2006: Culture and Customs of Zambia. Westport USA: Greenwood Press.

Reports

Berne Declaration (BD) 2011b: Specific Instance Regarding Glencore International AG and First Quantum Minerals Ltd. and Their Alleged Violations of the OECD Guidelines for Multinational Enterprises via the Activities of Mopani Copper Mines Plc. in Zambia. Press Folder, April 2011. <www.evb.ch> accessed on 02.04.2012.

Central Statistical Office Zambia (CSOZ) 2000: Main Zambia Census Report, Vol.10. Lusaka <www.zamstats.gov.zm> accessed on 28.01.12.

Central Statistical Office of Zambia (CSOZ) 2011: Zambia. 2010 Census of Population and Housing. Lusaka. <unstats.un.org/unsd/demographic> accessed on 17.07.2013.

Flütsch, Andreas 2012: Sambia legt einen Teil von Glencores Kupermine still. Available from <www.evb.ch/cm_data/9527484.pdf> accessed on 12.12.13.

Fraser, Alastair 2007b: Zambia: Back to the Future? Managing Aid Dependency Project, GEG Working Paper 2007 (30). Oxford: Global Economic Government Program <www.globaleconomicgovernance.org> accessed on 30.04.2012.

Fraser, Alastair and John Lungu 2007: For Whom the Windfalls? Winners & Losers in the Privatisation of Zambia's Copper Mines. Lusaka: Civil Society Trade Network of Zambia (CSTNZ).

Glencore International plc. 2010: Sustainability Report 2010. <www.glencore.com> accessed on 28.01.2012.

Gregow, Karin and Kenneth Hermele 2011: From Curse to Blessing? Africa and the Raw Materials Race. Stockholm: Forum Syd. Available from <www.ctpd.org.zm> accessed on 30.07.2013.

Hansungule, M., P. Feeney and R. Palmer: Report on Land Tenure Insecurity on the Zambian Copperbelt. Oxfam: 1998. Available from <www.mokoro.co.uk> accessed on 30.07.2013.

Hart Group 2013: Zambia Extractive Industry Transparency Initiative. Final Reconciliation Report February 2013. Oxfordshire: Hart Group. Available from <eiti.org/report/zambia/2010> accessed on 24.08.2013.

ICSG (International Copper Study Group) 2012: The World Copper Factbook. <www.icsg.org> accessed on 26.03.2013.

Mwambwa, Savior, Aaron Griffiths and Andreas Kahler 2010: A Fool's Paradise? Zambia's Mining Tax Regime. CTPD (Centre for Trade Policy and Development) <www.ctpd.org.zm> accessed on 16.6.2014.

Simpere, Anne-Sophie 2010: Project Mopani (Zambie). L'Europe au Coeur d'un Scandale Minier. Raport de Mission. Les Amis de la Terre France: Décembre 2010. <www.amisdelaterre.org> accessed on 28.01.2012.

Wanless, C., L. Tausky, C. Davis, C. Telfer, E. Hallock 2008: Can the OECD Guidelines Protect Human Rights on the Ground? A Case Study. The Evictions at Mufulira by First Quantum Minerals / Mopani Copper Mines. Toronto: The Umuchinchi Initiative. Available from <oecdwatch.org/publications> accessed on 01.05.2013.

World Bank 1998: Copper Smelting. Pollution, Prevention and Abatement Handbook. <www.ifc.org> accessed on 09.08.2013.

World Bank 2011: What Would it Take for Zambias Copper Mining Industry to Achieve its Potential? Report 62378-ZM. <openknowledge.worldbank.org> accessed on 11.07.2013.

Films

ARTE 2012: Afrika – Der ausgeraubte Kontinent, 29 Dec., available from: <www.youtube.com>.

Rensch, Christian 2013: Geld aus Afrika – Schweizer Profiteure der Glencore Millionen. SRF (23.01.2013).

TV5 Monde / RTS 2012: Glencore/Congo/Child Labour. 31 May, available from: <www.youtube.com>.

Schweizer Radio und Fernsehen SRF 2012: Reporter: Glencores Hinterhof. 25 Oct, available from: <www.youtube.com>

Odiot, Alice and Audrey Gallet / France Télévisions 2012: Zambia: Good Copper, Bad Copper. Available from: <www.youtube.com>.

Uranium Mining in Niger

Undermining Pastoralist Lifeworlds

Clare Oxby and Saskia Walentowitz[1]

1. Introduction

Long before French colonialization and the discovery of uranium, today's territory of Niger belonged to the vast Saharan region, which has been continuously inhabited by humans since the Holocene, despite dramatic climate change. Neolithic people, and those who followed in their footsteps, created complex cultural landscapes, as depicted in numerous cave paintings, rock carvings, and artefacts throughout the Sahara. These people proved able to maintain fragile ecosystems thanks to the invention of highly adaptive agro-pastoral economies and worldviews in which mobility is the key to access water and pasture (Claudot-Hawad and Lefèvre-Witier 2009). Today, Tuareg Berbers, as well as other pastoralists such as the Fulani, carry on the knowledge and institutions that make life possible in these inhospitable arid and semi-arid regions. However, it is only through the ongoing, annually renewed practices of transhumance and nomadic life that these places remain truly habitable for humans and livestock.

The present chapter focuses on the Tuareg and illustrates how this equilibrium has been systematically disrupted by colonial and postcolonial state powers. Uranium mining plays an integral part in this ever intensifying process of undermining pastoralist lifeworlds in northern Niger. It is an unprecedented case that demonstrates how remote environments, together with their sought after mineral resources, have played a major role in the development of the French nuclear-powered state and its Nigérien counterpart, both before and after colonization. For more than a century, these political and geo-strategical dynamics have led to a series of violations of human and environmental rights, that remain largely ignored. From "unruly nomads" to "Islamic terrorists", Tuareg people are repeatedly accused of standing against modern state building, education, and economics, while being subjected to systemic neglect by national and international bureaucracies and stakeholders (see Claudot-Hawad 2013).

Uranium mining in Niger, whose profits barely benefit the local people, is but one striking example of a process that is affecting many corners of the globe, as other chapters in this volume demonstrate. Local and international NGOs are denouncing the harmful effects of environmental contamination caused by uranium mining, on humans, water resources and livestock. By contrast, the French state-owned mining company Areva claims that radiation risks are well controlled and that local people's health problems are related to the local desert conditions and to general development deficiencies. In this chapter, we will contextualize this dialogue of the deaf within the broader political economies of France and Niger, and wonder for how long such policies against the interests of mankind and nature will continue to be pursued, especially in times of global climate change when a radical shift in attitude and practice is called for.

1 Saskia Walentowitz wrote large sections of Part 1, Clare Oxby wrote Part 2 while both authors made revisions and suggestions for the whole chapter; Tobias Haller contributed material to the chapter and drafted sections 1.1. and 1.3.

1.1 History and Current Context

Colonial History and Revolts

The pre-colonial history of the Sahara and the Sahel is a shared history of language communities that shaped each other in diverse ecosystems including river deltas, savannahs, coastal regions, deserts and highlands. These communities created specific, territory-bound agricultural, fishing as well as pastoral lifeworlds and cultural landscapes. They were sustained through mutual exchange of material goods and cultural assets, involving long-distance trade between the Mediterranean Sea and the great forests, as well as warfare and protection economies (see de Moraes Farias 2006). "Ethnic" boundaries and cultural identities were flexible among Sahelian and Saharan populations within highly regulated resource management institutions. These populations embraced Islam early. Political alliances and supremacy shifted in the course of major religious reforms that accompanied successive reorganizations of trans-saharan trading routes and networks (Meunier 1997). Between the 8th and the 19th century, today's territory of Niger has been part of several large sociopolitical landscapes. These comprise the empires of Ghana, Mali, and Songhay that arose alongside nomadic Berber and Bîdan Imamates and Emirates. Until today, the vast Azawagh Valley that stretches between the Adghagh Mountains in Northern Mali, and the Ayer (or Aïr) Mountains in northern Niger, counts several Songhay-speaking Berber communities. Many belonged to the former Tagaraygarayt confederation, itself born out of the fusion of an ancient Tuareg Imamate with parts of the Iwellemeddan during the Sokoto caliphate in the beginning of the 19th century (Walentowitz 2003). These historical dynamics across cultural and ecological boundaries were disrupted with the advent of French colonization.

In the early 19th century, rising European interests in these regions led to so-called exploration expeditions in search of the origins of the Niger River by Mungo Park, Heinrich Barth and others. During the "scramble for Africa" before and after the Berlin Conference of 1884, the territory of today's Niger was annexed to the French West African Territories which included today's neighbouring states such as Mali (French Sudan), Burkina Faso (French Upper Volta) and Chad (belonging to French Equatorial Africa). French West Africa was governed from a colonial office in Dakar; and what is today known as Niger, from the capital city of the Damagaram Sultanate, called Zinder. The French entered this city in the 1890s and fully controlled it by 1899 after conquering the territory of Damagaram. Zinder became the military capital from 1911 to 1926, while the village Niamey by the river Niger grew into the colonial administrative centre, because the French feared attacks from adjacent British Nigeria (Rivières 1965, Decalo 1979, Fuglestad 1983). The territories of the Tuareg confederations were not in full French control until the late 1910s. However, the Tuareg already suffered not only from severe casualties, but also from disordered access to land in the southern areas, and the disruption of trading and herding routes caused by the borders erected by the French and the British colonial powers (ibid., Krings 1995).

In the process of colonization, the five main Tuareg confederations known as the Kel Ahaggar and the Kel Ajjer (southern Algeria and Libya), the Kel Tademakkat (Central Mali), the Kel Ayer (northern Niger) and the Kel Tagaraygarayt (Central Niger) became increasingly isolated from each other. Thus marginalized, this situation fostered several upheavals. During the major, so-called *morti* revolt, some Tuareg groups preferred to surrender early in order to protect their people; while others, namely the Kel Ayer led by Kawsen, invented new forms of combat in order to resist French domination until their final defeat in 1917 (Claudot-Hawad 2012a). Following the divide-and-rule maxim that already separated "nomads" from "sedentary" people and disrupted vital pastoralist-agriculturalist symbioses, the colonial powers also unbound the larger part of the servile workforce (*iklan*), which until then had played a key role in Tuareg economy. The increasing political, social and ecological fragmentations, through ever more restricted access to water, pasture, food and other resources, laid the bedrock of crises, upheavals, and resistance in the following decades.

Postcolonial Politics

During the period 1950-1968, the Tuareg were faced with a major challenge, since colonial policy favoured farming activities on pastoral land previously managed by the Tuareg as common property (Krings 1995). In early post-colonial times, the government of Niger as well as international development agencies continued to prioritize agricultural production over pastoralism and neglect the territorial rights of the nomads. This seriously reduced hazard-minimizing strategies and, thus, the resilience of the Tuareg to cope with climate variability in the region, rendering them vulnerable to the drought periods of 1969-73 and 1981-1984 (Spittler 1989, Krings 1995). In consequence, many Tuareg were forced to move into larger cities in Niger and Mali, or were displaced to refugee camps. Many young men went into exile in Libya, where they sought work, but where many ended up serving as mercenaries (Krings 1995). These *ishumar* (a Berber version of the French word *chômeur*, "unemployed") became the Tuareg partisans of the late 1980s (see below). The Tuareg remained largely excluded from the clientelistic governments of Niger and therefore could not benefit from the wealth accumulated by the country's ruling elites.

After Niger's independence in 1960s, the country was governed by several military dictators and a few elected presidents, all of them more or less dependent on French interests. Indeed, Niger's history since World War II cannot be separated from French uranium politics (Granvaud 2012). Under President de Gaulle, France developed an increasing interest in nuclear power for civilian as well as for military purposes in order to take its place within the new world order. To finance the development of the nuclear sector, revenues from oil exploitation were tapped and a secret collaboration with the US-American government was established. The *Commissariat à l'Energie Atomique* (CEA) saw Niger as a strategic asset, being one of the few areas under indirect French control with large uranium reserves, along with Madagascar and Gabon (Granvaud 2012).

Hamani Diori became Niger's first President in 1960 with the support of the French government. By creating a one-party state under Diori, a socialist group called SAWABA and led by Djibo Bakary was prevented from political participation. The Tuareg remained largely excluded from political life as well. While Diori became a spokesperson for regional African affairs, he was accused by the opposition of corruption and of being too attached to French interests (Decalo 1979, Fugelstad 1983). France protected Diori against several *coup d'états* by the SAWABA group, until the 1970s, when Diori started criticizing France for not paying enough taxes for the SOMAÏR mine in Arlit. In 1974, Lieutenant Colonel Seni Kountché seized power, leading a repressive military regime. Interestingly, Kountché had held a higher position in the French Army before Niger's independence. After his death in 1987, the country experienced a more liberal phase under his successor Ali Saibo, who led Niger to its first democratic elections in 1993, won by Mahamene Ousmane. This president headed a coalition government that failed, however, to consolidate; and was ousted in 1996 by a *coup d'état* led by General Ibrahim Baré Maïnassara.

In 1999, Baré was assassinated and Commandant Daouda Malam Wanké took over. The latter called for free elections, which were won by Mamadou Tandja, the former interior minister who had led the particularly cruel military repression of the 1990s Tuareg revolt. Tandja's presidency in the 2000s coincided with a series of long-overdue decentralization measures, this time requested by external donor agencies. In the process, Tandja was again challenged by a Tuareg upheaval in 2007. At the same time, relations with the French government deteriorated, because Tandja sought to negotiate better deals for uranium exports (see section 2.5). Tandja's government ended with a *coup d'état* in 2010. New elections held in 2011 put Mahamadou Issoufou, the former opposition leader, into power as the new president. Issoufou's mandate has been marked by Mali's unprecedented political crisis since January 2012, when the National Movement for the Liberation of the Azawagh (MNLA) took control of the Malian North adjacent to Niger's North, in the aftermath of the 2011 Libyan civil war.

Tuareg "Rebellions" and the Sahara as a "Terrorist" Zone

Niger's political history and Tuareg opposition movements cannot be understood disconnected from mining issues. The northern part of Niger was, and continues to be, instrumental to French interests, which are in contradiction with those of the pastoralists living in this area since remote pre-colonial times. The persisting structural violence, repeated military repressions and massive human rights violations committed by the armed forces of Niger, continue to trigger revolts among the Tuareg, just as they did during the French colonization period. Krings described already in the mid-1990s how such processes of marginalization led young men to migrate to Libya, where one of the few options offered to them was to work as mercenaries for Gaddafi (Krings 1995). This is the context in which they received military training, while reviving ancient relationships between different Tuareg groups, thus fostering joint actions against the military forces of Niger and neighbouring Mali (see Claudot-Hawad 1996a).

Several massacres perpetrated by the military, notably in Tchin Tabaraden in 1990, set up an armed conflict, which escalated to civil war and mass arrests of Tuareg in 1992. By 1993, four different rebel groups were fighting the government: the *Front de Libération de l'Aïr et de L'Azaouak* (FLAA), the *Front Libération de Tamoust* (FLT), the *Armée Revolutionaire de la Libération du Nord Niger* (ARLNN) and the *Front Patriotique de Libération du Sahara* (FPLS), later amalgamated into the *Coordination de la Résistance armée* (CRA). After the 1993 democratic elections with Ousmane as president, imprisoned Tuareg were freed and negotiations started with the FLT fraction, supported by France. Meanwhile, the CRA published a memorandum claiming two-thirds of the territory of Niger as a pastoralist area, as well as 25% of the national revenues—mainly from uranium mining—for the North. Last but not least, they wanted the massacre of Tchin Tabaraden to be reinvestigated. This memorandum met stiff opposition from the government and from populations in the South of the country. Some members of the government even suspected France of being part of this plan, since during the 1950s, the French had considered establishing a separate Saharan state in order to better control mining activities there (Krings 1995).

Under the new government that led the country in the 2000s, structural violence continued. At the same time, Libya's influence in the region grew, and Tuareg were gaining increasing access to modern weapons and training in Libya. By the mid-2000s, corruption and bad governance had become a major issue for the Tuareg, who, however, still claimed national identities as Nigériens (Claudot-Hawad 2012b). An expression of this was the rebellion of the *Mouvement des Nigériens pour la Justice* (MNJ) in 2007 (see Keenan 2009 and section 2.4). Following the attacks of 9/11 in the USA, the "war on terror" also reached the area, starting with a conflict in Algeria and alleged links between local armed groups and Al Qaida. Since 2002, the Pan-Sahel Initiative—a US military assistance program to Mali, Chad, Mauritania and Niger—was set up, depicting the whole area as a "terrorist space" (see Keenan 2009).

A new chapter in the conflict area was brought about after the fall of Gaddafi in 2011, and the emigration of Tuareg from Libya back to Niger and Mali, some of them well armed with weapons looted from the disintegrated Libyan army. In early 2012, certain Tuareg groups in Mali, including the MNLA (*Movement National de Libération de l'Azawad*), took advantage of the political vacuum there to defeat the Malian army, and launched the creation of a new independent state called AZAWAD, which was announced to the international media on 6 April 2012. A new group called Ansar Dine took control shortly afterwards, having eliminated certain leaders of the MNLA, and established an Islamic religious state with a radical Salafist ideology (see Adeyemi 2015). According to Hélène Claudot-Hawad (2012a), it is unclear where this new group came from and whether it was actually linked to Al Qaida. The US and the EU strongly opposed these developments, using an "anti-terror" rhetoric; and France swiftly intervened by supporting ECOWAS forces in the area in order to defeat the separatists. The resulting situation firmly re-established the northern parts of Mali and adjacent Niger on the map of "Islamic terrorism" for the USA and Europe. For local people, the Mali war has led, once again, to increased militarization,

Fig. 1: Tuareg fighters of the MNJ. Source: Screenshot from "Niger : la bataille de l'uranium", France 5, 2008.

especially around the mining area of Arlit, and to the further undermining of the living places and futures of Tuareg and other pastoralists.

1.2 Pastoralist Life Worlds in Northern Niger

Since colonial times, modernization discourses depict the Tuareg Berbers as "archaic" and "unruly nomads", unable to adapt to a modern economy and nation-state organization (Claudot-Hawad 2012b). The list of stereotypes and prejudices is long. The reduction of complex pastoralist movements and land care practices to "erratic" nomadism legitimates, to this day, the denial of full territorial rights to Tuareg and other pastoralists in the Western Sahara and the Sahel. In combination with the enforcement of arbitrary nation-state borders, such de-territorialization processes have transformed well-managed pastoralist landscapes into a disintegrated no-man's land controlled by remote bureaucrats. According to Niger's postcolonial *Code Rural*, the land belongs to those who cultivate it, while the pastures and water resources of the *zone pastorale* remain state property that may be accessed by all. By contrast, former transhumant systems were rooted in common property regimes, while sophisticated social and political institutions allowed for interactive land and resource management that were the key to the annual regeneration of both land and life worlds. Colonial and neo-colonial policies conceptualized fragile Saharan and Sahelian ecologies as pre-existing, "natural" reserves, neglecting the fact that these cultural landscapes could only be sustained by interconnected mobile activities, such as caravan trade, that spanned huge territories reaching far beyond the *zone pastorale*.

In the following sections, we will further describe the linked development of people and place in these arid environments in order to better understand the crucial role that the uranium-rich plains, lying to the West of the Aïr Mountains, play in Tuareg pastoralism and socio-political organization. Through this description, it should become clear why unsafe mining practices are having harmful consequences for both people and land well beyond the immediate mining areas. We will also include a discussion of the challenges and limits presented by the recent land reforms that, together with decentralization policies, contribute to the decline of pastoralism rather than its protection. In the last section of Part 1, we will briefly remind readers of the extent to which Niger's national economy relies on uranium mining, while agricultural and pastoralist activities remain vital for the vast majority of the people.

Fig. 2: Watering pond in Eghazer. Photo: Collectif Tchinaghen (2008).

The Tuareg, a Berber People in the Sahara and the Sahel

After its repopulation in the beginning of the Holocene, the once green Sahara became a desert again some 5000 years ago. Despite this dramatic climate change, it continued to be inhabited thanks to the intensification of food production during the Neolithic period (Claudot-Hawad and Lefèvre-Witier 2009). Moreover the archaeological evidence from the Saharo-Sudanese Region, including sites in present day northern Niger, suggests that the Neolithic economy—characterized by domesticated livestock, pottery production, and seasonal settlements built by water sources, but without any evidence of agriculture—was based on herding and not farming[2]. Indeed, Saharan people are among the oldest herding peoples of the world. Pastoralist activities developed and fine-tuned over many millennia. In the course of desertification, herders became nomads, mountain dwellers, and oasis gardeners. The adoption of camels during the first millennium allowed them to navigate the entire desert and establish an extensive network of caravan trading routes that connected and shaped the various human ecologies of North-West Africa (Claudot-Hawad and Lefèvre-Witier, op.cit.).

The Tuareg (*imazighan*, singular *amazigh*) are the famous heirs of these ancient Saharans. Their number is estimated at over three million, including an important diaspora that formed in Saudi Arabia in the aftermath of the 1970s droughts. Most Tuareg live in Niger and Mali where they are considered as "minorities". Many others live in southern Algeria, Libya as well as in Burkina Faso. In precolonial times, Tuareg people organized into five major political confederations: The Kel Ahaggar and the Kel Ajer in the North, the Kel Tademakkat in the Ouest, the Kel Ayer in the East and the Kel Tagaraygarayt at the center of this vast Saharan territory. These groups were connected with each other through trade, warfare, political alliances, religious studies, and marriage. Tuareg society (*temust*) was enacted as a whole in annual rituals such as Gani (Claudot-Hawad 1993), and annual transhumance to salt pastures including the especially renowned ones in the vicinity of In Gall. During these vital summertime gatherings known as "resurrection" (*tenekkart* or *cure salée*), people and animals rejuvenated; and political assemblies were held with representatives of the major confederations, until colonial administrators prohibited them in order to prevent revolt.

2 These findings throw into doubt the mainstream theory according to which herding emerged as a later offshoot and subsequent specialization of agricultural production, and suggest the possibility in this region of a direct transition from hunter-gathering to an animal husbandry-based Neolithic in the absence of agriculture, see Arioti and Oxby (1997).

The Tuareg also maintained long-standing relations with many Saharan and sub-Saharan groups, including agriculturists, pastoralists and urban dwellers (Claudot-Hawad 1996a). This fact is still visible in the linguistic landscape: while most Nigérien Tuareg speak *tamajeq*, in either the Aïr (to the North and Central East of Niger) or Azawagh (to the West) dialects of what is a Berber language, some others speak a northern version of Songhay. Among the Kel Tagarayga-rayt in West–Central Niger, the Ayttawari Seslem and the Kel Eghlal n Enniger speak *tetserret*, a dialect that resembles the Berber spoken in Mauritania and Southern Morocco (Walentowitz and Khamed Attayoub 2001).

On both sides of the Sahara, Tuareg society materializes common cultural, economic, and political institutions in remarkably diverse ways (E. Bernus 1981). Besides pastoralism, some of the most significant traits are the emphasis on women's autonomy within a cognatic or bilateral kinship system (S. Bernus et al. 1986) in which opposite-sex siblingship acts as a primal relation (Walentowitz 2011); and a warrior ethos of honour and protection within a hierarchical society including craftsmen, religious specialists and servants. The principles of pluralism and fluidity are central driving forces in all fields of sociality and forms of exchange. There is no power without its counterpart, no living-together without the inclusion of others, be it among political elites, husbands and wives, kin and allies, masters and servants, or nomads and agriculturalists (Claudot-Hawad 2001). These principles shaped people and ecosystems over centuries, until colonial and postcolonial politics of division and standardization started to cut boundaries through cultural landscapes that cannot exist or persist, however, without the possibility of interconnection.

Pastoralist Landscapes: a Mosaic of Human Ecologies

Tuareg territories include a range of arid and hyper-arid habitats including a topography of highland, lowland, sub-desert and desert landscapes. In today's northern Niger, the Tenere Desert comprises *reg* (stony desert), *erg* (areas of shifting sand dunes) and ancient oases, while the Aïr Mountains offer clement mountain and valley habitats. Elevations range from 300-500 meters in the Tenere to nearly 2000 meters at Mount Tamgak in the Aïr highlands (Newby and Grettenberger 1986). To the West and South-West of the Aïr mountains, the valleys and flood plains of the Eghazer and Tamesna lead to the more humid Sahelian pastureland of the Azawagh valley that runs into central Niger.

Of all these regions, the Aïr presents the highest diversity (Newby 1992). Parametric climatic conditions are extreme; annual rainfall ranges from 0-100 mm in the North, to 100-200 mm where the Sahara meets the Sahelian zone. Temperatures can range from 0 to 50° C. In addition, rainfall is spatially and temporally highly unpredictable (Newby 1990); when it occurs it fills depressions (*wadi*), which then show favourable moisture and vegetation conditions. This harsh environment is home to a well-adapted fauna including Addax antilope (*Addax nasomaculatus*), Dama gazelle (*Gazella Dama*), slender horned gazelle (*Gazella leptoceros*) and ostrich (*Struthio camelus*), as well as predators such as fennec (desert fox), cheetah (*Acinonyx jubatus*), and Nubian bustard (*Neotis nuba*). Local flora contains dry resistant species such as wild olive (*Olea lappesinei*), wild sorghum (*Sorghum spp.*), wild millets (*Pennisetum spp.*), as well as a further 50 plant species, including shrubs and trees such as African mahogany (*Khaya senegalensis*), Albizia and tamarind among many others (see Le Houerou 1989). Many of the fauna and flora are endangered, which led to the creation of the *Aïr and Ténéré National Reserve* in 1991. The reserve includes the Aïr Massif and is Africa's largest protected area with 7.7 million hectares (UNESCO). It has been established, however, without the consultation of the local population, in the midst of an armed conflict, and forms the other side of the modernist coin and its paradoxical "politics of nature" (see Latour 2004): here pastoralist landscapes are conceived by outsiders as congealed and non-productive natural reserves to be protected from local pastoralists, while the productive land that unfolds to the West of the Aïr mountains may be openly accessed by all including mining companies.

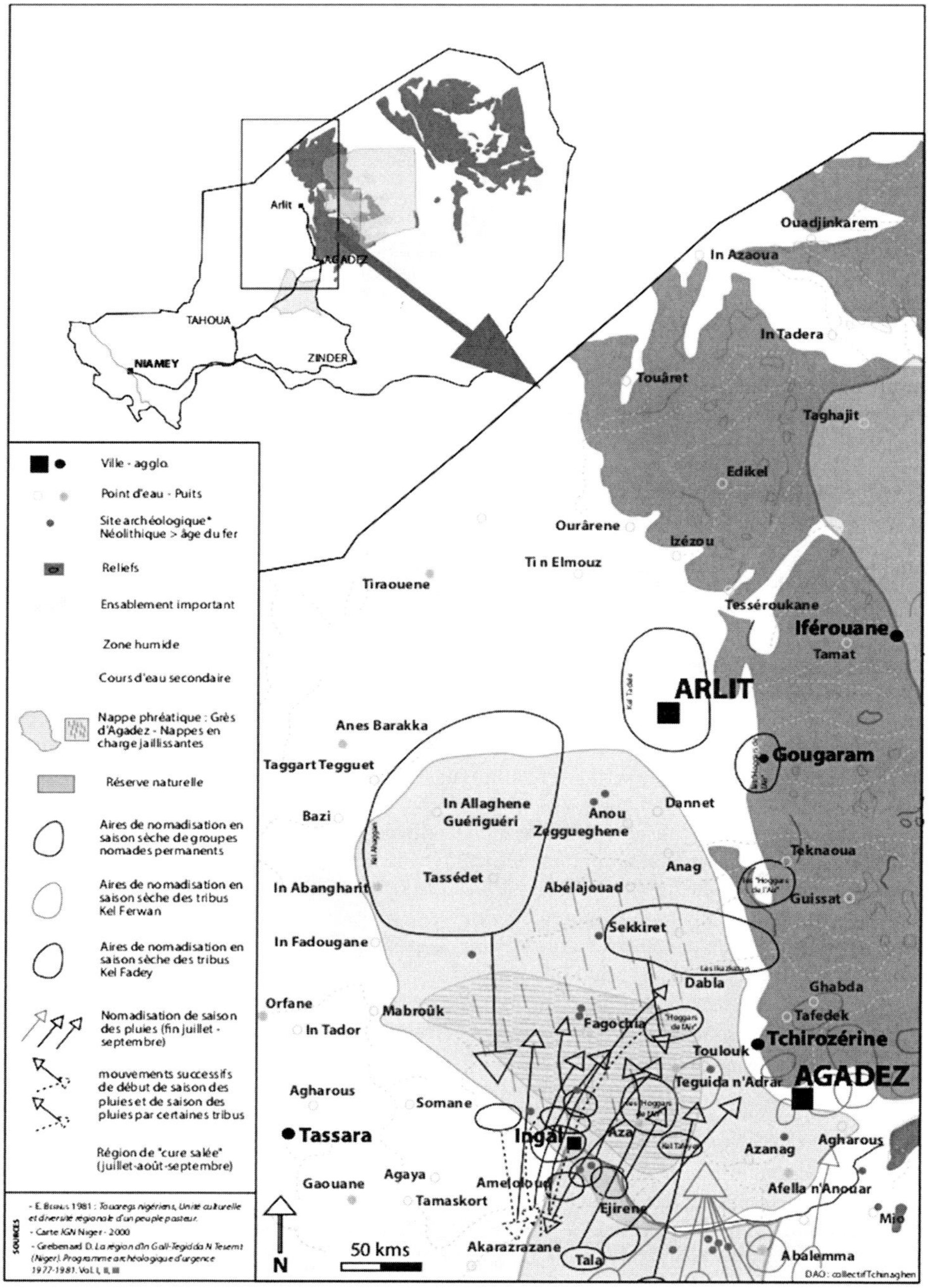

Map 1: "Tuareg way of life: a precious balance". Reproduced from Collectif Tchinaghen (2008: 6).

The Tuareg, by contrast, have their own environmental practices and categories based on dynamic relationships between people, animals, water and land. The Kel Azawagh and Kel Ayer Tuareg are mainly herders, raising camels, goats, as well as sheep and cattle further South in the Azawagh region. Before the heavy droughts, most Tuareg economies were centred on the production of milk, while small livestock was exchanged or sold for millet and sorghum (Bernus 1981). Together, milk and cereals form the basis of the daily diet. Caravans and the control of trans-Saharan trading routes played a crucial role in pre-colonial times, through the exchange of dates and salt for cereals, cloth and other manufactured goods. Mountain Tuareg such as the Kel Ewey are agro-pastoralists, having intensified oasis gardening after a huge loss of livestock during the French conquest. Although caravans became forbidden with the establishment of colonial border regimes, they maintained a triangular caravan trade between Bilma and Fachi in the Tenere, and Kano where their camels fertilize agricultural Hausaland each year after harvest (Spittler 1989).

Despite the diversity of Tuareg economies, all groups combined nomadic pastoralism with long-distance mobility, in order to cover the needs of humans and animals in these inhospitable environments and sustain their fragile ecosystems. Tuareg nomadism is grounded within well-localized territories used by various groups that shift, however, over time (Bernus 1981). Nomads follow an annual cycle of surface and underground water management. During the dry season, the different groups move shorter distances between wells, ponds and pastures; during the rainy season between June and September, all groups move longer distances towards the salty pastures that grow in the plains and valleys of Tamesna and Eghazer after the rains. Coming from the North, the South-West, East and South-East, the Kel Ahaggar, Kel Azawagh, Kel Ayer, Kel Ferwan and the Kel Gress, all converge on this region and other salt pastures at a similar latitude. People and livestock enjoy the annual salt cure and move home fully replenished after a long dry season that, often enough, ended with hunger. The rainy season, by contrast, is the time of marriage and the renewal of all kinds of alliances, a contra-point of abundance at the edge of the Saharan desert.

Nomadism, a Disrupted Dialogue between People, Animals, Water and Land

Time-honoured Tuareg ontologies of dwelling revolve around flexible territorial rights that are anchored in shared cosmopolitics of movement (see Claudot-Hawad 2012b). Homelands are actively built through pastoral activities and journeys that optimize the use of resources variously available in space and time. Wells, ponds and pastures are collectively attributed, managed by local elites and are always open for negotiation. Nomadism is a dialogue in which humans are the mediators between land, animals, plants and water. Neither nature nor culture, pastoral landscapes are the outcome of these mediations and must be regenerated each year thanks to the carefully established, partially overlapping movements of people and animals around fixed points that are constituted by wells. Major wells are preferably situated at the crossroads of various territories, which are redefined in such a way that no group may claim exclusive territorial rights. Interdependence is actively created in customary law, social norms and an education that valorizes the association of difference rather than the conservation of identity.

In the semi-arid and arid inhabited zones of Central and North Niger, access to land is still determined largely by rights to water, the supply of which varies seasonally and annually. Tree-lined dry valleys run through the sandy scrubland: in the short wet season rain water accumulates in ponds along these valleys, and in the long dry season the ponds dry up and people depend upon wells, which are sunk at intervals along the same valleys. According to community memories dating back to the period immediately after the 1917 revolt, the 'traditional' wood-lined wells of Central Niger were managed by the community leaders who had them dug and maintained by specialist well diggers, for the benefit of their kin and entourage. To this day the same situation largely holds with this type of well, which remains the most common type (even though resource use rights have subsequently been curbed as we shall see below); and the wells are usually known locally by the name of the relevant influential local person. The geographical features of the land thus coincide with a social landscape corresponding to members of a relatively wealthy elite who, in cooperation with their neighbours, manage access to water and the surrounding pasture, in particular during the hot dry season preceding the rains when these resources are most scarce (Oxby 2011: 59).

Since colonial times, nomadic pastoralism has changed dramatically, especially through the restriction of movement (*supra*) and ill-advised shifts in water management. New water points have been sunk along these same valleys as part of regional and national government programs, during the colonial period and increasingly after independence in 1960: cement-lined wells known as *ciminti*, and boreholes equipped with a motorized pump and reservoir, known as *fonfo*. The creation of deep wells has permitted a greater concentration of people and herds during the dry season, and has also facilitated the sporadic spread of crop cultivation, especially millet. Although the government negotiated the construction of the first deep wells with the so-called "customary leaders" who had been registered with the administration since the colonial period, these government wells were officially open to all users, and their association with local populations and their leaders was not formally recognized (Oxby 2011: 60).

In fact, local community leaders to this day remain largely removed from decision-making relating to the land they inhabit, and more generally feel unrepresented by a government which, apart from a few token individuals, does not understand or value their language, culture or way of life. Moreover, strategies for developing local natural resources are based on an outsider understanding of drought as an exceptional climatic occurrence, in ignorance of local views about seasonal land use and climatic variability (Oxby 2014: 123). More and more wells are dug, increasingly by well-off merchants and other urban dwellers who invest in livestock and *de facto* privatize the surrounding pastureland. This trend of privatization has been reinforced through the creation of shops near wells and has gone hand in hand with national and international sedentarization policies; in some areas, pastureland is even sealed-off in order to prevent access (Gangol and Afane 2010: 3). The land act or *Code Rural*, amended in 1993, paradoxically contributed to the decline of pastoralism rather than securing it (ibid.: 4). Even though additional amendments granted to pastoralists the right of usage and management of "traditional" dry-season *terroirs d'attache* (territorial bases), it did not define exclusive territorial rights, nor prevent the development of other activities such as agriculture.

In recognition of the failure of the 1993 *Code Rural* to protect adequately the interests of herders, whose suffering during the severe drought that unfolded during 2010 was publicized worldwide, the Niger government agreed a *Code Pastoral* (Pastoral Land Act) in May of the same year, the result of years of lobbying on the part of local and international NGOs representing herders. This *Code Pastoral* stresses the importance of human and animal mobility in maximizing productivity; in other words, it finally endorses the scientific importance of the age-old nomadic way of life, that had previously been so undervalued in the national perception. A recent analysis of the *Code Pastoral*, however, argues that it has been introduced too late, and in too diluted a form, to bring much support to mobile pastoralists, whose livelihoods are increasingly under threat, particularly during drought years and with the added factors of land alienation, demographic pressure, and climate change. Moreover this *Code Pastoral* is found to institutionalize rather than resolve a conflict of interest between two categories of pastoral land users: those who enjoy priority use rights to ancestral lands versus those who pass through such land and can exercise rights as temporary transitory users (Oxby 2011: 64). Finally, Niger's decentralization policies follow the same path: nomads are increasingly deterritorialized through the implementation of sedentary territorial logics, from the Nation-state borders all the way down to the limits of local communes (Gangol and Afane 2010: 7).

Global phenomena such as climate change demonstrate that the modernist split between a pre-given "Nature" and a human-oriented "Culture", that has underpinned the mismanagement of the *zone pastorale* ever since French colonization, is fast becoming a threat for mankind altogether.

Alternative ecologies (Descola 2013), including nomadic land husbandry, are no longer negligible rudiments of the past, but have become, once again, prerequisites for all of our futures. Although many of the pastoralist spatial practices in the Sahara and the Sahel have vanished, and those that remain are discouraged by sedentarization and development policies, it is important to record them, in order to understand that uranium mining constitutes the latest chapter in the colonial and postcolonial decline of nomadic pastoralism in Northern Niger. Indeed, following the same denial of territorial rights, mine excavation activities occupy an area that is crucial not only for local nomads, but for all major Tuareg groups during the rainy season (compare Maps 1 and 2).

The Tuareg refer to this destructive process that harms the society as a whole by the phrases *ahluk n temust* and *ahluk n tamattay*, meaning the "annihilation of a nation" or "people" through the avoidable destruction of elements, persons, resources or cultural assets, that are critical for its cohesion and survival. It is differentiated from *ahluk n aytedan* ("loss of people") caused by natural disaster, epidemics or famines. Violence, both structural and direct, against specific persons, groups, water and land resources can thus be considered as a form of collective violence threatening the "seed" (*iri*) of a society that must by all means be kept alive. When this local future becomes impossible, what is left for the younger generations, if not the modernist avatar of long-distance mobility, which increasingly fosters violence-based associations through the traffic of weapons and drugs along the Sahara's ancient trading routes. The mineral resource-addicted societies of the North rely on pastoralist territories just as much as pastoralists do. And so does Niger's national economy that, to this day, at the expense of its own citizens and common property, remains grounded in the nuclearization of France.

1.3 National Economy

The economy of Niger is focused on uranium mining, which constitutes between 60 and 80% of external revenues, and agriculture, which contributes about 40% of the GDP (CIA Fact Book 2012). Niger also has sizeable reserves of oil, and some oil production. However, the general indices of development do not look promising: with USD 800 GDP per capita, Niger ranks at 176th place in the world, and finds itself at the bottom of the Human Development Index (place 174 of 179 countries in 2012). Life expectancy is hardly higher than 50 years, and child mortality is extremely high with 163 per 1000 children. Two-thirds of the population lives under the so-called poverty line, one-third faces undernourishment (50% of the children), and most of the people do not have electricity. The country imports about double what it exports, leading to high external debt (1.3 billion USD in 2000) and dependency on International Monetary Fund (IMF) and World Bank loans. Foreign aid was nearly half of the annual budget in the 2000s (CIA World Fact Book 2012). Niger suffered heavily from several droughts in the Sahel zone between the 1970s and 1990s. Its food production is insufficient to cover domestic needs while at the same time the country exports cash crops (such as peanuts, cotton, cowpeas and onions). Only 15% of Niger's land is suitable for agriculture, the rest can only be used for livestock. Most people are either agricultural peasants or pastoralists engaged in a combination of subsistence and market production in order to meet cash needs (Maldonado and Gasarian 1998, Decalo 1997).

Drought and desertification since the 1970s, as well as population growth (some of the highest birth rates in the world, especially in the agricultural South of the country), are said to have undercut the economy. Future growth may be sustained by exploiting oil, gold, coal, and other mineral resources. The government has announced an expansionary budget for 2012, with anticipated revenue increases from uranium, oil production, and donor aid; and it has made efforts to secure a new three-year extended credit facility with the IMF, following the one that concluded in 2011 (CIA Fact Book 2012). These general figures do not show the historical and political-economic context of the crisis in Niger. The annual income of the mining company Areva is five times larger than that of the state of Niger. Areva gains two-thirds of the revenues of the SOMAÏR mine in the North of Niger, leaving one-third to the state. In the case of the COMINAK mine, Areva and the state of Niger hold 30% each. As Areva controls the whole chain from uranium production up to the production of electricity, the actual gains made from Niger's uranium are much higher. As Granvaud states, France owes much to Niger, as it depends up to 80% on nuclear power for electricity, while 25 to 40% of the uranium needed in France originates from Niger (Granvaud 2012: 63).

Historically, Niger has followed the "usual path" of development promises based on mining revenues that served as a basis for getting loans. As the promises vanished due to the diminishing relative prices for uranium coupled with rising oil prices and interest rates, the country's economy became trapped in debt payments. In addition, the devaluation of the national currency led to further losses (Granvaud 2012: 69). In the 1980s, Niger had to face a structural adjustment program with severe cuts and privatization of public services from water supply to agriculture and health, while 50 to 60 % of the annual budget was used for debt servicing. Together with the political instability and notoriously bad governance, the country serves as a prime example of the "resource curse" hypothesis (see Auty 1993).

Due to the economic crisis, the country qualified for enhanced debt relief under the IMF program for Highly Indebted Poor Countries (HIPC) in 2000, and concluded an agreement with the Fund on a Poverty Reduction and Growth Facility (PRGF). Debt relief provided under the enhanced HIPC initiative significantly reduced Niger's annual debt service obligations, freeing funds for expenditure on basic health care, primary education, HIV/AIDS prevention, rural infrastructure, and other programs. In 2005, Niger received 100% multilateral debt relief from the IMF, which translated into the annulling of approximately USD 86 million of debts to the IMF (IMF 2006, CIA World Fact Book 2012). When Niger became part of the program, the country was already unable to prevent the several hunger crises that followed the drought years (see Granvaud 2012: 70-72). The undermining of pastoralists' services of trade and transhumance had a negative effect on the availability and distribution of food, which was further aggravated by speculation on food in the 2000s. For Granvaud (2012: 68-70), the main problem remained, namely that Niger was never able to control its economy and invest the gains from uranium exports for a development more adapted to local realities.

Part 1 has laid out the various historical, ecological and political contexts in which mining takes place in the North of Niger. The details of the uranium mining operations, their impact on the local population, together with the response of local people, mining corporations and nation states, is the subject of Part 2 which now follows.

2. Uranium Mining: Impact and Response

France's uranium mining activities started during the race for uranium shortly after World War II. Domestic uranium resources were already being mined, but in the 1960s new public awareness of the dangers of nuclear power and the harmful effects of uranium mining, in combination with rising costs and decreasing resources, prompted France to further explore its ex-colonial territories. Uranium mining was found to be much easier and cheaper in poor countries of the 'South': minimal bureaucratic red tape, thanks to the political connections of colonial times; minimal interference by health and environment watchdogs; and conveniently remote locations where activities may proceed away from public scrutiny. As one commentator stated about the nuclear boom in Niger: "Getting a mine going in Texas takes two bookshelves full of authorisations... In Niger you give a shovel to a guy on $2 a day and you're mining uranium." (Greenpeace 2010: 14).

France as colonial and then as former colonial power has successfully defended its monopoly of uranium mining in Niger to this day: the corporation involved is the nuclear energy giant Areva, whose majority shareholder is the French government. With the rising demand in France for cheap energy, France has become increasingly dependent on Niger's uranium to meet electricity requirements: one out of every three light bulbs in France is lit thanks to Nigérien uranium (OXFAM - France 2013). As decades pass by, however, more and more doubts have arisen as to the possible negative impacts of uranium mining activities on human health and on the environment in particular. Substantial evidence in relation to Areva's activities in Niger gained international media attention in the aftermath of the change of government in 2010, which enabled a relaxation of the usual strict information control, as we shall see below.

The Greenpeace document 'Left in the Dust' is a product of this period, and is widely quoted in the following sections. Available in English and French (Greenpeace 2010), it neatly assembles the evidence available to their researchers in 2010, including previous findings by researchers from NGOs such as AGHIRIN'MAN, SHERPA and CRIIRAD (see website section of bibliography); this is supplemented by Greenpeace's own researchers' interviews and the results of the limited surveys they were allowed to carry out in Arlit in 2009-10. This evidence is supplemented by the work of other authors who have published over the decades since production began (Martin 1989, Grégoire 1999, Keenan 2008 and Grandvaud 2012, 2013, 2014).

A brief chronological summary of uranium mining in Niger now follows, before proceeding to the next sections which will deal with the environmental and health impact of uranium mining, and the response by local people, their organizations, the Niger government, and the international corporations.

Areva and its local subsidiaries have been mining uranium in Niger for over 40 years. In 1968, the holding company SOMAÏR (Société des Mines de l'Aïr), was created[3]; excavation began in 1971 at the open-pit mine, which varies from 50 to 70 metres deep, and is situated seven kilometres NW of the mining town of Arlit (see Maps 1 and 2). A second Areva holding company COMINAK (Compagnie Minière d'Akouta)[4] was set up in 1974; the location is about 2 km from the town of Akokan, itself situated some 5 km to the SW of Arlit; and production started in 1978. Unlike the SOMAÏR mine, COMINAK is an underground mine: with a depth of 250 metres and over 250 kilometres of galleries, it is among the largest underground mines in the world. Since the

3 SOMAÏR (Société des Mines de l'Aïr) was created in 1968. Its capital is held 63.6% by AREVA and 36.4% by SOPAMIN (Société du Patrimoine des Mines du Niger, the Nigérien agency that manages the State's mining interests). Source: www. areva.com, October 2014.

4 COMINAK (Compagnie Minière d'Akouta) was founded in 1974. AREVA holds 34% of its capital; SOPAMIN (Niger) 31%, OURD (Overseas Uranium Resources Development Company Ltd, Japan) 25%, and ENUSA (Empresa Nacional de Uranio S.A, Spain) 10%. Source: www. areva.com, October 2014.

beginning of production, SOMAÏR and COMINAK have excavated over 52 million tonnes of ore, from which over 100,000 tonnes of uranium have been extracted. On average, the mines produce over 3,000 tonnes of uranium and net €200 million in sales each year (Greenpeace 2010: 10).

A third mine, Imouraren, was originally planned to start production in 2013, projected to be the largest uranium mine in Africa and the second largest in the world, with an annual production capacity of 5,000 tonnes of uranium. Areva's revenues for 2008 were €13.1 billion, with a profit of €589 million. SOMAÏR generated €161.7 million of that revenue by producing 1,743 metric tonnes of uranium. COMINAK earned sales of €100.6 million for its supply of 1,289 metric tonnes of uranium concentrate" (ibid.). In May 2014, Areva, third global uranium producer in 2013, signed a new agreement with the government of Niger concerning conditions surrounding the exploitation of uranium in the north of the country: amongst other things it announced that plans for developing the new mine at Imouraren, about 60 km south of Arlit, were to be postponed, pending the improvement of the value of uranium on global markets.[5]

As already mentioned, Areva has until now enjoyed a monopoly over uranium exploitation in Niger. Since 2007, however, concessions for mineral prospecting in the Tuareg homelands of northern Niger have been granted to many other foreign nations including China and Canada (Collectif Tchinagen 2008 has a map identifying about twenty of the companies involved and the location of their concessions). Map 1 (page 162) shows the Tuareg land use in the areas concerned, and Map 2 (page 170) superimposes the location of the numerous concessions that had already been granted in 2008, also some others that were planned. The attempt by the Canadian mining company GoviEx to present their environmental impact survey to the public in Arlit is described in the next section. So far none of these other companies have begun exploiting uranium in Niger. The current competition between various global companies to break France's monopoly of uranium in Niger, combined with the unstable political situation characterizing the whole of the northern Sahel and Saharan region, pose a great threat to the people of northern Niger, who feel increasingly marginalized by the government. As outlined in Part 1, and will be further discussed in section 2.4 below, the use of force is being resorted to by some groups in northern Niger, in an attempt to redress the situation from the point of view of local inhabitants.

The mining of uranium in Niger is characterized by startling extremes: Niger is one of the wealthiest countries in the world if mineral resources are the yardstick, leaving France far behind; also, France is increasingly dependent on Niger's uranium resources in particular. On the other hand, Niger is amongst the poorest countries in the world according to standardized ways of measuring development and poverty worldwide: Niger consistently sits near the bottom of the Human Development Index which measures life expectancy, educational attainment and incomes, despite decades of mineral extraction which, rather than bringing wealth to the mining region, appear to have brought destitution and ill health, as we shall see in the following sections.

Section 2.1 deals with the impact of uranium mining on the environment; section 2.2 with human health impacts; section 2.3 with the impact on local livelihoods. Section 2.4 discusses the various reactive strategies of local people, including those of certain Tuareg groups that attempt to organize while facing the threat of "terrorist" labelling; and finally section 2.5 outlines the strategies of the states and corporations involved, in particular the way that the mining company operates in the shadow of the state of France and that of its local protégé, Niger.

5 Areva press release May 2014 published on www.temoust.org.

2.1 Environmental Impact

Uranium extraction is a risky and highly contaminating business not just for workers but for the environment at large, especially under dry and windy natural conditions. It exposes humans, fauna and flora to long-term radiation with variations in intensity and with substances such as radium, radon, polonium and others. This fact demands cautious and knowledgeable procedures in the production process for all actors in contact with these elements. It should be expected that dealing with these dangerous materials would lead to standard protective strategies by a company, irrespective of whether this is happening in Europe or Africa. However, despite reassuring assertions in the company's publications, the reality in the case of northern Niger fails to reflect best practice (see Granvaud 2012: 180ff.).

All the sources referred to in the previous section combine to represent the impact of mining in the Arlit region in a very different way from the statements of sound environmental and social management promoted by the mining company Areva. The available reports reveal that the local state control agencies neither dispose of the technology and the equipment to measure local radiation, nor do they possess adequate transportation means to monitor the area; therefore, one has to rely on NGO measurements for an impression. This section starts by drawing attention to the inadequacy, as far as the local population is concerned, of data provision on the local impact of mining; it proceeds with a review of the available evidence for the impact of mining on the soil, air and water. The consequences for human health follow in the subsequent section 2.2.

Data, Suppression of Data, and the Control of Information

On 14th May 2015, a meeting took place in Arlit, to which the mining company GoviEx had invited members of the public, in order to share details of its plans to develop a new uranium mine to the South-West of Arlit in sector Madaouela West.[6] According to locally agreed regulations, an environmental impact survey should be carried out and conclusions submitted to the public in advance of any such meeting, so that concerned NGOs and members of the local population groups can inform themselves of the likely impacts of the proposal, and prepare questions to the project promoters on how undesirable environmental impacts may be limited. Representatives of the NGO AGHIRIN'MAN complained at this meeting that no environmental impact survey had been made available to them or the public in advance of the meeting, and that the mining company was therefore in breach of regulations they had previously agreed to. Following a discussion about this at the public meeting, the Préfet (head of regional administration) of Arlit told the NGO representatives that they could no longer speak at the meeting, and ordered them out of the premises.

This incident sums up the inequality in access to environmental data between project promoters and representatives of the local population, and the way the Nigérien authorities side with the former against the latter. Environmental data collection is strictly controlled by the promoters in collaboration with the local administrative and military authorities; only a few independent researchers have managed to gain access and publish results relating to the mining locations, as we shall see in the review below. Although there are formal regulations in place to promote the inclusion of the local populations in decision-making concerning the environment, these are not necessarily implemented in the expected way, as shown by the above incident.

The evidence gathered for this section relates principally to the operations of the French-controlled company Areva, the only company yet to have engaged in uranium production in Niger. It comes from the work of NGOs such as: CRIIRAD (*Commission de Recherche et d'Information*

6 "AGHIRIN'MAN interdit de parole par le préfet d'Arlit" <uranium-niger.jimdo.com/les-actions/autour-de-la-gestion-des-sites-miniers>, accessed 2 June 2015.

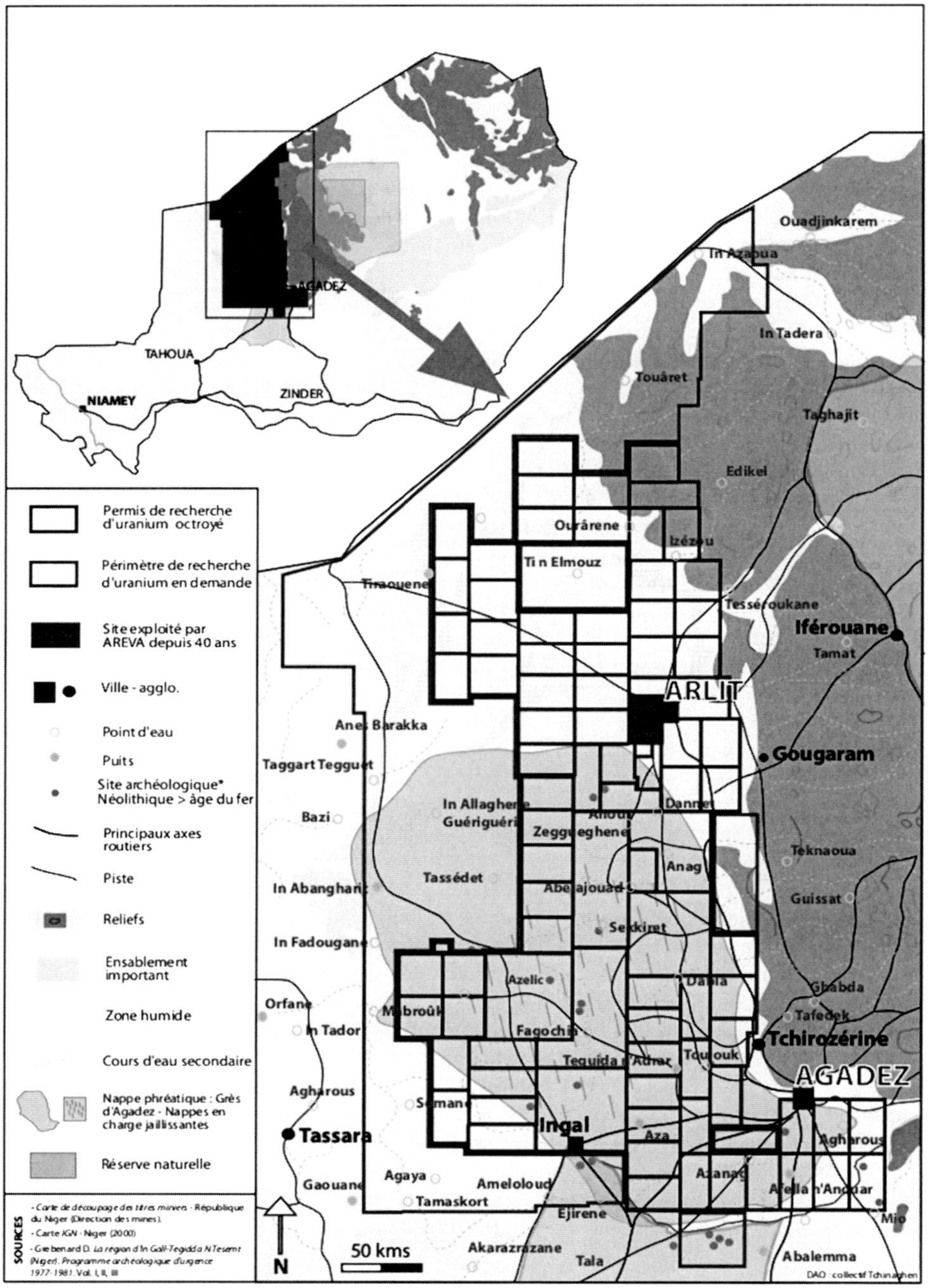

Map 2: "The Tuareg people threatened by uranium exploitation". Reproduced from Collectif Tchinaghen (2008: 7).

Indépendantes sur la Radioactivité), set up to improve information and protection of the public against ionizing radiation/radioactivity; SHERPA (a Paris-based association set up in 2001 to protect and defend victims of economic crimes); EJOLT (which supports collaborative research and action on environmental conflicts, see Chareyron 2014); and Greenpeace (especially 2010, also 2013). Another extensive overview of environmental and health-related problems in the area of Arlit was made by Granvaud, whose 2012 report summarizes research reports and information from the same NGOs, as well as from local actors and journalists. As years go by, however, the control of information by the French mining company seems ever greater, and the publication of impartial information on the impact of mining activities seems ever less. Although Chinese, Canadian and other companies holding research concessions for mineral prospecting in Niger have no doubt amassed their own data, this is not readily available either.

Soil and Air

In order to appreciate the impact of uranium mining, it is necessary to understand the working procedures of the mining itself: for the uranium proper—or yellowcake—to be reached, the subsoil has to be removed, and this removal process creates large amounts of soil contaminated with radiation: for every kg of yellowcake 335 kg of contaminated soil is produced. Reports estimate a total of 35 million tonnes of tailings, being the contaminated black soil that still contains high levels (85 %) of radiation (Greenpeace 2010: 40). These mountains of radioactive soils are exposed to strong winds and erratic—but nevertheless occurring—rains and by these means radioactivity is spread all over the area. Areva does sprinkle certain spots to bind the sand but wet soils leave the area as cars and workers move in and out. In 2007, CRIIRAD and AGHIRIN'MAN carried out radiological surveys in the town of Akokan, situated 2.5 km from the underground COMINAK mine. Directly in front of the Akokan hospital, they discovered shockingly high levels of radiation up to 100 times higher than normal background levels; this radiation was caused by radioactive waste rock that had been paved into the road during its construction (Greenpeace 2010: 41).

A further problem is air pollution by radon gas that escapes from the underground mine via the air ventilation system. The air vents of the ventilation system are freely accessible and thus dangerous for humans: the radon gas is then inhaled by humans or enters the food chain. In 2009, Greenpeace had aimed to install radon detectors near the mine vents, in order to study the transfer of radon into the environment, and to identify contamination of vegetation and milk from animals that graze in the area, but they were not allowed to because of 'security' restrictions placed on them. The radon measurements were therefore taken mainly from the urban zone, several km from the source. The measurement taken at the police station in Akokan, less than 2.5 km from the entrance to the COMINAK mine, shows a radon-222 concentration in the air 3 to 7 times higher than normal levels in the area. Such higher levels of radon-222 in the air could have been caused by open air storage of mining waste or the atmospheric emissions from the underground mine. The National Centre for Radioprotection is not permitted to measure radon and relies on information supplied by Areva (Greenpeace 2010: 29). Based on these figures, it is astonishing that Areva holds the certification ISO 14001, indicating sound environmental management (Collectif Tchinaghen 2008: 15).

Water

Amongst the most harmful effects of uranium mining in Niger is its impact on water resources (ibid.): in the semi-arid environment typical of northern Niger, access to water is the most crucial aspect of the environment and of local livelihoods, as already discussed in section 1.2.

Millions of litres of water are used daily in the mining operations, particularly in the leaching process to separate the uranium from the ore. The water is pumped from a huge groundwater reservoir—the Tarat aquifer—which is 150 meters deep. This is a fossil aquifer, and will take millions of years for it to fill up again. In 40 years of operation, a total of 270 billion litres of water

Fig. 3: Inside view of the underground mine in Arlit. Source: Screenshots from *World of Matter* / Ursula Biemann.

have been used in Arlit and Akokan (Greenpeace 2010: 21-22). Based on an old preliminary study from 1968, Areva claims that this represents about 20% reduction of the Tarat aquifer: a huge proportion, given that it is irreplaceable. About 35% of the water is said to be used for industrial use, while the other 65% is for drinking water production. Though Areva claims the use of water has declined in the last decade, there was an increase in water use in the SOMAÏR mine in the past years. When asked for an explanation, Areva said that more water is used in the new ore leaching process. Also, the water use is proportional to the increase in production: the more uranium produced, the more water used.

A consequent problem is the drying up of wells previously used by the local population for drinking water and watering livestock. A 2004 hydrogeological study reports: "In the section of the CK [COMINAK] mine, the water has been completely drained and its level has dropped to the wall of Tarat [aquifer] (decrease of 150 meters); this has sometimes led to an inability to operate the water wells dug near the mine, namely Comi_10, Comi_11 and Arli_987." (Greenpeace 2010 quoting COGEMA, Niger 2004).

But most problematic is the contamination of water previously used by the local population for drinking and watering livestock, creating enduring health problems for people and their livestock. The SOMAÏR open pit mine and the COMINAK underground mine both cut through the Tarat aquifer, which is opened up to the contaminants, that can spread throughout the water table. Where drilling has already taken place in Imouraren, the material extracted from the drilling is said to have five to nine times the level of gamma radiations than normal.

At certain places the groundwater is removed from the mines in order to mine the ore. The removed 'industrial' water becomes contaminated by the uranium and other radionuclides and can no longer be used as drinking water (Greenpeace 2010, quoting Areva). In addition, highly contaminating chemicals are used for 'lixivation'— the process of cleaning the yellowcake. Radioactive liquids are stored in ponds with dykes that sometimes leak or break. These leakages often occur close to settlements such as Arlit that are at a distance of only 3.5 km from the mine.

Despite information disseminated by the French company that they purify water and facilitate its access for local people, water samples collected in 2003 by NGOs show alpha radiation that is highly dangerous for human health if consumed. The 2010 Greenpeace study concluded that contamination of water had further increased during the last twenty years and that water also contained high levels of other chemical substances with higher levels of concentration than recommended by WHO. To make things worse, contaminated water is used widely for irrigation in the area. Water is being more and more depleted and polluted in this section of the groundwater table

Fig. 4: Dry tailings disposal site near Arlit, 2007. Photo: Collectif Tchinaghen (2008: 12).

that is being used by the 80,000 inhabitants of the Arlit town and the surrounding area. As clean water supplies are becoming scarce, Areva is considering the possibility of using other groundwater tables such as the one of Agadez in the near future (Granvaud 2012: 166ff.).

Climate Change

An impact of uranium mining that is often overlooked is the issue of climate change. The company portrays itself as providing a solution to the climate change problem, as uranium does not create CO2 emissions. However, locally the company operates with thermal power plants to produce electricity and for which, until now, 160,000 tonnes of coal have been burned producing high levels of CO2 emissions. In addition, gasoil and other carburant substances are needed for the production process. One tonne of uranium requires 9.7 tonnes of petrol and many other substances: sodium, ammonium, oils, magnesium, as well as cement and explosives. All of these CO2 impacts will rise exponentially if the new mine of Imouraren becomes operational.

2.2. Health Risks and Medical Evidence

Health Risks to Mine Workers

While the environmental impacts listed above are staggering, perhaps even more shocking is the way the mining company has failed to inform mine workers and citizens of the health risks of the mining operations so that appropriate protective measures may be taken. Areva argues that when the mines were installed in the late 1960s some of the health risks were unknown. However, one of the original incentives for the French company to prospect abroad was growing concerns over health in France. National and international standards for nuclear activity were already being put in place. "The white men, when they were here, they knew. The white men openly talk about the diseases caused by the mine," Fatima Daoui stated, referring to the 500 French expatriates who used to run the mines. It was not the same for her people, though; an air of secrecy stifled the Nigérien workers, who were not free to speak about health problems without fear of repercussions (Greenpeace 2010: 51).

In the early years of operation, the workers did not wear protective masks, while in France already in the 1950s such laws and regulations existed to protect workers' health. According to the 2008 Collectif Tchinaghen Association report: "The first individual safety outfits such [as] masks were placed at the workers' disposal in the middle of the eighties—nearly 15 years after the beginning of the mining. During the first years, the miners used to work with their ordinary clothes. So they used to come back home with the clothes they had worn to work, which were covered with ore dust." Only in the 1990s did Areva begin to inform the workers about contamination risks, during one-hour meetings. Meanwhile, workers were still being constantly exposed to a level of radiation more than double that allowed by European and French laws. Even when workers were provided with dosimeters, they were not informed of the dose they received. Furthermore, subcontracted workers, who are used frequently, still do not wear dosimeters. "Even today, we still see some guys who come from the area with their work clothes," said one ex-mine worker (Greenpeace 2010: 55). Another source of contamination are the accidents that occur while transporting uranium, which then expose the immediate environment to radiation (Granvaud 2012: 160-2).

The rules concerning work conditions in the mines only apply to workers formally employed by Areva and local mining companies: a high number of workers are employed by contract firms that provide the mines with cheaper work partly because they pay the workers less. Contract firms do not seem to feel that they are bound to national legislation and regularly violate security measures in order to increase profits (ibid.: 185ff.).

Health Risks to the Local Population

Work clothes covered with dust from the mines obviously pose a dangerous risk of exposure not just to mineworkers but also to their families and the wider population. The water and health facilities available to locally recruited mine workers and their families are generally of such a poor standard that they are constantly exposed to radioactivity. There has been no attempt to inform the wider public about the risks of uranium mining: for example, the particular dangers of using recycled uranium mine scrap wood and metal, materials that regularly find their way into settlements; such materials are used to make tools, cooking pots and for construction work, exposing people to constant radiation (Greenpeace 2013). In spite of being informed and alerted by French and Nigérien NGOs, Areva seems to downplay the impact of contaminated scrap metal on local health conditions. More seriously, local NGOs are said to be able to prove that contaminated scrap metal is sold on local markets (see Fig. 5).

This brings us back to the huge gap in formal scientific learning and understanding between local residents, and researchers advising the promoters, with the result that few members of the general public in Arlit would be able to understand the findings of an environmental or medical survey even if invited to a public meeting and presented with findings beforehand. There is a huge unmet need for education and publication of research findings in a way that local people can understand, using accessible terminology and in local languages. The promoters are clearly reluctant to fill this gap, which could fuel the dissatisfaction of local people, especially those excluded from the benefits enjoyed by mine workers.

On the contrary, the promoters are quick to dismiss any concerns and complaints by local individuals unless they are backed up by specialist scientific evidence. It is only thanks to a few dedicated locally based individuals, together with the Nigérien diaspora and their international NGO support networks, that any concerns about environmental impacts can be formulated and published in such a way that is taken seriously by the promoters (see section 2.4). A further barrier has emerged in that scientific evidence is being presented by the promoters more and more in English rather than French (CRIIRAD 2012), so that now even some members of the largely francophone support and advisory networks are finding themselves sidelined. Presumably, the Chinese companies produce their reports in Chinese, which makes them inaccessible even in Europe.

Meanwhile, most of the local population, of course, neither reads nor speaks either English or French, the most commonly used languages and scripts in the North of Niger being Tamajeq (Berber), Hausa and Arabic, each with their own specialized scripts. What local people do understand, however, is that their environment is being expropriated and transformed in a way that is detrimental to their usual and preferred livelihoods, and that many individuals living in the vicinity of the mines are suffering and dying as a result of ill health from new and unknown causes.

Medical Evidence of the Health Impact of Uranium Mining

This section looks at the medical evidence of ill health and the local service provision in response to these health risks. Areva has built two hospitals in the mining towns, one in Arlit and the other in Akokan. Areva's press pack states that "These facilities are open to everyone in the region, meaning that the population benefits from the best healthcare in Niger." However, it goes on to say that mine employees and their families account for two thirds of the patients treated, and the rest treated are from the general population. Statistically, in an urban area of 80,000 people, where the company only employs 2,400 individuals (not taking into account their families), the equal acceptance of all patients—given these figures—seems quite improbable. When questioned on health coverage, Areva representatives said that a worker is covered for three years following the end of his employment. The existence of coverage term limits also appears to conflict with the promise of access to free care for the entire population. Unlike formally employed workers, contract workers are unable to use the company hospitals. Local workers are very aware of the early death of many of their co-workers, and local NGOs are planning to investigate and compile a census of patients who have not been diagnosed in the company hospitals and arrived too late in the local hospitals where they were diagnosed with cancer and leukaemia (Granvaud 2012: 177ff.).

Fig. 5: Scrap metal dealer in Arlit. Source: Sreenshot from "Niger : la bataille de l'uranium", France 5, 2008.

The Areva-run hospitals do not staff any occupational health doctors, it is therefore impossible for someone to be diagnosed with a work-related illness. "There are surgeons, general practitioners and midwives. They are very good at their jobs! But there is not a doctor working to declare occupational diseases, as required by Nigérien law. Areva has done everything to avoid that, which explains the absence of occupational illnesses for 40 years. The problem we have with radioactivity here, these are low doses. Long term, low doses lead to cancer. Cancers that attack the kidneys, silicosis. There are many cases like this. That is why we require an independent radiological and epidemiological study to verify if these diseases are associated with radioactivity. We were confronted by Areva, which does not agree to this study." (AGHIRIN'IN MAN quoted in Greenpeace 2010: 55). According to an environmental impact survey carried out by COMINAK in 2000, the death rate due to respiratory infection in the town of Arlit, 16.19%, was twice that of the national average 8.54% (Greenpeace 2010: 30). However, in public documents, Areva continues to minimize and ignore the risks. "Allergies (concerning lungs and eyes), are some of the most common illnesses. They are found everywhere in the Sahara region and have long been recorded by the WHO as typical of desert areas. They are caused by sand irritating eyes and lungs and are unrelated to mining activities," (ibid.). Greenpeace researchers counter that, while it is true that respiratory ailments are common in desert regions, the COMINAK figures could indicate a pattern heightened by some other cause. Such evidence is potentially so controversial that the mining companies have prioritized the prevention of public scrutiny of their updated survey results, and of any alternative similar research being carried out whether by government or independent medical researchers.

In 2005, shocking allegations were made against one of the hospitals: that it was misclassifying cases of cancer as HIV. According to a 2005 report by SHERPA, a patient who suffers from lung cancer is never informed of that diagnosis. A former employee of one of the hospitals said that "the only confirmed cases of cancer involve people from outside the mining companies" and goes on to a very serious charge, "when these symptoms affect agents of corporations, one talks about malaria, AIDS..." (ibid.) As a former mine worker put it, "It is not possible to have access to medical records. Who is going to give them to you? We don't have the right. It is the company. ... Those in the company, they are afraid of being fired! If you [workers] write something, they dismiss you without cause!" (ibid.: 55)

In 2008, at the counter-event to the annual meeting of the World Economic Forum called 'Public Eye' and held in Davos Switzerland every year since 2000, both the People's Award and the Global Award were awarded to Areva, for the suppression of the true health conditions of uranium mine workers at subsidiaries SOMAÏR and COMINAK in northern Niger. In 2011, Areva announced that it would begin monitoring the health of thousands of workers at its uranium plants the following year, bowing to years of campaigning by advocacy groups. The new health scheme is modelled on one launched in Gabon the previous year after 1,000 workers fell ill there. The company, however, still denied any link between its mining activities and health problems amongst local people (Mark 2011).

To sum up, the health impact of the mining industry on the local population is largely unknown, because the mining company, that controls the collection of medical data, does not share it, and does not allow any independent medical survey even by the Niger government. The accumulated evidence suggests that mine workers and residents of the region do experience increased risks of mine-related health problems; and that many individuals and their families have and are continuing to suffer the deleterious consequences of this hazardous mining industry without due acknowledgement and care, let alone compensation.

2.3 Impact on Livelihoods

Livestock-based Livelihoods More Sustainable Than Mining

The mining sites crosscut transhumant migration routes and seasonal pastures (see Maps 1 and 2), some well water for livestock is now contaminated, and some areas where animals accessed mineral salts, crucially important for their health, are no longer accessible. For those Tuareg groups still depending on the herding of livestock, the mobility necessary for successful production is rendered increasingly difficult. Human and animal health risks and problems have further contributed to the demise of local livestock-based livelihoods, and will no doubt lead to far greater problems of desertification and poverty in the future. "The health of the animals is also in peril... The noise pollution will force many animals to leave the area... The exploitation of this whole area, as 159 permits [for uranium mining] ... condemns the pasture. This will force these animals to the south... and all this will aggravate and reinforce desertification," (Greenpeace 2010 quoting the president of the human rights organization FUSAD)[7].

7 FUSAD (United Front for the Safeguard of Democratic Gains), is a member of the national board of the *Réseau des*

Taking a long-term view of the semi-arid regions of Northern Niger, mobile and semi-mobile livelihoods based on animal husbandry are no doubt the most sustainable: these are highly specialized production systems evolved over many centuries, flexible enough to cope with huge variations in rainfall over space and time, and have enabled the population to inhabit and enjoy their challenging homeland environments (see Part 1). Such livelihoods are not isolated in that they include the trading of livestock on markets where food and other necessities are bought; they also combine seasonal horticulture and other productive activities with animal husbandry wherever and whenever conditions permit; and household incomes are supplemented by seasonal paid work both local and cross-border, together with remittances from household members based further away within the global Nigérien diaspora.

In addition, the negative economic labelling that pastoralism is not productive contrasts with the fact that agro-pastoral production systems represent 11% of Niger's national GDP, more than double the 5% for the uranium industry (Granvaud 2014, OXFAM-France 2013). Moreover, when compared with livestock-based livelihoods, earning a living through work in the uranium mining industry is highly precarious, in that contracts for local people are most often temporary or non-existent. And they are subject to fluctuating global commodity prices and other factors beyond the control of the local population: this is demonstrated by the recent decrease in value of uranium on world markets, causing the postponement of the start of production at the new Imouraren mine. Livestock-based livelihoods are therefore of paramount importance in assuring the long-term wellbeing, not only of the local population, but also of the national economy.

It is therefore shocking to see how the mining companies, supported by successive Niger governments, have disregarded the entitlement of local communities to live in their traditional homelands according to their chosen livestock-based livelihoods. Since the land rights of mobile livestock keepers are inadequately represented in national rural land legislation, there is a lack of formal legal evidence of community land use and ownership (Oxby 2011: 62ff), giving the mining companies carte blanche to move into Tuareg homeland areas, which the government treats as vacant land. This process of exclusion has already occurred in Imouaren in anticipation of the opening of the new mine there, and local herders have already been ordered to move elsewhere (Granvaud 2014). But where will they go? All the productive spaces are allocated to different social groups (see Map 1). These herders and their families will most likely end up in the shanty towns of Arlit and other cities in Niger and beyond its borders.

Because of the land alienation and breakdown of local livelihoods as described above, destitute ex-pastoralists have flocked to urban centers, especially Arlit, in the hope of earning a living from mining and its related service industries. The reality, however, is disappointingly different: housing in Arlit town is geographically and socially segregated into several types: first, accommodation for Areva workers: comfortable housing for the expats, with the cost of energy and water included; two-room accommodation for African employees, with lesser facilities; and dormitories for African workers, without water or electricity. Secondly, the *boukoki* or shacks of Arlit's shanty towns, whose quality is far below any of the housing for Areva workers (Granvaud 2014). Most of the 60,000 odd residents who live in these shanty towns are ex-herders from the North of Niger; most are without formal schooling and unqualified for jobs in the mines: "They say that their travels ended in Arlit because the mines have drained the natural water tables, leaving the area barren." (Sourt 2009). Such growing urban dependency may provide a convenient supply of contract workers for the mines; however, it is clearly no long-term solution to the region's development: agro-pastoralism continues to represent the most viable way of life for a large proportion of

Organisations pour la Transparence et l'Analyse Budgétaire – ROTAB (Network of Organizations for Transparency and Budget Analysis). See <www.frontlinedefenders.org>, accessed 11 Nov 2015.

the local population. What is urgently needed is that the mining companies should recognize this, and plan accordingly in collaboration with the government.

Mining Jobs to "Local People": Few Residents of North Niger Benefit

There is no doubt that many individuals and families of Arlit and more widely within Agadez region have benefited from employment in the uranium mines, as miners in the opencast and underground mines, and through related jobs and services. However, with the health risks of mining work, the physically demanding nature of the work, the frequent lack of employment contracts leading to job insecurity, the lack or inadequacy of employment and health rights—was it worth it? It is difficult to know the answer, since employees are generally silenced by the company they work for, for fear of losing their jobs. As we saw in section 2.2, however, a few ex-workers did speak out to Greenpeace and other NGOs, detailing their complaints and disillusion, sometimes their despair over lack of health care.

But whom have the Areva jobs benefitted, people from North Niger or from elsewhere? Ex-herders in Arlit complain that Areva does not hire them, that most workers do not come from the region but are brought in from southern Niger (Sourt 2009, and Orphans of the Sahara 2014, Part III, interview with local herder). Indeed the allocation of mining jobs to local people was one of the main demands of the Tuareg uprising of 2007, along with the devolution of power to northern Niger, an end to pollution and a better deal to the local population in relation to the uranium industry. The phrase 'jobs for local people' can be highly misleading in that what local inhabitants demand are jobs for residents of Arlit or Agadez region; whereas the foreign mining company may be allocating a certain number of jobs to Nigériens, many of whom may actually come from cities and regions far away from the North of Niger. It is only jobs allocated to residents of Arlit and Agadez region that can count towards providing alternative local livelihoods, and it is clear that most destitute ex-herders living in the shanty towns of Arlit have been disappointed. Until these demands by dispossessed local producers can be addressed, the main alternative livelihood opportunities in Arlit and Agadez region are petty smuggling and trading on the borders of legality, running for international drugs and people traffickers; and joining *Al-Qaida au Maghreb* (Sourt 2009, Granvaud 2012).

Bleak Future Livelihood Prospects for Northern Niger

When asked about the legacy Areva leaves to the area, the president of the local NGO AGHIRIN'MAN, said (in 2009, quoted in Greenpeace 2010): "Enduring pollution! Lack of water, because the ground water table is already at 70% (of what it was before mining activities began). As they (ground water stocks) fill every 100 million years, one can say they are not filling. The fauna has also disappeared. The flora has disappeared. It is a desert country, but there are trees... their roots cannot grow deeper than 60 meters! However, the water table is now at 300 meters: the trees cannot reach them. The heritage for us is enduring pollution." For a community relying on local water resources for humans and their livestock, flora for food collection and grazing for livestock, and fauna for game meat, this spells disaster.

Granvaud describes Areva's deliberately poorly publicized legacy to North Niger as "a social tragedy". This author ironically asks why, if Areva really is—as proclaimed—taking the needs of local herders into account, is there a need for fencing and military personnel to exclude them from the huge areas in question? "*Nous avons réussi à délimiter une surface d'exploitation qui ne risque pas de gêner les éleveurs*" (Areva 2009 source quoted in Granvaud 2014). "*C'est sans doute pourquoi les 200 km2 du site d'Imouraren sont sanctuarisés, clôturés, interdits d'accès et surveillés par des dizaines de militaires ou de para-militaires.*" (Granvaud 2014).

Now is the time to prevent the escalation of this social tragedy from Arlit to the new site of Imouraren, where the new mine is poised for Areva to start operating as soon as the value of uranium on world markets picks up. However, the outlook does not look bright: with increasing administrative and military protection of the concerned sites against any visitors and researchers other than those invited by the project promoters, research results appear increasingly unlikely to reach the public domain. Meanwhile, the dispossession of the local population continues apace, along with the depletion of crucial water sources, the pollution of the environment, the spread of radiation-linked disease to people and their livestock, the consequent breakdown of familiar secure livelihoods, and the inevitable increased reliance on desperate alternative survival strategies on the borders of legality.

2.4 Local People's Strategies

The people of northern Niger are not opposed to mining in their region *per se*. Most want a fair share of the revenue and better controls and regulations of the mining process to protect their environmental and health interests and concerns. Their strategies to address the problems outlined in previous sections fall into several categories as follows.

Migration Strategies

Following on from the traditional resource use described in Part 1, individuals and families have had to adopt a range of migration strategies in response to uranium mining in their homelands: these include coming in to the mining towns in search of paid employment; moving away from locations in their traditional homelands where mining has caused damage; and fleeing locations in their traditional homelands under threat from government or mining company administrative or military personnel attempting to reserve areas for mining activities. While some of these "migration" strategies may be described as voluntary, there is no doubt that others would qualify as forced migration as a direct consequence of uranium mining in the Tuareg homelands.

Representation through NGO Networks and International Advocacy

Certain dedicated individuals from a Tuareg heritage, based in the mining towns, have put themselves forward in order to represent the grievances felt by residents of northern Niger: they have created NGOs (non-governmental organizations) in collaboration with international supporters and researchers, in order to publicize at an international level the research findings concerning the impact of mining activities, and win international support with the hope of redressing the situation. Examples of such NGOs, already amply quoted above, are:

- AGHIRIN'MAN: based in In Gall, North Niger, publicizes information on the protection of the environment and the health of the populations of Arlit;
- Collectif Tchinaghen: *Paix et Solidarité pour le Nord-Niger*: environmental and human health watchdog, based in Agadez, North Niger;
- CRIIRAD: *Commission de Recherche et d'Information Indépendantes sur la Radioactivité*, set up in May 1986 in the aftermath of the Chernobyl disaster, in order to inform citizens about nuclear contamination risks;
- SHERPA: set up in Paris in 2001 to protect and defend victims of economic crimes.

These associations may link up with larger international NGOs in order to give further publicity to their claims, including: Greenpeace, Médecins du Monde, Survie (a watchdog on France's role in Africa), Terrains de Luttes (a francophone web-based information and research network concerning the situation of popular classes and the strategies of dominant classes), ROTAB (Network of Organizations for Transparency and Budget Analysis, see 2010), and EJOLT (Environmental Justice Organizations, Liabilities and Trade, see Chareyron 2014). This approach seems successful in publicizing grievances and research results: indeed, much of the evidence upon which this report is based comes from individuals working through NGOs according to such strategies.

Some of the representatives of the local populations of North Niger have travelled to international fora to present the situation and grievances of their people at high-profile conferences, including at the European Commission and the UN: for example, one individual from Arlit, member of the Indigenous Peoples of Africa Coordinating Committee (IPACC), and representing the local associations TUNFA and TIDAWT (Tuareg term for "together"), travelled to Geneva in 2011 to present a paper, on the right to water of indigenous peoples, at the United Nations Permanent Forum on Indigenous Issues, Tenth Session, Rights to Water. While this might make sense internationally, legal international treaties and conventions that deal with indigenous peoples are mostly not recognized by African states, unlike many countries in Latin America; so this type of approach seems to have yielded few tangible results so far.

Use of Force in the Context of Alternative Political Visions for North Niger

Certain Tuareg groups, seeing that peaceful research and lobbying methods have brought few effective results in redressing perceived injustices in Niger, have resorted to stronger means, in particular the taking by force of international mine workers as hostages, for longer and longer periods of time. The vice-president of the armed separatist group MNJ (*Mouvement des Nigériens pour la Justice*) declared in a 2007 interview that one of the movement's main objectives was to bring about a more just sharing of uranium revenues that would benefit the inhabitants of North Niger (*Le Monde* 13 May 2007). In April 2007, MNJ launched an attack on Areva's new Imouraren Base (Keenan 2008: 463); and abducted an executive of the Chinese uranium company Chino-U for 5 days in July 2007 (Keenan 2008: 451, 463). In June 2008, MNJ abducted 4 French mineworkers working for Areva, keeping them for a total of 6 months until their release in December 2008. A spokesperson for MNJ reported at the time that mining had damaged valuable pastoral lands, while mining revenues had failed to benefit local communities (BBC News 7 Aug 2008). In September 2010, following from the political alliance between local Tuareg militants and members of AQIM (al-Qaeda in the Islamic Maghreb, a group founded in 2006 in the Maghreb), an Areva employee and his wife were abducted from their homes together with 5 uranium mine workers working for Vinci, an earth-moving project under Areva supervision; 5 were French nationals, one Togolese and one from Madagascar. Although the non-French were released quite early on, the 5 French were held for over 3 years by AQIM, until their release on 24 October 2013.

These kidnappings have attracted global news attention to the mining sites of Niger and the companies involved. These and the threat of further actions have undoubtedly contributed to Niger's negotiating power with the mining companies, resulting in a larger government share of uranium revenues (see section 2.5). However, it is arguable to what extent these achievements have been devolved from Niger's capital Niamey to the mining region of North Niger, where residents still find themselves largely bypassed. Meanwhile, Areva has further militarized and secured its mining sites, and not only continues its mining activities in Arlit, but is planning an expansion with the huge Imouraren mine ready to start production.

2.5 Strategies of the Corporations and the State

Corporation Strategy: Stifle Publication of Any Local Evidence of Mining Risks

Areva has tried repeatedly to deny the environmental and health problems relating to its mining activities: its public stance on health issues caused by the mines, stated in its 2009 report, implies there is no problem at all. "Cancers are extremely rare. During 40 years of mining, not one case has been detected that was thought to have been caused by exposure to ionizing radiation. Cancer is an illness found mainly in Western countries with elevated pollution levels and high consumption of rich food, tobacco and alcohol (Greenpeace 2010: 55, quoting Areva). When denial fails to satisfy critics, the company minimizes the risks (ibid.: 30), or justifies the ongoing bad practice by referring to low standards at the start of the operation 40 years ago (ibid.: 39).

Certain claims, however, have been difficult to refute; for example, the presence of radioactive materials in the streets of Akokan (see section 2.1), resulting from a survey by CRIIRAD: the corporation publicly accepted these findings, and immediately set about cleaning the location in question. In October 2008, the local subsidiary COMINAK reported that all contamination problems had been addressed, and in September 2009 Areva confirmed to CRIIRAD that the clean-up was complete (ibid.: 42). When Greenpeace carried out further surveys in different locations, found further evidence of contamination, and published their findings in November 2009, Areva had to take action again. Soon after, the corporation informed Greenpeace that the seven spots indicated by Greenpeace had been cleaned up, and Areva drew up an action plan to check the villages. Greenpeace welcomed this partial clean-up of the village of Akokan by Areva, but emphasized the need for a comprehensive study so that all areas could be made safe for the community (ibid.: 44). Until now this has not happened.

In recent years, Areva has followed the CNRP (Canadian Nuclear Revitalization Partners) recommendations on laundry work. "We also asked that the washing of work uniforms will not be sent in town for cleaning, but done there [at the facility]; and they have taken steps in this direction," CNRP spokesperson said in relation to Areva (ibid.: 55). Such isolated clean up actions illustrate Areva's awareness of the health risks despite their denials; however, the response is partial, in response to specific published claims, rather than representing a responsible overall surveying and risk minimizing strategy. The corporation's main response to critical independent evidence is to further militarize the mining zones and prevent further access by independent researchers (Granvaud 2014).

Niger Government Strategy: Play Off the Mining Corporations

The importance of uranium to Niger as its number one foreign exchange earner cannot be overestimated; uranium contracts thus provide a crucial context for understanding national politics in Niger. As one observer put it: ... "les élites politiques font des affaires et les hommes d'affaire dirigent le pays." (Michel and Beunet 2008: 189). During the early decades of uranium mining in Niger, Areva enjoyed an uncontested monopoly. The world energy crisis in the early 2000s gave nuclear energy a new lease of life, and the price of uranium rose on world markets from earlier levels of USD 10 a pound (543 grams) to a record USD 136 in June 2007 (Keenan 2008: 454). There resulted a scramble by foreign corporations to acquire exploration rights in Niger, and for Areva to expand production with plans for a new mine at Imouraren.

In 2006, Niger awarded exploration rights to a group of Chinese companies: by 2007, it had granted around 90 exploration permits with a further 90 under consideration. France's dominant position in relation to Niger had been broken. It was perhaps not a coincidence that President Tandja's son Ousmane happened to hold the post of Commercial Attaché at Niger's Hong Kong embassy at the time and that the signature of the contract between Niger and the Chinese occurred on 14 July, a day of national pride for the French (ibid.: 177, 455). Rumors followed that France/Areva had supported the Tuareg rebellion of 2007 in order to frighten off foreign, especially Chinese, competition. Meanwhile, the MNJ were identifying weapons of Chinese origin during their raids in North Niger. In fact, Chinese engineers had been prospecting for uranium North of Agadez the day before the 8 Feb 2007 MNJ attack on the military garrison at Iférouane which resulted in 3 dead and 2 being taken as prisoners: the Chinese were rapidly taken out of the area. On 6th July 2007, a high profile Chinese engineer was taken hostage by MNJ for 10 days before being handed back to the International Red Cross with a statement from MNJ that this action served as a warning to the Chinese mineral prospection companies, specifically because they do not employ local labour and do not invest in local infrastructure (ibid.: 182).

The Niger government went so far as to accuse Areva of financing the rebels, and expelled Areva's head of operations in July 2007 in a move that provoked the direct intervention of President

Sarkozy and high-level talks in Niamey with President Tandja (ibid.: 458). Keenan argues that the Tandja government used Areva's alleged involvement in the rebellion to help break its monopoly on uranium production in Niger and thus gain a greater revenue stream by creating an expanded, internationally diversified and competitive market (ibid.: 459). As the intrigues escalated to higher and higher international political levels, the demands of the residents of North Niger fell into the background, while they and their representatives suffered the full weight of military repression following the MNJ attacks, with many fleeing into exile across international borders.

On 19 February 2010, a military coup ousted President Tandja. The military junta announced that it would audit all uranium exploration permits awarded in the month before the coup, since evidence had emerged that the permits enriched the ousted president and devastated the Tuareg population (CSM 2010). Meanwhile, Al-Qaida-linked AQIM took advantage of the unstable situation in Niger by infiltrating the country from its cross-border locations in the no man's land of the Niger/Algeria/Libya triangle, and consolidating its networks within Niger (Giraut 2010, RFI 29.09.2010).

The change in government power structure enabled a general airing of information about mining activities in North Niger, and some limited updated surveys, including the comprehensive report of Greenpeace (2010) that has been referred to throughout this chapter. This report even contained statements from newly appointed government ministers speaking out against the excessive control exercised by Areva: "As the AREVA-run hospitals do not staff any occupational doctors, it is impossible for someone to be diagnosed with a work-related illness, says Nigérien Environment Minister Issouf Baco. [...] The doctors who are there and who are hired by the company cannot do what the company does not ask. There must be an independent substitute from the state... There has been no independent study. Even less independent health care!" (Greenpeace 2010: 55).

Elections held on 31 Jan 2011 brought the opposition leader Mahamadou Issoufou into power. After months of negotiations, during which hundreds of people took to the streets in a series of demonstrations to call for the terms of Niger's 2006 mining code to be respected, and Areva threatened to close its two mines thus putting 5,300 people out of work, President Issoufou finally signed a new agreement with Areva in 2014. The Niger government succeeded in requiring that Areva should conform to Niger's 2006 mining code, give up a number of tax breaks, and increase royalties from the previous 5.5% to as high as 12% depending on profitability. On the other hand, the exploitation of the huge new mine at Imouraren, due to start producing 5,000 tons of uranium a year in 2015, was postponed pending an improvement in the value of uranium on world markets (Radio France Internationale, 26.05.2014). The contract for exploiting Imouraren went to the French company despite a rival bid from the Chinese, re-establishing France's firm dominance of Niger's uranium industry albeit with several strong rivals including China, US and Canada.

Although the need for environmental impact surveys is emphasized in the 2006 code, it remains to be seen how this will be implemented, along with addressing the health concerns of mine workers, ex-mine workers, and the residents of northern Niger in general. One thing is clear: the protection of the rights of the local populations, to enjoy their historical homelands and pursue their livestock-based livelihoods there in peace and safety, has not been a high priority for the government in its recent negotiations with the mining companies. Indeed, the lack of response to the needs of the population of North Niger appears to continue unchanged under the current government, with little difference from previous post-independence governments in Niger, which have repeatedly and progressively marginalized the inhabitants of the vast North of the country (Claudot-Hawad 2013); and have consistently channeled most foreign aid and development funds towards other more compliant regions further south.[8]

8 The current well-digging program of Swissaid, for example, has predictably been directed by Niger's central government

3. Conclusion: Keeping the Sahara Alive

The uranium tragedy in Northern Niger constitutes the latest episode in a long-standing colonial and post-colonial history characterized by the gradual but unrelenting undermining of pastoralist lifeworlds. But the predicament of the pastoralist way of life there represents more than just a localised subsistence economy under threat from predatory international mining companies. To this day, many Tuareg and other pastoralists continue to live their chosen lifestyles and make finely-tuned adjustments to it: it is thanks to the knowledge and skills of their ancestors, and the people who preceded them, that the once green Sahara and Northern Sahel has remained a *habitable place* over millennia[9], despite dramatic transitions towards extreme arid conditions some 5000 years ago. In stark contrast, the short-sighted, life-wrecking practice of alienating territory for the purpose of mining projects, together with the growth of alternative means of survival such as drug and people trafficking that this has facilitated, will ultimately transform this whole territory into a barren, *non-habitable space*.

This is a self-fulfilling prophecy based on the modernist ontological assumption according to which an objectified, exploitable nature legitimizes "politics without Earth". In our present time of the anthropocene and climate change, this assumption no longer holds: "We are now all farmers of Earth's troubled land" says Bruno Latour (2011: 74). Land and water grabbing for the exploitation of uranium, oil, and other non-renewable resources does not simply affect a remote nomadic minority, who follow a so-called 'archaic' lifestyle: these practices of the last few decades are contributing to bringing an ill-fated end to one of the greatest feats of human ecology and settlement of our planet. *This* has consequences for people and places throughout the Sahara and beyond, as the recent pro-autonomy movement escalating to war in Mali has demonstrated; more important, it has world-wide consequences for all of us, since insecurity in the Sahara has now become a crucial focus of global geo-political interests and strategies.

In December 2015, France hosts the next United Nations conference on Climate Change, COP 21. This conference has the potential to mark a turning point towards a liveable future for all mankind; and yet, how can the nuclear-powered nation of France, together with others, pretend to take action on climate issues, while continuing to undermine the lifeworlds of those who clearly supersede them in the art of mastering unpredictable and changeable environments?

to Dosso and Tillabéri regions in the vicinity of the capital Niamey, rather than faraway Agadez (see www.swissaid.ch).

9 Evidence of the occupation of Arlit and surrounding areas from the Paleolithic through the Neolithic, Iron age, pre-Islamic and then Islamic eras, including stone tools, rock art, tombs, and numerous copper and iron smelting sites, are documented with photos and maps in the rapid surveys undertaken just as the mining work was beginning in the 1970s, by a team of multi-disciplinary researchers concerned that so much human cultural heritage was being destroyed as a result of the uranium rush (Poncet 1983).

References

Adeyemi, A.D. 2015: Terrorism and Transnational Security Threats in West Africa: A Global Perspective. Xlibris Corp.

Areva (Hg.) 2011: Areva and Niger, a Sustainable Partnership. <www.areva.com/EN/operations-587/representation-exploration-and-mining-in-niger.html>.20.02.2012.

Areva 2008: *50 ans au Niger.* In: Areva.com. <niger.areva.com/FR/niger-94/actionnaire-principal-des-societes-minieres-somar-cominak-et-imouraren.html>. 25.02.12.

Areva ne fera pas la loi au Niger 2007: *Une région pillée, un peuple sacrifiée.* <areva.niger.free.fr/docs/communique.pdf>. 19.01.2012.

Areva: *Présence dans le monde.* France.<www.areva.com>.29.02.2012.

Arioti, M. and Oxby, C. 1997: From Hunter-Fisher-Gathering to Herder-Hunter-Fisher-Gathering in Prehistoric Times (Saharo-Sudanese Region). *Nomadic Peoples* 1(2): 99-118.

Auty, R. 1993: Sustaining Development in Mineral Economies: The Resource Curse Thesis. London: Routledge.

Bernus, E. 1981: *Touaregs Nigériens. Unité culturelle et diversité régionale d'un peuple pasteur.* Paris: L'Harmattan.

Bernus, S. et. al. 1986: *Le fils et le neveu. Jeux et enjeux de la parenté touarègue.* Paris: Maison des Sciences de l'Homme, Cambridge: Cambridge University Press.

Chareyron, B. 2014: The Radiological Impact of Uranium Extraction near Arlit and Akokan in Niger. *Factsheet 15*, Environmental Justice Organisations, Liabilites and Trade (EJOLT).

CIA 2012: The World Factbook. Niger <www.cia.gov/library/publications/the-world-factbook>, accessed 28.02.12.

Claudot-Hawad, H. 1993: *Les Touaregs. Portrait en fragments.* Aix-en-Provence: Édisud.

Claudot-Hawad, H. 1996a: La fragmentation touarègue ou le prix de la "paix". *Ethnies/Survival International* 20-21: 37-53.

Claudot-Hawad, H. 1996b: Identité et altérité d'un point de vue touareg. Éléments pour un débat. In: Hélène Claudot-Hawad (ed): *Touaregs et autres Sahariens entre plusieurs mondes.* Aix-en-Provence: Édisud / IREMAM: 7-16.

Claudot-Hawad, H. 2001: *Éperonner le monde. Nomadisme, cosmos et politique chez les Touaregs.* Aix-en-Provence: Édisud.

Claudot-Hawad, H. 2012a: Révolutionner l'opinion touarègue: les stratégies novatrices de Kawsen contre la colonisation du Sahara (début du XXe siècle). *ArOr/Journal of African and Asian Studies* 80(2): 243-258.

Claudot-Hawad, H. 2012b: Nomadisme (Touaregs). *Encyclopédie berbère* XXXIV: 5590-5602.

Claudot-Hawad, H. 2013: La "question touarègue": quels enjeux?' In: Galy M and Badie B (eds): *La guerre au Mali: comprendre la crise au Sahel et au Sahara: enjeux et zones d'ombre*, Paris: La Découverte, 125-147.

Claudot-Hawad, H. and Lefèvre-Witier, P. 2009: Désertification et civilisation nomade au Sahara: naissances conjointes. *Antropo* 18: 87-94.

Collectif Tchinaghen 2008: *La malédiction de l'uranium. Le Nord-Niger victime de ses richesses.* Agadez (Niger): Dossier d'Information, Collectif Tchinaghen, Paix et Solidarité pour le Nord-Niger.

CSM (Christian Science Monitor) 2010: Niger Coup: Good for Tuareg Rebels, Bad for Uranium Investors. Accessed on <www.temoust.org>.

de Moraes Farias, P. F. 2006: Touareg et Songhay: histoires croisées, historiographies scindées. In H. Claudot-Hawad (ed.): *Berbères ou Arabes?* Paris: Éditions Non-Lieu, 225-262.

Decalo, J. 1979: Historical Dictionary of Niger. London: Scarecrow Press/Metuchen.

Decalo, S. 1997: Historical Dictionary of Niger (3rd ed.). Boston & Folkestone: Scarecrow Press.

Denis, F., S. Manier and N. Siby 2008: *Niger. La bataille de l'uranium.* France 5 and Galaxie Presse (Film).

Descola, P. 2013: The Ecology of Others. Chicago: Prickly Paradigm Press.

Fuglestad, F. 1983: A History of Niger 1850-1960. Cambridge: Cambridge University Press.

Gagnol, L. and A. Afane 2010: Quand l'injustice est spatiale. Le nomadisme pastoral face à l'impératif territorial dans le Sahara nigérien. *Justice spatiale/Spatial justice* 2: 1-17.

Giraut, F. 2010: L'uranium, la Chine et Al-Qaida: le Niger au coeur de terribles enjeux. *Le Temps,* 01.03.2010.

Granvaud, R.. 2012: AREVA en Afrique. Une face cachée du nucléaire francais. *Dossier Noires* 24. Marseille: AGONE.

Granvaud, R. 2013: *Niger: le coût de la sécurité... et de l'uranium.* <www.survie.org>.

Granvaud, R. 2014: *Areva en Afrique.* (Article summarising the author's 2012 volume on <www.TerrainsDeLutte.org>).

Greenpeace 2010: Left in the Dust. AREVA's Radioactive Legacy in the Desert Towns of Niger. (French version: *Abandonnés dans la poussière. L'héritage radioactif d'AREVA dans les villes du désert nigérien*). Amsterdam: Greenpeace International, <www.greenpeace.org/international/en/publications/reports/Left-in-the-dust>, accessed 20.02.2012.

Greenpeace 2013: Nuclear History Repeats in Niger. <www.greenpeace.org>, 21.01.2013.

Grégoire, E. 1999: *Touaregs de Niger, le destin d'un mythe.* Paris: Karthala. (See especially chapter 2: *Exploitation minière et accumulation*).

Human Development Report 2007/2008. United Nations Development Program.

Keenan, J. 2008: Uranium Goes Critical in Niger: Tuareg Rebellions Threaten Sahelian Conflagration. *Review of African Political Economy* 117: 449-466.

Keenan, J. 2009: The Dark Sahara: America's War on Terror in Africa. London: Pluto Press.

Krings, T. 1995: Marginalisation and Revolt Among the Tuareg in Mali and Niger. *Geoforum* 36(1): 57-63.

Latour, B. 2004: Politics of nature. How to bring the sciences into democracy. Cambridge: Harvard University Press.

Latour, B. 2011: Politics of Nature: East and West Perspectives. *Ethics & Global Politics* 4(1): 1-10.

Le Houerou, H. N. 1989: The Grazing Land Ecosystems of the African Sahel. Berlin: Springer.

Maldonado and J. Gasarian. 1998: *Sécteur informel: Fonctions macro-économiques es politiques gouvernementales: Le cas du Niger.* Document de recherche S-INF-1-20. Département du développement des entreprises et des coopératives, Organisation internationale du Travail – OIT.

Mark, M. 2011: French Nuclear Group to Monitor Health at Niger Uranium Mine. *The Guardian,* 12.12.2011.

Martin, G. 1989: Uranium: a Case-Study in Franco-African Relations. *Journal of Modern African Studies* 27(4): 625-640.

Médecins du Monde 2011: *La Responsabilité sociale et environnementale d'AREVA est mise en cause par Sherpa, Médecins du Monde et la CRIIRAD.* <www.medecinsdumonde.org>, 04.08.2011.

Meunier O. 1997: *Les routes de l'islam.* Paris, L'Harmattan.

Michel, S. and M. Beunet 2008: *La Chinafrique. Pékin à la conquête du continent noir.* Paris: Hachette Littératures. (See especially chapter 7: *Ruée vers l'uranium au Sahara*).

Newby, J. and J. Grettenberger 1986: The Human Dimension in Natural Resource Conservation: A Sahelian Example from Niger. *Environmental Conservation* 13(3).

Newby, J. 1990: Niger: The Air-Tenere National Nature Reserve. In: Agnes Kiss (ed): *Living with Wildlife: Wildlife Resource Management with Local Participation in Africa.* Washington, D.C.: The World Bank.

Newby, J. 1992: Parks for People—a Case Study from the Air Mountains of Niger. *Oryx* 26 (1): 19–28.

Oxby, C. 2011: Will the 2010 Code Pastoral Help Herders in Central Niger? Land Rights and Land Use Strategies in the Grasslands of Abalak and Dakoro Departments. *Nomadic Peoples* 15(2): 53-81.

Oxby, C. 2014: Social Differentiation of Risk: Perceptions of the Future in Drought-prone Central Niger. *Journal des Africanistes* 84(1): 106-129.

OXFAM-France 2013: Areva in Niger: Who is Benefiting from the Uranium? 19.12.2013, <www.oxfam.org>.

Poncet, Y. 1983: La Région d'In Gall – Tegiddan Tesemt (Niger): Programme Archéologique d'Urgence 1977 – 1981. *Etudes Nigériennes* 47. Niamey (Niger): Institut de Recherches en Sciences Humaines.

Radio France Internationale English 29.09.2010: Sahel Intelligence Chiefs Meet to Plan Fight against AQIM. <www.rfi.fr> accessed 17.06.2014.

Radio France Internationale 26.05.2014: *Niger, entrée en vigueur du nouvel accord avec Areva.* <www.rfi.fr> accessed 17.06.2014.

ROTAB (Réseau des organisations pour la transparence et l'analyse budgétaire) 2010: *Bulletin d'information du réseau des organisations de la société civile pour la transparence* 19. <www.rotabniger.org/rotab2/publications>. 30.01.12.

Rivières, E. S. 1965: Histoire du Niger. Berger-Levrault.

Sourt, C. 2009: Niger and the Real Cost of Nuclear. *The Guardian,* 1 Feb. 2009.

Spittler, G. 1989. *Dürren, Krieg und Hungerkrisen bei den Kel Ewey (1900–1985).* Stuttgart: Franz Steiner.

UNESCO. Réserves naturelles de l'Aïr et du Ténéré. <whc.unesco.org/fr/list/573>, accessed 22.11.2015.

Walentowitz, S. 2003: Mukhammad Al-Jilani. *Encyclopédie Berbère* XXV: 3904-3905.

Walentowitz, S. 2011: Germanité, alliance, filiation: les dynamiques autopoïétiques de la parenté touarègue. In: P. Bonte, E. Porqueres, G. and J. Wilgaux (eds.): *L'argument de la filiation aux fondements des sociétés européennes et méditerranéennes anciennes et actuelles*. Paris: Maison des Sciences de l'Homme, 107-137.

Walentowitz, S. and Attayoub, A. K. 2001: La tetserrét des Ayttawari Seslem: un parler proche du berbère "septentrional" chez lez Touaregs de l'Azawagh. *Annuaire de l'Afrique du Nord* XXXIX: 27-48.

Images

Courtesy of Collectif Tchinaghen (2008); screenshots from the film 'Niger : la bataille de l'uranium' (2008) and World of Matter / Ursula Biemann (w.y.): "Arlit Uranium Mine", in <www.worldofmatter.net>.

Additional Information

Films

2008 'Niger : la bataille de l'uranium' by Nahan Siby, Frédérique Denis an Stéphane Manier. France 5 / Galaxie Presse.

2011 'Qui sème le Vent' by Fred Garson. ARTE.

2014 'Orphans of the Sahara' Three part documentary produced for and broadcast by Aljezeera in Jan 2014. See especially Part 3: Exile, www.aljezeera.com

Websites

www.temoust.org
www.medianiger.info
www.rotabniger.org
areva.niger.free.fr
www.airinfo-journal.com
www.criirad.org
www.cooperation.net/aghirinman

Artisanal Gold Miners Encountering Large-Scale Mining in Guinea:
Expulsion, Tolerance and Interference

Matthieu Bolay[1]

1. Small- and Large-Scale Gold Mining: Who Competes for What?

Most gold mining companies in western Africa have settled in areas where artisanal gold mining (ASM[2]) was already practiced before the companies came to the place. The relations between artisanal gold diggers and the mining companies are often conflictive, which is usually explained by a supposed competition for access to the same resources. However, it was proven that the independent artisanal gold miners and the large-scale multinational mining companies do not compete for the same gold (Hilson 2011). The first ones are technically constrained to limit their digging to surface reefs or alluvial placers, while mining companies extract large quantities of ore from deep reefs (Hilson and Yakovleva 2007: 102; USAID 2005: 3). Recent and recurrent incidents on gold mining fields in Guinea and elsewhere have shown that, despite their interest in different kinds of deposits, artisanal gold miners were continuously labelled "illegal robbers" when they were caught on the lease of an active gold mining company. The criminalization of artisanal mining is part of the everyday life of gold diggers in a non-formalized context of work, mainly due to administrative constraints and the high costs involved in making their activity enter the licensing system (Fisher 2007: 739). The informality of their practices is usually not much objected to by the state authorities as long as they do not interfere with the same spaces of the large-scale gold mining companies.

However, the topic has recently evinced some interest also from the media, mainly after several accidents taking place on the territory of gold mining companies that caused the death of many artisanal gold diggers.[3] In Upper Guinea (Siguiri), the company AngloGold Ashanti was forced to react and reaffirm its position as legal owner of the land vis-à-vis the "illegal robbers mainly coming from outside". These accidents have accelerated the deployment of new forms of control by the concerned company, mainly through its Corporate Social Responsibility (CSR) programs.

After having spent a total of one year with artisanal gold diggers in the region of Siguiri, where AngloGold Ashanti operates since 2004, in this chapter, I present how artisanal miners interact with the neighbouring spaces occupied by AngloGold Ashanti. I will describe three forms of typical encounters that appeared successively and analyse them with regard to the application of CSR

1 A Review of this chapter was undertaken by Prof. Katja Werthmann.

2 According to the global report on artisanal and small-scale mining (Hentschel et al. 2002), Artisanal and Small-Scale Mining (ASM) refers to informal mining activities carried out using low technology or minimum machinery.

3 In November 2008 and February 2009, 13 and 23 artisanal gold miners died in two shaft collapses that occurred on the lease area of the mining company AngloGold Ashanti in the region of Siguiri, in Guinea. The accidents were disseminated by media in and beyond the region (A.F.P. 13.11.2008).

programs. I rely on a political ecology approach, which seeks to understand how the control over specific resources is organized. This becomes particularly visible in situations where different stakeholders meet and/or negotiate their access to resources—in this case, gold ore. Therefore, I examine different forms of encounters that take place between a gold mining company and the surrounding gold diggers, who are either local residents or itinerant artisanal gold miners. The analysis of this interaction shall help in better understanding of how measures undertaken by a gold mining company, in order to control spaces of extraction, are perceived by artisanal gold diggers, and what consequences these measures have for the gold diggers.

First, the context of the case study will be described with regard to the history of gold mining and the related movements of people in the region. A special focus is placed on the social organization of artisanal mining at the local level. Second, I will briefly present the research methods and the theoretical framework. Third, I will describe three forms of typical interactions that I respectively call expulsion, tolerance and interference. Finally, the implications of CSR programs towards the social organization of ASM will be discussed, especially concerning the fragmentation of the social organization of artisanal gold mining.

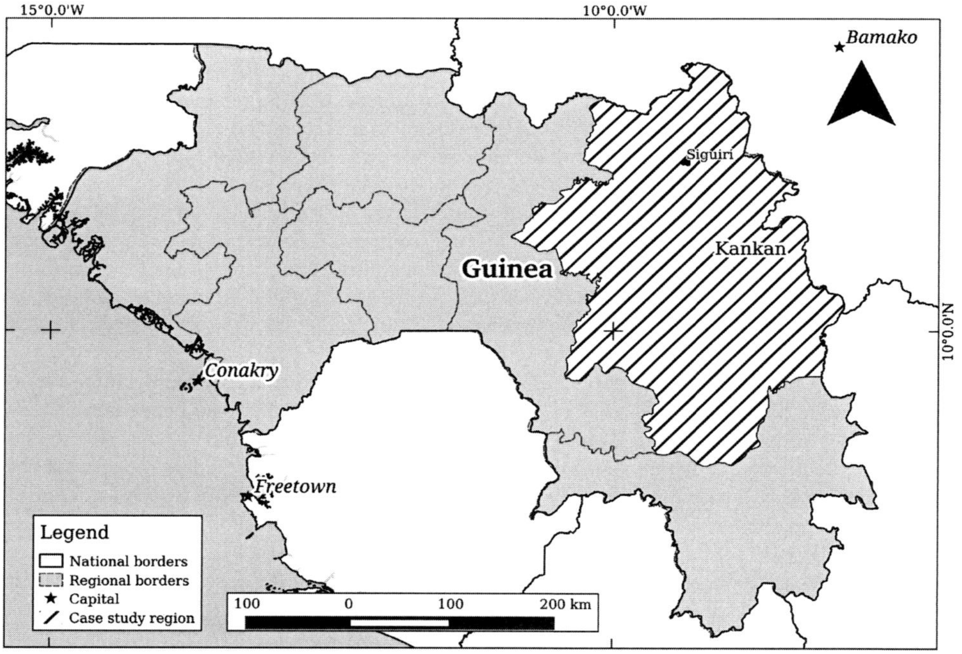

Map 1: Location of Siguiri in the Kankan Region of Guinea.

2. Gold Mining in Bouré: Long-Standing Practices and the Standardization of Encounters

Gold mining has always been powerful in attracting people from faraway places to the regions of extraction. The most known and widely documented examples are the gold rushes in the frontier-era in the United States. Similar dynamics can be found in contemporary Africa (Jacques et al. 2006 ; Werthmann and Grätz 2012: 11), where the concept of frontier has been adapted to grasp the interactions of fortune-seekers with local communities, with reference to migratory dynamics and the related social boundaries (Grätz 2004). According to historians, the region of Bouré in northern Guinea, where this study was conducted, was already a common venue for populations of northern arid areas during the dry season at the time of the Malian Empire. Between the 12th and the 16th century, the Malian Empire, which had exclusive ownership on the gold nuggets, already relied on a seasonal workforce (Ki-Zerbo and Niane 1991: 195). The activity was exclusively conducted during the dry season and strictly depended on the agricultural calendar. Whole villages or families left their places of origin for temporary settlements, some of which became permanent villages (Herbert 2012: 26-28). Contrary to other areas of Guinea where gold deposits have been discovered only recently, the region of Bouré has historically developed mainly upon gold mining and numerous waves of "transitory" migrants. Temporary workers were traditionally hosted through the institution of *tutorat*, which is a customary convention that gives its beneficiaries a grant of land tenure rights in exchange for continuing gratitude and allegiance towards the owner. Sometimes, the benefit includes the provision that the *tuteur* receives part of the production from his "guest" (Chauveau 2007: 66). The institution of *tutorat* became standardized through hosting practices embedded in migration narratives. These narratives still play an important role. First, they shape identities of the local populations through land claims and by reference to the first settlers. Second, they serve as a regulation for the immigrant or temporary settled populations (e.g. Werthmann 2006). Later, during colonial times, temporary workers kept on maintaining regular contacts with artisanal mining settlements where they had established prerogatives, in particular through *tutorat* relations. These dynamics have visible impacts on the current social organization, such as the commonality of multilocal households, but have also met major changes that I will address later.

Since more than ten years, artisanal gold mining has literally exploded in western Africa—both in terms of the number of people engaged in the activity and the extent of the exploited mining fields (Hentschel et al. 2002). On the one hand, the rapidly increasing gold price contributed to the attractiveness of the activity among populations that were traditionally not involved in mining (for example, urban dwellers and educated people). On the other hand, unemployment in urban centres is also a major cause of engagement in the ASM sector (Hilson and Potter 2005). Unemployed people often participate in the increasing urban-rural migrations (Beauchemin and Bocquier 2004: 2253) or contribute to "return" migrations even if they had never lived in their "place of origin" before (Potts 1995). Thus, in the tradition of long-standing circular movements, an increasing number of people engage in full-time artisanal gold mining by adopting a mobile way of life that allows them to move quickly from one mining field to another without any intention to settle (Jonsson and Bryceson 2009).

This chapter is concerned with encounters between multinational companies and the surrounding populations, comprising both local residents and the above-mentioned itinerant artisanal miners. Since the 1980s, these relations started getting increasingly standardized through CSR programs. According to Carroll (1999), the shift from mutual avoidance to a sustained and organized involvement in the social life of the local population was partly due to strong incentives from international institutions and NGOs. This position contributed to a rising critique of companies' environmental and social damages. Regular critiques were developed by the NGOs and academics about the meagre impacts of measures for the benefit of the local population compared to the benefits of legitimizing the presence of extractive multinational companies in developing countries (for a synthesis: Banerjee 2008). In the mineral extraction sector, Hönke (2010) showed how companies used their CSR programs to better control the territories where they were operating. Other studies deal with the consequences of these programs for the social life of the concerned communities such as conflicts between villages (Frynas 2005: 592) or among villagers (Luning 2012) when it comes to accessing benefits. Despite an obvious unequal distribution of power, artisanal miners and the concerned communities cannot be considered passive victims. Several case studies in the Melanesian context have shown how communities adopted CSR and the developmental discourses on land tenure, to benefit from compensations and CSR investments. Communities strategically organize in "customary land groups" or create a "landowner" status in order to get access to compensation funds of the mining companies or the state, as well as to benefit from quota-share through the Ministry of Mines (Filer and Macintyre 2006; Jorgensen 2004). Interactions between communities and the mining companies are hence increasingly standardized through CSR programs and the related discourses, agents and programs.

Therefore, I propose to analyse encounters between artisanal miners and a gold mining company, from the miners' perspective (both locals and itinerants), by taking into consideration the existing social boundaries among the gold diggers. The next section describes some specificities of the present case study, especially regarding how work is organized in the artisanal mining fields in proximity to the concession area of the company.

3. The Case Study

The present study was conducted in the region of Bouré, 40 kilometers from Siguiri town in Upper Guinea. Largely known for its important gold deposits, it is the first allocated concession in Guinea that was given to a large-scale mining company. First called Société Aurifère de Guinée (Ashanti Goldfields Corporation), it then merged with AngloGold in 2004 and became part of AngloGold Ashanti Ltd. In the region, the company is known as "SAG".

SAG started exploiting gold in 1989. The arrival of the company first evoked great expectations concerning employment perspectives in the region. Disappointment was equally great, as most hired workers were expats and the few available jobs for the local population were physically demanding and low-paid. A former employee in the artisanal mines reported: "Salary is so low, that with the rush, I finally went back to artisanal mining. Work is very hard too but there [with artisanal mining] you can be lucky and you are free."[4] As a consequence of the unsatisfying working conditions, relations between the company and the population in Siguiri and the villages of Bouré rapidly turned conflictive and resulted in frequent strikes by local employees and protests by the neighbouring population of Siguiri, among whom many are part-time or seasonal artisanal mine workers.

Around the active pits of the company, the soil is drilled by thousands of small shafts made by the artisanal gold miners. Those are usually young men who leave their home place to generate an income before settling down back home. However, the forms of engagement in the artisanal gold mining sector differ, and can be ideally presented by three specific types: First, some miners engage in long-term itineraries that sometimes last several years. This implies cutting off most family ties and a return is only possible when they acquire recognition in the sector and adequate savings to start living on their own. Such miners often refer to mining as a form of "adventure",

4 All the quotes in this chapter are from interviews conducted during the fieldwork in Guinea between 2009 and 2012.

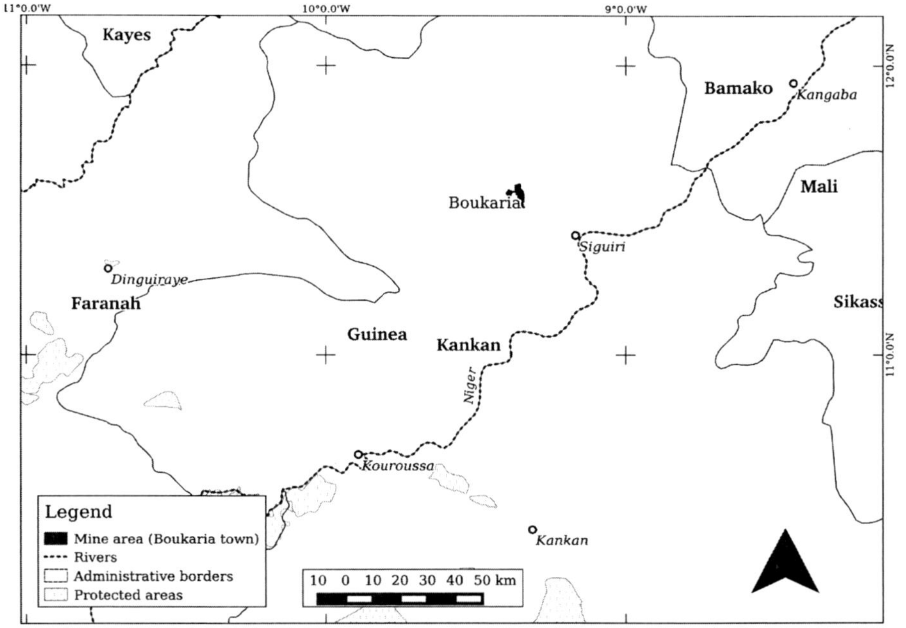

Map 2: Mining area and settlements near Siguiri.

which is a common reference in the migration narratives in western Africa (Bredeloup 2008). Second, there are miners who engage seasonally or temporarily, usually as part of the household strategies of securing livelihood. Such miners often consider mining as a complement to their main activity, which is farming. Third, local residents of mining areas usually depend exclusively and most often indirectly on gold mining activities. They mainly combine several income generating activities related to gold mining, except participating directly in the extraction process. They invest in mining activities either by owning a shaft, by providing mechanical support, by doing business with gold ore, or by offering services in the mining camps; for example, in small restaurants and coffee shops, or in stores selling clothes and extraction material. Their autochthonous status, besides owning a shaft, also allows them to be considered as part of the local militia.

Seasonal workers are often hosted by local residents, either through the institution of *tutorat* or through family ties. They usually maintain strong personal networks extending over several years. On the contrary, itinerant gold miners settle in temporary mining camps close to the mining fields. The camps can host up to several thousands of people during the dry season and thus become major hubs for trade and business for the whole region (Mbodj 2009). Different kinds of workers characterize working conditions on ASM extraction sites: First, a team of shaft diggers who associate with a local shaft owner, whose responsibility is the maintenance of the team until gold is reached. The shaft owner is hence considered a member of the team in the distribution of the latter extracted ore, which is divided among the initial stakeholders and washed individually. Second, when the ore is finally reached, experienced gold diggers, specialized in underground extraction, called *kalayantigilu* start joining the initial team and are asked to "take a turn" inside the galleries. The work in the tunnels is considered much harder and is usually done by *kalayantigilu*, who sell their workforce and experience from shaft to shaft. They can earn much more than the shaft owners and initial teams as these gold diggers multiply their operations in the best shafts.

In any operation, success can hardly be ensured. This pushes some gold diggers, be they itinerant or not, to try their luck in the neighbouring territories of the mining company, in particular when they face difficulties to meet their daily expenses. Because of the mechanized operations undertaken by the SAG, rich deposits are accessible easier there. In some places, abandoned ore can be conveniently rewashed through ASM and these small quantities can cover daily needs. Yet, these operations imply greater risks because of the soil fragility. From the miners' accounts, collapses can be anticipated by listening to and reading the warning signs given by the subsoil. However, such skills require time and experience and are associated with the so-called "masters" or "officers".

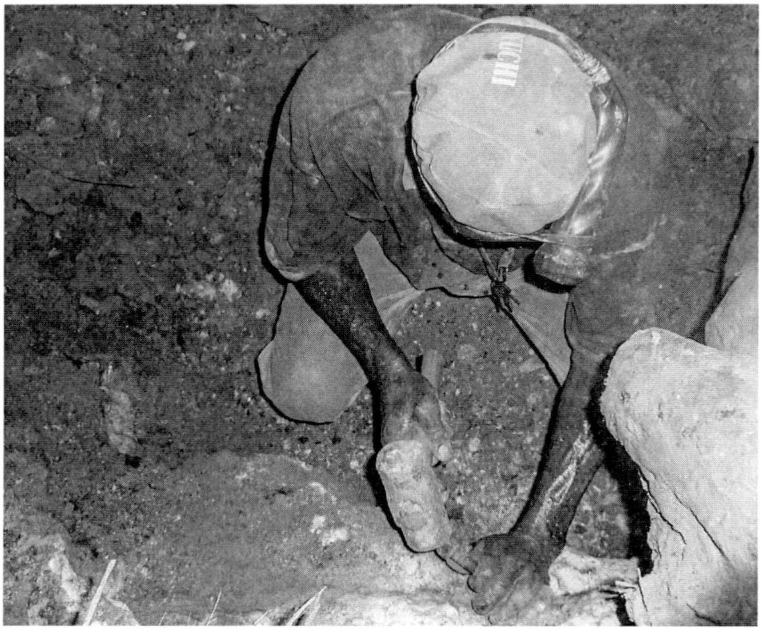

Fig. 1: Searching for gold in Kintinian tunnel. Photo: Matthieu Bolay.

Between 2009 and 2010, several accidents took place on the territory of the SAG. Shaft collapses caused the deaths of more than 30 artisanal gold diggers and were largely covered in the regional media and much farther. In response to these fatalities, the SAG accelerated the "strategic action plan" of its "community development" plan. The plan officially aimed at providing infrastructural support to the neighbouring communities and improving the relations with them.

After briefly describing data collection methods in the next section, I will then present how artisanal gold miners active in the area re-interpreted relations with the mining company by paying special attention to the successive forms of distribution of control over gold mining spaces induced by CSR.

Fig. 2: Artisanal miners at work in Kofilani Boure. Photo: Matthieu Bolay.

4. Methodology

The field research was conducted during a total of ten months in gold mining areas in Upper Guinea between 2009 and 2012. More specifically, I collected data in the mentioned region of Bouré, where the SAG is settled. Diverse anthropological methods were combined and several stakeholders' perspectives were taken into consideration. The first objective aimed at capturing and comparing access strategies to mining resources of itinerant and resident mine workers in different locations. Village-managed artisanal mining fields, abandoned artisanal mining fields and mining spaces upon the lease of the mining company constituted my main field. I first stayed in two miners' households for two three-month periods where I followed day-to-day activities on the neighbouring mining fields. These insights were complemented by 24 interviews with male resident and temporary miners. I also interviewed two women but did not include this data because of the exceptionality of their experience in the view of the other male interviewees.[5] Interviews mostly focused on miners' strategies of access to mining resources both formally and

5 Women make up half of the population living on the artisanal gold mining fields. The division of labor prevents them from doing any kind of subterranean work. This makes them less prone to operate in the concession area of the mining company.

informally, on their general perception of the mining company, as well as on their interpretation of the programs undertaken by the SAG. Complementary interviews were conducted with *tom bolomas*[6] and local state officials in order to obtain their perspective on relations among the miners, and between the miners and the mining company. Because of the "tripartite" context that this chapter addresses, informal discussions were also held with other local stakeholders such as village chiefs, soldiers, employees of the SAG, among whom some had executive positions, and two police officers. Those insights were useful to better understand how they interpret the social heterogeneity of the mining community, especially in their diverse relations with the SAG. I also collected copies of police statements for casualties in the area of Bouré through local informants, and reviewed public relations and official statements of the SAG to capture its official position toward the "mining community".

In my data, the perspective of the artisanal miners has obviously more weight than the one of the company, which was almost inaccessible through formal interviews. Some noteworthy studies approach the question from this perspective and recall that private firms should not be considered as monolithic entities. Strategies rely on personalities and political ties with local state institutions, and are thus often far from the officially stated position of the mining company (e.g., Davis and Franks 2011 ; Frynas 2005 ; Kapelus 2002). However, as the main concern of this paper is to understand how artisanal miners experience neighbourly relations with the mining company, its activities directed toward the "community" and the situations that arise from their encounters, the perspective of the mining company is less addressed.

The analysis of the material has been guided by a political ecology approach, which aims at identifying inequalities in terms of access to natural resources; in this case, to soils of high grade gold ore. Gold is not a renewable resource at human scale and the potential conflict should hence not only be considered "environmental"[7] but also include its economic and political dimensions (Escobar 2006). Natural agency is considered, however, as it obviously affects actions and/or reactions both of artisanal miners and of the mining company. For instance, shaft collapses provoke fatalities and attract media coverage and its potential occurrence influences choices of the gold diggers. In our case, socio-environmental conflict theory provides a framework focusing on conflictive situations in the access to and over resources (for example, high grade ore soils). With this framework, we can examine who controls what, who is allowed to decide what and who is expected to do what tasks (Little 2007; Robbins 2012). By asking these basic questions, we can better understand how the increasing scarcity in gold access due to appropriation by private companies and an increasing number of gold diggers leads to conflicts between social entities (local and itinerant artisanal miners, artisanal miners and non-miners, artisanal miners and the mining company). The three presented cases show that the more cooperative the approach of the company toward the "community", the more conflictive it turns within the targeted community. The argument here is that the conflict moves from the tensions between company and the miners to tensions between local populations and the miners. The main reason is that the local autochthony conventions are challenged by CSR programs and turned into effective social boundaries in the context of work of the artisanal gold miners, by following an approach based on autochthony through the interference of CSR programs.

6 *Tom bolomas* literally means "the ones that hold coercive power" in Maninkakan; they act as a local militia and control the artisanal mining fields under the mandate of the village authorities.

7 Contrary to the consequences of mining (e.g., soil degradation) that often lead to environmental conflicts.

5. From Mutual Avoidance to Interference

Different stakeholders are involved in framing the encounters between the SAG and the artisanal gold miners. This is visible at two different levels. First, the level of decision-making, and second, the level of decision-implementation. These decisions mostly concern control over territory and the application of programs towards the "communities". The following image refers to the main stakeholders and their relations concerning power in decision-making and control over working spaces.

The presence of the state is less dealt with in this chapter, because it is usually limited to episodic incursions of agents of the Ministry of Mines in the villages managing artisanal mining fields in order to collect taxes. In the everyday relations between the SAG and the gold diggers, the state is usually not directly involved, though its representatives at the local level (for example, the president of the local government) or the army (in cases of strikes and protests) occasionally can play an important role. I examine here three types of encounters between the SAG and the artisanal gold miners that appeared successively, and reflect the evolution of the company's strategy towards "unwelcome" gold diggers. The three examined cases appeared successively but should not be considered isolated steps as they add up and integrate with the previous ones.

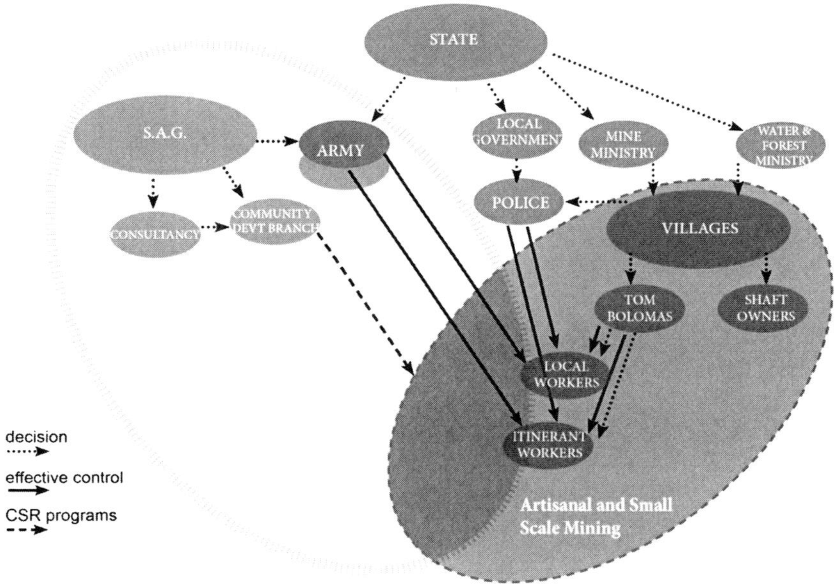

Fig. 3: Map of stakeholders.

Fig. 4: Miners' camp in Bouré. Photo: Matthieu Bolay.

5.1 Expulsion: Reactions to the Incursions of the Clandos

Concerning decision-making power, the common understanding in the past years was a simple territorial division. On the one hand, the SAG manages its mining lease area. On the other hand, the villages, through their authorities and local militia, control the artisanal mining fields. This clear divide illustrates well the position of mutual avoidance of the different actors and the "fortress strategy" of the SAG that they were applying worldwide before the broadening of the CSR (Caroll 1999). Despite this clear territorial divide, artisanal gold miners often cross these lines to access the pits of the company where the ore is more easily accessible. Digging the soil in the lease of the company is risky on two grounds. First, gold diggers expose themselves to important risks of shaft collapses because of soil instability largely caused by drilling and dynamite explosions. Second, they also expose themselves to military patrols in charge of controlling the territory; being caught results in heavy fines and often imprisonment by the local police. Intervention of the family and bribes are usually needed for release.

Incursions can be of two kinds. Initiators are often individuals or small groups of gold diggers and sometimes also massive movements of artisanal mine workers benefitting from the temporary lack of company control over their licensed areas. Experts in these practices are called *clandos* ("clandestines") because they operate outside the law and because they have been the main target of the patrols. An old man in Balato reports on the common view on *clandos*: "Ever since the SAG is here you have had people digging in their pits. For us (the *tom bolomas*), it is not a problem if they want to try their luck there. If they get caught it is their own responsibility; that is why they often go at night. Most of them are in the shafts in the *faro*[8] during daytime and at night they hide to get to the [company] pits." This statement illustrates well how responsibilities in controlling the neighbouring spaces of the lease and the *faro* are distributed. From the perspective of the local

8 The *faro* is the artisanal gold mining field.

population, entering the lease of the SAG to get gold ore is certainly a risky operation but is not morally condemned.

While *clandos* usually operate with maximum discretion, the territory of the SAG is, on certain occasions, massively occupied by all kinds of artisanal gold miners, especially when strikes and protest emanating from Siguiri town lead to the temporary closure of the mine. This was the case in 2009 when hundreds of unemployed youths blocked the road connecting the processing factory to the town for several days in order to demand more company investment in the provision of jobs and in access to services such as health care or electricity supply. A small summary of that day spent with local gold diggers in the pit will illustrate the classic expulsion strategy of the company towards the gold diggers.

> "We knew for two days that strike would start on Saturday. In Siguiri town, most youths at the coffee shop intend to join the march to Koron. In the neighbouring villages, plans are more concerned with occupying the pit than blocking the road. I hence decide to remain in Fatoya for the night in order not to be blocked by the protest. On Saturday morning, I join Seydoujan to go to the pit, which is already totally occupied when we arrive. Most people collect left-over ore on the edges of the pit, while others attempt to dig small trenches at the bottom of the pit. Women do instantaneous washing and fill bags with the ore that looks most promising. In a few hours, the whole population of the neighbouring mining fields of Fatoya, Boukaria, Kofilani and Kintinian seems to have rushed and replicated the organization of the *faro* in the unoccupied pits. No later than one hour after our arrival, Seydou spots three military vehicles approaching fast towards one of the edges of the pit. At this same moment, almost all trenches have already emptied and everyone is running in the opposite direction. Seydoujan advises not to run and to walk quietly out of the pit. It is just expulsion, he says, they cannot arrest everyone so they usually arrest no one. When we come back to Balato, Seydoujan explains that this same scenario, called *ratissage* by people, has been repeated several times during the year and is increasingly being coordinated with the protests from Siguiri town."[9]

Clandos incursions and *ratissage* do not rely on the same purposes and circumstances. Clandos incursions are part of the everyday life occurring on the fuzzy borders of the SAG territory, while massive movements of people in the pits depend on protests organized in the town. However, they both provoke the almost instantaneous expulsion of the gold diggers out of the concession areas of the company. "Expulsion" clearly affirms the property of the SAG upon a certain territory by controlling it with the support of contracted soldiers. Decisions are unilaterally taken by the SAG and intended to prevent any confrontation with the gold diggers by symbolically reaffirming the figure of the fortress.

5.2 Tolerance: the Illusion of Uncontrolled Zones

As a reaction to the numerous accidents that occurred on the lands managed by the SAG and which resulted in the exercise of its responsibility towards the *clandos* operating on the mining lease, the company developed a more cooperative approach towards the artisanal gold miners. Agents of the SAG started roundtables with the neighbouring village authorities and related *tom bolomas* committees in order to redefine the controlled territory of the company. According to members of the *tom bolomas*, the company used the following arguments: an increasing number

9 Adapted from personal fieldnotes.

Fig. 5: Abandoned mining site occupied by artisanal miners in Bougaria Bouré. Photo: Matthieu Bolay.

of people were caught on the mining lease and their imprisonment was being criticized by the local elites. Further, fatalities of artisanal miners on the company's territory would damage its image. The proposed solution was a withdrawal of the patrols from areas that had been fully exploited by the company so that artisanal miners would be free to access these zones. While the message was well received by the gold diggers, it posed several problems for village authorities. First, artisanal mining fields represent important incomes for village authorities and families owning the land. They could lose parts of their incomes if gold diggers started operating massively on spaces where *tom bolomas* would not be allowed to collect taxes. Consequently, the local militia could not control these new spaces dedicated to artisanal mining. Second, members of the local militia mentioned that the risk of shaft collapses was higher in places that had already largely been exploited by industrial means such as drilling and use of explosives.

This new approach of the company towards the artisanal gold miners hence suggests a shift in the control of the area and a redistribution of responsibilities. From controlling all the concession territories, the company decided to limit control to the active zones of exploitation. Villages, ruled by *tom bolomas*, remained in charge of controlling the artisanal mining fields. Yet, no responsibility was defined for the remaining zones located within the company's lease where artisanal mining was now tolerated. There are two consequences of this policy of tolerance in the day-to-day practices of artisanal gold miners. First, people in the most vulnerable positions started digging these areas, where ore was now easily accessible but in lower quantities. In particular, this activity became an option to find day-to-day basic incomes for young itinerant mine workers without either a stable network of peers or mining experience. Second, accidents in those remaining zones increased. Personal reports based on the local Red Cross and police reports show that out of 200 fatalities the majority occurred outside of the artisanal mining fields controlled by the local militia. Third, SAG did not report these casualties occurring on their licensed lands. This contributed to enhancing the company's image in its yearly report. Further, SAG could escape from accusations concerning human rights violations, as such incidents are unwelcome for a company that has subscribed to the Voluntary Principles on Security and Human Rights (AngloGoldAshanti 2010).

The question of responsibility for what could appear as a hecatomb is well summarized by one of the heads of the local militia: "Of course, it is too dangerous [to dig shafts in the remaining areas of the company]. We often come here and tell them to get out of the shafts, but the following day they are back. We already have to manage the village plots and decide where people can dig and not dig, but there it is the SAG which decides. We don't take taxes there; it is not for the village, so we cannot prevent their coming".

This unambiguous statement shows that the decision of the company to tolerate the venture of the gold diggers in abandoned pits involves the inability (and somehow unwillingness) of local authorities to extend their management upon these areas. Controlling space is a way to control access to resources, but it is also a way to regulate work in a hardly formalized environment. In consequence, the withdrawal of the SAG from the control of some segments of its territory fosters the appearance of unregulated spaces of work where the most vulnerable workers operate at greater risk in order to secure their livelihoods. Though not explicitly mentioned, practice shows that space is regulated by boundaries based on experience and locality in this case. Newcomers, who can neither count on their experience, nor on strong ties with local residents to access the good spots on the artisanal mining fields, are more vulnerable and more prone to turn towards remaining zones.

5.3 Interference: Intervention in the Social Life of Artisanal Mining

A third type of encounter between the mining company and the artisanal gold miners can be identified through the implementation of CSR programs in the "community". CSR programs (public awareness on environmental issues, investment in community relation improvement, investment in health care) extend the scope of influence of the SAG from the spaces of gold extraction to the neighbouring villages and mining camps. Following the "collaborative" trend initiated in 2009, the company, for example, invested in electricity infrastructure for the "communities". At the beginning, the question of who constituted the targeted "community" had to be solved. On the one hand, understanding the community as an equivalent for the "village" meant putting aside the miners' principal problem that most of the villagers were not local residents. On the other hand, focusing on the "community of artisanal miners" signified formal recognition of their presence and work also on the remaining areas that were legally still under the responsibility of the company. The 2009 consultancy study brought the following definitions of artisanal miners "groups" such as "two separate artisanal mining groups operating at Siguiri—the local, traditional *orpailleurs*[10] and the artisanal miners who come from outside the Siguiri area and, in most cases, are not Guinean. Specifically, [the report] highlighted the existing tension and conflict between these two groups who compete for the same mining sites".

Further, in the report, the so-called "traditional *orpailleurs*" are associated with regular artisanal mining in the villages' places, while in contrast, "strangers" are said to work "outside the traditional system (...) within the mine's lease area" (AngloGoldAshanti 2009). As discussed earlier, artisanal mining is historically embedded in practices of mobility, which standardized, through the institution of *tutorat*, the "multi-locality" category, and the networks of peers going beyond the single mining field. The provided definition contributes to the identification of a legitimate community to interact with, and consequently to the exclusion of non-native mine workers from the targeted group despite the close relations that were traditionally maintained between them.

10 *Orpailleur* is a French term for the gold prospectors.

Based on this definition of "mining community", the company, through its consultancy office, addressed the administrative and customary "natives" as representatives to manage the distribution. Unsurprisingly, electrification quickly turned into a divisive issue between recognized natives, whose access was guaranteed through CSR programs, and the considered allochthonous, who had to pay significant taxes to get access to electricity on the informal market. Thus, people who were established for long and gained their autochthonous status within the social space of the artisanal mining fields — for example, through the right to own a shaft, to participate in the plots sharing, or to integrate the *tom bolomas* team — were excluded from access to new infrastructure services. In other words, they were considered strangers again, which impacted upon their artisanal mining activities and meant their exclusion from decision-making processes in the management of corporate investment benefits, which only focused on the villages.

Hence, from the control of workspaces, the SAG, through its CSR policy, controlled specific infrastructures and interfered in the social life of both itinerant and local resident artisanal miners. Decisions were mainly taken by consulting local authorities and implementation occurred by transferring allocation tasks to local elites. In consequence, itinerant gold diggers were progressively deprived of some of their established prerogatives, which caused conflicts among the artisanal miners. For instance, ordinary energy providers were suddenly accused of cheating on the prices and of imposing an unequal competition. These artisanal miners often bring their own gas infrastructure from other regions and thus competed with the locals who were developing an informal energy market through their access to company-installed electricity lines. As these companies were from Burkina Faso, Mossi-speaking artisanal mine workers were quickly and violently expelled. The same occurred with the management of ore mills and automatic sluices that often came from Malian markets. Also other machine owners who were not residents of the place had to justify their linkage to "native" families in order not to be expelled. For instance, a *tom boloma* established for long in the region argued in reaction to the autochthony politics of village authorities: "They took me in the militia because I had been a student and I can write. First they were saying no because I was a stranger. For me, it is twenty years that I am living here; I know the mines, how to read, to write...of course, I can be a *tom boloma*. But now, again, some say I should not remain in the team. Although I do not even have a machine, some tell me to leave..."

Thus, the unequal allocation of access to energy resources based on autochthony has weakened the usual conventions regulating the presence of itinerant gold diggers. The application of an external categorizing discourse hence made social boundaries effective through the interference of revised property rules on the technical elements needed for the production of gold in the artisanal sector.

In this third type of encounters, the conflict of gold diggers opposing the mining company is externalized to the field of artisanal gold mining, where the company indirectly rules by controlling access to energy. By creating facilities in the extraction and treatment of gold through energy supply and by letting remaining zones be uncontrolled, the SAG provides a new framework for the surrounding artisanal mining fields where the issue of access to land is partly replaced by the one of access to energy. The transfer of allocation tasks to the natives' authorities also contributes to the autochtonization of work relations between resident and itinerant workers. Without controlling space *stricto sensu*, the company through CSR programs thus interferes in the social organization of artisanal mining and partially prevents the appearance of gold diggers in its active pits.

6. Discussion and Outlook

In this chapter, I have shown three successive typical forms of interaction between a gold mining company and the surrounding artisanal miners from their perspective. Each of these encounters highlights different ways of controlling spaces for gold extraction and the subsequent division of power in the control of working spaces. The three types of interactions appeared successively between 2009 and 2012 but should not be considered as isolated steps. Each new type integrates with the previous one(s) in such a way that the control over working spaces is progressively reshaped during those three years. The first type of encounter (expulsion) relies on a clear division of control between the area of the mining concession, on the one hand, and the artisanal mining fields, on the other. Coercive power is transferred to the army on the territory of the SAG and to the local militia on the artisanal mining fields. Artisanal miners caught on the licensed land are expelled in the best case or transferred to the police for imprisonment. The second case (tolerance) appeared after several controversies about the responsibility of the company concerning fatalities that occurred upon its lease. Control is here withdrawn from leftover zones, which remain free of access to the artisanal miners. Fatalities drastically increase in those places and often affect young, inexperienced and poor itinerant mine workers. In the third type (interference), the former division of control of working spaces remains the same. However, the energy supply to the "community" (the natives in the given definition) by the company through CSR programs allows the SAG to interfere in the social organization of artisanal mining. The consequence is a return to conflicts and autochthony-based claims and the expulsion of numerous itinerant workers from the artisanal mining fields.

From those three interrelated types, it appears that the company diversified the types of control over the territory, and even extended it to the surrounding gold diggers. Direct control is maintained and strengthened in the active pits. Besides, soft control is deployed beyond the licence area through energy supply, extending also to the spaces of artisanal extraction. Through this, village authorities gain power in the zones of artisanal extraction, upon which they reactivate autochthony boundaries that usually only concerned the living spaces (camps and villages through *tutorat*). Finally, the transfer of allocation tasks to the village authorities implies that the responsibility of the company in sensitive zones is removed and potential conflict is partially externalized to the artisanal miners. Thus, it appears that the more the company and the gold diggers interact, the more conflictive the relations among the gold diggers become. Also, the case study shows that the control over territories is gradually reshaped in order to maintain a strict coercive control in the active industrial mining areas without regard to the original lease of the company, while facilitating access to abandoned areas. Finally, the control of energy resources contributes to the autochtonization of work relations among artisanal gold miners despite a long history of established conventions of work and hosting between local and itinerant miners.

Finally, the study suggests that the common understanding of environmental conflict to grasp conflictive situations between large-scale mining and small-scale mining is not fully applicable to the case presented here. As mentioned by Escobar (2006), political and economic dimensions must be taken into consideration to understand better the motives for control in specific territories. While the issue is understood from both miners' and company's points of view, as a concern of access to and control over spaces of work, the motives to control them differ. From the miners' perspective, the question is essentially an economic concern of accessing gold ore, depending on the resources they have (for example autochthon status, stable network of peers and experience). On the other side, the mining company does not face effective loss of gold ore but acts at the local level out of political concern to maintain its image, freed from responsibility for the accidents occurring upon its land. Moreover, the CSR initiatives contribute in externalizing potential conflicts away from the responsibility of the company.

In contrast to other studies that focus on the relations between companies and the state, I have centred the analysis on the relations between a mining company and the surrounding collectives to understand how the evolving categorization of spaces dedicated to artisanal gold mining and control over spaces contributed to increased conflicts among the gold diggers. The extension of social boundaries based on autochthony conventions to the working sphere of the mining fields has in particular one main consequence. From the gold miners' point of view, be they locals or itinerants, the application of CSR programs de facto opens a new register for justification for the procurement of the needed resources such as energy for the extraction of gold (Boltanski and Thévenot 2008).

In a context of rising worker mobility—facilitated by networks in a process of professionalization despite the informality of the practices in legal terms—the current state of the art raises questions regarding the echoes of these modes of governance at the level of the state. For instance, while autochthon discourses are not new in western Africa, their political uses by state actors has often contributed to generate tensions (Bayart 2001; Ceuppens 2005; Chauveau 2000). Nevertheless, it is interesting to recall that, since the new Guinean presidency of Alpha Condé in November 2010, the official discourse on mining resources has changed: while companies are pressed to renegotiate their concessions and to give higher revenues to the state, the artisanal mining sector also becomes involved in terms of potential incomes. The first speech given by the Presidency to the national association of small-scale gold and diamond diggers highlights this new trend. After regretting the "lack of organization" of the gold diggers, the Presidency warned them against neighbouring countries taking advantage of the situation. This statement well illustrates how non-nationals are designated as responsible for all kinds of problems in the ASM sector at different institutional levels despite their historical involvement in artisanal gold mining and the massive presence of Guinean mine workers in the neighbouring countries. Thus, in most governmental offices concerned with artisanal mining, "foreigners" are also referred to as appropriating national wealth.

Far from considering the transnationality of the artisanal mining sector, the official position regarding ASM participates in the growing discourse based on autochthony at the level of the state to the point of extending this divide to the category of nationality. Thus, as long as the mobile component of gold extraction and the proper modes of organization of this exploitation system are not taken into consideration by decision-makers and local administrators, the results are growing tensions among the workers. Moreover, it is known that artisanal mining activities and the related movements of people constantly arose in the last decade (IOM 2012 ; UNECA 2003). The traditional seasonal displacements of workers from their home places to the mining areas are complemented by long-term displacements to more and farther mining fields spread beyond national borders. For most mine workers, adopting an extreme mobile way of life has become an inherent condition for work in the artisanal mining sector, and as such forms a necessary component of this socio-technical system. Therefore, the categorizations emanating from the CSR programs based on a generic understanding of belonging through the national-ethnic lens when assessing transnational work practices only partly reflect the issues of the workers on the ground. Rather, they generate supplementary conflict potential among them by externalizing the competition for resources.

References

AngloGoldAshanti 2009: Country Report: Guinea; and 2010 Country Report: Guinea.

Banerjee, Subhabrata Bobby 2008: Corporate Social Responsibility: The Good, the Bad and the Ugly. *Critical Sociology* 34(1): 51–79.

Bayart, Jean-François; Geschiere, Peter; Nyamnjoh, Francis 2001: Autochtonie, démocratie et citoyenneté en Afrique. *Critique internationale* 10(1): 177–194.

Beauchemin, Cris and Philippe Bocquier 2004: Migration and Urbanisation in Francophone West Africa: An Overview of the Recent Empirical Evidence. *Urban Studies* 41(11): 2245–2272.

Boltanski, Luc and Laurent Thévenot 2008: De la justification. Les économies de la grandeur. Paris: Gallimard.

Bredeloup, Sylvie 2008: L'aventurier, une figure de la migration africaine. *Cahiers internationaux de sociologie* 125(2): 188.

Carroll, Archie B. 1999: Corporate Social Responsibility: Evolution of a Definitional Construct. *Business & Society* 38(3): 268–295.

Ceuppens, Bambi and Peter Geschiere 2005: Autochtony: Local or Global? New Modes in the Struggle over Citizenship and Belonging in Africa and Europe. *Annual Review of Anthropology* 34: 385–407.

Chauveau, Jean-Pierre 2000: Question foncière et construction nationale en Côte d'Ivoire. Les enjeux silencieux d'un coup d'État. *Politique Africaine* 78: 94–125.

Chauveau, Jean-Pierre 2007: Transferts fonciers et relation de "tutorat" en Afrique de l'Ouest : évolutions et enjeux actuels d'une institution agraire coutumière. *Journal des Sciences Sociales* 4: 7–32.

Davis, Rachel and Daniel M. Franks 2011: The Costs of Conflict With Local Communities in the Extractive Industry. First International Seminar on Social Responsibility in Mining, Santiago.

Escobar, Arturo 2006: Difference and Conflict in the Struggle over Natural Resources: A Political Ecology Framework. *Development and Change* 49(3): 6–13.

Filer, Colin and Martha Macintyre 2006: Grass Roots and Deep Holes: Community Responses to Mining in Melanesia. *The Contemporary Pacific* 18(2): 215–231.

Fisher, Eleanor 2007: Occupying the Margins: Labour Integration and Social Exclusion in Artisanal Mining in Tanzania. *Development and Change* 38(4): 735–760.

Frynas, Jedrzej George 2005: The False Developmental Promise of Corporate Social Responsibility: Evidence From Multinational Oil Companies. *International Affairs* 81(3): 581–598.

Grätz, Tilo 2004: Les frontières de l'orpaillage en Afrique occidentale. *Autrepart* 30(2): 135–150.

Hentschel, Thomas, Felix Hruschka and Michael Priester 2002: Global Report on Artisanal & Small-Scale Mining. Mining, Minerals and Sustainable Development. IIED. 70.

Herbert, Eugenia W. 2012: Elusive Frontiers: Precolonial Mining in Sub-Saharan Africa. In: Werthmann, Katja and Tilo Grätz (eds.): *Mining Frontiers in Africa. Anthropological and Historical Perspectives.* Köln: Rüdiger Köppe Verlag, 23–32.

Hilson, Gavin 2011: "A Conflict of Interest"? : A Critical Examination of Artisanal/Large-Scale Miner Relations in Sub-Saharan Africa. In: Botchway, Francis N. (ed.): *Natural Resource Investment and Africa's Development.* Edward Elgar Publishing.

Hilson, Gavin and Clive Potter 2005: Structural Adjustment and Subsistence Industry: Artisanal Gold Mining in Ghana. *Development and Change* 36(1): 103–131.

Hilson, Gavin and Natalia Yakovleva 2007: Strained Relations: A Critical Analysis of the Mining Conflict in Prestea, Ghana. *Political Geography* 26(1): 98–119.

Hönke, Jana 2010: New Political Topographies. Mining Companies and Indirect Discharge in Southern Katanga (DRC). *Politique Africaine* 120(4): 105–127.

IOM, International Organization for Migration 2012: Global Eye on Human Trafficking. Geneva. 11.

Jacques, Eric et al. 2006: Artisanal and Small-Scale Gold Mines in Burkina Faso: Today and Tomorrow. In: Hilson, Gavin (ed.): *Small-Scale Mining, Rural Subsistence and Poverty in West Africa.* London: Practical Action135–147.

Jonsson, J. B. and D. F. Bryceson 2009: Rushing for Gold: Mobility and Small-Scale Mining in East Africa. *Development and Change* 40(2): 249–279.

Jorgensen, Dan 2004: Who and What is a Landowner? Mythology and Marking the Ground in a Papua New Guinea Mining Project. In: Rumsey, A and J Weiner (eds.): *Mining and Indigenous Lifeworlds in Australia and Papua New Guinea.* Wantage (UK): Sean Kingston Publishing, 68–100.

Kapelus, Paul 2002: Mining, Corporate Social Responsibility and the "Community": The Case of Rio Tinto, Richards Bay Minerals and the Mbonambi. *Journal of Business Ethics* 39(3): 275–296.

Ki-Zerbo, Joseph and Djibril Tamsir Niane 1991: Histoire générale de l'Afrique—Volume 4, L'Afrique du XIIe au XVIe siècle. Paris: Présence africaine/Edicef/Unesco.

Little, Paul Elliott 2007: Political Ecology as Ethnography: a Theoretical and Methodological Guide. Translated from *Horizontes Antropológicos*, Porto Alegre. 12(25): 85-103. <socialsciences.scielo.org/pdf/s_ha/v3nse/scs_a12.pdf> accessed on 20.6.2014.

Luning, Sabine 2012: Processing Promises of Gold: A Minefield of Company–Community Relations in Burkina Faso. *Africa Today* 58(3): 23–39.

Mbodj, Faty B. 2009: Boom aurifère et dynamiques économiques entre Sénégal, Mali et Guinée. *Echogéo* 8.

Potts, Deborah 1995. Shall We Go Home? Increasing Urban Poverty in African Cities and Migration Processes. *The Geographical Journal* 161(3): 245–264.

Robbins, Paul 2012: Political Ecology. Oxford: Wiley-Blackwell.

UNECA, United Nations Economic Commission for Africa 2003: Report on Selected Themes in Natural Resources Development in Africa: Artisanal and Small-Scale Mining and Technology Challenges in Africa. Addis Ababa: Third Meeting of the Committee on Sustainable Development, United Nations Economic Commission for Africa.

USAID 2005: Minerals and conflict: A toolkit for intervention. Office of Conflict Management and Mitigation. Washington DC: USAID.

Werthmann, Katja 2006: Gold Diggers, Earth Priests, and District Heads: Land Rights and Gold Mining in Southwestern Burkina Faso. In: Kuba, Richard and Carola Lentz (eds.): *African Social Studies Series.* Leiden: BRILL, 119–136.

Werthmann, Katja and Tilo Grätz 2012: "Introduction". In: Werthmann, Katja and Tilo Grätz (eds.): *Mining Frontiers in Africa: Anthropological and Historical Perspectives.* Köln: Köppe, 9–22.

Constructed "From Above" or Driven "From Below"?

Youth Agency, Mining and "Spaces for Change" in Sierra Leone

Roy Maconachie

Over the past two decades, neoliberal reforms, soaring commodity prices and heightened global resource demands have led to significant growth in extractive industry investment across sub-Saharan Africa. A surge of investment has triggered a variety of responses in mineral-rich communities—from outright rejection, to protest over labour conditions, to acceptance in anticipation of gainful employment. Drawing on recent research carried out in Kono District in Sierra Leone, this chapter critically explores these contrasting responses to mining activities, by focusing on how youth perceive and respond to extractive industry expansion.[1] In doing so, the analysis broadens understanding of why youth perceptions of mining investment differ, and illuminates the various factors underlying a diverse range of responses to the expansion of extractive industries. The focus on youth and its heterogeneity as a social category has important policy implications and the chapter makes a contribution to understanding the dynamics and diversity of youth livelihood strategies in resource rich developing countries. In making a distinction between autonomous spaces of community-led development "from below" and corporate controlled spaces of development "from above", the chapter reflects on how youth perceptions of extractive industry expansion may also be influencing the ways in which mining companies understand and fashion their business and social responsibility strategies.

1. Introduction

Across sub-Saharan Africa, pressures on natural resources are rapidly increasing in response to growing global demands for fuel and non-fuel minerals. Over the last two decades, the globalization of the extractive industries has led to dramatic technological, organizational and regulatory changes across the African continent, as many governments have adopted new mining codes, or revised existing ones, to stimulate a flood of foreign direct investment in mineral extraction (Bridge 2004). While Africa's natural resources have become increasingly important in global markets, with some sub-Saharan countries experiencing phenomenal growth (Weng et al. 2013), paradoxically, the continent remains the world's poorest region. The communities surrounding the locations of extraction are often the most marginalized and economically depressed, as serious social and environmental impacts displace indigenous populations from their land and transform local livelihoods.

1 The fieldwork upon which the chapter is based was funded by a grant from the British Academy (SG112077). The author would like to thank Tony Bebbington, John Rogan and Nick Cuba for their feedback on an earlier draft of the chapter, which was originally presented as a paper at the 2013 AAG Annual Meeting in Los Angeles.

Fig. 1: The town of Kayima in Kono District. Photo: Roy Maconachie.

In such contexts, as Bebbington et al. (2008) suggest, investments in mineral extraction have become a particularly contentious and ambiguous face of global capitalist expansion, often leading to a pronounced increase in conflict and social mobilization around the detrimental effects of mining and hydrocarbon projects. In many extraction-impacted communities, evidence also suggests that youth—understood here as a social category defined by a combination of age, social status and relative livelihood dependence—are in particularly disadvantaged positions. Even in locations of abundant natural resource wealth, young people often lack access to land, credit, employment opportunities and other assets. At the same time, however, they may be in better positions to mobilize, being more literate, more able to take advantage of new communication technologies, and more attuned to global narratives on the environment and rights.

Focusing on the case of Sierra Leone, this chapter critically explores contrasting youth perceptions of, and responses to, recent extractive industry expansion. Questions that concern the future of Sierra Leone's natural resource wealth, and of what is often referred to as a "crisis of youth", are both central to post-conflict development trajectories, government peace-building initiatives and security sector reform. Indeed, in recent years, there has been renewed interest in debates about the underlying causes of Sierra Leone's decade-long diamond-fuelled civil war of the 1990s, with considerable difference of opinion about whether the conflict was driven more by politics ("grievance") or exploitative economic interests ("greed") (e.g. see Collier and Hoeffler 2004). While it is true that some of the bitterest battles of the war were fought in the diamond mining areas of Kono District, the root cause of the conflict is now increasingly being linked to underlying questions that concern the marginalization of youth (Peters and Richards 2011; Mokuwa et al. 2011). A growing literature on the agrarian dimensions of Sierra Leone's war, and more specifically the intergenerational tensions that drove the conflict, suggests that large numbers of socially excluded young people were prompted to embrace the war in a desperate search for empowerment. These same concerns for youth marginalization remain high on the post-war agenda for donors and policy makers, largely driven by particular concerns that continuing youth alienation may be a potential source of further conflict.

While the "crisis of youth" may have become the master narrative of post-war reconstruction in Sierra Leone, youth concerns have also increasingly become a focal point for a host of new international mining companies who are keen to invest in a politically stable country. With 42% of Sierra Leone's population below the age of 15 years, and 34% between the ages of 15 and 35, young people represent the majority of the population (GoSL 2004). But at the same time, this very age cohort is most likely to be equipped with few employable skills and be inactive in the labour market (Peeters et al. 2009). Government estimates suggest that 70% of all youth are presently unemployed or underemployed (Peeters et al. 2009) and youth employment generation has understandably been designated as a central development pillar in the country's third Poverty Reducation Strategy Papers (PRSP), "The Agenda for Prosperity". Meeting the challenges of youth is a key element of the government's vision for achieving middle-income status within the next twenty-five years, a goal which also relies heavily on natural resources being the initial driver for rapid growth (GoSL 2012).

In this context, the objectives of this chapter are twofold: first, it seeks to explore how changing global-economic patterns and processes associated with extractive industry investment are shaping livelihood perceptions and opportunities for young people in resource-rich communities; and second, it aims to understand how youth responses to extractive industry expansion may be influencing the ways in which mining companies understand and fashion their business and social responsibility strategies. Drawing upon recent research carried out in diamondiferous Kono District, where there has been a proliferation of foreign interest in mineral extraction, the chapter critically explores heterogeneity within the social category of youth, and its relationship to social activism, mining and livelihoods. For young people in Kono District, the underlying agendas and motivations for engaging with mining companies vary considerably between different subgroupings of youth, as do the responses of companies to these different voices.

In trying to better understand how the agency of youth shapes and responds to extractive industry investment, different "spaces of participation" are explored and the distinction is made between "invited" corporate controlled spaces of development "from above", and more autonomous non-institutionalized spaces created "from below" through social activism. Discussions with young people in Kono District reveal that much youth participation in mobilizations driven by mining expansion has crystalized on the back of severe exploitation and marginalization by powerful corporate actors. But at the same time, evidence also suggests that some mining-based advocacy groups have been equally motivated by the opportunity of attracting external development funding from international actors. As poverty and hardship have become entrenched during the post-war period, some youth may be drawing on mining-focused social activism as a new livelihood strategy and an avenue to advance their causes.

Following this introduction, the first section of the chapter outlines a broad conceptual approach for understanding how different actors respond to the impacts of mining, and how community development initiatives are driven. Indeed, many mining companies now adopt the language of "participation" and "sustainable development" as a central platform of their corporate social responsibility (CSR) strategies. Drawing upon the work of Andrea Cornwall (2002a; 2004), it is suggested that a spatial lens is a particularly useful metaphor for exploring how power and participation are situated within the community development initiatives found in many mining areas. More specifically, a distinction is made between corporate controlled spaces of development "from above" and youth-led spaces of development "from below". This discussion sets the stage for section two, which focuses in further detail on youth perceptions of mining investment, and the factors underlying their diversity of responses to the expansion of extractive industries. In section three, the chapter reflects on how different youth responses to extractive industry expansion may be influencing the ways in which mining companies understand their social responsibility strategies. Ultimately, as is explored in the conclusion, a better understanding of how diverse interpretations of "sustainability" become shaped by conflicting interests and underlying agendas remains critical for ensuring that young people in Sierra Leone are included in development processes that have for so long been dominated by powerful actors.

2. Conceptualizing "Spaces for Change"

Over the past two decades, "participation" has become enshrined as a central tenet of development discourse and is now widely accepted as "standard practice" in development policy and practice around the world. Corporate approaches to community development are no exception to this norm, with some scholars arguing that "[t]he notion of 'participation' is integral to the discourses of CSR and development, both as a process and as an end in itself" (Kemp 2010: 203). While there is now a burgeoning body of scholarly work which has explored participation in a wide range of development contexts, there are a number of recent works which are especially germane to understanding how CSR initiatives are being designed and implemented in mining communities in Sierra Leone. Most notably, the sub-category of work which reflects on issues of power and difference in participation and development is particularly relevant to understanding how youth are being included, or excluded, from community development processes.

Recent critiques of participatory approaches to development argue that there is often a danger of participation being based on "over-optimistic notions of agency... combined with romantic ideas about groups and institutions" (Cleaver 2004: 271). Even when local people are included in decision-making processes, there is still often a failure to transform existing social, political and economic structures in ways that give voice to those who have been marginalized (Hickey and Mohan 2004; Cooke and Kothari 2001). Consequently, in exploring the processes that are underpinning social transformations in regions of extractive industry expansion, it remains important to appreciate that spaces of engagement are always mediated by relationships of power between actors, and by the modes of influence that different actors exercise. As succinctly summarized by Hamann (2003: 251):

> "While resisting the temptation to misuse their power advantage,... [companies] need to support communities' right to prior informed consent, and acknowledge and respect the culture gap between the corporation and indigenous communities. One implication of this is that CSR strategies cannot be taken 'off the shelf': they need to be established within the local African context and in continual dialogue with key stakeholders."

In trying to gain a better understanding of the dynamics of power in arenas of participation, Cornwall (2002a; 2004) argues that the concept of "space" is a particularly useful construct. She suggests that a spatial lens can help to convey the situated nature of participation, as well as the "bounded yet permeable arenas in which participation is invited, and the domains from within which new intermediary institutions and new opportunities for citizen involvement have been fashioned" (2004: 75). Accordingly, Cornwall (2002a, 2002b) examines different kinds of space in which participation can occur, distinguishing between two main contexts in which they are created. First, she notes that "invited" institutionalized spaces of participation are those that are most often created "from above" by powerful actors, such as government agencies, international NGOs or multinational mining corporations; and second, she suggests that "autonomous" non-institutionalized spaces are those created "from below" through collective action or social activism. In both cases, she points out that it remains important to remember that participation is shaped and defined by power relations. When considering the creation of "new spaces" of participation, Cleaver (2004: 247) further adds that questions remain as to how "old" sources of authority come to shape and legitimize the new rules and institutions of liberal democracy which claim to support citizen voice. For the purpose of this chapter, this distinction—between invited corporate-controlled spaces of development "from above" and autonomous community development "from below"—provides a useful lens when considering how youth respond to extractive industry investment. Recent fieldwork carried out in Kono District illustrates how the emergence of public spaces—both in the form of invited institutionalized spaces as well as more organically created, people-driven sites of political action—may be opening up new opportunities for post-war public engagement in resource governance, particularly amongst youth.

Fig. 2: Artisanal diamond mining in Kono district. Photo: Roy Maconachie.

3. Youth-Led Spaces of Development "From Below"

In exploring social transformations associated with extractive industry investment and, more specifically, considering how they may be impacting upon livelihood opportunities for young people in resource-rich communities, in-depth fieldwork was carried out in Kono District in the Eastern Province during the months of January and June 2013. The investigation focused on the Kono diamond mining centre of Koidu in Tankoro Chiefdom, a settlement with a long, rich history of diamond mining, and now a thriving town of 87,539 people. A broad range of qualitative methods was adopted in the fieldwork—including key informant interviews, focus group discussions, and structured observations at community meetings—to gain an understanding of the wide range of perspectives held by young people. Initially, a purposive sampling frame was used to identify key informants, such as youth political leaders and those working in youth-focused civil society organizations, who possessed specific knowledge and understanding of the mining investments that had taken place. But this eventually gave way to a broader investigation, whereby different self-generating groupings of youth were targeted and interviewed, including those working as casual labourers in industrial mining operations, those engaged in artisanal and small-scale mining (ASM) activities, those who had previously mined but had given it up to pursue other livelihood options, and those who had never mined but had rather focused their attention on farming. Following interviews, informants were asked to identify other potential interviewees, and also if they would be willing to join focus groups. Focus groups consisted of 4-5 people and allowed for an in-depth exploration of how, if at all, attitudes varied between different groupings of youth.

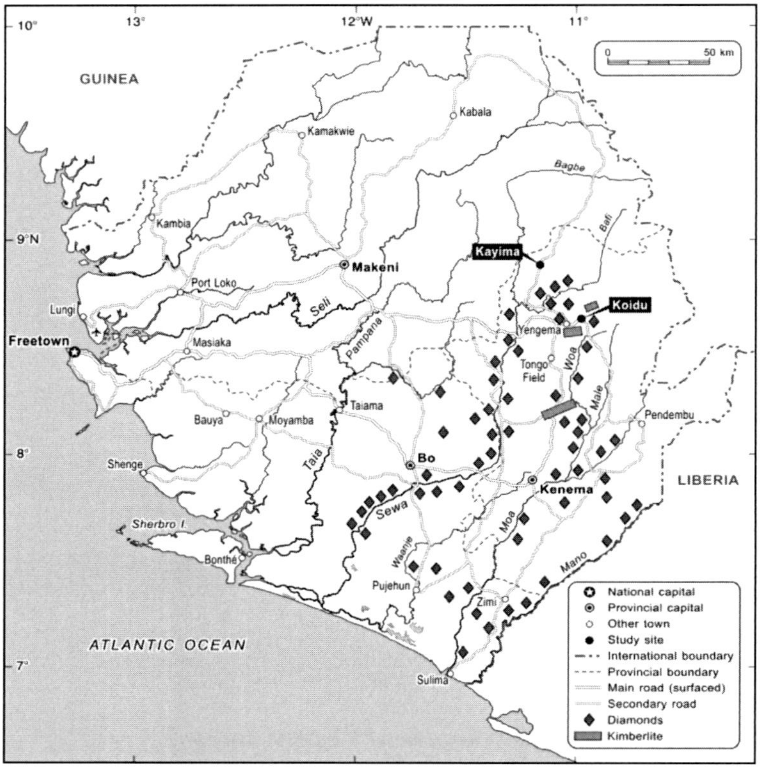

Fig. 3: Diamondiferous regions of Sierra Leone. Source: Author's fieldwork.

As anticipated, a wide range of perceptions were reported. For many youth who were employed by industrial mining operations, and to some extent those working within the ASM sector, there was a general consensus that diamond mining had always been, and would continue to be, an essential livelihood activity in the economy and society of Kono District. While most of the miners interviewed reported that the working conditions they were subjected to, the remuneration they received for their labour, and the uneven distribution of mineral rents associated with the sector were far from desirable, there was also a general feeling that they were fortunate to be employed at all, as livelihood opportunities were few and far between. This dichotomy of feelings was represented in a number of interviews with labourers at the country's largest Kimberlite diamond mining company, OCTEA Ltd., suggesting that the same ambiguous relationship between mining and development described by Bebbington et al. (2008), was also clearly present in Koidu. Many young miners felt that there was great potential for mining investments to be a more effective driver of local economic development, particularly if the government played a stronger role in negotiating more beneficial conditions in the agreements made with international mining companies. As one labourer employed at OCTEA Ltd. commented in an interview:

> "Artisanal mining is no longer paying off for most people. There are just not as many alluvial diamonds around any more, and a digger can spend years waiting to recover a winning stone. So many young people are turning their eyes to the companies who are mining intensively, although it is true that they cannot provide jobs for everyone. So you see the companies have a captive audience. They can pay what they want and young people will still come running. These companies are making big money, and so it only seems right that the authorities should be doing more to ensure that workers are paid a better daily wage" (personal communication, OCTEA labourer, Koidu, 05/06/13).

While some young miners remained ambivalent about mining investment, others displayed great indignation about the very uneven benefits that accrued to different actors within the sector. These discussions revealed that many youth-focused Civil Society Organizations (CSO) had mobilized in reaction to perceptions of severe exploitation and marginalization by powerful corporate actors. More specifically, there was widespread concern for both the labour exploitation that is characteristic of many mining companies, and for the dispossession of land and other assets that typically occurs during the process of extractive industries expansion. As one miner put it:

> "Internationally, people think the companies are creating employment for youth, but in actual fact it is very limited. OCTEA is doing very little for the local population and the jobs they offer are not secure. Recently more than 300 workers were made redundant and many of them had to go back to the mining fields to do artisanal mining. So the company and the local people are not on good terms. Many youth are now coming together and they are really shouting. We have had two riots so far, both of which have resulted in the loss of life" (personal communication, artisanal miner, Koidu, 03/06/13).

Such sentiments are, of course, not peculiar to Kono District and other scholars have noted how the un-inclusive and unsustainable nature of mining has also motivated social mobilization and conflict in other West African country contexts (see for example, Hilson and Yakovelva 2007; Pegg 2006). But such conflict may not always be detrimental, and as Bebbington et al. (2008: 893) point out, in some cases conflict may have the ability to create a "political pathway towards the construction of institutions that could foster more socio-economically inclusive and less environmentally damaging forms of mineral expansion" (2008: 5). Others have noted that resistance to mining, and particularly the struggles of youth, may not only be driven by the need to defend human and citizenship rights, but also the need to assert an identity (O'Brien 1996).

In Kono District, much of the present youth-led activism around the extractive industries is fundamentally concerned with citizenship rights. Ideas about rights and identity among Kono youth are strongly tied to livelihood security in whatever form it can be found, and this has increasingly driven social mobilization. More broadly, a recent World Bank study of youth employment reported that the post-war period has seen "a remarkable upsurge in self-organized social activism amongst young people" (World Bank 2007: 87-101) and a parallel revival of reciprocal labour companies, cooperative associations and rotational credit clubs have been noted in rural areas (Fanthorpe 2007). While these modalities of social cooperation are not new, they have acquired a new salience under conditions of post-war economic hardship and the decline of patron-client relationships. The mobilization of youth-dominated associational groups and civil society organizations, and the subsequent youth-led protests that have recently been staged against industrial mining operations in and around Koidu, are examples of how such spaces for citizen engagement in diamond governance may be opening up. These mobilizations have been driven by new forms of mass communication and consumption—including the use of mobile phones and the internet, and a thriving youth-orientated music industry—all of which have helped youth to shape political ideas in ways unimaginable a generation ago.

While new communication technologies have undoubtedly been an important instrument for holding mining companies to account, and for disseminating local grievances against companies to international audiences, it is worth noting that most diamond mining in Kono District has historically been associated with the ASM sector. In the past, large-scale and capital intensive mining operations have only accounted for some 10% of Sierra Leone's diamond exports by value (Diamond Industry Annual Review 2006). The bulk of the country's diamonds have come from ASM production, an activity demanding heavy inputs of unskilled (youth) labour. But a recurrent theme in the history of diamond mining areas, especially Kono, has been confrontation between, on the one hand, state agencies and concession-holding companies seeking greater control over diamond production and, on the other hand, peasants and artisanal miners seeking to maximise their own returns from the industry and defend their putative rights and entitlements. Confrontations of this kind are a signature feature of bureaucratically weak states, but the presence of diamonds has lent them a particular intensity in Kono District.

More recent interviews with Kono youth in and around the town of Koidu suggest that changes in the local economy are having bearing on both the nature of confrontation and the direction of youth activism. As many of Kono's prime alluvial diamond areas have become "mined out" and ASM operations have waned in favour of more capital-intensive modes of mechanized extraction (also see NACE Monitoring Report 2006), tensions between small-scale and large-scale mining have somewhat subsided. Consequently, youth-led activism has shifted its focus to a number of high-profile international mining companies (Fanthorpe and Maconachie 2010). However, many of the youth-focused CSOs in Kono that are presently lobbying against mining companies are not suggesting that mining should be halted altogether. Rather, they are advocating that the institutional arrangements should be put in place to ensure that communities are compensated with development investment and better employment opportunities. In fact, discussions with some youth revealed the perception that the permanent closure of large industrial mining operations, such as OCTEA, would be disastrous for the local economy. The main grievance was that more benefits needed to be channelled back into diamondiferous communities, predominantly the creation of more jobs for youth. But as mining has increasingly shifted to more capital-intensive modes of extraction, and is becoming more attractive to larger companies who aim to pursue deep-shaft mining, it is the youth who are ultimately losing out. While youths are no longer able to engage in ASM activities—a sector which has been notoriously popular because of its low barriers to entry—chiefs and village elites still continue to benefit from rent-seeking activities as new relationships are formed with companies, and income is derived from various corporate kick-backs. In short, declining demand for unskilled labour in the mining sector may thus emerge as one of the greatest challenges for overcoming the social and economic marginalization of youth and related security problems.

As the mobilization of Kono-based youth groups has been a vehicle for addressing the labour issues and employment conditions of predominantly larger-scale mechanized companies, the tension that this has created between companies and communities has, on occasion, led to antagonistic situations. The disturbance at the OCTEA (then called Koidu Holdings Limited) blasting area in December 2007, is a case in point. A group, predominantly composed of youths calling itself the Affected Property Owners Association, staged what was initially a peaceful protest outside the company blasting area, to demand a less intrusive schedule of blasting and better compensation for environmental degradation for local residents. These protests were supported by a number of CSOs, but they soon turned into a full-scale riot, resulting in the death of two individuals and numerous injuries.

In the aftermath of the disturbance, the government halted company operations pending a Commission of Enquiry into the events. The recommendations of the Jenkins-Johnson Enquiry resulted in a government White Paper that laid out new regulations for corporate social behaviour, but it remains unclear as to whether or not any real lessons were learned from the conflict that took place. For example, more recently in December 2012, a group of aggrieved mine workers at OCTEA staged a second peaceful protest, this time to demonstrate their frustration and dissatisfaction over poor working conditions and what they believed to be corporate deceit. The atmosphere that was borne out of the protest was, once again, tense and the situation quickly degenerated into an all out battle between the protesting mine workers and the armed police guards stationed at the mining concession, with two people being killed. Whether or not the Jenkins-Johnson recommendations have actually addressed any of the grievances of Kono youth therefore remains to be seen. As stated by the Co-ordinator of one local youth-focused CSO:

> "The Jenkins-Johnson document is a beautiful list of recommendations. If you read the document, its beautifully written on paper, but it's not doing anything. One of the most important recommendations was that resettlement of the community should take place before mining continued. But they're not going by it. So who should hold [the company] accountable? It should be the Government. But the Government is not doing anything towards that" (Personal communication, Co-ordinator of KoCEPO, Koidu, 11/01/13).

Fig. 4: Mechanized diamond mining in Kono district. Photo: Roy Maconachie.

4. Corporate-Controlled Development "From Above"

While the conflicts described above represent extreme cases where the tensions between youth and mining companies have escalated into tragic confrontations, the demands being placed upon companies such as OCTEA may be illustrative of the wider trends that are currently unfolding across sub-Saharan Africa. In natural resource-rich regions in all corners of the continent, it has become clear that mining and oil and gas companies are now increasingly being expected to secure a "social license to operate", and to fund and implement community development projects. In many cases, companies are expected to take on responsibilities for which governments have traditionally been liable. This was particularly the case following the initiation of long-term IMF structural adjustment programmes in the 1980s and early 1990s, as measures were implemented to reduce spending and the direct economic involvement of the state. While these policies were, in rhetoric, implemented to provide an "enabling environment" for private enterprise to thrive, the draconian "conditionalities" associated with adjustment most often left African governments impotent and unable to provide services and infrastructure for their citizens.

In the case of Sierra Leone, the devastation caused by a decade of civil war during the 1990s further exacerbated poverty, destroyed much of the country's infrastructure and reduced the capacity of the state to provide development for its citizenry. In many instances, extractive industry companies in Kono District have indeed embraced the opportunity to fill these development voids in some of the country's most impoverished areas, seeing it as an opportunity to enhance their international reputations and strengthen their social license to operate. But at the same time, it has often been challenging for mining companies to build constructive relationships with rural populations. This has particularly been the case because of a strong sense of community ownership over land and resources in rural Sierra Leone, which it has been suggested has led to an equally strong sense of entitlement to benefits from outsiders using "community land" for profit and self-enrichment (Fanthorpe and Gabelle 2013).

However, in recent years, a number of mining companies in Kono District have become much more focused on contributing to social development programmes for the youth population. There is now an increasingly strong "business case" for companies to actively support sustainable development initiatives that go beyond impact mitigation. This pro-business CSR stance has also been highlighted in recent academic literature, with proponents arguing that mining companies have the potential to effectively address poverty and contribute to human development, at the same time as improving financial performance. But the focus on youth programming has particular resonance in Sierra Leone, given the role that impoverished youth played in the country's resource-fuelled civil war. There are still strongly held fears that increasing youth marginalization may lead to insecurity and further conflict, an environment that is quite obviously not conducive to "good business" for international mining companies. As explained by Hamman (2003: 242): "Local community opposition, ranging from protest to sabotage, can cause costly delays in production or even termination of production. Such production delays are particularly problematic in the context of increasingly demanding customers in just-in-time supply chains."

In this light, a number of interviews with mining company officials suggested that they were now taking youth concerns much more seriously. OCTEA, for example, has recently launched a programme to promote reading and learning for youth in Koidu schools and the surrounding Kono communities, training teachers and providing reading material to more than 7,000 school children. In addition, the company also spends USD 100,000 each year supporting two local premier league football clubs, an investment that has received the widespread approval of the region's football-passionate youth population. It would therefore seem that questions around youth livelihoods and wellbeing have become of increasing concern to OCTEA's CSR strategy, perhaps having gained more significance since the forced temporary closure of operations following the December 2007 riots.

Since performance standards and community development initiatives are most often voluntary, a number of scholars have argued that good community relations are crucial for a company's reputation, and can also influence the nature of global debates surrounding sustainability (Auty 1998; Blowfield and Frynas 2005). In the case of Sierra Leone, Fanthorpe and Gabelle (2013: 52) point out that mining companies "...have a clear incentive to build and foster relationships with a wide range of stakeholders (notably, companies, trade unions and development partners) to ensure security and protection for their operations...". From this perspective, as was clearly articulated by the Public Relations officer at OCTEA, there is both an economic and a security incentive for companies to take CSR on board:

> "...when you come to operate, you get your mining licence from the government and for that you have an agreement. You have obligations that you are supposed to meet... and as long as these are met, you are fine legally. Nobody is going to take you to a court of law for meeting your obligations. But what you will not get is a social licence to operate, and that is why mining companies tend to want to embark on corporate social responsibility.... You have to embark on activities that will engender a very cordial relationship between your operations and the community, providing you with an enabling environment so that at the end of the day you are able to carry out your operations and maximise your profit" (OCTEA Public Relations Officer, personal communication, Freetown, 07/06/13).

But considerable concerns have been raised by a number of scholars who have demonstrated how elsewhere in sub-Saharan Africa, extractive industry companies often implement community development programmes with little knowledge or understanding of the socio-cultural contexts of the people's lives in which they operate (Hilson and Banchirigah 2009; Tschakert 2009). In extreme cases (e.g. see Gilberthorpe and Banks 2012), an over-emphasis on meeting global sustainability performance standards has led to inappropriate and ill-conceived development outcomes at the local level, which has resulted in further fragmentation and inequality. This also raises the question of how sincere mining companies actually are about their youth-focused development initiatives, and whether or not the pursuit of "blueprint" CSR schemes might play a role in decontextualizing local development needs.

As such, unequal power relationships play a central role in reinforcing the most commanding interpretations of "sustainability", as they become conditioned to serve particular interests. While young people in Kono District are generally in structurally disadvantaged livelihood positions, and are likely to find it difficult to effect meaningful change, powerful mining companies that are flush with resources are able to control any development that might take place, and define what "sustainability" entails on their own terms. As David Mosse (2005: 8) has suggested, "power lies in the narratives that maintain an organization's own definition of the problem...[and] success in development depends upon the stabilization of a particular interpretation...". Dauvergne and Neville (2010) add that in weak states such as Sierra Leone—where governments often lack the power required to influence the underlying agendas of transnational companies—an absence of state regulatory oversight and control makes it even more unlikely that voluntary corporate initiatives will lead to meaningful environmental and social outcomes.

In short, as Newell (2005) reminds us, power disparities between companies and communities have significant implications for corporate-community interaction, and the impact that CSR can have on poverty reduction and human development. In order for CSR to be both appropriate and effectively operationalized, and therefore a potential vehicle for positive change, corporate-led development initiatives must seek acceptability, credibility, and ultimately legitimacy, from multiple and competing sources. But perhaps being central to this task is a more explicit appreciation of the unequal power relations between different stakeholders in mining investments, and a recognition of their varying agendas, which should help to ensure that the needs of youth are more fully incorporated into CSR strategies.

5. Conclusion

In Sierra Leone's most recent PRSP, "The Agenda for Prosperity", the message is clear that the government aims to pursue an extractive-led development trajectory in the immediate future and the years to come. Indeed, the country is now poised on the verge of an unprecedented period of economic growth, driven principally by the discovery of vast new deposits of strategic minerals, and more recently, offshore oil deposits on the Sierra Leone-Liberia basin. Moreover, significant international investments in the exploitation of rich iron ore deposits at Marampa suggest that the management of natural resources will continue to assume a high priority on both political and economic agendas. In the wake of a pending "revenue bonanza" driven by natural resource exploitation, the question of how resource benefits will be channelled back to their communities of origin, and indeed the general populace at large, has returned to centre stage. Perhaps, at the heart of these concerns are the very issues raised in this paper that concern relationships between youth, companies and extractive industry investment.

In exploring these relationships, there are three main conclusions that can be drawn from this chapter. First, while it is clear that interactions between youth and corporate actors are taking place in a variety of different ways, and are being driven by different agendas, spatial metaphors remain helpful for understanding these relationships. The linkages between community-driven development and corporate-controlled development are often not well defined or articulated, leaving the possibility for significant misunderstanding to occur. Moreover, the issues of sustainability that arise in young people's day-to-day lives may be conceptualized in ways so different from those incorporated within business management plans that they simply become inconceivable within the terms conceptualized by the company. In mapping out youth participation in mining, and making sense of how companies respond to youth, it is therefore useful to make a distinction between autonomous forms of community-led development "from below" and corporate controlled development "from above".

Second, in the context of Sierra Leone it also remains clear that, numerically at least, the most important actor occupying these spaces of change is "youth" and this will continue to be the case given the demographic context. While young people do not have a single position on mining, and there are a wide range of youth responses to mining investment that can be explained by a variety of factors, pressures from these different positions of youth will affect the mining sector one way or another. In this sense, youth are playing an important role in opening up spaces "from below", and in doing so will influence the nature of spaces "from above".

Finally, the inverse side of the relationship between young people and mining companies is that the CSR spaces that are being driven "from above" also do not have a single position on youth—some seem more accommodating, while others appear to be more instrumental. Indeed, a multiplicity of agendas must be accommodated in order to insulate corporate actors from potential risks and to protect their "social license to operate". But whatever the case may be, these corporate spaces of engagement must respond to youth. As the government continues to pursue an extractive-led path to development and corporate investment in the mining sector is encouraged in the years to come, it will also to be the case that the physical spaces around mining sites in Sierra Leone are occupied by large contingents of youth. The youth occupying those physical spaces necessarily will affect the political spaces that emerge around mining investment, and companies would be wise to recognize this.

References

Auty, R. 1998: Resource Abundance and Economic Development: Improving the Performance of Resource Rich Countries. Helsinki: The United Nations University World Institute for Development Economics.

Bebbington, A., Hinojosa, L., Humphreys Bebbington, D., Luisa Burneo, M. and Warnaars, X. 2008: Contention and Ambiguity: Mining and the Possibilities of Development. *Development and Change* 39 (6): 887-914.

Blowfield, M. and Frynas, J.G. 2005: Setting New Agendas: Critical Perspectives on Corporate Social Responsibility. *International Affairs* 81 (3): 499-513.

Bridge, G. 2004: Mapping the Bonanza: Geographies of Mining Investment in an Era of Neoliberal Reform. *The Professional Geographer* 56 (3): 406-421.

Cleaver, F. 2004: The Social Embeddedness of Agency and Decision-Making. In: Hickey S, Mohan G (eds.). Participation: From Tyranny to Transformation? Exploring New Approaches to Participation in Development. Zed Books: London. 271–277.

Collier, P. and Hoeffler, A. 2004: Greed and Grievance in Civil War. *Oxford Economic Papers* 56 (4): 563-595.

Cooke, B. and U. Kothari (eds.) 2001: Participation: The New Tyranny? Zed Books: London.

Cornwall, A. 2002a: Making Spaces, Changing Places: Situating Participation in Development. *IDS Working Paper* 170. Institute of Development Studies: Brighton.

Cornwall, A. 2002b: Locating Citizen Participation. *IDS Bulletin* 33(2): 49–58.

Cornwall, A. 2004: Spaces for Transformation? Reflections on Issues of Power and Difference in Participation in Development. In: Hickey S, and Mohan G (eds.) *Participation: From Tyranny to Transformation? Exploring New Approaches to Participation in Development.* Zed Books: London. 75–91.

Dauvergne, P. and K. J. Neville 2010: Forests, Food and Fuel in the Tropics: the Uneven Social and Ecological Consequences of the Emerging Political Economy of Biofuels. *Journal of Peasant Studies* 37 (4): 631-660.

Diamond Industry Annual Review, Sierra Leone 2006: The Diamonds and Human Security Project, Partnership Africa Canada, Freetown, February 2006.

Fanthorpe, R. 2007: BRACE (Building Resilience and Community Engagement). Institutional Survey. Unpublished Report, Concern Worldwide, June 2007.

Fanthorpe, R, and Maconachie, R. 2010: Beyond the "Crisis of Youth"? Mining, Farming and Civil Society in Post-War Sierra Leone. *African Affairs* 109 (435): 251-272.

Fanthorpe, R. and Gabelle, C. 2013: Political Economy of Extractives Governance in Sierra Leone. Report for the World Bank. July 2013.

Gilberthorpe, E. and Banks, G. 2012: Development on Whose Terms? CSR Discourse and Social Realities in Papua New Guinea's Extractive Industries Sector. *Resources Policy* 37: 185-193.

Government of Sierra Leone (GoSL) 2012: The Agenda for Prosperity. Draft Zero. Ministry of Finance and Economic Development, Freetown, 12 October 2012.

Government of Sierra Leone (GoSL) 2004: Population and Housing Census: Analytical Report on Population Size and Distribution Age and Sex Structure. November 2006, Freetown. <www.statistics.sl> accessed on 11.12.2013.

Hamann, R. 2003: Mining Companies' Role in Sustainable Development: The "Why" and "How" of Corporate Social Responsibility from a Business Perspective. *Development Southern Africa* 20 (2): 237-254.

Hickey, S. and Mohan, G. 2004: Participation: From Tyranny to Transformation?: Exploring New Approaches to Participation in Development. Zed Books: London.

Hilson, G. and Banchirigah, S.M. 2009: Are Alternative Livelihood Projects Alleviating Poverty in Mining Communities? Experiences from Ghana. *Journal of Development Studies* 45 (2): 172-196.

Hilson, G. and Yakovlela, N. 2007: Strained Relations: a Critical Analysis of the Mining Conflict in Prestea, Ghana. *Political Geography* 26 (1): 98-119.

Kemp, D. 2010: Mining and Community Development: Problems and Possibilities of Local-Level Practice. *Community Development Journal* 45 (2): 198-218.

Mokuwa, E., Voors, M., Bulte, E., and Richards, P. 2011: Peasant Grievance and Insurgency in Sierra Leone: Judicial Serfdom as a Driver of Conflict. *African Affairs* 110 (440): 339-366.

Mosse, D. 2005: Cultivating Development: an Ethnography of Aid Policy and Practice. London: Pluto Press.

National Advocacy Coalition on the Extractives (NACE) 2006: Monitoring Report. Freetown, Sierra Leone.

Newell, P. 2005: Citizenship, Accountability and Community: The Limits of the CSR Agenda. *International Affairs* 81: 541-557.

O'Brien, D. 1996: A Lost Generation? Youth Identity and State Decay in West Africa. In: Werbner, R. and Ranger, T. (eds.) *Postcolonial Identities in Africa.* Zed Books, London. 55-74.

Pegg, S. 2006: Mining and Poverty Reduction: Transforming Rhetoric into Reality. *Journal of Cleaner Production* 14 (3-4): 376-387.

Peeters, P., Cunningham, W., Acharya, G. and Van Adams, A. 2009: Youth Employment in Sierra Leone: Sustainable Livelihood Opportunities in a Post-Conflict Setting. World Bank, Washington, DC.

Peters, K. and Richards, P. 2011: Rebellion and Agrarian Tensions in Sierra Leone. *Journal of Agrarian Change* 11 (3): 377-395.

Tschakert, P. 2009: Recognizing and Nurturing Artisanal Mining as a Viable Livelihood. *Resources Policy* 34(1-2): 24-31.

Weng, L., Boedhihartono, A., Dirks, P., Dixon, J., Lubis, M. and Sayer, J. 2013: Mineral Industries, Growth Corridors and Agricultural Development in Africa. *Global Food Security* 2 (3): 195-202.

World Bank 2007: Improving Opportunities for Sustainable Youth Employment in Sierra Leone. Draft Report, 28 June 2007. 87-101.

El Cerrejón Coal Mine and the Wayuu in La Guajira, Colombia

Claudia Howald and Karmen Ramírez Boscán[1]

1. Situating the Case Study

The biggest open-pit coal mine in Latin America is located in Colombia, and is called El Cerrejón. Since the mine was established in the 1980s, it has affected the lives of the local population living in the Guajira department in many ways. Many Wayuu, Afro-Colombians and peasant communities that previously lived on the land, where the mining corporation began its activities, were displaced and their way of life has ever since been deeply transformed. Currently, more than 30 years after the mining activities started, corporations have changed, as other companies have entered the Guajira Peninsula. However, displacements and other social and environmental impacts have not stopped. Nonetheless, local people have learnt to organize themselves, to mobilize and attract national and international attention to their claims. Together with support of their allies, they have been able to adopt new legal tools in order to face the mining company and protect their territories.

1.1 History and Current Context

Colombia covers an area of 1,138,910 km^2 and is inhabited by 46.5 million people. This makes Colombia the fifth largest and third-most populated country in Latin America (after Brazil and Mexico). Colombia is a culturally diverse country with several ethnic groups. According to official figures, which are based on how people identify themselves, indigenous peoples make up 3.4%, Afro-Colombians 10.6%, thus leaving 86% of Colombia's population belonging to the "non-ethnic group" identified as *mestizos*[2] (Hörtner 2006: 8; DANE 2005b; 2012).

Unusual for Latin America, Colombia has only a short-lived history of dictatorship. Colombia can indeed look back on an "enduring tradition of democracy". However, it is marked by excessive political violence (Haller et al. 2007: 415; König 2008: 7). The political scenario of the constitutional Republic of Colombia is characterized by two traditional parties, the *Partido Liberal* (PL) and the *Partido Conservador Colombiano*, both of which were established in the middle of the 19th century and ever since have played an important role in parliament. Two civil wars resulted from the bitter rivalry between the Conservative and Liberal parties: The War of a Thousand Days (1899–1903) and *La Violencia* (1946–1957). In July 1957, an alliance between former Conservative President Laureano Gomez (1950–1953) and former Liberal President Alberto Lleras

1 Claudia Howald has an MA in anthropology from the University of Neuchâtel, Karmen Ramírez Boscán is a founding member and consultant on international affairs of Fuerza de Mujeres Wayúu. The opinions expressed in this chapter are their own and are not necessarily consistent with those of the organization.

2 In Colombia, *mestizo* refers to people born of a mixed couple, i.e., Spanish descendants with indigenous people or Spanish descendants with African descendants. The term is more generally used to refer to people who do not belong to any other ethnic group.

Camargo (first presidency 1945–1946) led to the creation of the National Front, which established a power-sharing agreement between the two parties (Hörtner 2006: 62, 83f.). During the period of the National Front, it was first and foremost the liberal presidents Alberto Lleras Camargo (second presidency 1958–1962) and Carlos Lleras Resprepo (1966–1970) who tried to advance Colombia's overall and economic development through reforms. Their main concern, which met with little success, was to stop the enormous rural-urban migration through implementing agrarian reforms (König 2008: 148–151).

The Colombian armed conflict finds its origins in the oligarchic distribution of land and power. The 1960s saw the emergence of the two left–wing guerilla groups, the *Fuerzas Armadas Revolucionarias de Colombia* (FARC) and the *Ejército de Liberación Nacional* (ELN) (Hörtner 2006: 103f.). Until today, the FARC is the largest guerrilla group in Colombia. The history of the armed conflict is complex, involving armed confrontations between the national army, different guerrilla movements, and paramilitary groups that emerged in the 1980s (and united in the Autodefensas Campesinas Unidas de Colombia—AUC in 1997). Today, the conflict is sustained by multiple structural causes linked with economic, territorial, social and political issues (Eraso 2009: 5–8; Hoyos 2006: 2). Over the years, multiple peace initiatives and negotiations involving the Colombian state together with various armed actors have only resulted in isolated demobilizations and short-term solutions (Hoyos 2006: 4–5). The recent formation of neo-paramilitary groups, labelled BACRIM (emerging criminal gangs) proves the failure to demobilize paramilitary structures in the country (ASK 2011). Negotiations with the FARC and the ELN have continued under President Juan Manuel Santos' mandate (since 2010). The armed conflict leaves a trail of an increasing number of civilian victims, who are caught in the crossfire. The figures from CODHES (Consultoría para los Derechos Humanos y el Desplazamiento)[3] show that Colombia has almost five million refugees and internally displaced persons (El Espectador 09.11.2010).

The Constitution of 1991, replacing the one from 1886, is of great importance for Colombia, especially for its indigenous peoples. Politicians from all parties realized that it was necessary to have a new constitution with more participatory and decentralized elements. For the very first time in Colombia's history, indigenous peoples were involved in the process of law making. The 1991 Constitution defines Colombia as a unitary, decentralized, democratic, participatory, pluralistic and social constitutional state (König 2008: 165). The Constitution grants indigenous peoples autonomy over their own land and cultural institutions in the form of establishing *resguardos* (legally recognized indigenous reserves). Black communities (*comunidades negras*) are acknowledged as ethnic minorities and are entitled to collective property rights in the Pacific region. The Constitutional Court was created through the new Constitution and is responsible for interpretation of legislation according to the Constitution (Dalto 2011). However, in spite of constitutional reform, impunity for violations of human rights and political crimes is still rife. There is an evident discrepancy between the rights that exist on paper and the rights that exist in reality and practice due to the inefficiency of the legal system (Hörtner 2006: 169). In many regions of the country, the new legal means for improving the situation of ethnic minorities in the new constitutional framework has "most often only symbolic value" (König 2008: 166). Furthermore, corruption in the public sector is quite a central issue in Colombia.[4]

3 CODHES is a national NGO working on displacement cases since 1985.

4 Colombia is ranked 80 on the Corruption Perception Index (2011), with a score of 3.4 out of a maximum of 10. See: <cpi.transparency.org/cpi2011/results/#CountryResults> accessed on 09.06.2012.

1.2 Geography and Ecosystem

Colombia is located in northwestern South America, bordering Venezuela and Brazil in the east, Ecuador and Peru in the south, the Caribbean Sea in the north, Panama in the northwest and the Pacific Ocean in the west. The Atlantic Caribbean coast has important harbours and is the economically more important of the two coasts, also due to its historical development.

The country is crossed by the Andes relief, which is divided into three ranges: West Andes, Central Andes and East Andes. The most populated area is the Caribbean coast and the Central-Western range, where the main cities have developed, while the peripheries (mostly savannah and rainforest plains as well as mountainous areas) are more sparsely inhabited. This geographical and demographic reality goes hand in hand with the lack of social services, the harshness of the armed conflict and high poverty rates.

Map 1: The Guajira department in northern Colombia

Fig. 1: View on landscape of excavated "overburden". Photo: Stephan Suhner, Arbeitsgruppe Schweiz-Kolumbien.

The resource-rich country encompasses 18 different ecosystemic regions, with 65 different types of ecosystems,[5] stretching over an area of 1.14 million km². Following Brazil, Colombia is the second richest country as measured by biological diversity, which is astonishing, given the fact that it only makes up 0.8% of the earth's land surface. Colombia is home to 45,000–51,000 plant species, i.e. 15% of the world's total plant species, and home to 1752 bird species as well as 583 amphibian species (GEO and GTZ 2001; Grupo Semillas 2008: 8). The term biodiversity has been more and more associated with Colombia during the last few years and is nowadays very present in official, public and academic discourses about the country. This is especially true when it comes to the Amazon and Pacific regions of the country, known for being the places in the world with the highest biodiversity. This discourse stresses the importance of conservation of this "biodiversity", presented as a world heritage. At the same time, it also reflects the idea of biodiversity regions as wealth sources for Colombia, considering a possible biotechnological revolution worldwide (Restrepo 2013).

Colombia is located in the inter-tropical region, which means the greatest part of its territory has humid tropical climate. Nonetheless, the country also includes sub-humid, sub-dry and hot tropical climate (principally in Orinoquia and the Caribbean coastal plains), as well as arid and semiarid areas (Caribbean coast, principally the Guajira peninsula, and in Inter-Andean spots) (Márquez Calle, without year). The Guajira peninsula receives an annual rainfall of just around 1000 mm in the southern part (where the Cerrejón mine is located), and below 500 mm in the northern desert part. Rainfall peaks in September to November, with almost no rain from January

5 However, different ways to categorize ecosystems exist worldwide. For a description of the different Colombian ecosystems see Márquez Calle (without year).

to March. The mean annual temperature is above 27° C, with little variation throughout the year.[6] The main watercourse is the River Ranchería, which flows from the southern, relatively humid part, to the north-western, dry part of the peninsula.

1.3 National Economy and Legal Framework

According to the IMF valuation of country GDP, Colombia's economy is the fifth largest economy in Latin America after Brazil, Mexico, Argentina and Venezuela.[7] In 2011, it was ranked 87 in the Human Development Index.[8] Colombia is a free market economy with major commercial and investment ties. The Santos administration (since 2010) has been promoting Free Trade Agreements (FTA) with the United States and increasingly with other countries as well.[9]

In 1990, the administration of President Cesar Gaviria (1990–1994) initiated economic liberalization or aperture. This entailed tariff reductions, financial deregulation, privatization of state-owned enterprises and adoption of a more liberal foreign exchange rate regime. These policies eased import restrictions and opened up most of the sectors to foreign investment, although agricultural products remained protected. Colombia has considerable mineral and energy resources, especially coal and natural gas. Mining and energy-related investments have grown because of higher oil prices, increased demand, improved output and pro-business reforms. These reforms have significantly liberalized Colombia's petroleum sector, leading to an increase in exploration and production contracts from both large and small hydrocarbon industries (U.S Department of State 2011). Apart from mining, the new policies also favour export-related agribusiness, including oil palm, sugarcane and livestock. Additionally, emeralds, floriculture and textiles are considered as an important part of the national economy. During the last three decades, successive governments have widely promoted a policy of "investor confidence" to ensure private investments, and mining has been the main activity for attracting international capital (Idárraga Franco et al. 2010: 28). Colombia, like many of the countries in the global South, is a major supplier of raw materials for products that are later processed in Northern countries.

The Importance of Coal in Colombia's National Economy

Colombia is progressively becoming one of the most important energy producing countries within Latin America. It has significant mineral reserves, fossil fuel resources and also important "new and clean energy" developments. Out of the total of 99,069 short tons[10] of coal produced in Central and South America, Colombia produces 81,957 short tons, placing the country as the largest coal producer among Latin American countries and 13th in the world (U.S. Energy Information Administration 2012). However, its role as an exporter is much more significant: it is ranked 4th in the world in 2009, after Australia, Indonesia and Russia, and ahead of South Africa.

With measured reserves of coal at more than seven billion tons and with 73 million short tons being mined in 2010, Colombia has seen a significant expansion of its coal industry. The major coal producers are El Cerrejón, Drummond and Prodeco, companies that are expecting to double

6 Instituto de Hidrología, Meteorología y Estudios Ambientales (IDEAM): Atlas Climatológico de Colombia. <www.siac.gov.co> accessed on 10.04.2014.

7 IMF 2011; World Bank 2011: GDP Ranking. <data.worldbank.org/data-catalog/GDP-ranking-table> accessed on 25.06.2012.

8 UNDP 2011: Human Development Index (HDI)—2011 Rankings. In: Human Development Reports – UNDP. <hdr.undp.org/en/statistics> accessed on 25.07.2012.

9 Among these, Colombia signed a FTA with Switzerland in 2009, which entered into force on 1 July 2011.

10 The short ton is a unit of mass equal to 2,000 pounds (907.2kg).

their production and export capacity in the coming years (World Coal Association 2010: 7). These numbers seem to indicate that Colombia is a very rich country with high growth rates and a stable economy. However, the poverty rate is at 45.5% according to official numbers, which means that more than 20 million Colombians are poor (DANE 2010). The Guajira department ranks third in terms of its NBI (*Necesidades Básicas Insatisfechas*, Unsatisfied Basic Needs) with a high figure of 65.1%, with the departments of Chocó and Vichada ahead of it with 79.2% and 66.8% respectively (DANE 2005c).

Legal Framework for Mining

Regarding the legal framework, mining is regulated by the *Código de Minas* (Mining Law, Law 685) dating from 2001. This law creates a specific policy regarding mining at the national level. Before the advent of multinational mining corporations, mining activities were characterized by subsistence and small-scale mining (mainly by indigenous peoples, Afro-Colombians and unemployed, migrant miners) and medium-scale mining (developed by medium-sized entrepreneurs). The law, framed with the aid of Canadian consultants, facilitates the concession of licenses for exploration and exploitation, thus enabling competition with other mining countries in South America. Critics point out that mining corporations heavily influenced the process of formulating the law. Moreover, no consultation with indigenous communities took place when the national legislation on mining was framed, in violation of Law 21 from 1991 that establishes previous consultation according to the ILO Convention 169 in the Constitution (Idárraga Franco et al. 2010: 37). Due to this and additional reasons, the Constitutional Court declared the law unenforceable, in May 2011. However, it suspended this decision's effect for the next two years, so that the law remained legally effective while it had to be modified in the National Congress (El Tiempo 11.05.2011). Another fundamental point regarding mining in Colombia is the creation of the *Plan Nacional para el Desarrollo Minero Visión 2019* (National Plan for Mining Development Vision 2019) in 2005, which envisages the development of Colombia into a mining country. Among its objectives is the doubling of coal production, quadrupling the production of precious metals and augmenting the level of investments in the country by privileging mainly private investments. According to these goals, mining licenses increased significantly in the last few years (Idárraga Franco et al. 2010: 39f.).

1.4 Local Population: the Wayuu

According to official figures of the last national census of 2005, indigenous peoples make up 3.43% of Colombia's 47 million people, i.e., 1,378,884 individuals belonging to 87 different indigenous groups (DANE 2005a: 15; IWGIA 2011: 120). Indigenous peoples live in all departments of the country, among them La Guajira being one of the departments with the highest percentage of indigenous population. Official figures count a total population of 655,943 inhabitants in La Guajira (DANE 2005a: 95).[11] 48% of them self-identify as indigenous peoples, 32% affirm not belonging to any ethnic group and 14% identify themselves as Afro-Colombians (ibid.).[12] The inhabitants who do not identify themselves as belonging to any ethnic group are generally considered and identified as being *mestizos*. Almost a third of the national territory is formed of indigenous reserves, though a large part of them are occupied by different kinds of companies, specializing

11 The Departments of La Guajira, Cauca, Nariño, Córdoba and Sucre account for approximately 65% of Colombia's indigenous population (DANE 2005b: 29).

12 Figures from the national census of 2005 (DANE 2005a) are not clear with regard to the remaining 6% of La Guajira population.

Fig. 2: Wayuu women in the old Tamaquito community. Photo: Stephan Suhner, ASK.

in extractive economic activities (oil companies, mining companies, banana and palm oil monocultures, etc.) (IWGIA 2011: 120). The vast majority of indigenous peoples living in La Guajira are Wayuu. The Wayuu number 280,000, making them the most numerous indigenous group in Colombia (CNRR 2010: 39; Fuerza de Mujeres Wayuu 2011). La Guajira has 25 *resguardos indígenas*, mostly Wayuu.[13] Wayuu territories are among the most affected by extractive projects of different kinds.

Wayuu History

The literature on the Wayuu often highlights their historical resistance to Spanish domination and the violence occurring in the region. Various authors even assert that the Wayuu people maintained control over their territory and its economic activities until the entry of the El Cerrejón coal mine (Puerta Silva 2010: 156). The historical perspective on the relations of the Wayuu with other social actors since the *Conquista* (the Spanish conquest of America) is central for understanding the transformations and the mechanisms of sociocultural reproduction of Wayuu society.

According to anthropological studies, the Wayuu are an indigenous group of Arawak origin that migrated from the Amazonian basin and arrived in the Guajira region around 150 BC. This has been retraced through the study of *wayuunaiki*, the language spoken by the Wayuu, which belongs to the Arawak linguistic family. With their arrival, the Wayuu gradually took over the land and displaced other groups inhabiting the area. Since the 18th century, they controlled the

13 Here, we are only addressing the part of the La Guajira peninsula belonging to Colombia, not the part in neighbouring Venezuela, where also Indigenous peoples like the Wayuu are living. Haller et al. (2007: 235–282) address the Wayuu communities in Venezuela and their struggle against oil companies.

major part of the peninsula (Puerta Silva 2009: 13; Villalba Hernández 2008: 46). The Guajira Peninsula has been known since the beginning of the Spanish conquest for its legendary Cabo de Vela ("Cape of Sails"), one of the first points of reference concerning land in the Americas (ibid.: 47). To the Spanish authorities of the 18th century, the *indios guajiros* were literally an enemy nation, and their continued resistance meant that the region had not yet been incorporated into the Nuevo Reino de Granada. La Guajira was perceived by the conquerors as a wide empty desert space, a region beyond their control, and in which they had no interest. During the 18th century, as a result of the Spanish imperial modernization, interest in the region became stronger and it was finally incorporated under Spanish jurisdiction.

An important factor in the relations with the colonial powers was the presence of English, Dutch and French influences in the Caribbean. The Wayuu people interacted with these colonial powers through trade and exchange in the Caribbean Sea. This helped the Wayuu in maintaining control over their territory, since the English and Dutch provided them with horses and firearms that they could use against other colonial powers. The Wayuu were among the few indigenous groups in Latin America in possession of, and with the capacity to use, firearms. Though the Wayuu resisted colonization, they gradually lost vast areas of agricultural land and hunting areas, losses that were replaced by the introduction of herding. Indeed, goats became increasingly important for both trade and subsistence economy (ibid.: 48).

Social Organization

According to Claudia Puerta Silva, the principal characteristics of Wayuu society are the dominant matrilineal system, the poly-residential nature of their living, the polygyny and their strong sense of belonging to the territory where their ancestors were buried (Puerta Silva 2009: 13). The Wayuu are composed of different clans and lack a central authority. Ideally, the clans or kin groups are clustered and configure a locality composed of different *rancherías*, the domestic unities. The *rancherías* are usually composed of various houses and kitchens, a canopy, a fence for the animals and a vegetable garden. Nearby a *jagüey* (a water reservoir) and a cemetery are located. Each of these localities acts autonomously in its decisions and actions, the political system being decentralized (CNRR 2010: 40; Puerta Silva 2009: 14).

Already the Spanish referred to the weak leadership among the Wayuu when describing the *cacicazgos* (headmanship) in the colonial period. Authority is lost when prestige and wealth are lost (Puerta Silva 2010: 170). Status and social prestige are attained through participation in the dynamics of political and economic competition. Prestige in this sense is strictly linked to economic wealth and to potential alliances. Central to Wayuu social organization are the alliances, since they serve as socio-political regulators. Alliances can take the form of marriages and political coalitions. They are forged through gifts, services or the payment of bride price in the case of a marriage (ibid.: 17). Among the Wayuu, traditional authority is exerted by the *alaula*, the maternal uncle of the clan. This becomes visible in conflict resolution and maintaining social order. Another central figure is the *pütchuipü*, a traditional mediator also called *palabrero* because of his role as spokesman between the various *alaula*. A third central figure is represented by the *outsu*, who is generally a woman mediating with natural and supernatural entities in order to maintain harmony and health (Fuerza de Mujeres Wayuu 2011; Puerta Silva 2009: 16).

Mediation is a central part of the Wayuu social organization and its reproduction. Some anthropologists refer to *corredores*, intermediaries who built bridges between the *alijuna* and the Wayuu people (in *wayuunaiki, alijuna* is a term used to refer to people who are not indigenous). Intermediation has been rather focused on individuals, through representatives of clans or sectors of them. The role of such intermediaries has become increasingly important. Practices and discourses were improved and are now influencing decisions about the future of their communities. This evolution did not take place through an organizational process that was consciously collective, but rather individually, sharing with other individuals (Puerta Silva 2010: 168). Very few events generated mobilizations involving the entire community of the Wayuu people or at least a significant portion of them. The only instance of centralized organization to be mentioned is Yanama, a Wayuu organization defining social programs for the indigenous group, existing since the start of the El Cerrejón mining project. Many authors see the absence of centralized organizations as one of the main reasons why the Wayuu did not at first mobilize against the mining project on their territory. In other words, the lack of organizations and the internal divisions of the indigenous people was the cause for the disadvantageous agreements regarding land and compensations (ibid.: 169).

During recent years, important changes concern the role of women. Traditionally, women were more linked to the world of the dead through activities accomplished in the domain of observing the collective mourning of a community. Today, they are increasingly responsible for administrative and political aspects and are active in the mediation between Wayuu communities and other non-indigenous actors. The entire Wayuu society, and women in particular, is a victim of the armed conflict taking place in Colombia (Cortés Severino 2009: 220). Threats, homicides, massacres, and other human rights violations perpetrated by the armed actors cause the displacement of numerous Wayuu. The situation generates a climate of general terror and distrust among the population of La Guajira. Various Wayuu organizations and leaders (including women leaders) are denouncing this situation and try to attract national and international attention to it. Public institutions have largely refused to recognize the impacts of the armed conflict, especially the violence engendered by the paramilitaries, and rather attribute the violence in the region to traditional clan conflicts (Villalba Hernández 2008: 61–62).

1.5 Local Economy and Resource Use

The Wayuu were labelled by the Colombian society and by anthropologists either as transhumant or as seasonal pastoralists. Today, some of them are pastoralists, mainly possessing goats. Other traditional activities are fishing, which is still significant at present, and horticulture, the extraction of products such as salt and plaster or the production of charcoal. These productive activities are linked to membership of certain clans. Furthermore, trading activities play a very important socioeconomic role. As mentioned above, it is through trade and exchanges that the Wayuu had established strategic relations with non-indigenous people like French and Dutch traders at the time of the Spanish colonization. A short view of the history of the Guajira Peninsula reveals the constant presence of extractive economic activities that left few benefits for the inhabitants. Among those activities, scholars mention pearl extraction out of the Caribbean during the first centuries of Spanish domination; the exploitation of the *palo de brasil* (also called "brazil wood", used as red dye in textile manufacture) during the first years of the Republic; the *bonanza marimbera* ("marihuana prosperity") during the 1970s; and the more recent construction of the coal extraction complex in the 1980s that stretches up to the present. The planning of future projects in the region, for example, the Trans-Caribbean gas pipeline project, highlights a historical continuity in the economic dynamics of La Guajira. Despite the presence of such projects and economic activities, relatively few Wayuu dedicate themselves to seasonal or salaried work (Puerta Silva 2009: 13; Villalba Hernández 2008: 46).

The Wayuu communities are characterized by a profound attachment to the territory, indicating thus the centrality of the land in Wayuu society (Dover et al. 1997; Villalba Hernández 2008). In their worldview, everything around them, rocks, minerals, winds, etc., are living beings. The Wayuu speak of the "Great Mother" when referring to earth. They consider themselves to be "sons and daughters of the Earth" and the sole owners of everything on it. According to this worldview, water is a common good shared by all living beings. Water sources are rather scarce in Wayuu territories, which are mostly semi-desert savannahs (Conversation with a Wayuu leader, November 2011). There is only one permanent river, called Ranchería, and other smaller temporary rivers and watercourses. The Wayuu have developed their own systems of storing water and each locality is provided with *jagüeyes*, water reservoirs (Puerta Silva 2009: 13).

Access to land and sea is regulated by the clan system. Regarding land, property is assigned according to the burial place of the clan's ancestors. For the Wayuu, the dead people live next to them, since in their worldview, the life cycle does not end with death. Dead Wayuu are symbolically part of the conformation of the territory where they once lived, the link to the dead people thus becoming an integral part of their relation with the territory. As a Wayuu leader expresses it: "We Wayuu are from where our dead people are" (Cited in Cortés Severino 2009: 220, own translation). In this sense, cemeteries represent the materialization of their sacred deeds. The point is fundamental in order to understand the deep impact that economic projects such as the El Cerrejón mine complex can cause on Wayuu society. The issue of the cemeteries and their removal to new locations became crucial in the negotiations with mining companies (Montero de Daza/Mestra Narváez 2008: 148). Not only extractive activities, but also the armed conflict and the consequent displacements are further causes of social destabilization (Cortés Severino 2009; Ramírez Boscán 2007).

Legislation Regarding the Protection of Indigenous Peoples' Territories

The department of La Guajira is divided into areas with different legal status; among these are *resguardos* and *reservas indígenas*. The process of creation of *resguardos* legally started with Law 89 from 1890 and is implemented through Law 135 from 1961, which aimed at the integration of the indigenous communities into Colombian society on the one hand, and the preservation of their territorial and cultural patrimony on the other hand. Through this legal framework, several *resguardos* were constituted on national territory, and also in the Guajira Peninsula, where the first Wayuu *resguardo* was created in 1967. The *resguardos* are indigenous territories with collective land titles and their own legal jurisdiction. The 1991 Constitution protects ethnic and cultural rights and treats the *resguardos* as territories that are inalienable, imprescriptible and protected against seizure. *Resguardos* are considered as municipalities and, according to Article 330, indigenous communities participate in decisions related to natural resources on their territories. More specifically, the *resguardos* are regulated through Decree 2164 from 1995: "The resguardos are a special form of legal and socio-political institution conformed by one or more indigenous communities which own their territory through a collective property title and govern themselves [...] through an autonomous organization [...] and its own normative system" (Cited in Flórez Lopez/ Millán Echeverría 2007: 53, *own translation*).

In 1984, an important part of the peninsula was recognized as Resguardo de la Alta y Media Guajira, covering an area of 1,000,000 ha. The remaining land titles granted to indigenous communities all concern smaller portions of land. There are actually up to 25 legally constituted *resguardos* in the Guajira department, but many Wayuu live outside these territories and thus do not benefit from any legal protection. When Wayuu people speak about their territory, they usually refer to the ancestral Wayuu territory, which is larger than the sum of *resguardos* currently acknowledged (Villalba Hernández 2008: 57).

The Colombian legal framework regarding the ethnic and territorial rights of indigenous peoples is a further tool to be used by the indigenous communities. The Constitutional Court of Colombia has repeatedly referred to the close relation that indigenous peoples have with land and natural resources as being a "relation of vital dependency". Thus, indigenous territories and collective land ownership are essential for the survival, the worldview and the spirituality of indigenous peoples.[14]

14 Many sentences of the Court referring to territorial rights and natural resources are in favour of indigenous people (ST

In addition to national laws, Colombia has ratified all international agreements and conventions on human rights and on indigenous peoples' rights. Colombia ratified the 169th Convention of the International Labour Organization (ILO) with the inclusion of Law 21 under the ambit of the Constitution of 1991. This convention contains the basic principles for the rights of indigenous peoples over their territories and the responsibility of the state to respect them. One important point of this law regarding the exploitation of natural resources is the prior consulting procedures with the indigenous communities. However, different interpretations of this right for consultation exist. At the international level, indigenous peoples from Colombia are struggling for the inclusion of the Free, Prior and Informed Consent (FPIC) in interventions and decisions regarding their communities. Such an interpretation goes even further than the procedures of prior consultation actually recognized by the state and the corporations.

The 1991 Constitution also recognized the rights of other ethnic groups, such as the Afro-American population of Colombia; e.g., with Law 70 from 1993. The law legally recognizes the existence of Afro-Colombian communities through the constitution of *Consejos Comunitarios* (community councils) and assigns them collective land titles. The Afro-Colombian legislation has thus become somewhat similar to the indigenous legislation (Flórez Lopez/Millán Echeverría 2007: 83). Afro-Colombian people have been living in the Guajira Peninsula for some centuries, earning their living mainly by farming (Idárraga Franco et al. 2010: 76). On the contrary, Afro-descendants[15] inhabiting La Guajira are not legally constituted in *Consejos Comunitarios*, thus they are not legally recognized as Afro-Colombian communities with their own specific rights and their own collective territory (Conversation with Swiss NGO activist, December 2011).

428 of 1992; ST 567 of 1992; ST 188 of 1993; ST 257 of 1993; ST 380 of 1993; ST 405 of 1993; ST 001 of 1994; C-519 of 1994; T-342 of 1994; C-139 of 1996) and become an additional tool for the protection of their rights (Flórez Lopez/Millán Echeverría 2007: 60).

15 In Colombia different terms are used to speak about blackness (Wade 2009): black people, black community, Afro-Colombian, Afro-descendant. Each term has a history and is linked to different representations. Generally, the terms used before the 1990s were more ambiguous, while right after 1991 the term "black community" got hegemonic. This term is strictly linked to the Law 70 from 1993 and it refers mainly to the rural areas situated in the Pacific region (Flórez López/Millán Echeverría 2007: 73). This "ruralized" representation is disputed by the urban black movements, whose main issue is racism, and who prefer the terms "Afro-Colombian" and "Afro-descendant" as more inclusive and centred on the "African heritage" and on the idea of mestizaje (Wade 2009: 169; Flórez López/Millán Echeverría 2007: 72). In daily life, however, these terms are used as synonyms, showing a great flexibility. In this text, we will principally use the term "Afro-Colombian" in order to avoid the essentialization inherent in the notions "black people" or "black communities". These terms are used here to refer to black communities as defined in Law 70.

2. The El Cerrejón Mine

The El Cerrejón mine complex is the largest of eight major economic projects implemented in Wayuu territory. All of these projects are related to the exploitation of natural resources (water, coal, gas, wind, etc.). Some of them, like the Jepirrachi Wind Park or the Hydroelectric Complex on the Ranchería River, are related to the Cerrejón mine, for which they generate electricity (Idárraga Franco et al. 2010: 82). The El Cerrejón mine has its origins in the wake of energy crisis of 1973, when high oil prices favoured a reorientation of national and international markets towards coal. Coal exploitation in Colombia can be divided into 3 different phases: The first phase started in 1976 with an exploitation contract between the state-owned company Carbocol (Carbones de Colombia) and Intercor (a subsidiary of Exxon Petroleum Company). Infrastructure construction started in 1981, production in 1986. The second phase of transition is characterized by the selling of Carbocol in 2000 and Intercor in 2002. This was carried out under advice from the International Monetary Fund and resulted in the entrance of new multinational corporations into Colombia. In 2002, Carbones del Cerrejón Limited was created, marking the beginning of the third phase that has lasted until the present. This company belongs in equal parts (33.3% each) to three big transnational corporations: Xstrata plc (Switzerland), BHP Billiton plc (Australia) and Anglo American plc (South Africa).

Fig. 3: View on part of the El Cerrejón mine. Photo: Stephan Suhner, ASK.

Xstrata shares are traded in the stock market of London. The company is among the major producers of nickel and copper, and is the largest coal exporter on the international market.[16] BHP Billiton's stock is traded on the markets of London and Melbourne, its activities include extraction of iron ore, petroleum, coal and other products such as diamonds. Anglo American operates mainly in South Africa, Australia and in the Americas. The company focuses on the extraction of coal and metals such as copper and iron ore.[17]

The extensive area of the mine is divided into five extraction areas: *Zona Norte, Zona Central, área de Patilla, Oreganal* and *Zona Sur*,[18] located in territories within the jurisdiction of the municipalities of Albania, Barrancas, Uribia, Manaure and Maicao. In 2010, the mined area covered 11,488 ha, with reserves reaching 69,000 ha. Production was around 30 million short tons of coal a year, which contributed 41% of coal production in Colombia. The net operating income of El Cerrejón in 2010 was of 4,471,605 million Colombian pesos (about USD 2,520 million) (Cerrejón 2010: 9). The agreements signed between the Colombian government and the corporation regulate the concession of the mine until 2034 (Fuerza de Mujeres Wayuu 2011). The extraction areas are connected by railways to the company's port in Puerto Bolívar, 150 km from the mine, which is the most important coal port in South America, having enormous capacities for shipping and stocking. The train is exclusively meant for the transport of coal from the company. On average, 10 trains a day with up to 120 uncovered wagons transport up to 12,000 short tons of coal. The whole operation cycle of each train (loading, transporting to the port and return to the mine) is 12 hours.[19] In addition to the railways, a road was constructed to connect the mine to the port. Both cross Wayuu territories and *resguardos* and represent important changes in the landscape of the Guajira Peninsula. Also the port was constructed in a Wayuu territory called *Media Luna* (Puerta Silva 2009: 133f.).

Currently, new mining concessions are being granted by the state throughout the national territory, including the Guajira Peninsula. The further expansion of the mining area of El Cerrejón and the entrance of new companies in Wayuu territories are foreseen, such as the gas exploration activities of Pacific Rubiales Energy (Canada) and the construction of railways and a port by MPX (Brazil) (El Informador 10.05.2011).[20]

2.1 Environmental and Social Impacts

In more than 30 years of coal mining operations in Wayuu territory, not only the Wayuu people and their organizations but also Afro-Colombians, peasants and *mestizos* have been trying to cope with the complex impacts of mining on health, social organization, culture and environment. In this section, the environmental and social impacts of El Cerrejón are first presented according to the company's own terms, including the measures taken to reduce them. Then additional third party sources are considered, and finally this is contrasted with a Wayuu "inside perspective".[21]

16 In 2013, Xstrata was acquired by Glencore. See also case studies Las Bambas, Tampakan and Mopani.

17 Cerrejón: Shareholders <www.cerrejon.com> accessed on 09.06.2012.

18 Cerrejón: Mina. <www.cerrejon.com> accessed on 06.06.2012.

19 Cerrejón: Ferrocarril. <www.cerrejon.com> accessed on 06.06.2012.

20 Map of Properties. In: Pacific Rubiales Energy. <www.pacificrubiales.com/map-of-properties-construccion.html> accessed on 11.01.2012; Statement by tribes and communities affected by mining mega-projects in La Guajira, Colombia. In: RECLAME. (20 November). <www.reclamecolombia.org> accessed on 11.01.2012.

21 Based on direct experience and conversations with community members by co-author Karmen Ramírez Boscán.

2.1.1 Environmental Impacts: El Cerrejón's Environmental Programs

> "At Cerrejón we prevent, mitigate, and compensate for the environmental impacts of the mining, conserve biodiversity, and apply the best operational standards in the industry" (Cerrejón 2012)

According to the company, environmental management is a fundamental component of its sustainability strategy. Special programs have been established in order to ensure compliance with the legal framework (Idárraga Franco et al. 2010: 28). In its sustainability report of 2010, El Cerrejón advertises its "eco-efficient initiatives", specifying that these mean "producing more using less". Moreover, the chapter on "environmental management" promotes the prevention or mitigation of climate change and the conservation of biodiversity, based on the application of the best operational standards in the coal industry, in order to secure a "socially and environmentally responsible mining operation". The main environmental programs of El Cerrejón project are related to air-quality management, quality and efficient use of water, land claims, management of wildlife and biodiversity conservation, management of solid waste as well as climate change mitigation and eco-efficiency (ibid.: 72). For example, the program called "relations with stakeholders"[22] promotes environmental responsibility among mine workers through the "Education and Implementation of Environmental Management Programs", and communities are involved in the "Air Quality Oversight Committee", where they are responsible for the control of the air quality monitoring system in the sites where measuring devices are installed.

El Cerrejón also emphasizes the alliance with the environmental NGO Conservation International (CI). CI is working on issues related to "conservation, communication, and research into biological diversity within the area of influence and in the department of La Guajira in general" (ibid.: 73). El Cerrejón, in this strategic alliance with CI and with the participation of Corpoguajira[23], is promoting the establishment of protected areas, such as the Natural Park in Bahía Portete (located in a vast area of the Resguardo Wayuu de la Alta y Media Guajira) and in the Montes de Oca (a major mountain range in the South of the department).[24] The company claims that it acts according to international initiatives and standards. Explicitly mentioned are: the Universal Declaration of Human Rights; fundamental Conventions of the International Labour Organization (ILO); Social and Environmental Performance Guides 1, 5 and 7 of the International Finance Corporation (IFC) of the World Bank; the Voluntary Principles on Security and Human Rights; the United Nations' "Protect, Respect, and Remedy" and "Global Compact" frameworks; the Global Reporting Initiative; and the Sustainable Development Framework of the International Council on Mining and Metals (ICMM) (Cerrejón 2010: 28). Furthermore, El Cerrejón has received the Lazos Award 2010 for its environmental responsibility (ibid.: 72).

22 The Spanish name of the program, *relacionamiento con las comunidades* ("community relations") sounds more problematic from the Wayuu point of view: they see it as a strategy to gain the trust of the communities to negotiate favorable agreements for the company, while not guaranteeing their rights (Conversation with a Wayuu leader, February 2012).

23 Corpoguajira, Corporación Autónoma Regional de La Guajira (Regional Autonomous Corporation of La Guajira) is the environmental authority of La Guajira Department.

24 It is important to mention that CI has been criticized by indigenous communities in other parts of the world for promoting conflicts related to the privatization of nature and for plundering biodiversity; for example, in Papua New Guinea (Choudry 2003).

Fig. 4: Advertisement of a nature recovery program by El Cerrejón. Photo: Stephan Suhner, ASK.

Contrasting Cerrejón's Environmental Programs with Other Information

Most public information regarding its environmental impacts has been provided or influenced by El Cerrejón. Like other mining companies, El Cerrejón promotes studies about the risks of natural resource extraction. Companies thereby promote themselves with slogans like "green mining" or "green quarries" (Idárraga Franco et al. 2010: 7), although mining is inherently unsustainable. Information regarding the environmental impacts coming from sources other than the company itself contradict the self-presented image of El Cerrejón. Compared to agricultural activities where land can be used for generations, it can only be used once for mining. The yearly extraction of 30 to 35 millions of short tons of coal accelerates the growth of the size of the open mine pit, now at 12,000 hectares, with the whole deposit spread over 69,000 hectares. Mining, transport and shipping are done without adequate protection, which produces large quantities of harmful coal dust that affects air, soil and especially water quality vital for human beings, animals and plants. Affected water resources are the Ranchería River, the traditional water reservoirs (*jagüeyes*), and even the ocean coastline (ibid.: 72). The huge size of the open pits also affects ground- and surface water flows—most obviously with the plan to change the course of the Ranchería River over 26 kilometres to access the 600 million tons of coal that lie beneath its original route (see section 2.2).

Up to the time of writing, no exhaustive and independent environmental impact assessments have been carried out. Information provided by NGO researchers are only approximations and lack detailed measurements. Socio-environmental impacts of mining in El Cerrejón have been approached in a study by a Colombian civil engineer working on environmental issues, Jaime Salas Bahamón, who had previously coordinated the implementation of the Environmental Management Plan of El Cerrejón in 2002–2003. According to the study, the following environmental problems caused by El Cerrejón have reduced fauna and flora numbers, affect the communities and agricultural production and cause emigration from of the region (Salas Bahamón 2004: 11): Dust emissions are produced by transportation on the roads of the mine. Although the records of metering devices do not exceed national and international standards, the population is highly affected. Most affected are the people living in the Wayuu *resguardos* of Provincial and San Francisco that are located very close to the operating area, those in Media Luna community located in the Resguardo de la Alta y Media Guajira, nearby the coal port of Puerto Bolivar, and those in Barrancas and Hato Nuevo municipalities.[25] Moreover, in open-pit mines coal bed methane can burn spontaneously, which generates gases such as sulphur oxides, nitrogen oxides and other gas emissions which lead to acid rain. Noise and vibrations are caused by equipment and the explosives used in the mining operations. The vibrations even lead to cracks in houses. This is perceived as having a negative impact especially on the well-being of the people living in the *resguardo indígena* of Provincial and in the municipality of Albania.

Environmental Impacts Perceived by Wayuu Women

> "Wayuu territory is located in arid, in some regions, desert ecosystems. Pastoralism, fishing and commerce are part of the most important activities of the Wayuu people. (...) According to the worldview of the Wayuu people, all the components of nature are endowed with life and in that sense, a very precise distinction between living and inert elements of nature is not possible. Hence, rocks, streams, wind, rain... have spirit and are also considered alive, which means that they are components of biodiversity" (Ramírez Boscán 2008, *own translation*).

In June 2011, the indigenous organization Wayuu Women's Force (Fuerza de Mujeres Wayuu) organized a workshop with the aim of analysing environmental impacts from the Wayuu women's perspective.[26] From a cultural perspective, Wayuu indigenous women concluded that everything is alive and that *Wounmainkat* ("Mother Earth") is the biggest woman that gave life to all humanity. In this sense, all operations carried out to exploit natural resources constitute an attack and even a rape against *Wounmainkat*. Women participating in the workshop recognized that every human action, and in this case, every action of the multinational companies, produces multiple collateral effects on the natural and social environment. In this workshop, the women identified and classified some of the environmental impacts, including air pollution, deforestation, loss of flora and fauna, impacts on water and rainfall, noise pollution, ground vibration and climate change, which were mostly seen as persistent and irreversible.

Wayuu women have openly questioned the El Cerrejón Land Rehabilitation Program, which is promoting the rehabilitation of over 2,800 hectares of land (Cerrejón 07.12.2011). First of all, traditional land was occupied by the company without approval and consensus of its original inhabitants. Furthermore, a model of devastation of "Mother Earth" was imposed severely disrupting Wayuu social order. Thirdly, Wayuu women concluded that the rehabilitation of 2,800 hectares will never compensate for the damage done through the exploitation of more than 69,000 hectares. The recovery of the ecosystem is very slow, and a complete regeneration is almost im-

25 According to Curiel Arismendy, these dust emissions are a cause of many of the diseases found among the local communities, such as acute respiratory infections, parasitic diseases, acute diarrhea or skin diseases (2010: 8).

26 In La Guajira, there are many Wayuu organizations; e.g., traditional authorities' associations or foundations. This section is based on data produced during the mentioned workshop in June 2011 with only one Wayuu organization. However, because its members come from many different communities, it can nonetheless serve as an accurate summary.

possible. Wayuu people compare the area where the El Cerrejón operations take place with a sick woman who will never fully regain her health. Recovery programs promoted by El Cerrejón cannot outweigh the damage, destruction and devastation caused through its operations (Wayuu women workshop, June 2011). According to conversations with indigenous leaders, apart from the land ecosystem, sea resources are also affected by the mining activities. Wayuu communities living on the coast now have restricted mobility at sea and had to reduce their fishing activities. Leaders have indicated that, at present, many fish species are very difficult to find due to the pollution produced by operations in Puerto Bolivar. This impact has not yet been widely documented.

2.1.2 Social Impacts: El Cerrejón's Social Programs

The El Cerrejón mine causes various negative social impacts. The company asserts that "social responsibility" and the respect of human rights are a top priority, especially after the subscription to the United Nation Global Compact in 2008. However, national and international organizations supporting local communities have documented ongoing human rights violations (Ramírez Boscán 2006).

In its sustainability report of 2010, El Cerrejón explains what it understands as "Relations with Stakeholders". The company decides about inclusion or exclusion of people in the category of "stakeholders" depending on their relation with the mining activities, their needs and expectations. National government, departmental and local authorities, educational and healthcare institutions, local communities, employees and trade unions, mining sector companies, suppliers and contractors, NGOs, regional, national and international media and opinion leaders are all considered to be part of this stakeholder approach. The objective of these relationships, "[...] based on trust and transparency", is the mitigation of social impacts and the respect of human rights, according to El Cerrejón (2010: 31). Several activities and special programs are organized for the communities located in the area of influence on a regular basis, in order to ensure and inspire confidence in this kind of "relationship". Especially targeted are the most affected communities of the Wayuu (234,829 Wayuu people living in 253 communities and 226 *rancherías*) and other communities whose population is mainly Afro-Colombian, like Oreganal, Tabaco and Los Remedios. Regarding human rights, El Cerrejón developed an approach that aims at ensuring the rights of workers, contractors and communities, by strengthening the implementation of the Voluntary Principles on Security and Human Rights. A special complaint procedure was introduced as a pilot program, whereby the Complaints Office follows the principles and recommendations of John Ruggie, United Nations Special Representative for Business and Human Rights (ibid.: 51).

Discrepancies in El Cerrejón's Social Programs

According to Salas Bahamón (2004), there are several negative social impacts of the El Cerrejón mine that are easy to identify:

Resettlement: Due to further expansion of the mine pits, some communities have been relocated. Some of them were forcibly resettled, like Tabaco in 2001 (ibid.: 10). The sudden changes evidently caused social and cultural instability in communities where the construction of new houses and the payment of compensation does not compensate for the loss of their traditional attachment to the land.

Employment: It is estimated that only 1% of the people employed in the mine are indigenous, though they make up one-third of the population of the department (ibid.). Although El Cerrejón promotes special training programs to train staff at their convenience, most of the Wayuu men and women aspiring to work in the mine are said to be unqualified for higher positions. The indigenous people who are employed by the mine mainly work in unskilled jobs. Among the few professional Wayuu working with El Cerrejón, most are given the task of approaching local communities in order to promote the "Relations with Stakeholders" program, which results in stigmatization from the cultural and social perspective of many Wayuu (Ramírez Boscán 2006).

Immigration: People from the region, as well as displaced families from other parts of the country, have been moving to communities that have to be resettled, hoping to get a share of the compensation paid by El Cerrejón. This is particularly true for non-indigenous communities such as Patilla and Chancleta. The situation creates problems including an increased demand for public services, prostitution, insecurity and poor waste management (Salas Bahamón 2004: 10).

Impacts on Wayuu Communities

> "The damages caused by the displacement of our people are very few compared to what occurred with our dead when their cemeteries, their places of rest and preparation for the passage from this life to Jepirra were violated. Wounmainkat is crying."
> (Words of a Wayuu woman, 2011; *own translation*).

Through information based on interviews and personal communication with Wayuu people, we would like to point out some additional social impacts that have been neglected so far, especially concerning the damage to the territory, social disruption and health.

Damage to territory: Chomsky et al. consider that the primary negative impacts of the mining project on affected indigenous people are related to the loss of ancestral lands (2007: 53). This has caused irreparable changes not only in the environment but also in the social structures of the Wayuu people. At the beginning of the project, over 29,000 hectares were expropriated from the indigenous people by declaring it as being in "public interest". Agricultural and grazing lands were undervalued and houses and cemeteries simply removed from the places where they had been located since time immemorial. On the other hand, free mobility within the territory was restricted, due to the vigilance that has been mounted over the entire operation that includes the mine, railways and port. This is still continuing: according to testimonies of Wayuu people from Media Luna, to carry out the expansion of the port at Puerto Bolivar, El Cerrejón now intends to appropriate Wayuu territory that is under *resguardos* jurisdiction. It was left unoccupied by Wayuu communities because of the excessive pollution there. However, Wayuu inhabitants of Media Luna have firmly stated that they will not cede their territory, even if it is currently unoccupied.

Social disruption and conflicts: The high expectations of development generated by El Cerrejón and the strategy of individual negotiation with Wayuu communities have resulted in internal divisions of an indigenous people's structure that was once the basis to resolve conflicts. To give only one recent example: Testimonials of Wayuu people have reported a case occurring in the area known as "4-Vías de Maicao", where El Cerrejón, together with the local government, is planning to build a modern shopping centre in the middle of the communities' land. With this project, the company would displace Wayuu women selling food at the traditional market located next to the railway line. The land is owned by a Wayuu family and inhabited by eight more families, all living in one house and a ranch. Additionally, since 2007, the seat of Fuerza de Mujeres Wayuu is located there. The place is thus a centre for training activities and organization of protest demonstrations. The expectations of compensation promised by El Cerrejón are causing strong divisions, with some Wayuu families wanting to displace other residents in order to receive the compensation.

Health: Impacts on people's health have not been studied in depth. However, mining activities are known to affect the skin, respiratory system and eyesight apart from causing other diseases, affecting particularly children and elderly people. The only independent information about the mine's health impact was published in 2007 in the book *Bajo el Manto del Carbón* (Chomsky et al. 2007). Data is based on interviews with members of affected communities carried out by medical experts. The main pathological symptoms included respiratory problems, abdominal pain, diarrhea, fatigue, headache, skin problems, eye irritation and permanent cough. In addition, many people complained about a foul-smelling cloud that descends on communities during the rainy season. This "toxic cloud" is the result of coal combustion, which produces gases such as sulphur dioxide, hydrogen sulphate, bisulphate, carbon monoxide and volatile organic compounds (ibid.: 114, 123). A medical expert involved in the study, asserts "increases in mortality and morbidity in cardiovascular and respiratory diseases and cancer are associated with air pollution" (ibid.: 119).

Militarization and Human Rights Violations

In the Colombian context of an internal armed conflict, the state's strategy is to militarise areas where multinational companies are located in order to protect their interests (ACIN 2011). This is what has happened in La Guajira, where soldiers claim to protect the railroad rather than the people.[27] Also, there is a permanent proliferation of illegal armed groups in the department. Despite the amnesty declared in 2005 by the Colombian government for the demobilization of paramilitary groups, such groups still have control over vast areas of the region. On the other side, guerilla groups try to destabilize El Cerrejón's production and have launched several attacks against the railways (El Espectador 18.12.2011). Local community members declare that both official state forces and illegal groups are trying to recruit them. On the one hand, the government promotes the armed forces as a career option for the Wayuu. On the other hand, illegal groups try to pursue locals to join their ranks or, in the worst cases, practise forced recruitment. Militarization has intensified the armed conflict in the Guajira department, and several massacres have been reported since 2000. Best known is the Bahía Portete massacre of 2004 with at least 6 people killed (CNRR 2010). In total, more than 200 cases of killings and disappearances have been documented over the last 10 years (NotiWayuu 2011). In other cases, Wayuu women have been monitored by state security agencies and threatened by paramilitaries, leading to their forced displacement (Fuerza de Mujeres Wayuu 2010). All these facts have been condemned by the Inter-American Commission of Human Rights.[28]

2.2 Local People's Strategies

Concerning the mine's presence, the responses of the local population are quite diverse, sometimes contradictory, and changing over time. Also the relations between different sections of the Wayuu, the corporations and the state have changed a lot. During the last 30 years, new forms of policies emerged, such as the recognition of a multicultural national identity and of "corporate social responsibility" (CSR). Today, these policies play a central role among the Wayuu and other local communities. Especially since the new Constitution of 1991, indigenous and other ethnic groups have legal tools to defend themselves against abuses and economic projects that drastically impact on the communities' realities (Puerta Silva 2010: 152).

In the beginning of the 1990s, Wayuu people started to demand reparations for the damages caused by the mining activities. They argued that negotiations on sale of land which had taken place at the beginning of the mining project were unfair since the corporation profited from their lack of knowledge and preparation. Demands were largely economic, though cultural aspects were included, too. The Wayuu also claimed their right to decide their own future, which would inevitably be next to the coal mine (ibid.: 157). In this sense, according to the anthropologist Claudia Puerta Silva, the Wayuu never remained passive in the face of the mining project. Rather, they tried to find a way to render it beneficial to them. The Wayuu developed strategies that permitted them to continue reproducing their life in La Guajira. Some of these strategies are short-term and

27 From a conversation between a Wayuu woman and national army soldiers located in the area of 4-Vias in Macao, in 2009.

28 IACHR 2010: IACHR condemns murder of human rights defender in Colombia. In: Press release N° 73/10 (2 August). <www.cidh.oas.org> accessed on 24.08.2012; Servindi 2009.

more improvised; others more elaborate and longer-term. Over time, discourses were increasingly extended to include autonomous ethnic articulation and self-determination. Key points are the right to receive compensation for the negative impacts of mining and for the occupation of their ancestral territory, the recognition of their cultural difference, autonomy of decision making and of determination of their future and the ability to articulate with Colombian society on their own terms (ibid.: 175).

Especially since the 1990s, the most important strategy of the Wayuu organizations facing the El Cerrejón mine is the use of national and international laws and conventions. For this, they are investing in capacity- and knowledge-building programs about the international system of human rights and indigenous peoples' rights. The use of legal tools is facilitated by alliances that Wayuu and other local communities were able to build with various national and international institutions, NGOs and other actors such as the lawyers' collective *Colectivo de Abogados José Alvear Restrepo* (CCAJAR). The latter is a very central actor in supporting displaced communities (e.g. for the well-known case of the community of Tabaco). Networks such as the *Red Colombiana frente a la Gran Minería Transnacional* RECLAME (Colombian network against transnational large-scale mining) or organizations like the *Organización Nacional Indígena de Colombia* ONIC constitute significant spaces for articulating the interests of local people affected by the El Cerrejón mine. In this context, new communication technologies are becoming increasingly important for building and maintaining alliances as well as for gaining support from the international community. Websites, blogs and the production and diffusion of documentaries have become common tools for online campaigns (Fuerza de Mujeres Wayuu 2011).[29]

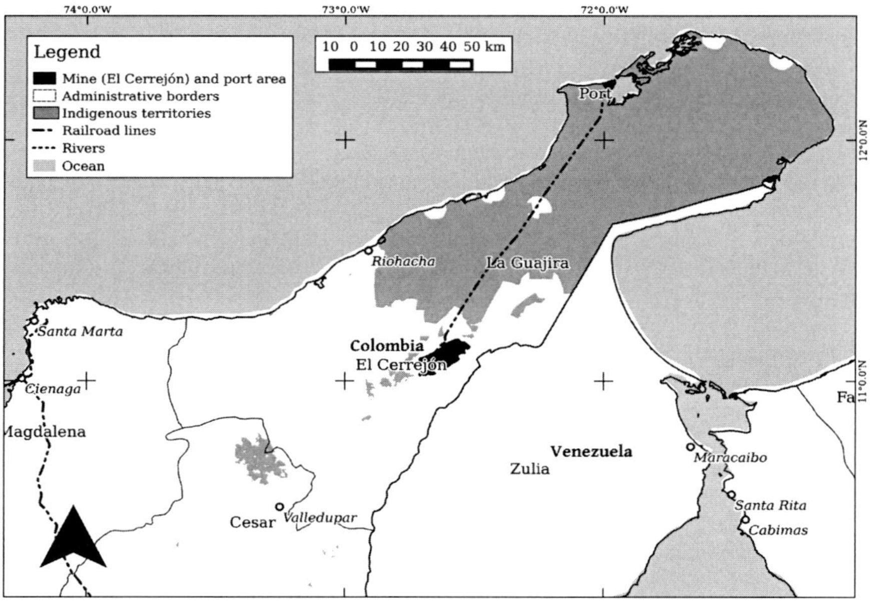

Map 2: Indigenous 'resguardos', the Cerrejón mine and transport infrastructure in La Guajira.

29 A good example of such a blog is <www.notiwayuu.blogspot.com> accessed on 12.01.2012.

The Wayuu increasingly emphasize their identity as an ethnic group and convert it into the axis of their claims and demands (Puerta Silva 2010: 167). Wayuu leaders developed identity politics as a negotiation strategy. The installation of the mining project coincides with a favourable context for ethnic claims, principally land rights claims directed at the El Cerrejón mining company. In fact, in Colombia, the easiest way to be granted land titles is by claiming to belong to some ethnic group. For ethnic groups like the Wayuu, collective land titles (like the *resguardos*) have become a legal form to protect their territories from economic megaprojects because it obliges the company to negotiate with the communities (ibid.: 157f.).

A crucial point in the strategies of the indigenous peoples is the struggle to enforce the application of Free, Prior and Informed Consent (FPIC) in any decision affecting them and their territories. Negotiations with the state and the corporations are focused on the right to FPIC, a procedure that is often ignored, bypassed or not fully accomplished. The ONIC, representing the national indigenous movement in Colombia, gives high priority to the recognition of the right to FPIC. The same can be said for many Afro-Colombian organizations that have placed FPIC on the top of their agenda. As a step to implement FPIC, Wayuu organizations are negotiating with the government to be included in the National Development Plan, which constitutes the basis for development in Colombia. This would put the Wayuu in a better negotiating position to protect their territory and to ensure their right to education, housing and food (Fuerza de Mujeres Wayuu 2011).

With regard to negotiations with El Cerrejón, only some of the local communities are participating, while others refuse to negotiate with the company. This means that the local population cannot be represented as a single actor negotiating with the corporation. Also, the Wayuu indigenous people do not constitute a homogeneous social actor. Negotiations and decisions are mostly taking place on a community level, with each community acting quite autonomously. But people have become increasingly aware of the strength of a united strategy and recently many Guajira inhabitants participated in creating new organizations trying to mobilize collectively against the mine. Such attempts gave birth to FECODEMIGUA (*Federación de Comunidades Afectadas y Desplazadas por la Explotación Minera en La Guajira*), and to the *Comité Cívico de la Guajira frente a la Gran Minería Transnacional*. Both organizations transcend the community level and the boundaries of particular ethnic groups, uniting interests of various communities—indigenous, Afro-Colombian and mestizo. In recent years, such organizations became more popular and gained visibility both locally and internationally.[30] They are protagonists of the increasingly bigger and more frequent mobilizations and protests that unite different sectors of the population. For example, a protest march was organized on 25 November 2011 in the town of Barrancas, opposing the project of expansion of the mine and diversion of the Ranchería River (see box below). The march united all the affected communities that had previously participated in a big meeting on these issues a few days earlier in the indigenous *resguardo* of Provincial.

30 Conversation with Swiss NGO activist, December 2011.

Resistance Against the Diversion of the Ranchería River

The most recent conflict regarding the Ranchería River is related to the planned change of the river's course over 26 kilometres for the expansion of the El Cerrejón mine. The Wayuu communities in this area were not informed about the planned activities of the El Cerrejón company, and they only came to know about it in April 2010, long after the project got all the environmental licenses granted by government institutions like the Ministry of Environment and Corpoguajira.

The dispute has an antecedent: in January 2006, a dam construction project was started in Ranchería water. The construction ended on 30 November 2010 (Inciarte Durán, without year). This dam project, spread over the municipalities of Fonseca, Distracción, Barrancas and San Juan del Cesar, was meant to provide water to the region and allow the installation of a central power supply. The dam flooded 638 hectares, affecting mainly the indigenous territory of the Wiwa living in the Sierra Nevada de Santa Marta, where the source of the Ranchería River is located. The dam is also indirectly affecting several communities, *resguardos* and settlements of Wayuu indigenous people, as well as non-Wayuu, by reducing water flows in the Ranchería River course.

Although very little information is available on the new plan to divert the river for the mine expansion, it is certain that a number of Wayuu communities, as well as peasants and Afro-Colombian communities would be affected. Hence, they have been in permanent resistance against the project (El Espectador 22.12.2011; El Heraldo 22.04.2012; NotiWayuu 2012). After the resistance initiated by the Wayuu communities, joint initiatives from different sectors of society were consolidated and the "Civic Committee to defend the Ranchería River and Manatial of Cañaverales"[31] was formed. In 2010 and 2011, multiple mobilizations were carried out against the river diversion, led by communities and accompanied by national and international organizations. Political parties, such as the Liberal Party and the Polo Democrático Alternativo, were publicly against diverting the river. The issue also became well-known thanks to internationally renowned artists like Manu Chao and Calle 13, who supported the resistance struggles of Wayuu women. On the 8th November 2011, a public report issued by El Cerrejón announced the postponement of its study to divert the Rancheria River. The company declared that the main reason for this temporary postponement was lower coal prices (El Espectador 08.11.2012). Still, it can be assumed that the local resistance also contributed to the decision.

The ability of the Wayuu to mobilize against the diversion of the Ranchería River was the result of a long process that started on 25th November 2006, when the organization Fuerza de Mujeres Wayuu decided to have an annual commemoration of the International Day of Elimination of Violence against Women. Fuerza de Mujeres Wayuu is an indigenous women's movement, from which men are not excluded. Born as a reaction to human rights violations against the communities, its aim is to call attention to the conflicts that threaten the Wayuu territory. As the Wayuu were previously not used to political mobilizations and campaigns, Fuerza de Mujeres Wayuu plays a central role in establishing open opposition against mining operations and human rights violations. Such mobilizations, together with the creation of local organizations, are very important to overcome the historical internal divisions of the Wayuu people. They permit the formation

31 The Manatial de Cañaverales is a water source which is at risk of disappearance because of coal concessions granted to other companies like the Brazilian CCX and the Canadian Pacific Rubiales.

of alliances at specific moments and for very precise goals, while respecting the social organization and the decision-making dynamics of the Wayuu people. For example, this kind of local resistance has played an important role in the case of CCX Colombia (formerly MPX), a Brazilian company which had obtained concessions to exploit coal deposits in the region. Still at the stage of ongoing consultations with some communities, the company announced on the 22nd January 2013 that it was stopping investments in La Guajira for three reasons: Problems in obtaining environmental clearance, low coal prices and the lack of agreements with indigenous communities.[32]

The presence of El Cerrejón led to the emergence of new intermediaries and leaders, who learned and specialized in the process of negotiation with the corporation and the state. According to Puerta Silva, the mining project constituted a "training camp" for such intermediaries, specializing in project management in the field of participation and development (Puerta Silva 2010: 170). In fact, the corporation and the state legitimize the mining activities through a discourse about development for La Guajira and its inhabitants. In response, local communities started to formulate their own discourses and visions on development and its form and content. In this sense, they do not reject development per se; e.g., in one case, a community co-financed together with the mining foundation an aqueduct and the construction of reservoirs permitting them easier access to water (ibid.: 174). Local leaders, communities and organizations thus adopted the vocabulary of development in order to enter into dialogue with corporation and state. However, they use it on their own terms and understanding. An example of such a reformulation of a vision of development is represented by the Wayuu women's organization Fuerza de Mujeres Wayuu (Fuerza de Mujeres Wayuu 2011).

Responses to the mine differ in some respects between Wayuu, Afro-Colombian and mestizo communities. According to a Swiss NGO activist working in the region, the Wayuu present more internal cohesion and clearer positions towards the mine than Afro-Colombian and other communities. He cites the community of Tamaquitos as an example of a clear and united organizational process (Pers. communication, December 2011). Tamaquitos is an indigenous community in the municipality of Barrancas, one of the communities most affected by the El Cerrejón mine. Though the company has no interest in buying its land, since it does not have any coal reserves, it already possesses the whole land around Tamaquitos, thus completely surrounding and isolating the community. Inhabitants of Tamaquitos lack the land needed for their subsistence according to their traditional way of life, therefore many have abandoned their homes (Idárraga Franco et al. 2010: 84). Other communities such as Roche, Patilla or Chancleta do not have the same internal cohesion and are divided in their approach to the mining corporation. The three are Afro-Colombian communities that sooner or later accepted relocation on El Cerrejón terms (ibid.).

Still, even in communities where the position and strategies are more consolidated, individual visions of the future are often in contrast with and contradictory to the discourse against the mine. For many, it seems that a job in the mine represents the only option to achieve a good life, as agriculture and other ways of life are slowly disappearing as alternatives. Local organizations and communities are usually more consolidated when facing the state than when facing the mine, due to the fact that the mining corporation has more local power and, in some cases, even substitutes the state. The mine, with its 30 years of presence in the region, has become a central part of the local dynamics. Consequently, there is no outright opposition to the existence of the mine itself, but rather a struggle against its negative impacts and a growing resistance against its expansion and new projects and mining concessions.

32 Portafolio.co: Batista saldría de proyecto de carbón en La Guajira, 22.01. 2013 <www.portafolio.co> accessed 8.2.2013.

2.3 Corporate Strategies and the State

In La Guajira department, El Cerrejón presents itself as the bearer of cultural and social well being. Models to protect and preserve the environment are part of the company's propaganda. Additionally, El Cerrejón claims to represent development and wealth for the country. Critics of the corporation's behaviour are often met with a discourse of "before and after". Mistakes made by previous owners of the company are emphasized and contrasted to the behaviour of the new owners.

In order to promote this benevolent corporate image, communication strategy has gained importance. Since 2005, El Cerrejón has published its Sustainability Annual Reports. These have become a mechanism to demonstrate progress in the promotion and respect of human rights as well as environmental protection. However, the publishing of these reports was only possible thanks to NGO pressure. In the reports, there is a general trend of analysing all issues from the perspective of "no-damage due to the coal operation". In other words, the company promotes investments in favour of the local communities or in favour of the Guajira department. At the same time, these investments allow them to keep the same communities and the government in a permanent state of control based on the promise of benefits. In its studies and reports, El Cerrejón highlights general positive effects of its operations, but barely refers to any positive effect the operation could have for the directly affected communities.[33]

Benefits offered by El Cerrejón are presented to Wayuu communities as part of the corporate social responsibility strategy. These benefits include, amongst others: housing improvements, watermills, special technical training programs, professional scholarships and productive projects. All programs and benefits are managed by the Cerrejón's Foundations System by means of agreements with local and national government institutions. This strategy means that budgets allocated by the state to implement programs for community health, education and housing are not fully utilized. In this sense, El Cerrejón lowers the costs for government agencies and institutions to provide social benefits for its citizens.[34]

Programs and projects that El Cerrejón offers to communities promote the strategy of maintaining good relations with the local population. Nevertheless, promises of social investments, projects and initiatives supported by the company are not offered to all communities. With this behaviour, the corporation is indirectly promoting patronage to gain support of some clans, families and even "false" leaders, who end up being the only valid interlocutors for any kind of negotiation or benefit. Strategically, the company prefers to deal with leaders and communities who do not object to their aid, thus avoiding consensus finding processes and creating divisions in a social structure, which is known for its decentralized manner of decision-making. El Cerrejón takes advantage of the lack of central authority and of organizational weakness while interacting with the communities. Many of the staff recruited by the company to carry out stakeholder dialogues are anthropologists, social workers and sociologists, among others. They have previously worked with local communities and have gained knowledge about them as well as their trust. In other cases, El Cerrejón offers very attractive salaries to well-known leaders, especially those who criticise mining projects. These offers are significant in a context where people live with precarious economic resources. Some leaders denounced these actions as leadership co-option.

33 Conversation with Wayuu community leader, January 2012.

34 On its website, the four Foundations of the Cerrejón Foundations System are presented as "[...] carrying out a strategic mission that independently drives a sustainable development 'shock' into the region. Indigenous Guajira aims to improve ethnic standards of living, Water to improve supply and quality of water in the region; Progress for sustainable development of the region and Institutional Strengthening for transparent and efficient use of mining royalties for community wellbeing" <www.fundacionescerrejon.org> accessed on 5.2.2013.

El Cerrejón, in its eagerness to show the best image of the company, constantly promotes relations with local, national and international media as well as with international NGOs. Media are used to highlight the support given by the corporation to big events and to highlight benefits offered by the company, but are also manipulated through restrictions on their access to reliable data. Additionally, El Cerrejón puts pressure on international NGOs by inviting them to special sessions organized at the international level. In these meetings, the company also invites Wayuu community members who are not empowered to represent the Wayuu communities at the international level ("false" leaders). This was the case during a meeting held simultaneously in London and Boston in 2011, where international NGOs were invited and consulted about the alteration of the Ranchería River's course. The consultation meeting did not include the legitimate representatives of the affected communities, who were only informed about the project months later.

El Cerrejón thus validates "false" leaders who do not follow Wayuu traditional ways of decision-making within communities. This strategy allows the company to ignore inherent rights guaranteed to indigenous peoples, such as the right to prior consultation established in Law 21 of 1991. In this sense, El Cerrejón respects neither prior consultation as established by law nor the FPIC defended by indigenous peoples in Colombia. Instead, the company only mentions "relations with stakeholders" with regard to local communities.

At present, concessions have been given to new companies in the south of La Guajira for the exploitation of other coal deposits. Companies like Pacific Rubiales from Canada and CCX (formerly MPX) from Brazil have already obtained environmental licenses to establish coal mining operations. In some instances, the three companies that form El Cerrejón have carried out separate consultation processes with local communities. This approach creates confusion and fatigue among the communities because they are forced to negotiate at all costs with many different companies.

So far no Wayuu community has initiated a case against El Cerrejón before the Constitutional Court of Colombia. This fact is important; since most of the verdicts of the Constitutional Court are in favour of indigenous peoples, they become an additional tool for protecting their rights. Not all cases reach the Constitutional Court though, as the processes are very long and fraught with many obstacles. Moreover, no company takes into account all the verdicts that the Constitutional Court has already delivered in favour of the protection of indigenous peoples' territories elsewhere in the country. Although there is no case involving Wayuu communities, legal precedents should be applied to any case in which a company, or the state, violates the right to prior consultation.

Finally, like many other companies exploiting natural resources in indigenous territories of Colombia, El Cerrejón declares itself a victim of the internal armed conflict, particularly of the FARC-EP guerrillas. These are said to have undertaken attacks against the company's railway last year. This fact justifies and promotes official militarization of the territory under the pretext of protecting "national interests", which in local practice means the interests of the El Cerrejón mine (Caracol Radio 2008; El Espectador 29.10.2011; El Mundo 19.12.2011; El País 29.10.2011). Similar cases of transnational corporations declaring themselves victims of the armed conflict are widely documented in the Colombian context. Well known are the cases of Drummond Company and Chiquita Brands, which are being investigated for links (mainly financial support) with paramilitary groups. Still, they affirm to be victims of the conflict (Romero and Torres 2011).

Role of the State

Both the companies and the state support each other through their strategies, as the Colombian government has an explicit policy of attracting international investments. Still, the role of the State differs in important ways from the companies. State activity is principally based on legislation, i.e., on the application or non-application of the national law. Despite the existence of a constitutional process which recognizes important international pacts and conventions, one of the main concerns of the indigenous movement is the systematic dismantling of national regulatory legislation (Houghton 2011). Legislation in Colombia increasingly benefits foreign investment instead of guaranteeing the protection of human rights. The importance of laws and decrees guaranteeing the right to prior consultation and protecting the rights of indigenous peoples is downplayed, while laws enabling the diversion forest, water and mineral resources to transnational corporations are prioritized. Although the Colombian Constitutional Court has issued an important case law on protection of human and territorial rights of indigenous peoples, the state often ignores existing jurisprudence (ibid.).

In this context, an important law is the *Código de Minas* (Mining Law, Law 685), adopted in 2001. Chapter XIV of the Código gives indigenous peoples and Afro-Colombians the possibility to create *zonas mineras indígenas* (indigenous mining zones) and *zonas mineras de comunidades negras* (Black communities' mining zones) in their territories. The announcement by the state in 2010 about the creation of more than 100 indigenous mining zones in Colombia seemed a very positive evolution, because the *Código de Minas* recognizes the right of priority (*derecho de prelación*) to ethnic communities for exploiting minerals in their territories (Ministerio de Minas y Energía, wy: 19). However, this law generated confusion among indigenous communities, who interpreted it as a new legal instrument permitting them to protect their *resguardos* from mining activities. Many indigenous communities constituted indigenous mining zones, without knowing that, according to the *Código de Minas*, if they do not start exploitation in the three years following the formation of the *zona minera indígena*, a third person can apply for licenses to explore and exploit the zone. Such a right of priority that induces indigenous communities to constitute mining zones and exploit minerals, only to then expire, is not considered a "real right" (Lagos Zapata 2010). Rather, the granting of such "rights" seems to be a strategic mechanism used by the state in order to implement its mining policy.

The issue of prior consultation with ethnic groups can not be separated from the extractive activities in indigenous territories. In Colombia, consultation is not binding for the state, even if the Constitutional Court offers some binding precedents through sentence T-769 of 2009 (Corte Constitucional 2009; Lagos Zapata 2010). It is interesting to note that a request for annulment was submitted against the mentioned sentence in 2010 by the then judicial director of the Ministry of the Interior and Justice and by the representative of a big mining corporation (Muriel Mining Corporation) (Acto 053 of 2012, Corte Constitucional 2012). Prior consultation, which should be guaranteed as a right, has been reduced to an administrative procedure in which indigenous peoples cannot decide over previous agreements between the state and multinational companies (ACIN 2011). In some cases, state institutions ignore the existence of indigenous communities in areas where extractive or infrastructure projects are planned. To do so, the areas are surveyed by air, instead of doing "on the ground" research (Amparo Rodríguez 2009: 64). Also, applications for the constitution of *resguardos* or for the titling of collective Afro-Colombian territories have been pending for 20 years in some cases. It must be noted that the majority of these applications concern territories rich in natural resources, where mining corporations hold concessions.[35]

35 A well-known example is COCOMOPOCA in the Department of Chocó, see Dueñas in El Espectador 18.09.2011.

The actual procedures for prior consultation are regulated by Decree 1320 of 1998, which has several important limitations. For example, Decree 1320 stipulates prior consultation only with communities that inhabit legally recognized territories (*resguardos* or collective land titles of Afro-Colombians) (ibid.: 59). Secondly, prior consultation, which is based on the right to participation in decision-making processes, does not give communities the right to take final decisions or to refuse projects. That is the reason why indigenous and Afro-Colombian people's organizations are struggling for the application of the Free, Prior and Informed Consent (FPIC) in decisions affecting them. Such a procedure would be based on the principle of self-determination and would allow communities to decide over their own future (ibid.: 71f.). The United Nations Declaration on the Rights of Indigenous Peoples of 2007 includes the right to Free, Prior and Informed Consent of indigenous peoples, making it legally binding for nation states. However, Colombia has not signed the Declaration yet: in April 2009, the Vice-Minister of Multilateral Issues of the Ministry of Foreign Affairs declared the unilateral support of the Declaration by the Colombian State.[36] Thus, with regard to the indigenous peoples on its territory the Colombian state shows itself open to signing and approving international conventions and treaties in order to maintain an image of democracy and respect for human rights. This is also the case with the United Nations Declaration on the rights of indigenous peoples: Colombia does not officially sign it, but supports it, without legally committing itself.

For the procedures of prior consultation, some traditional authorities are officially recognized by the Ministry of the Interior and Justice as valid representatives and interlocutors for a specific ethnic group, but others are ignored. Over the years, this has resulted in an increase in state-recognized traditional authorities to carry out procedures of prior consultation. However, these representatives have not always been recognized by the indigenous and Afro-Colombian communities themselves. This situation led to the dismantling of traditional authorities (Coronell 2011), which is even more accentuated among the Wayuu, because of the fact that their social organization lacks any centralized authority.

3. Conclusion

Mining activities in the Guajira Peninsula started more than two decades ago. The context in which the El Cerrejón mine is operating has changed over time. National and international legislative frameworks, national economic policies, international standards for multinational corporations, strategies of indigenous and local communities as well as those of the mining corporation have constantly evolved during the operation period. Colombian legislation protecting indigenous peoples and other legally recognized ethnic groups was created and improved since the 1990s. Indeed, the country is known to have "a progressive legislation with regard to the recognition of the rights of the indigenous and also with regard to environmental laws" (Haller et al. 2007: 448). When mining operations started in La Guajira department, such tools were not available; local communities learned and adopted these new tools and strategies through the process. Moreover, they learned to act through organizations in order to mobilize wider sectors of the affected population.

Colombia's new economic policy focuses on the mining sector, aiming at transforming the country into a *país minero* (mining country). By implementing this policy, the state prioritizes multinational corporations by creating favourable conditions for them. Some of these benefits

36 Ministerio del Interior y de Justicia 2009: 21 April 2009. Gobierno de Colombia anuncia respaldo unilateral a la Declaración de Naciones Unidas sobre los Derechos de los Pueblos Indígenas. <www.mij.gov.co> accessed on 29.07.2012.

run against the state's duty to protect and guarantee the rights of its citizens, indigenous peoples included. Consequently, legislation regarding indigenous peoples and ethnic minorities as well as environmental legislation is often ineffective. Nonetheless, international circumstances obliged mining companies to change their strategies. In the case of El Cerrejón, it is evident that the company's image has become crucial. In its new strategy, which follows international trends among multinational corporations, discourses are extremely important, as indicated by terms such as "corporate social responsibility" or "relations with stakeholders". The Colombian state contributes to this by legitimizing extractive projects through a discourse based on the ideas of progress, development and the creation of employment. This argumentation frames mining projects as highly important for national and public well-being, thus trying to present public national interests as a priority, above all other interests, such as those of indigenous or Afro-Colombian communities (Puerta Silva 2009: 84).

Despite the existence of progressive legislation and discourses on corporate social responsibility, local communities continue to denounce serious human rights violations and damage to their social and natural environment. Protests and lobbying activities are organized, aiming at international attention in order to stop negative impacts and new extraction projects as well as to get reparations for past damages and losses. Although the action of affected communities appears increasingly united, communities are aware of their fragility in terms of internal divisions. With its 30 years of presence in the region, El Cerrejón has become part of La Guajira, and also part of local communities' options. It possesses certain power on the local dynamics and in some areas even substitutes the state. Thus, there is no outright opposition to the mine, but resistance rather focuses against the negative impacts of mining activities and against further extractive interventions in the peninsula.

References

Amparo Rodríguez, Gloria 2009: El papel de la consulta previa en la pervivencia de los pueblos indígenas y demás grupos étnicos de Colombia. *El Otro Derecho* 40: 55-74, Instituto Latinoamericano de Servicios Legales Alternativos.

Chomsky, Aviva, Garry M Leech and Steve Striffler (eds.) 2007: Bajo el manto del carbón: pueblos y multinacionales en las minas de El Cerrejón, Colombia. Bogotá: Casa editorial Pisando Callos.

Cortés Severino, Catalina 2009: Scenari di memorie del conflitto armato colombiano da una prospettiva etnica e di genere. *Quaderni del ramo d'oro* 2: 204–225. <www.qro.unisi.it> accessed on 18.12.2011.

Curiel Arismendy, Iliana 2010: Salud de los Wayuu, en cuerpo y alma. *Informativo UnNorte* 64: 8–9. <www.uninorte.edu.co> accessed on 30.01.2013.

Dalto, Marzia 2011: Le juge constitutionnel colombien vis-à-vis du déplacement forcé. *Revista Derecho del Estado* 27: 47–74.

Dover, Robert VH, Marta Rincón and Maria Zapach 1997: Estudio: impacto sociocultural en los indígenas del sur de la Guajira, generado por la explotación minera del carbón. Bogotá: Universidad de Alberta and Censat AguaViva.

Eraso, Carolina 2009: Expériences et stratégies de vie des personnes déplacées en Colombie: approche critique d'une catégorie juridique nationale. Mémoire de licence en ethnologie, Université de Neuchâtel.

Flórez Lopez, Jesús Alfonso and Delma Constanza Millán Echeverría 2007: Derecho a la alimentación y al territorio en el Pacífico colombiano. Tumaco: Diócesis de Tumaco, Quibdó: Diócesis de Quibdó (etc.).

Grupo Semillas 2008: Biopiratería. Una amenaza a los territorios colectivos de Colombia. Estrategias y acciones desde la sociedad para enfrentarla. Bogotá: ARFO Editoriales e Impresiones Ltda.

Haller, Tobias, Annja Blöchlinger, Markus John, Esther Marthaler and Sabine Ziegler (eds.) 2007: Colombia. In: Haller, Tobias et al. (eds.): *Fossil Fuels, Oil Companies and Indigenous Peoples.* Münster: LIT Verlag, 413–453.

Hörtner, Werner 2006: Kolumbien verstehen. Geschichte und Gegenwart eines zerrissenen Landes. Zürich: Rotpunktverlag.

Hoyos, Cristina 2006: Frieden in Kolumbien? – Möglichkeiten einer Schweizer Beteiligung am Friedensprozess. *Schweizerisches Jahrbuch für Entwicklungspolitik* 25 (2): 211–218.

Idárraga Franco, Andrés, Diego Andrés Muñoz Casallas and Hildebrando Vélez Galeano 2010: Conflictos socio-ambientales por la extracción minera en Colombia: Casos de la inversión británica. CENSAT Agua Viva, Amigos de la Tierra Colombia. <censat.org> accessed on 30.01.2013.

König, Hans-Joachim 2008: Kleine Geschichte Kolumbiens. München: C.H. Beck.

Márquez Calle, Germán (without year): Curso Sistemas ecológicos y sociedad en Colombia. Programa de Maestría en Medio Ambiente y Desarrollo del Instituto de Estudios Ambientales -IDEA- de la Univesidad Nacional de Colombia. <www.virtual.unal.edu.co> accessed on 07.04.2014.

Montero de Daza, Judith and Luis Mestra Narváez 2008: Los Wayuu, la tierra y la política del carbón en la Guajira, 1980–2000: una visión desde la historia para reparar y no repetir. *Palobra* 9: 145–162.

Pacini Hernandez, Deborah 2007: Estudios socio culturales y ambientales. Impacto de la Mina de El Cerrejón sobre la Comunidad Wayuu 1993. In: Chomsky, Aviva, Garry M Leech and Steve Striffler (eds.): *Bajo el manto del carbón: pueblos y multinacionales en las minas de El Cerrejón, Colombia.* Bogotá: Casa editorial Pisando Callos, 41–50.

Puerta Silva, Claudia 2009: Les indiens wayuu et le projet minier du Cerrejón en Colombie: Stratégies et politiques de reconnaissance et d'identité. Doctorat, École des Hautes Études en Sciences Sociales, Paris.

Puerta Silva, Claudia 2010: El proyecto del Cerrejón: un espacio relacional para los indígenas wayuu, la empresa minera y el Estado colombiano. *Boletín de Antropología* 24 (41): 149–179, Universidad de Antioquia Medellín, Colombia.

Ramírez Boscán, Karmen 2005: La perspectiva de los Wayúu sobre los conocimientos y la biodiversidad. In: Escobar, Elsa Matilde et al. (eds.): *Las Mujeres Indígenas en los escenarios de la biodiversidad.* Bogotá: UICN, Fundación Natura Colombia, Instituto Colombiano de Antropología e Historia, 39–42.

Ramírez Boscán, Karmen 2007: Desde el desierto. Notas sobre paramilitarismo y violencia en territorio Wayúu de la media Guajira. Maicao.

Restrepo, Eduardo 2013: El giro a la biodiversidad en la imaginación del Pacífico colombiano. *Revista de Estudios del Pacífico Colombiano* 1: 171-199.

Romero, Mauricio and Diana Fernanda Torres 2011: Drummond, Chiquita y paramilitares: adaptación y negociación de ventajas en medio del conflicto. In: Romero Vidal, Mauricio (ed.): *La economía de los paramilitares. Redes de corrupción, negocios y política.* Bogotá: Debate, 149–190.

Salas Bahamón, Jaime Ernesto 2004: El Cerrejón y sus efectos: una perspectiva socioeconómica y ambiental. Bogotá.

Szegedy-Maszák, Ildikó 2008: Corporate social responsibility, the example of the mining project Cerrejón and its relation with the indigenous group Wayuu in Colombia. *Universitas* 117: 295–322.

Villalba Hernández, José A 2008: Wayúu resistencia historica a la violencia. *Historia Caribe* 5 (13): 45–64.

Wade, Peter 2009: Defining Blackness in Colombia. *Journal de la Société des Américanistes* 95 (1): 165-184.

Reports, News Articles and Internet Sources

ACIN 2011: Cuestionario sobre proyectos de desarrollo y extracción de recursos naturales dentro o cerca de territorios indígenas. Presentado al Relator Especial de la ONU sobre los derechos de los pueblos indígenas, Aportes de la Asociación de Cabildos Indígenas del Norte del Cauca ACIN Cxhab Wala Kiwe, abril de 2011.

ASK 2011: Jahresbericht 2010 der Arbeitsgruppe Schweiz-Kolumbien ask. <www.askonline.ch> accessed on 12.03.2012.

Caracol Radio 2008: El Cerrejón en contingencia tras atentado dinamitero. (27 May). <www.caracol.com.co> accessed on 06.06.2012.

Cerrejón 2010: Sustainability Report 2010. <www.cerrejon.com> accessed on 09.02.2012.

Cerrejón 2011: 7 December 2011. Cerrejón Presents Book "Towards Rehabilitating Lands Affected by Open-Pit Mining". <www.cerrejon.com> accessed on 09.02.2012.

Cerrejón 2012: 14 February 2012. Una realidad que comienza en grande: llega el primer camión del proyecto de expansión p40. <www.cerrejon.com> accessed on 06.06.2012.

Choudry, Aziz 2003: Conservation International: Privatization Nature, Plundering Biodiversity. Seedling (October). <www.grain.org> accessed on 10.02.2012.

CNRR — Grupo de Memoria Histórica (ed.) (2010). La El Cerrejón y sus efectos: una perspectiva socioeconómica y masacre de Bahía Portete: Mujeres Wayuu en la mira. Bogotá: Aguilar, Altea, Taurus, Alfaguara. <memoriahistorica-cnrr.org.co> accessed on 18.12.2011.

Coronell, Daniel 2011: Operación clonación. Semana (10 December). <www.semana.com> accessed on 31.07.2012.

Corte Constitucional 2009: Sentencia T-769/09. <www.corteconstitucional.gov.co> accessed on 29.07.2012.

Corte Constitucional 2012: Auto 053/12. <www.corteconstitucional.gov.co> accessed on 29.07.2012.

DANE 2005a: Análisis regional de los principales indicadores sociodemográficos de la comunidad afrocolombiana e indígena a partir de la información del censo general 2005. Bogotá.

DANE 2005b: La visibilización estadística de los grupos étnicos colombianos. Bogotá.

DANE 2005c: Necesidades Básicas Insatisfechas (NBI). Boletín Censo General 2005. <www.dane.gov.co> accessed on 10.02.2012.

DANE 2010: Resultados cifras de pobreza, indigencia y desigualdad 2009. MESEP, Misión para el empalme de las series de empleo, pobreza y desigualdad. Bogotá. <www.dane.gov.co> accessed on 10.02.2012.

DANE 2012: 6 June 2012. Reloj de Población. <www.dane.gov.co/reloj/reloj_animado.php> accessed on 06.06.2012.

Dueñas, Mauricio 2011: Territorios titulados a COCOMOPOCA en peligro de ser hurtados. El Espectador (18 September).

El Espectador 2010: Colombia registra 4,9 millones de desplazados por conflicto armado. (9 November). <www.elespectador.com> accessed on 01.02.2011.

El Espectador 2011a: Ataque en vía férrea del Cerrejón paraliza la exportación de carbón. (18 December). <www.elespectador.com> accessed on 20.12.2011.

El Espectador 2011b: Atentado contra mina de El Cerrejón deja dos obreros heridos. (29 October). <www.elespectador.com> accessed on 06.06.2012.

El Espectador 2011c: Trasteo de un río en La Guajira. (22 December). <www.elespectador.com> accessed on 09.02.2012.

El Espectador 2012: Cerrejón pospone estudios sobre la posible desviación del río Ranchería. (8 November). <www.elespectador.com> accessed on 31.01.2013.

El Heraldo 2012: Frente común para evitar desvío del Ranchería. (22 April). <www.elheraldo.co> accessed on 06.06.2012.

El Informador 2011: Complacencia en La Guajira por decisión de MPX de abrir oficina regional en Riohacha. (10 May). <www.el-informador.com> accessed on 11.01.2012.

El Mundo 2011: Paralizado Cerrejón tras atentado. (19 December). <www.elmundo.com> accessed on 06.06.2012.

El País 2011: Atentado contra mina de El Cerrejón deja dos obreros heridos. (29 October). <www.elpais.com.co> accessed on 06.06.2012.

El Tiempo 2011: Corte Constitucional tumbó Código de Minas. (11 May). <www.eltiempo.com> accessed on 16.12.2011.

Fuerza de Mujeres Wayuu 2010: Informe presentado a los expertos miembros del Foro Permanente de Naciones Unidas para las Cuestiones Indígenas (FPCI) en su misión a Colombia, 5 al 9 de julio de 2010, Valledupar 7 de julio de 2010.

Fuerza de Mujeres Wayuu 2011: Wayuu Indigenous Women: Raising Voices for Justice, Peace and Consultative Development. Presentation by Karmen Ramírez Boscán for the Training Program to Enhance The Conflict Prevention and Peacemaking Capacities of Indigenous Peoples Representatives. Geneva: United Nations Institute for Training and Research (UNITAR), June 2011.

GEO and GTZ 2001: Kolumbien. In: Internationaler Tag der Biodiversität 2001. <www.biodiversity-day.info> accessed on 15.01.2011.

Houghton, Juan 2011: Debates indígenas ante la reconquista minera del Cauca. Grupo de seguimiento a políticas públicas, Casa del Pensamiento de la ACIN.

IMF 2011: Report for Selected Countries and Subjects, Data and Statistics. <www.imf.org> accessed on 24.01.2012.

Inciarte Durán, Leonel (without year): Análisis de Conflictos socio ambientales represa El Cercado Rio Ranchería. Presentation of Corporación Autónoma Regional de La Guajira. <www.minambiente.gov.co> accessed on 09.02.2012.

IWGIA 2011: Colombia. In: IWGIA: *The Indigenous World 2011*. Copenhagen.

Lagos Zapata, Adriana María 2010: Zonas mineras indígenas y afrocolombianas. ¿Reconocimiento de un derecho o legitimación de una injusticia histórica? Razón Pública (22 November). <razonpublica.com> accessed on 29.07.2012.

Ministerio de Minas y Energía (without year): Cartilla didáctica para la asimilación del Capítulo XIV del Código de Minas. "Grupos étnicos", Comunidades indígenas. <www.simco.gov.co> accessed on 25.06.2012.

Ministerio de Minas y Energía 2001: Código de Minas. Ley 685 de 2001. Diario Oficial No. 44.545 de 8 de septiembre de 2001. Bogotá: Ministerio de Minas y Energía. <www.secretariasenado.gov.co> accessed on 18.12.2011.

Ministerio de Minas y Energía 2011: Colombian Mining Statistical Yearbook. <www.simco.gov.co> accessed on 09.02.2012.

NotiWayuu 2011: Itinerario de Víctimas Wayúu. <victimaswayuu.blogspot.ch> accessed on 24.08.2012.

NotiWayuu 2012: 21 March 2012. Colombia: Wayuu de Resguardo de Provincial dicen no a Consulta para desviar Río Ranchería. <www.notiwayuu.blogspot.ch> accessed on 06.06.2012.

Presidencia de la República de Colombia 2007: 8 July 2007. Desviado Río Ranchería para dar paso al proyecto más grande de adecuación de tierras. <www.presidencia.gov.co> accessed on 09.02.2012.

Ramírez Boscán, Karmen 2006: Derechos Humanos del Pueblo Wayuu. Informe presentado ante la Misión Internacional de Verificación sobre la Situación Humanitaria y Derechos Humanos de los Pueblos Indígenas de Colombia.

Ramírez Boscán, Karmen 2008: Pueblo Wayuu, Biodiversidad y Conocimientos Tradicionales: Cuatro notas para la reflexión. <www.indigenousportal.com> accessed on 12.01.2012.

Servindi 2009: 20 March 2009. Colombia: Indígenas Wayuu instan a CIDH interceder ante amenazas paramilitares. <servindi.org/actualidad/9347> accessed on 24.08.2012.

U.S. Department of State 2011: 13 July 2011. Colombia. In: Bureau of Western Hemisphere Affairs Background Note. <www.state.gov> accessed on 22.01.2012.

U.S. Energy Information Administration 2012: International Energy Statistics: Coal. <www.eia.gov> accessed on 09.02.2012.

World Coal Association 2010: Ecoal. Quarterly Newsletter of the World Coal Institute 72. <www.worldcoal.org> accessed on 12.02.2012.

The Marlin Gold-Silver Mine in San Marcos, Guatemala

Annick Vollmar, Joel Zumbühl, Maximilian Rau[1]

1. Situating the Case Study

With the Marlin Mine, metal mining returned again to the public arena of Guatemala after a lapse of two decades. The gold and silver deposits were discovered in 1998 in the department of San Marcos in western Guatemala. In 2003, the government granted the exploitation for 25 years to the Canada-based company Glamis Gold.[2] Open-pit and underground extraction began in December 2005 for an expected duration of ten to fifteen years (Zarsky/Stanley 2011: 6, 9).

Even though the Guatemalan government ratified international conventions and agreements such as the ILO Convention 169, the local population affected by the Marlin Mine was never properly consulted. However, a large majority showed its opposition to the mining project in January 2005 in a local referendum (COPAE 2007: 1). People rejected mining because they were forced to leave their land and due to the over-use and contamination of water sources, leading to health problems. Also, they feared increased social conflicts in their communities, as they had already experienced acts of violence and harassment against opponents of the mine. Since 2005, in 42 of Guatemala's 333 municipalities similar local referenda have shown overwhelming majorities favouring a ban on large-scale mining (Dougherty 2011: 3).

High gold prices and the low cost of production in Guatemala generate significant returns to shareholders, but the mine's economic contribution to local communities and the country are ambiguous. Human rights violations as well as negative environmental impacts have led to international furore. Many people from Guatemala and abroad, including exponents of the Catholic Church, politicians, journalists, and other representatives of civil society, have engaged in the debates (Castagnino 2006: 7, 11).[3] The case of the Marlin Mine is not only marked by asymmetric power relations and putting economic interest above local communities' rights, but also by on-going impunity, ethnically biased jurisdiction, and deep-rooted inequality in Guatemalan society.

1.1 History and Current Context

Guatemala has a long tradition of metal mining, dating back to precolonial times. During colonial times, mining was enforced to pay taxes to the Spanish Crown. In the mid 20th century, gold and silver production increased significantly with ups and downs depending on market fluctuations. In 1965, a mining code was approved with the objective of stimulating investment in the

1 The authors would like to thank Helen Gambon and Anna Leissing for their comments and collaboration on the completion of the chapter. Final revision: Helen Gambon, Madlen Kobi and Thomas Niederberger

2 In 2002, Glamis Gold merged with the Canada-based Francisco Gold Corp., partner of Montana Exploradora de Guatemala S.A., and owner of the Marlin Mine, and in 2006 with Goldcorp Inc.

3 Bickel Ulrike 2006: Menschenrechtsverletzungen und Umweltzerstörung durch Gold-Tagebergbau in Guatemala. In: Guatemalanetz. <www.guatemalanetz.ch> accessed on 24.11.2012.

industry. The same year an area of 410 km^2 in the departments of Alta Verapaz and Izabal was granted to the Canadian-based International Nickel Company (INCO). A rise in oil prices and a drop in the price of nickel on international markets in the early 1980s finally led to a shutdown of mining activities for almost two decades (Castagnino 2006: 8–10).

The highland of northwest Guatemala, Chiapas as well as parts of Belize and Honduras were settled by the Mayas between 1500 BC and 300 AD (Allebrand 1997: 32; Kurtenbach 1998: 53). At the peak of the Mayan culture, from 300 AD to 900 AD, numerous city states and extensive trade networks existed in the area. The Maya also developed a complex system of hieroglyphic writing as well as mathematical and astronomical knowledge (Weisshar 2008: 732). The time from 900 AD to 1523 AD is characterized by a significant decrease in population and a process of fragmentation that led to the emergence of the 22 Mayan languages which exist in Guatemala today. It is assumed that a combination of soil exhaustion, climate change, overpopulation and rivalry between the different empires and royal dynasties caused the collapse of the Mayan civilization (Kurtenbach 1998: 60; McKillop 2004: 12). Many cities were in ruins when the Spanish

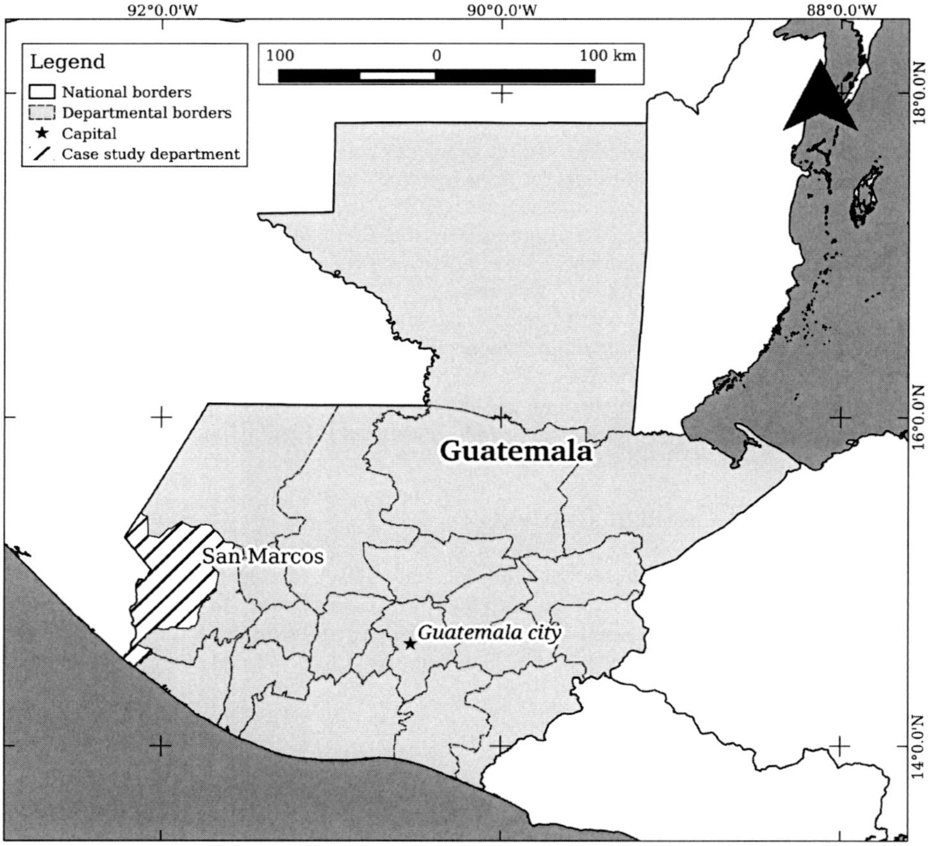

Map 1: The department of San Marcos in Guatemala.

conquerors arrived in the 16th century. The introduction of diseases further reduced the Mayan population by one-third. Colonial Guatemala developed no significant economic prosperity. From 1503 onwards, the Spanish Crown distributed land to the conquerors. With this, the indigenous population living on these lands was subjected to the so-called *encomienda* system. The justification was that the landowner should protect and convert them to Christianity. However, they were primarily exploited as slave labour for plantations and mines. Due to continuous misuse, the system was legally abolished in 1542. However, in view of the outrage of the Spanish settlers, it was restored in 1550 (Allebrand 1997: 56–57; Lateinamerika-Institut 2010). The introduction of new agricultural products such as sugar and wheat brought along forced labour, malnutrition and further population decline. Imported methods of animal husbandry led to erosion and depletion of the formerly fertile soil. Besides, different Mayan groups avoided contact between each other, which led to the establishment of enclosed communities (Allebrand 1997: 58–59; Molkentin 2002: 67).

The independent state of Guatemala was established in 1840. The years until 1871 were marked by cautious economic modernization under conservative regimes. At the end of the 19th century, economic modernization and integration into the international market accelerated under the impression of a shift to large-scale coffee and banana plantations, led by the United Fruit Company and other US-American corporations. The need for more land and agricultural labour again led to serfdom and the forced acquisition of indigenous common property land. The ethnic division within society further deepened, as most important positions in politics, economy and society were held by Ladinos, while the indigenous population was repressed (Allebrand 1997: 83; Kurtenbach 1998: 68).

Between 1945 and 1954, democratic structures such as voting rights as well as progressive labour laws were introduced, forced labour was abolished, and the education system was improved. The state expropriated unused lands from large landowners and redistributed it to the indigenous population. However, when the government of Arbenz Guzmán (1951–54), in line with the land reform, tried to confiscate part of the land of the United Fruit Company, politically right-leaning officers and the US intelligence service came together to carry out a military coup (Molkentin 2002: 87–95; Allebrand 1997: 84). The subsequent military dictatorship abolished the social reforms that had been achieved in the previous years. In 1960, a coup attempt by a left-leaning group of the military failed, marking the first step of the guerrilla movement and the beginning of a 36-year-long civil war (Viscidi 2004 , Spetzler 2009: 30). In this longest and bloodiest civil war of Central America, the military government and right-wing paramilitary organizations fought against left-wing insurgents organized in the Guatemalan National Revolutionary Unity (Unidad Revolucionaria Nacional Guatemalteca, URNG) (Molkentin 2002: 113–114). More than 200,000 persons, most of them civilians, were killed or disappeared and about 1.5 million were displaced (GHRC 2011: 2). The UN Truth Commission of Guatemala (Misión de Verificación de las Naciones Unidas en Guatemala, MINUGUA) calculated that 83% of the victims were indigenous persons. 93% of the investigated acts of violence were committed by the military and right-wing paramilitary. The army suspected indigenous people in different parts of Guatemala to be guerrilla allies and therefore directed random and massive aggressions against them. Today, the so-called "scorched earth operation" is recognized as genocide, since the use of force against indigenous people was not only aiming at destroying the base of the guerrillas, but rather at demolishing the cultural values and the very existence of the Mayan communities. The Truth Commission stated in its report that "[...] the violence was fundamentally directed by the State against the excluded, the poor and above all, the Mayan people, as well as against those who fought for justice and greater social equality" (CEH 1999).

In 1996, a peace agreement between the government and the URNG finally put an official end to the war. The agreement included an improved recognition of the rights of indigenous peoples within a multicultural state of Guatemala. The URNG became a political party and parliamentary and presidential elections took place (Molkentin 2002: 271; Hirschmann 2007: 89), won by Alfonso Portillo and his Guatemalan Republican Front. During the following term of neoliberalism-oriented President Óscar Berger Perdomo (2004–06), only minor improvements in the condition of the indigenous population were recorded. Since 2011, the presidency is held by Otto Pérez Molina, who was a commander in some of the most violence-ridden areas during the civil war. There have been accusations that troops under his command committed massacres among the indigenous population in the province of El Quiché.[4]

The country continues to be severely challenged by high poverty rates, inequality, deficient law enforcement and an increase in violent crime linked to the so-called *mara* youth gangs and drug cartels (GHRC 2011). Guatemala has a high rate of violent deaths (7 times the world average) and almost absolute impunity (98% of reported cases).[5] Land conflicts with forced displacement and assassination of peasants are still frequent, often linked to large-scale plantations for agricultural exports.

1.2 Geography and Ecosystem

Guatemala shares its borders in the North and West with Mexico, in the East with Belize and the Caribbean Sea, in the Southeast with Honduras and El Salvador, and in the South with the Pacific Ocean. With an area of 108,889 km², the country is inhabited by 14 million people. The narrow coastal plain, which lies between a volcanic mountain range and the Pacific Ocean, is one of the country's richest agricultural areas and concentrates three quarters of the population.[6]

Almost two-thirds of Guatemala is mountainous. The Central American Cordilleras divide Guatemala into two mountain systems: The Sierra de los Cuchumatanes in the North-West of Central Guatemala and the Sierra Madre in the South, dividing the densely populated volcanic area and the sparsely populated region of Petén to the North. The mountains define two major hydrographic regions, with rivers flowing into the Atlantic and into the Pacific Ocean. The Marlin Mine is located across two sub-watersheds, the Tzalá and the Quivichil, which drain into the Río Cuilco and then through Mexico into the Gulf of Mexico. Río Cuilco and its subsidiaries are important for the local population as a source of drinking water and irrigation.

Guatemala is located within the tropical zone and has an elevation from sea level up to 4000 meters. The climate is defined by altitude with average temperatures around 25° C in the lowlands and below 15° C with regular sudden frost between December and February in regions above 3000 meters. The dry season lasts from November to April. The trade winds from the Caribbean bring heavy rainfall for the rest of the year, with severe tropical storms especially during the months of September and October. Due to the high altitude, precipitation in San Miguel Ixtahuacán and Sipacapa in San Marcos is a bit lower than in other parts of the country (around 1000 mm per year with a relative humidity of about 85%). In the department of San Marcos, temperatures range on an average between 6° C and 20° C during the year.[7]

4 Comlay, Elinor and Mike Mc Donald 2011: Retired General Sweeps to Power in Guatemala Election. Reuters. 7.11.2011. <www.reuters.com> accessed on 26.11.2012.

5 ¡Fíjate! 2008: Gewalt und Bedrohung.<www.guatemala.de> accessed on 28.11.2012.

6 Encyclopaedia Britannica. Guatemala. Relief. <www.britannica.com> accessed on 28.11.2012.

7 Instituto Nacional de Meteorologia Guatemala 2012. <www.insivumeh.gob.gt> accessed on 11.04.2012.

1.3 Local Population

Guatemala is an ethnically diverse country. Estimates of the percentage of the Mayan indigenous people vary according to different sources and definitions. According to the 2001 census, 40.3% of the population belong to one or the other of the 22 linguistic groups that constitute the Maya.[8] 59.4% of the population belong to the so-called Ladinos and Europeans. Ladino is the term used for *mestizos* of mixed Amerindian-Spanish descent and indigenous people who designate themselves "westernized." Taking into account the genocide of the Maya population during the civil war and the on-going discrimination based on ethnicity, it can be assumed that the actual percentage of indigenous population is higher as many people currently prefer to self-identify as mestizos. According to IWGIA, indigenous peoples constitute a majority of 60%.[9]

Maya and Ladinos are ethnic groups mainly distinguished by criteria such as the language spoken or the clothes worn (Fischer/McKenna Brown 1996: 9). The Marlin Mine is located within the two municipalities of San Miguel Ixtahuacán and Sipacapa in the department of San Marcos in the western highlands at around 2000 m.a.s.l. Most of the residents of these two municipalities are indigenous Maya. Almost 90% of the Marlin Mine is located within the borders of three communities of San Miguel Ixtahuacán and little less than 10% within the border of one community in Sipacapa (Zarsky/Stanley 2011: 9). San Miguel Ixtahuacán, which consists of the municipal capital San Miguel, 17 villages and 43 hamlets, has an estimated population of 37,000 people according to the 2002 census.[10] There are two ethnic groups, the Mam, who represent more than 99% of the population, and the Ladinos with less than 1%; the predominant languages are Mam and Spanish. The municipality of Sipacapa consists of a municipal capital called Sipacapa, 12 villages, and 19 hamlets. According to the census in 2002, Sipacapa has about 14,000 inhabitants. There are two ethnic groups, the Maya Sipakapans whose share is over 75% of the population and the Ladinos with less than 25%. Sipakapan and Spanish are the predominantly spoken languages (Diagnostico Municipal 2006). The poverty rate in the two municipalities is very high, with 97.5% of inhabitants living on or below the poverty line, 80% of them in absolute poverty (Van de Sandt 2009: 22).

1.4 Local Economy and Resource Use

Guatemala has a highly inequitable distribution of land ownership: about 2.5% of the landowners occupy 65% of agricultural land, while 90% of farmers depend on only one-sixth of the agricultural land.[11] 60% of the population depends on agriculture for its livelihood. The majority of the population in San Miguel Ixtahuacán and Sipacapa are peasants, cultivating primarily corn, beans, vegetables and fruits on so-called *milpas* (traditional crop-growing system in Mesoamerica). Many households additionally keep livestock. Each family holds and uses one or several plots, but these form part of the common property of the community (*ejido*). So, although people have the right to sell their plots to another resident family, these contracts must be formally validated by the municipal mayor as well as by the community mayors. However, most of the small-scale farmers do not have enough land for subsistence farming. Available land per head is shrinking as rural families grow. Moreover, intensive land cultivation as well as effects of climate change have

8 The term "Maya" was introduced in the 1950s by scientists to describe a group of related idioms. Only one ethnic group, on the peninsula Yucatán, used the term Mayab for self-designation (Garbers 2000: 2; Weisshar 2008: 731).

9 International Work Group for Indigenous Affairs: The Indigenous World Yearbook 2011

10 Instituto Nacional de Estadistica: Guatemala. Estimaciones de la Población total por municipio. Período 2008-2020. <www.ine.gob.gt/np/poblacion/index.htm> accessed on 28.11.2012.

11 USAID Country Profile Guatemala. <usaidlandtenure.net> accessed on 24.07.2013.

led to soil degradation, especially on the steep slopes of the western highlands where most of the small-scale farmers live. Therefore, many farmers migrate to work as seasonal labourers in coastal coffee and sugarcane plantations in rural Guatemala or as wage labourers in Guatemala City, Mexico, or the United States. Remittance from family members living abroad is a major source of cash income in the region (Van de Sandt 2009: 21, 25).

Since the Mayan peasants have a strong spiritual tie with nature and a deep attachment to land, represented in religious rituals, the land has not only economic, but also spiritual and symbolic value. Land use shows the webs of kinship relations and therefore forms a part of their identity. Hence, if the Maya peasants lose their land, their livelihoods and their social and spiritual values too are affected. Maintaining the family lands and the traditions secure subsistence and some basic income, even if due to limited land many farmers cannot participate in large-scale cash-crop cultivation (Hamilton/Fischer 2003: 89–90).

1.5 National Economy and Legal Framework

Guatemala is characterized by high poverty rates and social inequality. The Gross Domestic Product (GDP) per capita was USD 5,300 in 2012. This is roughly one-half of the average for Latin America and the Caribbean and it is ranked 156 out of 228 countries in this respect.[12] After Haiti, Suriname and Honduras, Guatemala has, with 54%, the fourth highest poverty rate in Latin America.[13] The incidence of poverty is rising due to the recent economic crisis. In rural areas, about 70% of the people live in poverty, 75% of them are of indigenous origins. Almost half of the children under the age of five are chronically malnourished. This is one of the highest rates worldwide.[14] Guatemala has a highly unequal distribution of income, with a Gini Index[15] of 55.1% in 2007. The richest 10% of the population earn 42.4% of the total income whereas the poorest 10% earn 1.3%.[16] Guatemala has a Human Development Index (HDI) of 0.581 (rank 133 out of 187 in 2012), which is below the regional average of Latin America and Caribbean with a HDI of 0.741.[17]

As most of Latin America, Guatemala experienced economic growth during the 1960s and 1970s due to the export of agricultural products such as coffee and bananas. In the 1980s, with a collapse in the prices for coffee and other agricultural products on the international markets due to liberalization policies, low or negative economic growth and aggravated national debt followed (Galemba 2012: 723). The late 1980s brought a growth in foreign investment, tourism and the economy in general, but also continuing negative trade balance and external debt. With the objective of revitalizing the economy, the government promoted diversification and expansion of non-traditional export products such as snow peas and cut flowers and introduced free trade zones. Since the end of the civil war in the mid 1990s, major reforms had some success in macroeconomic stabilization.

Remittances from Guatemalans living abroad are the most important source of foreign income. They exceed the revenue of exports and tourism together and are nearly equivalent to 10% of the GDP. Agriculture still plays an important role, as 38% of the labour force is involved in it, but

12 CIA World Factbook: Guatemala. <www.cia.gov> accessed on 24.07.2013

13 Index Mundi 2012: Country Comparison. <www.indexmundi.com> accessed on 24.07.2013..

14 World Food Programme: Guatemala. <www.wfp.org/countries/guatemala/overview> accessed on 24.07.2013.

15 The Gini index measures the degree of inequality in the distribution of family income in a country.

16 CIA World Factbook: Guatemala. <www.cia.gov> accessed on 24.07.2013.

17 The HDI shows social and economic development, composed of health, education and income. International Human Development Indicators 2012: Guatemala. <hdrstats.undp.org> accessed on 24.07.2013.

Fig. 1: Indigenous women and children watching the open pit of Marlin Mine. Photo: Anna Leissing, Guatemalanetz Bern.

it only contributed 13% to the GDP in 2011.[18] Since the 1990s, the tourism industry has become increasingly important. Main attractions are the cultural sites and ruins of the ancient Mayan civilization. Tourism in the western Highlands, where the Marlin Mine is located, is however not well established and brings only limited income. Guatemala has a negative trade balance with exports of USD 10.09 billion against imports of USD 15.84 billion in 2012. Exports mainly include coffee, sugar, petroleum, apparel, bananas, fruits, vegetables and cardamom and imports include fuels, machinery and transport equipment, construction materials, grain, fertilizer and electricity. In 2013, its most important trading partners were the United States (with about 38% of overall exports and imports), followed by El Salvador, Honduras, Mexico, Nicaragua and Costa Rica.[19]

At the end of 2010, there were 274 valid mining licenses granted by the Ministry of Energy and Mining (MEM) of Guatemala. 199 of these sites were in active operation. In recent years, an increasing number of gold and silver mine projects are being established in the north-eastern and western regions. Since the opening of the Marlin Mine in 2005, gold and sliver have become the main components of mineral production, representing 97% of sales of mineral products in

18 CIA World Factbook: Guatemala. <www.cia.gov> accessed on 24.07.2012.
19 Ibid., accessed on 10.05.2015.

2011 (94% of total sales by Marlin Mine alone) (MEM 2011: 5, 9). However, mining is not yet an important economic factor. Due to the very low taxes and royalties, it contributes less than 1% to the national GDP (CIEN 2009: 34, 42). At the household level, mining can substantially contribute to family incomes: salaries for gardeners at the Marlin Mine are about 16% higher and the wages for heavy machinery workers (tunnellers and processors) about 87% higher than the minimum wage. This makes the mine an attractive working site for the local population (Zarsky/Stanley 2011: 15, 23).

It is generally attractive for foreign mining companies to invest in Guatemala. The mining law of 1997 allowed foreign companies 100% ownership of their projects and fixed a record low royalty rate of 1% (tax on non-renewable resources), to be shared equally between central and municipal governments (MEM 1997). This was changed in January 2012, when President Otto Perez Molina and the mining industry agreed to increasing the royalty rates. Gold and silver companies now voluntarily pay 4% of gross revenue as royalty. However, the distribution of the royalties remains the same.[20] In the case of the Marlin Mine, Goldcorp pays an additional one percent royalty. Thereof, 80% goes to the two municipalities of San Miguel Ixtahuacán and Sipacapa to be used "to implement the[ir] economic development plans." The remaining 20% is reserved for the Ministry of Energy and Mines and the Ministry of Environment and Natural Resources.[21]

Land Rights and Indigenous Peoples' Rights

Conflicts over land rights have a long history in Guatemala, dating back to the time of the Spanish conquest. The Spanish conquerors occupied most of the productive soil while the indigenous population was forced to change their residence to marginal and infertile areas in the Western Highlands. The independence in 1821 brought only minor changes to the fortunes of the indigenous population whereas an emerging Ladino oligarchy gained more and more control over land and labour and forced indigenous peasants to transfer their land to them. Many people were compelled to migrate in order to find work on the coastal plantations (Viscidi 2004: 1).

In 1952, an agrarian reform was initiated with the objective of redistributing land back to the indigenous population. However, after the military coup in 1954, the agrarian reform was reversed and the land returned to the previous owners. During the civil war, large areas were granted to military officers and rich landowners affiliated with the military. The peace agreement from 1996 stresses rural development and especially transfer of state land to indigenous communities (Viscidi 2004: 1). A constitutional amendment from 1993 recognises the rights of the indigenous population, including the protection of communal lands, the restitution of state-owned land to indigenous communities, and the provision of adequate health, security and welfare conditions for workers migrated outside their communities.[22]

As part of the peace agreement, Guatemala signed and ratified Convention 169 of the International Labour Organization (ILO) in 1996, which concerns the duty of the government to ensure consultation with the indigenous population before starting any projects that affect them or their territories, including mining. Article 15 passage 2 from the ILO 169 states that "[the] governments shall establish or maintain procedures through which they shall consult these peoples, with a view to ascertaining whether and to what degree their interests would be prejudiced, before undertaking or permitting any programs for the exploration or exploitation of such resources pertaining to their lands". Also the concerned people should receive a fair compensation for any

20 Kosich, Dorothy: Guatemala Introduces Sweeping Mining Law Reforms. In Mineweb. 25.06.2012. <www.mineweb.com> accessed on 15.10.2012.

21 Goldcorp Inc., News January 27, 2012: New Voluntary Royalty Agreement to Benefit Stakeholders of Goldcorp's Marlin Mine. <www.goldcorp.com> accessed on 15.10.2012.

22 Political Constitution of the Republic of Guatemala 1993: Section III, Articles 66-69.

damage that may happen due to resource extracting projects.[23] However, despite the articles in the constitution and the ILO Convention, the situation has not yet significantly improved, because of insufficient implementation. Land is still distributed highly unequally. The "coffee crisis" of 2000, when coffee prices plummeted throughout Central America, aggravated the problem. Many rural families which lost their employment in the coffee plantations were violently expelled from their homes by the landowners' private security forces and the police (Viscidi 2004: 2–3).

2. The Marlin Mine

The gold and silver mine "Marlin" occupies 6.5 km^2 of land in the municipalities San Miguel Ixtahuácan and Sipacapa (85% and 15%, respectively) in the department of San Marcos. Marlin is one of Goldcorp's most important and lowest-cost projects in Latin America, with an estimated total of 1,250,000 ounces of extractable gold. The mine is in operation since 2005 and had a projected life-span of ten to fifteen years. The ore is processed by crushing, grinding and cyanide milling. It started as a combined underground and open-pit mine. In 2012, the open-pit was exhausted and the focus put on the underground development.[24] Also a filter plant was built to dry tailings and use them as backfill for the open pit. In 2013, production was at 200,000 ounces of gold a year, out of almost 2 million tonnes of extracted ore (at a grade of 3.3g/t).[25]

In 1999, the Ministry of Energy and Mining (MEM) granted the exploration license for the project to Montana Exploradora de Guatemala S.A. At the end of 2000, the Canadian corporation Francisco Gold acquired Montana Gold Corporation, another Canadian enterprise and owner of Montana Exploradora, and continued with the exploration. Two years later, Francisco Gold was acquired by Glamis Gold, which started with the construction of the mine in 2003. In the same year, the government granted a mining license for 25 years. The financing of the mine was supported by the World Bank's private sector arm, the International Finance Corporation (IFC), with a USD 45 million loan.[26] In 2006, Glamis Gold merged with Goldcorp, the Canadian mining company which has been operating the mine since then (Dougherty 2011: 7). Goldcorp controls mines in North, Central and South America and describes itself as "one of the world's fastest growing senior gold producers."[27]

At the time of writing, the company estimated that the Marlin Mine would continue to operate for another five years.[28] The extent of long-term impacts are uncertain, but there is a high risk that environmental contamination will continue for years. Goldcorp's mine closure plans are considered by the 2010 Human Rights Assessment to be one of the weakest aspects of the mine's management (On Common Grounds Consultants 2010: 195). The closure plans were not publicly available, and the legally required guarantee deposit of USD 1 million, was considered insufficient for the closure of the mining site. An independent study estimated that the closure will cost about USD 49 million.[29]

23 ILO: C169 Indigenous and Tribal Peoples Convention. <www.ilo.org> accessed on 19.4.2012.

24 On Common Ground Consultants (2010: 6); Goldcorp Inc., News July 13, 2012: Dispelling the Myths of Marlin. <www.goldcorp.com> accessed on 20.10.2012.

25 Goldcorp Inc.: Marlin — Overview & Operating Highlights. <www.goldcorp.com> accessed on 02.02.2015.

26 International Finance Corporation (World Bank Group) 2004: Montana Exploradora: Marlin Gold Project. <siteresources.worldbank.org/INTOGMC/Resources/marlinfactsheetenglish.pdf> accessed on 19.04.2012.

27 Goldcorp Inc.: Mines & Projects. <www.goldcorp.com> accessed on 15.10.2012.

28 Goldcorp Inc.: Marlin — Overview & Operating Highlights. <www.goldcorp.com> accessed on 20.10.2012.

29 See Rojas, Álex: Cierre de la Mina Marlin Valdría US$49 Millones. In: Prensa Libre. 21.04.2012. <www.prensalibre.com> accessed on 28.10.2012. and Zarsky/Stanley (2011: 38).

2.1 Environmental and Social Impact

Gold mining is known to have far-reaching negative environmental, health and social impacts, particularly in combination with poor governance and management. However, it is not always easy to prove direct responsibilities and the outcomes of scientific studies often depend on who is measuring what, and who is the sponsor. In Guatemala, according to the Mining Act from 1997, the Ministry of Energy and Mining (MEM) is the administrative body that coordinates all kinds of mining activities. Also, the Environmental Impact Assessment that a company has to submit before receiving a licence is addressed to the MEM. However, the Ministry of Environment and Natural Resources (MARN) is responsible for actually monitoring and following up on compliance with environmental regulations (Castagnino 2006, 14). These intertwined responsibilities complicate the situation. The following section gives an overview of some of the documented impacts allegedly linked to the Marlin Mine.

2.1.1 Environmental Impact

In the case of the Marlin Mine, there are well-founded allegations about negative impacts on water quantity and quality. The Guatemalan mining law from 1997 allows the companies to use water without paying for it. For the leaching process with cyanide, the Marlin Mine uses an enormous amount of 250,000 litres per hour, which equals an average Guatemalan family's household water consumption over 22 years (McBain-Haas/Bickel 2005: 17). According to the company, there is no significant impact of the mine on water supply as it does not use the same aquifer source that supports the local wells. However, according to the local organization ADISMI (Asociación de Desarrollo Integral de San Miguel Ixtahuacán), over 40 wells have dried up in the surrounding communities since the beginning of the mining process.[30]

Water contamination poses another problem: over the years, several independent studies on water pollution have been published. A study of the El Salvador Centre for Research on Investment and Trade about the water quality of the Sipacapan Tzalá River has shown that the river downstream of the Marlin Mine is heavily contaminated. The content of copper was 80 times higher than the acceptable limit established by the Guatemalan law, aluminium 13 times and magnesium 2.5 times higher (Bianchini 2006: 4). Another report, published by the Pastoral Peace and Ecology Commission (Comisión Pastoral Paz y Ecología, COPAE) in July 2009, revealed that the Tzalá and Quivichil Rivers around the Marlin Mine contain high levels of iron, aluminium, manganese, copper and nitrate, and are thus not safe for drinking. The metal and nitrate content surpass the standard for open-pit mining established by the World Bank as well as the Environmental Protection Agency of the United States (US EPA), the Ministry of Environment and Natural Resources (MARN) and the Guatemalan drinking water standards (COPAE 2009: 7). The contamination of water sources around the mine has direct impacts on the daily life of the local residents: The rivers are not only used as drinking water sources for the people but also as water source for the cattle. Peasant farmers have reported several cases of dead livestock after drinking from water sources below the Marlin Mine.[31] A study of the health situation around the mine, published by Physicians for Human Rights in May 2010 in response to the accusation of human rights violations in the region, showed elevated levels of manganese in human urine samples. The examined residents living closer to the mine had higher concentrations of mercury, arsenic, copper and zinc in their

30 GHRC/USA (Guatemala Human Rights Commission/USA): Goldcorp's Mining In San Miguel Ixtahuacán. <www.ghrc-usa.org/AboutGuatemala/Goldcorp.htm> accessed on 15.10.2012.

31 FREDEMI: Guatemala. Rural Farmers Lose Livestock Due to Water Contamination by Marlin Mine. 26.11.2012. <upsidedownworld.org> accessed on 25.07.2013 and Russell, Grahame: Guatemala: Even The Cows Are Being Killed By Goldcorp's Mine In Guatemala. In: <indigenouspeoplesissues.com> accessed on 17.10.2012.

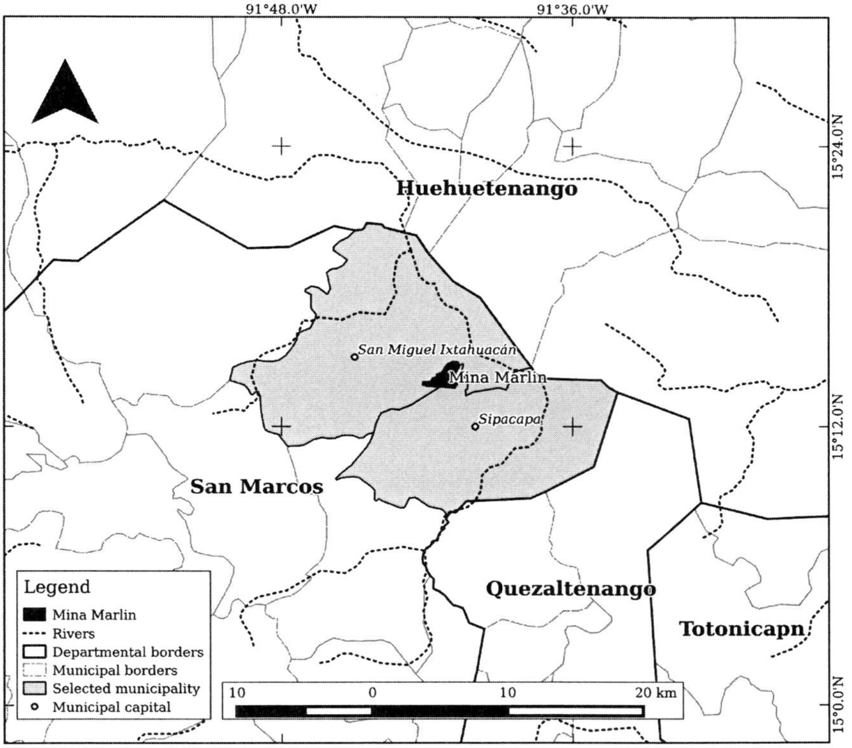

Map 2: Location of Mina Marlin in the department of San Marcos.

urine than people living farther away. Since the study dealt with metal pollution, chemicals like cyanide were not investigated. Nevertheless, the health complaints made by residents, like skin rashes and respiratory ailments, are likely to be a result of cyanide exposure (Basu/Hu 2010: 11, 15). The water held in the tailing pond must comply with the discharge standards before being released. A study conducted by E-Tech International (2010: 79) showed that water quality is not in accordance with the IFC/World Bank guidelines. Also, there are allegations that the design of the tailing pond is deficient as no impermeable liner was used to prevent leakage into the ground, and there were repeatedly problems with the handling of cyanide, including a spill during transportation and the lack of an import licence, for which the company was fined by the MARN (Dougherty 2011:8). Nevertheless, Goldcorp denied any responsibility for the contamination of water or negative impacts on people's health, referring to water samples taken by governmental institutions and the company-funded Asociación de Monitoreo Ambiental Comunitario (AMAC): "[t]o date, over a million data points have been analysed; all of them comply with internationally-accepted standards." Also, these test results do not "show any significant adverse impacts related to mining".[32]

32 Goldcorp Inc. 13.07.2012: Dispelling the Myths of Marlin. <goldcorp.com> accessed on 20.10.2012.

Fig. 2: View of the tailing pond of Marlin Mine. Photo: Anna Leissing, Guatemalanetz Bern.

Another problem often voiced by local people is the vibration caused by heavy machinery that damages houses. A study of cracked walls and floors in the houses near the mine concluded that "the most heavily supported cause of the structural damage in the villages [Agel, El Salitre, San José Ixcaniche and San José Nueva Esperanza] around the Marlin Mine is the Marlin Mine blasting and increased vehicle traffic of heavy mine supply trucks." This was the result of a comparison between the villages Escupija and Chininguitz, sharing similar climate and seismic conditions. However, different houses have different damages. Since the research team was not around before the mine started to operate, the cause of the damage cannot be determined with absolute certainty (COPAE/UUSC 2009: 49).[33]

2.1.2 Social Impact

A major problem in the area are the increased social divisions among the local communities and the resulting conflict and violence. There is a divide between mining proponents and families which resist selling their plots to the company. The latter receive threats and suffer acts of violence at the hands of people who profit from the mine, such as mining workers.[34] One example is

33 COPAE: Guatemala: Canadian Miner Accused of Damaging over 100 Homes. In: Mines and Communities, 11.11.2009. <www.minesandcommunities.org//article.php?a=9644> accessed on 15.10.2012

34 MiningWatch Canada 2010: Urgent Action. Shooting of Community Leader Opposing Goldcorp Inc.'s Marline Mine in Guatemala. <www.miningwatch.ca> accessed on 28.11.2012.

the assassination of the son of an anti-mining leader on 7 September 2009. The incident happened three months after his father, Arturo Mendez, denounced Goldcorp's activities in Guatemala in front of Goldcorp's CEO, a board of directors and shareholders. Before that, Mendez had received threats that one of his family members would be killed if he would not stop his engagement.[35] Another incident took place on 7 July 2010: Two young men showed up at the door of Diodora Antonia Hernández Cinto, who is part of the resistance movement, and asked for a cup of coffee. When she went to hand them the coffee, one of the men shot her in the head. Diodora Hernández survived the attack. Her husband stated that a company's official had told the mine workers one week earlier that there would be less work in the future because she had refused to sell her plot to the company. After the incident, the government provided the woman a security guard living with her family. The two men who committed the crime were released after questioning. One of the men had been working for Goldcorp's subsidiary Montana, the other for one of Montana's contractor companies.[36] Due to the high degree of impunity in Guatemala, none of the perpetrators have been held accountable for these crimes, leading to a further decrease in the local people's trust in governmental institutions.

2.2 Local People's Strategies

The anti-mining movement in Guatemala is linked to a transnational activist network including organizations from North, Central and South America and Europe. In the case of the Marlin Mine, the involved actors are mainly indigenous peasants from rural areas of the Western Highlands of Guatemala, supported by the local and diocesan Catholic Church, as well as local, regional, national and international NGOs and solidarity groups, including some large ones like Oxfam America. Dougherty (2011: 410) distinguishes four different tracks of resistance against mining in Guatemala: 1) judicial: use of national and international courts to address grievances; 2) transnational activists: solidarity campaigns to shame the Guatemalan state and the companies; 3) legislative: advocating for new a mining legislation; 4) incidents of civil disobedience.

In 2004, social and environmental organizations founded the National Front Against Opencast Metal Mining (FNRMM in Spanish), promoting a reform of the mining law and supporting environmental protection and the rights of the affected population. The following year, different events such as a Regional Conference of Indigenous Authorities and a Regional Mayan Legislative Meeting were held to debate the issue, whereby mining was clearly opposed by the participants (Castagnino 2006: 20). The arrival of Glamis Gold to Guatemala nourished the controversy around mining. Company and government agencies organized information campaigns in favour of mining, to facilitate the land acquisition process. However, there was no transparent information made available to the affected communities (Willems 2010: 45–46). According to the Guatemalan constitution and ILO Convention 169, the implementation of a mining project requires the prior consultation of the indigenous population. However, this did not happen in any meaningful way, as a member of an investigation by the Human Rights Ombudsman's office comments: "there are some people who say there [sic] were invited to coffee to learn about the project. They went and their information was taken down [...] they were required to sign, and this appears to be the document that was presented as proof of the town's approval" (cited in Castagnino 2006: 21; see also Fulmer et al. 2008: 101–104).

35 Goldcorp Out Of Guatemala 01.08.2009: Son of Guatemalan Anti-Mining Leader is Assassinated. <www.goldcorpoutofguatemala.com> accessed on 25.10.2012.

36 Indigenous Peoples Links 11.07.2011: Diodora Hernández Continues in Resistance to Goldcorp Inc. <int.piplinks.org/Diodora+Antonia+Hernandez+> accessed on 25.10.2012.

Local people were concerned about the potential environmental damage the mine could cause and felt insufficiently informed and consulted. The resistance against the mine took off during its construction. In November 2004, an opinion poll in San Miguel Ixtahuacán and Sipacapa, carried out by the company Vox Latina, showed 95.5% rejection of the mining project.[37] In December 2004, a convoy bringing a milling cylinder to the mining site was blocked for forty days by protesters, until the National Police and the Guatemalan Army intervened violently, killing one and injuring 16 protesters (Zarsky/Stanley 2011: 10). In January 2005, communities in Sipacapa, with the support of the NGO Madre Selva, submitted a complaint to the Compliance Advisor Ombudsman (CAO) of the IFC concerning potential environmental and health impacts and the lack of adequate consultation. In March 2005, the CAO visited the region to mediate between the local population and the company (Willems 2010: 48; Zarsky/Stanley 2011: 11). On the 18th of June 2005, the population of Sipacapa held a community referendum about the project. 2,500 people of 13 Sipacapa communities, about 45% of the registered population, voted in the consultation, of which 98% rejected mining on their territories. The referendum attracted considerable international attention. About 75 national and international observers were present to assure a democratic voting. Glamis Gold had tried to stop the referendum in advance and denied its legitimacy. It required an order from a Civil Court, which decided in favour of the company and requested the mayor of Sipacapa to suspend the consultation. At the same time, the request for an injunction was rejected by the Constitutional Court, based on the ILO Convention 169 and the municipal code. However, the municipal code has inconsistencies with regard to the required minimum participation necessary to make decisions binding, which made it easy for the company and pro-mining government agencies to ignore the results. Nevertheless, the consultation "set a rights-based example of resistance for indigenous communities in the Western Highlands of Guatemala" (Willems 2010: 49; Castagnino 2006: 22).

In 2008, Gregoria Crisanta Perez and seven other women damaged a power line to the mine, which was installed on Perez' property without her permission. After this act of sabotage, the mine was not able to exploit gold for over a month. Perez went into hiding, until the police caught her in February 2010. When the police wanted to transfer her to San Marcos, villagers blocked the road and liberated her.[38] In another incident, protesters set fire to a pick-up truck and an exploration drill rig of the company and justified this with the fact that the company had not obtained the rights to access the land.[39]

In December 2009, a coalition of community groups from San Miguel Ixtahuacán submitted a complaint to the Canadian Department of Foreign Affairs committee, according to the Guidelines for Multinational Enterprises of the Organization of Economic Cooperation and Development (OECD)[40]. They requested an investigation into the mine's impact on water quality and quantity, health problems, as well as structural damage of houses in the area. In May 2011, the responsible Canadian National Contact Point of the OECD closed the complaint, recommending the parties to "participate in a constructive dialogue".[41]

37 Ramírez, Alberto 2004: Opposition to Mining Activity in San Marcos. *Prensa Libre*. 04.11.2004. <www.minesandcommunities.org> accessed on 16.10.2012.

38 Russell, Grahame: Goldcorp Inc. Linked to Capture of Gregoria Crisanta Perez. In: Rights Action. 02.04.2010. <www.rightsaction.org/articles/Goldcorp_&_Mam_woman_020410.html> accessed on 16.10.2012. and Romero, C. and T. Keppeler: Schadstoffe in Minen in Guatemala: Gold macht Kinder krank. *Tageszeitung*. 20.07.2011. <www.taz.de/Schadstoffe-in-Minen-in-Guatemala> accessed on 16.10.2012.

39 Mines and Communities. 07.06.2009: Protesters burn equipment at Marlin mine, Guatemala. <www.minesandcommunities.org> accessed on 20.10.2012.

40 French, Cameron: Guatemalan Groups File Complaint on Goldcorp Mine. In Mines and Communities. 09.12.2009. <www.minesandcommunities.org> accessed on 16.10.2012.

41 Canadian National Contact Point: Final Statement on the Notification dated 9 December 2009. 3.5.2011.

Fig. 3: Women walking by a ruin in an abandoned village close to Marlin Mine. Photo: Anna Leissing, Guatemalanetz Bern.

Another strategy concerns the revaluation of the traditional indigenous peoples' municipal council, the *Alcaldía Indígena*, which dates back to the colonial era. In 2005, the government dissolved the *Alcaldía Indígena* in San Miguel Ixtahuacán, apparently to reduce resistance against the mine. But in 2008 it was re-launched by members of the resistance movement. They see the *Alcaldía Indígena* as a legitimate organ that represents the interests of the indigenous population better than the state institutions. In the meetings, concerns about mining impacts and possibilities for action are discussed (Spetzler 2009: 70). On several occasions, the resistance was expressed in larger demonstrations and gatherings. The "Central American Week of Resistance against Mining" in June 2006 included forums and community meetings, a march ending in front of the Canadian Embassy in Guatemala City, as well as a demonstration in Sipacapa against the Marlin mine.[42] Again, on 28 July 2010, a big demonstration was held in Guatemala City, led by the internationally known Bishop Álvaro Ramazzini and Nobel Peace Price winner Rigoberta Menchú Tum.[43]

<www.oecd.org/daf/inv/mne/48754883.pdf> accessed on 03.07.2013.

42 Mines and Communities 2006: Week of Anti-mining Actions in Guatemala. 13.6.2006. <www.minesandcommunities.org> accessed on 19.04.2012.

43 Cascadia Solidaria 2010: Maya Convergences in Guatemala City for Defense of Natural Resources and Territorial Rights. 30.07.2010. <cascadiasolidaria.wordpress.com> accessed on 20.10.2012.

The anti-mining protests also attracted the attention of international human rights bodies: in March 2010, the ILO requested the government to suspend the mining activities because of inadequate and untimely consultations. The same year, James Anaya, UN Special Rapporteur on Human Rights of Indigenous Peoples, focussed his visit in Guatemala on the Marlin Mine. His report stated an "alarming lack of legal protection for the rights of indigenous people over their lands and territories" and requested the government to adopt a law that defines and assures rights to consultation for resource development projects such as mining (Human Rights Council 2011). In May 2010, the Inter-American Commission on Human Rights (IACHR), a body of the Organization of American States, ordered the Guatemalan government to suspend mining operations as a precautionary measure "to prevent and protect against potential serious and irreparable harm to persons or groups of persons who are in imminent peril."[44] However, this never happened (see next section). But it did have consequences at other levels, such as the temporary removal of Goldcorp from the Dow Jones Sustainability Index (DJSI) in September 2011.[45] Finally, in July 2012 the Peoples' International Health Tribunal met in San Miguel de Ixtahuacán. After hearing testimonies from people affected by different Goldcorp mines in Central America (including Marlin), the expert committee of this unofficial court found the company and the governments of the host countries and Canada guilty for promoting mining investments with diverse negative impacts.[46]

Although the protests against the mine have not achieved its closure, resistance continues also after the extraction has moved from the open-pit to underground, with probably just a few years left to the definitive closure. Aniseto Lopez from the Front in Defense of San Miguel de Ixtahuacán (FREDEMI) once stated: "And cost us what it will, we will achieve our goal—because reality is on our side."[47]

2.3 Corporate Strategies and the State

Goldcorp presents itself as a sustainable company with particular emphasis on "responsible mining". On its website, Goldcorp states that it is "committed to being responsible stewards of the environment and to maintaining the highest health and safety standards possible" and that it tirelessly advocates human rights and maintains "a principled, conscientious approach to corporate citizenship".[48] The "sustainable prosperity" objective is to make positive and lasting contributions in the communities where it works. Goldcorp demonstrates commitment through the application of Corporate Social Responsibility standards, occupational, health and safety policies, as well as through the application of codes of conduct. It adheres to the Voluntary Principles on Security and Human Rights (since 2006), the Cyanide Management Code (since 2007, Marlin being certified in 2009), and is a member of the ICMM, the UN Global Compact and other CSR initiatives. Since 2005, it publishes annual sustainability reports, both for the company and for specific mines.

44 IACHR in: Organization of American States w.y.: Precautionary Measures. About Precautionary Measures. <www.oas.org> accessed on 28.10.2012.

45 The DJSI, launched in 1999, serves as benchmark for "sustainable investors" and corporate sustainability guidelines. See Dow Jones Sustainability Indexes 2012 : Index Family Overview. <www.sustainability-indexes.com> accessed on 16.10.2012.; and MiningWatch Canada: Goldcorp Removed from Dow Jones Sustainability Index. 21.9. 2011. <www.miningwatch.ca> accessed on 19.04.2012.

46 Some testimonies and the verdict are documented on <healthtribunal.org>; accessed on 15.05.2015.

47 The Esperanza Project. Indigenous Guatemalans to UN: No to Mining, Yes to Life. 21.6.2010 <www.censat.org> accessed on 23.9.2013.

48 Goldcorp Inc.: Responsible Mining: Creating Sustainable Prosperity. <goldcorp.com> accessed on 22.10.2012.

Considering the Marlin Mine project, the company "has endeavoured to build lasting partnerships with surrounding communities and strengthen families through increased health, economic, educational and social opportunities", as stated on its website.[49] To back up these lofty words, Goldcorp has established the Sierra Madre Foundation in San Marcos, through which it finances development projects such as the construction of schools, roads and sport fields, as well as the construction and equipping of a local hospital.[50] Goldcorp's strategy also includes payments to the local municipalities. For instance, it donated a total of eight million *quetzals* (roughly USD one million) to the municipality of Sipacapa in March 2010.[51] The company also itself conducts studies on environmental and social impacts.

Confronted with independent studies showing contamination and other problems at Marlin, Goldcorp's strategy is to selectively ignore them. Instead, it refers to studies made by agencies of the government of Guatemala and others which it commissioned itself. According to Mining Watch Canada, in 2011 none of these studies had been made public. Also, they "were conducted by agencies without the capacity to do them".[52] These studies contradict and delegitimize the findings of independent scientific studies and thus allow Goldcorp to avoid controversial topics.

Similarly, when the IACHR ordered the closure of the Marlin Mine as a precautionary measure in 2010, Goldcorp simply ignored it and the mine continued to operate. Since Guatemala is a member of the Organization of the American States, the IACHR's order was binding, but not enforced by state authorities. Instead, more than one year later, the vice-minister of Energy and Mines Ricardo Pennington stated that "[a]fter all the studies, analyses and participation of various [governmental] actors involved, we conclude that there are no grounds for the suspension of the mine."[53] In December 2011, the IACHR withdrew its previous order but requested the state of Guatemala to take the necessary measures to guarantee that the people living in the surrounding communities have access to safe drinking water.[54]

While Goldcorp on the one hand ignores unfavourable verdicts, on the other hand it makes use of criminal charges against protesters, for example after an encounter between local farmers and the mine's private security force in 2007. In Guatemala, victims of human rights violations are often subjected to penal prosecution by their aggressors, but powerful actors such as Goldcorp not only "enjoy impunity for their harmful actions but they are able to manipulate the justice system as a tool for repression".[55] Activists do not only have to fear legal persecution, but also death-threats, intimidation and defamation campaigns as well as frequent assassinations (Yagenova/García 2009: 76). The use of both private and state security forces has received substantial civil society criticism. In Guatemala, this is a particularly sensitive issue, considering the country's history of conflict. In the Department of San Marcos, fifteen massacres took place in the civil war between 1962 and 1996. Armed mining security forces therefore trigger a powerful collective memory (Holden/Jacobson 2008: 337). The Marlin Mine has so far contracted three private secu-

49 Goldcorp Inc.: Marlin — Community Initiatives. <goldcorp.com> accessed on 22.10.2012.

50 Goldcorp Inc.: Fundación Sierra Madre. <goldcorp.com/Responsible-Mining> accessed on 20.10.2012.

51 Rigalt, Carlos: Medidas Cautelares de Cierre de Mina Marlin sin Ser Acatadas. In: El Periódico. 17.02.2011. <www.elperiodico.com.gt/es/20110217/pais/190985/> accessed on 16.10.2012.

52 MiningWatch Canada: FAQ Sheet What You May Not Know About the Marlin Mine in Guatemala. <goldcorpoutofguatemala.files.wordpress.com> accessed on 20.10.2012.

53 Cited in: Macpherson, Andrew: Prospecting the Terrain of Struggle. In: The Dominion. 03.08.2011. <www.dominionpaper.ca/articles/4106> accessed on 15.10.2012.

54 IACHR in: Organization of American States w.y.: PM 260-07 — Communities of the Maya People (Sipakepense and Mam) of the Sipacapa and San Miguel Ixtahuacán Municipalities in the Department of San Marcos, Guatemala. <www.oas.org> accessed on 28.10.2012.

55 Ahniwanika Schertow, John 2007: Trial of "Goldcorp 7" Continues in Guatemala. 28.11.2007. <intercontinentalcry.org> accessed on 03.07.2013.

rity companies, of which at least one has a military background. On top of this, various units of the military are active in the region, some of them directly engaged in the protection of the mine (for example, a small contingent protecting explosive magazines). State security forces (military and police) intervened in confrontations between protesters and the mining company on at least five occasions. At the time of the violent breakup of a blockade in 2005, the company did not have any human rights policies and procedures with regard to the interaction with state security forces, and it is not clear in what ways it was co-responsible for the repression. But also after Goldcorp adopted the Voluntary Principles on Security and Human Rights in the Marlin Mine in 2006, several incidents ended with criminal charges against community members, and there was still no formal agreement with the state security forces to clarify the terms for their interventions (On Common Ground Consultants 2010: 160–171).

Role of the State

The Guatemalan government faces a dilemma of interests and obligations: it has an economic interest and seeks to attract foreign capital in line with the neoliberal economic model. At the same time, the state is required to protect human rights, including the rights of indigenous peoples, according to the Guatemalan Constitution and international conventions and agreements such as ILO Convention 169. On the one hand, government agencies use a discourse of human rights and sustainable development, expressed in a fairly impressive legal and institutional framework protecting human rights and indigenous people's rights, as well as in certain specific actions such as the provision of security guards after the threats against Diodora Hernández (see section "social impact"). On the other hand, the government acts (openly or in a hidden manner) in favour of Goldcorp, expressed for instance in the non-enforcement of the IACHR order to suspend mining at Marlin. Exponents of the *Alcaldía Indígena*, the indigenous assembly in San Miguel Ixtahuacán, complain about public authorities trying to anticipate and weaken the resistance against the mine. The official mayors are seen as acting in favour of their political parties, intimidated by superior (pro-mining) state powers, and therefore trying to keep local communities calm rather than defending their interests (Spetzler 2009: 70). As mentioned above, the criminalization of social protest can only happen with the complicity of state agencies. This is illustrated for example by an anonymous document titled "Organizations That Promote Social Conflict in Guatemala", prepared as information to the President, in which local organizations and the Catholic Church (particularly, Bishop Álvaro Ramazzini) are characterized as belonging to communist ex-guerrilla movements, allegedly connected to drug trafficking (Anonymous 2012). Such discourses obviously serve to undermine the legitimacy of the anti-mining movement and prepare the ideological base to justify their violent repression by state security forces.

3. Conclusion

Guatemala, marked by its violent background with a 36-year-long civil war, faces huge challenges such as a corrupt government, organized crime networks, land conflicts, as well as high rates of violence, especially against women and human rights defenders. The Guatemalan government has so far been unable and/or unwilling to protect indigenous peoples against human rights abuses. In case of the Marlin Mine, Goldcorp has taken the chance of operating in a weak governance context and assumes state-like functions in the municipalities where it operates. This is visible, for example, in infrastructure projects such as the construction of roads and schools. At the same time, the complicity of state and private security forces in the violent repression and criminalization of protest shows the close relationship between the extractive industry and the state. This collaboration is highly questionable in a fragile context such as Guatemala, where governmental institutions should rather be focussing on strengthening their accountability towards the population. Although the government emphasizes its efforts to protect the rights of indigenous peoples, it has failed to consult the affected communities adequately, as required by ILO Convention 169 and the National Constitution. This has clearly violated indigenous rights to participate in decision-making concerning the development of their territories.

Adding to the lack of consultation and grave health and environmental concerns are deep social divisions within the affected communities, further intensified by a high poverty rate. Most people living around the mine depend on remittances from migrated family members and on subsistence farming. Tourism or other alternative income sources are largely absent in the Western highlands. Those who profit from the few workplaces created by the mine thus often get into conflicts with the rest of the population. The power imbalance between Goldcorp and the local communities is one of the most striking features of the Marlin Mine. With its immense economic capital and good relations with the government, the company has the power to create its own institutions for monitoring environmental and health impacts of the mine, thereby producing investigation results that contradict the findings of independent studies.

The Marlin Mine has become the rallying point for a human rights-based movement well beyond the highlands of Guatemala, which has also attracted important international attention. A broad coalition of different actors from local to international NGOs, churches and farmers' groups supports the affected indigenous communities in their struggle against mining, with the right to consultation and to a healthy environment as the core issues. Besides the few people who directly benefit from the mine, the local population articulated the overwhelming disapproval of mining activities in a self-organised community consultation in Sipacapa in 2005. However, the government and company have both refused to accept this decision. Despite all this, mining is still going on. However, the last stage of the Marlin Mine is approaching fast. The mine closure process, expected to unfold in 2017, will change the situation dramatically and may set important precedents for the future viability of mining in Guatemala. If it turns out that, after the gold is exhausted, the local population is left with a toxic legacy and even the few economic gains made by them were just temporary, the opposition against new mines may become even stronger.

References

Allebrand, Raimund 1997: *Die Erben der Maya. Indianischer Aufbruch in Guatemala.* Bad Honnef: Horlemann.

Anonymous 2012: *Informe Preparado para el Presidente de la República y Autoridades Gubernamentales en Materia de Seguridad, Justicia y Desarrollo Energético. Organizaciones que Promueven la Conflictividad Social en Guatemala.*

Basu, Niladri and Howard Hu 2010: Toxic Metals and Indigenous Peoples near the Marlin Mine in Western Guatemala. Potential Exposures and Impacts on Health. Published by Physicians for Human Rights.

Bianchini, Flaviano 2006: *Estudio Técnico. Calidad de Agua del Río Tzalá* (Municipio de Sipakapa; Departamento de San Marcos). <blog.reportero.org/wp-content/estudio_de_agua_del_rio_tzal.pdf> accessed on 19.04.2012.

Castagnino, Vincent 2006: Metal Mining and Human Rights in Guatemala. The Marlin Mine in San Marcos. Peace Brigades International. < www.peacebrigades.org> accessed on 21.04.2012.

CEH (Comisión para el Esclarecimiento Histórico) 1999: *Memoria del Silencio. Conclusiones y Recomendaciones del Informe de la Comisión para el Esclarecimiento Histórico.* <srhrl.aaas.org > accessed on 4.9.2013.

CIEN (Centro de Investigaciones Económicas Nacionales) 2009: *Contribución de la Industria Minera al Desarrollo de Guatemala.* <www.cien.org.gt> accessed on 28.11.2012.

COPAE 2007: *Die Mine Marlin. Bilanz nach fast zwei Jahren des Bergbaus.* Guatemalanetz. <www.guatemalanetz.ch> accessed on 24.11.2012.

COPAE 2009: Second Annual Report on the Monitoring and Analysis of the Water Quality. Current Situation of the Water in the Tzala and Quivichil Rivers in the Influence Area of the Marlin Mine, Located in the Municipalities of San Miguel Ixtahuácan ans Sipakapa, Department of San Marcos. Guatemala. <www.ciel.org> accessed on 19.04.2012.

COPAE and UUSC (Unitarian Universalist Service Committee) 2009: Damaged Buildings near the Marlin Mine: Preliminary Investigation and Analysis of Building Damage in the Villages of Agel, El Salitre, San José Ixcaniche, and San José Nueva Esperanza, San Miguel Ixtahuacán and Sipacapa Municipalities, San Marcos Departement, Guatemala. <www.ciel.org> accessed on 19.04.2012.

Diagnóstico Municipal, Oficina de Planificación Municipal, 2006: Sipacapa. Fundacion Sierra Madre. <www.fundacionsierramadre.org/FSM17_sipacapa.htm> accessed on 28.11.2012.

Dougherty, Michael L. 2011: The Global Gold Mining Industry, Junior Firms, and Civil Society Resistance in Guatemala. *Bulletin of Latin American Research* 30 (4): 403-418.

E-Tech International 2010: Evaluation of Predicted and Actual Water Quality Conditions at the Marlin Mine, Guatemala. <www.etechinternational.org> accessed on 8.9.2013.

Fischer, Edward F. and R. McKenna Brown 1996: Maya Cultural Activism in Guatemala. Austin: University of Texas..

Fulmer, Amanda, Angelina Snodgrass Godoy and Philip Winter Neff 2008: Indigenous Rights, Resistance, and the Law: Lessons from a Guatemalan Mine. *Latin American Politics and Society* 50 (4): 91-121.

Galemba, Rebecca B. 2012: "Corn is Food, not Contraband": The Right to Free Trade at the Mexico-Guatemala Border. *American Ethnologist* 39 (4): 716–734.

Garbers, Frank 2000: Symbolische Kämpfe. Die Maya in Guatemala kämpfen für das Recht auf Ungleichheit als Weg zur Gleichberechtigung. *Lateinamerika Nachrichten* 315/316.

Guatemala Human Rights Commission (GHRC) 2011: Human Rights Analysis Report. <www.ghrc-usa.org> accessed on 26.11.2012.

Hamilton, Sarah and Edward F. Fischer 2003: Non-Traditional Agricultural Exports in Highland Guatemala. Understandings of Risk and Perceptions of Change. *Latin American Research Review* 38 (3): 82-110.

Hirschmann, Barbara 2007: *Vom Indio zum Maya. Identitätspolitik der Mayabewegung in Guatemala.* Diplomarbeit. Wien: Universität Wien.

Holden, William N. and R. Daniel Jacobson 2008: Civil Society Opposition to Nonferrous Metals Mining in Guatemala. *Voluntas* 19 (4): 325–350.

Human Rights Council 2011: Observations on the Situation of the Rights of the Indigenous People of Guatemala with Relation to the Extraction Projects, and other Types of Projects, in their Traditional Territories. UN Doc. A/HRC/18/35/Add.3. <www2.ohchr.org> accessed on 3.10.2012.

Kurtenbach, Sabine 1998: *Guatemala.* München: Verlag C. H. Beck.

Lateinamerika-Institut 2010: *Caminos – Eine Reise durch die Geschichte Lateinamerikas. Encomienda.* <www.lai.fu-berlin.de/e-learning/projekte/caminos/lexikon/encomienda.html> accessed on 28.11.2012.

McBain-Haas, Brigitte and Ulrike Bickel 2005: Open Pit Gold Mining- Human Rights Violations and Environmental Destruction. The Case of the Marlin Gold Mine. Heidelberg/Aachen: FIAN International/Misereor. <www.fian.org> accessed on 4.9.2013.

McKillop, Heather 2004: The Ancient Maya. New Perspectives. Santa Barbara: ABC-CLIO.

MEM (Ministry of Energy and Mines) 1997: Mining Law and its Regulation. <www.mem.gob.gt> accessed on 4.9.2013.

MEM (Ministry of Energy and Mines) 2011: *Anuario Estadístico Minero 2010.* <www.mem.gob.gt> accessed on 19.04.2012.

Molkentin, Gudrun 2002: *Kriegsursachen und Friedensbedingungen in Guatemala. Eine historische Untersuchung über das kriegsursächliche Gewaltgeschehen und die Herausforderungen an den gegenwärtigen Prozess der Friedenskonsolidierung.* Frankfurt a. M.: Lang.

On Common Ground Consultants 2010: Human Rights Assessment of Goldcorp's Marlin Mine. Executive Summary. Vancouver, BC, Canada. <www.hria-guatemala.com> accessed on 19.04.2012.

Spetzler, Julia 2009: *Megaprojekte und internationale Begleitarbeit in Guatemala. Fallstudie über die internationale Begleitarbeit von Menschenrechtsaktivisten im Konfliktkontext von Megaprojekten in Guatemala.* Masterarbeit. Luzern, Schweiz. Available from <www.peacewatch.ch> accessed on 21.10.2012.

Van de Sandt, Joris 2009: Mining Conflicts and Indigenous Peoples in Guatemala. CORDAID. <www.cordaid.org> accessed on 8.9.2013.

Viscidi, Lisa, 2004: A History of Land in Guatemala: Conflict and Hope for Reform, Americas Program. Silver City, NM: Interhemispheric Resource Center. <www.nisgua.org> accessed on 4.9.2013.

Weisshar, Emmerich 2008: Sprachen und Kulturen Guatemalas. In: Kurtenbach, Sabine, Werner Mackenbach, Günther Maihold and Volker Wünderich (Eds.): *Zentralamerika. Politik, Wirtschaft, Kultur.* Frankfurt am Main: Vertuet Verlag. 731-750.

Willems, Luis 2010: Mining and Indigenous Peoples in Guatemala. The Local Relevance of Human Rights. University of Gent. Available from <www.catapa.be> 30.03.2013.

Yagenova, Simona and Rocío García 2009: Guatemala. El Pueblo de Sipakapa Versus la Empresa Minera Goldcorp. *OSAL* 25. Buenos Aires: CLACSO.

Zarsky, Lyuba and Stanley, Leonardo 2011: Searching for Gold in the Highlands of Guatemala: Economic Benefits and Environmental Risks of the Marlin Mine. Medford, MA: Global Development and Environment Institute, Tufts University.

The Las Bambas Copper Mine Project in Apurimac, Peru

Annette Wallimann[1]

1. Situating the Case Study

The proposed Las Bambas copper mine is situated on the border between the provinces of Grau and Cotabambas in southern Peru. According to official statistics, these provinces in the Apurimac region are underdeveloped and extremely poor. Mining is therefore considered to be the only opportunity for economic development for the rural population with Xstrata, the Anglo-Swiss mining company,[2] as the major player promising the achievement of socially, economically and ecologically sustainable development to Cotabambas and the entire Apurimac region. In 2004, the company acquired the first concession to promote copper extraction in Las Bambas. For this undertaking, Xstrata provides the area with self-styled environment-friendly measures and social contributions to the local farming communities directly affected by the mining activities. In 2011, exploration was finished and construction of the mine had started.

The project is set to become one of the world's largest copper mining projects. Former President of Peru, Alan García (1985–1990 and 2006–2011) called it "The Mining Treaty of the Century" (El Comercio 2010) while renewing the contract and extending Xstrata's mining concession for exploitation on 2 September 2010. At the national and international levels, the project is seen as a model of socially responsible and sustainable mining. Xstrata and the Las Bambas project belong to a set of projects denominated as "new mining", according to the economist José de Echave[3]. The mining sector has made a huge effort to overcome its (earlier) negative reputation (De Echave et al. 2009b: 271). However, a number of social conflicts are associated with mining projects and have constantly increased since the mid-1990s. These conflicts have different causes and manifestations, but they are all linked to the social, ecological and economic effects of mining. According to De Echave et al. (2009b), over 50 per cent of social conflicts in Peru were directly related to mining problems in 2009.

This chapter is the result of a two-month field research in Cotabambas, carried out for a Masters thesis in social anthropology at the University of Bern in 2010 (Wallimann 2011). It gives an example of the dynamics which develop at the early stage of a mining operation and the impact of an international mining company like Xstrata on national and local policy and on the social fabric of the region, long before the first tonne of minerals has been extracted.

1 This chapter was translated from German by Annabelle Piguet and Andrea Müller.

2 On 2 May 2013, ownership of Xstrata was acquired by Glencore and the Las Bambas project was later sold to a Chinese consortium (see update at the end of this chapter).

3 De Echave is a former director of CooperAcción, a national level NGO.

1.1 History and Current Context

Since its inclusion in the Republic of Peru in 1873, the Apurimac department has been divided into seven provinces, marked by a continuous tendency of fragmentation. The regional government has been making every effort to achieve political unification. There are strong links and similarities between parts of Apurimac and adjacent regions of Cusco, Arequipa and Ayacucho. Lack of social cohesion and internal articulation are the main causes for the political difficulties of Apurimac's administration (PDRC 2010: 35). The processes of national integration, centralization and modernization in the early 20th century have brought negative consequences for this southern Andean region. During this period, political and economic power began to mainly concentrate in the capital, Lima. The mountainous and Amazon regions were seen as 'empty spaces' due to their insignificant economic activities, even if they had natural resources of great value. Since the establishment of the Republic of Peru in the early 19th century, the *haciendas* system had perpetuated the injustice of the Spanish colonial land regime with natives working for *hacienda* landowners under conditions comparable to serfdom. After the Second World War, thousands of people from Apurimac began migrating towards the coast, then a series of peasants' riots and land occupations followed in 1966 and 1967. Finally, in 1969, land reforms abolished the old feudal structures of the *hacienda*. This also brought in an advancing capitalist economy, transforming Apurimac from a rural area to the urban-rural region it is today (PDRC 2010: 37-40).

A social characteristic of the region is the climate of mistrust and social tension that emerged in the 1980s with the armed conflict between the state and the Maoist *Sendero Luminoso* ("Shining Path") guerrilla movement. This movement was characterized by a brutal strategy of terror that was directed against politicians and state security forces on the one hand, and against all members of civil society suspected of cooperating with the government on the other. At its peak, Sendero Luminoso consisted of over 70,000 guerrilla fighters who controlled huge parts of the rural areas in the centre and south of Peru, Apurimac being one of their strongholds. The population, particularly in the rural communities, suffered from the state's armed forces conducting violent counterattacks during the second half of the Sendero Luminoso period until 1992, when a majority of its leaders were arrested and the organization fell largely apart (Marti in NZZ 6 June 2009). According to official estimates, over 30,000 people were killed and about 600,000 people displaced in this conflict, most victims being civilians (Comisión de la Verdad y Reconciliación 2011). The structural conditions left after this long period of violence are considered to be a fertile ground for the present clientelism and corruption among local authorities (PDRC 2010: 38).

Mining is not a new phenomenon in Peru. Large parts of the country have a long history of mining, dating back to the Inca empire. From the colonial conquest in the 16th century to independence in 1821, Peru's gold and silver mines were central for the economy of the Spanish crown. However, the Apurimac region was not considered an important mining area until a few years ago, as there was just some informal artisanal mining going on. This changed at the turn of the millennium, when the central and regional governments began to actively promote large-scale mining. By 2009, over twenty major mining projects were under planning in the department of Apurimac. At the time of the research, in 2010, all these projects were still in an early evaluation and exploration phase with a high chance of being exploited in the near future (De Echave et al. 2009b: 150; PDRC 2010: 68).

Bebbington (2007) states that the mining industry has become a major player in Peru's development scenario. Since the mid-1990s, mining activities in Peru have rapidly and steadily increased. Today, Peru is one of the leading countries receiving foreign investments in the mining industry. Among all Latin American countries, it is listed in the second place after Chile (Alayza 2007: 13). Its attractiveness for mining investments as well as its leading role in the export of mineral resources were increased with the promotion of neoliberal economic policies and structural adjustments in the early 1990s under President Alberto Fujimori (1990–2000). With the aid of the World Bank, Peru's mining sector was made more attractive for foreign investors, providing better access for transnational companies (Bebbington 2007: 49–64). The World Bank presented this as an effective way to combat poverty and also supported the sector through loans (De Echave 2009: 108). While neoliberal reforms were introduced all over Latin America, Peru became one of the world's most liberal economies. Extractive industries were explicitly supported also under the presidencies of Alejandro Toledo (2001–2006), Alan Garcia (2006–2011) and the current president Ollanta Humala. An evident effect of this was the drastic increase of mining concessions given to international companies (Bebbington 2007: 49–64; Bury 2004: 78–80; CooperAcción: Comunicaciones).

1.2 Geography and Ecosystem

The Apurimac region is situated in the Andean Mountain Slope of south Peru. The copper deposits of Las Bambas are part of the larger "southern copper belt", which reaches from the cross-border region of northern Chile to southern Peru (De Echave et al. 2009b: 149). The capital city of Apurimac is Abancay. The department's political administration consists of a regional (department), provincial (province), and local administration (district). Most of the fieldwork for this research was carried out in the province of Cotabambas, which is divided into six districts: Challhuahuacho, Haquira, Coyllurqui, Mara, Cotabambas and Tambobamba. Xstrata's coordination centre for the Las Bambas mine is located north-west of Challhuahuacho, the capital of the district by the same name.

The Las Bambas operations are situated at 3300–4800 metres above sea level in the border area between the provinces of Grau and Cotabambas. These provinces belong to the two higher altitude regions called Suni (3000–4000 metres above sea level, with a moderate cold climate) and Puna/Jalca (4000–5000 metres above sea level, cold and dry climate) (MINAG 2011). These Andean highlands are characterized by steep mountain ranges with high peaks and plateaus, separated by deep valleys. The seasonal climate of the Andes shows significant temperature differences between day- and night-time. The rainy season lasts from October to April (Baier 1989: 17–19). All streams and rivers run into the Apurímac River, the biggest river in the department and lifeline of the Apurímac valley. Despite the diversity of hydrological resources, the mountainous topography leads to local water scarcity in some side valleys. In the dry season from May to September, rivers, lakes and moors drastically lose water. This creates a shortage and decline of water quality for human consumption and for agriculture.

Due to the different types of climates, we find a large diversity in the region's flora and fauna. Besides numerous tree species, there are a large variety of different sorts of fruits and berries, diverse herbs and medicinal plants, and a wide range of cultivated species. Woodcutting for fuel, agricultural land expansion and slash-and-burn agriculture contribute to the degradation of natural vegetation and soils, leading to desertification in certain areas and reinforcing climatic aridity. Regional wild animal species include pumas, Andean bears, deer, foxes, weasels, skunks, and various bird and fish species. Part of the landscape are also domesticated animal species belonging to the family of camels (alpacas, lamas and vicuñas), besides cows and sheep (PDRC 2010: 56–58).

1.3 Local Population

The social and cultural diversity of Apurimac's and Cotabamba's population is not documented extensively. The original inhabitants of Apurimac, before the Inca Empire, mainly belonged to the Quechua and the Chanca ethnic groups. The area was occupied at the peak of Inca expansion (around 1450 A.D.), followed by forced resettlements and immigration of other ethnic groups. There were no significant urban centres, the main town being Tambobamba, mainly used for transit and accommodation. Its name comes from the Quechua words *tambo* (place of recreation) and *pampa* (plateau). The area of Cotabambas was called Cuatanapampa or Cutac Pampa at that time. The Spanish colonization of Tambobamba forced most of the Incas and other population groups to live under serfdom (PDEC de Cotabambas 2010: 16). Today, the department of Apurimac has the highest proportion of Quechua-speaking people in Peru. Two variations of the Quechua language are being spoken: Quechua Ayacuchano and Quechua Cusqueño, the latter increasingly dominating in Cotabambas (Vega-Centeno 2005: 33–34).

The province of Cotabambas had a population of 45,700 inhabitants in 2007[4] with an annual population growth rate of 0.61 per cent. Twenty seven per cent of the inhabitants count as urban and 73 per cent as rural population (PDEC de Cotabambas 2010: 9–10). Besides the two biggest cities Abancay and Andahuaylas, there are only a few small urban centres, primarily the provinces' and districts' capitals. According to the regional development plan, 54 per cent of the 405,000 inhabitants of Apurimac live in rural areas. Crop and livestock farming are the primary economic activities in the province. Moreover, 71 per cent of the population speak the Quechua language, which indicates their indigenous origin. In total, there are more than 470 peasant communities in the rural areas of Apurimac (PDRC 2010: 41-42), mostly belonging to the indigenous population of the Peruvian Andes. However, the community members rather call themselves *campesinos* (peasants) and not *indígenas* (indigenous). The denomination *comunidades campesinas* (farming communities) has historically grown in the region as Hilda Rayme, a member of the NGO CADEP explains:

> "In accordance with the national law, they say *comunidades campesinas*, since the Agrarian Reform period. Before that, there were the *gamonales* and the *hacendados* [landlords] and the '*indio*' or '*indígena*' as they called the people who worked for them. So, historically the term *indio* or *indígena* had an excluding function [...] the connotation is someone who lacks education and does agricultural labour. And still today [...] one can feel a certain disdain when the indigenous topic arises [...] as if the *indígena* was a citizen of second category, someone without a voice, without knowledge. But in recent times, some have become conscious of what it means to be *indígena,* as being in a constant struggle [for the recognition of indigenous rights]."[5]

While the native people of the Peruvian Amazon area define themselves as "indigenous" since longer (Christian Aid 2005: 45), the people of the *comunidades campesinas* in the highlands are slowly becoming aware of the political power and international recognition that the concept of "indigenous peoples" entails. However, as Hilda Rayme observes, there has been an increasing interest in benefiting from this concept among the local population in Cotabamba, related to its promotion by NGOs and especially in connection with mining.

The mining sector enjoys a strong support in Apurimac due to its socio-economic situation and lack of alternative development opportunities. Apurimac and Cotabambas are among the poorest regions in the country (INEI: Peru en Cifras). In 2009, the poverty rate in Apurimac was at 70 per cent and even at 82 per cent for Cotabamba (PDRC 2010: 42). Life expectancy was at 71 years and literacy around 75 per cent (PNUD 2009). Additionally, the regional development plan of Apurimac department states that child malnutrition in 2010 was at 43 per cent (PDRC 2010: 42).

4 This number refers to the latest population census at the time of writing.

5 All the quotes in this chapter are translated from Spanish and taken from interviews conducted in 2010, if not indicated otherwise.

Fig. 1: School children in Choaqquere (note the pullover with "Las Bambas" logo). Photo: Annette Wallimann.

Social Institutions

The *comunidades campesinas* (peasant communities) are the primary legal organizational level of the rural population since the Peruvian land reform in 1969. Among other things, they are in charge of administrating land titles. The *asamblea general* (general assembly), where all the registered members participate, is the most important institution of the peasant community. It is led by an executive board named *consejo directivo* or *junta directiva*. Various committees under the board's authority are assigned different thematic tasks according to the situation and needs of the communities. The six executive board members are elected in community meetings every two years. The board represents the community vis-a-vis external actors like the government agencies or companies like Xstrata.[6]

The community refers to values of solidarity and reciprocity with the concept of *ayllu,* meaning that the village community is perceived as one family. Within the system of reciprocity, called *ayni*, labour, products, festivities, advices, favours, and loyalties are exchanged (PDEC de Cotabambas 2010: 34). The common law called *empadronado* is related to respect for rights and duties. For instance, community members are obliged to take part in the so-called *faenas*, the getting together to perform community tasks during the dry season. Non-compliance with these duties can lead to punishment by the general assembly.

Most of the communities in the Las Bambas area possess formally registered land titles. They are thus entitled to regulate and administer the arable lands. Traditional institutions of governing access to land are considered as being relatively well preserved in Cotabambas, with several coexisting forms: zoned pasture land used by individual families (*cercos*); productive areas distributed among the families and cultivated by them in rotation (respecting the long fallow periods) and collectively used communal and pasture lands (e.g. *laymes; pastizales*). The right to use and

6 Most of the information in this section was collected during the fieldwork period in 2010.

purchase land owned by the community can only be granted by a two-third majority of the general assembly (ibid. 34).

Even as community spirit is important in the *comunidades campesinas*, there are also heterogeneity and power asymmetries between families, community sectors and men/women, which can lead to conflicts (Bebbington et al. 2007: 128f.). The elected members of the executive board have a predominant position due their direct contact with external agents. It is apparent that the executive board members, being in constant exchange with Xstrata staff, are vulnerable to the company's strategy of individual negotiation, which also undermines their legitimacy among the other community members and easily leads to accusations of corruption.

Religion and Understanding of Nature

Religion is central to Andean culture and closely related to the understanding of nature as well as to the social structure. Catholic elements introduced by Spanish colonization have mixed with pre-Christian elements of nature worship like *pachamama* (i.e. "Mother Earth") and the *apus* (personalities of mountains or hills) (see Gouley 2005: 45; Rist 2000: 312). This syncretistic mix is expressed in the agrarian calendar and in festivities to honour various patron saints (PDEC de Cotabambas 2010: 19). The productive cycle is determined by a series of rituals. The cycle starts with the *Pago a la Tierra*, an act of sacrifice to the earth before sowing. It then continues with a large celebration of the sowing (*ccashuay*), and finally ends with the harvesting season and the rites of thanksgiving and veneration (*huancascca*).

These rituals are a space of intense cultural activity, reproducing a local identity marked by a strong interaction between humans and nature. People are connected to *pachamama*, the Mother Earth and goddess of fertility, in a reciprocal relationship. She offers her fruits to mankind in order to claim them back. Similarly important is the relation to the *apus*, the personality of the mountains, inhabited by the spirits of the ancestors. The coca plant plays a crucial role in that understanding and in its rituals, as it is considered to be a sacred plant that connects the inner world of *pachamama* and *apus* with the outer world of mankind (ibid. 2010: 21). The regional development plan states that there has been a marked decline in the rites connected with nature worship in Cotabambas. Some of the reasons may be the influence of new religious groups such as the evangelist churches, and the growth of the market economy.[7]

1.4 Local Economy and Resource Use

The mountainous topography of the Apurimac region has a diversity of production types due to extreme differences in altitude and climatic conditions. Three agro-ecological zones can be distinguished: the lowest zone includes the belt between 1000 to 2800 meters above sea level, the middle zone from 2800 to 3800 meters above sea level, and the upper zone from 3800 metres upwards. Whereas the permanent cultivation of fruits, pastures and some vegetables dominate in the lower zone, the upper zone is characterized by potatoes, millet, *quinoa* and various kinds of tuber crops. The middle zone offers favourable conditions for hybrid agro-pastoral activities. Here, in addition to potatoes, corn, barley, wheat, anise and some fruits are also produced (PDRC 2010: 55).

Las Bambas is situated between 3300 and 4800 metres above sea level and hence belongs in the middle and upper zones. Cultivation is characterized by a long average fallow period of six years. In all communities of the zone, potatoes are the most important food base. In higher regions,

7 Interview with Julia Cuadros, CooperAcción.

only potatoes resistant to drought can be cultivated. Only a small amount of the potato harvest is sold on the market while the largest part provides the peasants subsistence needs (Xstrata Peru S.A. 2004: 190). Livestock farming is at least as important as agricultural cultivation. Cows, bulls, sheep, horses, chickens and guinea pigs are reared both for subsistence and for sale on the local markets (ibid. 2004: 190–191). A household generally owns between one and five horses or cows and up to 30 sheep. Horses serve as a means for transportation. The occasional selling of a sheep guarantees a rapid cash income (Cuadros 2010: 30–31). In Cotabambas, the soil is more suitable for livestock breeding than for agriculture, and access to pasture and water are governed by common property institutions. However, a key constraint in keeping livestock is the growing water scarcity. Local peasants have noticed a steady decline of water resources over the last few years, and many sources have dried out completely (PDEC de Cotabambas 2010: 31, 34). While arable farming is organized communally, livestock is in most cases privately owned by individual households or by extended families. The average size of a household extends to four to five people. Possibilities for paid labour were very limited in the region (Xstrata Tintaya S.A. 2010a: 136–140). However, between 1993 and 2007 the proportion of people working in the agricultural sector has decreased while commerce, services, construction and mining sectors have significantly increased, most of all due to the mining project Las Bambas (PDEC de Cotabambas 2010: 13).

Mobility and Communication

The peasant communities in Cotabambas have very limited access to mobility. All of the interviewed peasants replied that there were no motorised vehicles in their communities. Most distances in daily life are covered on foot. Minibuses are the only existing public transport—they run twice a day, mainly connect the urban centres, and are often not affordable for the local villagers. Similarly with communication: apart from the communities located near urban centres, most neither have telephone lines nor mobile network connection. The only means of media communication are local radio stations, a rather one-sided way of broadcasting information. This lack of mobility and communication makes it difficult to be informed and participate in public events, workshops and meetings. Women have the lowest access to such meetings because they are often bound to their households and children.[8]

1.5 National Economy and Legal Framework

According to the Peruvian Private Investment Promotion Agency (ProInversión), mining accounted for 24 per cent of totally USD 22 billion of direct investments in 2011, making it the leading sector in foreign investments. 56 per cent of the investments came from Spain, the USA, South Africa, Chile and Great Britain. Switzerland had around 4.3 percent (ProInversión: Estadísticas).

As a result of the opening of the economy and the increasing access to international markets, export rates in Peru have been rapidly increasing since the 1990s. The export volume of mineral products reached USD 27 billion in 2011 and accounted for more than 50 per cent of total exports (Banco Central de Reserva del Peru: Series Estadísticas). The most important products were copper and gold, followed (at a large distance) by lead, zinc, iron, molybdenum and silver. Copper accounted for 39 per cent of all mineral exports, making Peru the second largest copper producer with 8 percent of global production. According to the Ministry of Energy and Mining of Peru (MINEM), production increased from 1,100,000 in 2005 to 1,235,000 metric tonnes in

8 Interview with Hilda Rayme, CADEP.

2010. Gold has almost the same importance for mining exports, with a 37 per cent share of all mineral exports. The statistics show Switzerland as the major importer of mining products from Peru, amounting to almost USD 5.8 billion in 2011. However, most of these products never reach Switzerland physically, but are just traded by Swiss companies.[9]

Indigenous Land Rights

The mining concession is the most important juridical element that binds the operations of mining companies to legislation in Peru. A concession allows a company to explore and extract mineral resources within a determined area during a specific time span, even without being the owner of the land. Therefore, a concession can be overlapping with the land ownership of indigenous communities. According to De Echave et al. (2009a: 15), concession areas for mining in Peru have increased from 2.26 million hectares in 1991 to 15 million hectares in 1997. For 2011, MINEM informs that 23.66 million hectares, which is 18.5 per cent of the total area of the country, have been claimed. This includes planned and definite concessions as well as areas in extraction (MINEM 2011:10). 54 per cent of the department of Apurímac and 60 per cent of Cotabamba Province were covered by such concessions (CooperAcción: Mapas).

Article 89 of the Peruvian Constitution of 1993 determines the rights of indigenous peoples of Peru. The article concedes autonomy for peasant and indigenous communities with respect to their communal form of organization and land use. The formulation of this article is based on the *Ley de Comunidades Campesinas Nativas* ("Law of Native Peasant Communities"), which goes back to 1974 when indigenous rights, and specifically communal land tenure, were legally recognized for the first time. With this law, indigenous land was declared "inviolable" (Haller et al. 2007: 274). As Pinto (2009: 90) points out, the right of communal land tenure is the strongest means for peasant communities to defend their territory against the neoliberal legislation of the Peruvian government.

In Peru, compulsory land acquisitions are only possible for purposes of "national security" or "public utility", and mining does not fall into these categories. Nevertheless, the inviolability of indigenous land ownership is limited by the Constitution as the state has sovereignty over Peru's natural resources.[10] This legal ambiguity can cause different conflicts (Alayza 2007: 37). The "General Law of Peasant Communities" defines the rights and responsibilities of peasant communities, regulates the communal rights and declares that the general assembly is the highest political authority of peasant communities. The law indicates that any activity on their land can only be realized upon approval by two-thirds of the community members. This approval is called "social licence" (*licencia social*).[11] The social licence implies that a company that aims at conducting extractive industrial activities in Peru must come to an agreement with the affected peasant communities. However, what happens is that consultations are typically characterized by vertical flows of information and compensation payments instead of a horizontal communication between the communities and the mining companies. According to Alayza (2007: 116), "[t]he conception that underlies the implementation of the Corporate Social Responsibility (CSR) is a vertical one; CSR programs come from above, that is from the company, and pass through different stages, always trying to achieve acceptance for their mining operations within the communities. [...] The experiences with the social licence show that this mechanism is weak in terms of participation of the local communities."

9 ProInversión Online: Oportunidades Sectoriales; The exception is gold, due to the role of gold refineries in Switzerland, as described in chapter 2 of Part II.

10 Constitution of 1993, Art. 66 (Tribunal Constitucional del Perú Online: Legislación Constitucional).

11 Ley General de Comunidades Campesinas (Archivo Digital de la Legislación en el Perú Online: Ley No. 24656).

According to Cuadros et al. (2005: 34), the concept of social licence is conceived differently by the players involved. While companies and government institutions reduce the acquisition process of social licence to the organization of informative meetings, NGOs and representatives of peasant communities have a different understanding. For them, the social licence follows the procedure of free, prior and informed consultation, as framed by ILO Convention 169. This Convention establishes the indigenous communities' right to information and participation with regard to projects affecting their territory and concerning decision-making. Peru ratified the ILO Convention in 1994 (ILO Online). The governments of the signatory countries are responsible for the implementation of the Convention, particularly of "prior consultation" (*consulta previa*). In 2011, the legal bill on prior consultation for indigenous communities was approved by the Parliament and the regulation came into force in April 2012.

Other disputes concerning the land rights of the peasant communities happened in the course of the liberalization reforms imposed by Peruvian president Alberto Fujimori (1990–2000) in the 1990s (Haller et al. 2007: 274; Pinto 2009: 89-95; De Echave et al. 2009a). Central among these reforms was the "Land Law"[12] of 1995. This law placed the importance of mining and energy projects above communal ownership. It consolidated the instrument of the "servitude" (*servidumbre*)[13] of individual and communal land ownership (Pinto 2009: 91). With the support of the general direction of the MINEM, the regulation of servitude allows mining companies to obtain the right to use a specific piece of land even if they do not come to an agreement with landowners regarding the compensation payments. In this case, the process of servitude is initiated and further negotiations between the different parties follow. If these negotiations are not successful, the MINEM has the right to impose temporary use rights even if this is against the will of the landowners or the affected communities. In this case, the MINEM also has the right to decide independently upon the amount of compensatory payments. The servitude instrument therefore turns out to be an instrument that helps companies who are supported by the government to enforce their claims for extraction permissions without being reliant on the direct approval of the peasant communities (Christian Aid 2005: 7).

Legal Framework for Mining Companies

Several reforms concerning the mining industry were introduced in the Peruvian legislation in 1992-93, related to the neoliberal government strategy of pushing economic growth through the attraction of foreign investments (see Bebbington 2007: 53). A key element in this was the centralization of all the tasks relating to taxation, employment and environmental regulation under the responsibility of the Ministry of Energy and Mining (MINEM), through the "General Law of Mining" of 1992[14] (Christian Aid 2005: 4). Moreover, several regulations concerning environmental protection in the existing environmental code[15] were degraded. Apparently to attract mining investments, specified environmental responsibilities of mining companies were replaced by generally formulated principles, thus reducing the regulatory effect. Simultaneously the handling of mining concessions was liberalized, the rights of concession holders were

12 Ley de Tierras (Law Nr. 26505).

13 The servitude regulates the rights of a neighbouring plot on a parcel of land. Servitude is given if the owner of a parcel of land has a partial right on the land ownership of his neighbour. Known examples of such processes concern "way-leave" or "water rights". The imposition of servitude does not mean a loss of land for an affected land owner, but rather the constraint on his own land to the benefit of a neighbouring land plot (Del Castillo and Castillo w.y.: 20). The servitude is regulated in Art. 7 of mentioned Ley de Tierras (Archivo Digital de la Legislación en el Perú Online: Ley No. 26505).

14 Ley General de Minería (Archivo Digital de la Legislación en el Perú Online: Decreto Supremo No. 014-92-EM).

15 The environmental code from 1990 regulates the protection and conservation of the environment and natural resources (Archivo Digital de la Legislación en el Perú Online: Decreto Legislativo No. 613).

improved, and the entrepreneurial role of the state in the mining sector was abolished (Christian Aid 2005: 8).

Nevertheless, Peru has adequate regulations to protect the environment, if properly applied, but the problem lies more in the surveillance and sanctioning of infringements (Haller et al. 2007: 374).[16] In 1993, a new requirement for companies to elaborate and publish an Environmental Impact Assessment (EIA) was established (Christian Aid 2005: 8). Companies have to investigate and define all possible environmental impacts of a proposed mining project and the envisaged prevention and mitigation measures in an EIA report to be handed in to the MINEM previous to the exploitation phase (see Pinto 2009: 96).

The responsibility for larger industrial mining projects are in the hands of the national mining ministry and thus of the central government. However, the state's presence at the regional and local levels is not sufficient to accompany and survey all activities in the mining sector. When the Ministry for Environment (MINAM) was founded in 2008, it was with the mission to oversee the compliance with environmental regulations. Its capacity to act in the field of extractive industries was limited, as it remained under the control of the MINEM. Therefore, the state so far has not been able to deal with the important challenges concerning environmental issues in the mining sector in a consistent way (De Echave 2009: 123). Several authors mention difficulties relating to state regulations on mining companies and the handling of the EIA (see Haller et al. 2007; Christian Aid 2005; Pinto 2009). One concern is that indigenous communities mostly do not have a possibility to participate in the preparation of the EIA. The companies themselves commission the preparation of the EIA. Thus, the independence, integrity and verifiability of the EIA remain questionable (Haller et al. 2007: 374; Christian Aid 2005: 40). The expected social impacts are also included in the EIA and not in a separate social responsibility assessment. Consequently, social problems are subordinated to environmental issues and are thus underrepresented. At the end of 2012, a law challenging the control over the EIA of the mining ministry was accepted. This law established a national system for the environmental certification of sustainable investments (SENACE), which should control all EIA of the extractive industry sector. However, experts have conflicting opinions whether this will lead to a real improvement of the EIA's quality with regard to the distortion of social and environmental impacts of large mining projects (El Comercio 2012).

16 Interview with Thomas Steeb, collaborator of the German Development Service and working as a consultant within the environmental department of the Ministry of the Interior of Apurimac.

2. The Las Bambas Mine

The planned copper mine in Las Bambas was one of several projects developed by Xstrata in Peru.[17] Xstrata was headquartered in Zug (Switzerland) and had a second registered office in London. Because of Xstrata's decentralized structure, Las Bambas was under the responsibility of the Xstrata copper unit with headquarters in Brisbane (Australia). During the field research in 2010, besides the Las Bambas project, Xstrata owned in Peru the mine Tintaya, the running operation in Antapaccay (Department of Cusco), and one-third of the Antamina mine in the Ancash Department. Tintaya is meanwhile closed and its infrastructure is used in the Antapaccay mine. The ore mined in Las Bambas was planned to be transported in a pipeline to the Tintaya-Antapaccay infrastructure for further treatment to obtain the product for export, a copper concentrate, that would still need to be refined for final use. Xstrata also announced at that time that a new project, named Coroccohuayco, was evaluated in the Cusco region (Xstrata Copper 2013).

In September 2003, the regional government, the Ministry of Energy and Mining (MINEM) and the Peruvian Private Investment Promotion Agency (ProInversión) decided to elaborate a mining convention to encourage private investment in Apurimac. Both national and local authorities began to stimulate the mining industry, considering it as the only chance for economic growth in the region. Furthermore, high tax revenues were expected due to a strong demand on the international market. In 2009, over twenty major mining projects were commissioned in Apurimac department. In 2010, most of them were still in an early phase of evaluation with prospects of exploitation in the near future (De Echave et al. 2009b: 150). In October 2003, the mayors of Grau and Cotabambas province, as well as their district governments, signed the "Declaration of Challhuahuacho". By this act, they ratified support to the ongoing process of exploration of copper resources in Las Bambas. The declaration states that the promotion of private investments in the Las Bambas project provides an important opportunity for development in Apurimac. In the words of the declaration: "the concession [of the Las Bambas project] signifies a major step forward in the region's development and in combating poverty" (quoted from Valderrama 2010: 7). Xstrata has announced that the total investment in the Las Bambas project will be around USD 4.2 billion with an expected output of around 400,000 tonnes of copper per annum (Xstrata 2013: Media Centre). In Las Bambas, copper is mostly obtained by open cast mining. This means that the ore is superficially removed from the rock. The duration of the project is formally divided into four stages: 1. Exploration, 2. Construction, 3. Exploitation and 4. Closure. The actual mining of copper is expected to start at the end of 2014 after the ongoing construction phase would be finished and the mine is supposed to be closed in 2031 (Xstrata Tintaya S.A. 2010a: 9). On 1 October 2004, Xstrata and the central government of Peru signed an option agreement providing Xstrata with a concession for phase 2 of the project (construction phase). Moreover, Xstrata receives the right to transfer the current contract into a concession right for the actual exploitation. Xstrata paid USD 121 million for the concession of an area of 33,200 hectares (De Echave et al. 2009b: 150–153).

The signing of the agreement and transfer of a first partial amount of USD 91 million symbolically happened in the directly affected peasant community Fuerabamba. At that time, President Alejandro Toledo as well as representatives from the central, regional and local governments were present at this event (Cuadros et al. 2005: 11–12). This first and optional contract regulated the following: the price for a later transfer into a definitive contract, the annual concession fees

17 In 2013, Xstrata merged with Glencore, which had important implications for the Las Bambas project (see editor's update at the end of this chapter).

and the taxes to be paid. It also referred to the social responsibilities of the company vis-à-vis the affected region. The latter includes the payment of financial contributions for the affected provinces of Grau and Cotabambas. These contributions are called *aportes sociales* in the Peruvian context. They include an initial social contribution of 50 per cent of the concession payment and several payments during and after the transfer of the optional contract into a definitive contract as well as payments in case these contracts are renewed. At the signing of the first contract, USD 45.5 out of 91 millions were paid into a state-administered social fund called Fideicomiso Las Bambas. In 2008, this fund was transformed into a private society and was renamed into Fondo Social Las Bambas (FOSBAM). It aims to benefit the two affected provinces of Grau and Cotambambas and was fed with another USD 13.5 millions with the later signing of the transfer contract at the beginning of the construction phase (ibid. 15). Besides the payment made in the framework of the *aporte social,* Xstrata was responsible for investing in the maintenance of "social conditions for the promotion of the mining project Las Bambas". This included the respect for local customs and human rights, the creation of labour possibilities for the workers from the region, the promotion of local firms, the implementation of infrastructure projects, the prevention of environmental impacts, or bringing information to the population about the project (Gouley 2005: Anexo 4).

Xstrata itself declared its commitment to make "social investments" in the form of productive and social projects in the directly affected communities, and that local residents would receive labour opportunities. According to Cuadros et al. (2005: 12), these conditions were not included in the formal optional contract with Xstrata and are therefore not legally binding. Implementation and compliance hence mainly lie in the hands of the company.

2.1 Environmental and Social Impact

In the Environmental Impact Assessment (EIA), Xstrata decided on their own criteria, as to which communities are located in a direct sphere of influence and which ones are not. The EIA distinguishes between zone 1 and zone 2. Zone 1 refers to communities where the most significant potential impacts have been identified. Six of the communities in zone 1 are directly located in the concession area for Las Bambas (Gouley 2005: 68). The community Fuerabamba is mentioned in particular here because of its location directly in the planned construction area. The community thus had to be resettled and was negotiating over the resettlement conditions during the period of research. Another important area located in zone 1 is the urban city centre of Challhuahuacho.

Zone 2 includes those communities where minor effects are expected. Among those are Chila and Choaqquere, both pressured to sell parts of their land to Xstrata in order to resettle the community of Fuerabamba. Due to their relevance for the communities in zone 1, they were also included in the study. A main part of the fieldwork for this study was carried out in the community of Choaqquere. According to Xstrata, the zones of influence 1 and 2 altogether include 20 local communities (Xstrata Tintaya S.A. 2010b: 25–26). Also the communities situated along the planned pipeline to Tintaya-Antapaccay belong to zone 2. This means that 24 additional villages were included in this study even if they had no direct contact yet with Xstrata in 2010. However, the company did not take into account indirect impacts, which would cover a much larger area. The non-profit organization CooperAcción estimates that 49 communities lie within the area of influence of the mine (those along the pipeline not yet included) (Cuadros et al. 2005: 53–54).

2.1.1 Environmental Impacts

There were no specific environmental impacts of the Las Bambas project until 2010. Although Xstrata had to pay a number of environmental fines during the research period (see Wallimann 2011: 101), no general conclusions can be made about the company's behaviour towards the

Fig. 2. Part of the exploration area at Las Bambas. Photo: Clothilde Laime.

environment. However, local residents and NGOs expressed concerns regarding the potential effects of copper mining on water, soil, and air. This is based on experiences from other regions that people had been reported to the people. Even if they lack direct personal experience, they have heard that a copper mine of this size can have significant impacts on the environment. Studies about Xstrata's copper mining project in Tintaya (see Fuentes 2012) and others show that these can be immense, even if the correlations are often hard to prove. While locals only had a vague idea about this, representatives of NGOs frequently referred to the Tintaya project based on more solid knowledge.

Xstrata itself has stated in the Sustainability Report of 2009: "We aim to preserve or restore the long-term health, function and viability of the natural environments affected by Xstrata's operations, so that they can sustain biodiversity, landscape functions and the needs of local communities, during our activities and over the long term once our operations close" (Xstrata Plc 2009: 55). The measures proposed in the EIA (see box) to mitigate environmental impacts were considered as relatively appropriate by different experts of NGOs working in the region. Nevertheless, all respondents presumed a long-term danger for water resources. This was seen as especially problematic, because it adds to the negative consequences of climate change, already noticeable in Apurimac, with glaciers disappearing and small rivers drying out. As the chemical processing of the ore to produce the copper concentrate was to be done in Antapaccay, the danger of chemical pollution was of less worry in the Las Bambas region. However, during exploration drilling itself, certain chemical substances can be released into surrounding waterways. The threat for water bodies is thus the most serious potential environmental impact—not least because the future mine is situated at the source of the Challhuahuacho River, which supplies numerous downstream communities (Wallimann 2011: 72–75).

The Environmental Impact Assessment (EIA) in Las Bambas

On 7 March 2011, the mining ministry approved the EIA for Las Bambas, a large document of several thousand pages (MINEM Online: Asuntos Ambientales Mineros). The study deals with the future Las Bambas mine and its three planned opencast mining locations. It includes the waste disposal, worker accommodations, the energy provision and a planned pipeline of 215 km that is supposed to transport the raw material from Las Bambas to Antapaccay for further processing. Moreover, the EIA lists the relevant legal regulations, defines the direct and indirect zones of influence, includes an inventory of the actual ecological and socio-economic situation and makes an evaluation of the possible ecological and social impacts (Xstrata Tintaya S.A. 2010a). During March 2009 and February 2010, Xstrata conducted official hearings in the direct zone of influence and in different districts to inform about the mining project and the planned pipeline. According to Xstrata (2010a: 31), these activities enabled the inclusion of concerns and wishes of the peasant communities in the EIA. Representatives of NGOs perceived these information events as one-sided. Susanna Anderegg, at that time employed as researcher in the local NGO CADEP, describes in a newspaper article that the information given to the residents was incomplete and biased to benefit the projects of the company (in *NZZ* 9 September 2010). During the time of this research, the timespan to send additional suggestions to the ministry after the publication of the EIA was restricted to 30 days. Members of peasant communities as well as experts from state institutions and NGOs had neither enough time nor the possibility to review the published EIA. Moreover, it is written in an expert language inaccessible to the people from the peasant communities (Wallimann 2011: 53).

2.1.2 Social Impacts

The most important social impact of the Las Bambas mine visible in 2010 was the growth of the urban centre of Challhuahuacho. This former peasant community had become an urbanized area within just a few years. In 2010, it was a trading centre where peasants from all around came to sell their products on the Saturday market. The district of Challhuahuacho had 1,811 voters in 2002, 2,200 in 2006 and 3,662 in 2010. An executive of the Ministry of Agriculture expected an increase to 20,000 inhabitants in Challhuahuacho by 2015, which in his opinion would cause supply problems. As a consequence of population growth and employment possibilities, the prices on the local market increased. A social issue often mentioned in interviews with local people concerning the mine is the high rate of alcoholism. Various interviewees perceived a close link between the high consumption of alcohol and the operation of the Las Bambas mine, but no study has so far delivered clear evidence. Furthermore, peasants, governmental authorities and NGOs feared an increase of drug abuse and the establishment of sex workers in the region, as it had been previously observed around the Tintaya mine in Espinar (Wallimann 2011: 65–66).

The perspective that the mine would bring development and the promised "social investments" of Xstrata in the form of job opportunities had caused high expectations among the local population in the Las Bambas region. According to Gouley (2005: 40), approximately 200 workers were employed by Xstrata Las Bambas in 2005, 40–50 per cent of them coming from the region. This number was expected to significantly increase during construction and exploitation phase. The EIA from 2010 expected an average of 3,800 jobs to be created throughout the operations, peaking to 6,500 during construction and decreasing to 1,350 during the exploitation phase (Xstrata Tintaya S.A. 2010a: 9). However, Susanna Anderegg, working with the NGO CADEP, estimated that only one-third of the workforce can be recruited locally, due to lack of practical experience and training (in *NZZ* 9 September 2010). A bigger potential is seen in indirect job opportunities due to the boosting of the local economy (Gouley 2005: 41).

By defining a direct zone of influence (zone 1), Xstrata determined those communities who had the right to join a development project in the framework of Xstrata's social investments (not be confused with the social fund FOSBAM). This situation led to a division between the communities of Cotabambas. While some communities directly benefited from the mining operations, others had not received any assistance until summer 2010. Hilda Rayme, project coordinator of CADEP, describes the situation as follows: "At the moment, the mine divides [the communities] in defining one part as 'directly affected'. But not only the 'directly affected' people, but also the others who live in the surroundings [will experience the impacts]. The influence of the money

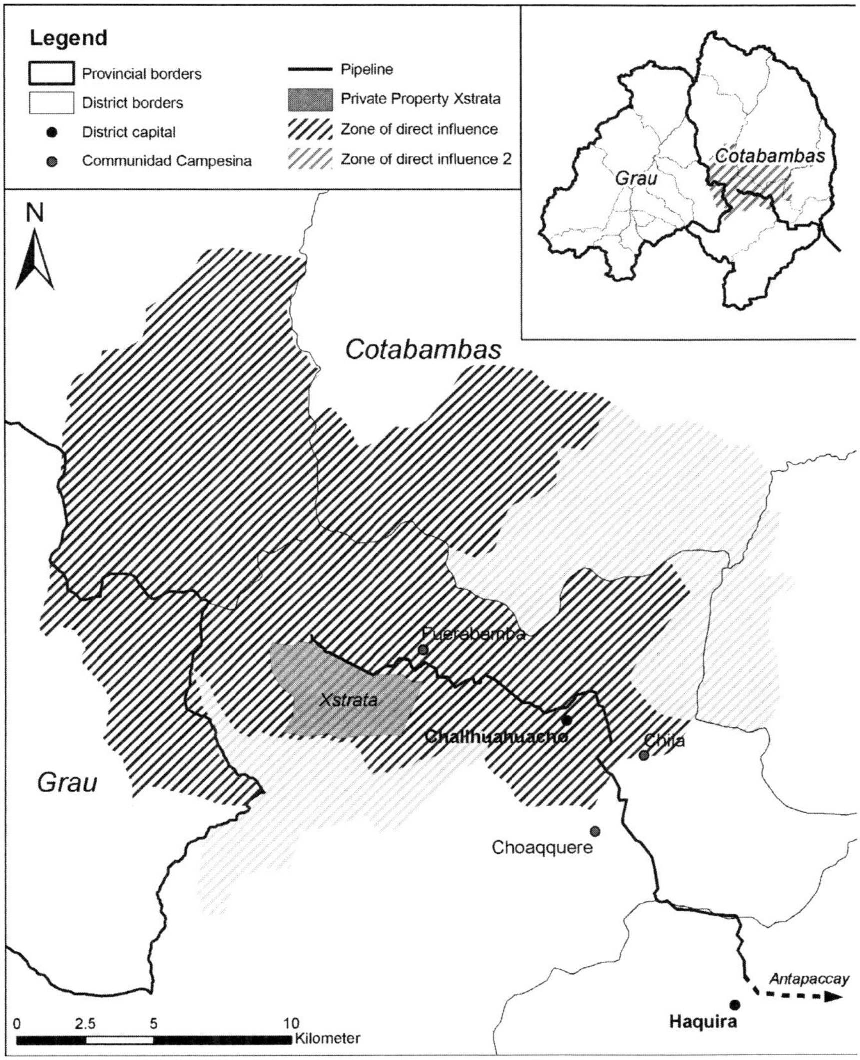

Map 1: The "zones of influence" 1 and 2 with the private land property of Xstrata, the route of the planned pipeline, and important locations for the research (Wallimann 2011).

given by the company to the 'directly affected' is splitting them. And the others [...] feel excluded, marginalized or are frightened."

Many people in the Fuerabamba community, for instance, were employed to do exploration work in the future mine. Others got paid for setting up different development projects in their still existing village community. Also locals from other communities in the direct influence zone got work opportunities in the mining area or in the already mentioned projects for their communities. The sudden availability of work and a certain financial income was a profound experience for the people involved and opened new perspectives to them. Moreover, the inhabitants of the affected communities hoped that projects for health care, education and agriculture would improve their living conditions. According to Xstrata, through their "social investment" between 2005 and 2010, over USD 13 million were invested in 41 social programs (see Xstrata Copper 2010). The company itself mostly administers the implementation of these projects. Of these 41 programs, five were implemented by external NGOs, five in coordination with local and regional governments and six with other state institutions (Wallimann 2011: 73). These efforts, though seen as a great chance among the locals, will result in dependency as well. Alicia, inhabitant of Choaqquere comments: "The company promises that we will go hand in hand with them as we are part of the direct zone of influence. They will support us by all possible means."

Under the impression of these financial benefits and new development prospects, community members who had emigrated earlier are returning to their villages. While the rural population of Cotabambas is often moving or sending family members to the bigger cities of Abancay, Cusco or Arequipa, the operation in Las Bambas attracts former migrants to return to their communities and to reclaim their communal rights. These migrations cause conflicts within the community because communal rights usually get lost after a certain period of absence. Hilda Rayme of CADEP reports that the community of Fuerabamba numbered around 300 registered community members before Xstrata's arrival. In 2010, the community numbered more than 1300 members. This is due to the fact that workers must live in the affected region to get access to unskilled work in the mine. Moreover, the affected people have placed a lot of hope on Xstrata's development assistance. This phenomenon of reverse migration can be observed in many of the communities (Wallimann 2011: 79–80; 84–85).

Other communities not located in this zone of influence feel excluded from the project, income opportunities and from the wage labour provided. The privileged treatment of the Fuerabamba community has caused feelings of resentment, alienation and envy. Valentín Roqque, President of the peasants' association, said in an interview that the relationship between the people of Challhuahuacho and Fuerabambas is no longer a friendly one. He sees one of the reasons for this in the fact that the inhabitants did not communicate with each other and instead conducted negotiations with Xstrata on their own. Therefore, the region's peasants at the time of research were basically divided into two groups: one group being in favour of the Las Bambas project and having a certain chance of receiving wage labour and access to development projects; and the other group being sceptical about the project and without prospects of receiving any support from Xstrata (ibid. 82–86). Moreover, it became apparent that the locals possess little knowledge and information about the mining operations and its general implications for Las Bambas.

There are several explanations for this situation: on the one hand are the previously mentioned limitations of disseminating information about the project in Cotabambas; on the other hand, state institutions do not get involved in providing information but leave this duty to Xstrata. The company on its part only consults those communities from which it needs approval to the project. Knowledge and perceptions about the mine vary considerably within the communities also because many refrain from exchanging information with civil organizations as soon as they are in regular contact with Xstrata's community relations staff.

2.2 Corporate Strategies and the State

According to Xstrata, the company takes into account international agreements elaborated by the United Nations, like the UN Global Compact or the UN Declaration of Human Rights. Further it refers to sectoral, voluntary initiatives like the International Council on Mining and Metals (ICMM), the Extractive Industries Transparency Initiative (EITI) or the Voluntary Principles on Security and Human Rights (Xstrata 2013). The contribution to sustainable development within the communities where Xstrata operates is stated to be a core issue of the company's Corporate Social Responsibility (CSR) strategy. The compliance with ethical principles is presented as part of Xstrata's overall corporate strategy: "We believe that operating to leading standards of health, safety and environmental management, contributing to the development of sustainable communities and engaging with our stakeholders in a two-way dialogue, regardless of our location, enhances our corporate reputation and is a source of competitive advantage" (Xstrata Plc 2009: 0).

As Xstrata points out here, sustainability is also to be understood as a source of competitive advantage. In implementing these standards on the ground, both Xstrata and the local population around the mining area are supposed to benefit. The company considers its presence in Peru as a win-win-situation for everyone and a great chance for the impoverished regions to benefit from the added value and development services of Xstrata. On several occasions, the company insisted that the operations should be done with respect towards and in collaboration with the local communities to bring development to the area (Wallimann 2011: 89–90). Thus, the company is acting not only in an economically profit-oriented way, but claims to be an agent of development, as the largest taxpayer and employer and in shaping the social life in Grau and Cotabambas provinces. The company thus offers development to local communities and the whole region of Apurimac in exchange of the right to exploit the region's mineral resources. With its "social investment", Xstrata also aims to compensate for potential negative impacts. In the case of the Las Bambas project, the company's Department of Community Relations (*relaciones comunitarias*) is in charge of communication, negotiations and sustainable development in the local communities.

In conclusion, Xstrata apparently involves only the parties and actors considered useful for successful negotiations. Besides occasional public hearings, Xstrata negotiates bilaterally with individual communities. Xstrata only works in partnership with NGOs recruited for its own projects. Other NGOs which are involved with mining issues are considered as opponents by the Xstrata staff and thus excluded from the negotiation processes (Wallimann 2011: 106–107).

According to Julia Cuadros from the NGO CooperAcción, who has been associated with the community of Fuerabamba over a long period, in its consultations with this community Xstrata chose a negotiation strategy that is not in accordance with the legal rights assigned to *comunidades campesinas*. Company officials first visited individual families, then representatives of each sector, then the community executive board. However, the land law (*Ley de Tierras*) states that matters concerning collective property must be approved by a two-thirds majority of the community assembly. Also in Choaqquere, the company staff visited all the families parallel to the community assembly and tried to convince them individually to sell. Enrique Quispe of CADEP, who had been working with the community at that time, describes: "They came to see us every day, going from door to door, trying especially to convince those who did not want to sell." With this strategy of individual negotiations and the creation of individual incentives, Xstrata's employees suppressed the possibility of collective action in the communities.

Furthermore, in public events as well as in community consultations, the company staff only provides specifically framed information. Many community members are conscious about this, especially when their own experiences contradict the provided information. Felipe, a peasant from Choaqquere, comments: "The company is always showing the positive aspects, never the negative, [otherwise] the population would not accept it, would they? That is why they are always talking that the mine is good, only good. Of course the mine is good, but there are certainly some negative aspects, which we also see [...]. There are plenty of cars now and that is why there is dust. Our animals eat this dust and die of diseases." Xstrata thus filters all the information it provides to the communities. Although the issue of negative environmental impacts is not completely omitted in its information events, negative social impacts are not addressed at all.[18] Neither are these specified in Xstrata's sustainability reports, even though it stresses the well-being of the communities in its sustainability strategy.

At the same time, in its public communication Xstrata continuously affirmed the willingness to dialogue and ensure participation and considers itself as exemplary in this respect. This emphasis is strongly linked to the demand for participation of the local population in industrial resource extraction projects and to the concept of informed participation as formulated in the ILO Convention 169. The importance of participation for the sustainability of Xstrata's activities is stressed by the company both in general and for specific projects. For example, the sustainability report 2010 states about the mining projects in Peru: "Our permanent processes of participative dialogue allow us to strengthen our relationship of trust with local communities. We are satisfied to see that our operations and projects in development directly contribute to sustainable development in our areas of influence" (Xstrata Copper 2010: 8; own translation from Spanish).

A good example for Xstrata's version of dialogue and participation was the presentation of the Environmental Impact Assessment (EIA) in a public hearing in Challhuahuacho on 15 July 2010. Xstrata later promoted this event in publicity leaflets as proof for the broad approval of the Las Bambas project, saying that 5,310 people had participated and that 166 oral and 466 written questions had been answered. In retrospect this event was presented by Xstrata as a kind of prior consultation (*consulta previa*) where the population could freely inform itself, voice concerns and doubts, and where Xstrata could sense the general approval for the project. According to Ramon Ginestet-Menke, consultant of the local ombudsman office for human rights in Abancay, Xstrata did in fact put much more effort into dialogue with the communities, compared to other companies, but always subject to its own interests. At the hearing he realized that: "Xstrata had planned this day meticulously, down to the last detail, focussed on a specific target. It certainly had its effect. This includes that critical voices had no chance of being heard. So Xstrata aims to encourage dialogue, but at the same time they control it for their own interests."

The interviewed representatives from NGOs and peasant communities agreed that this hearing cannot in any way be seen as confirming a free and informed consent to the project. Several participants observed that unknown people had been brought from outside who publicly stated their approval at the hearing, but local people with doubts felt they had no chance to speak up. Similar dynamics have been observed also in other public hearings. With this strategy of controlled dialogue, engineered by Xstrata's community relations staff, a broad consent is simulated without enabling actual participation (Wallimann 2011: 91-92). As Julia Cuadros of CooperAcción comments concerning the public hearing of the EIA: "5,300 people were present at the public hearing. This is not a participatory approach to me. A large crowd of people being present does not guarantee transparency, democracy, participation. It is just a gathering."

Xstrata also had an offensive promotion strategy, primarily perceived in the city of Challhuahuacho but also apparent in the rest of the province. It was present at public social events such as the weekly market (see figure 3) and visually manifests in the urban space, through posters, brochures, promotional gifts, or setting up signposts for upcoming projects (figure 4). Both Xstrata's Community Relations and the FOSBAM were part of this promotion strategy. Although the company stresses that the FOSBAM is mainly managed by state officials who work independently of Xstrata, most people from the communities noticed no difference between

18 Interview with Susanna Anderegg. 23 May 2011.

Field Study Notes: Establishing Contacts with Xstrata

In Cusco, a colleague and I went to see Xstrata Tintaya's well-guarded branch office. After a few efforts and by presenting a letter from the University of Bern about our research project, I succeeded in arranging a meeting with the then-executive officer in charge of the Las Bambas project. I introduced myself and told him about my research project, with the hope of asking him a few questions. He then questioned me about my research strategy, about who I was in touch with, and offered me a visit to the mining area with the company's responsible persons. I refused the proposal at that time because I was worried that I would then no longer be able to operate independently from Xstrata, but declared my interest to maintain good contact with the company. When he heard about my contact with the NGO CADEP, he knowingly nodded and explained that there were players who considered mining per se as negative and reminded me that I should try to stay objective. Without asking for permission he then took a picture of me with his mobile phone, saying: "this is for the people to recognize you and to keep them informed about your plans in Las Bambas", and warned me that the region could be dangerous because people in the countryside were narrow-minded and many of them had alcohol problems. He agreed to give me an interview at a later stage, when I was more familiar with the situation, and concluded by wishing me good luck for my master's thesis and mentioned that as an anthropologist perhaps I could work for Xstrata one day (field notes 21 and 27 July 2010).

When I tried to get in touch with the executive officer again after my stay in Cotabambas, he did not answer my requests. Due to his silence, I instead contacted the head of institutional relations of Xstrata Peru. He agreed to receive me and to give an interview in the offices of Xstrata Copper in Lima. There, he handed over a portfolio of Xstrata's latest Sustainability Reports and spoke for one hour about the Xstrata financed social programs in Las Bambas and the social fund FOSBAM without leaving me any chance to speak. When he was finally done with his speech, I asked again for an opportunity to interview him. Firstly, he agreed, but immediately changed his mind when I asked if I could record the conversation. Instead, he offered to respond to my questions by email. So, the conversation was suddenly over, when surprisingly the executive officer entered the room. Remarkably friendly, unlike in our first meeting, he asked about our experiences and gave us a textile handicraft of the Las Bambas region with the logo of Xstrata Copper. Both promised their full support when we left (field notes 16 September 2010). Back in Switzerland, I sent several emails to them with questions, which they never answered.

FOSBAM and Xstrata due to their similar advertisements and behaviour. Xstrata increasingly dominated the local radio stations in Cotabambas (see Wallimann 2011: 95-96), where the development efforts of Xstrata and FOSBAM were advertised in turn but with similar content. The logo of "Xstrata Las Bambas", sometimes together with the logo of FOSBAM, was spread over the entire province (Wallimann 2011: 92–96). In a region where people have few possessions, Xstrata reached out to the public by distributing small gifts, also in communities which had so far not been in direct contact with the company. The company's concept of development thus took the centre stage in the local public debate through diverse visual and acoustic promotion activities. By spreading its logo and linking it to the idea of sustainable development, Xstrata achieved a symbolic power superior to the one held by the weak state presence.

A further strategy that the company pursued was to tie development services in the communities to conditions of loyalty. Due to its multifunctional role in the region, Xstrata had enormous political and economic power. Julia Cuadros of CooperAcción clarified how the company attached its development services to conditions, which prevent the communities from making collective decisions: "The company offers things to them. People are being conditioned. In this

Fig. 3. A stand of Xstrata's "community relations" at the Saturday Market in Challhuahuacho. Photo: Annette Wallimann.

way, they will not let them speak up for themselves." The critical NGO voices mentioned in this chapter all coincide in that Xstrata makes use of its development projects and distribution of benefits to obtain certain requested actions from the peasant communities. This pressure forces them to opt for or against the company and therefore for or against development. Many people thus opted for cooperating with the company due to their material needs, sometimes even against their own conviction. Flor, a peasant woman from Choaqquere, described the dynamics of such difficult decision in her community: "The land negotiations were not easy. In the beginning, there were disputes and discussions, a majority was against and did not want to sell. I was neutral [...] I thought about my sons' future. One of them studies mining and the other metal construction engineering [...]. I could not oppose, because I [...] hoped that one of them might get the chance to find work there. So, if I opposed the sale, doors would be closed to them. Under these circumstances, I could not say 'no'. I could neither say 'yes', because most of the community did say 'no' [...]. You get into trouble by doing so."

In Choaqquere, community members have already internalized this dependency. Alicia, another peasant woman of Choaqquere said: "We still have a lack of organization. We are slowly beginning to organize ourselves. [...] but the company does not allow us to organize with other communities. We are under pressure." Xstrata thus generates opportunities subject to the condition of loyalty, which hinders people in expressing their concerns and opinions. As a consequence, the collective organization within and between the communities is harmed.

Role of the State

The Peruvian Government's strategy is generally to foster the mining industry in Apurimac and the Cotabambas province. Former President Alan Garcia spoke of "the mining contract of the century", and all the representatives of regional and local authorities interviewed consider mining to be the only opportunity for development in the region. They focus on the region's marginalization and see mining as a socially responsible way to combat poverty.

Before the arrival of Xstrata, the Cotabambas province had been largely excluded from public services because the state administration lacked the necessary financial resources. During an official integration celebration in the community of Chila, Pedro Sánchez from MINEM expressed to the gathered communities that "the beginning of this project is what you call a public-private-communal partnership between the government, private investors and the community. You know that the government does not have sufficient resources for all these investments. We therefore welcome the private investor to come here, to bring funds, to bring knowledge and to set up these institutions". Apart from specific development services, state authorities hope that mining will significantly increase tax revenues and thus the available budget. For these reasons, central and regional government representatives see themselves more as promoters of the mining industry and less as supervisory bodies. At the time of research there was little interest and there were no concrete strategies to examine environmental and social impacts and to ensure compliance with the obligations Xstrata signed up to in the contract for the Las Bambas project in 2004 (Wallimann 2011: 112–113).

2.3 Local People's Strategies and the Role of NGOs

Within the project's direct zone of influence in Cotabambas, many people consider Xstrata to be an entity that, unlike the local NGOs, not only offers capacity building but also material support (interviews with H. Rayme, CADEP and J. Cuadros, CooperAcción, 2010). The local communities were hardly accessible by road and their inhabitants have little chance to obtain jobs and their financial means are thus very limited. At the time of research, the new prospects provided by the company were stronger than the insecurity about the impacts of the future mine. As one of the peasants said during a visit in Fuerabamba: "We, who used to be poor, we all have money now. [...] We did not have any money in the past, we had nothing. [...] Now we are more or less satisfied, more or less independent and reassured. Sometimes we worry, because we live on top of the hill [projected for mining]. "

Even peasants who are critical about the mine would not turn down the chance of working for Las Bambas, because they see no alternatives. Ernesto, inhabitant of the community of Peccoy, describes his ambivalent expectations: "There are two types of change that will take place in the province. The positive and the negative. In this case, there will be more negative than positive changes [...] pollution of the environment and the social conditions, something that many of us might not perceive. Our way of living, the way we dress will not be the same like now." Still, he confirms that he would immediately work for Xstrata: "[...] because we are all interested in finding an employment, I think. If they gave me a job, I could be free. What a fool I would be, if I did not take up such an opportunity."

Many people in the communities feel historically marginalized and want to break out of this situation, says Gauchinho of the Choaqquere community: "[...] So we began to negotiate over the land in order to benefit our community. Mainly because of the fact that our community has never received any support in the past. Neither from the district authorities nor from the institutions. There was no support whatsoever." With the establishment of the mining industry in the province, the communities hope to benefit from the emerging infrastructure and economic growth. Thus, by 2010 almost all the communities within the zone of direct influence had decided to cooperate with Xstrata. The motivations voiced were to improve financial income, education, health and the general standard of living.

Community members not always focus primarily on the community's future in their deliberations, but also on family or individual interests. When the Choaqquere community agreed to sell land for the resettlement of Fuerabamba, the strongest support came from people already employed by Xstrata. Flor, a peasant women from Choaqquere, experienced the negotiations as tense and dishonest. She suspected that leading voices advocating the sale in the community had been bribed and suddenly switched to oppose it once they did not receive what they wanted: "Could it be that the workers of the community relations promised something to them? Perhaps, they were promised money, as a payment for their leadership to say 'yes'. Possibly, they did not give them the money, so from one moment to the next, they turned their back to the company."

In the discussions and interviews, many community members expressed visions of the future that increasingly placed their families, but not the community, at the centre. Felipe speaks of a division within the community (which was distancing itself from its president at the time of fieldwork), where everybody acted in his/her own interests: "If local authorities, the community's president, the people etc. would work together with the company, that would be nice. But now it seems to me that everybody goes his own way. The president is heading in one direction, the community in another [...]."

Like in the Choaqquere community, in all communities inside and outside the direct zone of influence, a shared strategy of how to deal with the mine did not seem to exist. As mentioned previously, most peasants had little knowledge about the mining project, its size, duration and details of the benefits promised by the company. According to Cuadros et al. (2005: 21), most communities had not elaborated a communal development plan (as it is promoted by local NGOs) and thus had no common strategy on how to deal with the challenges and opportunities when the project appeared. As a result, the communities depend on the company not only financially but also with regard to a development strategy.

In the communities outside the direct zone of influence, at the time of research, some exponents tried to get more information about the processes related to mining. They did not form any majority, but a small group of motivated people collaborating with NGOs, participating in trainings and organizing themselves across community borders. Some of them also intended to improve their negotiation skills, expecting that sooner or later they would be confronted with Xstrata because of the mine's expansion or other issues. The local organizations in Cotabambas that cooperate with NGOs had very limited power to mobilize against the mine. For instance, the local peasants' association had so far been the most important grassroots organization in the province concerning its position on the Las Bambas project. However, the number of representatives of directly affected communities actively supporting the association had decreased. This can be explained by the company's pressure strategy. As explained previously, people from the direct zone of influence often refused to work with NGOs, and similarly with the peasant's association. Hilda Rayme from CADEP comments: "[...] It has turned out that authorities or delegates of the communities concerned are not part of the peasants' association. Why? Because they feel pressured by the mining company; especially due to the diverse economic benefits that the company provides to them, they are being corrupted." The strategy of the peasants' associations was therefore unclear at the time of research. According to the association's president, it is necessary to wait until the company makes a mistake which would reduce the communities' acceptance of the mining project and would enable to mobilize them anew.

Another peasant explained a basic problem concerning negotiations with Xstrata that the communities have to deal with: "Here, we are only familiar with small amounts of money, very, very small amounts. I think this is one of our weaknesses. Apart from that, money is numbers. We are not as used to numbers. 20 Soles or 100,000 Soles? Most of us cannot read numbers. This will make the negotiation process very difficult." On the one hand, the communities had too limited knowledge of the forthcoming project and its impacts to be able to adopt specific strategies. On the other hand, there was a lack of space for articulation, to work out common strategies. According to Hilda Rayme, it was mainly the young people of Cotabambas who hoped to find employment in the mining sector and to change their living conditions. At the same time, especially members of the communities outside the zone of direct influence complained that they were excluded from decision-making processes concerning the Las Bambas project. In sum, the conditions to

Fig. 4. Xstrata/FOSBAM billboard for the construction of an irrigation system. The header says: "We work for the sustainable development of Grau and Cotabambas". Photo: Annette Wallimann.

act collectively and to defend common interests were lacking due to internal disagreements and divisions (Wallimann 2011: 82–89).

The Role of NGOs

The two NGOs mainly mentioned in this chapter are CooperAcción and CADEP (Centro Andino de Educacion y Promocion "Jose Maria Arguedas"). CooperAcción is organized at the national level and specifically supports communities affected by extractive industries, both politically and through development projects. CADEP is an NGO with its headquarters in Cusco that has traditionally operated in the provinces of Apurímac and Cusco and had recently focused on mining due to its increased significance in this region. As Las Bambas is one of the largest mines in Apurímac and the most important one in the Province of Cotabambas, both NGOs through their work have come face to face with the mining project and its operator Xstrata. At the beginning of the project of Las Bambas, CooperAcción had been partly collaborating with Xstrata to create a roundtable dialogue for all stakeholders involved in mining. After the failure of this dialogue, the cooperation between CooperAcción and Xstrata was terminated due to insurmountable differences.

Both NGOs collaborate with local communities, local governments, other NGOs and civil society organizations. They mainly work on capacity building of peasant communities. Both organizations stress that the communities should get the opportunity to independently determine their path of development and that they should at least get the opportunity to articulate their needs in local and regional development plans. In order to enhance the participation of local communities in mining, community members need to acquire certain competencies in environmental science, negotiation, law, institutions and other general knowledge. The NGOs therefore demand that Xstrata respects the ideas and priorities of local residents in the ongoing development processes. In the experience of both NGOs, it is difficult to grasp the visions and strategies of

Fig. 5. Community assembly in front of a Xstrata sponsored schoolhouse. Photo: Annette Wallimann.

communities according to their own endogenous concepts. One aim of these NGOs is to support local and regional grassroots organizations, like the mentioned peasants' association, but these organizations had become weak and were no longer able to build majorities (Wallimann 2011: 113–114).

According to representatives of CADEP and CooperAcción, there was an increasing tendency that the company referred to them as "anti-mining organizations". This complicated their work in the region and made their position appear to be extreme. However, both CooperAcción and CADEP do not consider themselves as mining opponents, but rather as civil organizations. They demand from the company that it maintains social and environmental standards and advocate the protection of the rights of the local population. Their main concern was that communities should be able to take a position on the Las Bambas project and related developments in a free and informed way, which they considered as becoming increasingly difficult. The company's policy of distributing gifts and projects to the communities in the zone of direct influence was partly subject to the condition that the people do not cooperate with CooperAcción. Both CooperAcción and CADEP were confronted with a further problem: while community members were paid for their work in Xstrata's development projects, the time they invested in NGO programs did not generate any income (Wallimann 2011: 101–103).

3. Conclusion

As has become clear, the interests concerning the realization of the mining operations in Las Bambas at the time of research varied widely. The representatives of the company, state authorities and important parts of the affected communities consider the project to be an opportunity for the development of Cotabambas and to combat poverty in the region. Also NGO staff generally think that mining has the capacity to change the communities' living conditions in a positive way.

However, they criticise Xstrata's current practices concerning the participation and consultation processes, the unequal treatment of communities in and outside the official "zone of influence" and the content of development projects, which are seen as subordinated to the company's urge to make profits. Xstrata neglects the possible negative social impacts of Las Bambas in its EIA as well as in the public discourse. Moreover, the company had so far not made any forecasts concerning these impacts. NGOs expected that the company's development efforts should be sensitive to present and future conflicts between and within the communities. Peasant communities within the zone of direct influence considered ongoing social conflicts to be a result of the compensation payments made by the company. Communities outside of the influence zone had not experienced such social impacts yet. Xstrata stimulated individualist interests and strategies within the local communities and at the same time weakened alternative development paths through the control of material and media power.

All actors were aware of possible environmental impacts. However, this awareness was of a rather latent character and so far not very visible. Xstrata as well as state actors either ignored the issue of environmental impacts in their discourse or described them as negligible and a remnant of previous, less sustainable mining techniques. Furthermore, the underdevelopment of the province Cotabambas and the whole region was constantly stressed. This discourse of underdevelopment and poverty served to legitimize the pushing forward of private investments and was the key to Xstrata's corporate social involvement. Xstrata's strategy was to provide generous compensation payments to affected communities and to promise economic development for the provinces of Grau and Cotabambas. Additionally, this development discourse reached the entire Cotabambas province and established mechanisms of inclusion and exclusion that forced people to adopt a position either for or against the Las Bambas project.

Because the development discourse of the company and its symbols are present all over the province, Xstrata has managed to define the concept of development in the area. Adopting mechanisms of inclusion and exclusion, the corporate discourse equals development with the acceptance of the project and implicitly penalises opposition with marginalization. The discourse of the company has a significant influence on the political, social, and economic integration of individuals and groups in Cotabambas province. Because of the weak presence of the state and its unilateral cooperation with the company, peasant communities increasingly depended on Xstrata. Because of historic and social fragmentation, communities have no possibility to mobilize and jointly formulate alternatives to the mining project. Through their corporate social responsibility activities to promote development and the implementation of "sustainable mining", Xstrata obtained a good reputation within the mining sector on a regional, national and even international level before the actual exploitation of any resources had started. This has created both legitimacy and preventive political security for the Las Bambas project.

It has become evident that two factors might lead to a stronger rejection of the Las Bambas project in the future: on the one hand, the emerging conflict over jobs claimed by the local population. Due to Xstrata's high demand for skilled workforce, employment of local people will remain limited. This conflicts with the communities' call for paid labour possibilities for the inhabitants of the two affected provinces. On the other hand, the decreasing water supply already taking place in some peasant communities has a significant conflict potential. Agricultural activities are highly dependent on the condition of the environment, and in particular on water resources. If water supply for livestock and fields is no longer guaranteed, even Xstrata's own productive projects implemented in several directly affected communities will fail, as it has happened in the area of the Tintaya mine before (see Fuentes 2012). Moreover, the resettlement of communities will radically change people's social situation and livelihoods in the province from a long-term perspective. The risk that the expectations held so far will be disappointed is there, and with this, also the attitudes toward the mine could change.

Update

After the 2013 takeover of Xstrata by Glencore, the Las Bambas project was sold in April 2014. The sale was demanded in exchange for the approval of the takeover by the Chinese Ministry of Commerce, out of concern about Glencore's possible domination of the world copper market. The mine is now being developed by the Australia-based mine operator MMG, a fully owned subsidiary of the Chinese Guoxin International. Further, the Chinese state-owned CITIC Metal Co. Ltd. holds 15.0% of Las Bambas. The transfer price was set at USD 5.85 billion plus reimbursement of the investments done so far by Xstrata, amounting to 7 billion in total (note that the concession was priced at USD 121 million in 2004).[19] In 2015, the mine's infrastructure was largely finished and production was planned to start in early 2016. The design has been changed: instead of transporting raw mineral slurry for further processing to the existing plant at the Antapaccay mine, the copper concentration and molybdenum extraction plants are now located in the Las Bambas concession area. The copper concentrate is then to be transported by road to the port of Matarani.[20]

By 2014, the new houses for the resettlement of 441 families of the peasant community of Fuerabamba, agreed upon in 2010, were finished. The new village features multi-storied concrete blocks with little space for animals and agricultural activities, and replacement lands spread out over a large area. It was still to be seen how the villagers would cope with the changes (MultiWatch 2014: 111; see also the case documentation on www.multiwatch.ch).

Just as this book goes into print, the situation at Las Bambas is changing dramatically: at the end of September 2015, a *frente de defensa* ("defence front"), formed by the peasants' association and a large number of the surrounding communities, mobilized to block the almost finished mine, after previous complaints remained unaddressed. The main reasons cited are the lack of consultation regarding the above-mentioned, fundamental changes in the mine design,[21] and the related increase of contamination risks and water consumption, as well as the dozens of heavy trucks that will cross through the villages. On the 28 of September, more than 10,000 people surrounded the mine and occupied parts of the installations. When the about 3000 strong police force started to intervene, the violence escalated, leaving three villagers dead and many wounded by gunshots. The government reacted by declaring a state of emergency—suspending fundamental rights of assembly and political expression—in Cotabambas and three other provinces of Apurímac, as well as Chumbivilcas and Espinar in Cusco. This indicates that the government fears an extension of the agitation to other areas affected by large-scale mining.[22]

19 Reuters.com, 01.08.2014: "Glencore reaps $7 billion from Las Bambas mine sale to China's MMG".

20 MMG: "Our Operations: Las Bambas" and "Las Bambas Factsheet", <www.mmg.com> accessed 18.2.2015.

21 See: CooperAcción y Observatorio de Conflictos Mineros del Peru: Caso Las Bambas, Informe Especial, Octubre 2015. <cooperaccion.org.pe>, accessed 16.11.2015.

22 Compiled from various sources: La Republica, 29.09.2015: "Explotó Las Bambas", and "Declaran estado de emergencia en Apurímac", <www.larepublica.pe>; La Mula, 19.10.2015: "El tema de Las Bambas aún no esta resuelto", <www.otramirada.lamula.pe>; and <www.multiwatch.ch/de/p97002113>; all accessed 24.10.2015.

References

Alayza, Moncloa Alejandra 2007: No pero Sí: Comunidades y Minería. Consulta y Consentimiento Previo, Libre e Informado en el Perú. Lima: CooperAcción.

Baier, Wolfgang 1989: Peru als Beispiel eines südamerikanischen Entwicklungslandes. Frankfurt a.M.

Bebbington, Anthony (ed.) 2007: Minería, Movimientos Sociales y Respuestas Campesinas. Una Ecología Política de Transformaciones Territoriales. Lima: IEP.

Bury, Jeffrey 2004: Livelihoods in Transition: Transnational Gold Mining Operations and Local Chance in Cajamarca, Peru. *The Geographical Journal* 170 (1): 78–91.

Cuadros, Julia 2010: Impactos de la Minería en la Vida de Hombres y Mujeres en el Surandino. Los Casos Las Bambas y Tintaya. Lima: CooperAcción.

Cuadros, Julia, Jaime Consiglieri and José de Echave 2005: Mapeo de la Situación Social e Institucional del Proyecto Las Bambas. Lima: CooperAcción.

De Echave, José 2009: Minería y Conflictos Sociales en el Perú. In: De Echave, José, Raphael Hoetmer and Mario Palacios Panéz (eds.): *Minería y Territorio en el Perú. Conflictos, Resistencias y Propuestas en Tiempos de Globalización*. Lima: Programa Democracia y Transformación Global etc. 107-131.

De Echave, José, Palacios Mario Panéz and Raphael Hoetmer 2009a: Introducción. In: De Echave, José, Raphael Hoetmer and Mario Palacios Panéz (eds.): *Minería y Territorio en el Perú. Conflictos, Resistencias y Propuestas en Tiempos de Globalización*. Lima: Programa Democracia y Transformación Global etc. 15–31.

De Echave, José, Alejandro Diez and Ludwig Huber et al. 2009b: Minería y Conflicto Social. Lima: CBC, CIPCA, CIES and IEP.

Fuentes, Golda Amanda 2012: Entwicklung? "Ja aber Nein": Comunidades Campesina-indígenas in Espinar, Peru im Widerstreit mit neokolonialen Entwicklungsdispositiven der transnationalen Bergbauunternehmung Xstrata Tintaya S.A. Bern: Masterarbeit, Geografisches Institut, Universität Bern. (unpublished)

Gouley, Clotilde 2005: Conflictos Mineros, Interculturalidad y Políticas Públicas: El Caso de Las Bambas, Provincias de Cotabambas y Grau, Departamento de Apurímac. Informe Final. Cusco: CBC.

Haller, Tobias, Annja Blöchlinger, John Markus, Marthaler Esther and Sabine Ziegler (eds.) 2007: Fossil Fuels, Oil Companies and Indigenous Peoples. Münster: LIT.

MultiWatch (ed.) 2014: Milliarden mit Rohstoffen. Der Schweizer Konzern Glencore Xstrata. Zürich: edition 8. (Available in Spanish and English as Pdf from www.multiwatch.ch)

PDEC de Cotabambas [w.y]: Plan de Desarrollo Estratégico Concertado de la Provincia de Cotabambas 2010 al 2021. Equipo Técnico, Municipalidad Provincial de la Provincia de Cotabambas.

Pinto, Vladimir 2009: Reestructuración Neoliberal del Estado Peruano, Industrias Extractivas y Derechos sobre el Territorio. In: De Echave, José, Raphael Hoetmer and Mario Palacios Panéz (eds.): *Minería y Territorio en el Perú. Conflictos, Resistencias y Propuestas en Tiempos de Globalización*. Lima: Programa Democracia y Transformación Global etc. 87-106.

Rist, Stephan 2000: Linking Ethics and the Market: Campesino Economic Strategies in the Bolivian Andes. *Mountain Research and Development* 20 (4): 310–315.

Valderrama, Escalante Gonzalo 2010: Recursos Comunales, Institucionalidad y Minería. El Caso de Challhuahuacho, Apurímac. Cusco: Centro Bartolomé de las Casas.

Vega-Centeno, Imelda 2005: Cultura, Identidad y Región en Cusco y Apurímac. *Cuadernos Decentralistas* 17. Lima: Grupo Propuesta Ciudadana.

Wallimann, Annette 2011: Entwicklung in privater Hand – Der Neue Bergbau und sein Entwicklungsdiskurs am Beispiel des Bergbauprojekts Las Bambas in der südperuanischen Provinz Cotabambas. Bern: Masterarbeit, Institut für Sozialanthropologie, Universität Bern (unpublished).

Xstrata Copper 2010: Desarrollo Sostenible en Las Bambas. Inversión Social 2005–2010. Lima

Xstrata Tintaya S.A. 2010a: Estudio de Impacto Ambiental Proyecto Minero Las Bambas. Resumen Ejecutivo. Lima: Golder Associates.

Xstrata Tintaya S.A. 2010b: Estudio de Impacto Ambiental Proyecto Minero Las Bambas. Volumen A: Generalidades del Estudio de Impacto Ambiental. Lima: Golder Associates.

Xstrata Tintaya S.A. 2010c: Estudio de Impacto Ambiental Proyecto Minero Las Bambas. Volumen F: Evaluación de Impactos Socio-Económicos. Lima: Golder Associates.

Online Sources

Anderegg Susanna 2010: Xstrata malt die Zukunft in rosigen Farben. Der Bergbaukonzern wirbt in Peru um die Gunst der Bevölkerung im Projektgebiet der Kupfermine Las Bambas. *Neue Zürcher Zeitung* 09.09.2010.

Archivo Digital de la Legislación en el Perú. <www.congreso.gob.pe> accessed in December 2012.

Banco Central de Reserva del Perú. <www.bcrp.gob.pe> accessed in January 2013.

Christian Aid 2005: Unearthing the Truth. Mining in Peru. < www.christianaid.org.uk/Images/unearthing_the_truth.pdf > accessed in June 2013.

Comisión de la Verdad y Reconciliación. <www.cverdad.org.pe> accessed in February 2011.

CooperAcción. <www.cooperaccion.org.pe> accessed in December 2013.

Del Castillo, Laureano and Pedro Castillo: Tierras Agícolas y Servidumbre Minera: CEPES. <www.cepes.org.pe> accessed in June 2013.

El Comercio 2010: Mandatario calificó de "Contrato del Siglo" al Convenio para Desarrollo de Proyecto Las Bambas. 02 September 2010. <elcomercio.pe> accessed in June 2011.

El Comercio 2012: ¿Cuál es el futuro de las evaluaciones ambientales con la creación del Senace? 03 December 2012. <elcomercio.pe> accessed in June 2013.

ILO. International Labor Organization. <www.ilo.org> accessed in January 2011.

INEI. Instituto Nacional de Estadística e Informática - Perú. <www.inei.gob.pe> accessed in March 2011.

LIPortal. Deutsche Gesellschaft für Internationale Zusammenarbeit (GIZ) Gmbh. <liportal.giz.de> accessed in January 2013.

Marti Werner 2009: Der Sendero Luminoso macht wieder Schlagzeilen in Peru. *Neue Zürcher Zeitung* 06.06.2009.

MINAG. Ministerio de Agricultura - Perú. <www.minag.gob.pe> accessed in February 2011.

MINEM. Ministerio de Energía y Minas - Perú. <www.minem.gob.pe> accessed in March 2011.

MINEM 2011: Annuario Minero. Reporte Estadístico. <www.minem.gob.pe> accessed in January 2013.

PDRC 2010: Plan de Desarrollo Regional Concertado Apurímac del 2010 al 2021. Abancay: Equipo Técnico Regional del Proceso de Actualización del PDRC. <www.regionapurimac.gob.pe> accessed in September 2010.

PNUD (Programa de las Naciones Unidas para el Desarrollo) 2009: Informe sobre Desarrollo Humano Perú 2009: Por una Densidad del Estado al Servicio de la Gente. <www.undp.org> accessed in June 2013.

ProInversión. Agencia de Promoción de la Inversión Privada - Perú. <www.proinversion.gob.pe> accessed in June 2013.

Tribunal Constitucional del Perú. <www.tc.gob.pe> accessed in June 2013.

Xstrata Perú S.A. 2004: Evaluación Ambiental. Proyecto de Exploración Las Bambas. Volumen 1. <www.lasbambas.com> accessed in June 2010.

Xstrata Plc 2009: Sustainability Report 2009. <www.glencorexstrata.com> accessed in June 2013.

Xstrata 2013. <www.xstrata.com> accessed in January 2013.

Xstrata Copper 2013. <www.xstratacopper.com> accessed in January 2013.

The Ranger and Jabiluka Uranium Mines in Northern Territory, Australia

Isabella Brom and Daniela Bucher[1]

1. Situating the Case Study

The uranium mines looked at in this case study are located in the Alligator Rivers Region in Australia's Northern Territory (NT). Two-thirds (20,000 km^2) of this region, also referred to as "Kakadu Region", is part of the Kakadu National Park (KNP), a UNESCO World Heritage site. Besides being ground for fishing, hunting and gathering, ceremonies, customary law, spiritual connection and socio-political systems of Aboriginal communities, the land is also subject to pastoralism, wildlife conservation, nature reserves, tourism and mining (GAC 2011e).

The main conflict in the uranium-rich Kakadu region relates to the strong focus of the government on international interests in uranium and monetary benefits instead of the well-being of the local people and the preservation of Aboriginal culture. The two major uranium mines that are subjects of this case study are Ranger, which started to operate in 1980, and the proposed Jabiluka project which was stalled in 1999 due to opposition by the indigenous landowners. Energy Resources of Australia (ERA) is the operating company for these mines and a central actor together with its major shareholder Rio Tinto and the Australian Government. The third relevant party is the local Aboriginal Mirarr community, who live in and adjacent to KNP and the mining areas. The Mirarr oppose mining developments on their ancestral land and have launched an international campaign to ensure the implementation of their formal rights through the Aboriginal Land Rights Act and to stop the development of the Jabiluka project.

1.1 History and Context

Some of the main protected features of the KNP in the NT are its floodplains, estuaries, tidal flats and high biodiversity rate as well as diverse aquatic and terrestrial ecosystems (Johnston 2000: 104). The Aboriginal people in West Arnhem Land and the NT, including the Kakadu region, are referred to as "Bininj" or "Mungguy", since the term "Bininj Gun-Wok" is the name of their highly polysynthetic language (Evans 2004: 71). The remains of ancient rock art, found at approximately 5,000 sites in the area, illustrate one of the world's oldest historical chronicles of indigenous people and are often used as proof for Aboriginal land claims (Australian Government 2011). In the 1960s, a regional national park was proposed in Kakadu in order to preserve its ecosystems and resources. Local Bininj were little involved in park planning, since most of them live outside the park perimeter and the government dismissed their position as traditional land owners (Wellings 2000: 120). Shortly after, a new Federal Government recognized the Aboriginal people as Australian citizens with full voting rights and improved the protection of their territorial rights.

1 Laura Rindlisbacher contributed to the first draft. After finishing this chapter, Daniela Bucher and Loretta Scherler (co-author of the Philippines case study) carried out field study in the Kakadu region for their master thesis (see Bucher 2014 and Scherler 2013). Some of their results have been added later on.

Fig. 1: The Ranger Mine. Photo: Dominic O'Brien.

In 1975, the Federal Government commissioned the Ranger Uranium Environmental Inquiry (RUEI) in response to concerns in the community and the worries of non-governmental organizations about Australia's involvement in uranium mining and the prospect of long-term mining activities around KNP (Wellings 2000: 121). In 1977, the RUEI concluded, "[t]he mining in the Alligator Rivers Region could proceed without having significant impact on the environment provided that a comprehensive system of environmental protection was implemented" (Johnston 2000: 103). It suggested that the Australian Government should not prevent the development of uranium mining despite the opposition of Aboriginal landowners (Johnston 2000: 103). Additionally, the granting of Aboriginal Land titles to a substantial part of the region as well as the creation of the KNP was recommended by the inquiry (Ling 2012)

At about the same time, the Aboriginal Land Rights Act of the Northern Territory of 1976 (ALRA NT) was established. It was the government's first attempt to legally recognize the Aboriginal system of land ownership (O'Faircheallaigh 2008: 33f.). When the Act was passed, the former "reserves" became Aboriginal land and unalienated[2] Crown Land, and Aboriginal people could claim land ownership based on the Act. However, a successful land claim requires the indigenous landowners to prove their traditional relationship to the land, which involves extensive research by anthropologists. The land claims are handled by Land Councils, which are representative bodies of Aboriginal people (CLC 2011). The ALRA also includes the principle of "Free, Prior and Informed Consent" (FPIC), which gives the Aboriginal landowners the right to veto mining exploration on their land and to negotiate agreements (Sosa 2011: 6).

2 Land which is assumed to be of no interest or not being used by other parties.

1.2 Geography and Ecosystem

The Alligator Rivers Region (approximately 28,000 km^2) is broadly defined by the catchments of four large rivers that run into the Van Diemen Gulf. The tropical climate in the Northern Territory is characterized by temperatures between 22 °C and 34 °C all year round, especially high humidity in summer and a mean annual rainfall of up to 1,565 mm. The wet seasons are marked by localized monsoon depressions, bringing heavy rain, winds and tropical cyclones (Press et al. 1995: 94f.). This climate is a determining factor for the surface water hydrology, vegetation and also the gradual development of the landforms of the park region (UNESCO WHC 2011).

Between 1979 and 1991, UNESCO declared the majority of the Alligators Rivers Region a national park with world heritage status based on its unusual combination of largely uninhabited areas with extraordinary biodiversity and a high concentration of Aboriginal rock art of unique significance (Johnston 2000: 103; Haynes 2009: 67; Aplin 2004: 156–158). The large diversity of landscapes, particularly the water bodies of Kakadu, is the environmental value of the park. Kakadu's rivers have produced large floodplains, swamps, estuaries, mangroves and mudflats. Its wetlands are known to contain the world's richest tropical breeding ground for water birds and are listed in the Ramsar Convention of Wetlands (Sweeney/O'Brien 2002: 2). The maintenance of water quality is essential for the stability of the wetlands and the survival of dependent organisms as well as to sustain clean groundwater, which is the primary source of potable water in Kakadu (Press et al. 1995: 98). The flora of KNP consists of distinct species of Eucalyptus trees and is one of the most diverse in northern Australia. KNP is home to 46 plant species considered rare or threatened and nine that have only been found within the park perimeter (UNESCO WHC 2011). Furthermore, it is home to diverse water birds, fish, invertebrates and small mammals and is their most important breeding habitat.

1.3 Local Population

In 2003, the Aboriginal population constituted a minority of 2% of the entire population of Australia. Reflecting a long history of colonization and discrimination, they continued to have a significantly higher mortality rate, a 45% lower average income and a 22-times larger likelihood of getting imprisoned than non-Aboriginal Australians (Crawley/Sinclair 2003: 362).

The fundamental paradigm of Aboriginal society is a belief system and intertwined philosophy referred to as "Dreaming", "Dreamtime", or "Story Time", or in many Aboriginal communities as "Tjukurpa" (Strang 2004: 209; Australian Government 2012b). Dreaming is the foundation of life, society and kinship and a fundamental religious and legislative conception among all Aboriginal groups in Australia (Morton 2005: 196). Its basic premise is that all aspects of the world, including humans, land, flora and fauna, and astral bodies are connected. Dreaming refers to the creation period when ancestral beings created the world (Kolig 1989; Bowie 2000; Lennon et al. 2001; Strang 2004). It is taught and passed on by oral traditions, ceremonies, singing, art and stories relating to specific sites and ritual dances and it is a cultural obligation to carefully transmit this knowledge between generations (Kolig 1989: 38; Rose 1996; Bowie 2000; Australian Government 2012b). Bowie explains that "Aboriginal cosmology is both practical and spiritual—there is no division between sacred and secular or between beliefs and the laws governing everyday life—the cosmology is The Law, entering every aspect of human existence" (1997: 238; in Bowie 2000: 141).

The social anthropologist Kolig (1989: 16) explains however, that due to Western impacts the Aboriginal worldview is no longer what it used to be in pre-colonial times. He claims that the traditional Dreamtime concept has undergone a functional change, finding itself in an area of conflict between traditional belief and rationalization (Kolig 1994; 2003; 1989: 16ff.). Rituals are inherited from Dreamtime ancestral beings and are a vital activity required for the maintenance of Aboriginal societies. On the one hand, through their intentional performances to influence the composition of society and its reconstruction or reformation, Kolig considers rituals a political tool. On the other hand, the performance of rituals is understood as a retrospective activity, which has already been acted out by the ancestors during the Dreamtime (ibid. 1989: 16f.).

Totemism is a major element of many mediating processes, bridging the gap between the human and the spiritual, the individual and the collective, and the living and the deceased (Morphy 1988: 249). Bowie (2000: 137) points out that this traditional belief system is neither universal nor exclusively shared by all Australian Aboriginal people. Kolig reckons that in today's declining traditional magical expectation of totem belief, rituals are losing their status as a politically efficient tool, leading to a process of gradual separation of politics, religion and magic (1989: 16f., 112, 157f.). Kolig (1994) further explains that, exposed to Western political and epistemic paradigms, Aboriginal people are adapting to Western ideology and the dominant political discourse (ibid. 1989: 156; ibid. 1994). However, despite the traditional concept getting increasingly exposed to scepticism and doubt, it remains intact to a large extent. Kolig (1989: 160) considers that there are two coexisting systems of thought in the Aboriginal society, which are intertwined and influence each other. An empirical orientation is shaped by daily experiences, whereas Aboriginal traditional belief is centred around a religious-magical system. Therefore, the spheres of politics, religion and magic are still not fully separated, resulting in a flowing transition between traditional ideology and new types of knowledge and experience. In addition, Kolig (1989: 107ff., 156) explains that Aboriginal people have increasingly adopted methods to improve their situation, such as the use of media to gain support in the wider public. Today, religious traditions can be used as a strategic resource in native title claims (ibid. 2003).

Also Altman (2012) discusses the "state of interculturality", in which indigenous people engage with two sets of values and neither exclusively inhabit their traditional nor a modern lifestyle. He claims that the Australian state does not take these changed circumstances into consideration and that this is at the core of the issues between mining companies, the state and the indigenous communities. In relation to ancestral land claims, Altman explains that the Australian indigenous people are trapped in Western legal definitions of authenticity. The concept of "repressive authenticity" (developed by Patrick Wolfe in 1999) is a state-authorized process used to recognize or deny the traditional ownership of land. The idea of "authentic" or "alleged" landownership however is based on two wrongly conceptualised, contrasting classifications of indigenous people: either they are categorized as undeveloped, "traditional" and inert in time and consequently considered legitimate traditional landowners. Or, they are seen as having adapted to modern circumstances and therefore do not qualify anymore for indigenous land rights. Following this concept of "repressive authenticity", the invasion of Australia needs to be seen as an on-going process, not a past event.

In addition, Altman (2012: 68) also refers to the notion of repressive authenticity to explain internal divisions within the Aboriginal population: only a particular subset of the indigenous population, the traditional owners, benefit and are empowered through mining companies. The wider Aboriginal group does not receive the same benefits, since only the consent of traditional owners is essential for mining agreements. Although Altman's academic authority is undisputed, it has to be noted that he maintains a close relation to the industry. As Justin O'Brien remarks with regard to Altman's above cited article of 2012: "[he] should have acknowledged [...] that he was a player in that debate and active agent [...], and that he shared the view of the Australian Government that the Jabiluka mine should proceed" (personal communication, 18.12.2012).

The land is the foundation of everyday life and the spiritual connection to their ancestors in Aboriginal belief systems (Queensland Studies Authority 2007: 1). The rich resources of the land provide nourishment for the body, mind and spirit, and it represents the connection between family members and generations (Rose 1996: 7; Lennon et al. 2001: 9; Queensland Studies Authority

2007: 1). Loss of land is equivalent to a separation from livelihood and the web of social relations that are embedded within the landscapes (Howitt 2001: 54f.). It is often emphasized that no amount of social welfare or compensation can cover the loss of land to Aboriginal people (Queensland Studies Authority 2007: 1). Jacqui Katona, former Executive Officer of Gundjeihmi Aboriginal Corporation (GAC), explained the relevance of land to her community accordingly: "Kakadu for our mob is our blood and our family. These lands give to us our identity, our history, and our future" (1999: 3). Her reflection on Kakadu illustrates the strong interrelation between the land and the social, political, cultural and economic dimensions of their life. Katona (1999: 3) highlights that the Mirarr's traditional system of clan relations in the region is based on co-operation, mutual obligation and respect for decision-making by traditional owners.

The Mirarr, the Aboriginal community central to this case study, is composed of 28 adult people and about 40 children. According to the Aboriginal Land Rights Act, the Mirarr hold freehold title to their country via the Kakadu and Jabiluka Land Trusts after successfully having claimed nearly all of their traditional territory (GAC 2011e). The Mirarr are sole "traditional owners" of a region comprising the Ranger and Jabiluka Mineral leases, Jabiru town, parts of KNP, the wetlands of the Jabiluka billabong country and the sandstone escarpment of Mount Brockman. The relationship of the Mirarr to their country has been confirmed by anthropologists to date back at least to 40,000 to 50,000 years B.C. (Butlin 1993: 9; Masterson 2010: 17; GAC 2011e). "Traditional owners" is a term created and introduced by European settlers but it is also a concept based in Aboriginal law (GAC 2011e). The traditional owners of an area have the collective responsibility to look after the land, hold the knowledge on how the land needs to be managed and maintained and have superior rights to others (Lennon et al. 2001; Strang 2004: 208ff.; Rose 1996: 44, 63). They can deny or grant permission to other people who wish to enter and use their homeland (Lennon et al. 2001: 9).

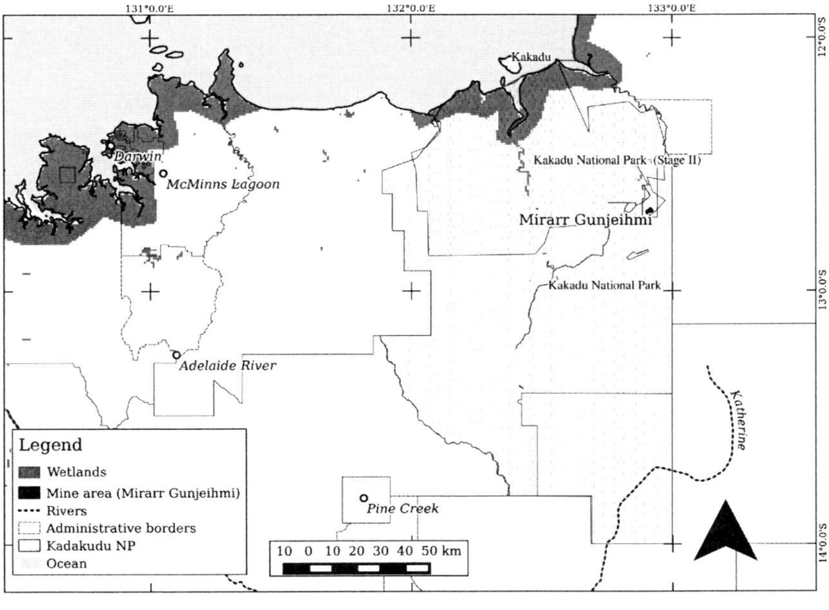

Map 1: Kakadu National Park and the Mining Site.

The tasks of the Mirarr as traditional owners include the protection of the land, maintaining sacred sites and taking care of the recognition of cultural rights (Lennon et al. 2001: 9; Strang 2004: 208ff.). The performance of traditional ceremonies is a primary means to ensure the country's continuing identity and fertility and to monitor the physical environment (Lennon et al. 2001: 9). According to customary law and tradition, Bininj are required to protect their land from possible destruction and violation of sacred sites. Mirarr possess, and are obliged to use, the right to speak for the land and for other Bininj (GAC 2011e). The Mirarr explain furthermore that they have the obligation to teach traditional law and responsibilities for the land as well as their traditional ways of resource use such as hunting, gathering and fishing to future generations (GAC 2011e). The capacity of the Mirarr to maintain their cultural traditions is hindered by the on-going mining activities on their land. This involves, in particular, restricted access to sacred sites or these places being dishonoured or destroyed (GAC 2001b: 32).

In the year 2000, approximately 1,200 Aboriginal people lived in Kakadu region and 450 to 500 inside the National Park (Wellings 2000: 119f.). This region contains two major townships: Oenpelli, an old Christian mission town, and Jabiru township. Jabiru was built together with the Ranger mine to accommodate the workers. The Bininj people arranged themselves in several organizations in order to deal with political, economical and particularly mine-related issues. The "Djabulukgu Association" consists of more than 90 people who are traditional owners and people affected by development on the Jabiluka lease. The association is the owner of several tourism enterprises within the Kakadu region. The "Gagudju Association" (Gagudju being the original word for "Kakadu") is the largest organization of the area, consisting of about 300 indigenous people associated with the Kakadu region. It was established in 1980 in order to manage the royalty money generated from the Ranger mine and has since established several small business enterprises such as tourist accommodations in the KNP (Collins 2000: viiif.; Rio Tinto 2005). In 1995, the Mirarr created their own institution called "Gundjeihmi Aboriginal Corporation" (GAC). It was established due to the unacceptable jurisdictional and institutional arrangements on the land of the Mirarr and aims to exercise their rights as traditional owners under the Aboriginal Land Rights Act.

The competing interests over the same land combined with close cohabitation of native Aboriginal communities and Western mining companies has become a reason for tension and cultural clashes. Howitt (2001: 253) reckons that overall it has been and still is the Aboriginal people's own endeavour to develop their own strategies to adapt to the changes created by the mining industry. Crawley and Sinclair suggest that, from an anthropological point of view, the optimal procedure for a successful collaboration of Western mining corporations, the Australian Government and Aboriginal people consists of learning from the latter, developing respect for indigenous cultures through multiple levels of interaction, including communication between corporate and indigenous leaders, the sharing of power and joint decision-making (2003: 372).

1.4 Local Economy and Resource Use

The Australian Aboriginal people traditionally cultivated an existence primarily based on hunting and gathering without permanent settlements. In the past, the communities used to continuously change their locations to places where resources were available. The knowledge of resourceful land was passed on verbally through generations (Australian Government 2012a). Hunting is still an integral part of Aboriginal traditional life despite considerable Western influence on Aboriginal culture. The Mirarr and other Bininj still consider hunting, gathering and fishing a part of their lifestyle and pursue it for economical as well as cultural reasons (Altman/Allen 1992: 117). GAC publications explain that berries, yams, roots and other bush fruits as well as the magpie goose, a seasonal specialty, are important food items for the Mirarr (GAC 2011d). However, through the influences of white settlers and globalization, the everyday life of the Aboriginal communities has changed and many do not maintain a hunter-gatherer lifestyle anymore but participate in modern economic activities (Australian Government 2012a).

In the 19th and 20th centuries, hunter-gatherers were understood as without sense of ownership or property and as incapable of advanced technological development. Sahlins (1972) argued, however, that they purposely seek a low living standard in terms of food, clothing or other physical capital, so that their needs were maintained and satisfied quicker. Sahlins (1972) and Keen (2010) understand the internal indigenous economy as a set of relationship categories, which have particular obligations. In his study, Keen describes Aboriginal economic relations as shaped by complex kinship obligations, relying heavily on reciprocity. Aboriginal communities maintain a network of duties and responsibilities, rights and credits; children learn the arrangements of obligations at an early age. In a broader sense, this system of kinship is the juristic model of rights and duties required by social statuses. These beliefs and values underlie all economic behaviour and relation and is also understood in terms of the concept of moral economy (Keen 2010: 4ff.; Sahlins 1972; Butlin 1993: 73).

The Aboriginal conceptions of Dreaming have remained resilient despite Western influence. The Aboriginal social identity remains founded on totemic associations. Every totemic clan is characterized by its specific ancestral tracks and story places, which provide its members inalienable and collective title to that particular land (Strang 2004: 210). The sacred sites or "story places" are a central part of this notion of totemism and the Aboriginal connection to the land. According to Bowie (2000: 142) and Kolig (1989), they are located where ancestral beings appeared, acted or reintegrated themselves with the landscape. The sacred sites located within Mirarr land require protection by the traditional owners. Access to them is often restricted, since Aboriginal belief predicts fatal and cataclysmic consequences if they are disturbed (GAC 2011e).

Fig. 2: Local rock art. Photo: Dominic O'Brien.

"Story lines" on the other hand, also known as "song lines", are the traces of the ancestors' journeys across the land in the physical features of the landscape. Through this creation and interaction with the land, the ancestral beings imbued all living creatures and natural phenomena with spiritual power (Bowie 2000: 14ff.; Kolig 1989: 36f.). Thereby, they transformed them into rocks, trees, waterholes, creeks and animals "like footprints in the sand" (Ingold 2000: 53). These features then became the "totems" of clans (Bowie 2000: 14ff.; Morton 2005: 197). The notion of totems further reinforces the Aboriginal system of land tenure, which is composed of both collective ownership and a traditional system of inheritance (Strang 2004: 210). However, it is not common among most Aboriginal land tenure to maintain strict borders between, for example, the country of the Mirarr and the estates of neighboring clans (GAC 2011e). The concept of ownership of Aboriginal cultures differs greatly from the Western institution of property, which prioritizes the commercial productivity of the land (Strang 1997: 253, 263ff.).

While in the last decades the Aboriginal spiritual attribution to sites and places has increasingly become a controversial political issue, a growing respect for indigenous cosmologies has been established in Australia and other parts of the world (Kolig 1989: 38; Bowie 2000: 144). Australia has developed a growing acceptance and application of indigenous ecological knowledge in ecological management (Rose 1996: V; Howitt 2001: 45). The incorporation of traditional ecological knowledge in resource and environmental management systems has become increasingly important (Howitt 2001: 43ff.). However, its application in scientific systems is often considered controversial and challenged by the dominant Western resource management paradigms. Australia's most prevalent examples of the recognition and application of Aboriginal ecological knowledge are the jointly managed national parks in the NT. The park management of Kakadu practices controlled burning in co-operation and consultation with Aboriginal traditional owners (GAC 2011d). The incorporation of the traditional owners as direct participants empowers their knowledge as a key contribution to conservation management (Howitt 2001: 44).

Overall, resource use of the Aboriginal people in the NT is determined by various factors. Due to extreme seasons, resource management varies widely across the region and the annual travelling and gathering cycles of the Aboriginal people are adjusted to the sharp contrasts between the extreme wet and dry seasons in Northern Australia (Strang 2004: 210). The availability of water resources and the aquatic life of the region further determines the patterns of resource use. Today, fishing remains one of the most significant economic and social activities. The landscape's role in providing a topographic, economic, social and spiritual orientation basically depends on water features (Strang 2004: 211). Consequently, Aboriginal people worry about the pollution of water resources. The impact of mining on the spiritual and social well-being of their community can be interpreted as a potent metaphor for the continuing colonization of their land, the destruction of their social and economic structures and the "cultural pollution" of their worldview, beliefs, values and identity (Strang 2004: 209).

The establishment of the mining industry has seriously affected the livelihoods of Australian Aboriginal people in the entire country. However, a partial integration of market, state and mine resources in their lives while still maintaining customary values and activities has led to a diversification of livelihood options. The economic hybridity in Kakadu allows indigenous people to remain connected to their ancestral land while at the same time using it for livelihood. Income-support from state, mining royalties and tourism influence this regional hybrid economy and the resulting strategies. The customary sector is thereby essential, since it connects modernized livelihoods and customary "on-land" activities and traditional values. "Caring for country" resource management activities are entirely in contrast to mining, supporting the protection and conservation of land, ancestral places and traditional ecological knowledge (e.g. as national park rangers). Altman (2012: 67) points out that the economic hybridity therefore allows flexibility and diversity, in accordance with the intercultural character of Aboriginal peoples' lives in the 21st century.

1.5 National Economy and Legal Framework

Australia's economy is one of the fastest growing in the world. In the past 20 years, it has experienced uninterrupted growth. In 2010, according to the nominal GDP, Australia was ranked the 13th largest economy in the world with a GDP of USD 1,532,408 (World Bank 2012). According to current figures, the mining sector contributes 5.6% to Australia's GDP (ABS 2012).

Australia's economy is well diversified with services (71.3%) and industry (24.9%) being the leading sectors of the economy in 2010. Characteristically, the services and financial industries are mainly based in the east of the country whereas western Australia encompasses and controls the majority of natural resources such as iron ore, gold, oil and natural gas. This east-west divide has led to differences in development plans for the country, with Australia's resources and commodities industries experiencing a boom period whereas its dominant financial and service sector came under pressure. It is expected that the growth of regions rich with minerals, oil or gas will soon exceed the economic progress of the densely populated states (EconomyWatch 2010).

Next to holding numerous free trade agreements with several countries, Australia is part of many economic organizations like APEC, the G20, WTO and OECD. China is Australia's biggest import partner and, since 2009, after surpassing Japan, also its largest export partner (EconomyWatch 2010).

Australia's Mining Industry

For the past ten years, the mining industry was a major stimulus for the country's economic growth, with around 35% of Australia's overall exports being minerals, coal, oil and gas. Mining in Australia has always been conducted by the private sector (McKay et al. 2000; EconomyWatch 2010). Australia is the world's largest exporter of coal, iron ore, lead, diamonds, zinc and zirconium and the second largest supplier of gold. With a total of 135,000 employees in 2009, the mining industry employs roughly 1.3% of the total labour force; only a small segment of Australia's workforce, considering the mining industry's large contribution to Australia's GDP and export market (ABS 2006). Indigenous employment by mining companies is a much-debated issue since 60% of Australian mines are located on or near Aboriginal land and the majority of Aboriginal people are amongst those living in poor socio-economic conditions. Only in the past decade, the mining industry has taken a more proactive approach towards increasing the number of indigenous employees, which since 2001, has increased from 1,390 to 2,488 employees (Brereton/Parmenter 2008: 68, 70).

Being home to 24% of the world's uranium deposits, Australia is the third-biggest uranium producer in the world after Kazakhstan and Canada although Australia itself is currently not using nuclear power. Uranium comprises approximately 35% of the country's energy exports. In 2009, it was responsible for 16% of worldwide uranium production and in 2010 for 11% (5,900 tonnes). After peaking in 2007 at 8,603 tonnes, Australia's uranium production is now declining (World Nuclear Association 2012). The exports mainly go to the USA, Europe, Japan, South Korea and Taiwan. In 2006, an agreement was signed with China, where the uranium was used to build up to 30 new nuclear power reactors. Allegedly, it may also be used for military nuclear programs, although the Australian Government claims that the uranium is only sold to be used for peaceful purposes. But even if that was the case, it would consequently free up other uranium (for example, Chinese) for other uses (Vincent 2006: 6f.).

Legal Framework

The Australian Government responded to concerns about Australia's involvement in the uranium industry and the prospect of large-scale uranium mining in Kakadu by establishing the Ranger Uranium Environmental Inquiry (RUEI) (Wellings 2000: 121) under the Environment Protection (Impact of Proposals) Act 1974 (Johnston/Needham 1999: 2; Howitt 2001: 252ff.). The establishment of the Ranger mine was enabled by the recommendations of the RUEI, which also considered—and recommended favourably—the establishment of the National Park and the recognition of land rights for indigenous people (Howitt 2001: 253ff.). Subsequent to the foundation of the RUEI, several environmental and collaborative acts were set up by the Commonwealth Government like the "National Parks and Wildlife Conservation Act 1975" (NPWC Act) under which the KNP is managed by the Director of National Parks and Wildlife through Parks Australia, an agency of the Commonwealth Government, in cooperation with the Aboriginal traditional owners (Johnston/Needham 1999: 3f.). The Commonwealth Government promotes uranium mining despite the fact that KNP is listed as a World Heritage site and was almost listed as a World Heritage site in danger (Aplin 2004). In order to monitor the social and environmental impacts of uranium mining, the statutory office of the Supervising Scientist for the Alligator Rivers Region was created by the Environment Protection (Alligator Rivers Region) Act 1978. Despite this, concerns remain regarding the achievement of environmental protection in the NT (Johnston 2000: 105; O'Connor 2011: 40f.). It came as no surprise that the assurances of scientists mean little or nothing to Aboriginal people (Johnston 2000: 111f.).

The Aboriginal Land Rights Act of the Northern Territory (ALRA NT) of 1976 provides indigenous "traditional landowners" the right to Free, Prior and Informed Consent regarding access to mineral resources. However, the ownership of the resources itself remains with the state (Altman 2012: 50). Being framed at the same time as the ALRA by a newly elected Commonwealth Government, the "Ranger Mining Agreement 1976" combined the recognition of indigenous land rights with providing corporations access to mine the uranium deposit. The approval of the Ranger mine was thus not only linked to a strengthening of environmental protection and monitoring and mitigation of impacts. Also, it included provisions for sharing economic benefits with the affected communities and for the regulation of indigenous land rights (Scambary 2009: 172f.). In addition, the creation of opportunities for education, employment and small business development was intended. Nevertheless, the mining town Jabiru, built in the 1970s, first became a restricted access area. With this step, which is even referred to as a "social contract", the intention was to minimize interaction between local indigenous people and the mining community (Scambary 2009: 173). It was only in 2000, that the Commonwealth Government announced a new indigenous employment policy targeted at Chief Executives. The program includes financial support for training and apprenticeship schemes for indigenous employees (Crawley/Sinclair 2003: 367). However, the Australian Government has not established a standard compensation regime for exploration or mining on land with native title rights (Stephens 2010: 12). This stands against the Australian judiciary's recognition of the native title since 1992 and the establishment of the Council for Aboriginal Reconciliation Act 1991 (Scambary 2009: 179).

2. The Ranger and Jabiluka Mines

The Alligator Rivers Region is home to several major uranium deposits. Two of the most important, situated close to each other, are Ranger and Jabiluka. The Ranger deposit was discovered in 1969 and first developed by a joint venture of two Australian mining companies and the state-owned Australian Atomic Energy Commission (AAEC), to satisfy a long-term delivery contract signed between the Australian and the Japanese governments. In 1978, the mining companies and the Northern Land Council (NLC) finalised the agreement to build the mine, against the will of the Aboriginal traditional owners of the affected land. The NLC, despite its responsibility to formally represent their opinion, did not act upon their instructions to prevent this development (Scherler 2013 and World Nuclear Association 2012). Mining started in 1980 and was originally planned to continue until 2006. When further ore deposits were discovered at Ranger, the government granted an extension of the operation period until 2020, again in disrespect of the traditional owners (GAC 2011b; ERA 2011b; SSD 2011).

Worldwide, Ranger was the fifth largest uranium producing mine in 2011 and "one of only three mines in the world to produce in excess of 100,000 tonnes of uranium oxide" (ERA 2011b: 3). Production varied between 3,258 (2002) and 5,760 (2006) tonnes of uranium yearly. By comparison, Australia's total uranium production was about 7,000 tonnes per year (World Nuclear Association 2012; ERA 2011b: 126). The mine area covers 78.6 km^2, which includes the open pits and the deposits for a yearly average of 2 million tonnes of waste rock. Mining operations at Ranger pit 3 were finished at the end of 2012, and the backfilling work had started. The total remaining ore deposits at Ranger are estimated to be approximately 140 million tonnes. However, these are of a considerably lower grade than the uranium processed so far. Mining will continue underground at the Ranger 3 Deeps deposit beneath pit 3, estimated to contain 34,000 tonnes of uranium oxide. In May 2012, ERA commenced work on the AUD 120 million Ranger uranium exploration tunnel. However, ERA has stated that they will not expand upon any planned mining operations without the expressed consent of the traditional owners (ERA 2011b: 3; Beaton/Tutty 2012; Atkinson/Thiebeault 2013: 5, 13, 15).

Jabiluka in comparison, similar in size to the neighboring Ranger deposit but planned as an underground mine from the beginning, was to produce an average of 1,000 tonnes per year over its planned 28 years of operation (GAC 1999a: 6). With an estimated 141,640 tonnes of uranium oxide, Jabiluka, discovered between 1971 and 1973, contains one of the world's largest known undeveloped uranium deposits (ERA 2011b: 16; Murdoch 2011). Located 22 km north of Ranger, construction work commenced in 1996 but was stopped in 1999 due to the resistance of the Mirarr traditional owners. ERA assures on its website that Jabiluka, in accordance with the Jabiluka Long Term Care and Maintenance Agreement passed in 2005, will not be developed without the approval of the Mirarr (ERA 2011b: 16). Yet, ERA is keen to mine the high-grade deposit, worth AUD 18.5 billion at spot prices in 2011 (Murdoch 2011).

Ranger and Jabiluka have relatively complicated ownership histories at the early stages, involving several medium sized Australian and Canadian mining companies, until the rights to both deposits were sold to Energy Resources of Australia Ltd. (ERA) in 1979 and 1991 respectively (Boukhari 1999; Deloitte Touche Tohmatsu 2008: 10, 23f.). ERA, a publicly listed company on the Australian Stock Exchange, does not own any other mines besides the two. First under majority control of North Limited, it was acquired by Rio Tinto in 2000, which since owns 68.4% of the shares (ERA 2012). ERA's revenue for 2011 was AUD 649 million, with a net loss of AUD 154 million. In 2012, it went down to AUD 395 million with a net loss of AUD 219 million, which was predominately because of depreciation of assets (Atkinson/Thibeault 2013: 9-10). Rio Tinto has operations in over 40 countries, making it the second largest mining corporation in the world and one of the most important uranium producers. Besides ERA, it owns the major Rössing uranium mine in Namibia and several projects in Canada.

Fig. 3: A meeting of the GAC. Photo: Dominic O'Brien.

2.1 Environmental and Social Impact

Since the discovery of the Ranger and Jabiluka deposits in the early 1970s, they have been central to the public debate about Australia's energy future and particularly its participation in the nuclear fuel cycle (Falk et al. 2006: 845). Indigenous peoples such as the Mirarr and environmental groups have pointed to the heavy impacts of uranium mining (Sweeney/O'Brien 2002: 2).

2.1.1 Environmental Impact

The main difference of uranium from other minerals lies in its radioactivity. In nature, it is kept relatively stable by the geological structure and hence has no radiological risk. Mining and milling uranium ore disturbs this natural balance and the efficiency of the rehabilitation of uranium mines is doubtful (Mudd 2006a: 2-5). In order to protect the Kakadu region from the effects of mining, extensive programs of chemical, biophysical, biological and social monitoring have been implemented, based on which proposals for or against the development of Jabiluka and Ranger were issued. The Ranger Uranium Environmental Inquiry (RUEI), also known as "Fox Report", claimed that if controlled and regulated properly, the development of uranium mines would not endanger the environment and suggested a system of environmental monitoring and research (Ling 2012).

The Supervising Scientist Division's (SSD) "Assessment of the Jabiluka project" (Johnston/Prendergast 1999) stated that the natural World Heritage values were not jeopardized by the proposed development of the lease. It concluded that the Jabiluka project initiated in 1996 could be pursued (Johnston 2000: 105) given that a high level of environmental protection as well as mining guidelines would be applied (Collins 2000). Likewise, the Kakadu Region Social Impact Study (KRSIS) (Collins 2000) was carried out by the SSD (Johnston 2000: 107). According to

GAC, the publications and advice issued by this organization need to be taken with caution, since the SSD is controlled and directed by the Minister for Environment and pursues an agenda based on the government's pro-uranium mining policy (GAC 1999a: 7). This indicates a collaborative relationship between the operating corporation (ERA), the NT government and the Commonwealth Government (Crawley/Sinclair 2003: 367). In contrast to assessments conducted by the SSD, the World Heritage Mission Report in 1998 found that Jabiluka poses serious threats to the cultural and natural values of the KNP and that development should not proceed (GAC 1999a: 8; UNESCO WHC 1998). A detailed Environmental Impact Statement issued by the Minister for Environment and Heritage and the Minister for Resources lists conditions of approval for the Jabiluka mine proposal. However, such investigations often do not take into account issues such as erosion, acid mine drainage threatening aquatic ecosystems or the long-term issues of radioactive waste (Mudd 2006a). They consequently cannot provide a holistic picture of the situation.

Key issues concerning environmental impacts of uranium mines in general are the management of the radioactive solid waste as well as water management and the controlling of chemicals and emissions generated from milling. Since the late 1970s, more stringent requirements have been placed on solid waste and water management at uranium mines in Australia, including reintegration of waste rock into the landform and operation with closed water circuits. Furthermore, strict regulations for rehabilitation after the closure of mine sites have been implemented (Mudd 2006a). The outflow of acidic water from metal or coal mines (Acid Mine Drainage, AMD) can contaminate aquatic systems and cause long-term environmental impacts when leaching out into the surrounding environment (Hamilton 1996: 3). The environmental impacts of uranium mining are intensified in the Kakadu region through seasonal monsoon and flooding. These increase the risks of AMD leaking from Ranger and Jabiluka into surrounding waterways and the wetlands of KNP, since rivers, waterways and airflows do not respect administrative boundaries.

ERA stated in 2010 that the Ranger mine has been operating since 1979 without harm to the surrounding environment (Janney 2010). However, during 2010, ERA was confronted with "challenging weather conditions and record rainfall" in the Kakadu region (ERA 2011a). Heavy rain caused water management problems and environmental risks. Various newspapers such as *The Age* or *ABC News* covered the issue of floods resulting in water contamination at Ranger mine (Murdoch 2010b; Masters 2010). In 2011, production at Ranger was suspended due to environmental risks posed by heavy rain (Statham 2011). This shows that the pristine natural and cultural values of KNP are put at risk by contaminated water regularly leaking into waterways and wetlands (GAC 1998: 9; Vincent 2006: 3; Beaton/Tutty 2012).

Both ERA and the Commonwealth Government's SSD initially underestimated these particular climate conditions. In the annual report of 2009-10, the SSD confirmed that the environment was not exposed to any substantial risk. Moreover, in 2010 and 2011, ERA's environmental management system's independent audits were successfully re-certified to ISO14001 (ERA 2011b: 74; ERA 2010a: 19, 49). ERA emphasized its priority on water management and environmental protection but still resorted to disposing off the uranium-contaminated water through an irrigation system at the Jabiluka site. The water management system of the Jabiluka project had already failed at an early stage and confirmed the concerns of traditional owners (GAC 2001a). However, ERA has conducted an environmental monitoring program through the entire operational period of Ranger, specified by the NT government and in consultation with the SSD. The report of that program claims to have provided an assessment of the extent to which the environment of the region, particularly KNP, has been protected during the operation of the Ranger mine (Johnston/Needham 1999: iii).

Despite the allegedly scientifically proven harmlessness of the uranium mines, the Mirarr and Australian environmental groups are convinced that mining is incompatible with long-term protection of the KNP and the surrounding region. This is based on hundreds of incidents of potential or actual contamination over 30 years since the opening of the Ranger lease, including leaks and spillages and the discharging of process water into the Gulungul and Magela creeks (Sweeney/O'Brien 2002: 3; Masterson 2010: 63). Apart from this, Australia's mining industry is the third largest energy user with 18% of the country's overall energy consumption and also has an enormous overall water intake. In the NT, it is responsible for a quarter of the territory's energy and high water consumption (ABS 2012; ABS 2011). The Ranger operation itself consumes extremely large amounts of energy and produces large quantities of greenhouse gas emissions from petrol, diesel and the processing machinery (ERA 2010c: 15).

2.1.2 Social Impact

The uranium mine turned the life of the local Aboriginal people upside down. With the mine also came the mining town of Jabiru, built for the workers. The close presence of the dominant western lifestyle especially impacts the indigenous youth, trapped between two contrasting visions. The undermining of traditional values, the destruction of sacred sites and transformation of the significance of family, community, ancestors etc. make it increasingly difficult to pass on the Bininj cultural knowledge to the younger generation. Problems connected to mining include alcohol abuse and disagreements within and between the communities (Margarula 2006: 8; GAC 2001a; Bucher 2014:148ff.). In a public statement addressing national and international audiences, Yvonne Margarula declared the Ranger uranium mine as well as the associated mining town Jabiru as the reasons of social problems in the Mirarr community. She considers her people, their culture and communal unity to be in a state of crisis and social breakdown and is convinced that another mining project in the area would jeopardize the cohesion of her people even more (GAC 1999a: 3). The Mirarr believe that their fundamental rights, such as the capacity to practice and enjoy cultural life, to access significant sites and to maintain the sanctity of ancient places are now denied (Sweeney/O'Brien 2002: 3). Sacred places such as "Malakanja II", an archaeological site that dates back by at least 50,000 years as well as ancient remains such as Aboriginal rock art are spread all over Mirarr country and are proof of one of the world's longest living traditions. These residues of the local Aboriginal culture are directly and severely impacted by the Jabiluka mining proposal. Proceeding with the latter would mean a continuing loss of the living cultural tradition of the Mirarr, which forms a major part of the unique cultural value of the KNP (GAC 1998: 4; GAC 1999a: 9).

Increased European immigration for mining jobs intensified the marginalization, the already low living conditions and cultural erosion of the indigenous people. In Jabiru, Aboriginal people are outnumbered approximately 7:1, by predominately white migrants. Aboriginal cultural practices in the town are disappearing and bad health and alcohol addiction are significant social problems (GAC 1999a: 7). Instead of acknowledging the impacts of mining on the local population in the Kakadu region, the Australian Government tends to portray them as unrelated to social problems. For instance, the Senate Uranium Mining and Milling Committee (1997) is assigning the failure, in part, to the adoption of a European model of living: "[...] many of the matters do not derive fundamentally from uranium mining but more generally from the impact of European society, economy and industry. This embraces pattern of life and schooling, relationships between parents and children, schooling and friendships" (ibid.). Furthermore, the committee states that many of the challenges and struggles witnessed in Jabiru and its surroundings are apparent elsewhere in the NT regardless of whether there is a mine in the vicinity or not. It admitted that the town is bringing a decline in Aboriginal culture, but Jabiru is portrayed as "a mixed blessing, providing Aboriginal people with access to services and amenities they may not otherwise have had" (ibid.).

Despite ERA's commitment to engage at least 25% Aboriginal workers, only five of its 350 employees were from that background, and none of them was ever employed in a senior position (GAC 1999a: 11f.). These figures slightly improved since 1998, with ERA recognizing the importance of the matter and taking action (ERA 2011b: 40). With 10% indigenous employment, the

NT shows a high percentage in comparison to other states. According to the census from 2006, NT mining companies employed 176 indigenous people in a total of 1,708 people in the mining workforce (Brereton/Parmenter 2008: 71).

A major impact of mining is the availability of significant compensation payments in the region. In 2011, ERA paid 4.25% of the net sales revenue in royalties to the Commonwealth Government. In 2009, ERA therefore contributed AUD 10.2 million, in 2010 AUD 26 million and in 2011 AUD 16 million in royalty money. These payments are ultimately distributed amongst NT-based Aboriginal communities, including the local Mirarr traditional owners (ERA 2011b: 46; ERA 2012). In total, the Ranger operation generated $220.9 million in royalties since the beginning of the project in 1980 (Rio Tinto 2005: 10). In the first four years (1994 to 1998) of its existence, the GAC received AUD 2 million in Ranger royalties out of a total of more than AUD 6.2 million paid over this period. In the same period, the Gagudju Association received AUD 4.3 million. In 1982, the Djabulukgu Association was founded to receive the future royalties from the Jabiluka mine (GAC 1999a: 10). As indicated by O'Brien, the Mirarr aim to terminate their membership with Djabulukgu (Interview Justin O'Brien 20/12/2012).

The royalty money has not made much difference to the poor living conditions of the local indigenous people (Katona 1999: 3; Sweeney/O'Brien 2002: 3; Margarula 2006: 8; GAC 2011a). Royalties are intended to be an exceptional reimbursement for destroying the traditional home country of the local indigenous people and are not supposed to act as a substitute for basic services and community infrastructure. Formally, the Australian Government is obliged to provide for this (GAC 1998: 11; Sweeney/O'Brien 2002: 4). As part of their opposition to mining development, the Mirarr demand recognition of citizen rights in order to receive basic government services like other Australians. In the current situation, their welfare is dependent on the mining company in return for their traditional country and sacred sites being excavated for a uranium mine (GAC 1998: 12). Following the approval of the Jabiluka project, the Mirarr would be dependent on the royalties for getting the most basic services for the next 30 years and the government would subsequently save money which otherwise would have to be spent on infrastructure and the well-being of the Mirarr people (GAC 1999a: 6).

The Mirarr are excluded from regional governance; their language, culture and traditions have been ignored and they had to endure extensive industrial development on their country against their will. In addition, their opposition to mining has made them subject to criticism, ill treatment and extensive pressure by government and mining interests (Sweeney/O'Brien 2002: 3). Seeing no future and having no hope, people turned to alcohol abuse, violence against their own community members and a constant feeling of disempowerment (GAC 1998: 12).

Field Study Results

During field research, it has become evident that the separation between environmental and social impacts is not very relevant from an emic perspective, as these dimensions are seen as interconnected on a deeper level. A predominant impact was identified in the colonization of the indigenous lifeworld through the presence of Western values, as exemplified in the mining town of Jabiru (Bucher 2014: 69ff.). This was however not seen as a problem separate from the mining impacts in a narrow sense. The forced presence of the mine and the associated town led to a deep cultural crisis and a feeling of depletion. The impossibility to compensate for the loss of cultural and spiritual values embedded in the landscape was confirmed (ibid.: 154; 174). An expression of this may also be the observed disregard towards community infrastructure financed with these compensations. An impressive example of the deep concern about the impacts of uranium mining could be seen in the reaction to the Fukushima disaster: the inability to prevent the extraction, and thus the failure to comply with the obligation to "care for country", has caused Mirarr elders to feel a shared responsibility for the disaster (Scherler 2013: 43; Bucher 2014: 119).

Fig. 4: Protest march against the Jabiluka mine. Photo: Dominic O'Brien.

2.2 Local People's Strategies

The Mirarr people have opposed uranium mining on their territory since the beginning of the development at the Ranger mine in 1971. Their disapproval of the Jabiluka project was very strong since the first proposal and even more so after the Fukushima nuclear disaster in Japan (GAC 2011a; GAC 2011c). That the small number of Mirarr people succeeded in stopping the Jabiluka mine already under construction is seen as an exceptionally unique triumph over powerful interests of multinational corporations, the state and as well as the internal divisions which often weaken the position of indigenous peoples (Altman 2012: 60). The Mirarr form a united front against the development of Jabiluka and appear to be respected by other Aboriginal people in the region as the sole decision-makers in this regard. Despite some individuals hoping to benefit financially from Jabiluka, the Mirarr urge their people to put country, community and traditions above royalty-money and jobs without a future (GAC 2011c; GAC 1998: 13f.). Their strategy includes legal measures, campaigns, education, peaceful protests and mobilization of national and international support through alliances with NGOs and civil society organizations. The resistance is based on prior negative experiences with close-by mining projects such Koongarra and Coronation Hill (Katona 1999; Murdoch 2010a), but most of all with the Ranger mine itself.

Both the Ranger and Jabiluka leases are situated within ancestral land, legally recognized as "Aboriginal Territory" (GAC 1998: 14). This gives the Aboriginal people the status of "traditional owners" (in this case, the Mirarr) and the power to negotiate agreements and measures to protect their cultural heritage before the government grants exploration or mining leases. They can also withhold their consent completely if no agreement is reached (O'Faircheallaigh 2008: 33f.). The Mirarr make frequent use of this veto right and also challenge earlier encroachments on their territory though native title claims, e.g. over Jabiru town (Mudd 2006b: 4) However, the legal protection proofed insufficient, as the government can make exemptions to its own legislation, therefore the Mirarr had to find alternative strategies.

The Ranger mine was approved by exempting the lease from the Aboriginal Land Rights Act and the following negotiation of an agreement between the companies and the Northern Land Council (NLC) against the will of the Mirarr. As the official representative body, the NLC has the task of communicating Mirarr interests and make their messages and positions public (GAC 2011a). In 1982, the NLC had signed an agreement to start production at Jabiluka with Pancontinental, the previous owner and original proponent of the mining lease. The Mirarr refused to accept this agreement, which was signed in a climate of extreme duress and pressure, without their official consent (GAC 1998: 5, 15). As the NLC had lost its legitimacy in view of the Mirarr, they founded the Gundjeihmi Aboriginal Corporation (GAC) as their self-constituted representative body, which became central both to the "Stop Jabiluka" campaign and to the negotiations with the mining company. Yvonne Margarula, the senior traditional owner, is the chairperson of the GAC (Margarula 2006: 8; GAC 2011a). The GAC assists the Mirarr to participate in decision-making processes and to communicate their perception on mining (GAC 1999a; Masterson 2010: 8).

With regard to the struggle for the protection of indigenous territories in Australia, Altman (2012: 71) considers three options: Firstly, indigenous representative organizations can act as political supporters. However, this can be problematic if they have been created by the state (such as the NLC) and therefore not only act on their own behalf but also depend on the government. A second possibility is to mobilize civil society and public opinion, which can be problematic if their interests do not align with the priorities of the indigenous peoples. Thirdly, he suggests the use of international instruments such as United Nations Declaration on the Rights of Indigenous People (UNDRIP) or the UNESCO Heritage Convention. While these are not legally enforceable, they have ideological force and can cause international embarrassment to the Australian Government (2012: 71f.). Justin O'Brien, executive officer of the GAC, points out that these options should not be considered exclusive. The GAC has used all three approaches in parallel, while emphasizing that the GAC is relying on independent means and therefore is not a part of the governmental control paradigm (J. O'Brien, personal communication, 18.12.2012).

As reasons for their strong resistance against uranium mining, the Mirarr cite their concerns about the environmental impact and the destruction of their cultural identity. Their territory contains many traditional places of significance and sacred sites, the most famous being Djidbidjidbi (Mount Brockman), adjacent to the Ranger uranium mine, and rock art findings (Masterson 2010: 31). Mirarr and other Aboriginal people put great efforts to achieve the registration of the Boyweg-Almudj Sacred Site Complex, which is partly located in the area of the Jabiluka ore body. The experience with the Ranger mine over the past 30 years proves their fears to be legitimate (GAC 2011a). The Mirarr have already suffered substantial negative impacts from the Ranger mine and are concerned that the long-term impacts of mining with all its aspects will destroy their society, natural environment and ultimately cause a cultural genocide. The fight against Jabiluka is at the same time about self-determination and to improve the general living conditions of the indigenous people (GAC 1999a: 5, 10). Being aware that without support it would be impossible to stop the Jabiluka project, the Mirarr launched a public campaign in 1996. In alliance with environmental groups and activists, this became "Australia's most high profile land rights and environment case" (GAC 2011a). Public forums in the NT organized by the GAC and a national speaking tour in 1996 received extraordinary response from the national and global media and the general public. Submissions sent to international bodies, alliances with other Aboriginal communities facing similar problems and a TV feature-documentary called "Jabiluka", broadcasted nationally in 1998 brought additional attention (ibid.). The campaign escalated in July 1998, when a nonviolent blockade of the Jabiluka mine construction site, organized together with environmental activists, was dissolved by the police. The mass arrest received extensive media coverage and definitely put the campaign on the national agenda.

By globalizing the "Stop Jabiluka" campaign, the Mirarr also underscored that they are not concerned solely with the impacts of uranium mining on their own territory. In 1997, they hosted a visit by survivors of the Chernobyl Nuclear Disaster and organized a forum called "Alliance Against Uranium" in Alice Springs, where they met with other Aboriginal people and a variety of support organizations (GAC 2011a). There are numerous NGOs, grassroots movements and environmental groups from all over Australia and the world which supported the campaign. In Australia, this includes for instance the Australian Conservation Foundation, Friends of the Earth, and a nation-wide "Students of Sustainability Conference". In Germany, "Stoppt Jabiluka" was initiated in 1998 by *infoe* (Institute for Ecology and Action Anthropology) after a visit by Mirarr representatives (INFOE 2004). Thanks to donations from supporters, the campaign was self-financing for some time (GAC 1998: 11).

The Mirarr also made use of the World Heritage Convention and took their struggle to the UNESCO World Heritage Committee in 1997, trying to get the KNP listed as "World Heritage in Danger". This embarrassment would have "required the national government to take action in order to maintain its international integrity" (GAC 1999a: 3, 8). UNESCO reacted with a risk assessment, which stated serious threats to the cultural and ecological values of the Kakadu World Heritage Area (UNESCO WHC1998). At the World Heritage Bureau meetings, all parties attended, including various Australian Government officials, lawyers and scholars, the mining company ERA, and many NGOs and other interest groups (Aplin 2004: 164). This forum raised awareness in the national as well as international media. The forum provided a platform for the demands of different actors to put pressure on the Australian Government to reconsider its policy (Aplin 2004: 171). Around that time, the Australian Government also received thousands of submissions urging that the project should not proceed. In response to the official approval of Jabiluka, in 1998, even the European Parliament passed a resolution calling for respect towards Aboriginal land rights and the status of the KNP as a World Heritage site (GAC 1999b: 3). International recognition of the Mirarr was also expressed when Yvonne Margarula and Jacqui Katona won the "Goldman Environmental Prize" (Goldman Environmental Foundation 1999).

Parallel to mounting public pressure, the Mirarr launched several court cases challenging different aspects of the approval of the Jabiluka mine, which were ultimately turned down (GAC 1998: 18). ERA still had to stop work at Jabiluka, since the Mirarr refused to give their consent for ore to be transported to the Ranger mine milling facilities. Building similar costly facilities at Jabiluka was not a profitable alternative for ERA (GAC 1998: 16f.)

In 1999, the combined pressures thus succeeded in stopping the construction at the Jabiluka site. In August 2003, rehabilitation work began on the site and the already extracted but not yet processed uranium ore was moved back into the mine shaft. But although ERA had confirmed not to develop Jabiluka without the agreement of the traditional owners, only the incorporation of the Jabiluka area into the KNP would definitely block future mining projects (Fitzgerald 2011). This same arrangement has been made with the Koongarra deposit south of Jabiru, which has been successfully protected by the sole traditional owner, Jeffrey Lee who requested it to be included in the national park through the Federal Government (Statham 2011).

In 2005, Rio Tinto/ERA signed an agreement allowing no further development of the Jabiluka lease without the consent of the Mirarr as the traditional owners. Since then, the relation between the two has become more cooperative. ERA assures to be in continuous and transparent communication with the GAC (Murdoch 2010b; ERA 2011b: 44). In 2011, ERA and GAC jointly developed a Cultural Heritage Policy. The latest achievement is the renegotiation of the Ranger mine agreement, which came to a conclusion in January 2013 after 14 years of negotiations (GAC 2013). According to the GAC, this agreement will provide the Mirarr with increased opportunities to participate in operations at Ranger, as well as create a new trust to fund socioeconomic initiatives within the Kakadu region. It should be noted that the agreement does not cover future mining beyond the existing Ranger Authority or underground mining (GAC 2013).

Even though the campaign against Jabiluka is thus on hold since several years, the Mirarr maintain a public presence, using traditional media as well as social mainstream media such as Facebook, Youtube and Twitter as well as their own website www.mirarr.net.

Field Study Results

The resistance strategy of the Mirarr can be seen as trapped between the two cultural responsibilities of caring for people and for country. In order to do the first, they need the compensations received for the destruction of the second. For instance, the provision of culturally appropriate education is a key element to empower the Mirarr youth, necessary to maintain the cultural base of the resistance. Even more, the GAC used royalty money from Ranger to fight against the Jabiluka mine. The adoption of Western tools for resistance was a strategic choice, apparently quite conscious about its risks, such as the fact that the Mirarr have to spend a lot of time figuring out legal strategies instead of caring for their people and country according to their own terms (Bucher 2014: 167ff.). The long years of struggle against the overwhelming power of state and corporation caused fatigue, which caused the GAC to change to more pragmatic tactics. It now has its focus on direct negotiations with the company, leaving the state on the side. The company has thus effectively replaced the state in many ways. The recent granting of Aboriginal title over Jabiru town and the renegotiation of the Ranger mining agreement finalised in 2013 can count as successful outcomes of this new approach. (Scherler 2013).

2.3 Corporate Strategies and the State

The operating company of the Ranger mine ERA, as well as its majority owner Rio Tinto, are good examples of the growing emphasis on Corporate Social Responsibility (CSR). Altman (2012: 54) highlights that "[i]mproved global communication and operation have made the activities of multinational mining companies subject to more critical gaze and transnational accountability." Thus, the development of formal and informal relationships between mining companies and some indigenous groups emerged due to public pressure (Crawley/Sinclair 2003: 361ff.). Mining corporations have become aware that they need to make efforts to meet the needs of Aboriginal people in order to continue operations profitably (O'Faircheallaigh 2009: 97). The stopped construction of the Jabiluka mine due to the pressure of a global campaign, centering on the demands of the indigenous Mirarr, illustrate this learning process.

According to ERA, the conditions for mining at Ranger and Jabiluka were framed by agreements made with the Northern Land Council on behalf of the Mirarr traditional owners under the federal ALRA (NT) 1976 (ERA 2011b: 80). However, according to the traditional owners, Ranger started to operate without their consent (Boukhari 1999). Since the ALRA (NT) was introduced, the mining companies are required to negotiate access to land with indigenous groups (Crawley/Sinclair 2003: 361ff.), but they only learned what this means in reaction to pressure.

Today, ERA portrays itself as contributing to the social welfare and economic prosperity of the indigenous communities and aspiring for a trustful relationship with them. The company acknowledges the Mirarr as the traditional owners of the land on which Ranger and Jabiluka are located (ERA 2012). ERA, the mining industry and both Federal and State Governments parade Ranger and Jabiluka as best practice examples (Friends of Earth 1999: 18). ERA provides health and safety programs for its employees, monitoring and rehabilitation programs as well as high operating standards in environmental protection. The company considers itself a fellow citizen which aims to engage with the "Mirarr Traditional Owners, local communities and governments [...], and to create cultural, social and economic development opportunities for our neighbors and future generations" (ERA 2011a). Being fully aware that it is operating in a "highly sensitive location", ERA emphasizes that Ranger and the Jabiluka lease are surrounded by, but separate from, the KNP national park (ERA 2010b; 2011b: 3). Consistent with Rio Tinto's trade union policy, ERA respects the rights of employees throughout its own operations and those of its business partners. Thereby, all employees have the right to choose to belong or not belong to a union and to seek to bargain collectively (Rio Tinto 2012).

As ERA's major shareholder, Rio Tinto is shaping the strategy, vision and practices of its subsidiary company. With annual net earnings of USD 14 billion in 2010, Rio Tinto is a "world leader in finding, mining and processing the Earth's mineral resources" (Rio Tinto 2012). Rio Tinto claims to adhere to the latest operations safety standards and to mine responsibly, contributing to the on-going global transition towards more sustainable development. Even more, Rio Tinto engages in developing new approaches to environmental and human development challenges such as climate change and poverty, according to the UN Millennium Development Goals (Rio Tinto 2012). The urge to be perceived as a "good corporate citizen" (Altman 2012: 55) further finds expression in codes of conduct such as the "Rio Tinto Cultural Heritage Guide", which aims to integrate appropriate cultural heritage management in the everyday work of Rio Tinto operations (Rio Tinto 2011). Overall, both Rio Tinto and ERA certainly have pro-active PR-strategies, including the publication of annual reports, media releases, information sheets etc. However, it would be too easy to dismiss this simply as "greenwashing", as there have also been remarkable changes on the ground in recent years.

ERA claims to be living up to its responsibility to return the land in good condition after temporary use (ERA 2011b: 79). It has designed elaborate rehabilitation programs for disturbed areas in order to restore the environment so that they can be incorporated back into the KNP after mining ends (ERA 2010b). Rehabilitation at the Ranger mine is continuously improved by including new real-time water monitoring points in local waterways and projects to control and divert surface and sub-surface water flows. Further programs include waste, greenhouse gas and weed management (ERA 2011b: 32-39). ERA understands itself as a well experienced mine operator (ERA 2012) with a strong focus on safety and high operating standards in environmental protection (Janney 2010). ERA claims that their norms and standards are consistent with those prescribed by the International Atomic Energy Agency, Australian Uranium Association, Australian Nuclear Science and Technology Organization, Australian Radiation Protection and Nuclear Safety Agency (ERA 2009b: 8). Further, it ensures that the uranium is only used for peaceful purposes and exports are subject to strict safeguards and non-proliferation conditions (ERA 2011b: 30, 47).

ERA holds continuous consultations with key stakeholders, particularly regarding environmental management. Any incident divergent from compliance with statutory requirements needs to be reported in a transparent manner, including incidents without any detrimental environmental impact (ERA 2011b: 47). The reports are also submitted to the GAC (ERA 2011b: 28, 47; 2010a: 28, 50). In addition, ERA appoints external independent auditors, rotating every five years, to assist the board of ERA in overseeing its financial and compliance obligations (ERA 2012; 2011b: 75, 121f.; 2010a: 50).

The standards of corporate governance, accountability and responsibility are specified in the ERA Code of Business Conduct (ERA 2012; 2011b: 75, 79f.). The company has developed an emphasis on indigenous involvement in mining operations, such as the enhancement of indigenous employment. With 17% Aboriginal employment rate at Ranger in 2011, ERA considers itself a leader in this field (ERA 2011b; 2010b). ERA's collaboration with the Mirarr is expressed in cross-cultural awareness training and a workplace diversity policy with regard to age, experience, gender, and background (ERA 2010a: 22ff.). Besides royalty payments, ERA makes a significant contribution to the local communities, to the wider territory and the national economy through development projects in collaboration with Rio Tinto (ERA 2012).

The Jabiluka Long-Term Care and Maintenance Agreement is an exemplary milestone of cooperation between the Mirarr, ERA and the NLC (O'Faircheallaigh 2008: 25f.). According to ERA's former chief executive Robert Atkinson, the relationship has made dramatic progress since the signing of the agreement (Wallace/Smiles 2007: 20ff.). The settlement of "Jabiru town native title case" in 2009 can be taken as a further step towards a better relationship between ERA and the Mirarr. Under the settlement, Jabiru is formally recognized as Aboriginal land and Mirarr traditional ownership, which the company sees as proof of its goodwill and support of the Mirarr's ambition to have their relationship with Jabiru formally acknowledged (ERA 2009a; 2009b: 9).

In the field study, the omnipresent role of ERA in Jabiru became apparent. The company sponsors the local collage, community events, and is generally perceived as replacing many government responsibilities due to its economic dominance in the region (Bucher 2014: 137ff.).

Role of the State

Despite the increasing Aboriginal resistance in the 1960s and 1970s, the Australian Government approved uranium mining in the Kakadu region (Wellings 2000: 120). It however set up a series of safeguards and balancing measures intended to minimize the risk to the Aboriginal people, as well as to protect the environment (Johnston 2000: 108).

Altman (2012: 71) considers the relations between the state, mining corporations, and indigenous people as binary, with the state and corporations allied in opposition to indigenous interests. Despite a recent controversy over increased taxation, the relationship between the state and mining companies has remained strong (Altman 2012: 70f.). In the last decade, the power inequality between mining and indigenous interests has further grown in favour of mining, mostly as a result of a move toward economic neoliberalism in government policy. This "hegemonic and narrow definition of development" leaves little room for the interculturality and hybridity of today's indigenous peoples (Altman 2012: 67).

The strong interests of the Australian Government in economic benefits from mining were demonstrated when it supported the mining industry at the UNESCO World Heritage meetings with its negative reaction to the definite "no-mining" recommendations of the World Heritage Committee (WHC), stating that the report "contains errors of law, fact and analysis, and recommendations that are flawed and unacceptable to the Australian Government" (WHC 1998, III, Mixed Sites, Item b.). The Australian Government's attitude toward the WHC was interpreted as "delaying tactics to defuse the issue" (Aplin 2004: 163). At the same time, the government launched a cross-departmental working group, together with ERA, in support of continuing work at Jabiluka. The group included representatives of the Department of Foreign Affairs and Trade, the Department of Science, Industry and Resources, Environment Australia, the Office of the Supervising Scientist, and the Department of Prime Minister and Cabinet, to co-ordinate an intensive worldwide lobbying effort and information campaign (Fagan 1999).

An important field of disagreement concerns the distribution of responsibilities and costs for the provision of services in mining areas. This has created friction between Federal and State Governments and the mining companies. While the latter do assume a certain role in "state building" as part of their CSR policies, they are generally reluctant to take over responsibilities from the state, "except when such action enhances prospects for access to Aboriginal lands and associated profits" (Altman 2012: 68). The underinvestment in physical and social infrastructure has resulted in poor health, low education and high rates of crime within many Aboriginal communities. Also, it has brought the mining industry to criticise the neoliberal state's shortcomings in delivering basic services such as education, health clinics, housing and water to rural areas. If all of these fall on the company instead, mining easily becomes unprofitable (Altman 2012: 70).

Indeed, the Ranger case shows that the payment of royalties by ERA resulted in the non-provision of government services on a fair-needs basis to indigenous people in the Kakadu region (Scambary 2009: 174). Jacqui Katona, the public face of the "Stop Jabiluka" campaign, states that "[t]he government didn't provide water, power, sewerage, and housing to Aboriginal people. The only way Aboriginal people could access those things was spending royalty money, and we felt, and continued to feel, that that's an exclusion to citizenship rights" (Wallace/Smiles 2007: 22).

Today's mining agreements are negotiated directly between the company and indigenous representative organizations, and usually include so-called "community benefit" packages with employment, training and business development programs, payment of cash compensation and heritage protection. However, these agreements mostly fail to overcome indigenous disadvantage and to deal with the diverse livelihoods of indigenous people, aggravated by a policy emphasis on mainstream economic development (Scambary 2009: 171f.). For the indigenous peoples, such agreements come with the risk that they "replace state dependence with mine dependence" (Altman 2012: 70). For the neoliberal state they are not only an easy relieve from its responsibility towards the Aboriginal citizens, but also play well into its interest to foster the mining industry: indigenous landowners who are dependent on mining for their basic welfare are likely to put less resistance against future mine developments.

3. Conclusion

The case study of Australia is particularly important to the subject of mineral resources, transnational corporations and indigenous populations. Australia is one of the few countries, where FPIC (Free, Prior and Informed Consent) is legally recognized and to a certain degree implemented. The ALRA (NT) is, even within Australia, a unique legislation, which grants the local indigenous people a veto in mining agreements and enables them to lodge land claims over their traditional homeland. Australia can therefore serve as an example for other countries and indigenous peoples, demonstrating the difficulties and opportunities such recognition can bring. However, the economic interests of the state can however lead to the disregard of these legislations, e.g., the exemption of the Ranger lease. In many respects the Mirarr are therefore in opposition to a united force of the state and mining companies.

This case study also illustrates exceptionally well the problems of representation and recognition within heterogeneous indigenous communities and interests. Since the start of mining development on their land, the Mirarr have been consistent and steadfast in their resistance with little internal division within their community. However, within the wider Aboriginal groups in Kakadu, ambivalence is prevailing and money and representative power foster divisions and conflicts.

With their effective and far-reaching campaign to stop the development of Jabiluka, the Mirarr attained international attention for their case and achieved the stoppage of construction works at the mine lease in 1999. This successful campaigning with broad alliance building leading to support of several national and international NGOs is another key factor that makes this case study unique and exemplary. Their active and progressive strategy was largely made possible through specific factors such as the UNESCO status but mainly through Australia's modern and democratic political system with the above-mentioned political and judicial settings, which regulate mining activities not only economically but also with certain consideration of environmental and social impacts.

The concept of CSR has emerged in Australia as a result of growing national and international public interest in the relationship between mining companies and indigenous people. Again, this case is exemplary due to the extremely sophisticated discourse of CSR and "sustainable mining" both ERA and Rio Tinto developed in response. Arguably, it is not only a discourse, but formal and informal relationships between ERA and the Mirarr have been established over the years. Australia, and in particular the Ranger/Jabiluka case, will possibly be taken as a model by others, having the ability to point out problems and contradictions within CSR.

Since many stakeholders with different intentions have a strong interest in the region, and cultural heritage cannot be sacrificed for economic-profit "[...] only initiatives that strive towards power sharing with Indigenous groups and strategies for broadening the organizational interface with Indigenous groups [...]" are likely to create a satisfying solution (Crawley/Sinclair 2003: 361). In a country with a violent history of culture clashes and colonization, indigenous communities with a strong connection to traditional customs have shown a strong opposition against unwanted developments. The struggle of the Mirarr against the Jabiluka mine can be an example for other regions where mining interferes with indigenous livelihoods.

References

Altman, Jon 2012: Indigenous Rights, Mining Corporations and the Australian State. In: Sawer, Suzana and Terence Gomez (eds.): *The Politics of Resource Extraction. Indigenous Peoples, Corporations and the State*. London: Palgraves. 46–74.

Anderson, Christopher 1988: Anthropology and Australian Aboriginal economy 1961–1986. In: Berndt, Ronald Murray and Robert Tonkinson (eds.): *Social Anthropology and Australian Aboriginal Studies*. Canberra: Aboriginal Studies Press. 125–188.

Aplin, Graeme 2004: Kakadu National Park World Heritage Site. Deconstructing the Debate, 1997–2003. *Australian Geographical Studies* 42 (2): 152–174.

Australian Institute of Aboriginal Studies (AIAS) 1984: Aborigines and Uranium. Consolidated Report to the Minister for Aboriginal Affairs on the Social Impact of Uranium Mining on the Aborigines of the Northern Territory. Canberra: Australian Government Publishing Service.

Boukhari, Sophie 1999: The Kakadu Affair Shakes the Heritage World. *The UNESCO Courier* 52 (10): 10–13.

Bowie, Fiona 2000: The Anthropology of Religion. An Introduction. Oxford: Blackwell Publishing.

Bucher, Daniela 2014: Impacts of Uranium Mining in the Kakadu Region. A Case Study about the Mirarr People in Jabiru and Surroundings. Unpublished Master Thesis, Institute of Geography / Social Anthropology, University of Bern.

Butlin, Noel George 1993: Economics and the Dreamtime. A Hypothetical History. Cambridge, New York and Melbourne: Cambridge University Press.

Collins, Bob 2000: Kakadu Region Social Impact Study Community Report. Report on Initiatives, November 1998-June 2000. Darwin: KRSIS Implementation Team.

Crawley, Amanda and Amanda Sinclair 2003: Indigenous Human Resource Practices in Australian Mining Companies. Towards an Ethical Model. *Journal of Business Ethics* 45: 361–373.

Evans, Nicholas 2004: Bininj Gun-Wok. In: Arnaud, Pierre. (ed.): *Le Nom composé. données sur seize langues*. Lyon: Presse Universitaires de Lyon. 71–99.

Falk, Jim, Jim Green and Gavin Mudd 2006: Australia, Uranium and Nuclear Power. *International Journal of Environmental Studies* 63 (6): 845–857.

Fox R. W., Kelleher G. G., Kerr C. B. 1977: Ranger Uranium Environment Inquiry Second Report. Canberra: AGPS.

Hamilton, Clive 1996: Mining in Kakadu. Lessons from Coronation Hill. A lecture to the Parliamentary Library 'Vital Issues' Seminar Series. Discussion Paper Number 9. Parliament House, Canberra.

Howitt, Richard 2001: Rethinking Resource Management. Justice, Sustainability and Indigenous Peoples. London and New York: Routledge.

Ingold, Tim 2000: The Perception of the Environment. Essays in Livelihood, Dwelling and Skill. Abingdon: Routledge.

Johnston, Arthur 2000: Environmental and Social Impact of Uranium Mining in Australia. Darwin: IAEA (362).

Johnston, Arthur and Bernard Prendergast 1999: Assessment of the Jabiluka Project. Report of the Supervising Scientist to the World Heritage Committee. Supervising Scientist Report 138. Canberra: Supervising Scientist.

Katona, Jacqui 1999: Aboriginal Rights in Kakadu: Breaking the Bonds of Economic Assimilation. *ANUNARU Discussion Paper* 16. Darwin: North Australian Research Unit.

Keen, Ian (ed.) 2010: Indigenous Participation in Australian Economies. Historical and Anthropological Perspectives. Canberra: ANU E-Press.

Kolig, Erich 1989: Dreamtime Politics. Religion, World View and Utopian Thought in Australian Aboriginal Society. Berlin: Dietrich Reimer Verlag.

Kolig, Erich 1994: Rationality, Ideological Transfer, Cultural Resistance, and the Dreaming. The Development of Political Thought in Australian Aboriginal Society. *Anthropos* 89 (1): 111–124.

Kolig, Erich 2003: Legitimising Belief. Identity Politics, Utility, Strategies of Concealment, and Rationalisation in Australian Aboriginal Religion. *The Australian Journal of Anthropology* 14 (2): 209–228.

Lennon, Jane, Michael Pearson, Duncan Marshall, Sharon Sullivan, Patrick McConvell, Warren Nicholls and David Johnston 2001: Natural and Cultural Heritage, Australia State of the Environment Report 2001. Canberra: CSIRO.

Masterson Andrew, O'Brien Justin, Kyle Geoffrey, GAC 2010: The Mirarr: Yesterday, Today and Tomorrow, a Socioeconomic Update. Jabiru: Gundjeihmi Aboriginal Corporation.

Morphy, Howard 1988: The Resurrection of the Hydra. Twenty-five Years of Research on Aboriginal Religion. In: Berndt, Ronald Murray and Robert Tonkinson (eds.): *Social Anthropology and Australian Aboriginal Studies. A Contemporary Overview*. Canberra: Aboriginal Studies Press. 241–266.

Morton, John 2005: Aboriginal Religion Today. In: Charlesworth, Maxwell John, Françoise Dussart and Howard Morphy (eds.): *Aboriginal Religions in Australia. An Anthology of Recent Writings*. Aldershot, Burlington: Ashgate Publishing. 195–204.

O'Faircheallaigh, Ciaran 2008: Negotiating Cultural Heritage? Aboriginal-Mining Company Agreements in Australia. *Development and Change* 39 (1): 25–51.

O'Faircheallaigh, Ciaran 2009: Effectiveness in Social Impact Assessment. Aboriginal Peoples and Resource Development in Australia. *Impact Assessment and Project Appraisal* 27 (2): 95–110.

Press, Tony, David Lea, Ann Webb and Alstair Graham (eds.) 1995: Kakadu. Natural and Cultural Heritage and Management. Darwin: Australian Nature Conservation Agency & North Research Unit.

Richardson, Graham 1987: Senate Hansard. 10 December 1987: 2882.

Rose, Deborah Bird 1996: Nourishing Terrains. Australian Aboriginal Views of Landscape and Wilderness. Canberra: Australian Heritage Commission.

Sahlins, Marshall David 1972: Stone Age Economics. Hawthorne, New York: Transaction Publishers.

Scambary, Benedict 2009: Mining Agreements, Development Aspirations and Livelihoods. In: Altman, Jon and David Martin (eds.): *Power, Culture, Economy. Indigenous Australians and Mining*. Canberra: Australian National University. 171–201.

Scherler, Loretta 2013: Development in a Mining Context: Aboriginal Corporations, Dependence and De-politicization. Unpublished Master Thesis at the Institute of Social Anthropology, University of Bern.

Stephens, Kevin 2010: Thirty-Fourth Annual Conference Perth, Western Australia. Session 6B. Native Title Agreements—A Quest for Certainty. 20 to 23 October 2010. Northern Territory.

Strang, Veronica 1997: Uncommon Ground. Cultural Landscapes and Environmental Values. Oxford/New York: Berg.

Strang, Veronica 2004: Poisoning the Rainbow. Mining, Pollution and Indigenous Cosmology in Far North Queensland. In: Rumsey, Alan and James Weiner (eds.): *Mining and Indigenous Lifeworlds in Australia and Papua New Guinea*. Wantage: Sean Kingston Publishing. 208–225.

Sweeney, Dave and Justin O'Brien 2002: Case Study: Jabiluka Uranium Project, Kakadu Region, Australia. Bali PrepCom IV. Australian Conservation Foundation and GAC.

Wallace, Paula and Vickie Smiles 2007: Uranium Mining and the Question of Corporate Social Responsibility. *ECOS* 138.

Wellings, Peter 2000: Uranium Mining and Indigenous Social Impact Issues. *IAEA* (362): 119–126.

WHC 1998: Item 5 of the Provisional Agenda: Report of the Rapporteur on the twenty-second extraordinary session of the Bureau of the World Heritage Committee (Kyoto, 27–28 November 1998). World Heritage Committee, Twenty-second session, Kyoto, Japan, 30 November– 5 December 1998.

Online Sources

ABS 2006: 6291.0.55.003 - Labour Force, Australia. Detailed, Quarterly, February 2006. <www.abs.gov.au/AUSSTATS> accessed on 25.12.2011.

ABS 2011: 1367.0 State and Territory Statistical Indicators, 2011. <www.abs.gov.au/ausstats> accessed on 18.05.2012.

ABS 2012: 1301.0 - Year Book Australia, 2012. <www.abs.gov.au/ausstats> accessed on 17.06.2012.

Altman, Jon and Allen Linda 1992: Living off the Land in National Parks. Issues for Aboriginal Australians. Aboriginal Involvement in Parks and Protected Areas. Report Series Australian Institute of Aboriginal and Torres Straight Islander Studies. Canberra, Aboriginal Studies Press. <www.anu.edu.au> accessed on 16.04.2012.

Atkinson, Robert and Steeve Thibeault 2012: 2011 Full Year Results. Presentation to the Financial Community. <www.energyres.com.au> accessed on 17.07.2012.

Atkinson, Robert and Steeve Thibeault 2013: 2012 Full Year Results. Presentation to the Financial Community. <www.asx.com.au> accessed on 13.03.2013.

Australian Government 2011: Aboriginal Rock Art. <www.environment.gov.au> accessed on 17.11.2011.

Australian Government 2012a: Aboriginal History and Culture. <www.environment.gov.au> accessed on 07.08.2012.

Australian Government 2012b: Uluru-Kata Tjuta National Park Note — Tjukurpa. <www.environment.gov.au> accessed on 07.04.2012.

BBC News 2008: Australia Apology to Aborigines. 30 January 2008. <news.bbc.co.uk> accessed on 07.08.2012.

Beaton, Cat and Justin Tutty 2012: The Radioactive Waste Creation and Effects on Rehabilitation of Mining 34,000 Tonnes of Uranium Oxide at the Proposed Ranger 3 Deeps Underground Mine, Kakadu, Northern Territory. A Report by the Environment Centre NT. <issuu.com/ecnt/docs/www.ecnt.org> accessed on 24.05.2013.

Brereton, David and Joni Parmenter 2008: Indigenous Employment in the Australian Mining Industry. Journal of Energy and Resources Law 26 (1): 66-90. <www.atns.net.au> accessed on 14.12.2011.

Bushfire Cooperative Research Centre (Bushfire CRC) and the Australasian Fire and Emergency Service Authorities Council (AFAC) 2009: Aboriginal Wetland Burning in Kakadu. Fire Note 36. <wikis.utas.edu.au> accessed on 17.11.2011.

Central Land Council (CLC) 2011: The Aboriginal Land Rights Act. <www.clc.org.au> accessed on 14.12.2011.

Deloitte Touche Tohmatsu (Insight Economics) 2008: Outlook for the Uranium Industry. Evaluating the Economic Impact of the Australian Uranium Industry to 2030. <www.aua.org.au> accessed on 18.04.2012.

EconomyWatch (2010): Australia Economy. <www.economywatch.com> accessed on 01.06.2012.

ERA 2009a: Settlement of Jabiru Native Title Case. Media Release. 25. November 2009. <www.energyres.com.au> accessed on 17.01.2012.

ERA 2009b: 2009 Sustainable Development Report. <www.energyres.com.au/documents/2009_ERA_Sustainable_Development_Report.pdf> accessed on 07.08.2012.

ERA 2010a: Annual Report 2010. <www.energyres.com.au> accessed on 13.01.2012.

ERA 2010b: Corporate DL. <www.energyres.com.au> accessed on 15.01.2012.

ERA 2010c: Sustainable Development Report 2010. <www.energyres.com.au> accessed on 15.04.2012.

ERA 2011a: Chief Executive's Message. <www.energyres.com.au> accessed on 17.11.2011.

ERA 2011b: Annual Report 2011. <www.energyres.com.au> accessed on 22.06.2012.

ERA 2012: ERA. <www.energyres.com.au> accessed on 15.01.2012.

Fagan, Matthew 1999: Local Strategies Jabiluka and Its Implications on International Law: 1-6. Jabiru. Unpublished Article, Submitted Australian Land Reform Commission. <www.mirarr.net/jabiluka> accessed on 04.10.2011.

Fitzgerald, Barry 2011: Jabiluka Still On, Says ERA. The Age 14.4.2011. <www.theage.com.au> accessed on 19.12.2011.

Friends of the Earth (FoE) 1999: A Review of Australia's Uranium Mining and the Proposed Jabiluka Uranium Mine: A Scientific Case for Placing Kakadu as World Heritage in Danger. A Technical Submission by Friends of the Earth to the World Heritage Committee of UNESCO: 18. <users.monash.edu.au> accessed on 24.05.2013.

GAC 1998: Mirarr Fighting for Country. Information Kit Module 1. Timeline. <www.mirarr.net/docs/timeline.pdf> accessed on 11.12.2011.

GAC 1999a: Mirarr Fighting for Country. Information Kit Module 2. Fact Files. <www.mirarr.net/docs/thefacts.pdf> accessed on 11.12.2011.

GAC 1999b: Mirarr Fighting for Country. Information Kit Module 3: Statements & Map. <www.mirarr.net/docs/statement.pdf> accessed on 11.12.2011.

GAC 2001a: Storm Clouds Gather for Kakadu Uranium Miner. Aboriginal Owners Refused Jabiluka Access for Inspection. GAC Media Release, 30.10.2001. <www.mirarr.net> accessed on 22.11.2011.

GAC 2001b: The Office of the High Commissioner for Human Rights Workshop on "Indigenous Peoples, Private Sector Natural Resource, Energy and Mining Companies and Human Rights". Submission by the Mirarr Aboriginal People. Kakadu, Australia. <www.mirarr.net/docs/GenevaSub.pdf> accessed on 07.04.2012.

GAC 2011a: The History of Binninj Opposition to Uranium Mining. <www.mirarr.net/history> accessed on 28.10.2011.

GAC 2011b: The Basis for Mirarr Opposition to Uranium Mining. <www.mirarr.net/jabiluka> accessed on 28.10.2011.

GAC 2011c: A History of Duress. A GAC Research Project. <www.mirarr.net> accessed on 28.10.2011.

GAC 2011d: Presentation to the NLC-AusAid Conference. Darwin. (Power Point Presentation, Geoff Kyle, GAC).

GAC 2011e: The Mirarr Gundjeihmi Clan. <www.mirarr.net/mirarr> accessed on 09.09.2011.

GAC 2013: Ranger Uranium Mine Agreement Updated. <www.mirarr.net> accessed on 24.01.2013.

Goldman Environmental Foundation 1999: Prize Recipient: Jacqui Katona & Yvonne Margarula. <www.goldmanprize.org> accessed on 22.07.2012.

Haynes, Christopher David 2009: Defined by Contradiction: the Social Construction of Joint Management in Kakadu National Park. PhD Thesis, Charles Darwin University. <espace.cdu.edu.au> accessed on 24.05.2013.

INFOE 2004: Kampagne Stoppt Jabiluka. <www.infoe.de> accessed on 07.09.2011.

Janney, Dan 2010: Energy Resources of Australia Ltd. General Manager Operations. 22 April 2010. <www.energyres.com.au/documents> accessed on 15.01.2012.

Johnston, Arthur and Steward Needham 1999: Protection of the Environment Near the Ranger Uranium Mine. Supervising Scientist Report 139. Canberra: Supervising Scientist. <www.environment.gov.au/ssd/publications> accessed on 31.07.2012.

Ling, Ted 2012: Commonwealth Government Records about the Northern Territory. Uranium Mining — Ranger Uranium Environmental Inquiry, 1975–77. National Archives of Australia. <www.naa.gov.au> accessed on 08.01.2012.

Margarula, Yvonne 2006: Look After Country, Look After People. In: Vincent, Eve (ed.): Yellowcake Country? Australia's Uranium Industry. A Beyond Nuclear Initiative (BNI) Publication: 8. <foe.org.au> accessed on 28.04.2012.

Masters, Emma 2010: Uranium Mine Leak '5400 Times Normal Level'. ABC News. Posted on 9 February 2010. <www.abc.net.au> accessed on 15.01.2012.

McKay, Bill, Ian Lambert and Shige Miyazaki 2000: Special Article. The Australian Mining Industry. From Settlement to 2000. Australian Geological Survey Organization.<www.abs.gov.au/ausstats> accessed on 12.06.2012.

Mudd, Gavin 2005: An Environmental History of Uranium Mining in Australia : A Scientific Review. Pres. Australian Uranium Conference 2005, Fremantle, October 2005. <users.monash.edu.au> accessed on 29.05.2013.

Mudd, Gavin 2006a: Uranium Mining. Australia and Globally. Fact Sheet 6. <www.energyscience.org.au> accessed on 28.10.2011.

Mudd, Gavin 2006b: Hot Spots. In: Vincent, Eve (ed.): Yellowcake Country? Australia's Uranium Industry. Beyond Nuclear Initiative (BNI) Publication 8: 4-5. <foe.org.au> accessed on 29.05.2013.

Murdoch, Lindsay 2010a: Mine Fears Grow as Pollutants Flow to Kakadu. The Age Online. 24 May 2010. <www.theage.com.au> accessed on 16.06.2012.

Murdoch, Lindsay 2010b: Kakadu Being Poisoned by Rio Tinto Mine, Group Warns. The Sydney Morning Herald. 24 May 2010. <www.smh.com.au> accessed on 15.01.2012.

Murdoch, Lindsay 2011: Jabiluka's Sacred Power 'Must Never Be Disturbed'. The Age Online. 7 April 2011. <www.theage.com.au> accessed on 19.12.2011.

O'Connor, Nigel 2011: Mine Companies Failing Tests. National Indigenous Times: 40-41. <www.mirarr.net/media> accessed on 15.01.2012.

Queensland Studies Authority 2007: The History of Aboriginal Land Rights in Australia (1800s–1980s). <www.qsa.qld.edu.au> accessed on 13.04.2012.

Rio Tinto 2005: Social and Environmental Report. <www.riotinto.com> accessed on 08.08.2012.

Rio Tinto 2011: Why Cultural Heritage Matters. A Resource Guide for Integrating Cultural Heritage Management into Communities Work at Rio Tinto. <www.csrm.uq.edu.au> accessed on 17.02.2013.

Rio Tinto 2012: Rio Tinto. <www.riotinto.com> accessed on 15.01.2012.

Senate Uranium Mining and Milling Committee 1997: Northern Territory. Canberra: Commonwealth of Australia. <www.aph.gov.au/senate/committee/uranium_ctte> accessed on 17.11.2011.

Sosa, Irene 2011: License to Operate. Indigenous Relations and Free Prior and Informed Consent in the Mining Industry. Sustainalytics. <www.sustainalytics.com> accessed on 13.06.2012.

SSD 2011: Uranium Mining in the Alligator Rivers Region. <www.environment.gov.au> accessed on 18.06.2012.

Statham, Larine. 2011. "Kakadu Owners Want Stop to Uranium Mining", in The Age 7.11.2011. <news.theage.com.au> accessed on 19.12.2011.

UNESCO World Heritage Committee (WHC) 1998: Report on the Mission to Kakadu National Park, Australia. 26 October to 1 November 1998. <www.environment.gov.au> accessed on 12.12.2011.

UNESCO World Heritage Center 2011: World Heritage List. <whc.unesco.org/en/list> accessed on 17.11.2011.

Vincent, Eve (ed.) 2006: Yellowcake Country? Australia's Uranium Industry. *Beyond Nuclear Initiative Publication* 8. <foe.org.au> accessed on 28.04.2012.

World Bank 2012: Gross Domestic Product 2012. <databank.worldbank.org> accessed on 22.12.2013.

World Nuclear Association 2012: Australia's Uranium. <www.world-nuclear.org> accessed on 08.01.2012.

The Ambuja Rawan and ACC Jamul Cement Plants in Chhattisgarh, India

Thomas Niederberger[1]

1. Situating the Case Study

This chapter is about the impacts of two cement plants and the associated limestone mines in India's central-eastern state of Chhattisgarh—with focus on the one operated by Ambuja Cements in the village Rawan of Baloda Bazar district (formerly part of Raipur district), but also including information from the ACC plant located in the industrial town of Jamul in Durg district. Even though they have their own respective histories, are operated by different companies and separated by a distance of 120 km by road, they have been linked to each other through the acquisition of both companies by the Switzerland-based Holcim Ltd., and the activity of a local union of contract workers.

This case study is not about a large-scale mining company, but about cement production which in the strict sense is an industrial category on its own, even though it involves mining. We decided to include this chapter for the sake of a better understanding of the background of the still-ongoing campaign, which involves a local union associated with the Chhattisgarh Mukti Morcha and an alliance of International and Swiss unions and organizations (including *infoe CH*). The campaign had made use of the OECD complaint mechanism and has succeeded in transmitting the voices of some of the most marginalized labourers and peasants of India to Holcim's corporate headquarters in Switzerland.

1.1 History and Current Context

Chhattisgarh was created as a separate state of India in 2000, by splitting off 16 eastern districts from the state of Madhya Pradesh. It is rich in minerals and has therefore become of major interest for India's fast-growing mining and related industrial sectors. At the same time, it remains a predominantly rural state with high poverty rates. Together with (parts of) the neighbouring states of Jharkhand, Orissa, Andhra Pradesh and Madhya Pradesh, it is part of the so-called "tribal belt", with a high percentage of *adivasi* (indigenous) population, concentrated in the extensive hilly forests of the southern and northern districts. Our case study concerns Chhattisgarh's central plain area, known as the "rice bowl" of Central India, which has many cement plants and abundant deposits of limestone.

The medieval history of Chhattisgarh is that of one of the various local kingdoms that remained at the margins of India's imperial forces until the arrival of the Marathas in 1741. Till then, the hill areas remained relatively autonomous. Chhattisgarh fell partially under the control of the British Empire in the early 19th century, and fully in 1854. The colonial agents administered it

1 Parts of the case study are based on a research paper by Sara Ryser and Lucile Thalmann and a draft article by Asgar Qadri who also visited the area and carried out several interviews. Reviewed by Shalini Gera and Yvonne Zimmermann.

from Raipur, as part of the "Central Provinces and Berar". Central India is famous for numerous uprisings by the so far largely self-governing *adivasis* against the imposition of feudal land taxation and colonial appropriation of forests. The British reacted by creating "excluded areas", that is, "institutional and legal arrangements [that] withdrew the application of colonial laws partially or fully in some areas" (Bijoy et al. 2010: 18). Also, they registered landholdings and carried out extensive anthropological surveys[2] of castes and tribes in order to help the administrators better understand the rather complicated country which they were supposed to manage (Sundar 2007: 157ff.).

With independence in 1947, India was constituted as a democratic, secular, socialist, federal Republic, proud to be the largest democracy in the world, remarkably resistant to all attempts at installing authoritarian governments. The Central Provinces became the state of Madhya Pradesh. Chhattisgarh now formed the hinterland of this mostly rural and economically weak state. The first prime minister Nehru, according to his policy of socialist planning and establishment of large dams and industries in "backward" regions, brought one of Asia's biggest steel plants to the small town of Bhilai in Chhattisgarh's district Durg: The Bhilai Steel Plant (BSP) was known as one of Nehru's "temples of modern India" (Parry 2008: 235f.). The subsequent industrialization of the region was concentrated in the 40 kilometres between Raipur and Durg, with BSP as the beacon, and some factories and mines elsewhere in the state. The wave of displacement that went along with it remained without much notice, as a "necessary sacrifice" for the nation.[3] Many displaced people have never received any compensation. There are only vague estimations about their total number: between 1947 and 2004 over 60 million people are estimated to have been displaced across India due to mega industrial projects (DRCWC/Samata 2010: 34).

Chhattisgarh had seen autonomist movements since India's independence. In 2000, it became a separate state, along with neighbouring Jharkhand. Possibly, the move had electoral and economic interests linked to it. Since 2003, Chhattisgarh is ruled by the Hindu-nationalist *Bharatiya Janata Party* (BJP). The separation came at a time of full economic boom and successive governments in Chhattisgarh pushed neoliberal policies and development through extractive industries to the extreme. But the state's impressive GDP growth rates of over 10% did not change the fact that it has rates of child malnutrition above 50% (Murali/Afifi 2012: 21).

Large-scale acquisition of agricultural land and forests for mining and industries increasingly face opposition from the local population. This takes various forms, from non-violent grassroots movements to armed rebellion. During the past decade, the *naxalites*—Maoist guerrilla fighters—have established their bases in the hilly forest areas of Central India. The most important one lies in the forests of southern Chhattisgarh and northern Andhra Pradesh. Their strongholds are in the very same *adivasi* territories targeted for mining projects (Chakravarti 2009: 34; Sundar 2007: 289). In 2005, a government-sponsored, armed "civil defence force" called *Salwa Judum* appeared on the scene, also made up by *adivasis* from the same region. Since then, the "internal security problem" definitively turned into a civil war: Within one year, some 150,000 people were displaced, 3000 houses burnt down and 500 to 1000 people were killed.[4]

In this conflict scenario, social activists, NGOs, unions and journalists come under double pressure: On one side, the Maoists have a strategy to set up legal organizations for the "formation of various tactical united fronts", following a logic of "worker-peasant alliance" and including

2 For an example regarding Chhattisgarh, see Russell (1916).

3 For BSP alone, land from 96 villages was compulsorily acquired—some of them disappeared completely to give way for the factory and related townships (Parry 2008: 236).

4 Numbers from Sundar (2007: 287); see also ACHR (2006) and various reports from "Campaign for Peace and Justice in Chhattisgarh" <cpjc.wordpress.com/reports-by-fact-finding-teams-on-salwa-judum> (accessed on 20.6.2013).

progressive sectors of the urban bourgeois class.[5] On the other side, "[...] those battling Maoism are desperate to 'expose' civil rights activists as enemies of the state" (Chakravarti 2009: 370), by accusing them of supporting the Maoist cause and using draconian anti-terror laws.[6]

1.2 Geography and Ecosystem

Chhattisgarh covers an area of 135,000 square kilometres and is situated in the central-eastern part of the Indian subcontinent. It is the ninth largest of India's 28 states (4% of total land area), with an extension of 640 km north-to-south and 330 km east-to-west. Chhattisgarh's central plain area covers roughly a fourth of the state and is divided into the administrative districts of Bilaspur, Durg, Mahasamund, Raipur, Janjgir-Champa, Dhamtari and the newly created Baloda Bazar.

The climate of Chhattisgarh is tropical, humid and sub-humid. It can be divided into three climatic zones: the northern hills, the Chhattisgarh plains and the plateau of Bastar in the South. Mean annual temperatures vary between 20 and 32 °C (Viswanath et al. 2000). Chhattisgarh is rich in mineral resources, especially iron ore, bauxite and coal, but also limestone, gold, diamonds and others. Also, it is rich in forests, which cover 44% of the total surface. However, this is concentrated mostly in the hills, while the forest cover in the plains is minimal, consisting mostly of degraded government-owned forests. The soil of the fertile Chhattisgarh plains are highly diverse and generally well suited for rice cultivation (Bhatt/Bhargava 2005: 180). The plains are watered by many streams and canals, which are tributaries to the Mahanadi River. The average annual rainfall is around 1400 mm. Rains largely fall in the monsoon season between July and September, with less than 10 per cent of the total annual rainfall left for the other months. From November to June there is almost no rain (Murali/Afifi 2012: 49). Sastri observes that due to global climate change the quantity of rainfall has decreased in some regions by 30–35% during the 20th century (2009: 156).

1.3 Local Population

The formation of a "Chhattisgarhi" identity over the centuries is usually described as a combination of its relatively secluded geographical location and the interplay between the rice growing plains, which came under the influence of the stratified Hindu caste society from early on, and the surrounding "tribal" forest areas. The main marker is Chhattisgarhi language, a dialect of eastern Hindi. Already in British times, the relatively emancipated role of women attracted attention: "[...] women [...] were to be seen going about in an open way, such as I had not seen elsewhere during my tour. [...] they are very independent, and their lords are not their masters except when it suits them" (Ball 1880: 630). Regional identity is marked by the numerous local gods and goddesses, their temples and the bathing places which form a "sacred geography" unknown to outsiders, and the practice of pilgrimage to these sites (Babb 1975: 11).

According to the Census of India,[7] the population of Chhattisgarh in 2001 was 20 million, which is roughly 2% of India's total population. About 80% live in the 20,000 villages, the rest in 95 towns and cities. By "religion", 95% name "Hindu", close to 2% "Muslim" or "Christian"

5 See "Urban Perspective Plan 2004 of CPI (Maoist), summarized in Chakravarti (2009: 127).

6 One example from Chhattisgarh is the case of Dr. Binayak Sen, who was released after 2.5 years in prison only after an international campaign and a Supreme Court decision (see: <www.binayaksen.net>).

7 Summarized in Bhatt/Bhargava (2005: 27ff.) and GoCh (2005: 10).

respectively. Under "Mother Tongue", almost 100 languages are recorded.[8] The category of "Scheduled Castes" (SCs, see Box 1) make up 11.6% of total population, or 2.4 million, out of which "Chamar" has by far the largest number (1.66 million). "Scheduled Tribes" (STs, see Box 1) make up 32% or 6.6 million. The statistic lists more than 40 different "tribes", with "Gond" as the most important (3.7 million).[9] There are important regional variations. In what was formerly Raipur's sub-district Baloda Bazar, now carved out to form a newly created district, SCs comprise 19% and STs 12.4%.

These numbers can serve as an illustration of India's famous cultural diversity, of which Chhattisgarh is no exception. Urban and industrial areas like Raipur city and Bhilai attracted immigrants from other states, from all levels of social status, and became "melting pots" to a certain degree (Parry 1999b). Also the rural areas are very heterogeneous. In large parts of central India it is difficult to distinguish a clear-cut "indigenous" population from colonizing settler communities. More relevant for the purpose of this chapter is the existence of a general aversion of the Chhattisgarhi "sons-of-the-soil" against "outsiders", as described by Parry (1999b). In the following section, the "conventional" caste structure at the village level is introduced with some elaboration on the Gond and the Satnami, which can serve as examples for the discriminated *adivasi* and *dalit* identities.

Social Structure and the Caste Hierarchy

Lawrence Babb did field research about popular Hinduism in Raipur district in 1966–67, in a village not too far from Rawan. The village showed the typical features of the *varna-jati* ("caste") system found in most regions of Hindu India: a hierarchical social order based on concepts of purity and pollution, where the "twice-born", upper-caste *brahmin* (priests), *rajput/kshatriya* (warriors) and *bania/vaishya* (merchants) were placed at the top. Then followed a series of *shudra* middle-castes (fishermen, farmers, herders and goldsmiths), lower-middle castes (Gonds, potters, blacksmiths and barbers) and low-castes (washerman and basketmakers). At the lowest rung, the "untouchable" *dalit* caste people were the Mehar (shoemakers) and 7 Satnami households, still cursed by their forefathers' occupation as leather workers. Outside of the village, ranked even lower and excluded from the community, was a settlement of Mehetar (sweepers) and Devar (swineherds and beggars). Also, the members of the lower castes strictly maintained the barrier to the "untouchables", by not intermarrying, not sharing cooked food and water, and adhering to other social taboos (Babb 1975: 16ff.). The largest community of 20 households were the Gonds, originally an *adivasi* "tribe", who were however fully integrated into the hierarchical order as a "caste" and had lost the Gondi language, but maintained certain specific religious practices (ibid.). Joyce Burkhalter Flueckiger's research on gender and folklore in the plains of Raipur district between 1980–1992 demonstrates the strong cultural influence of the *adivasis*, particularly the Gonds, who had immigrated long time ago from the hills to the plains and became integrated into the Hindu caste system. They occupied a relatively low position, mostly as agricultural labourers, but the performance of dance and songs by the Gondi women was regarded as a characteristic cultural feature of Chhattisgarh, and members of higher castes hosted it in front of their houses on certain occasions (1996: 8; 1987: 40).

8 The most spoken languages are 10.3 million for Chhattisgarhi and 1 million for related Surgujia, 2.5 million for Hindi, 1.15 million for Oriya and related Halabi and Bhatri. The main *adivasi* languages are Gondi (744,000) and Kurukh/Oraon (392,000). Many of the listed languages are actually near extinction. Important languages of immigrant communities include Telugu (134,000) Marathi (120,000), Sindhi (77,000), Bhojpuri (58,000), Bengali (53,000), Punjabi (51,000) and others (Bhatt/Bhargava 2005: 125ff.).

9 As listed by K.S. Singh, "Gond" includes 33 different tribes, spread all over central India (1994: 293–358).

Fig. 1.: Houses of a former village turned into urban area in Bhilai. Photo: Karin Scheidegger.

The Satnamis form a community specific to Chhattisgarh. Its origin goes back to a reform movement which was initiated by the Chamar farm servant Ghasidas around 1820 in the northern Raipur district. The Chamars were leather workers, associated with filth, bad odours and dead cows, and therefore were "untouchables" at the very bottom of the hierarchical order. Ghasidas called upon the members of his caste—by far the most numerous of all the untouchable castes—to refuse this position, stop eating meat and accepting to handle carcasses of cows. He founded a cult which rejected the Hindu gods and goddesses: The Satnami only believed in Satnam ("true name"), that is, a formless superior being. The movement spread and by the beginning of the 20th century the large majority of the Chamars had joined it.[10]

Jonathan Parry (1999b) studied the position of the Satnami community in two multi-caste villages in Durg in the 1950s,[11] at the time when "modernity" arrived in the form of the Bhilai Steel Plant (BSP). According to oral history, they lived in separate hamlets, had different water sources and bathing places, and were completely excluded from the caste Hindu society—which for the rest was largely structured as Babb (1975) describes it. Parry compares this to the present time and asks how reserved jobs for "Scheduled Castes" (for which Satnamis can apply) at the public sector BSP has changed this picture. Nobody dares to call them by their former name (Chamar), but they are still not considered part of the Hindu community. Their hamlets are now a sector of an urban slum settlement. They still bathe at their "old" places and rarely get invited to a Hindu marriage.

10 For an ethnohistory (1780-1950) of the Satnamis, see Dube (1998) and Bauman (2008).

11 The Satnamis were 15-20% of the population, by far the largest among the "untouchable" castes (categorized as part of the Chamar in the Census of 1961, which made 84% of the "Scheduled Caste" members).

However, some important changes have happened and led Parry to call Bhilai, and especially the BSP plant, a "melting-pot": among the mixed-up permanent workers of the factory, caste separations have largely been replaced by a "modern", cosmopolitan ethic, which has its influence over the whole town. However, the same is not true with respect to contract workers (mostly *dalit*, low caste and *adivasi*), and less in the private sector where reservations do not apply and employment tends to reproduce "primordial ties" (1999a, b). The important point to retain for this study is that even in urban settings like today's Bhilai, caste remains a factor of exclusion, also when it comes to modern institutions and companies.

Box 1: Indigenous Peoples, Scheduled Tribes, Castes and Areas
(Contributed by Asgar Qadri)

Indigenous Peoples: "The Government of India officially does not consider any specific section of its population as 'indigenous peoples'" (Bijoy and et al. 2010: 13). Therefore, the relevant United Nations Declarations and Covenants on Indigenous Peoples are not applied. The Indian government has usually favoured its own policies over international law. India has been a member of the International Labour Organization (ILO) since 1919, and has so far ratified 41 conventions out of 188, including ILO Convention 107 (indigenous and tribal population), which is in force since 1958. However, it did not sign ILO Convention 169. Bijoy et al. write: "In general, despite the existence of a plethora of complex institutional and legal arrangements for recognition and safeguarding of the rights of STs, the Indian situation remains very far from compliance with the requirements of international law" (2010: 57). Irrespective of the non-recognition as such, many organizations and activists in India self-define as "indigenous peoples", and/or they use the equivalent Hindi term *adivasi* ("original inhabitants"). However, towards the Indian state and judicial system, they can only use the recognized category of "Scheduled Tribes".

Scheduled Tribes (STs) and Scheduled Castes (SCs): Under the Indian constitution, discrimination due to religion, caste and sex is prohibited. However, members of STs and SCs are entitled to special protective provisions, a form of "positive discrimination". There is no clear definition as to who should be included in the category, and the "Scheduled" simply refers to a list which was notified and can be modified by the President of India. The state is obliged to promote the educational and economic interests of STs and SCs through reservation of seats in public institutions, public enterprises, government and administration at all levels.

Scheduled Areas: In district-/sub-district-levels, where the "Scheduled Tribes" form the majority of the population, there can be exceptions from the general laws and special legal provisions. A prominent example is the "Panchayat (Extension to Scheduled Areas) Act (1996)" or PESA, which gives additional powers to the self-constituted *gram sabha* village assembly. Notably, it has to be consulted for any development projects and land acquisition. However, this central law has not been properly implemented by the states (Bijoy et al. 2010: 71). In Chhattisgarh, where elaborate rules about the consultation procedure do exist, there are many reports of alleged "fake consultations", which involved massive threats and manipulation by police forces, authorities and industrialists (Bhushan/Zeya Hazra 2008: 136–140). Still, some *panchayats* have been able to give a voice to local communities. For instance, Sharma and Choudhry list 23 cases of *gram sabhas* in Chhattisgarh's Raigarh district, which passed resolutions or took legal action against land acquisition (Sharma/Choudhry 2011: 126; 130–31).

1.4 Local Economy and Resource Use

Since the earliest records, the centre of all economic activity in the plains of Chhattisgarh has been paddy (rice) cultivation. Other crops include different pulses and oilseeds, wheat, maize, groundnut and soya. Vegetables are mainly grown for subsistence needs. Cultivation is combined with rearing of livestock such as cattle, buffaloes, sheep, goats and poultry. Cattle and buffaloes are also used for transportation and ploughing. The production of cash-crops remains low and the technology is frequently termed "outdated" (Bhatt/Bhargava 2005: 196). Put differently, as there is little input of artificial fertilizers and mechanization, farmers are producing independently without much need for cash, mostly for subsistence and for small-scale trade at the local markets. Knowledge about farming techniques is passed on from generation to generation and varies from region to region (Rizwana/Lyaqet 2011: 384). In recent years, farmers have been shifting from long-duration local rice varieties to short-duration, high-yielding varieties, adapting to the decreasing rain quantity (Sastri 2009: 156).

Viswanath et al. consider that the Chhattisgarh farmers apply a traditional, highly adapted, agroforestry system in which different trees, especially babul, play an important role. An average farm has 20 babul trees of different ages, which provide fuelwood, brushwood for fencing, small timber for construction and implements, and gum and seeds, which can be easily stored and sold at local markets. The babul tree has a special significance as a "symbol of economic and ecological stability", especially for small-holding farmers (2000: 176).

The traditional method of paddy cultivation is called *byasi*: rice is sown in pre-ploughed fields in July, with the arrival of monsoon rains. About a month later, the field is ploughed again and the uprooted seedlings are bundled and transplanted, in order to receive a more uniform distribution The rice fields are surrounded by bunds, which are constructed manually with great effort. The bunds serve to maintain the fields covered with 10 to 20 cm of water during the initial stage from July to August, keeping the rice seedlings moist and reducing the growth of weed. On top of the bunds grow vegetables and dry crops as well as trees. In September, *kesri dal* (linseed) is sown in the rice fields as a relay crop. Also, fish can be caught easily at that time. The rice is harvested in October. For this, medium and large-holding farmers hire day labourers. In the following months, a second crop (called *rabi*) might be possible if irrigation is available. Otherwise, the harvest of babul gum and seeds become a major subsistence activity (ibid.: 160).

Irrigation

With only one-fifth of the total cultivated area under irrigation, the general level of irrigation in Chhattisgarh is rather low, but in Raipur district it is almost 50%. The main sources of irrigation are canals, followed by tube-wells and ponds (Bhatt/Bhargava 2005: 198f.). Chhattisgarh has an extensive network of irrigation canals, such as the Mahanadi Reservoir Project Complex (MRPC), established in 1927. However, many of them are in bad shape, and the distribution of irrigation water is conflict-ridden. Water from irrigation canals is used both for the rice crops at critical stages (e.g., erratic rains), and to fill up the village ponds, which serve for washing, bathing, animal consumption, fish cultivation and irrigation of small fields and gardens.

Since the 1990s, efforts have been made for devolution of powers from state authorities to local-level 'water users' associations' (WUAs, also known as *pani panchayats*), under the label of "participatory water management". The WUAs are constituted by all landholding farmers in the command area of a branch canal. They elect representatives at four different levels, who are empowered to make decisions regarding the distribution of irrigation water.

Marothia has studied the performance of 22 WUAs in Raipur district in 2001. Two years after introduction of the new system, he found that their performance was "mixed". Major challenges were the willingness of administrative authorities to cede powers to the associations, and the danger of WUAs becoming part of power struggles between political factions or getting captured by local elites (Marothia 2002). In their study about climate change and migration, Murali and Afifi report for several villages in the Raipur plains that the groundwater level had decreased considerably, tube-wells and ponds went dry during summer. Canal water was provided to most fields from July to September and to some also from January to March. However, locals complained about unjust distribution, suspecting that the authorities were giving priority to provide water from the canal to the coal-based power plants in the region (2012: 51f.) For the region in focus of this study, there was no information available regarding the performance of local irrigation water management.

Landholdings and Seasonal Migration

In the colonial period up to independence, the landholdings in the villages of Durg and Raipur were distributed highly unequally. The feudal system was controlled by the "all-powerful" *malguzar* ("landlord"). He was responsible for revenue collection, but also had a whole range of other functions of authority. In a village studied by Parry, the *malguzars* owned over 70% of the land in the 1930s, what he believes to be a "consequence of rights which [they] had acquired during the British period over formerly unbroken common land [...]" (1999: 17f.). Even though the landlord system was officially abolished after independence, the distribution of landholdings remained highly unequal. Most *dalit* were landless, while low-caste families held just 2 to 4 acres. Because few families could meet their subsistence from their own land, they had to work for large-scale farmers. In addition, seasonal migration was very common, e.g., to the tea-gardens of Assam or to early industrial centres like Calcutta (Parry 1999: 19; Bates 1985). Seasonal migration has continued and even increased until today, now often to construction sites in the urban centres (Murali/Afifi 2012: 81).

Sacred Landscape and "Occult Economy"

It is important to remember that "nature" is not perceived by everyone as simply a deposit of exploitable "natural resources". Popular Hindu beliefs, for instance, are linked to a "sacred landscape" and highly elaborated ritual cycles. For agriculture, the annual cycle with the three seasons of *garmi* (hot, March to June), *barsat* (rainy, June to October) and *sardi* (cold, November to February) is most relevant. Babb (1975: 127–28) shows how it is bound to numerous pilgrimage festivals. Mountain-top shrines, tree temples and holy waters have been interpreted as remainders of past animist or nature veneration, as they are still practised by less Hinduized *adivasi*. This can turn against industrial development projects because of the popular belief that the destruction of a certain shrine would enrage the deity who inhabits it, and there is a danger that the deity will take revenge. Therefore, especially local workers might refuse to do the job. Parry gives the example of Rajhara, where the land of the local deity Rajarav had been taken for iron-ore mines. Some workers from outside are said to have ignored the warnings of the locals, and after having destroyed the place, they fell sick and died one by one (2008: 241, 244).

Parry links this phenomenon to the "occult economy" of blood sacrifice. There are many stories about human sacrifices in Chhattisgarh of olden days. Animal sacrifice to plead with the gods for the fertility of the fields is still practised in remote areas, and was more recurrent at the time of the British. Also, in tribal myths, ironwork, mining and metallurgy had their origins in human sacrifice. The goddess Kali is associated with the forge of the smith, and therefore also with the giant furnaces of a steel plant (2008: 242). Kali is "the black one" who takes life if she is not properly respected. One should give sacrifices to her voluntarily, or she might take it forcefully at her will. For the enormous size of the factories, many Chhattisgarhis believed that only human sacrifice could be powerful enough to sustain them. The numerous workers who died at the construction sites were a proof of this—they were non-voluntary "sacrifices" taken by the goddess. Also, there were rumours that factory owners and managers were abducting helpless people to sacrifice them, preventively giving human lives in order to secure the goddess' cooperation. "Human sacrifice is just sensible risk management, and deploys the standard technology of ritual power. It is therefore no mystery that people suppose it a necessary precursor to major public works", summarizes Parry the popular beliefs at the time of arrival of the BSP, which made many shy away from working there in the 1950s. This has changed, but the rumours have not disappeared completely, and the "[...] idea that factories require victims persists [...]" (2008: 254, 249).

1.5 State Economy and Legal Framework

India has a highly sophisticated administrative, legal and judiciary system, largely inherited from the British, which often combines progressive laws with weak implementation. Laws of different epochs sometimes contradict each other, as do Central and State laws. The Supreme Court of India is highly respected and has issued many landmark verdicts, but is hopelessly lagging behind with tens of thousands of lawsuits in the waiting row. The same happens at the State level, where judges are, however, less trusted. Corruption is rampant in all sectors of the government, administration and judiciary.[12]

After the state-centred "socialist" development model was more and more difficult to uphold, first neoliberal reforms were introduced in the 1990s under the then Finance Minister Manmohan Singh and the ruling Congress Party. The World Bank actively pushed for "structural adjustments"—the well known mix of decentralization, deregulation and opening access for foreign investments (Breman 1999: 429). The reforms accelerated after 1999 under the BJP Government. The subsequent investment boom brought economic growth rates of around 8%, a high demand for better infrastructure, large-scale construction and intensified extraction of natural resources.

Since its creation in 2000, the state of Chhattisgarh has adopted various policies to foster the mining, energy and industrial sectors. The "Chhattisgarh Vision 2010" laid out the long-term development priorities of the state with a focus on economic growth, based on certain advantages for the industrial sector: abundant mineral reserves (coal, iron ore, limestone, bauxite, diamonds, a.o.), availability of surplus power and further potential for thermal coal power plants, strategic location, low cost of land and a "peaceful industrial workforce". The problem of land acquisition was identified as a major hindrance (Bhatt/Bhargava 2005: 141; 158). Nevertheless, Chhattisgarh's economy continues to be mostly rural (as does India as a whole), with over 80 per cent of the population depending primarily on agriculture and forest produce (ibid.: 192).

Mining and the Cement Industry

In India, offshore mineral resources are the property of the central government and everything underneath the solid ground belongs to the corresponding state governments, which have the authority to grant concessions for exploration and exploitation. Under certain circumstances, additional approval of the central government is needed (Ministry of Mines 2011: 47). In accordance with the economic reforms, the Government of India released a "National Mineral Policy" (NMP) in 1993, which guided substantial changes in the concerned laws and regulations, in order to attract foreign investment. In 2008, the NMP was revised again to "simplify the procedures for grant of mineral concessions" (Ministry of Mines 2011: 37). More powers were delegated to the state governments, and the "Mines and Minerals (Development & Regulation) Act, 1957" (MMDR Act) was amended to allow 100 percent foreign direct investment for exploration and mining (Nair/Majumdar 2006: 9). In 2004–2005, the value of Chhattisgarh's mineral production was 7% of India's total, whereby 94% was coal and iron ore alone. Limestone only accounted for

12 India ranks at position 94 (of 174) in the 2012 Transparency International Corruption Perception Index.

3.1%. The contribution of the mining sector to the state's GDP in 2001–2002 was 15%, making up for around 10% of the state's revenues (Bhushan/Zeya Hazra 2008: 120–122). For the production of cement, limestone is the main ingredient with about 80 per cent, complemented with clay, silica, fly ash and others. Limestone is relatively abundant and therefore cheap, so that royalty rates do not significantly affect the price of the end product. Limestone is generally mined with the open-pit method in mines that are in proximity to the cement plants and controlled by the same company (so-called "captive mines"). Cement production needs a lot of energy, which in India is mostly derived from coal fired "captive" power plants, often located directly next to the cement plant. India is the second largest cement manufacturer in the world after China. 137 large and 365 smaller cement plants annually produce more than 300 million tonnes, whereby the production capacity has doubled in just five years and is projected to reach 550 million tonnes by 2020.[13] Accordingly, also limestone mining increased and reached 229 million tonnes in 2009–2010. Chhattisgarh's contribution was at 8 per cent (Ministry of Mines 2011: 14), with estimated reserves at 3,580 million tonnes (Putul 2007: 48). The combination of abundant limestone and nearby coal deposits makes Chhattisgarh's central plain area a privileged location for the setting up of new cement plants. The Raipur district alone had 7 of Chhattisgarh's 10 major cement plants, and 30 of 63 proposed new plants. The expansion plans would mean an increase of the state's total production capacity from 13.8 to 100 million tonnes per annum (*Down To Earth* 2010: 20–21). In January 2012, a new district by the name of Baloda Bazar was carved out of Raipur, which is being projected as the "cement hub of the country" with four major plants in operation and two new projects coming up.[14]

Legislation

The establishment of large public sector industries after independence was combined with the development of strong trade unions and enactment of progressive labour laws. This also had its effect on the "formal" or "organized" part of the private sector, where the same laws apply. However, formal employment covers less than 10 per cent of the total workforce, and more than half of it is employed by public enterprises (Parry 2009: 180; Breman 1999: 407). This has created a privileged class of "labour aristocracy", which enjoys benefits like decent wages, health insurance and other bonuses, whereas most of the population depends on precarious informal work in the so-called "unorganized" sector, hardly protected by legal provisions. With the neoliberal reforms of the 1990s, permanent employment in the formal sector is under pressure, and even public sector companies have increased the number of temporary "contract workers" (Breman 1999: 427; Parry 2009: 190).

The major trade unions are affiliated to political parties and described by Parry as either "instruments of state hegemony and management control" (in the public sector) or as "tools of private capital" (in the private sector). "That is why most workers are militantly apathetic about them [...]" (2009: 180). They mostly represent the "privileged enclave" of the formally employed "company workforce [...] and defend the citadel walls [...] against a more radical union movement that has attempted to represent the interests of workers outside it." (2009: 201). Parry observed that contract workers at the Bhilai Steel Plant were mostly not unionized, liable to summary lay-offs and given the harshest, most dangerous jobs. Officially, they were paid more than the statutory minimum wage. However, the contractor through whom they were employed kept a

13 *Business Standard*: "India cement production to touch 320 million tonnes this year." Sept. 23, 2012. <www.business-standard.com> (accessed on 20.07.2013).

14 Krishna Das, R. 2012: "Balodabazar district in Chhattisgarh set to become cement hub." *Business Standard*, Jan. 19. <www.business-standard.com> (accessed on 20.07.2013).

Fig. 2.: The yearly Bhagat Singh Memorial March in Bhilai, organized by CMM, March 2014. Photo: Karin Scheidegger.

good part of the salary for himself, and sometimes paid a kick-back to the manager who had issued the contract (Parry 2009: 192–93). According to the "Contract Labour (Regulation and Abolition) Act" of 1970 and the "Industrial Disputes Act" of 1947, most of the contract work is technically illegal. If someone does a task that is "a permanent and perennial need", or after having worked continuously for 240 days for the same employer, he or she has to be employed on a permanent basis and receive all the corresponding benefits. This is obstructed by the companies by shifting workers, short-term lay-offs and falsifying or non-delivery of documents that would serve to prove the permanent nature of the work (Parry 2009: 192). For the cement sector, there are even stricter rules which were agreed upon in a tripartite agreement of 1983 between trade unions, the Cement Manufacturers Association (CMA) and the state. According to this so-called "Cement Wage Board Award", contract labour is allowed only in loading and packing. The agreement is still in force, but has been disrespected by the industry.[15]

Land Rights and the Panchayati Raj

Regarding land rights, there is a basic distinction between the privately owned and the public or "government" land. To acquire privately owned land for industrial or mining use, a company has two options: either it chooses to negotiate with the individual owners to buy the land, which often happens through middlemen (PUDR 2012: 9–10); the other option is through the "Land

15 According to S. Gera, the agreement was made when unions were strong, before the neoliberal reforms of the 1990s. ACC and Ambuja were both members of the CMA at that time. Although they left the CMA later, they are still bound by the agreement, which remains valid as long as no other agreement replaces the preceding one (personal com., Sept. 2013).

Acquisition Act (1894)". This colonial law has been used, especially for setting up large-scale industries, which often displace entire villages. It allows the state to forcefully acquire land to be used for whatever is considered a "public purpose". Under the guise of "economic development", state governments throughout India have termed the installation of mines and industries as such a "public" purpose (even if it is done by private investors) and passed on the forcefully acquired land to private owners at very favourable conditions. The law does not include any provisions for compensation, rehabilitation or resettlement of those who get displaced. After a wave of massive resistance against large-scale displacement, the Central government introduced a "Policy on Resettlement and Rehabilitation" in 2007 (Somayaji/Talwar 2011: 2). Also at the State level, different rehabilitation policies are in use, such as the "Madhya Pradesh Rehabilitation Policy (1983)", which was taken over by Chhattisgarh. It holds, amongst other points, that each family which loses land though forced acquisition should receive a permanent job for one of its members.[16] After years of debate, a new "Right to Fair Compensation and Transparency in Land Acquisition, Rehabilitation and Resettlement Bill, 2013" entered in force on 1 January, 2014.[17] The new law has no effect on rehabilitation of people who were displaced earlier.

In recent years, popular movements have achieved a number of legal revisions that aim at strengthening control of local communities over lands and forests, which officially count as "government owned". These are mostly the former village commons that fell under the control of the state in a process of de-commonization already under the British administration, and which has continued after independence. The focus, thereby, is on the *panchayati raj* local governance institutions, especially for the "Scheduled Areas". Different forms of village assemblies or councils (known as *gram sabha* or *gram panchayat*) had been functioning traditionally in many parts of India, but the officially recognized version came into existence only after a constitutional amendment in 1992, in a move to decentralize governance, especially in the rural areas. With the formalization, an electoral model was introduced. Madhya Pradesh was among the first states to pass a law in this regard, which was adopted by Chhattisgarh as Chhattisgarh Panchayat Raj Adhiniyam (1993). Still, it took until 2005 to hold the first elections. Chhattisgarh's three-tier *panchayat* system is made up of approximately 9,820 *gram panchayats* at the village level, 146 *janpad panchayats* at the intermediate level ("block") and 16 *zilla panchayats* at the district level. However, the power of the village-level *panchayats* is quite limited and their main function is to check if funds from social and development programs reach their supposed beneficiaries. To fortify this, the national parliament passed the "Right to Information Act" in 2005, which forces the administration to release specific information on demand, especially by the *panchayats* (Sohini 2006).

With regard to the setting up of major industries and mines, the need to consult and receive consent from the affected village assemblies in Scheduled Areas has become a contested issue, both legally and in the actual undertaking of "public hearings" for that purpose—often, the latter are either boycotted or obstructed by opposing villagers. However, in non-Scheduled Areas, there seems to be no clear obligation to hold such hearings. In any case, for small mines and the mere expansion of existing industries that existed before the new law, no consultation is necessary.[18] Though the present case study is not situated in a "Scheduled Area", the additional powers of the *gram sabha* according to PESA (see box 1) are relevant because they are frequently seen as a model to extend to non-Scheduled Areas also. In that sense, they are the closest to the principle of Free, Prior and Informed Consent to be found in Indian legislation.

16 Personal communication S. Gera, Sept. 2013.

17 *The Economic Times*, 16.10.2013: "The New Land Acquisition Act to come into effect from 2014." <articles.economictimes.indiatimes.com/2013-10-16/news> (accessed on 04.02.2014).

18 Personal communication S. Gera, Sept. 2013.

2. Holcim's ACC and Ambuja Cement Plants

In 2005, the Switzerland-based multinational Holcim entered the Indian market through an alliance with Gujarat Ambuja Cement Limited (Ambuja or ACL) and the acquisition of a majority of stakes in the ACC Limited. These two major Indian cement companies had a combined production capacity of 45 million tonnes of cement per year, making Holcim the second biggest player in the Indian market (Multiwatch 2012). In 2013, ACC had 17 integrated cement plants and 40 ready-mix concrete plants and Ambuja counted 5 integrated cement plants and 8 grinding plants, in different parts of India. In July 2013, Holcim announced a restructuring, whereby Ambuja would take over a majority share of ACC and then be merged with Holcim India.[19]

For this chapter, we focus on Holcim's operations in Chhattisgarh, primarily the plant of Ambuja in the village of Rawan, district Baloda Bazar, near the town of Bhatapara,[20] and are including some information from the ACC plant in Jamul, located in the urban area of Bhilai in Durg district. Both are among the major operations for their respective group of companies, and their capacity is constantly being increased. On its website, Holcim especially underscores the "serious growth potential" in its operations in Chhattisgarh.[21] Ambuja Rawan had its clinker production capacity expanded to 3.4 million tonnes and a new 33MW coal-based thermal power plant was installed. At ACC Jamul, production is to be increased from 1.58 to 4 million tonnes annually by 2015 (ACC 2011: 3). Both plants have their captive thermal power plants and limestone mines nearby. ACC Jamul sources additional limestone from a second mine at 22 kilometres distance (BWI/ICEM 2011: 3).

An important aspect of these kinds of integrated industrial complexes is that they include a township or "colony", a relatively secluded settlement for its permanent employees, with schools, medical facilities and other benefits provided by the company. The satellite view manifests a kind of "beauty and beast" image: a well-kept, planned town with green lawns, walled off from villages or slum neighbourhoods and just next to an industrial "monster" in dusty shades of grey, brown and black, including dumping sites (coal stock in black, slag heaps in grey).

In this chapter, mining of limestone is treated as an integral part of the production of cement. It is carried out by the same company that produces, distributes and sells the final product, under a well-known consumer brand name. In this sense, this case study differs from all the others included in this book, where mining is production of raw materials for a global market without direct relation to manufacturing and consumption, which usually happens elsewhere and is not under control of the mining company. The focus is thus not on mining itself, but on the impact of an entire industrial complex.

2.1 Social and Environmental Impacts

In January 2012, the *Pragatisheel Cement Shramik Sangh* (PCSS), a registered trade union representing contract workers in different cement plants of Chhattisgarh, filed a complaint against Holcim at the OECD contact point in Switzerland. The focus of the complaint is on the rights of contract workers to collective bargaining and on discriminatory practices against union members in the plants of ACC Jamul and Ambuja Rawan. Other issues, all concerning Ambuja Rawan, are

19 *The Economic Times*, 25.7.2013: <articles.economictimes.indiatimes.com>; see also IIAS: "Holcim, Ambuja and ACC Unlocking Synergies. Really?" <www.moneycontrol.com> (accessed on 11.02.2014).

20 The company refers to this plant in its documents as "Ambuja Bhatapara", but here we call it "Ambuja Rawan" for the village where it is actually located.

21 Indicating that the state holds around 9 percent of global coal reserves. <www.holcim.com> (accessed on 9.3.2013).

subsumed under "violation of human rights" and include non-compliance with rehabilitation policies of the state after the original land acquisition in 1983, destruction of livelihoods of landless agricultural labourers, excessive water use and dust pollution, discrimination against locals in employment, damages to houses due to vibrations from blasting in the mines and encroachment of village commons, including harassment of people who resisted the encroachment. The following section is largely based on the extensive documentation of this complaint,[22] including a document prepared by the "Tribal Welfare Society" (TWS) in 2008. Additional testimonies from Rawan have been collected by Asgar Qadri in July 2012.

2.1.1 Environmental Impacts

For an assessment of the environmental impacts of the cement industry, we would need to consider the flows of all the materials and energy needed for production, including land, water, limestone and a number of secondary ingredients as well as the coal used for electricity generation, which goes beyond the scope of this study. The Centre for Science and Environment carried out a rating of India's cement production in 2005, covering roughly 80% of the sector (Bhushan/Zeya Hazra 2005). It showed that in comparison to international standards, the production technology and consumption of energy and water were not particularly bad, and that environmental problems could not simply be attributed to outdated technology. Nevertheless, captive power plants for cement production made it the third largest user of coal, with a consumption of 11,400 million kWh in 2003–2004.[23] Water consumption was at an average of 0.3 tonnes per tonne of cement, totalling 39 million tonnes. The study found serious problems in the mining of limestone: 90% was extracted by blasting, which produces heavy dust emissions. The less impacting, but more expensive, 'surface miner' technology was used for less than 10%. The storage and transport of raw materials produced quite a high emission of particulates (330 gm/tonne cement). The production was concentrated in the poorer regions of the country, and 70% of the plants had communities living within one kilometre radius. The study reports widespread complaints from these communities regarding dust, pollution, cracks in houses from explosions, as well as negative impacts on agriculture due to a decline in the ground water table.[24]

In Chhattisgarh, the use of ground water is apparently not regulated and "a landowner has full freedom to harness water available under his/her ground". Also, there is no water resource plan (Bhatt/Bhargava 2005: 188). The Water Resources Department responsible for monitoring the ground water table has no overview of the complex cumulative impacts of all the running and planned mines and industries. The *Down to Earth* magazine estimates the annual water requirements of all the planned industries and mines in Chhattisgarh at 3,000 million cubic meters: This amount could meet the domestic water requirement of 274 million people (2010: 23). With climatic changes ever more affecting rain patterns, this produces additional stress, especially for small-scale subsistence farmers.[25]

22 The complaint includes an annex with testimonies by locals, including a *panch* (member of the *panchayat* village council) of Bhadrapali, a Gond agricultural labourer, the *sarpanch* (head of the *panchayat*) of Pausari, a *panch* of Rawan and others. Furthermore, it includes a letter of complaint to Ambuja signed in May 2010 by 19 representatives of various villages affected by the plant.

23 On global scale, cement production is estimated to contribute 5% of greenhouse gas emissions (Moody 2007: 115).

24 CSE India: "CSE's Green Rating Project releases rating of the cement industry" <old.cseindia.org> (accessed 18.12.2012).

25 For extensive documentation, including maps, about changing rains, poverty and migration see the case study from Chhattisgarh of "Where the Rain Falls" project: <wheretherainfalls.org/india> (accessed 18.12.2012).

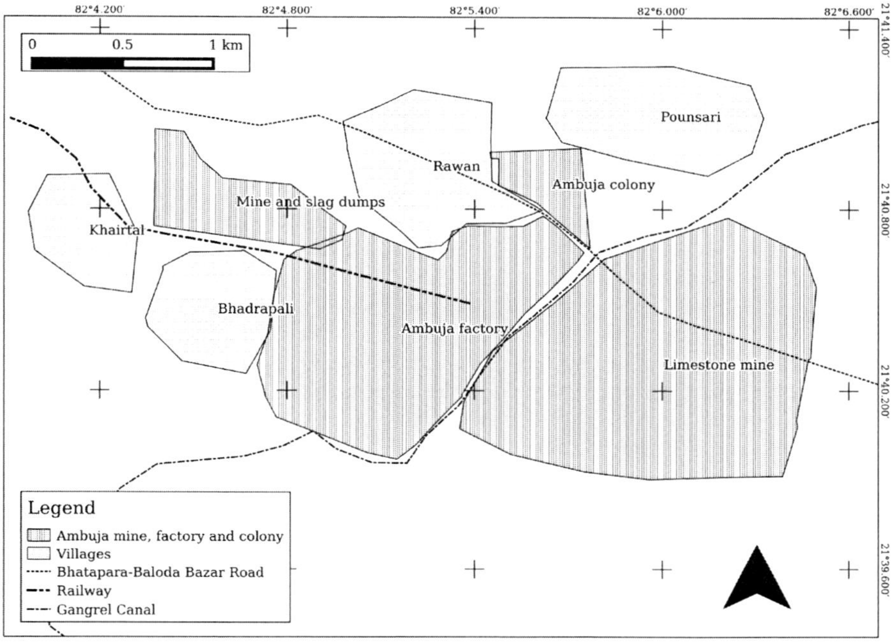

Map 1: Location of the Ambuja Cement plant, colony and limestone mines next to the villages of Rawan, Pounsari, Bhadrapali and Khairtal. Railway and road cut through from west, the canal from south-west flowing to north-east.

Water Use at Ambuja Rawan

According to the company,[26] the water requirement of the Ambuja plant in Rawan was 4912 m^3 per day in 2011. It had a permit from the Central Ground Water Authority to withdraw 6000 m^3 of ground water per day.[27] The company is implementing ground water recharge measures together with the Ambuja Cement Foundation (ACF) and is currently developing a long-term plan for water resource development. It has introduced a "water accounting", which shows that, for 2011, the water credit of the plant (2950 million litres) is 1.3 times the total water debit (2307 million litres)—which means that the water harvesting measures provide more water than needed by the factory.[28] ACF claims to have built 8 check dams, 94 hand-pumps for drinking water and undertaken deepening of 27 ponds in 14 villages surrounding the Ambuja plant. Interestingly, the document also includes a photo of a lake that formed in the abandoned mine pit at the west side of the Rawan plant as part of the company's water management efforts.

Apparently, the activities of Ambuja Cement and its foundation in water management are perceived quite differently by the local people. The TWS document of 2008 claims that a stop dam

26 Document provided by Peter Gysel, Deputy Head Corporate Communications of Holcim, per E-Mail, 26.11.2012.

27 5650 m^3/day through mine dewatering for mining below the water table and 350m^3/day from borewells (ibid.).

28 "Water credit or generation includes Ground Water Recharge (1411.82 million litres) through mine pits, structures constructed or contributed by the "Ambuja Cement Foundation" (ACF) (such as check dams, ponds and tanks), mine water harvesting for real-time use (1191.73 million litres) and Water Recycle/Reuse (346.70 million litres)." (ibid.).

built by the peasants at their own cost 30 years before the arrival of the company had been "face lifted" by ACF, including construction of a small canal, which overflowed into farm land and destroyed crops. Also, "it is well known that a few local officials of Ambuja Foundation have left the work half done and wasted lakhs [100,000s] of rupees [and] are submitting false information showing progress of the work on paper only [...]".[29] In the PCSS complaint to the OECD of 2012, several statements from local peasants and authorities allege that the excessive water use of the factory is affecting agriculture negatively. "Because of the arrival of the company, water problem has also become acute. Earlier, the water from the village ponds, and from the water tank of the big landlord (malguzar), used to be sufficient for the village" (Gaind Ram Dhruv, Rawan, Annex 44a); Ramu Sahu, *panch* of Bhadrapali, relates that "[i]n 2011, two villagers rented their borewells to the company. All through the summer, the company ran a pipe directly from the borewells and pumped water continuously, due to which the ground water level has reduced considerably", and that "[t]he village pond that was constructed by the villagers [...] is also being claimed by the company as its own." (ibid.: Annex 32A); Digeshwari Verma, the *sarpanch* of Pausari, confirms: "The company has dug several 100-feet-deep holes all over the lands, which act as water sinks and drain away ground water from the rest of the land. Around 8–10 years ago, the ground water level was 40–50 feet below the ground, but now it is 80–90 feet below the ground" (Annex 45A). From these statements, it seems that there is a competition between farmers' and the company's water needs, especially during the summer months. One incident regarding the canal from Gangrel dam,[30] used to irrigate fields and fill up the ponds, is especially indicative of that: "During the past summer, company dug up the canal [...] and diverted the canal water to the power plant in the company. Consequently, the canal water did not reach our village, Pausari, this past year and all the 40–50 villages downstream [...] also did not receive any canal water" (Digeshwari Verma, Annex 45A; see also Ramu Sahu, Annex 32A).

In a detailed response[31] to allegations by PCSS as summarized in MultiWatch (2012), Holcim denies any negative impact on the water availability. However, the Gangrel canal incident is not mentioned. Ethnographic experience shows that local peasants have a highly developed understanding of water conditions and their changes over time. Certain doubts remain, because the period from 1980s, when the factory and mine started operating, has generally been marked by changing climatic conditions and decreasing rain quantity due to global climate change. Nevertheless, the allegations indicate that the CSR program of Ambuja Foundation—focussing on improving water availability—is experienced by locals as a capturing of common water bodies. Thereby, it seems that the problem is less about total water quantity, but about its distribution at critical times. Ambuja's partial takeover of the management of local water resources could mean that whenever water is scarce, the needs of the company prevail over those of the farmers.

Pollution

To appreciate that dust pollution is a major impact of Abuja's cement plant in Rawan, a look at the satellite image is enough. Different parts of the factory, the coal dumps, the limestone mines and slag heaps all have their characteristic colour print from beige to black. However, information regarding air and water contamination remains scarce and independent scientific data are lacking. The so-called "co-incineration", i.e., the burning of toxic wastes, apparently has not been a big is-

29 PCSS 2012: Annex No. 27 p. 5-6.

30 Gangrel is a large dam south of Raipur, from where an extensive irrigation canal network covers large parts of the Chhattisgarh plains (see Marothia 2002). One of them serves Rawan and other villages further north.

31 Holcim.com: "Statements regarding MultiWatch allegations", April 27, 2012. <www.holcim.com/press-and-media> (accessed on 18.12. 2012).

sue so far in India.[32] Among the allegations that TWC had raised in 2008,[33] there are two cases of release of toxic liquids into small water bodies, both for Rawan and Jamul. In the PCSS complaint, we find various statements about dust and the vibrations from explosions: "Due to heavy blasting of the mines, the buildings are developing cracks", and due to air pollution from the cement plant and power plant "dust particles are settling on fertile land owing to which fertile soil is becoming barren [...]".[34] According to Digeshwari Verma, this negatively combines with the water shortage, leading to a decrease in agricultural productivity: "Earlier, the lands would give 16–17 quintal rice per acre, but now we get only 7–8 quintal rice per acre" (Annex 45A).

2.1.2 Social Impact

Industry- and mining-induced displacement has become a major problem throughout India. Especially vulnerable are *adivasis*, women, elderly people and landless agricultural labourers. While holders of ownership papers can at least negotiate some compensation, customary ownership is often not recognized. Also, there is no compensation for the loss of common lands and forest (Pandey 1998; DRCWC/Samata 2010). The impacts of displacement have been described as the loss of physical and non-physical assets like homes, communities, income-earning sources, subsistence, productive land, resources, social structures, cultural identity, networks and cultural

Fig. 3.: Village pond in Rawan, with Ambuja cement plant in the background. Photo: Karin Scheidegger.

32 However, problems have been reported from other cement plants of Holcim in Argentina and Mexico (see MultiWatch 2012; Moody 2007: 114ff.).

33 PCSS 2012: Annex No. 27.

34 Letter from leaders of village communities, 01.05.2010, PCSS 2012, Annex 51A.

sites (Downing 2002: 3).

In Rawan and surrounding villages, many families lost land when the cement plant was built in the 1980s. According to an open letter of 2011 by the union activist and lawyer Sudha Bharadwaj, only 20 of the 38 farming families of village Rawan which lost their land to the cement plant 25 years ago, have received some form of compensation. Since then, the company continued to acquire land in the villages of Rawan, Bhadrapali, Pausari and Kukurdih. Bharadwaj argues that the acquisition of land is, in many cases, illegal and by far exceeds the requirement of the company because vacant land is kept for future expansion of the plant capacity.

The account[35] of Hira Lal gives an impression of how the situation is perceived by many local people. According to him, the agricultural land on which the factory and limestone mine were built was bought at a very low price: "They took the land with the help of a local contractor, who was the richest man in the village. We were scared of him but even then refused to give our lands. However, they all came together—the government officials and contractor's men—and forced us to handover land papers to them." Lal thus started working as a contract labourer at the plant. In the beginning, he was paid about 8 rupees a day. When he retired after about 20 years of contract labour, he got 65 rupees a day (little over one dollar). As he puts it, "we were all comfortable farmers. They came and took our land. Today, they are rich and we have nothing."

Loss of Land and Village Commons

In PCSS' complaint to the OECD, there are several statements regarding land acquisition and encroachment of common lands: according to Ramu Sahu, *panch* of Bhadrapali, in his village of around 150 families, approximately 300 acres were acquired in 1982–83 from around 80 to 90 families (Annex 32A). Digeshwari Verma, *sarpanch* of Pausari, accounts that his village lost around 100 acres "to the mines": "Out of this, around 30–35 acres was government land, which was being used for the grazing of cattle [...] The remaining land was privately owned, which individual farmers have subsequently sold to the company over the years, as they faced into financial troubles" (Annex 45A).

The company thus acquired both private and common land, and apparently still continues to do so. While loss of private agricultural land affects the livelihood of certain families, the loss of commons affects the communities as a whole. In the letter of 2010, the village leaders allege that "[t]he company has illegally occupied the village commons, grazing grounds, roads joining one village to another, burial ground etc. of villages Rawaan, Bhadrapali, Pounsari, and Khairtal [...]" The mentioned road connects Rawan and Bhadrapali and its encroachment by the company forces the villagers to take a detour of 4.5 km (PCSS Annex 51A). The *sarpanch* Dhaneshwari Verma filed a case in the local court on behalf of Badrapali village, alleging that Ambuja Cement is illegally occupying government lands, which are shown in the records to be communal grazing lands.[36] The statement of Ramu Sahu, *panch* of Bhadrapali, confirms this: "The company has encroached upon agricultural land, in addition to what was acquired by it in 1983. The company is encroaching upon our village land, which is common land (government land) and is used for grazing our cattle. Our village has filed a case against the company for this illegal encroachment. The company started dumping material on the village land adjoining the village pond, which is close to the railway tracks, which is also land for grazing cattle. We also complained against that. At that time, the material was cleared from the land, but in 2009, the company again started dump-

35 Recorded by Asgar Qadri, Rawan, July 2012.

36 The application regards "Khasra (property) No. 134/1 and Khasra No. 149/1". It states that the lease deed signed on June 4, 1982, under which the company is claiming legal ownership of this land, expires within 3 years if the land is not utilized (application filed on October 10, 2010, under section 182 (2) (IV), in the court of honourable collector, Raipur). (Testimony and document collected by Asgar Quadri, July 2012).

ing material there [...]" (Annex 32A). Krishna Prasad Verma, *panch* of village Rawan, alleges encroachment also by the school that Ambuja runs for the children of its permanent employees: "The boundary wall of the school has encircled some 15–20 feet of the village common land. After I raised this issue, the company started harassing me" (Annex 50A).

Another case of encroachment, involving threats and physical attacks, is reported by Shankar Lal Yadav, a milkman and permanent employee of Ambuja. He refused to leave his house—built on government land 20 years ago—to make way for the company: "Company official, Mr. CR Verma, himself told me that I should demolish my house since the company wants this land. [He] had come along with some villagers to threaten me. The government land neighbouring mine has already been encroached upon by the company's Children's Park [...]. [...] the company has orchestrated 5 life-threatening attacks on me and my family members [...] These attacks were carried out by employees of the company or their children. [...] After the attack on 9.8.2009, when we reached the police station to lodge the complaint, Mr. Verma was already present in the police station in order to protect the attackers" (Annex 47A). This statement also points to the collusion of the local police with the Ambuja management (see also section 2.3).

According to numbers collected by PCSS, the total amount lost to the industry in the villages of Rawan, Bhadrapali, Pounsari and Khairtal is around 1070 acres of privately owned and 85 acres of communal land. Out of 225 families which lost some or all of their land, only 29 persons received a permanent job, and 115 were working as contract labourers.[37]

Precarious Livelihoods, Working Conditions and Discrimination

When the cement plant was built in Rawan in the 1980s, most of the necessary land was bought from villagers with a promise that one member of each family would receive permanent employment in the plant—a common "rehabilitation" measure. This went along with general promises of bringing prosperity to the area. This hope has not materialised. Today, most families are dependent on precarious wage labour, and the villages remain poor. Gaind Ram Dhruv, a Gond and landless agricultural labourer of Rawan, states: "Before the (Ambuja) company came to Rawan, around 40% of the families here owned land, and the remaining people earned their livelihood by working on other people's land. Most of the *adivasi* families did not own any land, but worked on other people's land. Before the company arrived here, most people found enough work for livelihood." Even if in this account the "pre-industry" past might be painted a bit too nicely, it gives a good example of the feeling of deception many seem to share. He also explains how the landless are especially affected: "[the company] has never accepted any of our demands to provide employment to those (landless) families who used to work on those lands which were acquired by it" (PCSS 2012, Annex 44A). Dhruv holds this responsible for an increased pressure on long distance migration in search of wage labour.

Non-provision of permanent employment to those who sold their land is mentioned in the letter of village leaders of 2010 (Annex 51A), as well as by Ramu Sahu of Bhadrapali: "From all these [80–90] families (whose lands were acquired), only 15–16 people have obtained permanent employment in the company. Around 85 people in our village work as contract workers in the plant. From my family itself, around 17 acres of land was acquired which had 5 brothers and 2 sisters as joint account holders. But [...] only my uncle obtained permanent employment in the company, and after he passed away, no one from his family has been given a permanent job." (Annex 32A).

The feeling of deceit is fortified by the observation that "outsiders"—migrants from other states—are getting most of the jobs. Gaind Ram Dhruv, for example, estimates that out of 600 permanent employees, just 50 are from neighbouring villages (Annex 44A). Because of not

37 Personal communication S. Gera, Sept. 2013.

obtaining permanent employment and the corresponding benefits, most people are dependent on the precarious, and often discriminatory, employment as contract workers. Here also the preference for "outsiders" is an issue. According to Dhruv, the middlemen who employ the contract workers prefer "outsiders", because "[t]hey believe that [locals] are more politically active". He estimates that just about a third of the contract workers are from the directly affected villages (PCSS Annex 44A).

Those who obtain a contract job in the cement factory or the mines are working under dangerous conditions, for meagre pay. According to PCSS, contract workers at ACC Jamul receive around 5200 rupees per month (approx. 84 USD, around 3.50 a day), just about the legal minimum wages fixed for Chhattisgarh. This compares poorly to the minimum salaries of permanent workers of 19,000 rupees per month (308 USD).[38] At Ambuja Rawan, until recently the wages were below the legal minimum, and in some cases still are (PCSS 2012: 10). Contract workers are often given the most dangerous jobs. An example reported by the human rights group PUDR is an accident on 31 January 2013 in the Rawan factory, where five workers were killed when an overloaded fly-ash recipient fell from the 5th floor to the ground. Apparently, the management first tried to deny responsibility for the accident, and infuriated villagers who demanded compensation were dispersed by the police.[39]

These allegations resonate with the general literature on working conditions in India. Breman, for example, states that giving preference to outsiders over local labourers is a standard practice—a major reason being the greater vulnerability of migrant workers—and that "without doubt, access to work is connected to caste membership [...] workers who perform the most humble and miserable forms of informal sector work are mostly recruited from the lowest ranks and are often from tribal and Dalit communities" (Breman 1999: 418; 413–14). Parry also notes that, different from the public sector with its reservation system, in the private sector, "primordial" ties of caste membership tend to be reproduced in the factories, with permanent workers and labour contractors, outsiders and contract worker Chhattisgarhis (Parry 1999a).

That caste discrimination is not explicitly mentioned in the accounts from Rawan does not mean that it is not existent. Being an ancient, and much internalized form of discrimination, it does not need to become explicit to function. But, more than for caste, we can suspect that a major reason why the locals feel dispossessed by the cement factory is that it continuously relegates them to a position of a "backward, illiterate village past". Therefore, they can only access the lowest, most precarious jobs, without recourse against the bone-breaking exploitation. The interest of the company always prevails over their own needs. On the other side, for the management and the better-off, permanent employees might count what Parry described for the BSP: they are marked by "[t]he idea of 'progress', of having embraced modernity and left rural darkness behind them, of belonging to a more 'educated' and 'civilised' world" (2008: 251). In Rawan, the modern industrial township, with its fully equipped middle-class "colony", forms a kind of "island of modernity"—but those living inside it have to continuously defend their privileges against the vast sea of the "village past" that surrounds them. Thus, the people of Rawan and the neighbouring villages have not just lost land. They have been "displaced by modernity".

38 Personal communication S. Gera, Sept. 2013.

39 PUDR 2013: "When criminal negligence passes for 'Industrial Accident'" <www.pudr.org> (accessed on 13.8.2013).

2.2 Local People's Strategies

India has a vivid history of social movements that mirrors its social and ideological diversity.[40] Local activists and groups take up different inspirations and ally with regional and state-level organizations and coalitions, where it seems useful to receive support and put pressure on the government, and sometimes also to link up at the international level. For the present case, we focus on the alliance that local people from the villages surrounding Holcim's Ambuja cement plant in Rawan built with the cement workers' union PCSS, and the way this succeeded in gaining public attention in Switzerland, with some reflections based on our own involvement through *infoe CH* and the MultiWatch coalition.

As there are no ethnographic accounts, we do not know enough to actually describe the complexities of collective action at the local level. However, it is remarkable that even though many of the affected people in Rawan could have rallied based on ethnic identity as *adivasi* or *dalit*, they chose a more inclusive "worker/farmer" identity to confront the company. The available documentation suggests that the antecedents and credibility of PCSS and its "mother organization", the *Chhattisgarh Mukti Morcha*, play a crucial role, and therefore this section is mostly about them.

From the Iron Ore Mines to the Cement Plants of Jamul

From its beginnings in the 1980s, the Chhattisgarh Mukti Morcha (CMM) was a hybrid between a workers' union and a social movement. Its symbol, the red-and-green flag, originally combined the red colour of the left and the green of the farmers. Later on, the green also gets an ecological meaning, with CMM taking up issues like the monopolization of water by the industry and promoting small-scale irrigation instead of big dams. CMM also played a crucial role in the movement for regional autonomy and cultural revival of Chhattisgarhi identity, taking up women's liberation and related issues like domestic labour (Omvedt 1993: 236–37; Parry 2009: 196). In 2009, CMM led a successful struggle against the Swiss multinational Syngenta's attempt to purchase an important seed collection, which contains many rice varieties bred by Chhattisgarhi farmers over generations (Lutringer 2009).

The origin of the CMM lies in the mining township of Dalli-Rajhara, where the iron ore mines of the BSP are located. From the 1950s, large numbers of unskilled manual labourers worked there on a precarious contract basis. They were peasants from nearby, many *adivasis* and many women—a sector that the major union AITUC, linked to the communist party, "never regarded as its natural constituency" (Parry 2009: 194). In the 1960s and 70s, there were major strikes by the contract workers demanding regularization. In the process, the AITUC leadership lost its credibility, the contract workers set up their own union and invited the charismatic Shankar Guha Niyogi to become its organizing secretary. From the late 1970s, Niyogi was at the centre of a real outburst of strikes and social unrest that spread all over the industrial belt of Chhattisgarh, under the umbrella of the CMM (ibid.: 194–196).

The struggle of the contract labourers for recognition and regularization remained a core issue of CMM. In 1989, the focus of the mobilizations shifted to the ACC Cement plant of Jamul (Parry 2009: 198–199), where the *Pragatisheel Cement Shramik Sangh* (PCSS) was founded as a CMM-affiliated union, focussing on cement industry contract workers. At the same time, first attempts to unionize contract workers also in Rawan at Modi Cement (now Ambuja) were successfully suppressed by the management (Chakravarthi/Gera 2011: 17). In Jamul, the agitation started off with

40 Broadly, we can distinguish three currents of "grassroots" movements: The *adivasi, dalit* and other movements against ethnic discrimination, the Marxist and the Gandhian movements. Actual organizations at any level often contain elements from all these currents, but may put emphasis on one of them. For a general picture on social movements in India, see Omvedt (1993).

a wild-cat strike at ACC's manual coal-unloading department, after the management had decided to reduce the number of workers. In that department, many people from the second generation of the original landowners of the industrial estate were employed. The strike thus came as an outbreak of "years-old resentment" (Nair 2009: 190–91). The social unrest spread fast and production in large parts of the industrial belt of Bhilai came to a standstill. The industrialists became increasingly nervous, and repression got violent. In 1991, Shankar Guha Niyogi was murdered by hired killers. In 1992, the police shot at CMM activists who blocked a railway line and killed 17 workers (Chakravarthi/Gera 2011: 17; Nair 2009: 193). This violent culmination marks the end of the most active phase of the Bhilai agitation.[41] Following this, the Government set up an enquiry committee, which came up with a report that acknowledged problems in the contract labour system and recommended that the disputes should be referred to the labour courts.

Methodology of the CMM

In reaction to the massive repression and the dismissal of many of its members from the factories, CMM and PCSS put more effort in building up its own institutions—including schools, cooperatives and a hospital for the workers (Parry 2009: 197). Also, "the union has increasingly become bogged down in lawsuits" (ibid.: 200). A team of lawyers is following all kind of litigations, with the case of the contract workers of ACC remaining the most prominent: In 2006, PCSS won a verdict of the industrial court, stating that 573 workers have to be regularized. The company appealed to the Chhattisgarh High Court, where a single judge confirmed the industrial court's ruling in March 2011 and also obliged the company to compensate about 300 workers who had been forced by the company to "voluntarily" quit their jobs. Again, the company appealed, and the case is now pending before the division bench of the Chhattisgarh High Court.[42]

The CMM systematically accompanies legal action with protests and direct action, such as a blockade of all the gates of the ACC factory and colony in 1995 (Nair 2009: 194), and another one in 2006 after the industrial court verdict: "to force the ACC management to obey the order, the CMM workers blocked the factory gate for eighteen hours and performed picketing. There was *lathi* fight with the 'goons' of the industrialists in which CMM activists, including women, were injured" (Nair 2009: 194). Apart from such spectacular actions, routine vigils have been maintained at the factory gates until today.

Manjusha Nair describes the CMM as using a "mixed repertoire" that corresponds to its hybrid character, acting on the one hand "as a radical trade union against the industrialists", and as "a social movement in relation to the state" on the other (Nair 2009: 180). The mix consists in a fine-tuned interplay of disruptive and non-disruptive actions, ranging from wildcat strikes and blockades to authorized mass celebrations, vigils and marches, and court actions. To force industrialists to obey the court orders, the harsher forms are *gherao*[43], a "harassment whereby workers detain their employers or managers on the premises, refusing to let them depart" (ibid.) and *bandh*, a total shutdown of the area. A softer form is the *dharna*, a vigil that can include hunger strike.

Nair observed the use of a "fierce rhetoric" against the state, but cautious actions and extremely disciplined demonstrations, which never resorted to any violence. The movement was not directed

41 A collection of documents is available from: "Labour Struggle in Bhilai 1990–2004 (Chhattisgarh Mukti Morcha)" <www.indialabourarchives.org> (accessed on 13.8.2013).

42 After that, it would go to the Supreme Court of India. If the company insists in going the full way, it will take years and many of the potential beneficiaries will be too old to take back their jobs (Chakravarthi/Gera 2011: 17; personal communication S. Gera, Sept. 2013).

43 The *gherao* is an intuitive form of collective action, widespread and popular in India. Parry describes how a factory owner is encircled by a crowd to demand compensation for the death of a youth in a work accident. He is effectively hindered from moving until he promises to pay (2008: 245).

Fig. 4.: A meeting of PCSS activists in Bhilai. Photo: Karin Scheidegger.

against the state, and not against any particular industrialist, but against the capitalist class (which had assistance from certain state authorities). "The repertoire that the CMM used against the state expressed this ambivalence: cautious when over-stepping the boundaries of what was "legal" and while resorting to protests that are "illegal", challenging but yet law abiding, and criticizing while co-operating with the state's attempts at reconciliation" (ibid.: 201). In all this, the CMM successfully takes up sentiments of regional identity and autonomy, in both urban and rural settings. Parry even considers the CMM practice of commemorating the "martyrs" of the movement, foremost its founder Niyogi, as linking up with the ancient notions of "sacrifice" (2008: 250). He also points out that the dependence on strong leader figures, typical for social movements in India generally, makes them vulnerable to criminalization, political murder and (allegations of) corruption (2009: 202).

In an interview with Sudha Bharadwaj,[44] a senior lawyer and activist of CMM and PCSS, she says she is reflecting on their fight within the context of increasing land right struggles and the civil war between the Maoist guerrillas and the state armed forces: "the Maoists have a strategy, they believe in the overthrow of that system. But democrats like us, we are confused." She clearly takes a "system change" approach, without believing that elections might change a lot. "We're not against elections and use different methods of contesting or opposing a particular candidate or supporting a particular candidate depending on what strengthens our organization." The CMM

44 Justin Podur, April 2013: "The Bastar Land Grab: An Interview with Sudha Bharadwaj". <kafila.org> (accessed on 13.8.2013).

is consistently trying to build alliances with other mass movements, e.g., under the banner of *Chhattisgarh Bachao Andolan* ("Movement to Save Chhattisgarh"), hoping to "have critical mass one day." Continuously under threat of being targeted by the government as "Maoists" under anti-terror-laws, she concludes: "Still, we keep trying to work legally, democratically, and do mass mobilization, express solidarities, try to widen the circle."

The Movement at Ambuja Rawan

Since the construction of the cement plant at Rawan in the 1980s, a large part of Rawan's adult population has either worked or is currently working as contract labourers for the plant and the mines. In many cases, they belong to the same peasant families who had lost (part of) their land to the company, or who had been working on these lands as agricultural labourers. This inevitably links the struggle of Rawan's farming community with that of the contract workers, as it is told by Bhagwati Sahu. He had led a local farmers' organization named *Udyog Prabhavit Kisan Sangh* (UPKS; i.e., "Organization of Farmers Affected by Industries"), which gave voice to concerns about land encroachment and water shortage. "In 2008, some local panchayat leaders tried to take up the cause of farmers and workers with the Ambuja management, but they either got corrupted or could not stand the pressure from the company."[45] Part of this first attempt was a high-level dialogue meeting with Ambuja, ACC and Holcim, facilitated by the NGO "Tribal Welfare Society" (TWC) and Holcim in July 2008,[46] which according to PCSS did not bring much results (2012: 9).

In 2010, Sahu was elected as a member of the local *panchayat*, and again took up the issue of alleged encroachments by the company. Around the same time, contract workers at the Ambuja plant were also trying to organize. The farmers of UPKS thus decided to join the PCSS cement workers' union, to link up both struggles. Sahu made contact with some experienced CMM activists, and the PCSS stepped in to support them with a series of written submissions to the management and administrative and political authorities. This was accompanied by a series of protests, pickets and strikes. Besides, there were many spontaneous actions in *gherao* style, e.g., for compensation after workplace accidents.[47] In just a few months, the movement grew to 400 to 500 members in the communities around the Rawan plant and the cooperation brought some quick results, as Sahu notes: "We succeeded in winning festival bonus for the contract workers for the first time in 30 years".[48] Also, the company agreed to pay minimum wages and stopped deducting the cost for personal safety gear from the worker's salaries (Chakravarthi/Gera 2011: 17f.).

However, by 2012, the movement had come to a standstill after several leaders ran into legal troubles and Sahu faced imprisonment (see section 2.3). In the meanwhile, PCSS was pushing to make the situation known on the international level. As the management of Ambuja plant refused to enter into negotiations, claiming that PCSS did not represent any workers in Rawan, the union's strategy was to directly address the top level, the corporate headquarters of Holcim in Zurich, Switzerland.

The Conflict reaches Switzerland

The local agitation first came to the attention of an international public in 2008, when the NGO Tribal Welfare Society nominated Holcim for the "Public Eye Awards". This shaming award is given yearly on the occasion of the World Economic Forum in Davos, Switzerland, and receives

45 Interview with Bhagwati Sahu; Asgar Qadri, Rawan, July 2012.

46 TWC organized 8 speakers: trade unionists and farmers from both ACC Jamul and Ambuja Rawan. Holcim was represented through Peter Gysel, deputy head of Corporate Communications. See PCSS 2012: Annex No. 27.

47 Personal communication S. Gera, Sept. 2013; see also PUDR 2012.

48 Interview with Bhagwati Sahu; Asgar Qadri, Rawan, July 2012.

considerable media coverage.[49] About two years later, PCSS approached the ICEM (International Federation of Chemical, Energy, Mine and General Workers' Unions) and BWI (Building and Woodworkers International) for support. In June 2011, they visited the ACC plant in Jamul and came up with a report, which upheld PCSS' main allegations and recommended bringing the case to the OECD (BWI/ICEM 2011). Following this, the issue reached the responsible functionaries for Holcim at the major Swiss union UNIA (a member of both ICEM and BWI). UNIA is also a supporter of the two main groups which later launched the "Holcim campaign": Solifonds, a small "solidarity fund for the liberation struggles in the Global South", and MultiWatch, a coalition of different NGOs, solidarity groups, unions and organizations of the global justice movement.[50] MultiWatch describes its aim as "to observe and make public human rights violations of Switzerland-based corporations."[51] A major challenge in this endeavour is the question on which one to focus, given that Switzerland has an impressive concentration of such corporations. In this case, the decision was nevertheless almost unanimous: news on bad practices had flown in also from other countries, including Argentina, Colombia, Guatemala, and there were some "open bills" for Holcim's crucial role in Apartheid South Africa. Another important reason was the fact that Holcim was about to celebrate its centenary jubilee in 2012, making it a perfect target for a public shaming campaign.

The campaign that MultiWatch set up circled around a "birthday present": a small, collectively authored book which described different cases, including two chapters on Chhattisgarh (MultiWatch 2012). This was accompanied by a "manifesto", signed by 100 prominent first signatories and then turned into an online petition, which basically demands that Holcim complies with its CSR claims.[52] The Campaign and the book were launched at Holcim's general assembly on 17 April 2012. Two dozens of Swiss activists distributed flyers to the shareholders—hundreds of old-aged Swiss—who arrived at the huge event hall in Zurich. Shalini Gera of PCSS, who had been invited by MultiWatch, confronted the Holcim management during the question round inside the assembly. This action caused considerable anger among shareholders and the management, but did not receive much media coverage. The actual media coup came later with the polemics that evolved around the Holcim-sponsored photo exhibition at the Museum of Arts in Bern, described in the next section. Starting off with a single person, who staged a scandal at the opening event, a debate on the independence of the public art institution evolved. MultiWatch took up the thread with a small action: twelve people, carrying photos that Gera had brought from Chhattisgarh, tried to hand over a letter to the director of the museum, demanding that "the other side of the picture" should be shown as well. Also here, the reaction was far from friendly and the group was quickly kicked out from the entry hall under threats of calling the police.[53]

By then, PCSS had already compiled a detailed complaint accusing Holcim of violations against the OECD guidelines for multinational corporations. The complaint was sent in January 2012 to the Swiss OECD contact point, located at the State Secretariat for Economic Affairs (SECO). The SECO applies a very rigid interpretation of the complaint procedures and demands absolute abstention from any public information by the parties in an ongoing dispute. Even the information that a complaint was accepted cannot be disclosed. PCSS decided to comply with the strict

49 e.g. Ryser, Hansjörg: "Nachhaltigkeit: Holcim top und doch am Pranger." *Bilanz*, 25.01.2008.

50 MultiWatch functions in campaign-centred working groups, bringing together activists and staff from different organizations, coordinated by one part-time employee. The editing organization of this book, infoe CH, participates in the coalition on an informal base.

51 MultiWatch: "Über uns". <www.multiwatch.ch> (accessed on 7.3.2013).

52 MultiWatch: "Das Holcim Manifest." <www.multiwatch.ch> (accessed on 7.3.2013).

53 Personal observation T. Niederberger, April 2012; see also: Der Bund: "Protest gegen Holcim-Ausstellung im Kunstmuseum." 16.4.2012. <www.derbund.ch> (accessed on 7.3.2013).

Fig. 5.: Pictures of the solidarity actions in Switzerland, displayed at the PCSS office in Bhilai. Photo: Karin Scheidegger.

regulations, in order not to give any pretext for the Holcim management to abstain from attending a possible dialogue meeting. This also implied that the public campaign in Switzerland became rather quiet.[54] For the time being, the case is "on hold", but the links that were created will certainly continue to have their effects in the future.

2.3 Corporate Strategies

Holcim was founded in 1912 in the Swiss village of Holderbank, as a factory producing cement and construction material. The company rapidly began to seek access to markets outside the small Swiss home market—first in neighbouring European countries, but already in the 1920s it had set foot in Egypt, Lebanon and South Africa. After World War II, the expansion continued in North and South America, and in the 1970s to the Asia-Pacific region.[55]

Today, Holcim is the world's leading and globally most diversified cement producer (alongside the French Lafarge) with a turnover of 21.6 billion Swiss francs in 2010. Between 2002 and 2007, the turnover has doubled and the number of workers has grown from 50,000 to 90,000. This aggressive expansion was especially pushed in the "emerging markets" where cement consumption grows rapidly—most notably in China and India. In India, Holcim controls 25% of the market

54 Personal communication, Yvonne Zimmermann of Solifonds, Sept. 2013.
55 Holcim.com: "History" <www.holcim.com/about-us/company-profile/history> (accessed on 7.3.2013).

through Ambuja and ACC Limited (MultiWatch 2012: 66–76). Interestingly, the move to enter India happened just at the time when Holcim sold its major position in South Africa, where it had made good profits during the Apartheid regime. The proceeds from this sale were just about equal to the amount needed to finance the new investment in India (ibid.: 95).

Holcim is not one of those synthetic corporations that float somewhere in the financial space and seem to have lost all contact to real live people. Its single largest shareholder, with around 20%, is Thomas Schmidheiny, grandson of the company's founder Ernst Schmidheiny. Valued at 4.5 billion Swiss Francs, this makes him one of the top 30 of the "richest Swiss". After many years, first as CEO and later as chairman of the board, he still is a member of the board, but has taken a more quiet role.[56] The Schmidheinys are part of the Swiss bourgeoisie, fund various cultural foundations and some of their kids go to public schools. It is thus no surprise that they care for their image as good citizens, which is intrinsically linked to the image of Holcim.

Creating an Image

Holcim presents itself as a pioneer in sustainability for the cement sector. The company is a member of the World Business Council for Sustainable Development (WBCSD) since 1999 and a founding member of the Cement Sustainability Initiative. It forms part of the UN Global Compact and the Corporate Support Group of the International Committee of the Red Cross. Since 2007, it has a high-level partnership with the International Union for Conservation of Nature (IUCN), and so forth.[57] The image as pioneer, however, must be attributed to Thomas' brother, Stephan Schmidheiny, who inherited the other part of the family empire: the asbestos cement company Eternit, famous for the thousands of cancer victims among its workers. Apart from dealing with this unlucky heritage, Stephan had played a crucial role in the setting up of the WBCSD during the UN Summit on Sustainable Development in Rio de Janeiro in 1992—a move that critics have described as a (successful) strategy to derail attempts to introduce legally binding rules for transnational corporations at the UN-level, by pushing for "voluntary initiatives" instead (Moody 2007: 116–117).

In 2012, Holcim assigned the PR-agency Krieg Schlupp Bürge (KSB) to create a concept for its centennial celebration. The agency engaged the famous photographers Hiepler/Brunier and Marco Grob, who usually does iconic portraits of Hollywood stars and Washington politicians, to travel to Holcim plants around the world and collect images of workers and buildings. The resulting photo book went as a gift to each one of the 80,000 regular employees.[58] The culmination point was the already mentioned exhibition "Industrious" at the Museum of Arts in Bern. In an interview, the responsible art director at KSB explained that their task was to find stories of workers, "who graphically demonstrate the heart of the corporation". The image that the project should create was that "Holcim shows esteem and respect for its workers."[59] The museum had received an important donation to show the Holcim exposition. Nevertheless, its director put down critics saying that "nobody would claim that these pictures are publicity for Holcim".[60]

When MultiWatch started its campaign at the Holcim general assembly in April 2012, the top management's first reaction was denial. CEO Bernard Fontana threatened to lay off the contract workers ("we do not need them"), and chairman of the board Rolf Soiron said that the "little red book" published by MultiWatch was full of "half-truths and incorrect factual presentations".[61]

56 BILANZ: "Die 300 Reichsten 2012". <www.bilanz.ch/die-300-reichsten-2012> (accessed on 7.3.2013).

57 Holcim.com: "Sustainable Development"<www.holcim.com/sustainable-development> (accessed on 7.3.2013).

58 Centennial.holcim.com: "Images". <centennial.holcim.com/images> (accessed on 7.3.2013).

59 Persoenlich.com: "Mit Holcim im Kunstmuseum Bern". <www.persoenlich.com> (accessed on 7.3.2013).

60 Di Falco, Daniel 2012: "Der Gaul riecht aus dem Maul". *Der Bund*, 24.4.2012 (own translation).

61 Personal Communication Yvonne Zimmermann of Solifonds, Sept. 2013. See also "Minutes of the 100th Ordinary

Later, in an open letter to all the first signatories, the allegations were denied point by point. After the management had denied an invitation by PCSS to talk just days earlier, they now offered a dialogue with UNIA and MultiWatch (but still excluding PCSS).[62] The angry reactions of the Holcim management (and of the director of the Arts Museum) suggest that they really wanted to believe the image they created. But finally, this means facing the difficulties of being consistent with one's own standards: "Our business has local roots and is geared to the conditions and needs of each specific market where we operate; yet, we have Group-wide standards and systematic benchmarking that allow us to enhance our efficiency. Therefore, our success lies in striking a balance between local responsibility and global leadership."[63]

The CSR-Policies of Ambuja and ACC

Holcim's image as a pioneer for CSR is well represented by its Indian subsidiaries. Both Ambuja and ACC portray themselves as concerned about the environment and the people affected by their mines and factories. ACC, for example, acknowledges "[...] its abiding concern for preservation of the ecological balance and safeguarding the health of the community".[64] However, the sustainability report of ACC lacks any details about the ongoing expansion works at Jamul, nor does it address any conflict there, but generally claims that "[t]here is no identified operation in which the right to exercise freedom of association/collective bargaining is at significant risk" (ACC 2011: 58). On its website, Ambuja presents itself in a kind of nation-building mission: "The cement industry is literally the building block of a nation. A country's cement output is considered one of the primary indicators of its level of economic development. As one of India's major cement companies, Ambuja Cements plays a key role in India's development and its blueprint for the future." The sustainability report of Ambuja boasts a lot about the positive impacts of the company's CSR policy, with 2.58% of net profit spent on community development projects (ACL 2011: 11; 43). The report is compliant to the Global Reporting Initiative (GRI) criteria, which ensures that conflictive issues should be addressed. However, if we look for hints about the conflicts in Chhattisgarh, we do not find much. The report claims that in 2011 "no cases of human rights violation" were recorded (ibid.: 44). The total workforce is indicated as 5591, of which about a third is covered by collective bargaining. The number of contract workers is not indicated, but it is stated that "wage payment to its contract employees [was] in accordance to the minimum wage rules" (at 11,418 rupees, more than double the statutory minimum of Rs. 4,940 per month) (ibid.: 83–85). In the report of the previous year, the numbers given were "3762 full-time employees and 2615 sub-contracted" (ACL 2010: 64).[65] Further statements include that Ambuja promoted "equality and diversity amongst employees and there were no incidents of discrimination in the reporting period"; that Ambuja "encourages collective bargaining to maintain harmonious industrial relations"; that Ambuja "adheres to the law while engaging labour. [...] The labourers engaged by the contractor are governed by the Contract Labour (R&A) Act" (ACL 2011: 85); there were "no strikes or lockouts at our mines" (ibid.: 71). The report also claims that there are no operations located "adjacent to Indigenous Peoples' territories", and that "a consultative approach to address concerns of local communities, if any, involving all the stakeholders" was followed, whereby "no significant disputes related to resettlement" appeared (ibid: 71).

Annual General Meeting of Holcim", April 17, 2012: <www.holcim.com> (accessed on 7.3.2013).

62 Personal communication Yvonne Zimmermann of Solifonds, Sept. 2013.

63 Holcim.com: "Local Management — Global Standards". <www.holcim.com/about-us> (accessed on 7.3.2013).

64 ACC Limited: "Corporate Social Responsibility". <www.acclimited.com> (accessed on 21.01.2012).

65 The second number is obviously too low, seen that PCSS alone represents around 1000 contract workers in a single plant in Jamul. S. Gera suspects that the number refers to directly employed "casual workers", and that the actual contract workers are omitted. Also the indicated (high) salary points to this conclusion (personal communication, Sept. 2013).

Ambuja has its own foundation, the Ambuja Cement Foundation (ACF), which works as its "CSR wing [...] committed to engaging rural communities in and around Ambuja Cement's manufacturing locations" (ACL 2011: 14). ACF presents itself as just another NGO with programs in the field of education, health, women's empowerment, and improvement of agriculture. The largest share, almost a third of its USD 10 million budget, is dedicated to water resource management (ACF 2011/12: 83). The focus on water issues can at least be taken as an indirect admission that the problems denounced by local farmers are not without merit. That ACF is simultaneously an independent foundation and part of Ambuja's corporate strategy poses many questions that should be studied at the local level: do the beneficiaries of ACF programs make a difference between the foundation and the company? What role does the foundation's ground staff play in case of conflicts, e.g., between farmers' and the cement plant's need for water? The ACF report remains silent about such questions; it is basically an "everything fine" paper, devoid of any conflicts and contradictions.

Divide-and-Rule and Criminalization

To appreciate the "back-stage" picture behind the CSR claims of Holcim and its Indian subsidiaries, we have to again look at its unresolved labour conflict in Chhattisgarh. Parry gives us a good general understanding about the divisions between regular staff and contract workers in the industrial estates of Bhilai, including those of ACC, and how they are maintained in a joint effort between industrialists, major trade unions and governmental policies. Generally, he estimates that in the private sector, only one-third of the workforce is directly employed, the rest being recruited through contractors or middlemen, which "[...] shields the company against the requirements of the law—which are needless to say never met" (2009: 183). This is supported by the very weak administration of the Labour Commissioner, a "toothless bureaucracy". Parry describes a divide-and-rule strategy whereby employers privilege certain unions and marginalize others: "[they] ensure that the majority of their workers belong to the most compliant union available (generally one that they sponsor themselves). They then refuse to talk to anybody else" (2009: 186). The degree of corruption and the separation of the privileged union leaders from the common workers lead to a situation where most describe them as "thieves". The large unions INTUC, AITUC and CITU almost exclusively represent the minority of formally employed workers, leaving the contract workers to the CMM, as described in section 1.5.

When the CMM started its campaign at the ACC plant in Jamul with a huge drive in 1989, "employers retaliated with intimidation and lock-outs; there were murders, assaults, and threats of rape" (2009: 198). According to Parry, the management summarily dismissed around 4000 CMM-affiliated contract workers and replaced them by others, who were recruited largely through the compliant AITUC union. This caused outbreaks of violence between the two camps of workers, "with the police and district administration extending a fraternal hand to the combination of Communist union and their capitalist sponsors" (2009: 199). Frequently, in these divisions, the lines are along caste and origin, more than class—with the CMM representing mostly local Chhattisgarhis, and other unions bringing in migrant labour, sometimes from the respective union leaders' state of origin.[66]

In linking the interests of industrialists and the state, caste membership seems to play a role. Manjusha Nair states that "[...] workers used to point out that the industrialists and the members of the provincial state cabinet had the same last names (suggesting that they belonged to the same

66 S. Gera insists that the CMM is using a constructivist, class-based notion of identity, whereby "Chhattisgarhi" is "someone who sweats for Chhattisgarh". This is illustrated by many CMM members (and leaders) with origins in other states, esp. Bihar (personal communication, Sept. 2013).

castes) and hence had closer relationships" (2009: 201). Her account of the crackdown against the CMM is confirmed by Parry: "ACC management retaliated with violence" (2009: 191), mobilizing support from the local administration to suppress the movement, making use of "the rhetoric of national interests and employment generation to protect their own interests" (2009: 199). Thereby, the AITUC union[67] played an important role, accusing the CMM of "anti-national activities" and links with the *naxalites* (Nair 2009: 193; see also Parry 2009: 202). The murder of CMM's leader Niyogi in 1991, the police killing 17 people one year later, alongside curfews, closure of the union office and numerous arrests (Nair 2009: 193), mark the culmination of violence. Since then, the management of ACC (and other companies) has tried to avoid any dialogue with the CMM-affiliated union PCSS, claiming that PCSS did not represent any of its workers (which is not completely wrong, given that many contract workers, known to be PCSS members, had been fired—in Ambuja Rawan, this is true for all the PCSS members). Still, occasional negotiations happen, when under pressure. The journalist Sudeep Chakravarti reports an accidental conversation with a human resource manager of ACC on the flight to Raipur in 2006. He was on the way to negotiate in a strike by contract workers in Jamul "[SC:] What does he [the union leader] want – Permanent positions for contract labour – Will you do it – No. We'll [...] make some permanent, but we can't afford to do it with all. It's the contractor's responsibility, too... it's also the responsibility of the state. [...]" (2009: 41–42). With regard to the court verdicts of 2006 and 2011, forcing ACC to reinstate and compensate PCSS-affiliated contract workers who were kicked out in 1991 (see section 2.2), the strategy of the company is simply to appeal to the higher courts, knowing that it will take many years to reach a final verdict.

The pattern described for ACC Jamul resembles what Chakravarthi and Gera write about the events at Rawan: after local farmers and workers started voicing demands, the management gave in to some of them and agreed to pay minimum wages to contract workers. Other demands, like proper documentation and inclusion in health insurance coverage, were turned down. Then, the management filed cases against 20 activists, accusing them of disrupting "industrial peace", and suspended 52 workers in 2010. It refused to talk to PCSS, claiming that the union does not have any members working for Ambuja, and ignored orders of the Assistant Labour Commissioner to end the lock-out, which he found to be illegal (Chakravarthi/Gera 2011: 18).

In March 2011, the local union leader Bhagwati Sahu, who had been a key person in building up pressure on the Ambuja management, was arrested after two of Ambuja's security guards had accused him of an assault and theft of a cell phone and some cash. Even though the allegation was baseless, he was released from prison only after thirteen months. Seven other PCSS members who were accused in the same case—including the senior union organizer Lakhan Sahu—were still hiding from the police by the end of 2013. The imprisonment of Bhagwati Sahu spread fear among other union members. According to union member Sunil, "after Sahu was sent to jail, the union lost its focus, the entire movement got derailed. The company succeeded in breaking the morale of the union."[68] Nevertheless, local activists acknowledge that after the international campaign, Ambuja's attitude toward the contract workers has become somewhat more careful.

The image that Holcim created for itself takes a new turn with the visit of Swiss photographer Karin Scheidegger to Rawan.[69] She had heard about the controversy over the photo exhibition at the Museum of Arts in Bern, and took the opportunity of a stay in India to take her own pictures

67 Which union is allied with the management (also known as "yellow union") is subject to change: Currently, at ACC Jamul INTUC is "yellow", but in Ambuja Rawan INTUC is collaborating with PCSS and AITUC is "yellow" (personal communication S. Gera, Sept. 2013).

68 Interview by Asgar Qadri, Rawan, July 2012.

69 Personal communication Karin Scheidegger; see also Blumer, Florian: "Im Reich des Zementriesen", *Surprise Magazin*, Nr. 303, June/July 2013.

of Ambuja's workers.[70] In the early morning of 17 April 2013, she and her local assistant Ajay T.G. took a photo in front of the factory compound entry gate of a sign board saying "Ambuja Cement Heartily Welcomes You for a Safe & Pleasant Visit". After they were sent away by the security guards, they drove out of the village to take a picture of a field with the factory in the background. Again, security guards arrived and started to hassle and then beat up her assistant. After the attackers had left, they went to the Baloda Bazar police station to file a complaint against the guards—one of them turned out to be the same "notorious security guard" who had sent Bhagwati Sahu to jail. Local journalists arrived, and finally the complaint was accepted. However, Ambuja also filed a complaint against Ajay T.G. for "trespassing" and "obscenities". Later, she learns that the husband of a local woman leader who had assisted her at the police station lost his (contract) job at Ambuja. The security guard in question is reported to have "resigned voluntarily" shortly after the incident.

Confronted by Scheidegger about the incident, the Holcim management in Switzerland officially upheld the version put forward by the Rawan security guard. Still, informal exchanges indicate that the incident might have had some positive effects on the local situation and that the local management and security guards have become somewhat more careful.

3. Conclusion

Holcim and its Indian subsidiaries Ambuja and ACC use a highly sophisticated, proactive CSR strategy. Still, they systematically downplay the problems around contract labour and land acquisition. Backstage, the anti-labour tactics used to suppress the Bhilai agitation in the 1990s continue, including divide-and-rule, threats and criminalization. The logic behind refusing to recognize the PCSS union could be quite simple: to negotiate with this union, which represents contract workers, is a clear admission that Holcim does employ many contract workers, which is legally problematic. The case fought by PCSS could set a precedent for all of Holcim's factories in India, and even for the cement sector as a whole. But PCSS' main demand remains recognition as a union of contract workers which has the right to collective negotiation, or to put it in CSR-language: It demands a "dialogue". For the workers and their families, this would mean recognition of their existence, as a first necessary step to be able to defend their rights and negotiate agreements with the powerful corporation.

A simple explanation for the "image vs. reality" contradiction could be that the CSR strategy of the corporate headquarters does not correspond to local branch managements' actions, because rapid growth has been given priority over integration of CSR standards after the takeover. Another is that CSR is just there as a media strategy to cover up the harsh reality. Probably, things are more complicated. It is well possible that the involved managers believe what they say. But, at a deeper level, under the guise of modernity, age-old discriminations against a "backward" rural population and its poorest sections, marked by the stigmas of caste society, are still much stronger than any well-meant CSR strategy—even more so, if by maintaining the discrimination one can profit from a seemingly endless source of very cheap labour.

Regarding the CSR programs in Rawan, it would be interesting to study exactly how far they actually function as additional benefits to Ambuja's permanent employees, rather than addressing the needs of the general population, including the contract workers and their families. A school, a playground for children and other installations that form part of the company's "colony" were allegedly built on village commons, which raises questions from that angle. Is the "colony" con-

70 A choice of Scheidegger's photos illustrates this chapter. Her personal website is <www.karinscheidegger.ch>.

tinuing what was called "colonization" earlier? Put differently, there are indications that the "decommoning", practised by the British colonial administration and independent India, now continues under private corporations. The state now takes a secondary, assisting role, which reflects the global power shift towards "government by corporations". The provision of apparently "public" services and development programs by a private company would thereby serve to disguise or supposedly compensate for the appropriation of common goods. Even if such programs sometimes actually reach the poor, they always function as a paternalist "good deed" in expectation of gratefulness, not as entitlements. And, as many philanthropic NGOs and development agencies have noticed before, just pouring money from the top does not guarantee that it is reaching the bottom.

Holcim is caught in the contradictions of the increasingly globalized flow of images: in Chhattisgarh, the "modern" cement company reproduces age-old discriminations against the "backward" world of the village and its "sub-humans", who are today relegated to the role of contract workers. At the international level, the company uses a sophisticated discourse on CSR standards, which definitely does not leave any space for the bone-breaking exploitation of "sub-humans". So far, Holcim's choice was to deal with this contradiction by simply ignoring their existence. But as networks of social movements are more and more capable of transmitting the human face of the exploited to large audiences, this strategy becomes difficult to uphold. Holcim's PR fiasco with the Bern photo exhibition illustrates this well: CSR, after all, is directed at a public. It needs public display, where it can be confronted with alternative representations ever more easily, as communication flows become denser.

This is not to say that "the Internet will solve it all". With more information accessible, it becomes even more important that "real people" make use of it. It is remarkable that the voices of farmers and workers of Rawan and Jamul have reached the global stage through "old-fashioned" trade union internationalism. This is an effect of the strategy of the local movement to link environmental and (workers') rights issues, rural and urban struggles, and to give equal recognition to the importance of fair salaries, clean water and right to land. There are certainly contradictions to this. To demand regular employment in a destructive industry and at the same time opposing its environmental impacts is a complicated stand. Also, in the fight for regular employment, only a few (the healthy, educated, well-behaved) can succeed. The demands that were voiced through PCSS are very pragmatic, alternative visions for "a world beyond cement" were not put forward. Nevertheless, in the age of climate change and multiple social and ecological crises, this case can teach a lot about effective alliance building. Especially, trade unions in post-industrial countries like Switzerland may take inspiration from it.

At the time of this writing, we are still waiting to see real improvements for the affected people in Chhattisgarh. In its own interest, Holcim should better not wait any longer, speculating on the eternal delay of the Indian judiciary. In the background lurks an armed conflict. As Nair (2009: 202) points out, the disaffection with the legal system and democratic activism plays into the hands of the Maoist guerrillas—and could bounce back badly at an industry which is very vulnerable to civil unrest. On the other hand, social movements like the CMM are an effective counterbalance against the drive to armed struggle, by opening the space of democratic debate, inclusion and negotiation for otherwise marginalized people. If they are suppressed, others might step in, proposing guns instead of words.

References

Babb, Lawrence Alan 1975: The Divine Hierarchy: Popular Hinduism in Central India. New York: Columbia University.

Ball, V. 1880 [reprint 1985]: Tribal and Peasant Life in Nineteenth Century India. New Delhi: Usha.

Bates, Crispin N. 1985. Regional Dependence and Rural Development in Central India: The Pivotal Role of Migrant Labour. *Modern Asia Studies* 19.

Bauman, Chad M. 2008: Christian Identity and Dalit Religion in Hindu India, 1868–1947. Michigan: Eerdmans.

Bhatt, SC, Gopal K. Bhargava (Eds.) 2005: Land and People of Indian States and Union Territories. Delhi: Kalpaz Pub.

Bhushan, Chandra and Monali Zeya Hazra (ed.) 2005: Concrete Facts: The Life Cycle of the Indian Cement Industry. New Delhi: Centre for Science and Environment.

Bhushan, Chandra and Monali Zeya Hazra 2008: Rich Lands, Poor People: The Socio-Environmental Challenges of Mining in India. New Delhi: Centre for Science and Environment.

Bijoy, C.R; Shankar Gopalakrishnan, Shomona Khanna 2010: India and the Rights of Indigenous Peoples. New Delhi: Asia Indigenous Peoples Pact (AIPP).

Breman, Jan 1999: The Study of Industrial Labour in Post-Colonial India – The Informal Sector: A Concluding Review. *Contributions to Indian Sociology* 33.

Chakravarthi, Indira and Shalini Gera 2011:. Commentary. Repression of Workers' Struggles in Chhattisgarh. *Economic & Political Weekly* 46(18): 16-19.

Chakravarti, Sudeep 2009: Red Sun — Travels in Naxalite Country (Revised Edition). New Delhi: Penguin.

Downing, Theodore E. 2002: Avoiding New Poverty: Mining-Induced Displacement and Resettlement. *Mining, Minerals and Sustainable Development* 58: 1-29.

Dube, Saurabh 1998: Untouchable Pasts: Religion, Identity, and Power among a Central Indian Community, 1780-1950. New York: State University of New York Press.

Flueckiger, Joyce Burkhalter 1996: Gender and Genre in the Folklore of Middle India. Ithaca: Cornell University Press.

Lutringer, Christine 2009: Acting on Institutions to Preserve Agricultural Biodiversity: The Syngenta Controversy in Chhattisgarh. In: Basile Elisabetta (ed.): *The Changing Identity of Rural India.* Anthem Press, India.

Mahajan, Malati 2000: Chhattisgarh. A Study in the Culture and Historical Geography. Delhi: Sharada Publishing House.

Marothia, Dinesh K. 2002: Institutional Arrangements for Participatory Irrigation Management: Initial Feedback from Central India. In: Brennan, D. (ed.): *Water Policy Reform: Lessons from Asia and Australia.* Proceedings of an International Workshop held in Bangkok, Thailand, 8–9 June 2001. ACIAR Proceedings No 106.

Moody, Roger 2007: Rocks and Hard Places: The Globalization of Mining. Zed Books.

MultiWatch 2012: *Zementierte Profite – verwässerte Nachhaltigkeit. 100 Jahre im Zementgeschäft – ein Blick auf den Schweizer Konzern Holcim.* Zürich: edition 8.

Murali, Janakaraj and Tamer Afifi 2012: "Where the Rain Falls" Project. Case Study India: Results from Janjgir-Champa District, Chhattisgarh State. Bonn: United Nations University, Institute for Environment and Human Security.

Nair, Manjusha 2009: Mixed Repertoire of an Indian Labor Movement, 1990-2006. *Journal of Historical Sociology* 22(2): 180-206.

Omvedt, Gail (1993). Reinventing Revolution: New Social Movements and the Socialist Tradition in India. London: East Gate Books.

Pandey, Balaji 1998: Displaced Development: Impact of Open Cast Mining on Women. New Dehli: Friedrich Ebert Stiftung.

Parry, Jonathan P. 1999a: Lords of Labour: Working and Shirking in Bhilai. In: J. P. Parry, J. Breman and K. Kapadia (eds): *The Worlds of Indian Industrial Labour.* Special Issue of Contributions to Indian Sociology 33 (1&2): 107-140.

Parry, Jonathan P. 1999b: Two Cheers for Reservation: The Satnamis and the Steel Plant. In: Ramachandra Guha and J. P. Parry (eds): *Institutions and Inequalities: Essays in Honour of André Béteille.* Oxford University Press: New Delhi.

Parry, Jonathan P. 2008: The Sacrifices of Modernity in a Soviet-built Steel Town in Central India. In: João de Pina-Cabral and Frances Pine (ed.): *On the Margins of Religion.* New York / Oxford: Berghahn.

Parry, Jonathan P. 2009: Sociological Marxism in Central India: Polanyi, Gramsci and the Case of the Unions. In: C. Hann and K. Hart (eds): *Market and Society — The Great Transformation Today.* New York: Cambridge University Press.

Putul, Alok Prakash 2007: No Man's Land. In: Khalshian, Rakesh et al.: *Caterpillar and the Mahua Flower. Tremors in India's Mining Field.* New Delhi: Panos South Asia.

Rizwana, Malik, Lyaqet 2011: Traditional Knowledge and Use in Paddy Cultivation in Raipur District, Chhattisgarh. *Indian Journal of Traditional Knowledge* 10(2): 384-385.

Russell, R.V. and Rai Bahadur Hira Lāl 1916: The Tribes and Castes of the Central Provinces of India. London: Macmillan and Co., Limited St. Martin's Street. [www.gutenberg.org/files/20583].

Sastri, A.S.R.A.S. 2009: Long Range Regional Climate Fluctuations/Changes and Their Impact on Agriculture — A Case Study for Chhattisgarh State in Central India. In: S. Panigrahy, S. Shankar Ray, JS Parihar (ed.): *ISPRS Workshop Impact of Climate Change on Agriculture.* December 17-18, 2009. Ahmedabad, India.

Sharma, Ramesh C. and Sheeba Choudhry 2011: From Rice Bowl to Industrial Grazing Lands: Compulsory Land Acquisition in Chhattisgarh. In: Sakarama Somayaji and Smrithi Talwar (ed.): *Development-Induced Displacement, Rehabilitation and Resettlement in India.* London: Routledge.

Singh, K.S. 1994: The Scheduled Tribes. New Delhi: Anthropological Survey of India / Oxford University Press.

Somayaji, Sakarama and Smrithi Talwar (ed.) 2011: Development-Induced Displacement, Rehabilitation and Resettlement in India. London: Routledge.

Sundar, Nandini 2007 (second edition): Subalterns and Sovereigns — An Anthropological History of Bastar (1854-2006). New Delhi: Oxford University Press.

Viswanath, P. K., R. Nair, P. K. Kaushik , U. Prakasam 2000: Acacia Nilotica Trees in Rice Fields: A Traditional Agroforestry System in Central India. *Agroforestry Systems* 50: 157–177, 2000. Kluwer Academic Publishers.

Reports

ACC Limited 2011: Sustainable Development Report 2011.

Ambuja Cement Limited (ACL) 2010: Sustainability Report 2010.

Ambuja Cement Limited (ACL) 2011: Sustainability... Our way of life — Sustainability Report 2011.

Ambuja Cement Foundation (ACF) 2011/12: Sustainability - Annual Report. <www.ambujacementfoundation.org>.

Nair, K.N. and Saumen Majumdar (ed.) 2006: Chhattisgarh Infrastructure Report 2006. <www.iipa.org.in>.

Asia Center for Human Rights 2006: The Adivasis of Chhattisgarh: Victims of Naxalite Movement and Salwa Judum Campaign. New Delhi. <www.achrweb.org>.

Bharadwaj, Sudha et al., 16.3.2011: Persecution of Contract Workers Union Pragatisheel Cement Shramik Sangh by the Multinational Corporation, Holcim in the Indian State of Chhattisgarh. Letter on behalf of CMM.

BWI and ICEM 2011: Mission Report about ACC (HOLCIM), Plant Jamul Town, Durg District, Chhattisgarh State, INDIA. 16th June 2011. <www.multiwatch.ch>.

Dhaatri Resource Centre for Women and Children (DRCWC) and Samata (Eds.) 2010: Contesting Women's Rights within the Political Economy of Mining in India.

Down To Earth 2010: Industrial Jungle. 1.-15. September, New Delhi, pp. 20-29.

Government of Chhattisgarh (GoCh) 2005: Human Development Report. CHiPS (Chhattisgarh Infotech and Biotech Promotion Society). <chhattisgarh.nic.in/book/hdr.pdf>

Government of Chhattisgarh, Commerce and Industries Department: Industrial Policy 2009–2014. <chhattisgarh.nic.in>.

Ministry of Mines India. 2011. Annual Report, pp. 9-49.

Pragatisheel Cement Shramik Sangh (PCSS) 2012: Violation of OECD Guidelines by Holcim Group's Companies in India.

Peoples Union for Democratic Rights (PUDR) 2012: Working Against Odds: Condition of Workers in the Cement Industry in Chattisgarh. Delhi. <www.pudr.org>.

Sohini Paul 2006: The Right to Information and Panchayati Raj Institutions: Chhattisgarh as a Case Study. Commonwealth Human Rights Initiative.

The Tampakan Copper-Gold Mine Project in Mindanao, Philippines

Irina Wenk and Loretta Scherler[1]

1. Situating the Case Study

This case study is concerned with the planned Tampakan Copper-Gold Mine Project by British-Swiss mining giant Glencore Xstrata and its Philippine partner company Saggitarius Mines. The mining project is in its exploratory phase since two decades with the beginning of actual operations being regularly postponed due to various conflicts and unresolved issues in the mine development area. The project site is primarily located in Tampakan, a municipality of South Cotabato Province, on the conflict-ridden southern Philippine island of Mindanao.

The beginning of actual mineral extraction has been postponed repeatedly due to a ban on open-pit mining issued by the provincial government of South Cotabato in 2010 in direct response to the proposed copper-gold project (Sarmiento 2012; Malaya 2013). This ban caused a delay in the issuance of the required Environmental Compliance Certificate (ECC) by the responsible government agency. Meanwhile, exploration activities continued and the mining company has been active in building up infrastructure and implementing considerable community development and social service programs. Were the Tampakan Copper-Gold Project finally approved and consented to by all stakeholders involved, it would become the fifth largest copper mine in the world in terms of production and the largest direct foreign investment in the Philippines (SMI/Xstrata 2011: 8).

Despite the persistence of the South Cotabato provincial ban on open-pit mining, the Department of Environment and Natural Resources' (DENR) Secretary Ramon Paje signed the ECC in February 2013. The ECC is bound to conditions that remain to be fulfilled, among others the attainment of the resident indigenous population's free, prior and informed consent (Reuters 2013; Hamm/Schax/Scheper 2013: 21). Even prior to the opening of the mine, the company's presence and activities are causing community displacement and environmental damage. The security situation has worsened with the advent of the proposed mining operation, and the deployment of both military and paramilitary forces in the area causes distress for the local population. In reaction, the indigenous people on-site, the Bla-an, are said to have declared a *pangayaw*, i.e., a tribal war of revenge, against the company. Mining in the Tampakan area is perceived by civil society groups and experts, the Philippine Catholic Church, various fractions of the local population and downstream communities as having severe social and environmental impacts. Approximately 870 households or 3,500 to 4,000 people—both indigenous Bla-an as well as Christian Filipino settlers living in the area—would need to be resettled once mineral extraction begins, and thousands more would be affected as the environment, especially water sources, would suffer (see AECOM/SMI 2011: 20 and Tomas/Celebrado 2008: 74).

1 A first version of this case study was originally drafted by Loretta Scherler and Michelle Pfister and then completely redrafted by Irina Wenk and Loretta Scherler.

Experts further warn of the high risk of mineral extraction in the area due to the fact that the Tampakan mine site sits on a number of faultlines and a strato-volcanic complex. The active volcano Mount Matutum is situated only twelve kilometres from the proposed mine site (Goodland/Wicks 2012: 7).

Current large-scale mining in the Philippines takes place under considerable tensions between two national laws: on the one hand, indigenous peoples' rights are supposedly protected under the Indigenous Peoples' Rights Act of 1997 better known as IPRA. The IPRA is a comprehensive law that includes significant provisions for collectively claiming title to Ancestral Domain and for indigenous groups to say "no" to outsiders' activities within such territories by way of withholding their Free, Prior and Informed Consent known as FPIC. On the other hand, the Mining Act of 1995 encourages large-scale foreign investment in the country's mining sector and, based on it, the government issues mining permits covering indigenous-owned or -claimed land.

The IPRA is a groundbreaking statute that provides for a wide range of rights, such as the right to Ancestral Domain, the right to self-governance or the right to cultural integrity. Based on the IPRA, organized indigenous groups can collectively obtain a so-called Certificate of Ancestral Domain Title known as CADT, which is a private, yet collectively held, land title that officially recognizes an indigenous group's right of possession and ownership over a delineated territory (Wenk 2012). With regard to projects within ancestral domain territories by non-indigenous external actors—and this holds whether an ancestral domain is officially titled or not—, the IPRA and its Implementing Rules and Regulations require the FPIC of all members of a claimant or title-holding group as officially recognized owners of the land. Within the designated mine site at Tampakan, there are at present three such collectively titled ancestral domains (CADTs) and one ancestral domain claim (CADC), all held by Bla-an communities. These ancestral domain areas cover roughly two-thirds of the proposed mining area (see Map 2). The Tampakan mining project, consequently, does not concern state-owned land granted to a mining company for mineral extraction but primarily indigenous-owned land that has been recognized as such by the government.

Challenging the hard fought gains of indigenous peoples' rights as chartered in the IPRA, is the Philippine Mining Act of 1995, a neoliberal scheme, which contains various incentives encouraging mining. Among these are a four-year income tax holiday, tax and duty-free capital equipment imports, value-added tax exemptions, and tax deductions where operations are posing losses. Most significantly, the Mining Act provides for two main types of production agreements: the Mineral Production Sharing Agreement (MPSA) and the Financial Technical Assistance Agreement (FTAA) (Holden 2005: 419). Both of these agreements are entered into with the Philippine Government, which grants mining rights or lease to companies under the Mining Act. The FTAA is larger in terms of area covered and in contrast to the MPSA, which provides for Filipino companies to run operations, the FTAA allows fully foreign-owned companies to do so and is therefore much criticized (see Dalupan 2000: 10, 64f., 67f.). SMI/Xstrata currently holds an FTAA for the Tampakan Copper-Gold Project covering 23,571 hectares across four provinces. In what follows, we discuss how the tension between the IPRA and the Mining Act plays out in practice and how the most important stakeholders deal with it.

Sources for this Case Study

For this case study, we drew on a range of literature and other sources, some of which were easy to find and access, such as the various news sources on the Tampakan mining case, while others were very difficult to find. Academic studies on the case are either rare or non-existent, with the exception—as far as we know—of Tomas/Celebrado (2008), on which we draw extensively. This, however, is an unedited and unpublished case study commissioned by three Mindanao universities and carried out by the Mindanawon Initiative for Cultural Dialogue. It is based on interviews and focus-group discussions with locals of the five main communities affected by the Tampakan mining project as well as with other stakeholders such as local governments, over a period of two months. A comprehensive research brief on Tampakan, with a particular focus on the Environmental Impact Assessment process was published by the Ateneo de Davao University Institute of Anthropology just after we had finished writing our study (Gatmaytan et al. 2013). Unfortunately, we were not able to include it or contact its authors as we drafted this case study. Most other sources concerning the local situation stem from civil society organizations involved (e.g., Christian Aid/PIPLinks 2004; Hamm/Schax/Scheper 2013), from experts on the effects of mining (Goodland/Wicks 2008 and 2012), or from the media. Goodland and Wicks (2012) provide a critical response to Xstrata/SMI's Environmental and Social Impact Assessment (ESIA), highlighting the serious risks involved in this mining endeavour. Robert Goodland is former senior environmental advisor to the World Bank and was the technical director of the Extractive Industry Review (EIR). Clive Wicks is the former head of the WWF UK International Programme covering Asia, Africa and Latin America and until recently, he was a Vice Chair of IUCN-CEESP. He does consultancies on conservation and development. Both have many years of experience on agriculture and environment and are members of the UK-based Working Group on Mining in the Philippines (WGMP) (Tampakan Forum 2011: 17). A great collection of news articles was accessed through the Mines and Communities (MAC) Website, which provides articles concerning the Tampakan project from various Philippine newspapers such as MindaNews, Bulatlat, Asian Correspondent, Sun Star and others. MindaNews, for example, is a very reliable news source based on a cooperative of independent and professional journalists. The online forum Indigenous Peoples Issues provided a further selection of articles, as did the Mindanao-based research institution Alternate Forum for Research in Mindanao (AFRIM). The websites of all NGOs mentioned in this study provided us with information and have been helpful (see a collection of links at the end of this chapter). The authors were also in personal e-mail contact with Clive Wicks, the Mindanao-based NGO Alyansa Tigil Mina (ATM), the Davao-based Undangon Ang Mina, and the Facebook-group "No to Mining in Tampakan, South Cotabato, Philippines".

We were also glad to have found a journalistic assessment of the situation in Tampakan (Romero 2008) and an overview of mining in the Philippines more generally (Landingin 2008) in a Philippine *NewsBreak* special issue on mining. Insights into the general problems of mining in the Philippines were also gained from the articles of Holden (2005) and Holden/Jacobson (2006). Holden is an environmental scientist/geographer and expert on mining in the Philippines. From the side of the company, aside from the business partners' websites, we primarily reviewed the Xstrata/SMI Sustainability Reports of 2010 and 2011 as well as its commissioned Environmental Impact Statement EIS, cited as AECOM/SMI (2011). For an assessment of the Mining Act in relation to the IPRA, we were able to draw on Dalupan (2000), who put forward a comprehensive and valuable legal study on the tensions between the Mining Act/the mining industry and the IPRA. Said study was supported by the Philippine Exporters Confederation in cooperation with the United States Agency for International Development through a project that supports improvement of the policy and regulatory environment in the Philippines.

Ethnographic material published on the Bla-an is very scarce. The earliest ethnographic account of some detail is by Cole (1913) in his "Wild Tribes of Davao District", with a small section on the "Bila-an". More recent publications considered for this study include those of Hyndman/Duhaylungsod (1998), who have mainly written about the T'Boli but also some on their neighbours, the Bla-an, the historian Rodil's account (1999) that was a commissioned work for the Danlag Bla-an ancestral domain claim in Tampakan, and an unpublished document by the Sal Naong Tribal Council in 1996 for their respective ancestral domain claim, of which Irina Wenk was able to secure a copy during former fieldwork in Mindanao. To date, there is no comprehensive monograph of any Bla-an people.

Throughout this study, we furthermore make reference to the yet unpublished dissertation of Irina (Wenk 2012). Her knowledge and writing is based on fieldwork in a province neighboring

South Cotabato and among people related to the Bla-an. The dissertation provides a lot of necessary background information and comprehensive analysis of some of the issues highlighted in this case study, such as, for example, the in-migration of settlers, the IPRA, as well as post-titling management of an ancestral domain in the face of outside parties. What this presented case study on mining ultimately lacks, however, is a view directly from the field. Much of the data presented here and the arguments developed need to be verified, elaborated on or challenged through thorough future ethnographic fieldwork.

1.1 History and Current Context

The Philippines is an archipelago of 7,107 islands with a population of 96.6 million. The first people to settle on some of these islands were the so-called Negrito, known in the Philippines as Agta, Ayta or Ata. In a later wave of settlement came Malayo-Polynesian groups who today constitute the majority population. Indigenous Peoples—an official legal category in the Philippines since 1997, comprising the Agta populations but mainly Malayan upland groups—are estimated to constitute around ten percent of the national population. Another eight percent is of Indian, Spanish, Arabic, Chinese and US-American origin. There are more than 150 languages spoken, with Tagalog (Filipino) and English as official languages and Cebuano and Ilokano as lingua franca in the south and north respectively. Ninety-one percent of the population is Christian (mainly Roman Catholic), five percent Muslim, and a remaining four percent adhere to Buddhism, Taoism, Hinduism and animistic beliefs (IWGIA 2012: 272; Reese/Werning 2006: 17f.).

The southern Philippines, comprising the Sulu Archipelago and Mindanao, has historically been influenced and ruled by Muslim groups with strong ties to neighbouring sultanates in Brunei, Ternate-Tidore and Makassar, in what today belongs to Indonesia and Malaysia. With Spanish colonization the spread of Islam northward came to a halt in the 17th century. The Muslim population of the southern Philippines is known as Moro, while those who did not adopt the Islamic faith are today referred to as Lumad or Indigenous Peoples. The latter, more interior groups on Mindanao were for centuries connected to the Islamic coastal powers through trade of forest products (see Rodil 2003; Majul 1999; Hayase 2007; Junker 2000).

Long Spanish colonization (1565–1898) affected Filipino culture and society profoundly. During the 20th century, historical events accelerated in a dramatic sequence: revolution, war against the United States of America (1899–1902), American colonization (1898–1946), Japanese occupation, severe World War Two devastation, independence, the imposition of martial law, the rise of "People Power" to overthrow the Marcos regime in 1986 and the search for a stable Republic since. Philippine populations have emerged from these varied encounters to engage in a quest for a unified national identity or, in some parts of the country, in separation, and they are caught between cultural influences from diverse indigenous origins, Muslim heritage, Spanish colonization, Americanization, and more recently modernization and globalization (Parkes 1999: 11).

Mindanao, the large southern island in focus here,[2] is today inhabited by what is referred to in the Philippines as "tri-peoples". The majority population in Mindanao consists of Christian Filipinos, who migrated there from other islands of the archipelago over the centuries but mainly so after World War II. This settler population is locally referred to as either Bisaya or Dumagat—the first identifier pointing to their principal place of origin, the Visayas region, and the language they speak, the second highlighting where they came from, seen from the perspective of those already

2 Mindanao is two and a half times the size of Switzerland and covers a land area of 104,530 km^2. The total Philippine land area is 299,764 km^2.

there (Dumagat, "those coming from the sea") (see Wenk 2012: 51). The descendants of all non-Islamized Mindanawons, combining a number of different ethno-linguistic groups across the island, are collectively referred to as Lumad, which means "native" or "grown from that place".[3] The Moro constitute Mindanao's third major population group. Their designation dates back to the Spanish colonizers who, upon their arrival in the 16th century, referred to Mindanao's coastal people as Moro, just as they had towards the "Moorish" Muslims from Northern Africa who had occupied Spain for nearly 800 years until 1492 (Rodil 2003: 3). Although a pejorative connotation has been associated with the term throughout the centuries, it has recently—with the emergence of a national consciousness and claims to nationhood (Bangsa Moro) among Mindanao's Muslim populations—become a designation of pride rather than shame.[4]

Mindanao and the Sulu archipelago was an inhospitable and strange land for the early Spanish colonizers and they never succeeded in conquering and subjugating the Moro or making their way inwards to establish control over the interior's inhabitants (Magdalena 1997). While the Moro remained hostile to giving up their sovereignty, the Lumad remained largely uncontacted. It was only with the entry of the Americans in the early 20th century that colonial control over Mindanao's populations was gradually established (see Wenk 2012: 55f.). From the 1930s onwards, roads were built into the interior and attempts made to convince the hill people living in a scattered manner to settle in confined villages, accessible for colonial administrators and tax collectors.[5] Along the Davao coast and its hinterlands, the establishment of such "tribal wards" was inextricably linked to the early development of a plantation economy under the control of US-American Governors (see Tiu 2005: 29). Facing labour shortages on their plantations, the colonial authorities induced the region's inhabitants—the Bagobo, Kulamanon, Bla-an, Tagacaolo, Mandaya and Mansaka—to become low-paid labourers on American-owned plantations through resettlement into centralized ward villages near the coast and the collection of an annual 10-peso *cedula* (head) tax. In cases of non-compliance with the tax, forced labour was imposed (Hayase 2007: 168ff.). Through these measures as well as the usurpation of their land for the expansion of hemp plantations, local people's traditional resource use system of swidden cultivation was seriously disrupted.

In the course of the 20th century, Mindanao became a typical agricultural frontier and was flooded by lowland Filipino settlers. It was in the late 1930 that the first Filipino president of the Commonwealth of the Philippines under U.S.-American colonial rule, President Manuel L. Quezon, coined Mindanao as "land of promise" to boost the island's settlement and agricultural development. An allegedly empty and fertile, yet "waste", land was to be cultivated and developed through settler-colonizers. Of the three major settlement sites opened up nationwide during the first half of the 20th century, two were located in Mindanao, namely the Koronadal and Alah Valleys of Cotabato. The Koronadal Valley, as a case in point, was thinly populated but inhabited and used by Muslim Maguindanawons and such ethno-linguistic groups as the Bla-an, Manobo and T'boli. The Koronadal and Alah Valleys received over 40,000 settlers within just five years between 1936 and 1941 (Crystal 1982: 96; Hayase 2007: 157f.; Wernstedt/Simkins 1965: 93).

The internal colonization of Mindanao increased drastically after World War II, when land-hungry settlers sought to restart their lives in Mindanao as agricultural pioneers. Post-war settlement of Mindanao was promoted as a state- and nation-building measure. The "Mindanao Problem", i.e., the continued resistance of the Moro against political domination from Christian

3 This pan-tribal term is linked to the emergence of the Philippines' indigenous peoples' rights movement and established itself during the 1980s.

4 For reasons elaborated elsewhere, only the non-Islamized groups of the Philippines have come to conceive of themselves as Indigenous Peoples (see Wenk 2012).

5 see Gatmaytan D.B. (1989) on colonial and post-colonial concentration policies.

Manila, was decisive for promoting mass-migration southward. From a government's perspective, to ensure that Mindanao remained part of the Philippine state, the Muslim population needed not only to be assimilated but a significantly larger, productive and government-loyal population created. Hence, both colonial and postcolonial governments placed Christian settlers and soldiers into the heart of Muslim territory in Cotabato and Maguindanao (Rodil 2003: 110; Magdalena 1997: 10). While the country's agricultural core areas in the Visayas and Central Luzon suffered from a growing landless population and agrarian unrest, Mindanao was to function as a safety valve, forced to absorb 1.25 million settlers between 1948 and 1960 alone. The provinces of Cotabato and Davao alone received sixty per cent of these settlers (Wernstedt/Simkins 1965: 83, 89ff.).[6]

During this process of internal colonization, Mindanao's original inhabitants were denied any legal rights to the lands they lived on and made use of through swidden farming as well as hunting and gathering. Mostly considered "abandoned" or "uncultivated", land was classified public, i.e., state-owned. The government could thus allocate such land and issue titles, despite it being inhabited and used by local people, to whomever it deemed eligible. Through such land acquisition processes, an institutional superiority of settlers' private rights over those of Mindanao's former inhabitants was established.[7] This situation only found redress at the national level with the passing of the Indigenous Peoples' Rights Act in 1997. Locally, however, ownership over such land continues to be contested between settlers and longer-term locals and this even more so in the presence of a mining company.

1.2 Geography and Ecosystem

The Philippines are located at the so-called circum-Pacific belt of fire, an area where the earth's tectonic plates collide and volcanic processes are most active. Volcanism and the convergence of tectonic plates resulted in the formation of important minerals throughout the islands, in particular non-ferrous metals such as copper, gold, lead, nickel, silver and zinc. Some of the copper and gold deposits are said to be particularly large (see Holden 2005: 418). Mining such deposits is attractive but is combined with great risks. Mount Matutum is a registered active strato-volcano found within only twelve kilometres from the proposed Tampakan mine site. Strato-volcanoes are filled with explosives and toxic gases of great destructive potential and experts consider it extremely risky to establish a mine site of such proportions in the vicinity of Mount Matutum. That said, the proposed mine site is also situated above a number of faultlines that cross each other, virtually under it there is an allegedly extinct strato-volcano complex, and it sits atop a shallow aquifer (Goodland/Wicks 2012: 1).

Designated by scientists as one of the world's top 20 "megadiversity" countries, the Philippines are rich in marine biodiversity. The archipelago boasts 500 of more than 800 known coral species, over 2,000 fish species and over 40 species of mangrove plants. Unfortunately, only five per cent of the country's coral reefs remain in excellent condition while mangroves and sea grasses have shrunk to less than one quarter of their original extent. Mining poses a threat to the marine environment due to its releases of mine wastes that result in the killing of fish, and coastal pollution. The Philippine Government seeks to conserve the country's biodiversity through the establishment of protected areas. Hence, approximately eight per cent of the country's land area is classified as Protected Areas and is to remain off-limits to mineral development. However,

6 See Wenk (2012) for an overview and discussion of settler colonization in Mindanao.

7 See Umehara (2004, 2009) for an analysis of this process in the Koronadal Valley.

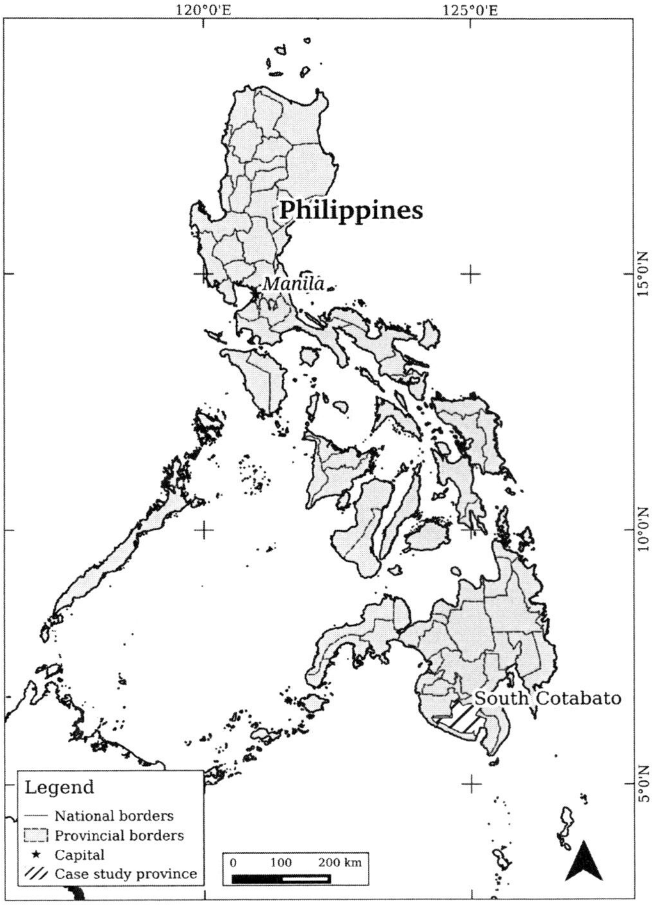

Map 1: The province of South Cotabato in the southern Philippine island of Mindanao.

according to the Philippine Biodiversity Priority Setting Program, only 41 percent of protected areas retain their original vegetation and the protected areas system fails to include some areas of high biodiversity (Miranda et al. 2003: 21).

The Glencore Xstrata copper-gold mine project site is located in the mountainous uplands of south central Mindanao, about sixty kilometres away from the city and harbour of General Santos. The mining area reaches across parts of two administrative regions, four provinces, four municipalities and nine *barangay* (village-type administrative units) (AECOM/SMI 2011: 39). The agricultural land in and around Tampakan is enhanced by Mount Matutum and the densely forested Matutum National Park that provides fertile soil and moisture all year round for agriculture down from the watersheds (Goodland/Wicks 2008: 106). According to a report produced by the Tampakan Forum (2011: 5), the mine site supports high plant diversity and thus high biodiversity value. Old growth forests mark the area, as well as a wildlife habitat, which is home to endangered species. The mine site accounts for a high percentage of wildlife that is endemic to Mindanao, which indicates the considerable environmental conservation value of the Tampakan project site.

Considering the hydrology and hydrogeology of the area, the Tampakan mine site straddles the headwaters of six catchments: the river Mal in the east and northeast, Manteo in the south, Altayan and Taplan in the west, Dalul in the north, and Mainit in the east. One part of the projected mine site is located within the Mal, Altayan and Taplan catchments. Water from these rivers is to be used for downstream irrigation of the project site (Tampakan Forum 2011: 27).

1.3 Local Population

The Tampakan mining area is inhabited by both local Bla-an and so-called Bisaya of settler origin. How each of these (non-homogenous) population groups is portrayed in the process of interaction with the mining company and how they each perceive themselves to be treated by the company derives from a complex inter-ethnic set up that can only be alluded to here.

The following table shows what the company calls "host communities" within the proposed mine site as well as those (indigenous) communities which lie outside the project site boundaries but are considered to be indirectly impacted by the proposed mining operation (AECOM/SMI 2011: 13f.).

Province	Municipality	Barangay	Directly affected host communities	Indirectly affected communities
South Cotabato	Tampakan	Danlag	Danleg	Danleg (Kawayan)
		Pula Bato	Fulo Bato	Fulo Bato (Gumamela)
		Tablu	S'bangken	S'bangken (Kolondatal)
Sarangani	Malungon	Malabod	Malabod	Malabod
		Bla-an	Bla-an	Bla-an
Davao del Sur	Kiblawan	Kimlawis	Kimlawis	Bongmal Kimlawis
		Tacub	Tacub	Tacub
		Bololsalo	Bololsalo	Bololsalo
Sultan Kudarat	Columbio	Datal Blao	----	Salnaong and Datal Saub

Table 1: "Host communities", according to AECOM/SMI 2011.

It is important to note that the mining area perimeter does not follow *barangay* (village units) boundaries but crosses them. Hence, portions of one *barangay* come to lie within, others without, the project area and, consequently, one community becomes "host" while its neighbour becomes "indirectly impacted". The single largest area included in the mine site is Barangay Tablu of Tampakan. For the others, only a relatively small portion of the *barangay* area lies within the projected mine site. The total population in the nine host *barangay* is 16,943, while those directly affected are said to be 870 households or 3,500 to 4,000 people. Of these, approximately 67 percent are Bla-an (AECOM/SMI 2011: 20). All of these 4,000 people will need to be resettled once extraction begins. Bla-an live in all of the nine host *barangay*, both within and without the project area, but form a majority in only five of them, namely in Kimlawis, Tacub and Bololsalo of the Municipality of Kiblawan, and in Tablu and Danlag of the Municipality of Tampakan. Within the project site, there are said to exist 582 Bla-an households, comprising 2,620 persons (AECOM/SMI 2011: 13). The rest are mostly Bisayan settlers and their descendants as well as smaller numbers of members of other indigenous groups and Maguindanawon, i.e., Moro people. The percentage of the Bla-an population in the host *barangay* as a whole ranges from 30 percent in Tampakan to 63 percent in Columbio to 84 percent in Malungon and Kiblawan areas.

The Bla-an are one of the largest indigenous groups in Mindanao and comprise approximately 240,000 people.[8] Their settlement area is located between the mountains west of the Davao Gulf and the Cotabato Cordillera. Prior to the advancement of settlers, Bla-an people occupied large portions of the lowland coastal area around the Sarangani Bay and the fertile Koronadal Valley (Kolon Nadal in Bla-an; *kolon* means plains) and Columbio (Kolon Bia-o). But, they retreated uphill and into their settlement areas of today in the mountainous interior, as did many other indigenous groups across the island, in the wake of the above described internal colonization process. That they formerly also settled in the lowland and coastal areas finds expression in how they continue to classify themselves into three main groupings: people of the mountains/higher regions, people of the sea and the Davao area, and people of the lowlands/plains (see Rodil 1999: 15f.; AECOM/SMI 2011: 3, 14; Hyndman/Duhaylungsod 1998: 87).[9]

According to Rodil (1999: 19), the Bla-an are organized into what is often referred to in the Mindanao context as "clans", called *saflanek* in Bla-an. A clan is composed of several extended families. The company's EIS reinforces this depiction of local organization into clans when it states, based on a recent census conducted for the EIS, that the 582 Bla-an households living within the projected mine site comprise 204 clans (AECOM/SMI 2011: 13f.). These, in turn, settle in so-called *banwu*, which is the people's definition of a community and the most important social unit defining a person's identity and origin. The concept of *banwu* includes not only people but their clearings and other land, rivers and creeks, hunting grounds and all other geographic or human features found within one community's territory. As such, it comes very close to how Ancestral Domain, as a recent land classification, is now officially defined in the Philippines (see IPRA, Ch. II, Sec. 3a; Wenk 2012: 8ff.).

Danlag, for example, is a mostly upland *barangay* and one of thirteen in the municipality of Tampakan. It lies both within and without the mining area boundaries, thus its population is directly affected, where it must be resettled and indirectly affected where it will bear other consequences of eventual mining operations. Danlag's Bla-an population is concentrated in the settlements of Purok Kawayan and Sitio Talfuk, with about 18 households each in the 1990s. Together, Kawayan and Talfuk form the Bong Banwu, the "big settlement" of Danlag but they each constitute an autonomous *banwu*, i.e., a community. Danlag means the act of searching for a clearing and used to be a communal hunting ground where the people from Pula Bato further south came to look for places to make their swiddens. It was later settled under one particular leader and his followers and made a *banwu* (Rodil 1999).[10] Each *banwu* has territorial boundaries known to the local people and it is an autonomous entity based on a community of close relatives (Rodil 1999: 18, 20, 22). Before settler colonization and various community organizing efforts that followed by government agencies and various companies, there was no "tribal" entity beyond the *banwu*.

8 See Ethnologue, Languages of the World at www.ethnologue.com (viewed Nov. 4, 2013). Data is based on the years 2000 and 2007 respectively.

9 See Umehara (2004) on the agricultural colonization of the Koronadal Valley and its impacts on the local population and the environment.

10 This leader died in the early 1970s, so the settlement/community of Danlag must have formed in the 1960s or even earlier.

As in most of non-Islamized indigenous Mindanao, each community had its local headman who was a respected and knowledgeable leader and an indispensable mediator in dispute settlements. Among the Bla-an, these leaders were, and are, referred to as Fulong, Bong Fulong, Bong Lebe or Datu.[11] Legitimacy to positions of power in local political organization, whether historically or recent, derives from tradition, custom, and the acknowledgement by the people of someone to be a leader. A leader must have proven wisdom and experience in dispute settlement, material wealth to come up with the necessary compensation payments to settle cases, a reputation of bravery, and vast social connections that allow him to acquire goods in times of need and to forge alliances.

In the ethnographic section of the EIS (AECOM/SMI 2011: 16), one finds an interesting statement on the present coexistence of a traditional leadership (*datu* or *fulong*) and a new formal leadership structure established, allegedly, under the IPRA. This new "formal" leadership operates within Bla-an ancestral domain areas and its leaders attend to dealings with the government through the relatively recent institutions of Tribal Councils and Chieftains. The "traditional leadership" in turn deals with the concerns "of the tribe"—such as marriage, dispute settlement, advising people on various issues, etc. Both forms of leadership are often composed of the same people, i.e., a *fulong* is also the modern "chieftain" of the Tribal Council and he may even sit in the Barangay Council, the government's smallest administrative unit to locally represent the State. In response to this depiction of parallel forms of leadership and local political representation as having only emerged with the IPRA, it must be highlighted that the formation of Tribal Councils across Mindanao is the consequence of various interventions by state agencies in the 1980s and 1990s. In the particular case of the Bla-an in and around Tampakan, such community (re)organization interventions were furthermore largely promoted and financed by foreign companies, i.e., Xstrata's predecessors in the area; a point we shall come back to later.[12]

1.4 Local Economy and Resource Use

Farming is the local population's main activity and source of income. This is supplemented by vending, occasional wage labour jobs or *barangay*-related work by some. People mostly grow subsistence crops, especially the Bla-an. For the various *barangay* as a whole, the major crops presently grown are corn, coconut, banana and tropical fruits (AECOM/SMI 2011: 6, 9, 10ff.). The Bla-an population's main subsistence practice is dry cultivation in swidden fields, referred to in Mindanao as *kaingin*, where they grow rice, corn and millet. This is according to the company's own recent assessment (AECOM/SMI 2011: 3). Further crops are sugarcane, banana, yam, sweet potato and cassava/manioc (Rodil 1999: 15). The agricultural cycle among the highland Bla-an includes two corn harvests per year and one dry-rice harvest. Rice or corn is the first crop planted in the annual cycle, followed by sweet potato and other root crops. In the past, the main crop consisted of sweet potato but it has been replaced as staple food by corn and rice. With the advent of logging companies in the 1970s, subsistence practices were altered and previously rare goods such as salt, clothing and canned goods introduced to most upland Bla-an families (Sal Naong Ancestral Domain Claim 1996: 5). Yet, still today, the above crops are supplemented by food gathering and hunting (AECOM/SMI 2011: 14). Settlement patterns correspond to the practice of swiddening and consequently, Bla-an houses are scattered along a family's swidden fields and not confined in clustered villages. To what degree this has changed in recent years is unknown to us.

11 See Rodil (1999: 28ff.) on leaders' special functions and duties.

12 This is not to deny that today's more hierarchical "tribal" structures are also the consequence of NGO interventions for ancestral domain claim-making purposes, for which local people are required to be organized and have identifiable leaders. For a detailed ethnographic account of this process among a neighboring indigenous group, see Wenk (2012).

Concerning production patterns of resident settlers, not much data could be found, but we assume that they are the main producers of the various crops found at local and regional markets. Where they produce rice, they usually do so under irrigated cultivation as opposed to the indigenous dry-rice production for limited home consumption. The province of South Cotabato, where many settlers live, is one of the finest agricultural areas in the country, supplying Manila and the rest of the Philippines.

Within the project site, there is also, and has been for a long time, evidence of small-scale mining (AECOM/SMI 2011: 2). Whether this is done by the indigenous population or settlers, or by both, is not mentioned in the company's EIS. But the Bla-an are traditionally known to be metal workers or smelters. They smelt bronze, copper and lead (Kinoc 2002: 21) and are thus—just as their neighbours the T'boli—familiar with the metals found in their territories and which now attract international extraction companies.

With a view to the local population's economic performance, the EIS provides figures for the average household income in the host *barangay*, said to be PhP 4,933 monthly (= 113 USD, conversion date August 8, 2013). 79 percent of households live below the official Philippine Poverty Line, which for 2007 was said to be at PhP 6,195 monthly for a family of five (AECOM/SMI 2011: 2f). In some specific areas within the project site, up to 96 percent of people live below this poverty line. It is only in Columbio, Sultan Kudarat, where the inhabitants seem to be better off with only 6.4 percent below the poverty line. Why this is so is not elaborated in the EIS. Host communities in general are portrayed in company documents as having low levels of education—among the Bla-an especially—, low living standards, low literacy rates and a predominantly rural subsistence lifestyle "in accordance with their traditional cultural customs." (AECOM/SMI 2011: 3, 12f., 16). Hence, the company can make itself useful through "community development" programs.

Settlers are generally materially better off and better educated in the formal school system than are the local Bla-an. The former perform more intensely in local and regional markets and are said by the company to have "assets" such as houses, motor vehicles, farmland and work animals. And they run family-owned businesses (AECOM/SMI 2011: 2). Such may be, for example, a sundry store known as *sari-sari* or an eatery (*carinderia*) along a major road or in the vicinity of a bus stop, or a local rice and corn mill. What is conspicuous in the company's EIS is that under the subtitle "properties within the project site", there is no mention of any "assets" held by the local Bla-an. Do the Bla-an not own anything, i.e., have they no assets to lose and be compensated for, once mining begins? What about land?

Traditional land ownership among the Bla-an emanates from occupancy. The acknowledged head of a clan gives permission to outsiders for the establishment of an *almigo* (a clearing/swidden field). Members of the clan itself did/do not have to seek permission, for any member is allowed to use land within the confined territory of the *banwu*, the boundaries of which are known to those both inside and outside. Those who cultivate a piece of land own the land in the sense of primary usufruct rights. Land can be acquired through marriage (the Bla-an are uxorilocal and the husband acquires land in the locality of the wife), the payment of debts, the customary exchange of goods in a conflict settlement or other social transactions (horses, water buffaloes and land were/are the most valuable goods exchanged), through inheritance, or by clearing a portion of unoccupied territory (Rodil 1999: 20ff.). Land is handed down from generation to generation and the basis of occupancy is "physical habitation", "actual maintenance of swidden fields" (whereby leaving a field to lie fallow is a form of maintenance), and the "reverential use of sacred places, regular tapping of the resources of mountains and rivers" (Rodil 1999: 27).

At present, there are three Certificates of Ancestral Domain Title (CADT) held by the local Bla-an, as well as one Certificate of Ancestral Domain Claim (CADC) from the former DAO 02 scheme. CADT number 108 covers Barangays Danlag, Pula Bato und Tablu in Tampakan with a size of 4,533 hectares and a "beneficiary population" of 2,867 people; CADT number 102 covers the community of Bong Mal in Barangay Kimlawis and more settlements in the Municipality of

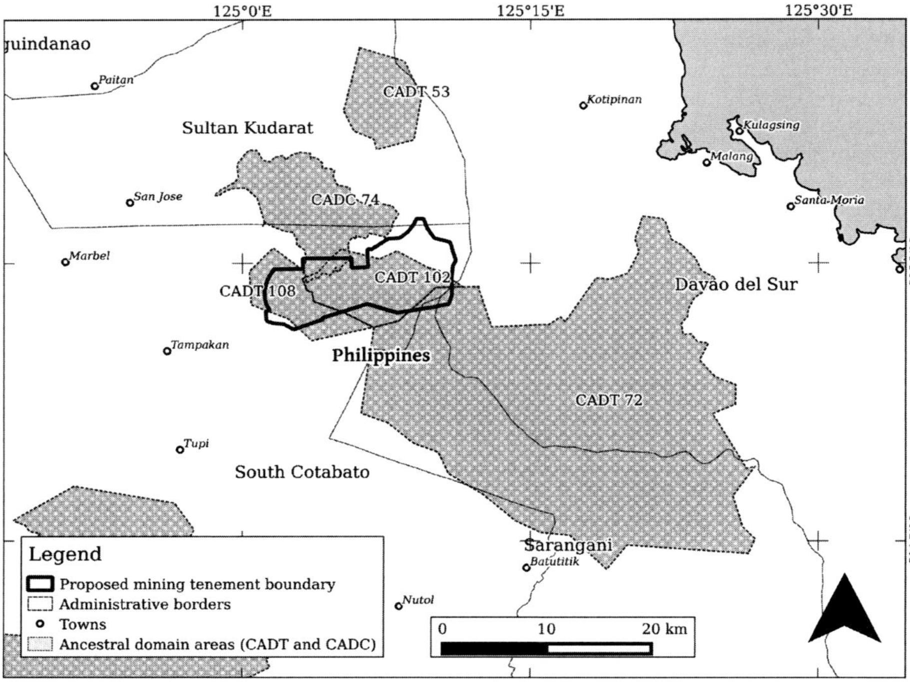

Map 2: The proposed mining area and overlapping ancestral domain areas in Mindanao.

Kiblawan covering a land area of 6,560 hectares and including 1,984 people; CADT number 72 covers a vast area of 74,804 hectares and includes 52,110 people, both Bla-an and Tagakaolo, in the municipality of Malungon in Sarangani Province. Only its north-westernmost tip is included in the proposed mine site, while of CADTs number 108 and 102 large portions are included. All three CADTs were approved in 2008 and 2009 respectively (NCIP 2013). Further north, with again only a small portion included in the mine site is CADC number 74. The CADC stems from IPRA's predecessor scheme known as DAO 02 and is not yet a land title but the government's certification of an indigenous group's claim to ancestral territory.

Consequently, most land within the projected mine site is officially recognized ancestral domain, meaning the land is privately held by Bla-an groups, who through the CADT are officially registered as an entitled collective of land owners. The EIS and hence SMI and Glencore Xstrata seem to recognize this legal fact in that they state that ancestral domain land cannot be sold and only be used by outside parties on the condition that processes stipulated in the IPRA are recognized and followed (AECOM/SMI 2011: 2, 17, 20). This includes obtaining all communities' FPIC (see Chapter 2.3) on the mining operation and the presence of the company in their area, a process the company is still engaged in as we write this (2013).

The communal land titles held by indigenous groups in the Philippines today are the result of decades of struggle and lobbying efforts by advocates of indigenous rights on regional, national and international level. They are a victory on the road towards self-determination, as demanded by an internationally active indigenous movement. However, they do not at present resolve the intricate issue of individual land ownership within titled domains. Yet, it is this that becomes a major concern to the local population as soon as a company comes in and issues of compensation and/or relocation emerge (see Wenk 2012). So too is the case of Tampakan. Who holds what rights to a single piece of land is increasingly contentious, especially so between Bla-an and settlers. The EIS states, for example, that in Barangay Tablu, 47 percent of households own land privately while in Barangay Blaan in Malungon, there is little privately owned land. Land in the latter case is considered by the company to be of little value, i.e., as having low rental rates (AECOM/SMI 2011: 20). With "privately owned land", the company means land held under individual title. What it fails to recognize—and consequently does not value correctly—is that a CADT likewise signifies private ownership of land, albeit not in the name of an individual but that of a collective.[13] So once again, as during the period of in-migration of settlers, the "assets" of the local Bla-an weigh less or nothing in comparison to those of the newcomers to the area.

Yet, settlers too have problems with their claims to the land because most have acquired land from Bla-an locals some decades ago and hold no formal land titles. Some hold deeds of sale that certify the transfer of rights to the land. Some hold individual stewardship certificates issued by the DENR, while again others hold CLOAs from the Department of Agrarian Reform (Tomas/Celebrado 2008: 21). A CLOA is a Certificate of Land Ownership Agreement issued under the most recent agrarian reform. In many cases, local Bla-an have lost control over land to settlers in so-called *prenda* arrangements, whereby land was/is mortgaged to a settler and could later not be redeemed by the original Bla-an owner/user. Where the land cannot be redeemed, it remains with the settler and is considered his (see Rodil 1999: 21, 23 and Wenk 2012). Most of these arrangements and forms of land acquisition leave settlers without a formally recognized land titles. Consequently, in order to protect their particular interests in the face of the mining company, settlers have formed their own associations or "landowner groups" (Tomas/Celebrado 2008: 21).

1.5 National Economy and Legal Framework

With its past of double colonization, the Philippines are continuously beset by political turmoil and economic problems. A country in ongoing political transition, its political economy ails with an elite's domination of politics and economics, unrealized land reform, and widening income inequality (Caballero-Anthony 2007: 11). The country's economic development and performance, understood generally to be poor and dismal (Severino/Salazar 2007), is hampered not only by fiscal deficits but a primarily negative image abroad, where the Philippines are equated with political uncertainty, instability and volatility, corruption, cronyism, weak infrastructure, an inconsistent policy environment as well as "peace and order" problems. Foreign direct investment has come in at very low levels but increased after the country's economic "under-performance" was addressed in the 1990s through economic policy liberalization (Sicat 2007: 143ff., 154, 163). Under a neoliberal framework, a series of policies were introduced to address issues of social justice to reduce conflicts, to make the country more attractive for foreign direct investment, to increase exports, and to reduce debt and fiscal deficits. New policy frameworks were, among others, the Local Government Code of 1991, the Philippine Mining Act of 1995 and the Indigenous Peoples' Rights Act of 1997. With decentralization, economic liberalization and privatization, and the lifting of restrictive provisions with regard to natural resources exploitation, foreign direct investment flows increased, although not lastingly (Sicat 2007: 167ff.). The Tampakan Copper-Gold Project and the conflicts related to it are to some extent a result of these policies of economic liberalization.

13 See Wenk (2012) for elaborations on the notion of "private yet communal" land as it has been legally adopted in the Philippines.

Poverty and social inequality in the Philippines is much higher than in neighbouring countries such as Thailand, Taiwan or Malaysia. In 2005, 27.9 percent of the population lived in absolute poverty and 15 out of 100 families could not afford enough food. This means that 22.5 million people are affected by hunger. Every third person has no secure access to drinking water, 47.2 percent have no electricity and 31 percent no sanitation. Only slightly more than half the population has a steady roof over their heads (Reese/Werning 2006: 54f.). The Philippine total population was 94,013,200 in the most recent census of 2010, with a population growth rate between 2005 and 2010 of 1.95 percent. The country's foreign debt was USD 64,856 million, and GNI per capita was at USD 2,470 by 2011 and 2013 respectively (NSCB 2013; Fischer Weltalmanach 2011: 542 and online).

Filipinos and foreign observers alike often hold the view that economic growth and development have bypassed the country's south. Yet, this is a false conclusion. Far from being isolated from the national economy, Mindanao has been a primary contributor to the country's productive capacities; it is just that the southern resource wealth and the income generated from the island's ceaseless exploitation have not benefited the majority of its people. Instead, poverty indicators for Mindanao are higher than the national average (Tadem 1992: 7).

Mindanao is the Philippines' most important producer of coconut and wood products. Moreover, it is estimated that over half the country's commercial fish catch comes from Mindanao waters. Mindanao rice lands account for 26 percent of the national total while corn lands occupy a 47 percent share. Mindanao is a *palay* (unhusked rice) surplus producing area, with central Mindanao considered the "rice bowl" per se and one of the top two rice-producing regions in the country. However, productivity is lower than in Luzon and the Visayas and prices received by Mindanao rice and corn farmers are lower than those earned by their northern counterparts. Most industries in Mindanao are export-oriented and dependent on the vagaries of international trade over which local producers have little to no control. Price instability and uncertainty has thus surrounded Mindanao's major products such as coconut and wood, banana, pineapple, minerals and fish (Tadem 1992: 14–18, 25).

The Philippine mineral endowment is among the highest of any country in the world. It ranks second in the world for gold, third for copper and sixth for chromate, the latter being the only natural source for chromium. There are also numerous deposits of nickel, coal, limestone, iron ore, silver and uranium. Hence, the Philippines are the fifth most mineralized country in the world and as such highly attractive for mining investments (Christian Aid/PIPLinks 2004: 6; Go 2008: 3). The Philippine Government believes that the extraction of minerals across the country leads to sustained economic growth and mining investments and operations are therefore actively encouraged. This is a common argument, but as studies on Philippine mining policy and performance show, the country is not ready for "socially responsible mining" and instead has a record of abandoned mines, tailings waste pollution, disasters and severe human rights violations connected to mining (see Go 2008: 3; Landingin 2008; Christian Aid/PIPLinks 2004; Hamm/Schax/Scheper 2013). The Philippines' mining performance supports the resource curse hypothesis, which suggests that natural resources wealth and its extraction can exacerbate conflict, corruption, weak governance and poverty (see Ballard/Banks 2003).

A comprehensive journalistic assessment of the Philippine mining industry in 2008 found that the "mining industry's economic impact remains negligible", with jobs created constituting only 0.4 percent of total national employment and revenues being less than one percent of total annual state revenues (Go 2008: 3).[14] The number of people employed in mining as a percentage of the

14 Mining's contribution to total employment rose to 0.7% by 2012, with a total number of 252,000 people employed in the sector (Mines and Geosciences Bureau 2013).

total Philippine workforce has been declining since the mid-1980s, particularly so in large-scale operations, due to new technology. But what is worse, mining tends to inhibit or damage entirely other industries such as farming, fishing, or tourism, which offer much more sustainable jobs. Mining has thus so far neither contributed to national development or overall poverty reduction, nor has it improved local livelihoods and job security in any way other than a very short-term and short-sighted manner. Experts have furthermore concluded from the industry's meagre contribution to overall employment and state revenues that the Mining Act of 1995 "(...) is heavily stacked against the state and in favour of the [foreign, added by Wenk] mining companies." (Landingin 2008: 6).

Large-scale mining in the Philippines can be traced back to the Spanish colonizers, who opened the first commercial mines in Benguet, which subsequently led to the establishment of the Benguet Corporation. The Mining Act of 1905 contained the provisions, that "all public land in the Philippines (is) to be free for exploration, occupation and purchase by citizens of the United States and the Philippines" (cited in Christian Aid/PIPLinks 2004: 6). Mining reached its peak in the 1960s and 1970s, when copper and gold prices were at an all-time high. Then President Marcos had a direct interest in the mining industry and owned half of the Marcopper mine on the island of Marinduque (see below) (Christian Aid/PIPLinks 2004: 6). Even after he was ousted in 1986, the succeeding governments' support to the mining industry continued.

In the late 1980s, mining experienced a decline. Responding to this downward trend, the Philippine Government tried to encourage new foreign investments with various strategies such as the liberalization of legal frameworks with less restrictive regulation, the assurance of government support, the provision of lower duties and tariffs, and tax breaks and holidays. The World Bank, whose influence over Philippine economic policy has been extensive ever since the country's serious external debt problems of the 1980s, encouraged these measures. In consultation with the World Bank and various international mining companies, the Philippine legislature eventually drafted the Philippine Mining Act of 1995 (Christian Aid/PIPLinks 2004: 7).[15] The Mining Act grants different types of mining rights or so-called leaseholders. These are exploration permits for corporations as well as individuals, the Mineral Production Sharing Agreement (MPSA), the Financial Technical Assistance Agreement (FTAA), and quarry permits. The centerpiece of the Mining Act and most contested is the FTAA, which is open not only to Filipino, but also fully foreign-owned, corporations (Dalupan 2000: 10, 65f.).[16] The Mining Act thus liberalized foreign investment, a step that has since caused much opposition and protest. An FTAA is entered into by the President him-/herself with a foreign-owned company, while the MPSA is approved by the DENR. The rights thus acquired last up to 25 years in both cases, with the possibility for renewal for another 25 (Dalupan 2000: 64). The MPSA is much less contested as it restricts foreign ownership of a mining project and requires the contractor to be at least 60 percent Filipino-owned.[17]

15 See Rovillos, Ramo and Corpuz (2005:185–200) on the role of the World Bank in Philippine mining policy and the developments in the mining sector over the past decades.

16 "An FTAA may be entered into for the exploration, development and utilization of gold, copper, nickel, chromite, lead, zinc and other minerals although no FTAAs may be granted with respect to cement raw materials, marble, granite, sand and gravel and construction aggregates. The phases under an FTAA include an Exploration Period of two (2) years, extendible for another two (2) years, a Pre-feasibility Study Phase of two (2) years, a Feasibility Study phase of two (2) years and the development, construction and utilization phases which take up the remaining years of an FTAA." (Dalupan 2000:65)

17 In the Mines and Geosciences Bureau Primer on mineral agreements (2005), the MPSA is defined as "(...) an agreement between a Contractor and the Government wherein the Government grants to the Contractor the exclusive right to conduct mining operations within, but not title over, the contract area. Mining operations that are allowed under Mineral Agreements include development/construction and utilization of mineral resources including the continuance of exploration works during the conduct of development/construction/utilization activities."

A foreign company, through the FTAA, can lay claim to an area of up to 81,000 hectares onshore or 324,000 hectares offshore, while Filipino-owned companies that hold an MPSA are by contrast restricted to 8,000 hectares in one province and 16,200 hectares across the country onshore and 40,500 offshore (Christian Aid/PIPLinks 2004: 8; Dalupan 2000: 64).

Despite much criticism on the Mining Act due to its liberal outline, it must also be said that unlike past mining laws in the Philippines, it contains extensive provisions on environmental protection, community development, and indigenous peoples' rights (Dalupan 2000: 63). It is also worth noting that the Mines and Geosciences Bureau (MGB) of the DENR, the agency in charge of implementing the Mining Act, lists the Mining Act of 1995 together with the IPRA and the Act Creating A People's Small-Scale Mining Program of 1991 as being the most relevant legal frameworks to mining in the country.[18] The Mining Act's Implementing Rules and Regulations (DAO 96-40), furthermore, introduce the concept of sustainable development in mining that is supposed to meet the needs of the present generation without compromising future generations' ability to meet their own needs (Dalupan 2000: 10). The Mining Act establishes an "environmental management framework" for mining operations and with regards to "host" communities, it requires that the holder of a mining agreement "(...) shall allocate a minimum of one percent (1%) of the direct mining and milling costs for community development activities (...)" (Dalupan 2000: 10f.). Concerning indigenous peoples in particular, there are provisions to ensure that no mineral agreements or permits are granted without their prior and informed consent. Once such consent has been given, an agreement on royalty payments must be concluded and these payments "(...) may not be less than one percent (...) of the gross output." Royalty payments by the company, a point we will come back to, are to form part of a trust fund established for the socioeconomic well being of the indigenous host communities and these trust funds are to be managed and utilized by the local people themselves (Dalupan 2000: 11). Dalupan further elaborates on numerous issues where the Mining Act and the IPRA conflict with and contradict one another. Eventually, she raises the sensitive legal issue of "whether and to what extent the Mining Act has been repealed by the IPRA" (Dalupan 2000: 77). This is a serious matter that requires further legal studies and has impacted any effective implementation of both laws.

To conclude this section, it is crucial to highlight that both the IPRA and the Mining Act have survived legal assaults on their constitutionality. The case against the IPRA was closed in December 2000 when the Philippine Supreme Court upheld the act's constitutionality.[19] Four years later, in December 2004, the Philippine Supreme Court also upheld the constitutionality of the Mining Act (Go 2008: 3; Landingin 2008: 6; see below).

In response to the Mining Act, foreign investment in mining increased by more than 50 per cent between 1995 and 2001, from USD 980 million to USD 1.5 billion (P55.2 billion to P84.5 billion) and the Philippines has experienced a mining boom ever since. Yet, production is still well below what it was during the peak years of the 1980s. Due to the newly created FTAA, the number of

18 See MGB: LAWS AND REGULATIONS (Republic Act), www.mgb.gov.ph, accessed on Sept. 5, 2013.

19 Background to the case: shortly after the IPRA was passed, a legal challenge was filed with the Philippine Supreme Court. The petitioners, a former Justice and an Attorney with ties to the mining establishment, claimed that the IPRA violates the Philippine Constitution for giving indigenous peoples control and rights of ownership over ancestral domains and ancestral lands, including the natural resources found therein. In the 1987 Constitution, most of the land claimed under the IPRA is considered "public" and consequently owned by the State (which in turn enables responsible government agencies to issue mining permits and other permits for the exploitation of natural resources to interested third parties). With the IPRA, however, all land claimed by indigenous groups ceases to be state-owned and is classified Ancestral Domain, i.e., it is no longer public but private land. As such, vast areas with extensive mineral resources are in significant ways removed from direct state control and therefore more difficult for mining corporations to access. This is why in order to mine an area that is titled or claimed ancestral domain, any mining company must now seek the local indigenous groups' free, prior and informed consent (on this Supreme Court case, see Leonen 2000; Malayang 2001; Castro 2000).

foreign mining companies increased by a dramatic 400 per cent between 1994 and 1996. Approved mining exploration and production permits increased from 294 in 2003 to 447 in 2007. By 2012, there were 6 FTAAs—the one in Tampakan having been the first to be issued—and 339 MPSAs. Another 28 FTAAs are pending applications. One single large company can make for half the foreign mining investments in the country, as was, for example, expected to happen between 2008 and 2010, when Xstrata was to invest USD 2 billion through its local affiliate SMI. However, actual investments were not quite as high and the figures remained almost 50 per cent below target (Landingin 2008: 6; Holden 2005; 420; MGB 2013). Despite these figures of growth, the industry's overall economic impact, as discussed, remains marginal. And with the boom came mining-related conflicts (Landingin 2008: 6).

2. The Tampakan Mine Project

Mining is just one of several natural resources extraction activities going on in southern Mindanao. It advanced along the settlement frontier, together with logging, agri-business and cattle ranching. In the late 1980s, a gold rush first reached the T'Boli homeland, then that of the Bla-an. As was common at the time, resource extracting agents and those looking for land did not consult with the local population before going about their business. As a T'Boli Mayor stated: "They didn't even consult us or inform us. They just acted as if we didn't exist. This is bad for democracy, we think" (cited in Hyndman/Duhaylungsod 1998: 95). The more than 25,000 non-local miners who invaded the T'Boli area west of the Bla-an in 1990, were followed by the Australian mining firm Western Mining Corporation or WMC. Having prospected the T'Boli area for gold, WMC later moved on into Bla-an territory, got a prospecting lease in the area of Mount Atmurok and later applied for an FTAA (Hyndman/Duhaylungsod 1998: 98ff.).

The entry of WMC into Bla-an land in the early 1990s exacerbated existing conflicts with settlers who had entered the area since the 1970s. The arrival of settlers and the advent of massive logging had seriously reduced the land base of the Bla-an and violent clashes in the 1970s and 1980s were common. With WMC's entry, the local population experienced additional harassment and restrictions. This, for example, when the DENR put up sign boards prohibiting the locals from farming and hunting within the MWC prospecting area (Hyndman/Duhaylungsod 1998: 99f.). The company hired goons under the control of prominent settlers and opponents of the local Bla-an to provide security for its prospecting operation, thus contributing directly to the escalation of the conflict. As Hyndman and Duhaylungsod describe (1998: 103), and as was common in much of Mindanao at that time and until today (see Wenk 2012), the local population armed itself in order not to be overrun at the increasingly contested settlement and resource extraction frontier.

The FTAA under which WMC later operated, was approved in March 1995 and covered an area of 99,387 hectares. Since then, the Agreement was subject to statutory reductions to currently cover an area of 23,571 hectares. Apparently, this first FTAA under the new Mining Act was approved by the Office of the President without consultations with local governments and much less with the affected communities (Tomas/Celebrado 2008: 9).

The local operating entity today is Sagittarius Mines, Inc. (SMI), a Philippine-incorporated company. SMI acquired exploration rights from WMC in 1999, the latter ceasing operations in the area that year. First known as Tampakan Mineral Resource Corporation, SMI formally obtained the FTAA in August 2002. SMI initially partnered with Indophil Resources NL, an Australian company but in 2007 Xstrata Copper assumed management of the project by acquiring a 62.5 percent share from IndoPhil. The latter retained a 32.5 percent interest in the project (Tomas/Celebrado 2008: 9). In June 2015, the project ownership changed again, with Glencore (the successor of Xstrata) selling its stake (see editor's update at the end of this chapter).

The main reason why WMC withdrew from explorations in the late 1990s was a legal battle that questioned the constitutionality of its FTAA and certain provisions of the Philippine Mining Act. The petitioners came from the Bla-an Sal Naong group, one of the affected communities in Columbio, Sultan Kudarat, which initially put up strong opposition against the presence of the company (see below). Organized into the La Bugal-B'laan Tribal Association and supported by Philippine legal advocacy organizations such as the Legal Rights and Natural Resources Center, the petitioners filed a case in the Philippine Supreme Court in 1997 arguing against the foreign ownership of a resource extracting company (see La Bugal-B'laan vs. Victor O. Ramos). The Supreme Court consequently declared WMC's FTAA unconstitutional in January 2004 but then reversed its decision in December that same year (Tomas/Celebrado 2008: 20).

The FTAA originally signed by WMC and later acquired by SMI/Xstrata (now SMI/Glencore Xstrata) is already eighteen years old. Yet, the mine is still not in operation. The Bla-an have engaged with these various corporate mining actors for over twenty years. There is consequently a long history of engagement that did, and does, include opposition but is just as much characterized by accommodation and (mutual) dependence, which has decisively shaped local people's development, decision-making and community formation processes over the years.

2.1 Environmental and Social Impacts

The actual and potential impacts of exploration and mining discussed in this section are by no means comprehensive. We also want to point out that environmental and social impacts are, in their actual effects, inseparable. Whatever environmental impacts occur has immediate and long-term effects on the social, political and economic life of the people affected.

2.1.1 Impacts on the Environment

So far, the environmental impacts of the exploratory activities at the Tampakan mine site are relatively small as far as we could assess from the existing literature. The most obvious interventions are the drill holes—numbering 250 holes by August 2007—but these are confined to certain localities and apparently the company does rehabilitation at these sites immediately after drilling (Tomas/Celebrado 2008: 35). Consequently, the severe environmental impacts the Tampakan mine is expected to cause are as yet potential. Some are serious warnings by experts, others apprehensions of various affected local people based on what they have heard might or has happened at other mine sites across the country and internationally.

Tomas/Celebrado (2008) have assessed what local and downstream communities fear most in terms of environmental impacts of the mine in 2007 and provide a comprehensive overview of several issues. And Goodland/Wicks (2012) have analysed the potential environmental risks of the mine. Firstly, host and downstream communities are worried about the impacts of mining on their water sources. Drinking water and rivers might be polluted and may cause illnesses and skin disease. Also, the rivers near the proposed mine provide drinking water to the people in Koronadal City and other towns and villages along them. They also supply tens of thousands of farmers downstream with water for their fields and fishponds, and they feed into Lake Buluan.[20] This fresh water lake in turn provides water for 40,000 people living in its immediate vicinity. Damaging or polluting the lake could result in further social unrest and violence in this already very unstable part of Mindanao. More generally, the company will acquire water rights, hence placing it in a

20 Lake Buluan is the third largest lake in Mindanao with a surface area of 61.34 square kilometres. Its primary inflow is the Marbel River (www.wikipedia.org, viewed March 22, 2013). Mining is understood as a serious threat to the lake.

position of full control over the water (Goodland/Wicks 2012: 5).

According to Xstrata's Environmental Impact Assessment (EIA),[21] 40 per cent of the total project area or 3,750 hectares consists of rainforest vegetation. This remaining forest cover is crucial as a primary source of water for the river systems that feed agricultural land and ponds downstream. A part of this upper mountain rainforest is old growth and protected by law (Goodland/Wicks 2012: 8; Sarmiento 2012). Most of these forested areas onsite are expectedly going to be cleared within the twenty year life span of the mine in order to expand the open-pit area, build access roads, and construct the various facilities including crushers, ore stockpiles, waste rock storage and more (Tampakan Forum Report 2011: 21). The destruction of forest and the ways water is going to be used for the mine pose a serious threat to agricultural production in the wider area of the mine. As Goodland/Wicks (2012: 5) state, there are 80,000 farmers and 200,000 hectares of farm land in South Cotabato that will be directly or indirectly affected by the mine. In reaction to this threat that does not halt at provincial boundaries, farmers and an irrigators association in Davao del Sur have asked the Aquino administration to stop the Tampakan project, arguing it would destroy the watershed that feeds into 13,000 hectares of mostly rice farming land downstream (Alave 2012).

Secondly, the cutting of the forest is only one of the expected transformations of the physical landscape in which people live and orient themselves. People are aware that open pit mining will result in many more such drastic changes. The proposed open-pit-method will obviously leave its long-term marks on the mountains of Tampakan alongside constructions such as roads, the waste rock storage facility, concentrator (processing facility), tailings storage facility, and a fresh water dam (Sarmiento 2012).[22]

Thirdly, people fear that landslides and floods will occur in consequence of mining. Local people as well as local governments complain that they are not fully informed about the technical details of where the mine tailings will be disposed (Tomas/Celebrado 2008: 35f.). Experts such as Wicks and Goodland hold that the risk to the aquifers and other water resources of having a toxic mine void 800 meters deep and covering 500 hectares is serious and the storing "forever" of thousands of tons of toxic water unacceptable. 2.7 billion tons of waste rock (to be piled over 300 meters high) and arsenic-containing tailings with a high potential for Acid Mine Drainage are great environmental threats in this critical water catchment area (Goodland/Wicks 2012: 7). If an accident that releases tailings into the water were to happen, the resultant pollution in subterranean water flows, aquifers, and lakes cannot be cleaned up and would severely affect agricultural production (Alyansa Tigil Mina 2012a).

Although SMI/Xstrata articulates its commitment to best practice and claims to intend to minimize negative environmental impacts, a draft response to the company's Environmental and Social Impact Assessment (ESIA) by Goodland and Wicks criticizes the company heavily for downplaying or omitting some of the expected negative impacts (2012: 3f.).

Due to its geological location, experts consider the Tampakan project to be one of the potentially most dangerous mining endeavours in the world (Sarmiento 2012).

21 SMI has completed an Environmental Impact Assessment (EIA) for the proposed Tampakan project. The EIA consists of various technical studies, which in turn were used to produce two reports: 1) the Environmental Impact Statement (EIS) and 2) the Environmental and Social Impact Assessment (ESIA) (SMI/Xstrata 2013). See Gatmaytan et al. (2013) for an assessment of the EIA process and content.

22 What it means for the sense of self to have one's familiar physical landscape altered so severely is discussed in anthropological writings on the meaning of place in studies on disaster, resettlement, displacement and environmental management (see, for e.g., Feld/Basso 1996; Carrier/West 2009; Oliver-Smith 2009; Leemann/Nikles forthcoming).

Fig. 1: Forests are still plenty in the proposed mines site. But these will have to be destroyed once operations begin. Photo: Bob Timonera/Fastenopfer.

Fig. 2: One of the springs in Bong Mal. Photo: Bob Timonera/Fastenopfer.

Fig. 3: These mountains will either be destroyed in the open pit mine or submerged in the tailings pond.
Photo: Bob Timonera/Fastenopfer.

There is a high risk of induced seismic activity, which can occur when miners drill or mine near fault lines and active volcanoes. Xstrata was warned about these risks by the technical working group Tampakan Forum but has not responded. The company also allows itself to ignore warnings from Philippine geologists, even though in its ESIA, the company's own engineers acknowledge that "the Tampakan mine has a high potential for loss of life and high environmental damage if a failure of dams or rock storage facilities occurs" (in Goodland/Wicks 2012: 7). SMI/Xstrata intends to counter such mega-disaster risks by simply designing dams that are supposed to hold the toxic materials in perpetuity. Technical experts of the Tampakan Forum find this approach "unbelievable".[23]

To conclude, two further impacts shall be briefly mentioned. First, the mine will produce greenhouse gases, which amount to less than 0.4 per cent of all greenhouse gases emitted in the Philippines. While a small figure as such, this is a massive amount of emissions for one mine only. And second, the ore will be transported to the coast through a pipeline, in the form of a slurry, which bears the risk of breaking and spilling toxic waste onto the land and into waterways (Goodland/ Wicks 2012: 7, 8). Facing all these (potential) impacts on the environment, local livelihoods are seriously at risk. With the removal of forest vegetation, the soil becomes degraded, water sources threatened, and landslides as well as siltation common. This in turn will affect the agricultural activities of the local population. As Tomas/Celebrado (2008: 65) hold, "long-term environmental effects of open pit mining are always irreversible."

23 Expressed during a public forum held in Marbel in South Cotabato on September 23, 2011. The great fear of a devastating mine tailings' spill is not without precedent in the Philippines: An accident in 1996 known as the Marcopper tailings spill incident illustrated to the Philippine public the extent of environmental harm mining can cause, with 1.5 million to 3 million m^3 of toxic tailings released into the Boac River (Holden/Jacobson 2006: 191f.; AFRIM 2011).

2.1.2 Social Impacts

The above-described environmental impacts will affect the ability of the local communities to sustain themselves and will consequently increase their dependence on the company and other outside actors. As also mentioned, the potential impacts of mining on the environment are expected to cause instability and conflict. But one does not have to wait for these environmental impacts to fully unfold to see the security situation in the area seriously at stake, for, with the entry of every company into a rural Philippine area comes militarization, an increase in armed parties and the proliferation of firearms, as well as human rights violations. Several incidents in Tampakan highlight this development: On January 1, 2008, the armed wing of the Communist Party of the Philippines, the New People's Army (NPA), attacked the SMI base camp in Tampakan and destroyed facilities worth about 12 million Philippine Pesos (USD 278,000, conversion date October 7, 2013). It also attacked the nearby military detachment in Barangay Tablu. This incident put the allegedly good relations between the company and the local population seriously into question and the company lost its sense of security, which its officials had previously enjoyed. We do not know in detail how the local Bla-an are linked to the NPA but Romero states that the locals of the area could have warned the company had they wanted to (2008: 52f.). People in Danlag were aware of the presence of the NPA and chose not to report its activities to the authorities and the company. This does not necessarily mean that the Bla-an are allied with the NPA but it means that at this particular moment, they might have endorsed the guerrillas' actions more than those of the company.[24]

More recently, on October 18, 2012, the Armed Forces of the Philippines' 27th Infantry Battalion passed the house of anti-mining activists Daguil and his pregnant wife Juvy Capion. The pregnant Juvy and her two sons, aged 8 and 13, were killed by the Armed Forces while her husband managed to get away. The incident has become known as the "Tampakan Massacre". Both Daguil and Juvy were among the Bla-an leaders openly opposing the mine and Daguil is said to lead an armed group of Bla-an against SMI/Xstrata. Capion was wanted by the authorities for the murder of three construction workers of an SMI contractor, for which he had admitted responsibility (see Sarmiento 2013b; Alyansa Tigil Mina 2013). Then, on January 29, 2013, Bla-an members Kitari and Diyo Capion were killed in an "encounter" with men identified as members of the KITACO Task Force, an official provincial armed force consisting of military and paramilitary personnel (Tampakan Forum 2013; Hamm/Schax/Scheper 2013: 30). Due to these incidents and the general harassment people experience from the presence of the Armed Forces as well as paramilitary troops, at least fifty individuals, mostly women and children, fled the Tampakan mine site in early February 2013. Some of them sought refuge in a shelter run by the Social Action Centre of the Diocese of Marbel (see below). On August 23 that same year, a Bla-an leader named Anteng Freay was killed by soldiers pursuing an armed tribal group in the area. Freay was the head claimant and *bong fulong* of the ancestral domain of the Atbol Bla-an (Sarmiento 2013c).

Violent encounters and consequent displacements are already happening at the exploratory stage, but the organized displacement of the local population at large is yet to come. The entire population within the project site will need to be resettled (AECOM/SMI 2011: 20). These are an estimated 870 households (582 indigenous and 288 non-indigenous), comprising an estimated 3,500 to 4,000 people (AECOM/SMI 2011: 2, 20).[25]

24 In a press statement, the Communist Party of the Philippines' spokesperson said that the NPA's "(...) raid against SMI is an important milestone in the effort to put a stop to the firm's destructive and plunderous mining operations, defend the ancestral domain of the B'laan tribe; protect the environmental balance of the Liguasan Marsh and the water supply of South Cotabato, North Cotabato, Sultan Kudarat and General Santos City areas; and resist the Arroyo regime's campaign to auction off the country's natural resources to big foreign capitalists" (cited in Tomas/Celebrado 2008:41).

25 A list with indigenous and non-indigenous households per *barangay* is given on page 20 of the EIS.

Fig. 4: A Bla-an village in Bong Mal that needs to be displaced for the mine. Photo: Bob Timonera/Fastenopfer.

While the Tampakan Forum (2011: 30) claims that the company's Environmental Impact Statement (EIS) has not determined any resettlement areas, the EIS version we were able to get hold of provides a map of "potential resettlement sites" as of October 2010. Here, 18 resettlement sites are identified, many of them in the immediate vicinity and border areas to the mining area. The EIS also states that a Resettlement Action Plan (RAP) will be developed. This plan is said to include extensive consultation with the people to be resettled as well as with the "receiving communities". The assessment of the value of land within the project site for compensation is part of the on-going resettlement program (AECOM/SMI 2011: 20). SMI/Xstrata claim their resettlement plan is in accordance with the Philippine Government's regulatory requirements and the policies of their managing shareholder, Xstrata Copper. Apparently, these policies comply with international standards, in particular with the standards set by the World Bank's International Finance Corporation (IFC) and the International Council of Mining and Metals (ICMM) (Xstrata/SMI 2011: 47). Considering that by 2011 the company was only about to "develop" its RAP while operations were expected to start within just a few years, to seriously worry about the actual implementation of this human displacement seems quite justified. This even more so since the (forced) resettlement of local populations in consequence of large-scale development projects has more often than not turned out to be a social disaster (see Oliver-Smith 2009).

We have here elaborated on just two major social impacts. There are several others to consider and we mention some of them here in the form of questions to encourage future research:[26] What internal conflicts will arise due to compensation measures? How will land and natural resources rights of local people be assessed in areas held under collective title (CADT)? What changes in land ownership and land use will occur within and without the project area? Since the presence

26 This compilation is in parts based on a listing in the EIS of potential impacts and options for mitigating these (see AECOM/SMI 2011: 40ff.).

of a mining company affects the economic set-up of an entire region, how do the different people, i.e., Bla-an and resident settlers as well as new migrant workers, cope with the transformations they face? How will local governments as well as local people deal with the enormous amounts of money in the form of royalties, taxes, compensation payments, and project budgets?

2.2 Local People's Strategies and Anti-Mining Activities

At the outset of this subchapter, two things are important to note: the first is not to conceive of "local people" as a homogeneous group; and the second is not to confuse the local population in the mining project area with "civil society", i.e., to assume that the NGOs which oppose large-scale mining automatically represent, act on behalf of, or are even synonymous with the local population. This is why we have added "anti-mining activities" to the predetermined title "local people's strategies". It is tempting to conflate local people and vocal anti-mining groups as it makes for nice stories of resistance but this obscures a more complex picture that emerges in terms of the reactions from local and extra-local interest groups. The approach taken here, consequently, is twofold: we first try to get a grasp on the reactions of local people towards the mining company and in a second step discuss the activities of the Catholic Church and civil society groups involved in Tampakan and their relations with the local population.

Local (Bla-an) People's Reactions and Strategies

Different sources (journalists, academics, the company's EIS) seem to agree that the directly affected communities today—both Bla-an and Bisaya—show basic support to the presence and activities of the mining company. Yet, as Tomas/Celebrado (2008: 40) write, "they continue to hold a critical stance in supporting the mining activities practicing their own protests when their interests are not met. Their shifting support is tested against the gains they can negotiate from the mining company." It must also be kept in mind that the basic support apparently found today has not always been there. Hyndman/Duhaylungsod (1998: 103f.), for example, report that there used to be broad-based opposition against WMC's presence in the first half of the 1990s but that the company succeeded in buying out local leaders through direct payments as well as promises of development. Irrespective of the general support for the mining company and its (future) operation, many people are concerned about particular issues affecting them. Thus, there are serious tensions between the local population and SMI/Xstrata. The most important of these concern employment, the assertion of Bla-an self-governance within their ancestral domains—particularly, with regard to the management of the Tribal Foundations (see below)—, overlapping land claims, compensation packages, the determination of royalty payments and other benefits, and resettlement (see Romero 2008: 53 and Tomas/Celebrado 2008: 8f.). People seek to have their interests met by the company through both protest and negotiation.

In 2008, Bla-an leaders and people blocked the access to the SMI base camp in order to address what they consider unfair treatment and discrimination in SMI's employment policy. Jobs such as driver for the company or contractual jobs in the drilling program are usually given to non-Bla-an because most local Bla-an do not fit the job requirements, which ask for a college degree and long work experience. SMI in response and to lift the blockade promised to look into its hiring policy and offered PhP 15,000 a month (USD 348, conversion date October 27, 2013) to the communities to keep the roads "barricade-free" (Romero 2008: 53). By that time, only "a few hundred jobs" had been created and the company apparently did not have enough jobs for the Bla-an graduates of its own scholarship program. Out of forty Bla-an SMI scholars who completed college, only two found a job with the company. Consequently, there was great frustration among the locals, who say that SMI promised to absorb its scholars once they graduated, but now that there are Bla-an BA graduates, the company asks for Masters degrees (Romero 2008: 53).

Fig. 5: A Bla'an woman walking past the SMI "Pioneering Camp". Photo: Bob Timonera/Fastenopfer.

To address the growing grievance among the local population with its employment policy, SMI introduced an employment system whereby contractual jobs are rationed to members "of the tribe" through weekly rotation. Each *barangay*, *sitio*, tribal council and landowner association is given a particular number of slots, which the leaders of these respective bodies then distribute among their members/constituencies. This is supposed to give everyone an equal chance to work. Yet, this system has again been heavily criticized, as it gives leaders a powerful role over their constituents in choosing who gets to work. Leaders have also been accused of diverting parts of the workers' salaries (Romero 2008: 53; Tomas/Celebrado 2008: 32). SMI, in turn, claims this hiring scheme was established after community-based consultations. Many Bla-an say the Bisaya are privileged in getting the better, longer-term jobs from the company.[27] In turn, SMI/Xstrata prides itself to be hiring primarily "Filipinos", neglecting inter-ethnic rivalries and unequal conditions on the ground. By December 2011, the Tampakan Project engaged 393 employees (97.2% from the Philippines) and 888 contractors (99.3% from the Philippines) (SMI/Xstrata 2011: 35).

According to the Mining Act, indigenous groups on whose land mining operations take place should receive royalty payments once mining is in full operation. The royalty fee of a minimum of one percent of the gross income should go to a trust fund for the benefit of the affected communities. However, the communities are not informed how much money this one percent actually is, i.e. about the gross income as such. The company does not provide this information (Tomas/Celebrado 2008: 51). Generally, the lack of information on crucial issues such as this is a serious problem and means that local people's decisions are hardly informed, i.e., based on comprehensive and understandable (in both language and content) information.

27 In 2007, there were about 360 rotational workers weekly. They were paid a daily wage of P224.50 (USD 5.20) for eight hours of work. Deductions for social security and PhilHealth amounted to P100 per week. There were another 320 workers hired through service agencies (i.e., local and outside labor contractors). These were mostly non-Bla-an. Regular company direct-hired workers were totalling 115, of which 48 came from Tampakan and only 19 were Bla-an (Tomas/Celebrado (2008:32).

In order to strengthen local knowledge about the extent and possible impacts of the proposed mining project, Kail Zingapan of the Philippine Association for Intercultural Development (PAFID) recently worked in the Tampakan area to produce an impressive 3-dimensional model of the Koronadal Valley and the Tampakan watersheds in close collaboration with the local indigenous communities.[28] The people involved showed Zingapan where their lands and fields are located and Zingapan in turn was able to show to them the extent of the mine development area and how it overlaps with their ancestral domain—for example, the location of the 500-hectare large open pit, with a depth of 785 meters (Hamm et al., October 19, 2013). Once the affected residents were able to see and consequently realized (probably, for the first time!) where exactly in their territory constructions such as the tailings dam and the open pit would be located, they were shocked. From this, Zingapan concluded that many people were mis- or uninformed regarding the future mining operation and its possible risks. PAFID then explained to them that based on the IPRA it remains their decision whether they want mining on their land or not. This information was used to confront the company in a public consultation in September 2011 (Alyansa Tigil Mina 2011; Tebtebba 2012).

With regard to the issue of information (not) provided by the company, people complain that if someone from SMI talks to "the people", it is actually only the leaders being talked to (e.g., Tribal Council members) (Romero 2008: 54f.). However, indigenous leaders and local government representatives are not 'the people' or the often referred to 'community'. As "owners of the land", those affected by the project demand regular reports, want to know how many tons of gold and copper have been found on their land, which land exactly will be mined, etc., but when they confront SMI community relations personnel with such questions, these reply that they are not in charge. SMI in contrast insists that since Xstrata assumed management, they have increased "stakeholder engagement" and "community consultations" and even established a formal grievance mechanism (Romero 2008: 55).[29]

Yet, it must not be forgotten that dependence is mutual and Xstrata cannot afford to overlook the local population—it requires its support for the Mining Development Plan. The leaders especially know that the company needs their consent and, even more, their help in containing people's discontent. And the price they ask in the form of financial payments has increased greatly over the years (Romero 2008: 54, 55). This relationship of mutual dependence and strategizing on both sides, developed over two decades of interaction, makes many local leaders profit from the presence of the company. Romero (2008: 62) recounts how the Bla-an leaders she met when visiting the mining area were extremely busy meeting SMI officials. One leader for example, together with the Fulo Bato Tribal Council, was in the midst of negotiating for an annual 5 million Pesos (USD 116,225) royalty payment from the company. This compared to the 1.7 to 2.7 million Pesos the other four Tribal Councils get annually. The Fulo Bato leaders justified their demand with the explanation that Fulo Bato has higher grade ores than the other areas in Tampakan.

With a view to the role of local leaders it is noteworthy that the Tribal Councils as well as *barangay* and municipal officials sent petitions to the provincial board of South Cotabato when the latter was about to vote on its new Environmental Code—asking the provincial board to drop the provision that bans open pit mining (Romero 2008: 55).

Positions toward the company and its activities also depend on the level of intimidation, abuses and discriminations suffered or expected by those taking an opposing stance. This can be well illustrated in the case of the Sal Naong people. The Sal Naong Bong Banwu, led by Majon Malid

28 See Rambaldi (2010) on principles and applications of 3-dimensional mapping, Zingapan/de Vera (1999) for a mapping case in the Philippines, and Wenk (2012) for an analysis of 3-D mapping as a strategic tool in indigenous land claims.

29 See Wenk (2012) for a description of a similar dynamic in neighboring Bukidnon and Cotabato when an agri-business investor began operating on titled ancestral domain land.

as Tribal Chieftain during the 1990s, were the one group of Bla-an who opposed WMC and later SMI staunchly and for the longest period. When all four other Bla-an "host-communities" signed their first agreements with WMC in 1997–98 (see below), the Sal Naong continued to oppose exploration activities in their area and underlined their stance with a position paper. Consequently, the community was harassed for years and left out completely of the social services and benefits the consenting communities were regularly receiving. Thus, brought down on their knees, the Sal Naong finally signed their so-called Principal Agreement with SMI in August 2004 (Tomas/Celebrado 2008: 22, 26, 41f., 44; see also Sal Naong Tribal Council 1996).[30]

Still, today, there is certainly a considerable number of people who wish the company would leave. They are worried about their birthplaces, the transformation of their way of life, the physical destruction of the environment and the coming of outsiders. Others reject large-scale development for ideological reasons. Tomas/Celebrado (2008: 40) found a "silent minority" that opposes mining. But there are also those actively and violently opposing mine "development", such as the tribal armed groups claiming to lead a *pangayaw*, a war of revenge, against the company. The press frequently reports about encounters of tribal armed groups and (para)military units in and around Tampakan (see, for example, Sarmiento 2013a; Galdu 2012; Alyansa Tigil Mina 2013), but it is impossible to assess the actual situation from afar. Questions remain with regards to who is involved in opposition, what these people's motivations are, and how much support they have among the local population.

To conclude, it is crucial to note that the five directly affected Bla-an communities, i.e., the Tribal Councils of S'bangken, Danlag, Fulo Bato, Bongmal and Sal Naong, do not act jointly on the many issues that concern them (Tomas/Celebrado 2008:77). This may have to do with the decentralized political structure of *banwu* organization, yet unilateral actions are encouraged by the mining company as it negotiates different contracts with each Tribal Council. Competition and divisions among the five directly affected groups are in the company's interest to push its operation through.

Settlers and Civil Society

Tomas/Celebrado (2008: 37ff.) provide an analysis of stakeholders of the Tampakan mining project and divide the various interest groups into core and critical stakeholders. Core stakeholders are those whose means of subsistence, livelihood, and existence are affected positively and/or negatively by the presence of the mining company and its activities. This includes the so-called "host communities", both Bla-an and settlers, downstream communities, host municipalities and *barangay*, as well as the mining company itself.

The critical stakeholders, in turn, are those with social, political, or economic concerns related to mine development in the area. They include the various civil society groups opposed to mining and a company-supported pro-mining advocacy group, various state institutions such as the Office of the President, the Mines and Geosciences Bureau, the National Commission on Indigenous Peoples (NCIP), and provincial governments. The media and academe, as producers of knowledge and information about mining, likewise form part of this outer circle of stakeholders, as does the military, the NPA, and various other armed groups. In what follows we introduce further some of the core and critical stakeholders involved.

Those residents at the Tampakan mine site who are of lowland Visayan origin hold grudges against the Bla-an as well as the company because they feel they are being treated as less im-

30 The Sal Naong under Majon Malid only agreed for 229 hectares of their ancestral domain to be opened for the mining company's activities. This area they identified themselves. Should the company later wish to expand its area of operation, the consent of the traditional occupants must be sought in a community assembly (Tomas/Celebrado 2008:36).

portant than the indigenous population. While the Bla-an make increasing demands and are able to ground these on the IPRA and their ancestral domain rights, resident settlers and *barangay* officials complain about being overlooked and getting less from the company than the "IPs" (indigenous peoples) (Romero 2008: 62). Much resentment concerns land rights. Settlers as directly affected locals hold no separate agreements with the company but have been subsumed in the agreements between the single *barangay* and the company. They consequently feel their rights to land not adequately considered. Claiming to be "landowners" too, non-indigenous residents have formed five recognized landowner groups, which run under local names as "farmers" and "landowners" associations. These associations seek to have their own negotiations with the company. To have their own Principle Agreements with SMI/Xstrata has become a means of clarifying their legal status within the ancestral domains, in which many settlers now happen to reside. The aim is to negotiate directly with the company about proper compensation for all the investments done on the land. To be able to negotiate their own terms, Bisayan land claimants have previously made their grievances heard through public protests in the form of road blockades and barricades (Tomas/Celebrado 2008: 21f., 49f.).

The Bisaya also increasingly face the return of previous Bla-an owners to the land they now occupy and consider theirs. Bla-an return to assert that the land they once "sold" or gave to the Bisaya for a very small amount is actually still owned by themselves since it is within the ancestral domain—which, in turn, enables them to access compensation payments from the company. This trend of returning and re-claiming land triggered conflicts between Bla-an and Bisaya, the latter questioning the rightfulness of Bla-an ancestral domain claims. The settlers in Tampakan seem to have difficulties understanding the IPRA and the rights of the Bla-an that emanate from it. They complain about the increased confidence of the Bla-an in dealing with settlers ever since the first ancestral domain claims (CADCs) were recognized (Tomas/Celebrado 2008: 48f).

Viewed from a different angle, it must be noted that particularly those settlers claiming to be landowners are very often also *barangay* officials (Tomas/Celebrado 2008: 73). As such they benefit considerably from the company's presence and payments and are much better off than the average Bla-an. Both *barangay* and municipal governments are generally supportive of the ongoing mining project. The local government units perceive of the presence of the industry as a means to advance the local economy in their areas of jurisdiction, expecting value-adding businesses, employment and small-scale investments to flow from the mine. Much more critical seem to be the downstream as well as neighbouring Muslim communities, both voicing their concerns about mining in Tampakan, arguing that the potential environmental impacts will seriously affect their livelihood (Tomas/Celebrado 2008: 40).

The main critical stakeholders, aside from the government bodies discussed in the following subchapter, are civil society groups and the Catholic Church, which both take a clear anti-mining position, in particular the Social Action Center (SAC) of the Diocese of Marbel and Soccsksargends Agenda.[31] Philippine civil society is very lively, vocal and active, and internationally well connected.[32] The Catholic Church is a vitally important part of present-day life as well as the past of the Philippines and it has been crucial in shaping civil society in the country since the time of Marcos in the 1980s. The Church's commitment to social justice was expanded to include indigenous peoples' concerns. With a view to Tampakan, the SAC, which forms part of the national

31 Soccsksargends Agenda (South Cotabato – Cotabato Province – Sultan Kudarat – Sarangani – General Santos – Davao del Sur Alliance for Genuine Development) is composed of peoples' organizations and NGOs as well as church-based groups and campaigns against development aggression in general and large-scale mining in particular (Tomas/Celebrado 2008: 51ff.).

32 See Silliman/Noble (1998) for a detailed analysis of the Philippine NGO scene. With a particular focus on Mindanao and indigenous peoples' struggles, see Gaspar (1997) and Rood (1998).

Fig. 6: Campaign poster "Scrap the Mining Act".

network Alyansa Tigil Mina (ATM), is one of the strongest opponents to the mining project; its anti-mining stance is non-negotiable (Plantilla 2012; Tomas/Celebrado 2008: 57). The SAC of the Diocese Marbel also forms part of the Tampakan Forum,[33] which supports the local struggle against the Tampakan mining project inter alia through legal assistance. In its report "Mining in Tampakan: Intensifying Conflict, Danger in Perpetuity", for example, the Tampakan Forum (2011) assesses Xstrata/SMI's EISA, critiques the company's EIS, and elaborates on its own legal evaluation of the Tampakan mining project and the environmental and social impacts the project is likely to cause.

Tampakan is a crucial case for the national anti-large-scale-mining movement, which lobbies for the repeal of the Mining Act of 1995 and related policies (Tomas/Celebrado 2008: 51ff.).

33 The Tampakan Forum is a technical working group on the Tampakan mine which includes the Philippine Misereor Partnership Anti-Mining Inc. (PMPI), Alyansa Tigil Mina (ATM), the Philippine Association for Intercultural Development (PAFID), the Legal Rights and Natural Resources Center-Friends of Earth Philippines (LRC-KSK), London-based Philippine Indigenous Peoples Links (PIPLinks), the International Working Group on Mining in the Philippines, IUCN CESP-SEAPRISE and others (Galdu 2012).

This movement aims to "scrap" the Mining Act and promotes an alternative mining policy for the country, a People's Mining Act that is to be pro-Filipino, pro-poor and less damaging to the environment. Their demands include the cancellation of all FTAAs and other related contracts that permit large-scale mineral exploitation as well as the declaration of a moratorium on open-pit mining.[34] Their forms of protest include mass mobilizations such as caravans and rallies, fact-finding missions concerning human rights violations, educational campaigns about the environmental and social impacts of large-scale mining, and community-level organizing in direct impact areas such as the "strengthening" of Tribal Council leadership. To give two examples: on September 23, 2011, the SAC Marbel and the office of the governor of South Cotabato co-organized a public forum at the South Cotabato gymnasium to bring SMI/Xstrata consultants together with experts from the anti-mining groups to enable discussions about the social and environmental impacts of the Tampakan mining project. Clive Wicks and Dr. Robert Goodland presented their critique on SMI's EISA and EIS. After that, Bishop Dinualdo Gutierrez of the Diocese of Marbel said: "No mining project will ever be good for us. Let us continue to support the South Cotabato Environmental Code that bans open pit mining" (Tampakan Forum 2011). In December 2012, people from Mindanao staged a 10-day-action called "Manilakbayan" in Manila, visiting mining companies' headquarters. The delegates of the "Manilakbayan" were organized and supported by environmental organizations linked to the Kalikasan People's Network for the Environment (Kalikasan PNE) and Panalipdan!-SMR[35] (Salamat 2012).

Civil society actors opposed to large-scale mining increasingly also make use of the Internet and new social media like Facebook to support their struggle. There is a Facebook-page entitled "No to Mining in Tampakan, South Cotabato, Philippines," which provides regular information about current mining issues in the Philippines and lobbies actively against the Tampakan project. The group is based in General Santos City and involves mainly students. Also releasing their latest news on Facebook and taking a strong stance against the killings of the Capion family is one of Mindanao's Lumad umbrella organizations, the Mindanao Alliance of Indigenous Peoples (Kalumaran). Aligned with the left-oriented National Alliance of Indigenous Peoples' Organizations in the Philippines (KAMP), Kalumaran declared that the tribal war—*pangayaw*—staged by some Bla-an against SMI/Xstrata was "just" (Cahayag 2012; see also Nawal 2012b).

It is crucial to recognize that the civil society organizations mentioned are not necessarily locally based or even well connected. They too, much like pro-mining parties such as the company itself, must engage in "community organizing" to find local allies and get local people and their leaders to adopt a general anti-mining position. This proves difficult and, if achieved, often short-lived, given the lures of the company. During the 1990s, when opposition to the presence of the mining company was still strong, especially among the Sal Naong, the SAC and Soccsksargends Agenda were highly influential and provided valuable advise and support to those resisting (Tomas/Celebrado 2008: 57). However, with all Principle Agreements in place (see below), their influence has waned (Tomas/Celebrado 2008: 57). This finding is supported by Daniel Hostettler, who visited Tampakan for a fact-finding mission in January 2013 and found "local resistance" primarily organized by the Catholic Church, with hardly any visible local presence aside from a few anti-mining slogans.[36]

34 See Reckordt (2012) on developments with regard to an alternative mining law in the Philippines.

35 Panalipdan!-SMR is a multi-sectoral alliance for environmental awareness and protection in Southern Mindanao. Its Secretary General is Benedictine Sister Stella Matutina, who has played an active role in the struggle against large-scale mining in the Philippines (Olea 2012; Basa 2011).

36 Email communication with D. Hostettler from the Swiss Catholic development NGO Fastenopfer, March 11, 2013. Contrary to this, at a Philippine Studies conference at the University of Lucerne (October 19, 2013), a colleague of Hostettler insisted that there is very strong and broad-based local opposition against the mine in Tampakan.

Romero (2008: 55) underlines this observation when she states that the Bla-an have largely rejected civil society groups' approaches. In positioning themselves not necessarily in opposition to the mining company but in what might be called actively negotiated dependence, the locals expect to benefit more than by taking a non-negotiable anti-mining position. Anti-mining NGOs hold little alternatives in store for the local population in comparison to the company's offers of livelihood programs, jobs, and payments. Bla-an leaders and Tribal Council members, *barangay* and municipal councillors have learned to engage with the company, and mutual dependence between SMI and these leaders allows each side to pursue its particular goals. How much the general population benefits from this arrangement remains an open question.

2.3 Corporate Strategies and the State

In its annual Sustainability Reports, SMI/Xstrata demonstrates its project plans and the possible benefits of mining in Tampakan. The 2011 report has a lot of community and environmental programs lined up and one can gather a number of strategies applied to push the Tampakan project. First of all, Xstrata/SMI[37] calls itself a "Filipino company with a commitment to the highest standards of sustainable development (SD)" (Xstrata/SMI 2011: 19; see also Goodland/Wicks 2012). It claims that the approach in Tampakan complies with international standards.[38] The company monitors and evaluates its sustainable development (SD) performance through an internal, SMI-run Sustainable Development Committee, which holds monthly meetings and regularly reports to the Xstrata Copper Executive Committee and Board (SMI/Xstrata 2011: 20).

SMI/Xstrata has reacted to the heavy criticism on their proposed mining operation in Tampakan by proactively addressing many of the issues criticized and by seeking contact with civil society organizations.[39] In response to the strong criticism on increased militarization and the company's involvement with paramilitaries and security task forces (see Cahayag 2012; Sarmiento 2013a; Sarmiento 2013b), the company claims that the SMI private security guards underwent screening on human rights and a National Bureau of Investigation background check. Voluntarily, a Human Rights Security Assessment was developed as part of the company's Social Impact Management Plan and all employees of SMI undergo "cultural induction training" in order to understand and appreciate Bla-an culture and history (SMI/Xstrata 2011: 28).[40]

37 We refer to this business partnership simply as "the company", albeit realizing that it is itself a complex arrangement of diverse parties.

38 Such as the International Council on Mining and Metals Sustainability Principles and Guidelines (ICMM 2013), the United Nations Global Compact, Voluntary Principles on Security and Human Rights, and relevant International Organization for Standardization (ISO) standards, the United Nations Universal Declaration of Human Rights and International Labour Organization conventions (SMI/Xstrata 2011: 19, 27).

39 Including Germany-based Misereor Development Policy Department, the Fastenopfer Philippine Desk in Switzerland, and Philippine-based legal assistance NGO Saligan. The company also claims to have sent response letters to local organizations such as the SAC of Marbel, the Columbio Multi-sectoral Environment Movement, the Diocese of Kidapawan and Catholic Church leaders (SMI/Xstrata 2011: 45).

40 We doubt the success of such "cultural induction". To give an example: The company's EIS lists 23 "cultural heritage/ archaeological sites" that were identified according to World Bank guidelines (AECOM/SMI 2011: 16f.). These are listed and mapped and there is a protocol about how the company is to manage burial sites and possible archeological finds. In the case of a burial site the protocol reads: "Leave the area undisturbed and make an estimate of the possible range wherein burial practice is feasible (...). If skeletal remains have been exposed, carefully collect them (...). Preserve the site and take appropriate measures to prevent further damage (...)." Then, "Inform the Municipal Health Officer and the Municipal Sanitation Officer or otherwise bring him to the site. It will be very useful, especially for proper determination of the site, if old local informants who had been there ever since will be asked to substantiate the finds. If it is established that the burial site belongs to the ancestors of the IPs in that area, then the local NCIP chapter should be informed. They

In its Sustainability Report 2011, SMI/Xstrata also addresses security challenges as evidenced in the fatal shootings of three suppliers in Bong S'bang in March 2011 and the killing of its Offsite Land Acquisition Superintendent in Davao City in September 2011. Further issues discussed are the distribution of project benefits, the company's Resettlement Action Plan (RAP), and water management. As mentioned earlier, the RAP is still being developed with consultations involving the NCIP, Local Government Units (LGUs) and the Mines and Geoscience Bureau (MGB). In order to start resettlement, SMI requires the Free, Prior and Informed Consent of all indigenous people affected (SMI/Xstrata 2011: 22). In this connection, Xstrata CEO Mike Davis stated at the Xstrata General Assembly in Zug on November 20, 2012: "We have not commenced any relocation activities and we will not do so unless we have received the Free, Prior and Informed Consent of the affected indigenous people—that is required by Philippine law (...). We expect that the process will be underway next year. Activities will be in partnership with the Philippine government's National Commission of Indigenous People."[41]

The IPRA defines the FPIC as the consensus of all members of the indigenous peoples to be determined in accordance with their respective customary laws and practices, free from any external manipulation, interference and coercion, and obtained after fully disclosing the intent and scope of the activity, in a language and process understandable to the community (IPRA, Ch. 2, Sec. 3g). It provides that a project cannot proceed in the absence of such consent. In order for mineral extraction to actually commence, therefore, SMI/Xstrata need the consent of all indigenous groups residing in the Tampakan Project area (Xstrata/SMI 2011: 48). In fact, experts insist that the right to give or withhold (!) the FPIC should not only be recognized with regard to the directly affected "host communities" but with regard to those living downstream as well. Tomas/Celebrado (2008: 65) call this the requirement to "follow the waste stream" of mining.

Despite SMI/Xstrata's alleged adherence to the IPRA's provisions, the FPIC and earlier consent acquiring processes conducted in the Tampakan area thus far seem to have been seriously flawed (Goodland/Wicks 2012: 4). And despite the provisions for FPIC in both the IPRA and the Mining Act, evidence suggests that the government agencies in charge of mining-affected areas, i.e., the Mines and Geosciences Bureau of the DENR and the NCIP[42], have failed to implement these provisions with a view to the people affected.

The crucial agreements in place with local stakeholders are the so-called Principal Agreements or PAs. In its EIS, the company defines PAs as "voluntary cooperation agreements between the proponent and host communities that outline reciprocal rights, commitments, and obligations of the parties within the context of the project exploration phase" (AECOM/SMI 2011: 36). Basically, PAs establish the consent of the local communities and other locally relevant stakeholders to the mining activities. Note that the first round of PAs under WMC was written in English and drafted solely by the company. For the Bla-an who signed, this was/is an unfamiliar language, even more so since such agreements are very legalistic and do not follow any culturally familiar norms for parties to come to terms with one another. The PAs were signed without previously providing the local people with information about the actual plans for and implications or risks of mining. Rather, the material benefits of the company's presence and activities were highlighted (Tomas/Celebrado 2008: 22).

could give advice as to the conventions regarding such cultural feature." (AECOM/SMI 2011: 19). Note how a disturbed burial site is foremost a health and sanitation issue, how old locals are only useful to substantiate the find, and how then yet another government agency—the NCIP—and NOT the local people whose ancestors' remains have been unearthed are to advise the company on what to do with these "features".

41 Livestream of the Xstrata General Assembly in Zug on 20.11.2012; downloaded from <www.xstrata.com> on 22.11.2012.

42 The NCIP was created through the IPRA to implement the statute. It is currently engaged in undertaking the FPIC process for Xstrata/SMI (NCIP 2013).

There are “Bla-an PAs” in place with five communities (Danleg, Fulo Bato, S’bangken, Salnaong and Bong Mal), “Barangay PAs” with five *barangay* (Danlag, Tablu, Pula Bato, Datal Blao and Kimlawis), and “Municipal PAs” with three municipalities (Tampakan, Columbio and Kiblawan). All of these, with the exception of Sal Naong, were executed by SMI predecessor WMC, mostly in 1997 and 1998. SMI adopted these agreements when it took over in 2004. In the same year, the Sal Naong PA was signed. In 2009, the project site area was expanded, including new “host communities”. SMI since seeks to update its agreements with affected communities and local governments—some have been newly concluded, some are in on-going negotiations (see list in AECOM/SMI 2011: 36f.; Xstrata/SMI 2011: 48; Tomas/Celebrado 2008: 22).

According to Tomas/Celebrado (2008: 44), the *barangay* councils, as local representatives of government, signed their PAs not quite voluntarily. They were pushed into signing because the municipal governments had signed before them, thus putting a stamp of approval on the mining exploration at a higher government level. The *barangay* councils consequently signed their PAs even though they did not fully understand the agreements’ content and terms, given they were written in English and presented to them by company lawyers together with OSCC (Office of Southern Cultural Communities) personnel,[43] who both highlighted the financial benefits of the company’s presence. Like the Bla-an communities at the time, the five *barangay* signed without having any legal advisers and without being allotted enough time to understand, let alone negotiate the terms they agreed to.

With the IPRA in place and taken notice of, agreements must nowadays be written in a language comprehensible in word and meaning to the local people concerned. Whether this is actually being done should be closely followed in the near future. Based on the IPRA, the mining company is supposed to now conclude a Memorandum of Agreement (MoA) with each affected community as it moves towards extraction. A MoA “should clearly state the consent of the affected indigenous community, and the commitments and obligations of all parties concerned”. In order to enter into a process to obtain a MoA, the community must give its FPIC. Whether such MoAs existed at the time the EIS was written, i.e., in 2010–11, is not stated in the EIS but the recognition of the process is there.

But who is “the community” that gives (or withholds) consent? The issue of representation is crucial and usually (often willingly) overlooked not just by the company and state agencies such as the NCIP but by leaders (and people) themselves as well. The IPRA requires that consent be given by the entire indigenous population within the ancestral domain. In practice, this has come to mean that the leaders who negotiate with company representatives are actually the ones giving consent (Tomas/Celebrado 2008: 65).[44] To what degree all the Bla-an people are, or have in fact been, involved in the conclusion of agreements with the company must be subject of further (field) research.

In order to have the proper “community” representatives in place, companies as well as the government—through its agencies in charge of administering indigenous (formerly “native”) people(s)—engage in what is called “community organizing”. In the case of Tampakan, these interventions have been considerable. In the second half of the 1990s, the company facilitated the organization of Tribal Councils (TCs) in the area. The formation of these was at the time promoted and implemented by the OSCC and the company saw to it that the process was rapidly wrapped up and served its purposes. WMC needed these institutionalized bodies as official representations of the “tribal communities” in order to negotiate the terms of its presence in the area. In return for

43 The OSCC was the NCIP’s predecessor agency in charge of the affairs of then so-called ”Cultural Communities“ in Mindanao.

44 See Wenk (2012) on this problem in post-titling ancestral domain management.

the "communities"' cooperation, WMC helped process the various CADC applications (except in the Sal Naong case), provided monthly financial assistance of up to PhP 30,000 (USD 695, conversion date 29 October, 2013), and facilitated basic services demanded by the TCs and/or the single communities. After the communities had signed their first PAs, the company's financial assistance was increased to an annual one million Pesos (USD 23,163) (Tomas/Celebrado 2008: 44).

The Certificate of Ancestral Domain Claim (CADC) processes, which were to give local Bla-an groups tenure security according to the logic of community-based forest management en vogue in the 1990s, were consequently enabled by WMC through the provision of financial and technical assistance. Required personnel such as an anthropologist, an ethno-historian, and a project coordinator was financed or hired directly by the company (Tomas/Celebrado 2008: 21, AECOM/SMI 2011). For the ethno-historical study on the Danlag people and proof of time immemorial occupancy that formed part of the claim, a well-respected Mindanao scholar was materially supported by WMC through its Community Relations Program (Rodil 1999: 14, 29). Two of the approved CADCs were later converted into CADTs under the IPRA, meaning they were turned into actual communal land titles. This again happened with the financial assistance from the mining company, this time SMI/Xstrata (Tomas/Celebrado 2008: 21).[45] That a company so actively intervenes and finances a people's claim to territory and self-government—the territory it then intends to mine and transform forever—is highly problematic. Not only because this goes against all of the law makers' intent of enabling people's self-government but also because it creates dependencies and hierarchies that largely serve the company's purposes.

In addition to the Tribal Councils and company-financed ancestral domains, WMC and later SMI/Xstrata promoted the formation of so-called Tribal Foundations (TF). At present, there are five such foundations: S'bangken, Danlag, Fulo Bato, Bongmal, and Sal Naong. The Sal Naong Tribal Foundation was formed only in 2004, after the tribal leaders gave up their opposition and finally signed their Principal Agreement. The TFs are initiated, recognized and maintained by the mining company and constitute the institutional arrangement through which the community development fund of Xstrata and other assistance is managed. They are also supported by the company—under the heading of "capacity building"—in their strategic planning with regard to the management of the ancestral domains. Each TF has a Board of Trustees (BoT) that is composed of three Tribal Council members, two NCIP representatives, one DENR-MGB representative and three mining company representatives. Beside the BoT, there is an executive officer, an accounting clerk and a clerical person (Tomas/Celebrado 2008: 26f., 33, 44). An important point of contention between the local population and SMI is the perceived control of SMI over the money allocated to and channelled through the TFs. Due to the equal representation of tribal council and company representatives, the SMI representatives in the BoT are able to control not only how TF funds are utilized but when they are released. The Bla-an leaders involved insist that they should be allowed to control the allocation of the funds in the name of self-government and because they are the ones most familiar with the needs of the people (Tomas/Celebrado 2008: 44, 47).

The company's stakes are high: it is about to develop one of the biggest copper and gold deposits in Southeast Asia and intends to invest up to USD 2 billion for this. And it needs the locals' consent at large to be able to proceed. The money that annually flows to the Bla-an, therefore, is considerable. SMI spends 27 million Philippine Pesos a year in financial assistance to the Bla-an communities (USD 626,000, conversion date 24 Oct. 2013), the *barangay* and the municipal governments. To each of the five Tribal Councils (TCs), it allots PhP 1.7 to 2.7 million, to the

45 The involvement of the company in the CADT processes was confirmed by Kail Zingapan of PAFID, one of the main NGOs doing land delineations for ancestral domain claims. Zingapan says that the delineation of the issued CADTs as well as the census undertaken were funded by SMI and are therefore contested (personal e-mail communication, April 1, 2013).

barangay units 2 million, to the Tampakan town government 5 million PhP. Forty percent of the 1.7 million given to each TC go into the scholarship program, less than one-fifth to livelihood, and another two-fifths to the TCs operational and administrative expenses. One million Pesos go to each community for livelihood programs and emergencies. It is the company that controls how the latter money is spent (Romero 2008: 54).[46]

Xstrata is required to pay a number of taxes to the local governments. These include an income tax and real estate tax as well as an annual occupation fee of PhP 10 per hectare (USD 0.23). According to Tomas/Celebrado (2008: 33), however, the *barangay* units at the time of their research in 2007 were not aware of this occupation fee payment or how they were supposed to benefit from it. The authors further report that Xstrata sets aside a minimum of one percent of group profit before tax each year to fund social programs that aim at sustainable development for the communities (Tomas/Celebrado 2008: 25f.). Such programs are implemented through the company's Community Relations and Development Department (CRD), which, apart from "community development", is in charge of a "heritage survey", the grievance mechanism initiated, as well as "community advocacy". In their heritage survey, the CRD identifies sacred sites before drilling and determines claimants; its grievance mechanism spells out procedures for conflict management; and through community advocacy, the local population is to be transparently informed and "educated" about project-related activities. For these purposes, CRD personnel are stationed in the field to regularly interact with the local population, i.e., to undertake "community dialogues" and assemblies, and to attend *barangay* and tribal council meetings.

In its Partnership Development Framework (PFD) initiated in 2007, Xstrata claims to pursue a "community demand-driven approach to sustainable development" with "community organizing" as its core strategy. With this, the company claims that it is in fact the "communities" that lead the initiated process of "development" and portrays itself as the benevolent well-meaning patron, who helps them in their efforts to improve. In fact, however, the "community development programs" are based either on company initiatives or form part of its community development commitments based on the signed PAs, which obviously are also company-initiated (Tomas/Celebrado 2008: 27). The initiatives cover programs in the fields of education, health, agro-forestry, capacity building and technical support in basic infrastructure, with education and health clearly being the company's flagship programs. And they are directed foremost at the local Bla-an, not the Bisayan inhabitants. Under its Corporate Social Responsibility (CSR) programs, SMI/Xstrata account for "community investment activities" called Corporate Social Involvement (CSI). Developed together with Local Government Units (LGUs) and development organizations, such activities are said to be in line with the UN Millennium Development Goals. Some of the programs that the SMI CSI contains are the Learning Values through Education (LIVE) program, which offers financial support to Tampakan families for school fees. There is also the Creating Access to Development Education Program (CADEP) and the Early Childhood Education Program in Malungon (SMI/Xstrata 2011: 43).

It is noteworthy how outside actors such as mining or agribusiness companies put education at the forefront of their community development. They—representing the well-educated—bring the 'right' kind of knowledge to the illiterate hill people, while completely disregarding and disrespecting existing local knowledge/education, and destroying the territorial basis on which such knowledge rests. The illiterate uplanders must be educated in schools, they must be educated in mining, and then their lives will improve. 30 to 60 percent of the funds going to the Tribal

46 The livelihood programs, says Romero, are not very successful and hardly reach all of the people. She gives an example of the TCs lending corn seeds to Bla-an farmers through the Tribal Foundations. This benefited the farmers as they did not have to buy the seeds. However, the corn seeds were a loan very few of the forty beneficiaries were able to repay, which made the program not last very long.

Foundations are, consequently, allocated for education in the form of scholarships from elementary to college levels. The company, through the *barangay* councils, further assists in the salary to public school teachers and the construction of school buildings (Tomas/Celebrado 2008: 27-29).

Health, the second flagship, includes the provision of basic health services, especially in remote settlements, which includes immunizations, circumcision, consultations, maternal and childcare, family planning, sanitation and distribution of iodized salt. These services are delivered through regular medical and dental missions involving further actors such as local governments and provincial medical/dental clinics (Tomas/Celebrado 2008: 30).[47]

The company also provides assessments of the basic services available to the local population—on which it then bases its projects (AECOM/SMI 2011: 3). Further community development programs concern technical assistance for land use and development planning, and for the development of a "sustainable growth plan" to local government units (Tomas/Celebrado 2008: 30). To minimize environmental impacts, the company has initiated a number of projects such as a "community-based environmental monitoring program" or a "multi-sectoral reforestation project". The latter intends to plant 250,000 trees annually in denuded areas around the project area (Tomas/Celebrado 2008: 23f.).

For its allegedly innovative and highly generous performance in Tampakan, Xstrata/SMI has received numerous CSR awards from government agencies, other mining companies and from an emerging CSR industry (Romero 2008: 53).[48] To what degree such projects are successful and from whose perspective is difficult to assess without being in the field. To us, the kinds of CSR initiatives initiated seem a mitigating measure to keep people busy and distract them from the serious environmental impacts of mining yet to come. Very hypocritical appears to be the establishment of a "field trial and research station" in Sitio Maticura in Tablu, Tampakan in 2003, which is envisioned as a "center for conservation" and for "experimentation on potential income earning opportunities for the local communities". It houses things such as a butterfly and bee farm (Tomas/Celebrado 2008: 24). While a people's habitat is planned to be destroyed in its entirety and for long-term, the people affected are being "educated" about conservation.

We conclude here with a discussion of a newer development in the Tampakan mining project, namely the issuance of the Environmental Compliance Certificate (ECC) to SMI/Xstrata in early 2013. The ECC is an important step towards full operation as it allows the company to start with actual mine development and hence to proceed from exploration to extraction. The ECC, issued by the DENR, assures that the mining operation, its installations and facilities do not have a significant impact on the environment and that the mining operator has complied with all the requirements claimed by the company's EIS. In the case of Tampakan, the ECC was initially denied to Xstrata/SMI in January 2012 but a year later, in February 2013, the DENR gave in and issued it nonetheless (Kalikasan People's Network for the Environment 2013; Reuters 2013; Malaya 2013).

The initial denial of the ECC by DENR secretary Ramon Paje was largely based on the South Cotabato ban on open pit mining, despite the fact that the Environmental Management Bureau's (EMB) Impact Assessment Review Committee had recommended the issuance of the ECC to

47 It is not mentioned in the texts available but highly likely that the Armed Forces are usually also involved in these medical missions to the mountains. As I have argued elsewhere based on my own observations, medical missions not seldom serve as entry points for the military into upland areas under NPA or tribal armed groups' influence and as such form part of the Armed Forces' counter-insurgency efforts throughout Mindanao (see Wenk 2012).

48 Including: The Presidential Mining Industry Environmental Award in 2006; the New Zealand Chamber of Commerce Corporate Social Responsibility Leadership Award. Further awards from the Philippine Department of Environment and Natural Resources (DENR) for its reforestation efforts and leading communication initiatives (Tomas/Celebrado 2008: 25; SMI/Xstrata 2011: 9). Note that this all happens in the exploratory phase where it is still relatively easy to put on a green jacket.

SMI. The Tampakan Forum hailed the denial of the ECC, yet criticized the agency's reluctance to bring forward other crucial reasons for withholding the ECC in 2012, such as the risk to life, property, and old growth forests (Alyansa Tigil Mina 2012b). The DENR as responsible government agency at the forefront of these developments as well as the Aquino government have expectedly been heavily criticized when the ECC was issued a year later. The role of the DENR hereby is problematic. While, according to its mandate, the DENR "shall be the primary government agency responsible for the conservation, management, development, and proper use of the country's environment and natural resources, specifically forest and grazing lands, mineral resources, including those in reservation and watershed areas, and lands of the public domain" (DENR 2013), the position of the agency is generally understood to be pro-mining. The DENR is further said to frequently issue mining permits or certifications such as the ECC without properly undertaking consultations and FPIC processes with the affected local populations (see Holden/Jacobson 2006: 192f.).

In all of this, the provision in the South Cotabato Environment Code of 2010 that bans open-pit mining constitutes an unusual stance against the general trend of stomping towards full mineral extraction in Tampakan.[49] It was a groundbreaking day on June 29, 2010, when South Cotabato Governor Daisy P. Avance-Fuentes signed the South Cotabato Environment Code into law. This virtually meant putting a halt to the Tampakan mining project. Fuentes signed the law one day before she stepped down as governor. When asked why she had signed the Environment Code, Fuentes cited the study "Philippines: Mining or Food?" by the Working Group on Mining in the Philippines (Goodland/Wicks 2008) and underlined her responsibility as environmental steward (Sarmiento 2012). In the May 2013 local government elections, Daisy P. Avance-Fuentes was re-elected as provincial governor and she is determined to uphold and implement the controversial Code, which her opponent and predecessor sought to lift. The ban on open pit mining will thus likely remain in place at least until the next elections in 2016, providing a major hindrance for the company to proceed to commercial extraction.

3. Conclusion

Multifaceted interaction between the local population in the projected Tampakan mining area and the successive mining companies has been going on since the early 1990s. And yet, mineral extraction has yet to begin and is currently projected to commence in 2019. The positionings of those involved that have emerged over the past two decades are varied and complex, have changed over time, and continue to be subject to intense negotiations. Concluding, we want to highlight three findings that seem the most relevant and outstanding also beyond this particular case. The first is the existence of a vocal and active regional, national and international opposition against the mining project in question that seemingly finds little endorsement among the local population. The second is that in the case at hand, the mining company has successfully made use of legal provisions formulated to protect indigenous peoples' rights against encroachment and displacement to achieve what indigenous rights are here to prevent: gain access to indigenous lands and eventually resettle the entire population. And this with the (bought) consent of the people concerned. The third finding concerns the transformation of inter-ethnic dynamics within the project site and increasing competition for land.

While the international campaign that condemns mining in Tamapakan is active and strong and admittedly has very good reasons for warning and calling for a stop to this mining endeavour,

49 The Environment Code can be viewed at <www.southcotabatonews.com>, accessed on October 27, 2013.

we are able to show, based on the available data, that large parts of the local population support the mining company's presence and activities in principle. Yes, there is opposition—some even violent—against future mineral extraction and the current activities of SMI/Xstrata among the local population, and there used to be strong opposition in particular project areas in the past. But the strategy the local people at large seem to pursue is one we propose to call actively negotiated dependency. This is to say while there are serious power inequalities at play between the company and the local population, the latter nowadays actively negotiate their terms and costs for the company to be in the area. Mining, for the host communities, both Bla-an and Bisaya, is, as Tomas/Celebrado (2008: 67) note, recognized as a means to avail of immediate and short-term benefits in the form of employment, the provision of basic services, education and health (note that all of these things the state is supposed to provide or help create but in the case at hand fails to do so). Most leaders have learned how to make demands and use their power to get short-term benefits out of a situation that is far from ideal. As in many other comparable situations, indigenous leaders' influence has been bolstered through their role as intermediaries between their own constituents and the company. In fact, many a leader's legitimacy to hold their present positions of power would cease to exist were the company to leave.

The existence of the IPRA adds a very crucial dimension to this dynamic of company-locals interaction: because the Bla-an in this area are the rightful owners and claimants of several ancestral domains, they are entitled to give or withhold consent to the company's activities and have the right to decide how mineral and other natural resources within their territories are being used and managed. It is this negotiating power, we argue—and the fact that the company depends on the Bla-ans' consent—, that leads local representatives at large to embrace the possibility of development the company offers.

People's general consent to the presence and activities of the mining company does not mean there is no contention with the company. People deeply worry about the impacts and effects of the mine on their ways of life and the environment. Yet, their contentions are not aimed against the company or mining as such but focus on particular issues such as employment, compensation and community development packages, the terms of which they seek to actively shape instead of resist. This is also because single individuals and communities have paid a heavy prize for opposition. And certainly, the amounts of money the company pays to the communities, the *barangay* and the municipalities have greatly helped in securing and maintaining people's consent.

The above said leads to our second finding, which is that the mining company has successfully made use of progressive legal provisions formulated to protect indigenous peoples' rights to land, territory and natural resources to achieve its goal of securing people's consent. From an indigenous peoples' rights perspective, this is exactly what should not have happened. Legal statutes such as the IPRA were designed to keep companies off indigenous lands, not to provide access to them. What SMI/Xstrata and its predecessor WMC were able to do, however, was to support the indigenous population in its claims to ancestral territory—providing an air of self-government—and then negotiate with the designated leaders the terms of land use and the price of people's consent to mining and to their own eventual resettlement. The people in turn seem to have largely embraced the company's support for their ancestral domain claims as well as their position of being able to make money out of being the owners of the land. The assertion of indigenous peoples rights is thus not realized in opposition to an encroaching party (be this a company or government), but instead happens in collaboration with a resource extracting party. How this was possible and what this means not only for local assertions to territory and resource rights but for indigenous rights advocacy at large should be guiding questions for future research. Thirdly, looking beyond company-indigenous interaction and relations, it is crucial to keep in mind that processes of transformation concerning the social, political, economic and religious life of the local Bla-an did not commence with the company's entry into the area but were already ongoing due to settlers, loggers and governmental interventions at the time WMC entered in the early 1990s.

The company further complicated ongoing dynamics of conflict and accommodation. Today, one finds resentment against one another on both sides, among Bla-an and Bisaya alike. A number of issues have been mentioned in this respect: the Bla-an, constituting a majority population in the designated mining area of approximately 67 percent, feel discriminated when it comes to employment. Those of Bisaya origin, in turn, feel discriminated against because they are not getting as much attention as do the indigenous Bla-an. Having been the stronger and government-backed party at the settlement frontier for decades, settlers nowadays face an empowered indigenous population and leaders, who voice their demands and are being considered first hand by the company.

The presence of the company, furthermore, leads to a renewed interest in land within the designated mining area in order to benefit from the mining company's compensation packages, both by Bla-an and settlers. The question of landownership and the problem of land disputes due to overlapping claims thus becomes an issue of primary importance and concern for local residents as well as the company. The legal situation is complex: With a CADT in place, the Bla-an are the rightful owners of the land, even where settlers have previously acquired land from individual Bla-an through "sale", prenda arrangements and the like. For one, such transfers were often unrecorded and hardly ever resulted in settlers acquiring actual individual title. Secondly, even where settlers formerly acquired titles, most of this land could never legally be owned as private land because upland areas such as the one in focus were classified as forestland and therefore state-owned. It is only with the creation of ancestral domain as an official land classification under the IPRA that such areas can be privately owned—by indigenous peoples only. Consequently, when it comes to land rights, settlers stand to lose vis-à-vis the indigenous Bla-an.

However, this is not quite the whole story, because the company seems nonetheless to value settler claimed or "owned" land higher than that of the Bla-an. The latter's land is said to be of little value because it is not individually titled. In the end, settlers might thus still be getting more compensation for their land than do the Bla-an. The dynamic triggered through the company's system of land valuation for compensation purposes is, consequently, creating a land market, in which settlers and Bla-an, residents and non-residents, increasingly compete.

Update

After finalizing the takeover of Xstrata by Glencore in 2013, CEO Ivan Glasenberg announced a radical shift away from Xstrata's strategy of developing expensive 'greenfield' mines, towards investing foremost in cheaper 'brownfield' expansion. It was estimated that about half of Xstrata's 88 projects would not proceed, or be sold off, after reviewing them.[50] The Tampakan project was put on hold: SMI drastically reduced its staff and activities in the project area, with rumours about its imminent sale appearing in the press. In January 2015, Glencore sold its shares of Indophil Resources to the Philippine-based Alsons Group.[51] Finally, in June 2015 Glencore issued a short press release announcing a "share sale agreement with an affiliate of Indophil for the sale of its entire remaining interest in the Tampakan Project",[52] without naming further details. Glencore's withdrawal was commented on by an NGO observer as "a victory of sorts", though the practical implications have yet to be analysed.[53]

50 James Wilson and Neil Hume: Ivan Glasenberg shows the 'Glencore way' to cut Xstrata 'fat'. *FinancialTimes,* 10.09.2013.

51 <www.indophil.com>, accessed 18.2.2015.

52 Glencore Press Release, 24 June 2015: "Sale of Glencore Stake in Sagittarius Mines Inc and Tampakan Project".

53 Per Email, 24 June 2015.

References

AECOM/SMI 2011: Environmental Impact Statement of the Tampakan Copper-Gold Mine Project. Sagittarius Mines, Inc., 1 April 2011, Document No. R10-033.

Ballard, Chris and Glenn Banks 2003: Resource Wars: The Anthropology of Mining. *Annual Review of Anthropology* 32: 287–313.

Caballero-Anthony, Mely 2007: The Philippines in Southeast Asia. In: Severino, Rodolfo C. and Lorraine Carlos Salazar (eds.): *Whither the Philippines in the 21st Century?* Singapore: Institute of Southeast Asian Studies, 1–17.

Carrier, James G. and Paige West 2009: Introduction: Surroundings, Selvers and Others: The Political Economy of Environment and Identity. *Landscape Research* 34(2): 157–170.

Castro, Nestor T. 2000: Three Years of the Indigenous Peoples' Rights Act: Its Impact on Indigenous Communities. KASARINLAN, *Philippine Quarterly of Third World Studies* 15(2): 35-54.

Christian Aid and PIPLinks 2004: Breaking Promises, Making Profits: Mining in the Philippines. London: Christian Aid and Indigenous Peoples Links. <www.piplinks.org> accessed on 30.10.2013.

Cole, Fay-Cooper 1913: The Wild Tribes of Davao District, Mindanao. Chicago: Field Museum of Natural History, 170.

Crystal, Rey 1982: Overview of Land Settlement Schemes in the Philippines. In: *Population Resettlement Programs in Southeast Asia.* G. W. Jones and H. V. Richter (eds.). Canberra and Florida: The Australian National University, 93–103.

Dalupan, Cecilia G. 2000: A Discussion Paper on the Mining Industry and the Indigenous Peoples Rights Act. Not officially published consultancy Paper, <www.sdsg.org> accessed on 23.7.2013.

Feld, Steven and Keith H. Basso 1996: Senses of Place. Santa Fe: School of American Research Press.

Gaspar, Karl M. 1997: The Mindanao Lumad Social Movement. *Mindanao Focus* 15(1).

Gatmaytan, Dante B. 1989: The Civil Reservation: The Concentration Policy as Land Acquisition Scheme. *Philippine Natural Resources Law Journal* 2(1): 1–9.

Gatmaytan, Augusto B., Maricel P. Hilario-Patino, Rosalinda C. Tomas, Leah H. Vidal, Virgilio G. dela Rosa, Jessie B. Manuta, Ronald Ryan G. Olarte and Lourdes R. Simpol. 2013. Research Brief Mining and Water Governance, August 2013. Davao: Ateneo Institute of Anthropology and Ateneo TropICS (Tropical Institute of Climate Studies).

Go, Miriam Grace A. 2008: First, Please Clean Up. Newsbreak Special Issue "The Big Dig: Mining Rush Rakes Up Tons of Conflict". Quezon City: Public Trust Media Group Inc., July/Sept. <www.newsbreak.com.ph>.

Goodland, Robert and Clive Wicks 2008: Philippines: Mining or Food? <www.piplinks.org/miningorfood>, accessed on 06/01/2012.

Goodland, Robert and Clive Wicks 2012: Working Group on Mining in the Philippines, DRAFT Response to Xstrata/ Indophil/SMI's ESIA of the Tampakan Mine Project in Mindanao, Philippines, Revised Version 8th February 2012.

Hamm, Brigitte, Anne Schax and Christian Scheper 2013: Human Rights Impact Assessment of the Tampakan Copper-Gold Project. Aachen, Germany and Luzern, Switzerland: Bischöfliches Hilfswerk MISEREOR und Katholisches Hilfswerk FASTENOPFER.

Hayase, S. 2007: Mindanao Ethnohistory Beyond Nations: Maguindanao, Sangir, and Bogobo Societies in East Maritime Southeast Asia. Quezon City: Ateneo de Manila University Press.

Holden, William N. 2005. Indigenous Peoples and Non-Ferrous Metals Mining in the Philippines. *The Pacific Review* 18(3): 417–438.

Holden, William N. and R. Daniel Jacobson 2006: Mining Amid Decentralization. Local Governments and Mining in the Philippines. *Natural Resources Forum* 30: 188–198.

Hyndman, David and Levita Duhaylungsod 1998: Ancestral Domain for Large-Scale Mining. The Dynamics of Land Rights and Resource Rights in Southern Mindanao, Philippines. In: Iorns Magallanes, Catherine J. and Malcolm Hollick (eds.): *Land Conflicts in Southeast Asia.* Bangkok: White Lotus Press., 83–118.

IWGIA 2012: The Indigenous World 2012. Copenhagen: International Work Group for Indigenous Affairs.

Junker, Laura Lee. 2000: Raiding, Trading, and Feasting. The Political Economy of Philippine Chiefdoms. Quezon City: Ateneo de Manila University Press.

Kinoc, Datu Antonio P. 2002: Blaan. KATUTUBO: Profile of Philippine Cultural Communities. Manila: National Commission for Culture and Arts.

Landingin, Roel 2008: Unearthing Strife. Newsbreak Special Issue "The Big Dig: Mining Rush Rakes Up Tons of Conflict". Quezon City: Public Trust Media Group Inc., July/Sept. <www.newsbreak.com.ph>

Leemann, Esther and Nikles, Brigitte 2014: Giving Up Fallows in Indigenous Swiddens in the Times of Global Land Grabbing. In: Cairns, Malcolm (ed.): *A Growing Forest of Voices.* UK: Earthscan (forthcoming).

Leonen, Marvic Vaio Victor F. 2000: Weaving Worldviews: Implications of Constitutional Challenges to the Indigenous Peoples Rights Act of 1997. *Philippine Natural Resources Law Journal* 10(1): 3–44.

Magdalena, Federico V. 1997: Population Growth and Changing Ecosystems in Mindanao. *Philippine Quarterly of Culture and Society* 25(1/2): 5–30.

Majul, Cesar A. 1999 (1973): Muslims in the Philippines. Quezon City: University of the Philippines Press.

Malayang, Ben S. 2001: Tenure Rights and Ancestral Domains in the Philippines: A Study of the Roots of Conflict. Bijdragen tot de Taal-, Land- en Volkenkunde, Leiden, *The Philippines Historical and Social Studies* 157(3): 661–676.

Miranda, Marta et al. 2003: Mining and Critical Ecosystems: Mapping the Risks. Washington: World Resources Institute.

Multiwatch 2011: Eine Kupfermine als regionales Vorzeigeprojekt? Das Tampakan Kupfer- und- Goldprojekt von Xstrata auf Mindanao, Philippinen. In: Multiwatch (eds.): *Nachhaltiger Bergbau durch Multis? Ein Dossier zum Schweizer Konzern Xstrata.* <www.multiwatch.ch> accessed on 04.11.2013.

Oliver-Smith, Anthony (ed.) 2009: Development & Dispossession: The Crisis of Forced Displacement and Resettlement. Santa Fe: School for Advanced Research Press.

Parkes, Carl 1999: Philippines Handbook. Emeryville, California: Moon Publications.

Rambaldi, Giacomo 2010: Participatory Three-Dimension Modelling: Guiding Principles and Applications. ACP-EU Technical Centre for Agricultural and Rural Cooperation (CTA). Online at: <pgis.cta.int>.

Reckordt, Michael 2012: Wie Rohstoffe in Südost-/Ostasien zur Entwicklung beitragen könn(t)en: Das alternative Bergbau-Gesetz der Philippinen. Köln: Philippinenbüro e.V. im Asienhaus.

Reese, Niklas und Rainer Werning (eds.) 2006: Handbuch Philippinen: Gesellschaft, Politik, Wirtschaft, Kultur. Bad Honnef: Horlemann Verlag.

Rodil, B. R. 1999: The Danlag Ancestral Domain Claim in Tampakan, South Cotabato. *The Mindanao Forum* 14(1): 13–51.

Rodil, B. R. 2003: A Story of Mindanao and Sulu in Question and Answer. Davao City: MINCODE.

Romero, Purple S. 2008: Fostering Dependence. Newsbreak Special Issue "The Big Dig: Mining Rush Rakes Up Tons of Conflict". Quezon City: Public Trust Media Group Inc., July/Sept. <www.newsbreak.com.ph>.

Rood, Steven 1998: NGOs and Indigenous peoples. In: G. S. Siliman and L. Garner Noble (eds.), Organising for Democracy: NGOs, Civil Society and the Philippines State. Hawaii: University of Hawaii Press.

Rovillos, R. D., S. B. Ramo and C. Corpuz Jr. 2005: Philippines: When "Isles of Gold" Turn to Isles of Dissent. In: M. Colchester and E. Caruso (eds.): *Extracting Promises: Indigenous Peoples, Extractive Industries and the World Bank.* Baguio City, Philippines: Tebtebba and Forest Peoples Programme, 185–222.

Sal Naong Tribal Council 1996: Sal Naong Ancestral Domain Claim. Datal Blao, Sultan Kudarat, Mindanao, Philippines. Unpublished Claim Book.

Severino, Rodolfo C. and Lorraine C. Salazar (eds.) 2007: Whither the Philippines in the 21st Century? Singapore: Institute of Southeast Asian Studies.

Sicat, Gerardo P. 2007: Philippine Macroeconomic Issues and Challenges. In: Severino, R. C. and L. C. Salazar (eds.): *Whither the Philippines in the 21st Century?* Singapore: Institute of Southeast Asian Studies, 142–179.

Siliman, G. S. and L. Garner Noble 1998: Organising for Democracy: NGOs, Civil Society and the Philippines State. Hawaii: University of Hawaii Press.

Steinberg, David J. 2000: The Philippines: A Singular and a Plural Place. Colorado: Westview Press.

Tadem, Eduardo C. 1992: The Political Economy of Mindanao: An Overview. In: M. Turner, R. J. May and L. R. Turner (eds.): *Mindanao: Land of Unfullfilled Promise.* Quezon City: New Day Publishers, 7–30.

Tiu, Macario D. 2005: Davao: Reconstructing History from Text and Memory. Davao: Ateneo de Davao University/ Mindanao Coalition of Development NGOs (MINCODE).

Tomas, Rosalinda C. and Donnabelle G. Celebrado 2008: Contestations in the Mountain of Gold: Digging the Cracks in the Tampakan Copper-Gold Project — The South Cotabato Experience. Unpublished Case Study Working Paper. Davao City, Philippines: Mindanawon Initiatives for Cultural Dialogue, Ateneo de Davao University.

Umehara, Hiromitsu 2004: Agricultural Colonization and Environmental Change in Koronadal Valley, South Cotabato. In: Umehara, Hiromitsu and Germelino M. Bautista (eds.): *Communities at the Margins: Reflections and Social, Economic and Environmental Change in the Philippines.* Quezon City, Philippines: Ateneo de Manila University Press.

Umehara, Hiromitsu 2009: Koronadal Valley: Half a Century After Land Settlement in South Cotabato, Mindanao. *Philippine Studies* 57(4): 505–541.

Wenk, Irina 2012: Ancestral Domain: Land Titling and the Conjuncture of Government, Rights and Territory in Central Mindanao. Unpublished Dissertation, University of Zurich, Switzerland.

Wernstedt, Frederick L. and Paul d. Simkins 1965: Migrations and the Settlement of Mindanao. *Journal of Asian Studies* 25: 83–103.

Zingapan, Kail and Dave de Vera 1999: Mapping the Ancestral Lands and Waters of the Calamian Tagbanwa of Coron, Northern Palawan. Philippine Association for Intercultural Development.

News and Online Sources

AFRIM 2011: "Leak at Marcopper mine site again raises fears" <afrim.org.ph> accessed on 20.03.2013.

Alave, Kristine L. 2012: "Farmers oppose open-pit copper-gold mine in Davao del Sur", Philippine Daily Inquirer, 27 May 2012. <www.minesandcommunities.org> accessed on 04.04.2013.

Alyansa Tigil Mina 2013: Tampakan Forum Press Statement: "B'laan Tribe Members killed in Alleged Military Ops". 5 February 2013. <indigenouspeoplesissues.com> accessed on 4.11.2013.

Alyansa Tigil Mina 2012a: "Tampakan Mining Project Cause of Conflicts in B'laan Tribe - Communities want SMI out", Alyansa Tigil Mina Press Release, 24 February 2012, <alyansatigilmina.net> accessed on 22.02.2013.

Alyansa Tigil Mina 2012b: "Philippines: Mr. President, Tampakan is not fit for Mining", Press Release, 22 August 2012. <hronlineph.com> accessed on 22.02.2013.

Alyansa Tigil Mina 2011: "Philippines: International and Local Experts Lambast SMI Consultants; Say Tampakan Mining Project has High Potential for Loss of Life and Environmental Damage". <indigenouspeoplesissues.com> accessed on 22.02.2013.

Basa, Mick M. 2011: "Courageous nun engages military for red-tagging", Davao Today News, 11 May 2012: <davaotoday.com> accessed on 22.12.2012.

Cahayag, Warren 22.10.2012: "Outrage sweeps Mindanao over Tampakan massacre". <www.davaotoday.com> accessed on 15.02.2013.

DENR 2013: Mission, Vision and Priority, Thrusts & Programs. <www.denr.gov.ph> accessed on 4.11.2013.

Fischer Weltalmanach 2011: Der Fischer Weltalmanach 2011. Frankfurt am Main: Fischer. Online version 2013 <www.weltalmanach.de/staaten/details/philippinen/> accessed on 4.11.2013.

Galdu 2012: "Increased militarization in Tampakan", <www.galdu.org> accessed on 18.02.2013.

Kalikasan People's Network for the Environment 2013: "Green Group denounces ECC approval of Tampakan copper-gold Mine Project", Kalikasan PNE Press Release, 19 February 2013, <www.kalikasan.net> accessed on 04.03.2013.

Malaya 2013: "DENR to rule on Tampakan ECC", <planpr.weebly.com> accessed on 17.03.2013.

Mines and Geosciences Bureau (MGB) 2005: Primer on Application for mineral agreements. <mgb.gov.ph> accessed on 4.9.2013.

MGB 2013: Mining Industry Statistics. (Release Date: 30 July 2013). <www.mgb.gov.ph> accessed on 3.9.2013.

Nawal, Allan 2012a: "Canadian Mining firm in Zamboanga wants to develop new sites", Mindanao Inquirer. <globalnation.inquirer.net> accessed on 27.03.2013.

Nawal, Allan 2012b: "Tribal war feared over killings of Blaan members", Mindanao Inquirer. <www.minesandcommunities.org> accessed on 4.11.2013.

NCIP (National Commission on Indigenous Peoples) 2013: <addis.ncip.gov.ph> accessed on 21.6.2013.

NSCB (National Statistics Coordination Board) 2013: Population Projections. <www.nscb.gov.ph> accessed on 4.11.2013.

Olea, R. 2012: "Sister Stella Matutina: Vilified for defending the environment", Bulatlat. <bulatlat.com> accessed on 22.12.2012.

Philippine Misereor Partnership Inc. 2012: "Philippines: Mining, Conflict and Death of Indigenous Peoples in Tampakan", 30 April 2012. <indigenouspeoplesissues.com> accessed on 22.02.2013.

Plantilla, Anabelle E. 2012: "Tampakan Mining project and the B'laans", Manila Times. <www.manilatimes.net> accessed on 18.02.2013.

Reuters 2013: "Philippine clears Xstrata's $5.9bn Tampakan mine", <www.minesandcommunities.org> accessed on 15.03.2013.

Salamat, M. 2012: "Mining-ravaged Mindanao folk deliver 'closure orders' to mining TNCs, Chamber of Mines", Bulatlat. <bulatlat.com> accessed on 16.12.2012.

Sarmiento, Bong S. 2012: "Special Report: The Tampakan Project: Battle over Southeast Asia's largest Copper-Gold Reserve" Mindanews, 11 Nov. <www.mindanews.com> accessed on 18.02.2013.

Sarmiento, Bong S. 2013a: "Tribal members flee Tampakan mine site due to heavy military presence", MindaNews, 3 Feb. <www.minesandcommunities.org> accessed on 18.02.2013.

Sarmiento, Bong S. 2013b: "Under cover of darkness, soldiers deployed to Tampakan mines site", MindaNews, 17 Jan. <www.mindanews.com> accessed on 18.02.2013.

Sarmiento, Bong S. 2013c: "2 dead in fresh violence at Tamapakn site – military." MindaNews, 23 August. <www.mindanews.com> accessed on 24.08.2013.

SMI/Xstrata 2013: Environmental Impact Assessment. <www.smi.com.ph> accessed on 4.11.2013.

Tampakan Forum 2011: Mining in Tampakan: Intensifying Conflict, Danger in Perpetuity. <de.scribd.com> accessed on 03.11.2013.

Tebtebba 2012: "Community Maps Can Empower Indigenous Peoples to Assert Land Rights." <www.tebtebba.org> accessed on 4.11.2013.

TVI Pacific 2013: North Zamboanga Projects. <tvipacific.com> accessed on 03.11.2013.

Xstrata/SMI 2010: Tampakan Copper-Gold Project. Sustainability Report 2010. <www.smi.com.ph> accessed on 08.01.2012.

Xstrata/SMI 2011: Tampakan Copper-Gold Project. Sustainability Report 2011. <www.smi.com.ph> accessed on 02.07.2012.

Legal Sources

La Bugal-B'laan Tribal Association, Inc., et al. 1998: Petitioners versus Victor O. Ramos, Department of Environment and Natural Resources, et. al., Respondents. Case Status and Summary of Final Memoranda of Parties. Philippine Natural Resources Law Journal 9(1):48-67.

Republic Act No. 7942, 1995: The Philippine Mining Act: An Act Instituting a New System of Mineral Resources Exploration, Development, Utilization, and Conservation.

Republic Act No. 7160, 1991: The Local Government Code of 1991. Department of the Interior and Local Government, Philippines.

Republic Act (R.A.) No. 8371, 1997: Indigenous Peoples' Rights Act: An Act to recognize, protect, and promote the Rights of Indigenous Cultural Communities/Indigenous Peoples, creating a National Commission on Indigenous Peoples, establishing implementing mechanisms, appropriating funds therefore, and for other purposes

NCIP Administrative Order No. 1, Series of 1998: Rules and Regulations Implementing Republic Act No. 8371, otherwise known as "The Indigenous Peoples' Rights act of 1997".

The Environment Code of the Province of South Cotabato 2010.

Links

Alternate Forum for Research in Mindanao (AFRIM): afrim.org.ph

Alyansa Tigil Mina: www.alyansatigilmina.net

Department for Environment and Natural Resources (DENR): www.denr.gov.ph

Kalikasan People's Network for the Environment (Kalikasan-PNE, Quezon City): www.kalikasan.net

LRC — Legal Rights and Natural Resources Center / Friends of the Earth Philippines: http://lrcksk.org

National Commission on Indigenous Peoples of the Philippines (NCIP): www.ncip.gov.ph

Philippine Association for Intercultural Development (PAFID): www.pafid.org.ph

PIP Links: www.piplinks.org

PART III
Case Study Comparison

Part III: Introduction

Thomas Niederberger and Tobias Haller[1]

The third part of this book provides an overview of the case studies, comparing their characteristic aspects and situating them in the larger debates on mining conflicts. In the sense of a reference handbook, certain repetitions cannot be avoided, and are even intended in order to facilitate targeted cross-reading of specific sections.

As outlined in the introduction of Part I, we are dealing with a wide variety of cases (the unit of comparison), which we have subsumed under large-scale mining carried out by transnational corporations (TNCs). To analyse them in relation to each other, we develop a qualitative comparison framework, informed by three principal sources: 1) the previous *infoe* book on fossil fuels (Haller et al. 2007; see Part I, Chapter 1) served as a model and is referred to from time to time, considering similarities and differences with the extraction of oil; 2) a range of theoretical debates from different fields, if possible, specifically related to mining. In line with Haller et al. (2007), the underlying theoretical frame refers largely to New Institutionalism;[2] 3) an extensive list of comparison criteria was elaborated, based on students' presentations of their first draft essays during the research exercise that marked the first step to this publication. This served as a checklist for the drafting of the case study chapters.

The comparison is structured in three steps, corresponding to the Chapters A, B and C: in *Chapter A*, the larger context and the specific characteristics of the different mines are discussed; in *Chapter B*, we analyse the strategies of the involved companies; *Chapter C* is about the reactions of local people confronted by mining and the forms of collective action (or "local people's strategies") that emerge from it. Each chapter starts with introducing the aspects of comparison and how they are related at a general level. Second, we provide short summaries of findings from the case studies to gain an overview. Third, the findings are discussed as a base for the labels given in the respective *Tables A, B* and *C*. Here, we also mention additional findings from the chapters that do not treat the entire range of aspects and therefore are not fully included in the comparison (i.e., the chapters 2 on gold refining in Switzerland; 3 on the explorer company Uranio AG; 6 on artisanal gold mining in Guinea; and 7 on youth and mining in Sierra Leone).

In analogy with Haller et al. (2007), the comparative analysis is guided by the assumption that the time of operation and the intensity of related impacts are crucial factors for both corporate and local people's strategies. Therefore, we treat the case studies in a historical-chronological order, starting with the oldest mine (Mopani, Zambia) and finishing with those in planning (Las Bambas, Peru) and exploration stages (Val Medel, Switzerland).

1 Reviews and comments were provided by Madlen Kobi, Stephy-Mathew Moozhiyil, Helen Gambon and Amadea Tschannen. Cover: detail from The True Cost of Coal, courtesy by The Beehive Design Collective, creative commons.

2 This interdisciplinary approach is related to institutional economics, political ecology and others, whereby theories of collective action, institutional change and the management of common-pool resources assume crucial importance (i.e., land ownership, access to natural resources such as water-bodies, forests, grazing lands etc., as well as the fora of collective decision-making themselves); For a similar approach, see the comparison of mining conflicts in Peru by De Echave et al. (2009). For an introduction to the anthropology of mining and related literature, see Part I – Chapter 1.

Referring to the theoretical debates on collective action, local resistance can be linked to increasingly negative social and environmental impacts over the time of operation. This relates well to an assumption that heavy impacts would be met by strong resistance, while minor impacts or the uncertain potential of future impacts (during exploration activities) would find less resistance. Starting from this hypothesis, we explore variations and identify other factors that are involved as well; for instance, global trends such as the introduction of social, economic and environmental policies and guidelines, or macro-economic and political dynamics, such as changing world market prices and the strength or weakness of state institutions. Discussed from a New Institutionalism angle, changes in relative prices of minerals lead to changes in bargaining power between different actors. The institutional arrangement of opening common and state property for mining offers an "easy way" to increase both state revenues and privatized profits, thereby aligning the interests of state and corporate actors, while overpowering those of local actors. This choice is then legitimized ideologically with discourses of "development" and CSR. However, when these expectations are not met, and at the same time serious environmental and social costs occur, resistance from the local level follows as an obvious reaction.

The comparison follows the same two-dimensional model as the case studies—contrasting corporate to local people's strategies and somewhat side-lining state actors—instead of the widespread "state–corporation–community" triangle (see Part I – 1 for reasons and implications of this choice). We suggest looking at mining conflicts as enacted by parallel networks that reach from local to global scale. The "local population", on the one hand, and "the corporation", on the other, form two different vantage points to analyse strategies which are local and global, horizontal and vertical, at the same time.[3]

One of the guiding questions throughout these chapters is how the turn to Corporate Social Responsibility (CSR) responds to resistance, while conditioning the possibilities for collective action at the same time. The interplay between corporate and local people's strategies, and the tendencies in resistance/negotiation exhibited by the latter, is discussed in the final conclusion (*Chapter D*). The focus thereby is on the role of prior experience and knowledge exchange, alliance building and the importance of values and worldviews. We argue that the question of *what way of life* is at stake is crucial: for groups that do not embrace the dominant ideology of development as "expectations of modernity" (see Ferguson 1999), mining impacts may be not just perceived on the immediate level of loss of land and related resources, but on a moral type of order, linked to collective identities embedded in a cultural landscape (see also Escobar 1999, Haller 2007). To understand the underlying drivers, it is therefore not sufficient to frame mining conflicts as a confrontation between corporate-driven, capitalist appropriation of common resources and local people's struggles for the legal protection of livelihoods. Rather, there is a third dimension involved, forming a "triple movement" of "commodification, social protection and emancipation" (Fraser 2013). This last dimension can subsume a variety of discourses and practices that strategically counter the de-politicized ideology of "development"—notions of well-being and "the good life", based on values that point to the quality of relations between humans and their environment.

Finally, it has to be noted that the employed comparative approach is itself a strategic choice: on the one hand, this is a reaction to the recurrent argument that problems at specific mines, if not deniable, are only exceptions to the rule, or "bad apples" within an industry that generally applies the best standards; on the other hand, it is also an attempt to adjust the exclusive focus on "worst cases" by many NGOs and media reports—and to counter the associated dangers of simplification born from the urge to show "heroic struggles" of local communities against "evil corporations".

3 This takes up a thought of Ferguson (2006) that the widespread "bottom-up versus top-down" framework not only describes, but tends to perpetuate power disparities, and is often inaccurate to understand globalized realities.

PART III: Comparison

A) Context

In this chapter, we compare the contexts of the case studies from Part II of this book. First, the aspects for comparison are introduced on a general level; second, we provide short summaries about these aspects based on the case studies, and third, we discuss the findings. An overview of the comparative aspects is given in *Table A* (see page 425). They are grouped in two blocks, and cover the following:

1) the national-level context includes the background of *(a) colonial and post-colonial histories, (b) the dependency of the national economy* and the state on mining sector revenues (including the vulnerability to price changes); the legal and institutional framework in place, especially considering *(c) protective legislation* for collective land ownership and indigenous peoples' rights, *(d) neoliberal reforms* and legislation that enables access for TNCs, and the general condition of the civil society is mentioned with regard to past and ongoing *(e) periods of authoritarianism and armed conflicts.*

2) the mine-specific aspects include *(a) the type of the mine, its product(s) and the stage in the mine-cycle, (b) the ecosystem and cultural landscape* where it is situated, and the rough characteristics of its *(c) environmental and social impacts.*

Context aspects can be termed as exogenous factors, in the sense that they tend to be perceived by local actors as a "given". However, they are not static, and may be changed or challenged by strategic action. How they are interpreted and strategically emphasized in different ways will be elaborated in the following chapters on corporate (B) and local people's (C) strategies.

1. Introduction

As a preamble, there is a need to discuss the larger contextualization of the political economy of the state–company interactions, taking up again the debates outlined in the introduction and Chapter 2 of Part I. Thereby, the dependence of a state on the extractive or mining sector and changing relative prices are central interrelated variables. One of the drivers is a specific change in prices of minerals, representing a general trend of relatively low prices since the mid-1970s and a rise since around 2000 onwards, lasting until recently. Theoretically, this indicates that relying on mining was also responsible for the external debt crises and following structural adjustment programs observable in many countries discussed in this book. The neoliberal reforms that downsized the state on the one hand promoted decentralization, while on the other hand gave potential room for increased civil society activity and legal protection of indigenous peoples' rights. The latter includes in some cases the obligation to obtain Free, Prior and Informed Consent (FPIC) from indigenous peoples for developments that affect them, or at least the duty to consult them beforehand, in line with the ILO Convention No. 169. Neoliberal reforms often proved to have a negative impact on development indicators and further undermined basic general infrastructure. When the relative prices for minerals and the interest in mining started to rise again in the 2000s, political structures and local-level constellations, rights and expectations had changed: because civil society had gained considerable strength internationally and in most local and national contexts, it has become difficult for companies to not address the discourse of environmental and social responsibility. Nevertheless, to operate in some countries may have become easier after structural adjustments and decentralization processes, as companies can to some degree bypass the state and negotiate directly with local groups. At the same time, due to the long-term investments and large profits made, a retreat of a company from an area is often not feasible, which makes it susceptible to pressures from governmental and non-governmental actors. In these global/local (or "glocal") settings, bargaining power oscillates: while decentralization provided more bargaining power to local people, this is lost in terms of demanding services from the state; and while local people's bargaining power increased based on the corporations' efforts to improve their reputation, companies partially substitute the government at the local level and are able to manipulate deals, thus raising their bargaining power again. How this field of power struggles is structured in the specific context is not only of central importance to this chapter, it will also influence other discussions as well as theoretical debates in the anthropology of mining and related fields.

1.1 National Context and the Global Political Economy

As mentioned in the introduction of Part I, mining companies can enjoy enormous leverage to pressure for policy adjustments. However, they are still dependent on governments to access mining licences (Ballard/Banks 2003: 294), and often for indirect subsidies such as for the construction of roads and infrastructure and for low-cost water and energy supply. The dependency of states on income from the mining sector and changes of prices in the international market have important macro-political implications, which can find their concrete expressions in state policies that privilege corporate interests against local resistance (Haller et al. 2007; 2010). The change in global market prices determines phases of expansion and retraction of mining at the global level, as well as the profit rates and the ability of governments to pressure for a larger share of revenue. Commodity market prices fluctuate in "boom and bust" periods of a few months to years, with underlying long-term "super-cycles" that can span over several decades. Although specific metals and minerals have their own market dynamics, depending on their use and characteristics, these do correlate with super-cycle developments (e.g., as shown by Jacks 2013). The last such super-cycle was at its peak in the 1960s and '70s, decreasing in the 1980s and reaching its bottom point in the mid-1990s. In the early 2000s, the current cycle took off with a general price boom for the whole range of commodities (ibid.), fuelled both by speculation and by rising demand from emerging economies, most notably China. Parallel to the rise of metal prices, the investments in exploration and new mining projects massively increased since the early 2000s, as did the value of shares of publicly traded mining companies. The global financial crisis caused a massive but short crash in 2007 to 2008. After a peak in 2010-11, prices are currently decreasing and while writing this, the debate is going on as to whether this is to be considered the end phase of the super-cycle.[1]

Due to the long time-span from investment decisions under the influence of high prices until new mines actually go into production, the increase of total metal production shows a time-lag of several years. Similarly, when prices decrease, dis-investments take about two years until they result in lower production (Larsson/Ericsson 2014). An obvious effect of lower market prices is the reduction of state revenues (for Zambia as an example, see Gibson 1999; Ferguson 2006; Haller 2013). The work of Haller (Haller/Merten 2010, Haller 2013) illustrates that such price changes have repercussions not only for mining locations, but for the hinterland as well, where

1 See for e.g.: Otaviano Canuto: "The Commodity Super Cycle: Is This Time Different?" World Bank Economic Premise 150, June 2014. <siteresources.worldbank.org> accessed on 15.12.2014.

Fig. 1: Copper Plates in Mopani Mines, Zambia. Photo: Meinrad Schade.

they concern the use of other resources, as unemployed workers from mines migrate back to their villages or to other hinterlands and look for informal cash alternatives by the commodification of common-pool resources such as fisheries, wildlife, pasture or forestry and water, which have been transformed from local common property to badly enforced and de facto open access state property. Such resources are then overused or negatively impacted over a large area. In mining contexts, specifically water, and also other mobile resources are mainly in danger of becoming scarce and overexploited.

In Chapter 2 of Part I, we have introduced the relevant regulatory framework for mining at a general level. The proliferation of voluntary agreements associated with Corporate Social Responsibility (CSR) will be discussed as part of the corporate strategies in Chapter B, while here we focus mostly on the national law. International conventions for the protection of indigenous peoples' and ethnic minority rights have made important improvements since the 1990s, which are partially reflected in laws at the national and sub-national level. Also, environmental legislation has become a common feature in regulating mining; for instance, by demanding Environmental Impact Assessments (EIA) as a condition for permits. However, the 1990s are also marked by a general "neoliberal turn": with so-called "Structural Adjustment Programmes", International Finance Institutions (IFIs) pressured developing countries to privatize state-owned companies, lower taxes, duties and royalties, and introduce mining laws that raise investment attractiveness (Ballard/Banks 2003: 294, Bebbington et al. 2013: 11). Between 1985 and 2004, more than 90 countries have adopted new mining laws or revised existing ones (Bridge 2004). A contrary trend, fuelled by high prices during the last decade, is what mining industry analysts call a resurgence of "resource nationalism": increase of taxes and regulations, up to (partial) nationalization of mining companies (often citing the left-wing governments of Bolivia, Ecuador and Venezuela as examples) (e.g., Wilson/Cervantes 2013).

Contradictory tendencies in the regulatory framework—improved recognition of land rights and decision-making by indigenous peoples and the simultaneous facilitation of access for mining corporations—have been characterized as a problem of "legal and institutional pluralism" (see Part I – 2). At stake are the quality of monitoring and implementation, the checks and balances between different levels of government and the judiciary, and related problems of corruption and nepotism. How legal and other standards are actually implemented depends a lot on the work of civil society organizations and independent media, which can act as "watchdogs". However, there is a challenge to this process: In authoritarian regimes and in countries affected by armed conflicts, political and civil liberties are systematically curtailed and activists and critical journalists are often denied access to the area, threatened or even killed. Mines are often focal points in armed conflicts: especially, when located in territories which are remote and difficult to control, they can become a source for financing armed groups by extortion and kidnapping of staff, for weapons captured from security guards, and for dynamite supply. In regions that are or have been affected by armed conflicts, the presence of army and paramilitary forces, police, private security companies and irregular armed groups leads to a general proliferation of violence. Be it as a side-effect or as a main objective, militarization and anti-insurgency programs can serve to repress potential resistance of local people as well as civil society groups against resource extraction.

The trauma, distrust and fear that result from war and violence can mar the lives of entire populations for years, even for generations. It is therefore necessary also to consider the context in a longer historical perspective of anti-colonial struggles and resistance against exploitation, lack of confidence in state institutions and memories of state violence, which can trigger opposition against mining projects from the outset (Kirsch 2012: 212, Bebbington 2012: 4). We can assume that the "long memory" of colonial exploitation and ongoing injustices will create a climate of suspicion or outright refusal, if mining is seen as a neo-colonial endeavour in continuity with past exploitation. Where a colonial history is directly related to mining, this may be an obvious but underestimated motive in contemporary mining conflicts. Furthermore, post-colonial expansion as well as "gold rush" periods accompanied by a sudden influx of settlers and migratory labourers need to be taken into account. As small- and medium-scale exploitation of metals and minerals remains superficial, their presence is an indicator for commercially exploitable deposits, and new large-scale projects are, therefore, often located in areas that have seen some form of mining, and related conflicts, before.

From the aspects introduced so far, we draw a first initial hypothesis: a regulatory framework which guarantees the recognition of land and related rights to common-pool resources of local and indigenous peoples is likely to motivate companies to enter into negotiations with them, at least pro forma. Contrarily, the continuity of colonial exploitation and non-recognition of rights, macro-economic dependency on mining rents, vulnerability to changing world market prices, and neoliberal reforms facilitate easy access to corporations, without paying attention to these rights. At the same time, they are likely to spur the conflictive imposition of mining projects even despite local resistance.

1.2 Landscape, Mine Type and Lifecycle

The second part of this chapter compares the project or mine-specific context of the case studies related to their location in the landscape, mine type and lifecycle, and corresponding impacts. The geological qualities, size and dissemination of the ore body, as well as the "materialities" of the target metal or mineral "[...] have causal powers and effects of their own" (Bebbington et al. 2013: 13) regarding the impacts, employed techniques and the economy of extraction. The surrounding ecosystems and cultural landscapes cannot be treated separately from the engineering problems, as the following statement of a mine operation manager illustrates: "[…] if you can pick where you put a mine, [the best is] an arid environment with no people... The worst is lots of rain and lots of people" (cited in Welker 2014: 216).

Open-pit mining needs substantially more land, water and energy but fewer workers than underground mining. High prices and improved technology thereby allow for the extraction of ores with very low metal content, in regions where it was not commercially viable two decades ago (Dougherty 2011: 5; see also Part I – 3). Generally, the size of a mine matters with respect to profitability, and there has been a trend towards extremely large mines. In South America, the term "*mega-minería*" is used to distinguish a new type of super large-scale open-pit mine from conventional mining (e.g., Svampa/Antonelli 2009). A further distinction is made between "greenfield" (new mines in not yet exploited areas) and "brownfield" projects (expansion or re-opening of old mines). To reduce risks regarding profitability, mining companies may prefer brownfield projects, which are much cheaper than the construction of completely new ones. In 2013, the construction costs for an average greenfield mine were about USD 800 million, twice as much than for brownfield mines (Larsson/Ericsson 2014). For the largest greenfield "mega-mines", the investment costs can reach several billions. On the other hand, operational and maintenance costs of an old mine are also considerable and under conditions of low prices, old depreciated mines can be closed quickly, either temporarily or completely, to save costs.

Additional costs arise related to the evaluation and mitigation of social and environmental impacts, which today covers an entire scientific and economic field of activity. There are well-established standards framed in Environmental and Social Impacts Assessment (EIA and SIA), included in CSR-policies and/or made mandatory by national legislations in many countries. Still, great challenges remain as actual impacts are often underestimated in such assessments (Kuipers et al. 2006). Technology can mitigate and prevent some of the environmental impacts; for e.g., by filtering and treatment of waste-water. However, this comes with additional costs and will affect the profitability of a mine. Once construction and operation has started, continuous monitoring is needed to check contamination risks.

The ecosystem, climate and topography have important effects on the impacts of mining in a specific setting: in a desert, contamination might be spread by wind, in wet regions by run-off. Mining in high altitudes can affect glaciers and river sources. There is an observed tendency of mining to push the frontier of extraction further into particularly vulnerable areas that so far were not touched by industrial development (Ballard/Banks 2003: 296, 288). These areas are often rich in biodiversity and wildlife and might even be inside or close to protected nature reserves. The extension and geographical distribution of impacts can contribute to major conflicts, not only within the concrete parameter of a mine but also beyond it. Typical cases are river systems where the communities nearby the mine receive compensations while people downstream receive only the pollution (e.g., the riverine tailings disposal at Ok Tedi in Papua New Guinea, see Kirsch 2014). Similarly, more distant communities might be affected by roads, power plants and supply lines, sourcing of water, and other impacts related to the extended infrastructure and inputs needed for mining.

In all this, what matters for different local actors is how their relation to the environment is affected; that is, the cultural landscape on which subsistence activities such as agriculture, fishing, herding or hunting and gathering are carried out in interrelation with practices of socio-cultural and spiritual significance (see Haller et al. 2013). Not all impacts are easily measurable, especially if the monitoring is done without participatory research procedures. Conflicts are triggered by the impacts local people perceive or are afraid of, rather than by what seem to be scientifically proven (or not). Natural sciences and related interdisciplinary sciences (e.g., epidemiology) tend to neglect livelihood strategies of different actors and their local knowledge and thus fail to anticipate the whole range of impacts in advance. For instance, the impact on poor nomadic women may not be part of an assessment if they are not recognized as "stakeholders". Socio-cultural, identity-related impacts are especially difficult to assess, or are simply unpredictable. From an emic perspective, land and landscape are commonly a base for kinship relationships, identity formation, and last but not the least, the place where the ancestors are buried, thereby inhabited by

Fig. 2: Ok Tedi Mine in Papua New Guinea. Photo: Wikimedia commons.

a spiritual world, to which local people perceive a vital relationship for their material and mental well-being. Fundamental changes in the local subsistence economy and the relation to environment and landscape will be perceived as a heavy impact (see Haller et al. 2007). In regions which are not yet strongly integrated into a market economy, the sudden and massive influx of cash and labour immigration can have disastrous impacts on the local society (Ballard/Banks 2003: 304).

While the evaluation of all these interrelated aspects in the context of specific mines is difficult, the stage of a mine in the mining lifecycle has obvious implications for the conflict timeline: in a simplified succession, we can assume that the exploration stage has limited physical impacts, with mostly test drillings and the construction of access roads; heavy social and economic impacts start in the project set-up stage, with consultations for land acquisition and resettlements, the formation of mining towns, accompanied by important environmental impacts once construction of the mining infrastructure starts (deforestation, change in water flows, etc.). During the active operation stage and mine expansion, social and environmental impacts are at their highest level, with a gradual increase in contamination and resource depletion and possibly incidents like pipeline leaks up to full-scale disasters. After mine closure, heavy socio-economic impacts can be expected from the sudden loss of incomes, and environmental contamination can continue for a long time.

Summing up the second set of context-related aspects, we draw a second hypothesis: the specific setting (ecology, geology, cultural landscape) and characteristics of a mine (its type, "materiality" of extracted resource, technology) have complex impacts, which change over time along the mine-cycle. For a mine that appears as "problematic" for its type, setting and stage, preventive investments in CSR are more likely than for an "unproblematic" mine. Apart from overall strategic management decisions (see Chapter B), a further aspect that can motivate CSR investments, and may contribute to a setting being perceived as "problematic", is the potential of local resistance; especially, at critical stages in the mine-cycle, as discussed in Chapter C.

2. Case Study Summaries

In this section, we give short summaries of the case study contexts, described in detail in the previous chapters. We move loosely from the general and historic to the specific and recent, touching the above-introduced aspects where relevant, with more or less stress according to how important they appear to be for the cases. In the third section, this will be further discussed, based on the overview in *Table A* (see page 425).

Mopani Copper Mines, Copperbelt Province, Zambia:

1) Zambia's colonial history is intrinsically linked with mining, carried out mainly under the British South Africa Company. Colonial expansion drove northwards along the railways to the Katanga mines in Congo, bringing white settler farmers to the area. In the 1920s, significant copper deposits were discovered in the "Copperbelt" area that had been inhabited predominantly by the Lamba people. As large-scale copper exploitation started, new mining townships attracted large numbers of immigrant workers. The organization of these workers into strong trade unions turned into one of the major political forces in the country, forming the base for the anti-colonial movement. After independence in 1964, Zambia became a model of African socialist development with the nationalized copper mines as the corner-stone of a state-controlled welfare economy. However, the global economic crisis of the mid-1970s led to a drastic fall of copper prices and Zambia plunged into an external debt crisis, high unemployment rates and decay of the mine infrastructure. From the mid-1980s to the late 1990s, the country became a testing ground for IMF / World Bank "structural adjustment" measures. In 1991, the government implemented a decidedly neoliberal program which specifically included the privatization of mines. The sell-out happened under very favourable conditions for the companies, framed in undisclosed "development agreements". The neoliberal reforms also brought a massive reduction in mining jobs and state services, weakening trade unions and state institutions. In recent years, the late President Sata tried to increase royalty rates in view of the massive price hike of copper and new Chinese companies moving into Zambia. Mining is again being seen as an issue of national importance. However, the condition of the original inhabitants of the mined land is not part of the national debate and minority or indigenous rights are not protected.

2) *In operation*: The Mopani Copper Mines (MCM)—the crown jewel of the Zambian copper industry—were sold to Glencore and First Quantum in 2000. Among the largest producers of copper and cobalt in Zambia, it comprises an entire mining-industrial complex, with two centres in the Copperbelt province towns of Kitwe and Mufulira that include four underground mines and some open-pits, a refinery and a smelter, with a declared yearly production capacity of 250,000 tonnes of copper and 2400 tons of cobalt in 2012. The mines date back to the 1930s, but have been continuously renewed and extended. After privatization, the new owners made major investments to increase life-span and production capacity and introduced the controversial "in-situ leaching" technology.

The major towns Kitwe and Mufulira depend economically almost completely on mining. This "mining town" situation means that all social problems, including precarious employment and living conditions, and sociopolitical conflict are directly or indirectly linked to the mining activity. In the 1930s, the Lamba, who are the original inhabitants of the area, had been dislocated and reduced to reservations, without any compensation for the loss of their land. Their system of collective land ownership and subsistence agriculture has largely broken down. But also many of the immigrants never obtained official land titles in the mining townships, creating legal uncertainty. The case shows that underground mining also can have severe negative environmental and health impacts, especially through the leaching and smelting processes. Reported are heavy sulphur dioxide emissions, as well as soil and water contamination, both chronic and through accidental leaks.

The Somaïr, Cominak and Imouraren Uranium Mines in Arlit and Northern Niger

1) The history of Niger is part of the larger colonial Scramble for Africa and the role of French Imperialism. The vast Sahara desert was thereby cut into arbitrary sections and the free movement of the Tuareg curtailed by policies that privileged sedentary groups against the supposedly "unruly" nomads. Also, after the formal independence in 1960, a strong neo-colonial dependency on French interests subjects people and space to external interests, which are not just shaped by business interests—most of all, the abundant uranium reserves—but get intermingled with geostrategic interests. Significantly, the period pre- and post-independence is when these deposits were discovered (in the 1950s) and developed (1960s, with extraction at Somaïr starting in 1971). This happened in a context of Cold War driven nuclear arms race and of nuclear energy being installed as a core development strategy of the French state.

The governing elites of Niger—repressive military regimes, with few exceptions—are mostly based in the south of the country. The marginalization and neglect of the Tuareg of north Niger has been systematic since several decades, and their rights to ownership of common property or autonomous institutions within the areas they roam since ancient times have never been respected. The "empty table" created by ignoring these rights, centring on essential watering places and pastures, was the precondition for the installation of the uranium mines and the sedentarization of many Tuareg in precarious settlements, including the mining town of Arlit. In recent years, following several armed uprisings and in the context of the extension of the "war on terror" into the Sahara region, the Tuareg are increasingly being subjected to a labelling as "terrorists". The militarization and authoritarian suppression of dissent create a scenario where civil society actors have very limited possibilities, and indeed, many activists working for the rights of the Tuareg have fled to exile in France and elsewhere.

2) *In operation / planning:* The case covers three different uranium mines: Somaïr is an open-pit mine 7 kilometres northwest of Arlit, in operation since 1971; Cominak is one of the largest underground mines in the world, located south of Akonak, in operation since 1978; Imouraren is a project for a major open-pit mine, located 60 km south of Arlit, that is currently on hold due to low prices for uranium. All three mines are controlled by the French major Areva. The state of Niger holds important shares in Somaïr and Cominak. In recent years, Areva lost its earlier monopoly over Niger's uranium, with numerous exploration concessions ceded to companies from other countries, including China.

The main environmental impacts of uranium mining concern the spread of radioactive particles through dust and rainwater, which accumulate in the few water bodies so precious for humans and animals, and secondly, the enormous amount of water consumed in the extraction and concentration processes. The water is in part pumped up from fossil aquifers. The groundwater table has decreased dramatically and many water sources and ponds have dried up. This, together with other restrictions and the loss of land, puts the livelihoods of local nomads and subsistence peasants in serious risk, making them more vulnerable to droughts. Furthermore, higher radiation levels cause illnesses for the mine workers and the general population, although the extent of this is unknown due to deficient documentation.

Ranger and Jabiluka Uranium Mines, Northern Territory, Australia

1) The colonial and post-colonial history of Australia is defined by the expropriation of common lands and forced assimilation programs that created a deep rift between the majority settler population and the indigenous Aboriginal population, today a minority comprising 2% of the population. Since the 1970s, there have been important steps to recognize and correct historical injustices. Most importantly, indigenous land ownership was restored in many cases. The Aboriginal Land Rights Act of the Northern Territory of 1976 (ALRA) is one of the earliest legal provisions worldwide that includes Free, Prior and Informed Consent (FPIC), giving indigenous "traditional land-owners" the possibility to veto mining projects. However, Australia is also a major mining

country, home to some of the largest mining TNCs (including Rio Tinto, majority shareholder of Ranger/Jabiluka) and ranked as the world's number three uranium exporting nation. The exploitation of the Ranger mine was only possible due to explicit exemption from the ALRA. Although subject to a well-functioning legal system and strong civil society, this contradiction remains a source of conflict. The adoption of neoliberal policies in the 1990s transferred social responsibilities from the government to corporations. Compensation and benefits from mining, directly negotiated between indigenous communities and corporations, increasingly replace publicly funded social programs and create new dependencies.

2) *Ranger: in operation; Jabiluka: suspended.* The Ranger/Jabiluka case covers two uranium mines close to each other: the Ranger mine started production in 1980 as an open-pit mine. Among the most important uranium mines worldwide, its most accessible deposits are now exhausted, but extraction could continue underground for several years. Jabiluka is an underground uranium mining project of similar size, for which construction started in 1996 but was stopped in 1999 due to the opposition by the indigenous Mirarr people.

The mining areas are surrounded by Kakadu National Park, listed as a UNESCO World Heritage site. Mining activities affect the large flood-plain area of the Alligator River, home to the Mirarr and other indigenous communities. Heavy rains make it difficult, almost impossible, to prevent contamination, showing that even the best technical standards cannot fully prevent harm from mining, especially with such a problematic element like uranium. In addition, the mines affect subsistence activities and spiritual aspects of indigenous land use, including sacred sites. The town of Jabiru, built together with the Ranger mine to accommodate the mine workers, represents the dominance of Western culture, and is seen as a major cause for the crisis of Aboriginal lifeworlds and associated problems such as alcoholism and drug abuse.

El Cerrejón Coal Mine, La Guajira, Colombia

1) In Colombia, the conquest by the Spanish brought the decimation of indigenous peoples and forced immigration of African slaves. Both of them are minorities in the actual context of 86% *mestizo* or mixed population. Since independence in the early 19th Century, Colombia's history is marked by political violence, oligarchic control and land disputes. Some indigenous territories in La Guajira have a special status since the 1960s, and the new Constitution of 1991 has improved recognition of indigenous and minority rights, including compulsory prior consultation for projects affecting indigenous territories, following more or less ILO Convention No. 169 and specified in a decree from 1998. However, in case of conflicting claims between a company's mineral exploitation permits and indigenous territorial rights, the former prevail in most cases. Neoliberal reforms, including a new mining law from 2001, facilitate access for TNCs. The coal mining sector is thereby of major importance. Colombia was ranked as the 4th largest coal exporting nation worldwide. Corporations have considerable power to influence policies and developments, both at the national level and in the area of operation, where they also benefit from protection by state security forces against potential guerrilla attacks.

The dependencies from colonial times as well as socio-economic disparities led in the 1960s to the ongoing civil war between left-wing guerrillas and the army and paramilitary groups, which produced thousands of civil victims and about 5 million internal refugees. The war also affected the Guajira region with about 200 killings between 2000 and 2011. Since 2005, the paramilitary groups have been officially dissolved and civil-war-related violence has decreased, but informal armed groups continue to be active in the area. A new wave of grassroots movements, supported by NGOs, has reopened space for non-violent action against corporate power.

2) *In operation, expansion:* El Cerrejón was developed since 1976, and started to operate in 1986. With a production of 30 million tons (fully dedicated to export), accounting for 41% of the total national production in 2010, it is the largest coal mine in Colombia. The mine affects an extensive area of the Guajira Peninsula, with several open-pit extraction sites covering more than

11,000 ha and reserves of 69,000 ha. Extraction is planned to continue and expand for many years, with current concessions valid until 2034. The installations include a company-owned 150 km railway-line, which cuts through almost the full length of the Guajira, and a port.

The area in southern Guajira, where the mine is located, receives about 1000 mm of rain, concentrated in a few months. The northern part consists of a desert-like landscape; its major water source is the Ranchería River, which flows through the mining area. The most obvious impacts are the damage to the local cultural landscape and dust emissions, noise and vibration from blasting and machinery, as well as air, soil and water pollution. A major potential impact that sparked resistance was the plan to divert the Ranchería River for mine expansion. Although this could be prevented, the change of water flows and increasing water scarcity, together with resettlement of villages and restrictions on movement imposed by the installations and the company's security guards create existential problems for the pastoral and small-scale agricultural lifestyle of the inhabitants of La Guajira. On the socio-cultural level, local Wayuu indigenous people indicate that the area as a whole is perceived and lived in as sacred, making reference to *Wounmainkat* (translated as "Mother Earth"). The mining operations therefore also affect local people's religious and cultural values.

Ambuja Rawan and ACC Jamul Cement Plants, Chhattisgarh, India

1) The colonial era has left many unresolved issues in the Indian state of Chhattisgarh, such as the expropriation of common property resources and the marginalization of subsistence-based communities. Mining became more important after India's independence in 1947, whereby state-promoted industrialization caused massive displacement of lower caste and *adivasi* ("indigenous" or "tribal") people. The profoundly hierarchical structure of Indian society gives privileged sectors access to the benefits of industrialization, while the vast majority is excluded.

India is a democratic, federal state with a highly developed legal system. However, many laws remain unimplemented, access to justice is highly unequal, and corruption rampant at all levels. This is balanced to some degree by relatively strong and diverse social movements and civil society groups. In the 1990s, the gradual opening of the markets under the influence of neoliberalism allowed for the entry of TNCs to the so far protected Indian market. The Swiss Holcim controls a large share in the cement market since the acquisition of ACC and Ambuja in 2005. Chhattisgarh has been eagerly pushing neoliberal reforms after its separation from the state of Madhya Pradesh in 2000, with particular focus on mining and related industries such as steel, cement and coal-fired power plants. This has further raised its dependency on a mono-sectoral, extractive economy. Chhattisgarh's forest-covered and hilly areas with a large majority of *adivasi* population have been the venue of a conflict with left-wing Maoist guerrillas, which escalated to open civil war in 2005. Although the central plains region, where the case study is located, is not directly affected, the conflict does have important repercussions on the entire state: basic freedoms have come under pressure and allegations of supporting the guerrillas are frequently used to criminalize social activists.

2) *In operation, expansion:* In this case study, limestone mining is one aspect of the larger industrial complex of cement production. Both the ACC Jamul and Ambuja Rawan plants source their own limestone from relatively small open-pit mines close-by and are powered by integrated, coal-fired power plants. Production capacities are being significantly increased in the context of fast growing demand on the domestic market. As the necessary inputs are available in large quantities in the state, there is no projected end of production.

The plant of ACC Jamul is located in the urban-industrial area of Bhilai. Ambuja's plant in Rawan, on which we focus here, is surrounded by villages in a densely inhabited rural area, where rice cultivation is the main activity. Annual rainfall is concentrated in the monsoon season, followed by a long dry season. Therefore, water ponds, wells and irrigation canals are of critical importance for the peasants. According to their observations, the ground water level has fallen and scarce irrigation water is diverted to the factory. Dust and air pollution have reduced the productivity of their fields. The social impacts concern precarious working conditions and extensive use of sub-contracting. This adds to a feeling of local peasants that they have not been properly compensated for past land acquisitions, and there are allegations of ongoing encroachment on private as well as common land by the company.

Marlin Gold-Silver Mine, San Marcos, Guatemala

1) Guatemala is a country rich in cultural diversity, with a 40 to 60% share of indigenous peoples, divided in 22 linguistic groups of the Maya language family. Its history is marked by some of the most brutal expressions of colonial exploitation, dispossession of indigenous lands, forced labour in plantations and mines and outright ethnocide. Guatemala continues to have some of the highest rates of inequality, poverty and violence worldwide. Although formally independent since 1840, it was never really de-colonized and was ruled mostly by oligarchs, military dictators and US-American interests such as the notorious United Fruit Company. Mining played an important role in Guatemala's economy during colonial times as well as in various waves from the 1920s to the early 1980s, when the oil and debt crises hit the country and mining almost ceded. Since the establishment of the peace agreement in 1996, democratic elections were held and important constitutional and legal improvements were made, finally recognizing indigenous peoples' cultural rights and territories; e.g., through adoption of the ILO Convention No. 169. However, there is a lacuna in implementation and indigenous people are still being forced off their land by private and state security units.

Guatemala's elected governments have introduced and defended neoliberal policies, including a new mining law in 1997. This gives numerous incentives to foreign investment in mining and sets an extremely low royalty rate of just 1% of net profits, which has created a "gold rush" for aggressive explorer companies, but not leaving many benefits to Guatemala, whose economy remains dependent on remittances from emigrant labour in the USA and some agricultural export products such as coffee. Under massive pressure from a strong anti-mining movement, the government renegotiated the mining law in 2012 and increased royalty rates to 4%, partly to the benefit of the affected municipalities.

The civil war between the US-sponsored right-wing military government and leftist guerrillas, which lasted from 1960 to 1996, left 200,000 people dead and 1.5 million displaced, most victims being indigenous peasants. Many indigenous and human rights activists, journalists and trade unionists were killed, leaving a weakened civil society and a traumatized population, though recovering fast in recent years.

2) *In operation:* The Marlin gold and silver deposit was discovered in 1998 and its exploitation started in 2005 as a combined open-pit/underground mine, with a total estimated lifetime of about 15 years. In 2012, the open-pit was exhausted and it continued as an underground mine. In 2013, almost 2 million tons of ore were extracted, to produce 200,000 ounces of gold a year (at a grade of 3.3g/t). With less than USD 200 per ounce, production costs are only about half that of average North American mines. The typical "crushing, grinding and cyanide milling" technique of modern gold mines is used, producing large amounts of toxic tailings which are stored in an artificial pond on site.

Situated in the two municipalities of San Miguel Ixtahuácan and Sipacapa in the tropical hilly region of western Guatemala at 2000 meters above sea level, the climatic conditions change between a dry season and months of heavy rainfall. The indigenous population of the surrounding villages lives on subsistence agriculture and animal husbandry, carried out on land held in common ownership.

Environmental impacts are caused by the high water consumption of about 250,000 litres per hour, which affects water sources in the region and leads to shortages in the dry season, and by the alleged contamination of the local rivers (used as drinking water sources). Locals report health problems (related to skin and respiration) and cattle deaths, which they attribute to contamination, supported by several studies showing elevated levels of toxic metals in the water. Other concerns were vibrations and noise from blasting, incorrect handling of cyanide, and uncertainties about the closure and rehabilitation of the mining area, together with social divisions and acts of violence against opponents of the mine.

Tampakan Gold-Copper Mine Project, Mindanao, Philippines

1) The history of the Philippines has seen Spanish (1565-1898) and US-American (1899-1946) colonial periods, but the interior hill regions of Mindanao came under direct colonial rule only in the first half of the 20th Century. The main driver for the colonial expansion at that time was the introduction of large-scale plantations, (forced) labour and taxation, coupled with the establishment of frontier-settlements. Swidden cultivation and hunting-gathering, the traditional land use of the local indigenous peoples, were ignored and the considered "unused" land was put under state ownership. After independence in 1946, mass-migration from densely populated northern Philippines to Mindanao was sponsored by the state as a nation-building measure. On a general background of inequality, high poverty rates, internal colonization and cultural discrimination against the hill people, movements for regional autonomy and low intensity warfare between various guerrilla groups and state security forces remain a factor in Mindanao's political landscape until today.

Mining has been a factor for the Philippine economy since the Spanish colonization, but to varying degrees. The peak was in the 1960s and '70s, during the authoritarian rule of Ferdinand Marcos, who personally owned the Marcopper mine. In 1986, Marcos was ousted by a popular movement, involving diverse civil society and indigenous organizations. This was followed by progressive legal reforms, including the Indigenous Peoples' Rights Act (IPRA) of 1997—arguably one of the best laws worldwide in recognizing and protecting indigenous territories and the principle of Free, Prior and Informed Consent. However, at the same time, neoliberal reforms were also carried out under pressure of the 1980s debt crisis, leading to the new Mining Act of 1995. This law grants various tax reductions and exceptions and facilitates access to mineral deposits for the TNCs through direct agreements with the central government. Though it is partially contradictory to the IPRA, it does include provisions for "sustainable mining" such as the need to consult local communities and frame benefit-sharing agreements with them. Tampakan is the first and biggest of a series of similar large-scale mining projects. However, their importance for the Philippines economy is still marginal, accounting for 0.4 percent of national employment and one percent of state revenue in 2008. The construction of the Tampakan mine, estimated at USD 2 billion, would make it the biggest single foreign investment project in the Philippines.

2) *Permits and planning:* The Tampakan gold-copper deposit is located in a municipality of the same name in Southern Cotabato, Mindanao. It was explored as part of a controversial agreement made in 1995, which originally opened almost 100,000 hectares to potential mining, later reduced to 23,500 hectares. Considered to be of world-class size, the mine pit is supposed to reach a depth of 800 meters, and its potential production is estimated at a yearly average of 375,000 tonnes of copper and 360,000 ounces of gold during a life-span of 17 years.[2]

Apart from the destruction of a rich cultural landscape, 40% of the mine site area of about 10,000 hectares is covered by humid tropical forest with high biodiversity. The Bla-an indigenous people who inhabit the area depend on shifting agriculture for subsistence. Downstream, potentially affected water sources such as Lake Buluan are used by 80,000 farmers to cultivate 200,000 hectares of land, including highly productive rice fields.

2 Numbers from SMI Website: "Project Description", <www.smi.com.ph> accessed 10.03.2015.

So far, the physical impacts have been limited to the exploration drillings and the projected mine is now stuck at the planning stage due to various uncertainties. Officially, production was planned to start in 2019. However, it is situated close to an active volcano, in a landslide-prone hilly area, which drains into an important river system, all of which make the construction of a mine challenging at best. Toxic tailings are to be stored in an artificial lake on site, while the mineral concentrate would be transported in the form of slurry by pipeline, for further processing near the seashore. The most immediate impact would however be the resettlement of 3500 to 4000 individuals from several communities, mostly the indigenous Bla-an.

Las Bambas Copper Mine Project, Apurimac, Peru

1) At least since the time of the Spanish conquest of South America in the 16th and 17th Century, the history of Peru is deeply entangled with mining, but not all regions were similarly affected. The Apurimac department came into focus for large-scale mining only in the 1990s. This economically marginal Andean mountain region had been suffering under a feudal landlord system until the land reform of 1969, which recognized the land rights of "peasant communities" (*comunidades campesinas*). From the 1980s to 1992, Apurimac became one of the strongholds of the Maoist *Sendero Luminoso* guerrillas. Both the guerrillas and the state repression against them were extremely brutal, killing an estimated 30,000 people and displacing 600,000 in the rural areas of southern and central Peru.

In 2011, Peru's mineral exports (mostly copper and gold) were valued at USD 27 billion, which was more than 50 percent of the total exports. The world-scale Las Bambas copper mine project will increase the output considerably and enjoys high priority support by the central and regional governments. With 8 percent of the global production, Peru is the second largest exporter of copper. Further, it is South America's second most important target for foreign mining investments after Chile.

Peru has a vivid civil society sector with combative indigenous and peasant movements, numerous NGOs and freedom of press. The country has seen a large number of mining conflicts in recent years, with several major projects blocked by popular resistance. Although local and indigenous peoples' territorial rights are protected by the ILO Convention No. 169 as well as constitutional and legal provisions, mining concessions overlap with community owned lands and the neoliberal policies crafted in the early 1990s are prioritizing access for corporations, which can be enforced by the state. The metal price boom has led to a dramatic increase in concessions, covering almost 20 per cent of the total area of the country and 54 per cent of Apurimac in 2011. The environmental laws, which regulate mining, are considered to be relatively strict; however, their implementation is obstructed by the central government.

2) *Permits and planning:* The copper mine project Las Bambas dates back to 2004, when Xstrata acquired the 33,200 hectares concession for USD 121 million in a public tender. Since 2011, the project is under construction, with production planned to start in 2015-16. Now owned by a Chinese consortium, the project is to be among the biggest copper mines worldwide, with investments estimated in 2013 at USD 4.2 billion and average production at 400,000 tonnes per year in three to four separate open-pits, with a life-span up to 2031. Originally, it was planned that the ore should only be milled in place and then be transported through a 215 km pipeline to the Xstrata-owned mine Antapaccay for further concentration, but the project design was changed and now includes a concentration plant on site.[3]

The mine is located on an altitude between 3300 and 4800 meters above sea level in the two provinces of Cotabambas and Grau in the rural Andean highlands. Climatic conditions are quite

3 Las Bambas: "Proceso" <www.lasbambas.com/acerca-de-las-bambas/proceso.html> accessed 11.09.2015.

cold and dry, with major rainfall between October and April. The local streams flow into the Apurimac River, the major river of the department. Climate change has brought a decline in glaciers, and peasants observe that many water sources are drying up. The main subsistence production is of potatoes, together with livestock rearing.

The impact assessment commissioned by the company defines 20 communities being in the "zone of direct influence", 6 thereof had land in the concession area and one had to be resettled completely. The NGO CooperAcción, however, considers that 49 communities are affected. The local urban centre of Challhuahuacho, as well as villages which can expect compensation payments and jobs, were receiving a lot of immigration, partially from people who had left earlier due to lack of employment and are now returning. This creates a "mining town" situation with social problems such as widespread alcoholism. Possible environmental problems mostly concern excessive water usage for the mine, as well as water pollution.

Val Medel Gold Exploration, Grisons, Switzerland

1) The general context in Switzerland is at first view very different from the other cases. Though the extraction of metals and minerals is not absent from its history, it was abandoned almost completely after World War II. As part of the "dirty industries", mining is generally considered incompatible with high standards of environmental protection. It has no economic importance and no political lobby, and indeed there is no specific mining law or policy at the federal state level.

The political structure of the federal state of Switzerland is highly decentralized. Its citizens enjoy far-reaching rights of co-determination through mandatory and legally binding "direct democracy" instruments such as the popular vote or referendum at the municipal, cantonal and federal levels, whereby the first two are autonomous in many respects. In this context, the Medel case is especially unique as an example of a mining law adopted and modified through several popular votes by the local population, in the absence of specific higher-level regulations. This makes it interesting in relation to the concept of FPIC, which aims at strengthening the decision-making powers of indigenous peoples at the local level.

Some issues that come up in Val Medel resonate with other case studies: depopulation and rural exodus, lack of employment in a "rural periphery" in the context of economic liberalization and cuts in public services. These are serious concerns from an emic perspective, even though, compared to other countries, we cannot call it a problem of poverty or of marked ethnic discrimination.

2) *Early exploration:* As the case study is about a project in the earliest stage of exploration, no environmental impacts can be observed yet. So far, the company claims to have found samples with very high gold content in comparison to international standards. It seems that the project is mostly speculative at this stage. If developed, it would be situated in an ecologically sensitive, mountainous cultural landscape characterized by high-altitude animal husbandry and rich water resources in the catchment area of the important Rhine River, inside the buffer zone of a proposed national park.

CASE	**Mopani**	**Arlit**	**Ranger/ Jabiluka**	**El Cerrejón**	**Rawan**	**Marlin**	**Tampakan**	**Las Bambas**	**Val Medel**
Timespan	1930s – ?	1960s – ?	1970s – ?	1970s – 2040s?	1980s – ?	2000 – ca. 2020	1995 – 2030s?	2004 – 2031	2008 – ?
Region, Country	Copperdistrict, Zambia	Arlit, Niger	NT, Australia	La Guajira, Colombia	Chhattisgarh, India	San Marcos, Guatemala	Mindanao, Philippines	Apurimac, Peru	Graubünden, Switzerland
1. General Background									
a) Colonial past	Yes, strongly mining related	Yes	Yes	Yes	Yes	Yes	Yes	Yes, mining related	No
b) Economic dependence	Very high (national)	Very high (national)	High (national)	High (national)	High (state level)	Low – increasing	Low – increasing	Very high (national)	No
c) Legal protection	Weak	Weak	Strong (FPIC in NT)	Medium (ILO 169)	Weak – medium	Medium (ILO 169)	Strong (FPIC)	Medium – strong	Strong (referendum)
d) Neoliberal reforms	Highly relevant: Privatisation	Relevant	Relevant	Relevant	Relevant	Highly relevant Mining law 1995	Relevant Mining law 1995	Highly relevant	Not relevant for extraction
e) Armed conflict	No	Yes, ongoing	No	Yes, ongoing	State, ongoing	Civil war (-1996)	Yes, ongoing	1980s – 1992	No
2. Mine Specific Aspects									
a) Type	Underground, smelters	Open-pit/ underground	Open-pit/ underground	Open-pit	Open-pit and cement plants	Open-pit → underground	Greenfield open-pit	Greenfield open-pit	Not yet defined
Product	Copper, Cobalt	Uranium	Uranium	Coal	Cement	Gold, Silver	Copper, Gold	Copper	Gold
Current Stage	In operation, expansion	In operation; new project suspended	R: In operation J: Suspended	In operation, expansion	In operation, expansion	In operation	Permits and planning	Permits and planning	Early exploration
b) Ecosystem; cultural land-scape	Urban	Desert, erratic water sources (*wadi*); nomadism, herding	Humid-tropical; UNESCO site, heritage and sacred landsc.	Dry, river; subsistence agro, herding	Subtropical; rice cultivat-ion	Tropical hills; subsistence agro and herding	Volc. mountains, tropical forest, rivers; swidden and rice cult.	High mountains, semi-dry; subsistence agro and herding	Mountains, water sources; national park, herding
c) Impacts	Pollution, health, mining town related social problems	Pollution, wa-ter use, health, livelihoods, armed conflict	Pollution, loss of sacred sites, mining town	Landscape, Water use, pollution, land loss, mobility	Groundwater, pollution (dust), land loss, labour	Water use and contamination, land loss	Resettlement, deforestation, pollution, disasters	Resettlement, mining town; potential: water use / pollution	Yet unknown / undefined

Table A: Overview of context aspects from case study summaries.

3. Discussion

In this section, the context aspects of the case studies as introduced above are discussed on the basis of the overview provided in *Table A* (above). We again take up and refine the hypothesis outlined in the introduction, on how the general and mine-project-specific context may influence corporate and local people's strategies; more concretely, the question is how the context fosters a tendency towards imposition or negotiation on the one hand, and towards resistance or negotiation on the other.

3.1 National Context

The quest for precious metals and other resources has been a major driver for colonial expansion (*Table A – 1a*) from its very beginning, and it comes as no surprise that this is also evident in the discussed cases. Most famously, the Spanish conquest of South America in the 16th and 17th Century was foremost about its gold and silver deposits (see the cases of Peru, Guatemala), similarly illustrative is the more recent British colonization of Southern Africa for copper (see Zambia). In other places, mining was more a side effect of colonization and became important only later on. That most of today's dominant mining sector corporations are based in Australia, Canada, the USA and UK/South Africa shows historic continuities with the colonial past and the ongoing exploitation and expropriation of land from indigenous peoples. But, on a more regional level, mining may have been limited to small- and medium-scale extraction and short gold-rush episodes (Apurimac Peru, Mindanao Philippines), or not relevant at all until quite recently (La Guajira Colombia, Arlit Niger, central plains of Chhattisgarh). How the colonial background influences the current setting needs to be considered as specific, path-dependent developments together with the post-colonial histories, such as periods of authoritarianism and armed conflict, as well as the state's dependency on the mining sector.

State dependency on mining sector revenues and price changes (*Table A – 1b*) can even concern individual mines, due to the enormous investments and output of modern large-scale mines. Our most extreme case is Niger with an economy dominated, over a long period, by just two mines, operated by the same corporation. In Zambia, the mining sector is more diverse but strongly focused on copper, which makes the country highly vulnerable to market price fluctuations of a single commodity. Peru is very dependent on mining as well, though the sector covers a broader range of metals and numerous mines and companies. In Colombia, coal exports contribute an important share to the GDP, and almost half of it comes from the El Cerrejón mine. In Australia, the historically important mining sector continues to play a major role, but has lost its predominance in a diversified economy. In all of these cases, one can expect that mining policy is of national importance and that the interests of state and corporate actors coincide in facilitating access to the sub-soil resources, if necessary even against the will of the local population. Somewhat different is the situation in Guatemala and the Philippines, where mining does not so far contribute to a large extent to the GDP but is said to have the potential to do so in the future. Hereby, the role of the national elites who present their own interests as "good for the nation" may be more important than the actual benefits to state revenues and the general economy. For instance, in India, mining exports are marginal, but the sector is pushed by the political and economic elite of the state of Chhattisgarh as a specialization within the domestic economy.

Where mining is the dominant economic sector, the distribution of benefits becomes a central question in political struggles. The nationalization of the entire copper sector by the newly independent Zambia in the 1960s is our most compelling example for this. However, for the last decade, a trend towards "resource nationalism" cannot be confirmed from the case studies, except for minor shifts such as slightly increased royalty rates in reaction to voters' demands (Zambia, Guatemala). Rather, the impression is that the negotiation position of national governments versus mining TNCs remained weak. For a rough illustration of the economic power of the involved TNCs and states, *Table 1* compares company sales to national GDP. Although this cannot be directly translated into concrete situations, the ranking does indicate power disparities; for instance, in Zambia, with the global sales of Glencore alone being about 10 times higher than the national GDP.

The relation between price levels and national policy can be observed best where it concerns a single metal such as in Zambia. Being implemented under the pressure of a long period of historically low copper prices, the deal which the government negotiated when privatizing its mines in the late 1990s was terribly bad. Again, when trying to somewhat improve its share after the record high profits of the early 2000s, the massive drop in 2008 weakened its position, as does the current downward trend. In Niger, the threat to close the mines was used by Areva to raise bargaining power in negotiations with the government.

Regarding the legal framework (*Table A – 1c*), the case studies align quite well with the general trend outlined above. All states introduced environmental laws, though of varying quality, and several have considerably improved the legal recognition of indigenous peoples' and minority rights over the past two decades.

Concerning the recognition of local people's rights to participate in decision-making, we make a simple qualification in Table A from "weak" (or absent) to "medium" to "strong". Zambia and Niger clearly fall into the first category, arguably also India. In the second, which roughly corresponds to an ILO 169-type recognition of "prior consultation", we place Guatemala, Colombia and Peru, with the last one tending towards the third group thanks to a strong protection of collective land ownership for "peasant communities". The third group is formed by the Northern Territory of Australia and the Philippines, which both have legally mandatory provisions for FPIC, and by Switzerland with its far-reaching direct democracy instruments such as the local referendum.

All mentioned countries have, to some degree, been subject to neoliberal reforms over the last two decades (*Table A – 1d*), often expressed in new mining laws or reforms somewhere in the mid-1990s (highly relevant especially in the case of Zambia, Peru and Guatemala, but also in Australia, India, Colombia, the Philippines, and Niger). Neoliberal reform packages, as promoted by the IMF and the World Bank, often included "decentralization" or the devolution of responsibilities to regional and local governance levels, which opened space for policies of "empowerment" and citizen participation, sometimes also better recognition of indigenous rights. However, in our examples, this does not apply to mining, which mostly remains controlled by central state agencies. Exceptions are Australia's Northern Territory and India's Chhattisgarh, both being part of federal states. In Peru, there is some devolution to the local level through covenants that regulate the share of benefits for the local and regional levels. The only example where mining legislation is fully decentralized (at the municipal level) is Val Medel, being an extreme case even within the federal state of Switzerland. However, this has historic reasons and cannot be linked to neoliberal reforms.[4] The contradictions arising from legal pluralism are an evident background in those cases which combine legal recognition of indigenous land ownership with the privatization of underground mineral resources (Colombia, Guatemala, Peru, Philippines, and Australia). The observed prevalence of the second over the former is mirrored by the predominance of Ministries of Economy and Mines over those for Environment or Indigenous Affairs.

4 This is not to say that neoliberal reforms are not relevant in Switzerland: at the level of low taxation and other policies framed to attract TNC headquarters, including miners and commodity traders, they certainly are (See Part I – 1: Box: State of the Debate in Switzerland).

Table 1: Comparison of national GDP of case study countries with global sales of corporations in 2010.

Country	GDP 2010 (Billion USD)	Corporation	Sales 2010 (Billion USD)
India	1708		
Australia	1141		
Switzerland	581		
Colombia	287		
		Glencore Xstrata (post-fusion)	232.6 (in 2014)
Philippines	199		
Peru	148		
		BHP Billiton	50.2
		Rio Tinto	41.8
Guatemala	41		
		Xstrata (pre-fusion)	22.7
		Anglo American	21.2
		Holcim	20.4
Zambia	20		
		Areva	12.2
Niger	5.7		
Guinea	4.7		
		Goldcorp	2.7
Sierra Leone	2.5		

Source: "GDP" from <data.worldbank.org/indicator>; "Sales" from Forbes Magazine "Global 2000 List" for 2010 and 2014, <www.forbes.com> accessed on 12.03.2015.

The mining industry has its own ratings for the regulatory framework and its implementation in different jurisdictions, and for good reasons: according to a survey of the Canadian Fraser Institute, the perception of Canadian mining managers is that 40% of the decision to invest in one project rather than another is related to policy issues, with 60% left to the mineral potential of a deposit (Wilson/Cervantes 2013). It is interesting to compare our tentative rating of "protection of local people's rights" above (*Table A – 1c*) with the detailed *Mining Policy Perception Index* (PPI) of the Fraser Institute.[5] For 2011-12, the case study countries figure in the (lower) middle range (Zambia 50, Peru 56, Colombia 64, Niger 68, out of 93), or at the very bottom (Philippines 88, India 89, Guatemala 92, out of 93). Only Australia's Northern Territory, with position 11, re-

5 The Fraser Institute Policy Perception Index is based on questionnaires of Canadian mining sector managers and evaluates these factors: "uncertainty concerning the administration of current regulations, environmental regulations, regulatory duplication, the legal system and taxation regime, uncertainty concerning protected areas and disputed land claims, infrastructure, socioeconomic and community development conditions, trade barriers, political stability, labor regulations, quality of the geological database, security, and labor and skills availability" (Wilson/Cervantes 2013: 5).

ceives a high rating. Switzerland was not included in the survey (Wilson/Cervantes 2013). Placing Zambia and Niger relatively high, the PPI shows a bias on easy access to minerals, while jurisdictions with strong anti-mining movements (e.g., Philippines and Guatemala, but also Romania, Greece and several Argentinian provinces) score at the bottom. A more independent rating is the 2013 *Resource Governance Index*:[6] Here, three case study countries figure as "satisfactory", with Australia (West) at position 4 (of 58), followed by Colombia (9; for oil), and Peru (11), while India (12; for oil), Zambia (17), and the Philippines (23) still score relatively well (Guatemala, Niger and Switzerland are not covered).

Most of the case study-countries can today be considered as imperfect but fairly liberal democracies with multi-party elections and constitutional rights such as freedom of speech, assembly and association guaranteed at least in principle. However, with the exception of Australia, Zambia and Switzerland, all case studies show a background of violent conflict or outright civil war (Philippines, Peru, Guatemala, India, Colombia, and Niger) (*Table A – 1e*). Although the motives are diverse, all of them are or were conflicts over the distribution of resources, based on deeply rooted social inequalities, dating back to the colonial era. Mining directly plays into the regional-level conflict dynamics in Niger, and to a lesser extent also in the Indian State of Chhattisgarh, Colombia's La Guajira and the Philippines' Mindanao. In Niger, India's Chhattisgarh, Colombia and the Philippines, state armed forces are directly involved in protecting mines against "terrorists" and left-wing guerrillas. Until quite recently, the Philippines (-1986), Peru's southern rural areas (-1992) and Guatemala (-1996) were ruled by authoritarian leaders or were under (de facto) martial law. In Colombia and Chhattisgarh, political violence and repression against dissent remain at high levels. Zambia introduced multi-party elections only in 1991, and Niger has seen its last military coup in 2010. Additionally, the case study regions could be characterized as extraction- (or resource-) "frontiers", associated with state policies which push for the development of peripheral regions through the extraction of resources (Geiger 2008: 97), for Niger, Australia, Colombia, Guatemala, Peru and most of all the Philippines.

For civil society, recovering after prolonged attacks under authoritarian rule or violence related to armed conflicts needs time. Nevertheless, we see that in all the cases relatively active organizations, critical media and social movements are factors which can have an influence on national politics and on the implementation of legal provisions—although with strong reservations for Niger, where most critical voices are in exile due to government repression, and for Zambia, where civil society (most of all its "backbone", the mine workers' unions) suffers from the profound social breakdown that came with the economic crisis. The prevalence of a history of armed conflict and authoritarianism, similar to colonial continuities, seems to be an ambivalent aspect: while it may weaken resilience of local populations against external shocks and fast socio-economic changes, it may also contribute to pre-existing organizational networks and a focus on human rights discourses that can provide a base for resistance movements. It may also make a certain region appear as "problematic" for mining companies and thus motivate preventive investments in CSR as a security strategy, together with other mine-site specific factors as discussed in the following section.

6 This index is compiled out of 50 indicators grouped into "Institutional and Legal Setting", "Reporting Practices", "Safeguards and Quality Controls" and "Enabling Environment", and published by the Natural Resource Governance Institute. See <www.revenuewatch.org/rgi> accessed on 20.07.2015.

3.2 Landscape, Mine Type and Lifecycle

After having explored the larger context in which the case studies are situated, in the following, we discuss basic aspects of the mining operations with regard to their setting in the cultural landscape. In the *second part of Table A*, we gave a rough qualification of expected and observed impacts, which will be taken up again in Chapter C, where we look at the reasons and demands that local people voice in mining conflicts and the changing dynamics over the mine life-cycle.

The case studies show a wide range of different types of mines and extracted metals and minerals. The cases cover different stages in the mine life-cycle, from exploration to planning, construction and operation, some of them approaching closure with follow-up projects in planning or construction. Considering such ongoing extensions, all the operations show a long-term character, with the shortest planned time-span from exploration to closure being 20 years (Marlin, Guatemala). The majority involve large-scale open-pit mines (El Cerrejón, Colombia and Tampakan, Las Bambas), while several are combined with underground mining or mines (Marlin, Cominak Niger, and Ranger, Australia). Cement production in India and the Mopani underground mines and smelter in Zambia are two examples where the most obvious impacts are caused by an entire industrial complex, while the open-pits are comparatively small. However, Mopani shows that technical innovations such as "in-situ leaching" can also change the impacts of "old-style" underground mining. The cases cover both brownfield (Mopani) and new greenfield mines (Tampakan, Las Bambas, Marlin). However, there are mixed cases where an open-pit is exhausted and is to be replaced by a new one nearby (Ranger/Jabiluka), combining and linking up new and old elements of a mining complex even over large distances (Las Bambas; possibly also the Imouraren project in Niger). Once a mining infrastructure is in place, the exploitation of the ore body demands continuous expansion of the mine pit on ground and/or into the underground, possibly requiring more land than originally expected. The most land-intensive mine in our sample, El Cerrejón, is slowly eating up an entire landscape; therefore, land acquisitions and resettlements are not a one-time intervention only at the beginning, but a permanent feature of the ongoing extraction.

Regarding contamination, Acid Mine Drainage is a frequent problem, hard to contain and destined to continue long after mine closure (see Part I – 3). Furthermore, the processes for chemical concentration are especially problematic. The processes used depend on the geology and the type of metal. Uranium (see Ranger/Jabiluka, Arlit Niger) comes with radiation risks, copper with sulphur dioxide emissions (see Mopani), and gold mining with cyanide leaching (see Marlin). Such processes may liberate heavy metals, arsenic or radioactive metals naturally contained in the rock. Modern mines increasingly exploit "polymetallous" ores, whereby a number of secondary metals are also isolated in several steps, possibly in several plants in different locations. For example, the product to be exported from Las Bambas is a copper concentrate (containing other metals), which needs further refining elsewhere.

The range of different ecosystems and climates includes desert to humid tropical and high altitude mountain regions. The affected areas can be quite large, though it is often hard to estimate how far diffuse impacts like contamination or the depletion of ground water resources reach. The case of Tampakan shows a typical situation with unequal distribution of compensation benefits and unaccounted risks for further downstream communities.

The cultural landscape aspect is central to all the cases, though in different and complex ways. In Niger, for example, uranium mines contaminate essential water bodies for herded animals. A spiritual significance of the landscape is potentially relevant for El Cerrejón, Ambuja Rawan, Marlin, Las Bambas, Tampakan, and most of all Ranger/Jabiluka. For the latter, the mines are located within a UNESCO heritage area, which means that both cultural and biodiversity aspects are highly recognized and protected. The Val Medel exploration lies within the buffer zone of a planned national park. In the introduction of this chapter, we quoted a mine manager declaring that the main challenges for a mine are people and water. This seems accurate, though in complex ways. The "arid environment without people" he prefers does not exist in our cases, and probably nowhere. Even deserts are not empty, and with mining, the population often increases due to immigration attracted by new economic opportunities (see Niger, Las Bambas, El Cerrejón). Hence, people are a given, and to treat them as a "problem" may be itself an underlying ideological conflict driver. The condition of aridness, which makes operations and pollution control easier, becomes an existential problem for local people if the enormous water consumption for mining competes with their needs. Water (consumption and contamination) is a key problem of mining (Kemp et al. 2010) and is crucial for understanding many of the conflicts presented here. Local peasants and herders will observe when age-old water sources go dry, or new-born animals increasingly show strange deformations, but an occasional outside visitor will not. The general picture, however, is that the impacts of different types of mines show a wide variety, much more for instance than the conventional forms of fossil fuel extraction discussed in Haller et al. (2007).

3.3 Context: Conclusion

Summarizing the first bundle of context aspects, a first hypothesis was outlined in the introduction in view of corporate and local people's strategies and the dynamics of conflict between them (see *Figure 3*): in a democratic setting with strong legal protection, monitored and acted upon by a vocal and diverse civil society, local people have a strong bargaining position in relation to mining companies, possibly even to veto an unwanted mining project. On the contrary, high dependency of the national economy, elite sectors or the general population on mining revenues (partly subject to shifting prices), and preference for corporate access through neoliberal policies potentially weaken the bargaining position of local people. The first set of factors should enhance a corporation's readiness to enter into meaningful negotiations with local people, while the second favours imposition and unilateral expropriation. Correspondingly, conflicts can be expected to be negotiated rather in courts and through non-violent political mobilization if the first set of factors predominate, and show escalated forms and/or passive resistance if the second set of factors does so. Strong colonial continuities and recent and ongoing periods of authoritarianism and armed conflicts are ambivalent aspects, which may weaken local resistance, but also motivate strong and immediate reactions, and at the same time cause companies to preventively invest in CSR measures for conflict mitigation and security.

In part, this argument resonates well with the "good governance" discourse *en vogue* in the last two decades. Thereby, capitalism combined with the rule of law in a liberal multi-party democracy, efficiently managed by a transparent and lean state, is presented as a necessity to attract foreign investments, and thus, for economic growth and development. Both, the neoliberal reforms and the recognition of minorities and common property rights are justified with varieties of this discourse. However, the reality turns out to be more complex: as James Ferguson (2005) shows for the African context, some of the worst-governed and war-ravaged countries like Sudan, Angola or the DR of Congo scored the highest rates in foreign investments and economic growth. This was driven almost exclusively by the extractive sector, especially oil. Journalist Naomi Klein (2007) explores a similar phenomenon under the title of "Shock Doctrine", reminding us that the first neoliberal laboratory of the so-called "Chicago Boys" was the Pinochet dictatorship in Chile. The shock of the military coup against president Allende in 1973, followed by the killing and expulsion of thousands of trade unionists, journalists and left-wing activists prepared the necessary "empty table" to introduce radical economic liberalization measures, which would have been unthinkable otherwise. The Chilean model of this "disaster capitalism" was then exported worldwide, creating fast growth and tremendous profits where people are under shock from war

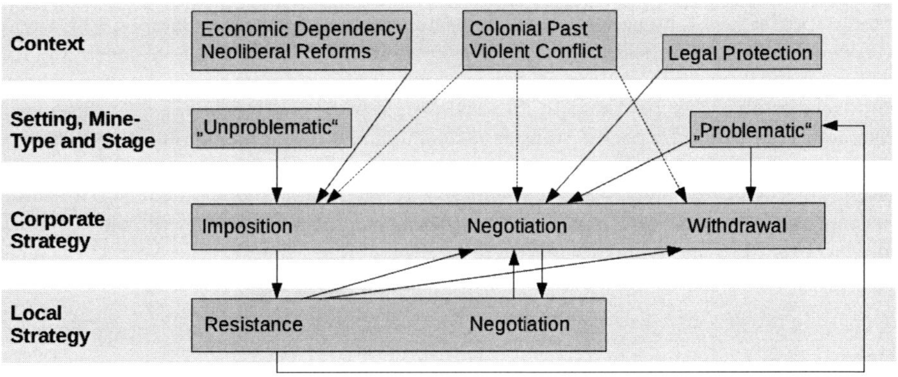

Fig. 3: Context factor influence on corporate and local people's strategies.

(e.g., Iraq) or natural disasters (e.g., New Orleans).[7] The pivotal role of private security contractors in this economic transformation, stressed both by Klein and Ferguson, is evident in the mining sector as well (see Chapter B).

On the background of a society affected by detrimental factors, the impacts of a mine can itself constitute a major shock. Looking at the strategies of corporations and local people, we should check for moves, which strengthen resilience against external shocks on the one hand, or weaken it on the other hand. As Klein maintains, the "shock effect" wears off with the years and creates a learning effect, immunizing local populations against future shocks.

Returning to the case studies, these observations are useful to describe certain underlying similarities in the general contexts: the combination of societies weakened by violent conflict, authoritarian rule and/or economic shock with the neoliberal reforms in the 1990s. Certainly, the type of disaster varies, from a not mining related civil war in Guatemala to a copper-price induced economic shock in Zambia, where "structural adjustments" meant that "[a]ll exchange controls, tariff barriers and food subsidies have been dropped in the shock-treatment switch-over to rampant capitalism."[8] Yet, we find strikingly similar policies favouring investments by mining corporations being introduced just at the time when governments and civil society organizations were, in retrospect, in the worst position to pressure for strict regulations and a fair share of future profits: at the bottom point between two "commodity super-cycles". This seems highly relevant for Guatemala, Zambia, Colombia, Peru and the Philippines. Similar scenarios regarding the negotiations between states and corporations may appear in the current context of decreasing metal prices and the possibility of a new, prolonged super-cycle low bottom.

Meanwhile, the improved recognition of indigenous and local people's rights in the last two decades, contemporaneous with neoliberal reforms, can be seen as a success of strengthened civil society and indigenous peoples' organizations, albeit a partial one. A cynical interpretation would

7 See Fletcher (2012) for an adaption of "disaster capitalism" to climate change.

8 Donald McNeil: "For Sale: Zambia's Rich Copper Mines, All of Them." *New York Times*, Aug. 11 1996. <www.nytimes.com> accessed on 17.12.2014.

be that it served merely as a tranquillizer, to keep them busy "at the surface", while quietly the (under)ground was prepared for the next big wave of corporate expropriation of natural resources. For instance, Lindroth (2014) identifies the improved recognition of indigenous rights as a tactic of cost-effective neoliberal governance, combining an improved "governability" of unruly sectors with the devolution of government responsibilities to the local level, in accordance with the doctrine of the "lean government". However, one can also see it simply as two opposing trends which clash in contested and contradictory settings as described in the case studies.

The specific characteristics of the mines we discuss can be summed up in a hypothesis with regard to corporate strategies (see *Figure 3*): if a mine project is seen as "highly problematic" for its setting and stage in the mine-cycle (put simply: a greenfield project in an area with "lots of people", plus a history of conflict, e.g., Tampakan), we can expect high investments in CSR and prevention of negative impacts. To the contrary, a brownfield mine in a setting which is regarded as relatively "unproblematic" would see less such investments (e.g., Mopani). This needs to be checked also in relation to the regulatory framework and changing market prices, as well as to the potential for local people to set up resistance at critical moments in the mine-cycle.

Pro-Negotiation Aspects	**Low**	**Medium**	**High**
Legal protection; Civil society	Niger, Zambia, India (Chhattisgarh)	Guatemala, Colombia, Peru	Philippines, Australia, Switzerland
Problematic setting	Niger, Zambia, India (Chhattisgarh)	Guatemala, Colombia, Switzerland	Peru, Philippines, Australia
Problematic mine-cycle stage (planning, expansion)	Zambia, India (Chhattisgarh)	Guatemala, Colombia, Switzerland	Peru, Philippines, Australia, Niger (Imouraren)
Pro-Imposition Aspects	**High**	**Medium**	**Low**
State dependence	Zambia, Niger, Peru	Australia, Colombia, India (Chhattisgarh)	Philippines, Guatemala, Switzerland
Neoliberal reforms	Zambia, Guatemala, Peru, Niger	Australia, Colombia, India, Philippines	Switzerland
Mixed-Impact Aspects	**Highly Relevant**	**Not Relevant**	
Colonial history	Niger, Zambia, India, Guatemala, Peru, Colombia, Philippines, Australia	Switzerland	
Armed conflicts	Niger, Colombia, India (Chhattisgarh), Guatemala, Philippines, Peru	Zambia, Australia, Switzerland	

Table 2: Context aspects favouring imposition or negotiation.

In *Table 2*, the cases are grouped according to the factors discussed so-far that foster corporate strategies of "negotiation" on the one hand, and "imposition" on the other. Though the categorization is a rough estimate and the factors are not weighted, it allows us to draw a spectrum on which the cases can be placed tentatively in relation to each other (see below).

Negotiation: *low* Imposition: *high*	*high* *low*
Niger–Zambia–India–Guatemala–Peru–Colombia–Philippines–Australia–Switzerland	

In the following chapters, we will discuss how far the strategies of corporations and local people correspond to this pattern. Concerning the comparison of local people's strategies, the mine-specific factors can be used to further specify the "impact–resistance hypothesis" from where we started: the resistance potential against mining thereby is expected to change in relation to the perceived impacts, and thus to the mine-cycle. At first, the most direct and visible impacts such as land loss and economic transformations dominate. Later, the long-term, "diffuse" impacts such as health problems caused by pollution become more important, while increasing economic dependence lowers the resistance potential. This means that protests can be expected to emerge with the project set-up, gradually increasing during construction, further intensifying and peaking during the first years of the active mining stage, and then decreasing as the new socio-economic realities take root and more people become dependent directly or indirectly on mining incomes. The above-mentioned background aspects, together with changing world market prices, thereby also play into the possibilities for local people to search for agreements, rather than to opt for outright resistance.

PART III: Comparison

B) Corporate Strategies

In this chapter, the strategies of the main companies involved in the case studies are compared. The first part introduces the three main angles of comparison: 1) "corporate typology" and ownership structure of the different mines (*Table B – 1,* see page 451); 2) "corporate impression management", i.e., the coherence between corporate "front-stage" and "back-stage" performance (*Table B – 2*); and 3) "corporate-community relations", the specific interactions of companies with local people regarding the core issues of consultation and consent, compensation and agreements (*Table B – 3*). The second part presents short summaries of the case studies with regard to these aspects, to be discussed in the third part. The results will be further compared in relation to local people's strategies in the following Chapter C and the final conclusion (D).

1. Introduction

Corporate strategy involves the strategic investment of different forms of capital to produce profits in a market-driven economy, while keeping costs low to achieve a competitive advantage over other companies in the same sector. The high cost risks of mining conflicts are increasingly taken into account in corporate decision-making and investment choices (see Part I – 1; Wilson/Cervantes 2013; Davis/Franks 2014). In the following, we identify major qualitative aspects of corporate strategies at play in mining conflicts.

1.1 Corporate Typology

In the introduction of Part I, we outlined the necessity but also the difficulties in opening up the "corporate black-box", that is, how internal differentiations and tensions affect the decisions taken within corporations, in reaction to various external pressures. Even though the general aim for maximizing profit and "shareholder value" underlies corporate decision-making, this is subject to improvization and guessing on short- and long-term costs and benefits in specific situations.[1] Welker's ethnography at the headquarters of Newmont Mining Corporation shows how specific decisions relevant to local communities are taken at the headquarters of a major TNC (2014: 40). Though such detailed insights are largely lacking for the case studies in this book, it is possible to differentiate between types of corporations and to evaluate how specific companies try to distinguish themselves from their competitors on a strategic level, according to the information they provide to the public.

To draft a typology of corporations (*Table B – 1*), a first basic distinction can be made between private, state-owned and publicly traded companies. In this book, we primarily deal with com-

1 For specialized news services for up-to-date mining and metal market developments, see: *Engineering and Mining Journal* (<www.e-mj.com>), <www.miningnews.net>, <www.mineweb.com> and <www.mining.com>.

Fig. 1: "Global Operations" of Glencore: "150 assets and more than 90 offices". Source: Screenshot of <www.glencore.com/our-world/global-operations>, Oct. 2015.

panies publicly traded at stock markets. They are accountable to and legally obligated to disclose information to their shareholders. The details of the disclosure rules depend on where the shares are traded, which can have important implications on the transparency of corporate decisions.[2] Contrarily, privately-owned companies are accountable to a very small number of owners and information about them is hardly available to outsiders, which is often the case for commodity trading companies (the example covered is Glencore prior to its stock market launch). A further, mining-specific differentiation is between junior, mid-tier and senior (or major) firms. This book mainly deals with senior firms, i.e., the so-called "transnational corporations" (TNCs), which own numerous mines and projects in different countries. Mid-tier miners develop or run one or several mines and are funded by a mix of risk capital and actual mining profits. Junior miners are usually active in exploration only and run on risk capital—today, they are mostly based in Canada thanks to special tax exemptions. A company of any size can have subsidiaries (full ownership or majority control over another company), minority participations, and enter into "joint venture" agreements with other companies. If the latter are state-owned, this can be a way for a host country to participate in mining benefits. Other corporate actors involved can be (external) consultants and advocates who provide services such as the set-up of offshore companies. Similarly, geological exploration, Environmental and Social Impact Assessments, CSR programs and other elements of a mining operation are regularly "outsourced" to specialized consultants, firms and NGOs. Substantial parts of the construction and operation of mines and related infrastructure and services, such as transport and security, can be commissioned to sub-contractor firms.[3]

All in all, ownership, corporate structure and legal accountability behind a single mine can be very difficult to decipher. Furthermore, a mine is usually not developed by the same company

2 For more detail, see the "mining investors guide" of MiningWatch Canada (Kuyek 2007). An open access database and guide on mining financing is <moneytometal.org>. See also Part I, Chapter 2.

3 A database of mining contractors and consultancy services is provided by <www.infomine.com/suppliers>.

through all project stages. Shares in mines and mining projects, as well as in entire companies, are constantly bought and sold. Smaller companies tend to be taken over by larger ones, but it also happens that a major company sells a mine project to a smaller one, which is ready to take more risks. The ownership history thus can have important implications for understanding current conflicts (Dougherty 2011). Summing up, the major, publicly traded, transnational mining corporations can be understood as focal points in a wider field of corporate actors. Thereby, we can assume that their impressive size and the financial power they can mobilize have important implications on their CSR strategies, which are the focus of the next two sections.

1.2 Corporate Impression Management

The discourse on CSR and a wide range of regulatory frameworks and voluntary standards developed parallel to the increasing power of TNCs.[4] CSR has become indispensable to positively influence the perception of a company by different stakeholders (most of all, shareholders), even more so in the context of new communication technologies, which have massively accelerated the global spread of information.

Corporations are sometimes depicted as "psychopaths" (i.e., legal persons without a consciousness), and mining in particular serves as prime example for "corporate crimes".[5] Or, as PR consultants may say, mining companies have an image problem due to bad communication, misinformation and lack of trust. What we here call "corporate impression management" (*Table B – 2*) constitutes in our view a core aspect of CSR strategy, as an attempt to react to and prevent criticism.[6]

The introduction of stringent CSR policies throughout all the relevant sectors of a corporation and down to contractor companies requires a transformation of corporate culture against considerable internal resistance. This process involves internal campaigning by specialized CSR staff, e.g., screenings of critical documentary films, to raise awareness among the staff. The attitude of members of the top management can be a crucial driver to push CSR. Further, one of the main argumentation lines for introducing CSR in Newmont was a "soft is hard " narrative: to counter the common perception of CSR as a "soft" discipline that only generates costs and no profits (unlike "hard" disciplines such as geology), the CSR staff stressed that social issues can create much higher costs, but also "high value opportunities" by potentially granting a "socially responsible" company access to contested fields where competitors fail because of their bad reputation. Therefore, CSR is ultimately contributing to the generation of profits (ibid.). Adapting these observations, we qualify CSR strategies in *Table B*) on a spectrum from "weak" to "strong", whereby the first extreme refers to a "soft" interpretation of CSR that mainly relegates it to the realm of public communication and occasional philanthropy, while the second corresponds to a realization of the "soft is hard" narrative, whereby CSR is understood as a core issue for the functioning of the corporation. The variations of CSR strategies covered in this book point to a field of competition about reputation and credibility between and within different corporations. Given the mentioned internal resistance and the time needed for any change in corporate culture, two major internal factors make a difference: the moment when a coherent CSR strategy was introduced and the support it receives from the top-level management.

4 The most relevant CSR instruments for mining and current debates are introduced in Chapter 2 of Part I.

5 E.g., Bakan (2004) and the film *The Corporation* directed by Mark Achbar and Jennifer Abbott (2003).

6 The terms "impression management" and "front-stage/back-stage" are from Erving Goffman (1959). Though we do not use them to describe "the presentation of self in everyday life" on an individual level, we think they can be freely adapted to the "Social Life of the Corporate Form" (see *Current Anthropology* 52 Special Issue 3).

CSR competes with other considerations about the profitability of a specific mine under current and expected prices, compared to other mines operated by the same or by other corporations. If a hypothetical company would do everything correctly, it could gain a lot of credibility, but the cost of its product would risk being so high that it would vanish soon due to competition by opportunistic free-riders.[7] We can assume that it is impossible that a company complies to all its commitments at all times—for instance, the best available technology for treatment of waste water can be so expensive that it is a big competitive advantage to avoid introducing it as long as possible. Taking this into consideration, we gain an impression on the quality of CSR strategies by discussing the (in)coherence between "front -stage" and "back-stage" performance of the main corporations in question. Is the "front-stage" communication about problematic issues proactive or rather a hiding strategy? What references to specific norms and standards are made? Is there a strategy of alliance building with NGOs, universities and local government bodies? If certain issues are denied or ignored, this may point to problems of "back-stage" performance that a company tries to cover up: non-compliance with laws and standards, manipulation of public opinion, criminalization and targeting of activists and the use of divide-and-rule tactics for "managing activism".[8]

To further qualify the CSR strategies in the case studies, we adopt and adapt the model of Benson and Kirsch (2010a; Kirsch 2014: 182ff., 185). This model distinguishes three phases of CSR in response to public critique between which corporations move back and forth: the first phase is denial and the proliferation of doubt in order to maintain low transaction costs, maximal externalization of negative impacts and highest possible profitability. The second phase is an acknowledgement that a problem exists together with gestures of accommodation such as the institution of audit culture and partnerships with certain NGOs, divide-and-rule tactics and co-optation of discourses. This means to assume certain costs to prevent that the responsibility for the full costs of damage must be incurred. The third phase is reached when the loss of legitimacy and financial risks threaten the survival of the corporation: as a "last resort", it is forced to actively engage in a "strategic management of critique and the establishment of a new status quo". For the mining industry, the third phase is relevant mostly at the level of individual projects. If the environmental damage and potential costs for closure and remediation of a mine become too problematic, a strategy may be the "transfer of liabilities to a third party" (e.g., a public enterprise which benefits from state subsidies). A strategy that may be placed somewhere between phase two and three is the re-branding and change of ownership to dissociate a "new" company from past failures, or the active reinvention of a corporate history.[9]

The Benson/Kirsch model is a useful tool to describe the top level of a corporate strategy, moving back and forth between the first two phases of response to critique, in order to avoid the third. But how corporations move along these phases depends not only on critique, but also on the commodity market prices and the general legal and political context, as well as on the profitability and technical challenges (see Chapter A). Relating to the corporate typology introduced above, one can assume that the larger and more diversified a company is, the less it depends on

7 This can be interpreted as a problem of collective action, whereby the lack of monitoring and sanctions lowers the motivation to ensure compliance with established standards. See for e.g., Kirsch (2014: 171-72).

8 See Deegan (2001). A summary of her advice of "divide and conquer" is given by Stauber: "First identify the 'radicals' who are unwilling to compromise [...]. Then, identify the 'realists'—typically, organizations with significant budgets and staffs working in the same relative area of public concern as the radicals. Then, approach these realists [...] start a dialogue and eventually cut a deal, a 'win win' solution that marginalizes and excludes the radicals and their demands. [...] Result: industry may have to make some small or temporary concessions, but the fundamental concerns raised by the 'radicals' are swept aside." PR Watch, Second Quarter 2002, <www.prwatch.org> accessed 14.03.2015.

9 As described by Rajak (2014) for the corporate history of Anglo American in the transition from the Apartheid regime to democracy in South Africa.

the success or failure of a single project. On the one hand, this means that a major TNC may stall or sell a controversial project to save its overall reputation, if targeted by international campaigns. Small and medium companies may depend fully on the success of a single project. On the other hand, major TNCs have more financial power to negotiate special treatment for taxes, subsidies and environmental standards.

Fig. 2: Billboard advertizing the construction of a cattle watering place by Xstrata, close to the Tintaya mine, 2010. Photo: Golda Fuentes 2010.

1.3 Corporate-Community Relations

Questions of consultation and compensation are at the core of what we call "corporate-community relations" (*Table B – 3a, b*). The insights gained from the case studies are limited on these aspects because they mostly lack the long-term ethnographic observation needed to understand the full dynamics of such processes. Nevertheless, we have some elements for a general evaluation and a base for further research.

With regard to consultation, current debates almost unanimously stress that local and indigenous communities should participate in the decision-making process on developments that affect them. However, there are very different views about what this means (see Part I – 1 and 2). Many CSR documents include statements on "dialogue", "participation" and "consultation" of stakeholders,[10] as a means to prevent conflicts and to ensure what miners call the "social licence to operate" (Coumans 2011: 30; Prno 2013, Owen/Kemp 2013). According to the concept of

10 See Voss/Greenspan (2012) for a collection of statements on community consultation and consent.

Free, Prior and Informed Consent (FPIC),[11] indigenous peoples are not to be considered merely as "stakeholders", but as "rightholders": as the custodians of their ancestral territories, they should have the right to ultimately give or withhold consent on any development that affects them. Arguably, companies have an interest to avoid reference to FPIC, as it gives indigenous communities much more power in negotiations, and therefore prefer vague terms like "stakeholder dialogue" or "community participation", which do not imply a veto right (see Part I – 2.3).

With the category of "consultation", we here refer to negotiation processes in which local residents cede access to the ore for a company in return for certain benefits or "compensation". Possibly, this is framed in a contractual agreement that is negotiated directly with local representatives. This process can be used to increase legitimacy of mining by linking it to (potential) benefits such as profit-sharing, development projects and the generation of local employment. However, several back-stage options exist for a company to cut costs by reducing the number of people who are recognized as "affected" by a project (Moody 2007: 173): non-recognition of traditional land ownership and related complex primary or secondary user rights (especially, where concrete records are missing), exclusion of impacts which affect larger areas, and neglecting elements related to quantity and quality of water available for farmers living downstream. Compensation payments and other benefits can have unintended results linked to the sudden influx of cash. Often, local views and power relations are not well understood and conflicting interests of local stakeholders and their strategies are not anticipated. To prevent compensations doing more harm than good, current debates stress the importance of participatory, fair and transparent negotiations prior to the development of a mine. The results should be fixed in direct agreements with organizations legitimately representing the interests of the indigenous and local people.[12] However, without a clear understanding of the heterogeneity and the power relations within local communities, a participatory process cannot be reached. This needs considerable research prior to the negotiation process, which can lead to a better representation, a reduction of negative impacts and an effective "management of expectations" about future benefits for the local population. On the other side, it can be used strategically to undermine resistance.

In Chapter A, we formulated two related hypotheses regarding the national-level context of mining operations and their specific characteristics and setting in the cultural landscape, identifying factors that either favour the imposition of mining projects under conditions that do not correspond to the aspirations and needs of the local people, or motivate companies to enter into negotiations with local actors and preventively invest in CSR. In the previous section, we added the corporate headquarters' CSR policy as another factor that may foster the potential for negotiations at the local level. With regard to the Benson/Kirsch model of three phases of corporate response to critique, we can adapt this by assuming that a "phase one" (i.e., denial) position would correspond to denial of local people's rights and the imposition of a mine against resistance; for "phase two" (i.e., partial acknowledgement), we expect at least pro forma recognition and some form of negotiation with certain local groups, while "phase three" is marked by substantial (and costly) compromises or even a full withdrawal from an unwanted mine project.

11 FPIC has been endorsed in the UN Declaration on the Rights of Indigenous Peoples and in some national Constitutions and laws. See Part I – 2; Doyle/Cariño (2013); Doyle/Whitmore (2014).

12 The negotiation of so-called "Impact and Benefit Agreements" is becoming a standard practice in Canada and Australia (See O'Faircheallaigh 2008, Gibson/O'Faircheallaigh 2010).

2. Case Study Summaries

In this chapter, we give short summaries of the case studies regarding the aspects introduced above. We move from a short description of the involved corporations to a general assessment of the front-stage/back-stage coherence of CSR-strategies and end with the "corporate-community relations", focussing on questions of compensation and consultation. In our analysis, we follow the structure of *Table B*, where 1) refers to the corporate type; 2) to the corporate impression management; and 3) to the corporate-community relations.

Mopani Copper Mines, Zambia

1) The Mopani Copper Mines (MCM) were fully state-owned through Zambia Consolidated Copper Mines (ZCCM) from the 1960s to the year 2000, when they were privatized and sold to the Switzerland-based Glencore (which holds the majority share of 73%) and the Canadian First Quantum Minerals (holding 17%). The Zambian state keeps the remaining 10% through ZCCM. The sale was made at only USD 43 million when copper prices were at a very low level. Glencore was ranked as number 9 copper producer worldwide in 2010.

2) In Zambia, Glencore was able to invest in an already existing mining industry under very favourable conditions thanks to neoliberal "structural adjustments" policies. Thereby, the company ensured that it could not be held responsible for damages prior to the acquisition.

Front-stage: As long as Glencore was a privately owned company, it was (un)known as "[t]he biggest company you've never heard of" (*The Guardian*, 19.05.2011), a perfect example of a "keep quiet as long as you can" strategy. Apart from not being obliged to release information to shareholders, it was also thanks to Glencore's opportunistic brownfield investment strategy (acquiring old and defunct mines at low prices) that a proactive CSR strategy was less needed. Mopani did not publish any specific CSR documents, nor did it have an own website. Glencore's first Sustainability Report is from 2010, but only in the 2014 report there is detailed information regarding Mopani. At the time of writing the case study, there were just a few general statements available regarding the participation in the EITI process[13] and issues of health and environment, social impacts and taxes. Investments in a new smelter were presented as a measure to reduce sulphur dioxide emissions, which was financed with a controversial USD 50 million loan from the European Investment Bank. Glencore presented employment generation and investments in production capacity as positive contributions to the Zambian economy. On taxes and financial transparency, it provided data to the 2010 EITI Report for Zambia and denied allegations of tax avoidance through a "transfer pricing" scheme.

Back-stage: The available sources indicate that there are severe impacts on the environment and health of the local population, particularly due to massive sulphur dioxide emissions, leading to respiratory and other diseases. Media and NGO reports hint at "cutting corner" strategies (non-compliance with legal norms in the absence of enforcement); especially, concerning waste deposit and the "in-situ leaching" technology. About the alleged tax avoidance, many questions remain open. The complex off-shore structure of the company seems to be designed to evade transparency of financial flows, what might well be legal, but does not legitimize it in the eyes of many Zambian citizens.

3) Consultation of local communities seems not to be an issue, in the absence of any legal requirements on minority or indigenous peoples' rights. For the expansion of underground mines, there is little need to acquire land. The only documented case was in 2001, when the company was challenged before the OECD contact point in Canada because it had threatened to call for

13 For the Extractive Industries Transparency Initiative (EITI), see Part I, Chapter 2.

the army to forcefully displace an informal settlement for the construction of a new mining shaft. In consequence, Glencore agreed on holding a dialogue process with the "squatter" community, which was started but later apparently abandoned without results.

Nothing is known about compensations, be it for environmental damage, land acquisition or other socio-environmental impacts. Likewise, there is nothing known about any direct agreements with local actors on the distribution of benefits. Mopani pays some taxes to the municipal authorities in the two cities where it operates. A retirement scheme and health programs were highlighted as CSR efforts; however, there are indications that they function rather as benefits for regular employees, while the sub-contracted workers and the general population do not benefit. The most relevant benefit for the local population is thus the employment of about 17,000 workers. An estimated 50% of the workforce was sub-contracted through other companies, with salaries just one-third of the fully employed workers, lowering social welfare costs and increasing the company's flexibility in hiring and firing depending on price fluctuations on the metal markets.

Arlit and North Niger (Areva)

1) Areva is specialized in nuclear energy production, from uranium mining up to the construction and operation of nuclear power plants. The company is controlled by the French state and plays a crucial role in providing the country's electric energy, dependent up to 80% on nuclear power, which is again largely fuelled by uranium from Niger (with an estimated share of 25 to 40%).

2) The neo-colonial setting of dependency of Niger on France, interlinked with France's dependency on nuclear power, means that the strategies of Areva can hardly be separated from those of the French state. Indeed, countless rumours and accusations concern the more or less hidden intervention of the French state to defend its interests in the region; that is, the interests of Areva. This concerns the protection of loyal military regimes against opposition, but also power play through supporting rebel groups if a government tries to increase its share in the uranium business, or to frighten away competitors (e.g., from China). In the absence of independent media and civil society observers, it is however mostly impossible to substantiate such claims.

Areva's strategy seems to be to hide from public scrutiny and to prevent criticism by obstructing independent research and the collection of data—for e.g., concerning the health conditions of workers and residents of Arlit. However, this is mostly not done directly, but rather through state security forces which criminalize critical local people and organizations, and whenever possible, prevent foreign NGO workers and journalists from entering the area. The context of armed conflict and "anti-terror" measures provide a perfect excuse for this, while the company stays in the background. In reaction to independent reports issued by NGOs around 2010 (in a temporary political opening of space for critique), Areva set up some programs to check the contamination, albeit only addressing certain "hot spots". Meanwhile, sources from local hospitals claimed that there was a systematic under-reporting of cases of cancer, especially for mine workers, and even a practice of wrongly diagnosing cancer as AIDS or other illnesses not related to radiation. The company seems to invest very little in social services and local development programs, with few benefits provided mostly to "expats" and regular staff.

An obvious example for Areva's power play happened when a new government challenged the company in 2011, demanding the implementation of the mining code of 2006 and an increase in royalty rates: Areva used the threat to close its mines, leaving thousands of people unemployed, to increase its leverage in the negotiations.

3) The case study mentions one incident of a consultation meeting (although for new concessions not related to Areva), whereby a local NGO was expelled for asking critical questions. Otherwise, there is nothing known about consultations or compensation agreements. The regular employment opportunities (several 1000s in total) do not seem to benefit many local residents.

Ranger/Jabiluka, Australia

1) The Ranger Mine in the Australian Northern Territory was discovered in 1969 and first developed by a joint venture between two Australian mining companies and the state-owned Australian Atomic Energy Commission (AAEC). In 1979, the publicly listed Australian mid-tier miner Energy Resources of Australia Ltd. (ERA) became the majority owner of the mine, which started operating in 1980. In 2000, Rio Tinto acquired a 68% share of ERA. The nearby Jabiluka deposit, discovered in the early 1970s, was acquired by ERA in 1991. ERA does not operate or develop any other mines except these two. Rio Tinto, the second largest mining corporation worldwide with presence in over 40 countries, has its headquarters in London and Melbourne (Australia). and is a leading uranium producer.

2) Ranger/Jabiluka is an exemplary case for the changes in the relationship between corporations and indigenous peoples over the last four decades. The company ERA had the opportunity to learn a lot from the resistance of a small group of Mirarr people, and can now proudly claim to have one of the best developed CSR strategies of the sector.

Front-stage: Ranger and Jabiluka are probably the best-documented mines included in this book. Since the 1970s, several extensive impact studies have been prepared. Although these reports highlight a number of risks and difficulties, they always concluded that the uranium extraction should proceed. ERA had to put in place expensive technical measures to prevent the spilling of radioactive water into the environment; it installed a relatively transparent monitoring system, and designed rehabilitation programs for disturbed areas. The company claims to comply with the highest technical standards in the sector. Apart from relevant laws, it follows its own codes of conduct, including the "Rio Tinto Cultural Heritage Guide", and carries out cross-cultural awareness trainings for its staff. It provides a lot of information to the public and shares reports with indigenous representatives. Last but not the least, ERA claims to ensure that the extracted uranium is used only for power plants and not for weapons.

Back-stage: The environmental damage caused by the Ranger mine will remain for a very long time, and leaks of contaminated water cannot be completely prevented under the difficult climatic conditions (heavy rains). That the Ranger mine never received the consent of the Mirarr landowners makes this heritage more problematic. The mine brought development and income to the area, which however did not change the fact that many Aboriginal people continue to live under conditions of poverty, health problems and drug abuse; or even, many of these problems resulted from the implications of the damage to the cultural landscape on the Aboriginal identity and values.

3) At the time of the development of the Ranger mine in the 1970s, the recognition of indigenous peoples' rights over their territories was just emerging, and the mine and mining town areas were explicitly excluded from it, so that the company could advance its project. By the 1990s, when ERA tried to install the Jabiluka mine, it was confronted and finally blocked by a strong movement. ERA has since then changed its strategy and invested considerable effort in building trust through involvement of indigenous representative organizations. Royalty payments (4.25% of net sale revenue in 2011) are redistributed to different indigenous associations of the region through a state-controlled fund. In the landmark "Jabiluka Long-Term Care and Maintenance Agreement" of 2005 between the Mirarr and ERA, the latter acknowledges that the Jabiluka project will remain stalled as long as it lacks consent from the indigenous landowners. Since then, several other direct agreements have been negotiated between the two, including one that restores ownership of the Mirarr over the area of Jabiru town.

El Cerrejón, Colombia

1) The El Cerrejón mine in Colombia was developed in the 1980s by a joint venture between the state-owned Carbocol and Intercor, a subsidiary of the US oil-major Exxon Petroleum. Since 2002, it is operated by Carbones del Cerrejón Ltd., jointly owned by three major TNCs which hold 33.3% each: Xstrata (now Glencore, Switzerland/UK), BHP Billiton (Australia/UK) and Anglo

American (South Africa/UK). The state has no ownership participation anymore.

2) With the takeover in 2002, the current owners profited from a relatively safe investment in an already established mine. At the same time, they can deny responsibility for problems and damages caused prior to the take-over.

Front-stage: El Cerrejón is another example of a company with a highly developed CSR policy and a proactive media strategy. As one of the largest producers of coal and thus greenhouse gas emissions worldwide, it even has a "climate change mitigation" policy. It claims to respect international standards such as the UN Global Compact, the Voluntary Principles on Security and Human Rights, and others. It installed community-based air quality monitoring systems and a complaint mechanism which follows the Ruggie recommendations[14]. El Cerrejón's CSR policy is not limited to "greenwashing" statements, but includes a complex system of community relation programs. Several company-controlled foundations promote regional development, education and training, as well as environmental and biodiversity protection. The company cooperates extensively with state agencies, a range of civil society organizations (e.g., Conservation International) and researchers.

Back-stage: Although the environmental impacts of El Cerrejón apparently do not exceed legally permitted levels, the huge open-pit mine causes obvious damages to the landscape and threatens the subsistence-based lifestyle of inhabitants over a large area. Dust pollution and restriction on movement not only affect communities close to the mine, but also fishing near the port. Locals report respiratory and other diseases that they attribute to air and water pollution. The rehabilitation program is seen as insufficient. The injustice at the time of land acquisition and forced resettlement of several communities remain unaddressed. To expand the mine, indigenous and Afro-Colombian communities are forced to accept insufficient conditions for resettlement. The civil war related militarization of the area continues, with company and state security forces closely collaborating to protect the mine, leading to allegations of joint responsibility for human rights abuses.

3) The CSR programs of El Cerrejón stress the importance of "community relations", whereby consultation is understood as "dialogue" in a non-binding way. This is promoted through a number of programs carried out with the help of social scientists, anthropologists and indigenous Wayuu key persons, which serve to build trust and loyalty among local communities. In the absence of an established umbrella organization which could represent all the affected communities, consultations involve individual communities and leaders, whereby those who take a critical position are frequently co-opted or side-lined. This leads to allegations of installing "false leaders", who do not enjoy the trust of their communities, using the community relations program for systematic divide-and-rule. Also, the negotiation of direct agreements concerns only individual communities, mostly regarding the terms and conditions of the resettlements. The compensations which were paid for earlier land acquisitions are now considered insufficient. A permanent benefit-sharing with local organizations is not in place, except through the ones controlled by the company itself. The latter offer some benefits for housing, education, training and setting up of small enterprises, effectively replacing state functions. Only few Wayuu and local people are employed by El Cerrejón, mostly in low-skilled jobs, with a few exceptions of jobs in the mentioned CSR programs.

14 See Part I, Chapter 2 on the UN Guiding Principles on Business and Human Rights and the recommendations of the UN Special Rapporteur John Ruggie.

Ambuja Rawan, Chhattisgarh, India

1) The cement plant of Ambuja Rawan in India's state of Chhattisgarh was built in the 1980s. Ambuja is a major cement producer, operating several other large-scale cement plants in the country. In 2005, the publicly traded, Switzerland-based cement company Holcim acquired a majority control over Ambuja and ACC cements, making it the number two player in the Indian market.

2) Both Ambuja and Holcim are well-established brands and directly distribute and sell their product to consumers, therefore their visibility is higher than in the case of mining companies.

Front-stage: Holcim is a founding member of the Cement Sustainability Initiative, member of the UN Global Compact and the World Business Council for Sustainable Development, and has partnerships with the ICRC, IUCN and others. Holcim is associated with well-known individuals of the Swiss high society, which may contribute to the need of portraying the company as a "good citizen". For the centenary celebrations in 2012, the company assigned a PR agency to create a photo book and exhibition of workers' portraits. The fact that the result was shown in the Arts Museum of Bern was criticized as an abuse of a public art institution for corporate communication.

All the involved companies have well-developed CSR policies and publish their own Sustainability Reports, containing variants of "stakeholder dialogue" claims. Ambuja explicitly mentions that collective bargaining is encouraged "to maintain harmonious industrial relations" and that it takes a "consultative approach" with regard to the concerns of local communities. In 2010, it spent 2.6% of the net profit on community development programs. These are carried out by the Ambuja Cement Foundation (ACL), which runs projects in education, health, women's empowerment and agriculture in the proximity of Ambuja's plants. The main focus is on water resource management. The company carries out "water accounting", which shows that its "water credit" (ground water recharge, water harvesting and recycling) exceeds the quantity needed by the factory. Although the Sustainability Report of Ambuja is compliant with Global Reporting Initiative criteria, it does not mention the problems at its Rawan plant.

Back-stage: The unresolved pre-history of the conflict at Rawan dates back to the workers' struggle at ACC Jamul in the late 1980s. The company ACC appealed against several court decisions that ordered permanent positions for contract workers, benefitting from the long delays in the Indian judicial system. Against members of the combative cement workers' union PCSS, summary lay-offs, divide-and-rule and criminalization are used. When the contract workers started to organize at the Ambuja Rawan plant in 2008, the respective management reacted in very similar manner. In both cases, there is an evident complicity with police and state authorities, and an outright refusal to negotiate with the union. At the Ambuja Rawan plant, we also find indications of "corner-cutting" behaviour regarding waste management and water supply, because effective monitoring is absent. Faced with these allegations, the Holcim headquarters chose to defend the management of its subsidiary companies. Although it offered to seek dialogue with Swiss campaign groups, it only agreed to negotiate directly with the workers' union under pressure of a complaint to the Swiss OECD contact point.

3) In the case of the contract workers struggle, there has been a break-down of negotiations for many years and the main interaction seems to happen at the judicial level. The general picture is of a discriminatory approach towards unions, whereby the ones who represent permanent employees and cooperate with the management are favoured while those who take up the concerns of contract workers are excluded. When local farmers at Rawan in alliance with contract workers organized resistance, some of their leaders were criminalized though false accusations by Ambuja security guards.

No proper documentation is available to understand how large-scale land acquisitions happened at the Ambuja factory in Rawan more than 20 years ago. Local people are now complaining about unjust or even complete lack of compensation. The acquisition was reportedly carried out through middlemen who promised that the company would provide one job for each family.

These jobs, even if they materialized, mostly turned out to be precarious contract work. Smaller land acquisitions seem to continue in a way which is perceived as unfair. Certain "community development" projects rather seem to benefit the permanent employees living in the factory colony, although they were allegedly built on common lands now walled off from the village. Likewise, though the problem of competing water usage is addressed pro-actively, it seems that local farmers perceive this as a take-over of a commonly-owned irrigation system, in order to ensure water supply for the factory at critical times.

Marlin Mine, Guatemala

1) The exploration licence for the Marlin mine in Guatemala was obtained in 1999 by Montana Exploradora de Guatemala S.A., a subsidiary of the Canadian junior Montana Gold, which was acquired shortly afterwards by another Canadian junior, Francisco Gold. The latter was acquired in 2002 by the somewhat larger Glamis Gold, which one year later, received the mining licence for Marlin. In 2006, Glamis was bought up by Goldcorp. This Canadian mining company operates about a dozen gold mines, fully or partially owned.

2) Goldcorp claims to be "one of the fastest-growing, lowest-cost senior gold producers, with operations and development projects in politically stable jurisdictions throughout the Americas", and having a focus on "sustainable growth".[15] In Guatemala, mostly the "low-cost" part seems to be confirmed.

Front-stage: Goldcorp's "responsible mining" approach includes commitments to various CSR standards, extensively mentioned in a special section of its website, such as the Voluntary Principles on Security and Human Rights, the Cyanide Management Code, and membership to the ICMM, the UN Global Compact, and others. Its first sustainability reports date from 2005. Most of its commitments were made after the acquisition of the Marlin mine in 2006, presumably in response to the mounting pressure from the anti-mining movement at Marlin. In reaction to critical studies and allegations about environmental and health impacts, Goldcorp commissioned reports which resulted in contrary findings. The 2010 precautionary order of the Inter-American Commission on Human Rights to suspend mining at Marlin was simply ignored.

Back-stage: Several independent studies indicate that Marlin does contaminate the local water bodies and has negative effects on the health of local villagers. In part, this seems to be the result of conscious decisions to cut costs, such as the failure to use an impermeable liner to prevent seepage of toxic effluents from the tailing pond into the ground. The massive use of water, for which the company apparently does not pay anything, causes shortages for local farmers. The cyanide leaching method, known to be highly problematic in general, seems to be handled deficiently, as indicated by several fines of the Ministry of Environment in this regard. Marlin has brought a proliferation of social conflict to the area. The mine's private security cooperates with state security forces in criminalizing and repressing protest. Even more, there were several incidents of death-threats and violent attacks (some of them deadly) against leading figures of the anti-mining movement, some of which were purportedly carried out by "pro-mining" individuals and employees of contractor companies.

3) Though required by the ILO Convention No. 169 and the Constitution of Guatemala, and specified in the Municipal Code, the consultation process during exploration, land acquisition and project development was flawed to such a degree that it seems fair to say that there was no consultation at all. Basic information such as the Environmental Impact Assessment was not made properly available and community consent was reportedly forged on paper only. This happened under the responsibility of Glamis Gold, previously to the takeover by Goldcorp and apparently

15 Claim from <www.goldcorp.com> accessed 03.03.2015.

under high pressure to bring the project close to production as fast as possible. Goldcorp thus acquired the mine already at an advanced stage of conflict, but failed to assume responsibility and address the earlier shortcomings. Even more, it tried to legally obstruct a self-organized community referendum in 2005. The company (and the government) then failed to acknowledge the result of over 90% rejection.

Arguably only in reaction to the opposition, Goldcorp strengthened its CSR commitments. The company funds a foundation called "Sierra Madre", active in the communities surrounding the mine. This foundation runs numerous projects in the fields of agricultural and economic development, health, education and professional trainings, as well as "community and institutional strengthening", including the setting-up of health centres, training of school teachers, and other services which are usually in the state's responsibility. Apart from this, the company also provides important financial contributions directly to the municipalities. Together with generating some local employment, mostly in contractor companies, this is sufficient to create a local pro-mining faction, and thus a division in the communities.

Tampakan, Philippines

1) In Tampakan, explorations were started in the 1990s by a mid-tier Australian firm called Western Mining Corporation (WMC). WMC sold its exploration rights in 1999 to the Philippine-based Sagittarius Mines Inc. (SMI), a subsidiary of the Australian Indophil Resources NL. In 2007, the Switzerland/UK-based major Xstrata acquired a 62.5% majority stake of the Tampakan project from Indophil. Together, Xstrata and Indophil had a 40% controlling share of the project developer SMI, while the 60% non-controlling shareholders of SMI were the Philippine-based Tampakan Mining Corporation and Southcot Mining Corporation. Since 2014, Indophil is a wholly-owned subsidiary of the Philippine-based Alsons Group; a major part of the interests in the project is thus controlled by Philippine investors. In June 2015, Glencore (which had acquired Xstrata) announced a sale agreement for its participation in the Tampakan project, without giving further details about the buyer.

2) The case shows a strong and well-crafted CSR-strategy at an early stage of mine development. Still, the company was so far not able to secure all the necessary permits.

Front-stage, SMI/Xstrata claim to comply with all the relevant laws and standards (including those of the World Bank/ IFC, ICMM, ISO and ILO, and the Voluntary Principles on Security and Human Rights), and boasts of numerous CSR awards. SMI publishes extensive CSR reports (since 2007) and has an internal "Sustainable Development Committee". Some problematic aspects of the project, such as security, are addressed both pro-actively and in response to critique by civil society organizations. A "Human Rights Assessment" was included in its "Social Impact Management Plan", and employees receive awareness trainings on Bla-an culture and history. Its programs for community development and even "community advocacy" cover a range of activities, with focus on education and healthcare. However, it seems that some critical issues are downplayed or ignored, such as the ban against open-pit mining by the South Cotabato province, the high risk of mining in the area of an active volcano, or the possibility that communities might withhold their consent once they have to be resettled. Though it acknowledges the need to receive FPIC from indigenous communities, as mandated through the IPRA Act, this seems to be interpreted as a procedural obligation, rather than as a right to veto.

Back-stage, many questions on the alleged co-responsibility of SMI/Xstrata in killings of several anti-mining activists by state security forces (partially funded by SMI) remain open. The militarization of the area also caused people to take refuge elsewhere. Up to 80,000 potentially affected downstream farmers are largely excluded from consultation and benefits.

3) The process which led to the initial "Principal Agreements" between the previous project owner WMC, three municipalities and four Bla-an communities in the late 1990s was flawed: the document was unilaterally drafted by the company, not translated into the local language, and no

independent advice was available. Instead, the company simply stressed economic benefits and the fact that higher government levels had already signed. SMI/Xstrata seems to recognize that these "agreements" are not fulfilling the character of "free" and "informed" consent as demanded by the IPRA act, and thus continues to negotiate with indigenous leaders and representative bodies to obtain a "Memorandum of Understanding", which would also cover questions of land compensation and resettlement. The IPRA Act defines FPIC as a consensus of all affected community members, according to customary laws and practices, fully informed and without pressure or interference.

What is most striking about the case is the observation that WMC together with government agencies had been facilitating the setting up of Tribal Councils in the 1990s. It even assisted communities to obtain the statutorily recognized communal ownership over their territories—except one community, which refused to sign the mentioned agreement. WMC provided financial assistance to the compliant communities and sponsored experts, including anthropologists, to carry out the necessary studies for the ancestral domain claim. Today, it is difficult to assess how far indigenous representative institutions, leadership and indeed, the "communities" themselves, were co-created by the company. For instance, it seems that current indigenous leaders have built up their authority largely by negotiating benefits from the company and redistributing them to the community. This has its structural expression in "Tribal Foundations" that disburse development funds, whereby indigenous leaders and company officials along with government officials have equal representation. Again, the initially opposing community was included in this system only after giving up its resistance. Already by 2007, over half a million USD was distributed to the communities and local government bodies, including Tampakan town. Apart from education programs, especially scholarships, and healthcare missions, a good share of the sum was used for administrative expenses of the Tribal Councils themselves, or freely distributed by company staff for "livelihood and emergencies". This suggests that the money is used to buy support for the company through patronage relations rather than empowering local people. The local government is also entitled to receive regular taxes from the company, though from the case study it is unclear how much was already paid and how much would be paid once the extraction would start. Similar uncertainties exist concerning local employment.

Las Bambas, Peru

1) The Las Bambas mine project started in 2004 when the Switzerland/UK-based Xstrata acquired the concession in a public tender by the national and regional authorities. It is a rare example of a project being controlled by a major corporation from the very beginning. In the process of Xstrata's takeover by Glencore in 2013, the project had to be sold due to conditions set by China's Ministry of Commerce, which feared that Glencore would otherwise control too much of the global copper market. Since 2014, Las Bambas is a joint venture project between the Australia-based operator MMG (62.5%), a wholly owned subsidiary of Chinese Guoxin International Investment Co. Ltd (which holds another 22.5% directly) and the Chinese state-owned CITIC Metal Co. Ltd (with 15.0%). For the transfer, Glencore received USD 7 billion.

2) Xstrata, being a relative newcomer among the major mining corporations, decided to push a proactive CSR-strategy since the early 2000s, considering its leadership in the field "a source of competitive advantage". Even more than Tampakan (see above), the Las Bambas copper mine project is a showcase model for a massive "sustainable mining" scheme.

Front-stage: Apart from the reference to a wide range of both general and mining-specific standards already mentioned in the Tampakan case, what calls most attention in Las Bambas is the elaborated development discourse. In a region presented as extremely poor and underdeveloped, with a marginal state presence and minimal employment opportunities, Xstrata arrived with the promise of a new era of progress and well-being. The company's obligations were framed in a contractual agreement that forms part of the mining concession, but it also went beyond that, with additional projects directly administered by Xstrata's Community Relations Office, in close cooperation with regional government bodies. On the one hand, there was a massive information campaign through public hearings in the affected communities and local radio stations. On the other hand, the funding of local authorities, a social development fund and some allied NGOs, together with its own projects, produced a boom of training programs, infrastructure construction and other projects.

Back-stage: The information provided to local peasants by Xstrata was criticized by observers as one-sided and partial, emphasizing development, while downplaying possible negative impacts, such as water shortage and contamination. Relevant documents, most of all the Environmental Impact Assessment, were written in an expert language barely accessible to the local peasants. At the same time, the loyalty networks established by the community relations staff discouraged people from exchange with NGOs labelled as "anti-mining", as this could mean loss of (potential) benefits. This prevented the existence of a space for public articulation outside the influence of the company and effectively curtailed the expression of critical voices and NGOs which could point out to possible problems based on experience from mines in other areas.

Communication and development projects were focussed on what the company defines as communities in the "direct zone of influence", most of all those needed to cede land for the mine and secondary infrastructure, while communities outside of this zone felt left out. This created envy and latent conflict between communities, obstructing the possibility of common action and weakening regional-level organizations such as the peasants' association.

3) Regarding the need to consult the affected peasant communities about land acquisition, the case is contradictory. At first sight, it looks as if Xstrata's community relations office made every effort to inform and consult all stakeholders in accordance with ILO 169 and related national law. A series of public hearings with large participation were held throughout the region. However, these were framed in a way that made local people feel that they could not freely express their opinion. Moreover, it seems that the land law, which demands a 2/3-majority consent of the community assembly for the transfer of common lands, was effectively side-lined. By negotiating door-to-door with individual persons, promising and providing jobs, training opportunities and distributing small gifts, individual short-term expectations were pushed and collective consensus-finding processes obstructed. Agreements were first made with government authorities, then with key individuals, and only after that, with the executive councils of individual communities.

The mining project has brought large amounts of cash to an area previously dependent on subsistence farming, and provided salaried jobs to a number of peasants. For the distribution of the development funds with an initial sum of USD 45 million that form part of the concession contract, a special foundation was set up. Employment was already at about 200 workers in 2005, half of them from the region, and was expected to rise to 6,500 during construction and then fall back to 1,350 during the operation stage.

Val Medel, Switzerland

1) The gold explorations in Val Medel and the Surselva started in the mid-1980s, but were soon abandoned again. The concessions passed through different small explorer companies until 2011, when the newly founded company SwissGold Exploration took them over. This company, consisting of two individuals who were also involved with earlier exploration attempts, is registered in Val Medel as a fully owned subsidiary of NV Gold Corporation, itself a junior explorer based in Vancouver, Canada (traded at the TSX Venture Exchange), which owns one more exploration permit in Nevada, USA.

2) Given the minimal capitalization of the company, its actual operations were so far limited to a collection of samples in 2011. There is very little documented communication from the company, making it impossible to distinguish a front-stage/back-stage pattern. The few elements at hand are an announcement of positive results to its shareholders, and some interactions with the

municipal council to prolong its concessions. During the latter, the company representatives also suggested changes in the municipal mining law, which raised suspicions that the mayor of Medel, who publicly took a pronounced position in favour of further explorations, may have economic interests in the company. In fact, it is mostly the mayor and the municipal council who appear as promoters of the proposal, while the only public appearance of a company representative was on the occasion of an information event. At this event, organized by the municipal council, the company presented vague information in English, but was unable to answer most of the questions of the audience. Both the company and the council presented a potential mine as mostly underground, using the best standards for environmentally less destructive technologies (excluding cyanide leaching).

The lack of anything like a CSR strategy from the company was partially compensated by the municipal council and an ad-hoc support-committee around the mayor of Medel. The arguments they put forth centred on the need to stop the depopulation of the valley and the creation of several hundred jobs, which would give new opportunities, especially for young people and families, to stay or return to the valley. Also, the generation of tax income for the maintenance of public services now under threat by spending cuts was mentioned. After the controversies leading to the negative vote on the proposal by the residents of Val Medel in 2012, the municipal council took a more careful stance and admitted that it was an error to seek advice from the company rather than from independent experts. Now, it argued on the need for further explorations to consider all options for the future development of the valley, while the ultimate decision on a potential project would be taken, again, by the people. With this argumentation, a revision of the municipal mining law was made with a narrow majority in 2014, albeit with a very low turn-out of voters, enabling the council, again, to grant an exploration concession to the company.

CASE	Mopani Zambia	Arlit/North Niger	Ranger/ Jabiluka, Aus.	El Cerrejón Colombia	Ambuja Rawan, India	Marlin Guatemala	Tampakan Philippines	Las Bambas Peru	Val Medel Switzerland
1. Corporate Type									
Company type (sector)	Majors (mining)	Major (energy)	Major (mining)	3 Majors (33.3% each)	Major (cement)	Mid-tier → Major (mining)	Major (mining), mid-tier	Major (mining)	Junior explorers
Owner(s), (Seat), (Share)	Glencore (CH/ UK) (73%), First Quantum (CAN) (17%), ZCCM (ZAM) (10%)	Areva (FR) (64% / 34%), OURD (JP) (31%), ENUSA (SP) (10%)	Rio Tinto (AUS/UK) (68%)	Anglo American (SA/UK), BHP Billiton (AUS/ UK), Xstrata (CH/UK)	Holcim (CH)	Glamis Gold (CAN) → Goldcorp (CAN)	Xstrata (CH/ UK) (62.5%), Indophil (AUS) (37.5%), local companies	Xstrata (CH/ UK) (100%)	NV Gold (CAN)
Operating Subsidiary	Mopani Copper Mines	Somaïr; Cominak	ERA	Carbones del Cerrejón Ltd.	Ambuja, ACC	Montana Exploradora	Sagittarius Mines Inc.	Las Bambas	SwissGold
State Share	ZCCM 10%	Yes (ca. 1/3)	Initial stage	Initial stage	No	No	No	No	No
Ownership history	State owned 1960s until 2000	Stable ownership	Several takeovers, last 2000	State–corp. joint venture (– 2002)	Majority control since 2005	Junior, Glamis, Goldcorp (2006)	Several takeovers	sold to MMG in 2006	Several other explorers
2. Corporate Impression Management									
CSR strategy	Weak	Weak	Strong	Mixed (–strong)	(Weak–) mixed	Weak (–mixed)	Mixed (–strong)	Mixed (–strong)	None
3. Corporate Community Relations									
a) Consultations									
Involved actors	None; except "squatter" case	–	Traditional owners	Individual, loyal clan leaders	Allied unions, individuals	–	Community leaders, Tribal Councils	Community councils 'direct zone'	Resident population
Excluded actors	–	–	Non-owner clans	Opposing groups and leaders	Contract worker unions; critical leaders	Self-organised local referendum	Opp. community; downstream pop.	Critical NGOs, 'indirect zone' communities	–
Relevance of FPIC	Not mandatory, no claim	Not mandatory, no claim	Ranger: No Jabiluka: Yes (incl. veto)	Not mandatory, no claim	Not mandatory, no claim	Ambivalent (ILO 169)	Mandatory, claims to obtain FPIC	Ambivalent (ILO 169)	Mandatory referendum
b) Compensation									
Development Projects	Few	Few	Through indig. associations	Various, own foundations	Various, own foundations	Various, own foundation	Various, own foundations	Various, own foundations	No
Local Employment	17,000; 50% sub-contracted	Few	Few	Few	Mostly sub-contract labour	Few	Few	Relevant	Not yet
Direct agreements	No	No	Yes, e.g. Jabiluka closure agreement	Individual communities	No	No	Yes, with directly-affected communities	Concession contract	No

Table B: Overview of corporate strategy aspects from case study summaries in Chapter B.

3. Discussion

In this part, we first highlight the multiplicity of corporate structures, to avoid a too monolithical reading of "the corporation". Second, "corporate impression management" compares how corporations compete on a credibility market, adopt CSR standards and engage in strategic alliances and public communication to build a positive "front-stage" image of their activities. Third, we discuss the "corporate-community relations": to what extent are corporations ready to enter into negotiations with local actors? Considering the specific contexts of the cases, we align the cases on a spectrum from "imposition" to "negotiation".

3.1 Corporate Typology

As described in the introduction of Part I, the choice of case studies was focussing on major, publicly traded transnational corporations (TNCs), with a certain bias on linkages with Switzerland. Nevertheless, the resulting sample (*Table B – 1*) turns out to cover a wide variety of corporate structures; especially, if we include the entire ownership history of each mine.

Major mining corporations which have important stakes in the case study mines are Anglo American, BHP Billiton, First Quantum, Glencore, Goldcorp, Rio Tinto and Xstrata, all listed in "traditional" places for the mining sector like London (UK), Toronto (Canada), Johannesburg (South Africa) and Australia. Anglo American, BHP Billiton and Rio Tinto are among the top 5 diversified miners, which produce a wide range of different minerals and metals all over the world. First Quantum and Goldcorp are somewhat smaller and more focussed on specific metals (copper and gold), but still belong to the top 15 mining companies by market value (see *Table 1*).[16]

Xstrata, listed in Switzerland and London and prominent in this book with four cases, is an example of an extremely decentralized, transnational corporate structure. Until 2013, it was run from the twin corporate headquarters in Zug (Switzerland) and London by just a few dozens of top-level managers who coordinated the different branches of the operating divisions such as Xstrata Copper and Xstrata Coal, based in Australia, or Xstrata Nickel in Canada (see EvB/BD 2012: 158; MultiWatch 2014). With the take-over of Xstrata in 2013,[17] Glencore has acquired a unique position as an integrated, diversified mineral producer and commodity trader, underscored by its high, trading-related sales of over 200 billion USD (see *Table 3* [18]), as well as its expansion into the agro-industrial sector. El Cerrejón (Colombia) is an interesting case of three competitors at the global level (Xstrata, BHP Billiton and Anglo American) that cooperate at the level of a specific operation. The sample shows that also corporations of the cement (Holcim) and energy (Areva) sectors carry out mining as a collateral activity, with different implications than in "pure" mining sector companies. Holcim sells the final product under the same name and is thus exposed to more visibility than a mining company. Areva's core business is the construction and operation

16 In this category, we can also include the South African company AngloGold Ashanti (see Bolay on Guinea, this volume), with 19 mines in operation. (www.anglogoldashanti.com). A mid-tier company with only a hand full of participations is BSGR, headquartered in Guernsey (UK), of which OCTEA diamonds is a 100% owned subsidiary; see Maconachie on Sierra Leone (this volume). and www.bsgresources.com.

17 Xstrata's growth to become a major miner was largely due to its link with the major commodity trader Glencore previous to the take-over. Though the move was first presented as a merger and for some time the brand "Glencore Xstrata" was used, the name Xstrata was deleted by 2014 and most top-level posts in the new company were taken by Glencore's executives.

18 The specific ranking for the mining sector of 2014 is not comparable for 2010. Glencore is not included in 2010, as it was not yet a public company. See "Global 2000 List" for 2010 and 2014 on <www.forbes.com> accessed 03.03.2015.

of nuclear power plants, which depends largely on long-term strategic investments by governments. Consequently, it is also our only example where governments (France and Niger, as well as Spain) have a major stake, and where we find long-term ownership stability. Minor state participations are relevant in Zambia's Mopani mine, which was fully state-owned until 2000, and in the first stages of El Cerrejón (Colombia) and Ranger (Australia). The almost negligible role of state ownership reflects the general trend after two decades of neoliberal reforms.

Finally, the sample also covers a junior explorer company based in Canada and its Swiss subsidiary with legal seat in Val Medel. Junior miners fiercely compete for access to promising mineral deposits and are under extreme pressure to cut costs, which means that they "typically under-perform in terms of social and environmental responsibility" and focus on "low-cost, environmentally risky extraction in policy environments with few constraints" (Dougherty 2011: 3, 12). What differentiates this company from hundreds of similar ones at the Canadian TSX Venture Exchange is thus its atypical field of investment. The rather shady nature of many of these explorers who run an "economy of appearances"[19] is illustrated by the explorer Uranio AG (see Niederberger, this volume). With a letterbox headquarters in Basel (Switzerland) and subsidiaries in Argentina and the Central African Republic as part of a complicated network dispersed over several jurisdictions, this case reminds us that "transnational" does not have to be equivalent to "large". In general, "letterbox" or "mailbox" companies which are set up to strategically conceal ownership, financial flows and legal responsibility make it difficult for researchers to "follow the money", as illustrated by the case of Glencore's Mopani mines in Zambia. The (incomplete) database of www.opencorporates.com lists 222 companies registered in over 60 jurisdictions as belonging to Glencore. [20]

In short, the typology above has only touched the tip of the iceberg. We mainly focused on publicly traded, major mining TNCs as central players within a larger field of corporate actors, like junior explorer companies, contractors, consultancies, NGOs and certain state bodies which are integral to, or at least involved in, corporate strategies to a certain extent. However, it is important to remember, that "corporate decisions are made by people engaged in negotiations and struggles over complex issues rather than by a metaphysical corporate actor that unerringly acts in the interest of profit-making" (Welker 2014: 28). The urge to hold a clearly identifiable company responsi-

Company (Metal and Mining)	Mining sector rank 2014 (market value)	Overall rank Forbes 2000 in 2014 (2010)	Market value in Billion USD 2014 (2010)	Sales in Billion USD 2014 (2010)	Profits in Billion USD 2014 (2010)
BHP Billiton	1	44 (63)	182.3 (192.4)	67.7 (50.2)	14.8 (5.9)
Rio Tinto	2	109 (68)	103.8 (118.3)	51.2 (41.8)	3.7 (4.9)
Glencore Xstrata (2010: Xstrata)	3 -	406 (202)	67.5 (46.87)	232.6 (22.7)	-7.5 (0.66)
Anglo American	5	551 (138)	35.3 (48.2)	29.3 (21.2)	-0.9 (2.46)
Goldcorp	11	1160 (872)	19.9 (28)	3.8 (2.7)	-2.7 (0.24)
First Quantum	15	1253 (-)	10.7 (-)	3.6 (-)	0.5 (-)
other sectors:					
Areva	Div. Utilities: 2	906 (358)	9.9 (16.58)	12 (12.2)	-0.7 (0.8)
Holcim	Construction: 2	328 (231)	27.3 (21.8)	21.3 (20.4)	1.4 (1.4)

Table 1: Ranking of Corporations from Forbes Magazine "Global 2000 – The world's biggest public companies".

19 See Tsing (2000) on the explorer company Bre-X, owner of an imaginary gold deposit in Indonesia.

20 The platform <www.opencorporates.com> tries to cover the legal registries of all companies worldwide. The numbers are from February 2015, when it had reached at 75 jurisdictions (Switzerland not yet included). Rio Tinto had 371 and Goldcorp 90 entries, each spread over about 40 jurisdictions.

ble for what happens at a certain mining site may sometimes motivate simplified accounts, which cannot give satisfactory explanations for specific decisions taken at specific times. Still, there is undoubtedly a need to identify underlying patterns structuring corporate decision-making and the transformation of corporate power.

3.2 Corporate Impression Management

In this section, we analyse how "corporate impression management" is at play in the case studies (corresponding to *Table B – 2*). By evaluating if "front-stage" claims are consistent with "back-stage" situations, we can estimate whether CSR is used merely as a "greenwashing" strategy or whether it does reflect actual changes in corporate decision-making.

From the case studies we can confirm that image does matter for mining companies, and that CSR indeed has become a common feature of all major, publicly traded mining corporations. As mentioned in Chapter A, mining is often a matter of national politics, and mining companies are to a large extent dependent on support from governments for permits, tax exemptions, subsidies and infrastructure. Delays or failures to get approval for new projects are very costly, thus there is "[...] the challenge of having to pacify communities and a sceptical public until operations generate a profit", therefore, "[t]he pressure to embrace CSR is far greater in the mining industry than the oil and gas sector" (Hilson 2012: 135). That there are high transaction costs is also confirmed in comparison with the previous *infoe* study on fossil fuel extraction (Haller et al. 2007; see also Kirsch 2012: 211): the input of capital is much higher for mining projects, which have a longer time horizon and cause geographically concentrated and highly visible impacts. Mining needs more natural resources in terms of land, water and energy, making it an issue in "land grabbing" and "water grabbing" debates, too. Oil can be transported relatively inexpensively via pipelines, while mining needs solid roads for heavy trucks and thus more constant capital input. Additionally, mining needs more workers, even if extraction is heavily mechanized. However, these are jobs for which local people are often not sufficiently qualified, or they qualify only for the badly paid and most dangerous jobs. Therefore, the discourse of (national and regional) development and CSR-related issues such as benefit-sharing and compensations are more important and conflict ridden. A good example for the appeal to a discourse of "development" is the case of El Cerrejón: the long life-span of the mine makes it important to ensure a socio-political setting which allows for an undisturbed continuation of the exploitation. The magnitude of its operations, its economic dominance and political influence are such that El Cerrejón resembles a "state within the state" in the Guajira region. In the cases of Las Bambas and Tampakan, the companies nurture expectations of development in order to obtain the necessary approvals by the local people and the authorities to secure multi-billion investments. Elsewhere, development discourses serve to immunize running operations against challenges from local groups and NGOs (e.g., Chhattisgarh, Guatemala). More ambivalent is Mopani, which used to be the beacon of national development before privatization and where it looks like Glencore tried to avoid nurturing expectations and stick to a narrow "extraction enclave" model instead (Ferguson 2005; see Part I – 1).

Fig. 3: "Supporting Indigenous capacity-building through partnerships". Source: Screenshot from the Rio Tinto Sustainable Development online Report 2013.

In addition to a discourse of "development", in almost all our examples "sustainability", "sustainable mining" or similar terms have been adopted in corporate communication, in what can be understood (in line with Kirsch 2014) as a strategy to co-opt the critical discourse of NGOs and social movements. The mimicking of NGO language is evident in CSR reports and on special sections for "responsible mining" on corporate websites. CSR reports often feature images of "happy locals", preferably indigenous people (see, for example, *Figure 3*[21]). From icons of NGO campaigns, they are now lending the image of authenticity to corporate environmental and social standards. To build credibility, claims are frequently underscored through alliances and financial contributions to civil society actors, e.g., to conservation NGOs that set up protected areas in an attempt to compensate for damages caused by mining. This shows that corporate development does not necessarily stand in contradiction with conservation environmentalism—rather, these can appear as two sides of the same coin (see Walter/Martinez-Alier 2010; Haller and Galvin 2011). Similarly, social development projects often involve local NGOs or foundations specifically set up for this purpose. In several cases, we find "community relations" departments well-staffed with locally respected mediators and experts such as social anthropologists, carrying out the "stakeholder dialogue", which has become a key element of CSR strategies. However, the discourse of sustainability and related "credibility shopping" can open new entry points for contestation, as it happened with Holcim's photographic representation of its workers in alliance with the Arts Museum of Bern (see Chapter C). Also at the academic level, corporate commissioning of expert reports and sponsoring of higher education institutions or scholarships constitute an expansion of corporate power to the public sphere, which has not gone unchallenged (see Kirsch 2014 and Part IV).

21 Screenshot from <www.riotinto.com/sustainabledevelopment2013/case-studies> (accessed 10.10. 2015); the report was produced by the London-based communication consultancy Black Sun (www.blacksunplc.com).

As an aspect of corporate alliance building, corporate strategies involve, to some degree, local and state actors, which act in the interest of the company, even if the latter denies responsibility. This is frequently the case when state security forces are sponsored by a company to protect their installations in regions marked by armed conflicts (see Chapter A)—mirrored in the case studies by allegations against mining companies of having a co-responsibility in armed confrontations or indirectly benefiting from violent repression and criminalization of dissent. The case studies of Niger, Colombia, Chhattisgarh, Guatemala and the Philippines illustrate this issue.

The impression gained about the relation between front-stage and back-stage performance from the case studies (Table B – 2) can be divided into three categories: the first ("strong"), with ERA/Rio Tinto as example, has a highly developed CSR strategy at the communication level and complies to it relatively well in the specific case; in the second ("mixed"), with El Cerrejón, Holcim and Xstrata (Tampakan and Las Bambas), there are major inconsistencies between a "best of class" CSR communication and serious doubts on the actual performance in the specific cases, situated in difficult contexts involving violent conflicts and weak enforcement of environmental and other legislation; the third ("weak") are the "late arrivals" with Glencore, Goldcorp and Areva, whose CSR was apparently just stepped up lately in response to major problems, thus enjoying little credibility. For this third category, Glencore is the most telling example. Actually, its front-stage profile had been remarkably consistent with the back-stage situation in Zambia and elsewhere, thanks to the fact that the company did not pretend to be more than a commodity trader plus mine owner who seeks maximum return on investments. There were few claims that the well-being of local communities or the general development of the nation was the company's business. After Glencore went public and the subsequent take-over of Xstrata, its corporate communication was unprepared and overwhelmed by a wave of media attention and public criticism. The reaction was a mix of denial of problems and the hurried installation of a makeshift CSR strategy. A complete lack of CSR policy can only be observed in the initial stage of mine development at Marlin in Guatemala and among the junior explorer companies (Uranio AG, Val Medel).

3.3 Corporate-Community Relations

In this section, we look on how different CSR strategies are expressed in the interaction between companies and people at the local level, regarding key conflict issues of consultation, consent and compensation (*Table B – 3a, b*).

As mentioned, companies as well as allied governments prefer weak interpretations of consultation and consent, and where mandatory, avoiding to understand it as a "right to veto". There is plenty of space for interpretation regarding the conditions of "free, prior, and informed" even in the strong FPIC concept. Relevant documents such as EIAs are written in an expert language inaccessible to the local people, benefits are exaggerated while risks are side-lined. The question of timing in the mine life-cycle is of crucial importance: Companies seem to prefer the notion that a one-time consent or consultation before starting the construction of a new mine is sufficient. Where a mine is running for decades already, consultation does often not take place at all, or it becomes a one-sided conversation masked as a "dialogue" that serves mainly to keep resistance low by granting some occasional benefits to key persons identified as "community leaders". This finding is in line with the study on oil companies (see Haller et al 2007). The historical liabilities of old mines and projects taken over from other companies remain unaddressed. The "prior" in FPIC may thus be misleading, or at least needs close attention. It has to be remembered that traditional decision-making in indigenous communities is usually based on long-lasting consensus processes, rather than on a one-time majority vote.

But the bigger challenge lies in the adjective "free", especially with regard to compensation. For companies, the first priority in negotiations with individual landowners or communities with recognized collective ownership is on acquiring the land needed for mining. The problem here lies in the definition of affected people in the context of complex ownership structures. (Post-) colonial states and globally driven rights systems have created a legal and institutional pluralism. To receive compensation, direct private ownership is often required, while traditional systems with complex webs of shared ownership, usage rights and obligations are ignored or misunderstood, which has direct and indirect negative effects on local livelihoods (see Haller ed. 2010, Haller 2013, Haller et al. 2013). In addition, one needs to consider in most cases colonial and post-colonial processes that created asymmetric power structures and relations so that the issue of representation of local interest groups also needs a close analysis (see for e.g., Haller ed. 2010 for an illustration of the colonially fabricated traditional leaders in African contexts).

A frequent, albeit vague form of compensation, is the promise of employment opportunities and general development due to investments, tax revenues and CSR programs. Many questions remain uncertain, and in the absence of concrete and enforceable contracts often depend fully on the company itself. How many jobs are actually created for the local population, what is the quality of these jobs, and how stable is the employment? Which development projects are benefiting the population according to their own priorities? How much revenue goes to the local government, and what is it used for? These questions can be clarified through direct agreements; however, local and indigenous organizations need considerable knowledge and resources to contract their own experts and lawyers in order to effectively challenge a company if it does not comply with its contractual obligations. Only in the Jabiluka case this condition seems to be met to a considerable degree. For further studies, it would be interesting to conduct an emic cost-benefit analysis over a longer period, covering expectations before the start of mining activities and the outcomes a few years later.

In the following, we consider the background of the case and mine-specific contexts as discussed in Chapter A, where we drew a spectrum from aspects that we assume would encourage a corporation's readiness to enter into negotiations with local groups. For order of discussion, we move from cases we considered above as showing a less developed or "weak" CSR strategy, passing through "mixed" to "strong" examples.

Fig. 4: Integration festivities for the resettlement of Fuerabamba community at the Las Bambas mine, Peru, 2010. Photo: Annette Wallimann.

In the first category, the case of Mopani in Zambia has a historical context of colonial destruction of collective land ownership with non-existent protection of minority rights, an old-age, underground brownfield mine in a highly mining-dependent setting. Apparently, under these conditions the mine management did not consider it necessary to set up preventive consultation procedures. The main benefits are some social services and development projects, minimal tax payments and some employment, though few compared to the total affected population and to earlier employment levels.

In Chhattisgarh, in the background of weak regulations, land acquisitions were and are directly negotiated with landowners, using strong-arm tactics and promises for individual compensation. Collective bargaining is limited to docile unions, while excluding the contract workers, and development projects are perceived as an expansion of corporate control over common property resources. In Guatemala, the Marlin project was imposed on the population without proper consultation, although legally mandatory. At the same time, a referendum showing widespread opposition was ignored. Compensations mostly appear as late attempts to curb the resistance.

Somewhere in between the "weak" and "strong" extremes is the El Cerrejón case in Colombia: the company has set up a permanent consultation scheme and negotiates agreements with communities affected by resettlements to expand the mine, as made mandatory by law and in accordance with CSR policy claims. However, there never was a consultation at the onset of the mine. The current practice is marked by manipulation and co-optation, and the extensive development programs and services seem to serve to control the opposition through divide-and-rule.

The examples of Las Bambas, Peru and Tampakan, Philippines tend from a "mixed" towards a "strong" position. In the critical stage of acquiring permits and access to land, these mega-scale mine projects both show a problematic setting (including the need for resettlement of entire communities) and will have profound impacts on an entire region. A high level of consent from local communities is legally mandatory and the company chose to comply pro-actively through extensive dialogue programs with selected key persons and communities, combined with the granting of generous benefits that distort the decision-making process. In both cases, the focus is on a small number of "directly affected" (and legally privileged) communities, while the large majority of the "indirectly affected" population is excluded.

A relatively "strong" CSR position is exemplified by the Ranger/Jabiluka case in Australia, situated in a context of strong legal protection and rule of law. Although the Ranger mine was built without proper consultation, the follow-up project of Jabiluka was blocked by the indigenous landowners. This is a rare example where a full FPIC is legally mandatory and, after massive resistance, was finally also recognized by the company as including a right to veto. The company negotiated several agreements with the legitimate representatives of the landowners, which include compensations and environmental safeguards for the running and closure of the Ranger mine. In the case of Val Medel in Switzerland, although the company has virtually no CSR policy, the municipal council stepped in readily in order to promote potential investments. The legal requirement for a binding local referendum at the earliest moment in the mining cycle was confronted with lack of information and vague expectations on possible benefits in the form of employment and tax revenues from a mine project that is still very hypothetical. Nevertheless, this example is outstanding for its pre-established democratic decision-making process and can be seen as a "best-case" scenario in so far as the involvement of local people was effectively guaranteed before any relevant investment decisions were made.

In sum, the cases show a wide range of possibilities and many deficiencies. Certainly, the level of legal protection, and especially the legally binding obligation to obtain FPIC from affected local people, is a major motivation for a corporation's readiness to negotiate, which confirms that CSR is most effective where it is legally mandatory, monitored and enforced. However, the legal framework in place and the quality of CSR strategies are not sufficient to explain the variations in the cases, especially if we consider their entire histories. The elements gathered so far allow us to place the cases along the three phases of corporate response to critique (the Benson/Kirsch model introduced in B – 1.2), from which we draw a spectrum from imposition to negotiation to withdrawal.

The positions on which we placed the cases in the model reflect tendencies, rather than stable conditions. In some cases, there have been important shifts over time. Most case studies, which we characterized as having a "mixed" to "strong" CSR strategy, are in phase two: they pro-actively manage the impression given to the public, introduce audits and collaborate with NGOs and other third party actors in order to confront criticism. This includes an increased readiness for consultation and compensations (a position of "negotiation"). In the cases where we consider the CSR strategy as "weak" (Mopani, Areva Niger, Ambuja, Marlin), a minor shift towards phase two came late, and only in response to insistent and strong criticism, perceived as involving high costs by the management. Before that, and arguably to a large extent until now, these are examples of a "phase one" denial or "imposition" strategy. But also the other cases show elements of phase one on specific issues, such as the proliferation of doubt about negative environmental and health impacts in the proclaimed absence of "scientific proof", or the denial of co-responsibility in acts of violence and criminalization of protest. The reason why we do not have an example that corresponds fully to phase one is the result of the choice being biased on controversial cases, but arguably also reflects the general trend in corporate "best practice". A phase three or "withdrawal" scenario is relatively rare: it means that external pressure and specific risks play together in a way that a company decides to partially or fully abandon its previous plans. The closest examples we have covered are the stalled deviation of a river at El Cerrejón in Columbia, and most of all, the stopping of the already initiated construction of the Jabiluka mine in Australia. The need to avoid a "phase three scenario" is certainly a prime motivation to invest in preventive CSR. This is strongly linked to the potential of local people to put up resistance at critical stages of mine development, as discussed in Chapter C.

3.4 Corporate Strategies: Conclusion

The results confirm that CSR discourses cannot be reduced to "greenwashing" alone (understood as merely a shift in corporate communication), but have created new norms and actual changes in the interaction between corporations and local people. In most cases, questions of consultation and compensation are given high priority. While, on the one hand, this may strengthen local people's bargaining power, on the other hand, it enables corporations to co-opt critique and pacify opposition. Moreover, one should not ignore that as part of larger strategies that give preference to negotiation, crude forms of imposition are still frequently practiced as a last resort against opposing individuals and groups within a divided social setting.

As stated by the "resource curse" literature, mining generally has "a dismal empirical track record to date in poverty reduction" (Pegg 2006; see also Part I – 1). Evaluating whether mining-generated benefits to the local level have increased over the last years in the case study regions would need more research on the emic evaluation of costs and benefits. Although the sums may remain marginal compared to the volume of investments and profits, for communities with high poverty rates and lack of public services they make a difference. Overall, the investments for a major new mine can be high enough to completely restructure the local economy from a subsistence-oriented rural setting to a mine-labour and market-oriented urban setting. Where a company considers it necessary to acquire a "social licence to operate", this may be experienced as corrupting entire regions, thereby creating new tensions and conflicts between different sectors of the local population. In most cases, we have shown that consultation and compensation are used by companies for divide-and-rule, thus perverting the original idea of the demand for fair treatment of local communities. By providing biased information, inciting high expectations about jobs and

Fig. 5: Anti-riot vehicle operated by the security contractor G4S in South Africa. Source: G4S corporate brochure.

development and distributing benefits on a one-by-one basis to select individuals, the outcome of a consultation process is manipulated. Later on, development projects, scholarships and community employment initiatives are used to maintain patronage networks through local leaders and loyal sections of the population that can be mobilized against "security problems" and protests if needed. Considering the highly sophisticated mechanisms of manufacturing consent described in some of the case studies, it seems more appropriate to speak of "co-optation" strategies instead of "negotiations". But even in the absence of any hidden agenda, well-intended CSR measures have unintended outcomes: ineffective projects and waste of resources, over-boarding administration, nepotism and corruption. And finally, CSR often covers fields which are usually considered state responsibilities such as health, education, security and basic infrastructure. Therefore, it has the potential to transform the relationship between local people and the state, and indeed, even the character of the state itself. As Himley observes, "the global mining industry itself has become an active institution builder", whereby new governance mechanisms are both "products of sociopolitical struggle [...] as well as instruments of power that have effects on the character and trajectory of that struggle" (2013: 398). The dynamics which arise in this diverse and contradictory field can be deployed strategically by local actors and their allies through targeted interventions (see Welker 2014: 217), which will be further treated in Chapter C.

We have found a kind of corporate state-building in several cases where a major mine project is situated in an area with little state presence and the company assumes the responsibility to "bring development". On the other side of the spectrum, one could propose the "extraction enclave" (or "oil-platform") characterized by Ferguson (2005), where a company avoids entering into contact with the local population and walls off all its installations in a bubble. This second model of resource extraction may be linked to the "imposition" stage mentioned above, while "negotiation" may be associated rather with "corporate state-building", whereby a company spends important amounts for its "social licence to operate". Consequently, "withdrawal" could correspond to a situation where the pressure to invest in corporate state-building in the first setting, or the costs to maintain the extraction enclave walled off from the surroundings in the second, affects the profitability of a mine to such a level that a company chooses to close it down.

However, this does not reflect the full picture. The examples where we find corporations building a "state within the state" show an extension of corporate control over the physical, social and discursive territory, from the capturing of state agencies and international institutions, NGOs and media, down to the co-optation, or even creation, of community leadership through patronage relationships. The corporation substantially frames the terms of references under which it carries out consultations and negotiates about compensation, using discourses of modernity and development as a front-stage legitimation strategy. Thereby, it often has not only the power to select the most "tame" participatory process at hand, but even to design an entirely new one, to suit its interests. This scenario thus constitutes a much more profound level of imposition than the crude expropriation we called "imposition" so far.

Is corporation-driven capitalism continuing the work of the state as an "agency of homogenization, uniformity, grids, and heroic simplification", as Scott proposes in "Seeing like a State"? (1998: 7-8; cited and contested in Ferguson 2005). Arguably, corporations which engage in state-building are not only confronted with similar problems like development agencies, but also with the problem of "imperial overstretch" encountered by colonial empires, that is, a dilemma of decreasing marginal returns on investments in territorial control. Indeed, the phenomenon is not new and has its historic precedent in the corporate-driven colonization in India and Southern Africa (Kapferer 2005: 290). What is new is the integration of critical concepts of sustainability and participation into a powerful discourse of corporate "development", a kind of corporate developmentalism that acts as an "anti-politics machine" (Ferguson 1990) and as a strategy against criticism. Thereby, corporate notions of "consultation" embrace the fluent and negotiated character of the corporate–state–community relationship. Demands for compensation are turned into a "management of expectations" that incites desires for concrete material goods, grounded in depoliticized ideologies of "development" and "progress", while ensuring that they stay "realistic" and focussed on those sectors most relevant to corporate interests.

Different to other forms of emergent corporate state-like formations based on consumption and the commodification of substitutes for religious, kinship and family values which are de-territorialized (see Kapferer 2005), the mining industry does need territorial control over entire landscapes and extensive transport and supply infrastructure. Indeed, a question for further research, which was not sufficiently addressed here, concerns the crucial role of security contractors in maintaining the territorial borders between the mining site and its surroundings (see e.g., *Figure 5*[22]).[23] But as a few dozen armed guards alone cannot provide security, the tacit or explicit support of relevant numbers of local people is needed, too. And as it would be too costly to fully control all elements and actors at play, the resulting pattern resembles shifting circles of varying levels of control, combining elements of corporate state-building with the extraction enclave. Thereby, participatory processes, which outwardly look like a serious engagement with local social and environmental conditions, are defined by corporations and mostly not by the affected populations and their diverse interest groups themselves. The engagement of companies in the establishment of CSR standards finally underlines their power vis-à-vis local populations and states.

22 The global security contractor G4S is a major actor in the mining industry. Image from the corporate pdf brochure "Case Study – Mining – Aquarius South Africa", available from <www.g4s.com> accessed 08.03.2015.

23 Or, more generally, the linkages between mining companies and the "rise of the privatized military industry" (Singer 2003), which links to the observations in Chapter A on "disaster capitalism" (Klein 2007). For example, the US-American engineering and construction company Bechtel—accused of profiteering from reconstruction contracts in post-war Iraq—is also a major provider to the mining industry (<en.wikipedia.org/wiki/Bechtel> accessed 08.03.2015). For the historic co-development of aluminium and the war industry, see Padel/Das (2010).

PART III: Comparison

C) Local People's Strategies

In this chapter, we compare the strategies of local people who are confronted with the impacts of mining, as described in the case studies. The aim is to identify hindering or enabling factors for successful collective action—understood as the negotiation of better conditions for compensations, preventing or mitigating negative impacts, or direct resistance against an unwanted project. We first introduce the focal aspects, then give short summaries of the case studies regarding these aspects and discuss them in the third section. The aspects are grouped in three parts, which are again divided into sub-issues (see *Table C,* page 477): First, we ask about the *actors and actor-coalitions*, providing a rough characterization of *(a) the local population* which is (potentially) affected by a mine; *(b) organized groups* which resist, voice demands or enter into negotiations; and the efforts to *(c) build alliances* and the resulting coalitions at the regional, national and international scales. Second, the *(d) reasons and demands* voiced in collective action are summarized, with special attention to the relevance of *(e) prior experience* and *(f) knowledge exchange* about mining impacts or relevant organizational skills, as well as *(g) possible alternatives* to corporate mining development. Third, we treat the *(h) means and methods* used.

1. Introduction

To identify something like "local people's strategies" is considerably more difficult than the previous analysis of corporate strategies. Even though the latter turned out to be more complex than anticipated, we could depart from identifiable actors with vertical structures of command and authorised communication, and an assumption of profit-maximizing as a strategic base-line. By contrast, even seemingly homogeneous local or indigenous communities are divided along age, gender, kinship etc.; additionally, the presence of elites and patron-client relationships, immigrants and settlers etc., might complicate the situation. We therefore assume that different sectors have diverging interests, expressed in possibly contradictory strategies which we try to locate on a scale from "resistance" to "negotiation". A third possible pole, "acquiescence", is treated rather implicitly as the absence of collective action and conflict.

While remaining attentive to diverging interests, identities and ideologies at play within the local population, we aim at identifying changes in the resistance/negotiation-dynamics over time along the stages of the mine-cycle, which will be taken up again in the concluding Chapter D.

1.1 Actors and Coalitions

The first question to ask is "who acts with whom?" What kind of local interest groups and organizations can be identified, and how are they linked to larger alliances and coalitions? Referring to collective action theories (e.g., Olson 1965, Ostrom 1990, as used in Haller et al. 2007), one can expect a number of endogenous factors which support the framing of collective strategies, such as a relative homogeneity of interests, a positive attitude towards subsistence economy and the spiritual importance of territory and nature, respected leadership and institutions for conflict resolution. Contrarily, a history of oppression, lack of information on possible impacts, divergent

Fig. 1: Anti-Mining Protests in Makati City, Philippines, 8 March 2013. Photo: www.alyansatigilmina.net.

interests due to dramatic socio-economic changes and further divisions induced by divide-and-rule tactics of companies such as selective bribes and gifts and negotiations with individual leaders and communities, are common hindrances to collective action. To overcome these hindrances, the "transaction costs" of organizing need to be in a good relation to the expected benefits, or the costs of not doing it. Also, the problem of free-riding (individual benefits in disregard of externalized costs) has to be addressed (ibid.: 567ff.).

In a comparison of 18 indigenous groups confronted with oil extraction, Haller et al. (2007) identified 5 that had successfully resisted an unwanted project. Thereby, the role of the regional indigenous organizations was decisive to overcome the "problem of collective action" and framing unified strategies. These organizations were capable of informing their member communities about the risks of oil extraction early on and prevented divisions by surveillance and sanctioning of individual leaders who would succumb to the temptation of accepting bribes in violation of the common interest. Traditional conflict resolution, sometimes together with histories of tribal warfare, was an important base for this. These indigenous organizations managed to build broader coalitions and mobilize support and resources at the national and international level, connecting their issue to larger political debates and thus putting pressure on the companies. Also, the (financial) support allowed them to deliver certain services and benefits to the communities. Against the lure of fast money offered by the companies, accompanied with high long-term environmental costs, these organizations credibly advocated for the defence of a neo-traditional lifestyle, while making careful adaptations to address the most pressing needs (ibid.: 576).

Though the criteria of the present comparison are informed by this previous study, there are significant differences due to the choice of including more mixed social settings, referred to as "local populations", rather than indigenous peoples with their representative organizations (see Part I – 1). Not all cases covered in this book involve populations that self-identify as "indigenous peoples" and claim related rights (as introduced in Part I – 2). Yet, a feature of almost all the cases are continuing discrimination and exclusion that have their origin in more or less long colonization periods, sometimes together with recent or on-going incidents of violence or civil war that may weaken local people's resilience against external shocks (see Chapter A). Poverty, low literacy and lack of public services may result in high obstacles for collective organizing, be it for simple reasons such as the difficulty to cover the costs of transportation for meetings and protest actions. The perceived isolation can be exacerbated in mining conflicts by powerful discourses that label opponents as "anti-development", "backwards", "eco-fundamentalists" etc.[1] At the same time, a long memory of struggle against oppression and exploitation, within which current events are to be located, can provide a powerful common ideological base for resistance.

Apart from local people and communities, there is a type of actors labelled as "Non-Governmental Organizations" (NGOs), involving a wide variety of organizational forms. NGOs need to build access to and legitimacy among local people in order to carry out their stated purpose, which they often do through alliances with other organizations that already have a longer presence in a certain area. This can evolve into complex coalitions that reach from the very local to the global scale. At the same time, NGOs need to legitimize their actions to their donors. This tension can give rise to allegations of NGOs benefitting from the knowledge created by local people's struggles without sufficiently contributing to it. Many scholars stress the need to better understand how "frictions" and mutual misunderstandings that arise in such processes can be overcome in broad coalitions, to construct a common ground (e.g., Tsing 2005). Kirsch mentions inspiring mestizo-indigenous alliances in Peru, but a lack of examples where peasants ally with workers' unions (2012: 209–10, 213). There is also a need to remain attentive to the emergence of new social movements in unexpected milieus that open up new possibilities for alliances (e.g., among "frontier" settlers in the Brazilian Amazon, see Campos 2008). This involves questions of identity and the construction of a common agenda regarding the reasons, demands and means of action.

1.2 Reasons and Demands

The reasons and demands, or the "why and what for?" of collectively organized activities, condition the possibilities for alliance building and can serve as indications for the resistance/negotiation-dynamics. Among their manifold expressions, one can distinguish between immediate and direct opposition against a mining project, its continuation or expansion, because of its (presumed or perceived) social and environmental impacts, and demands relating to the conditions under which it is carried out. The second can include demands to mitigate environmental impacts or improve compensation and benefit-sharing (including the availability and quality of employment). Conflict triggers can be high company profits (linked to high market prices), unmet or differently perceived promises and a sense of unfair compensation. In their comparison of 50 mining conflicts worldwide, Franks et al. (2014) identified 21 "issues in dispute", which are regrouped in:

1. "Social and cultural change": community health and safety, culture and customs, and the presence of vulnerable and marginalized groups (relevant in about 50% of the cases);
2. "Economic change" with the distribution of benefits as major issue (ca. 70% of cases);
3. "Socio-environmental change": pollution and access to resources (ca. 70%–80%);
4. "The process of change": consultation/communication and consent (ca. 40–50%).

1 E.g., in Argentina, the label "eco-terrorists" appears in the mainstream media referring to anti-mining groups (irrespective of their completely non-violent tactics). Thereby, the use of anti-terror legislation against road blocks and the surveillance of activists by the secret service are justified, leading to a general criminalization of the movement (Antonio Gustavo Gómez, presentation at University of Bern, 20.3.2012).

The framework we use covers these major aspects.[2] Special attention is given to issues of consultation and consent, or process-related demands, in accordance with their crucial role in discourses of CSR (see Chapter B) and indigenous peoples' rights.[3] Where the local population is not identifying itself or recognized as indigenous, similar demands can take the form of a local or regional referendum. Although mostly not mandatory, nor legally binding, this can be a powerful instrument to show that there is no "social licence to operate", bringing together reasons and demands (the right to decide) with a means to express opposition.[4]

As underlying reasons to which people may refer to in collective action, we already mentioned a background of anti-colonial struggle, as well as memories of state violence. More specifically, direct prior experience with similar projects in the region, and/or knowledge transfer from exchange with other regions might trigger early resistance (Haller et al. 2007). In Argentina for example, a spill-over effect from the convergence of negative experiences with the Alumbrera mine in Andalgalá and mobilizations in Esquel generated a wave of anti-mining movements throughout the country, leading to the prohibition of large-scale open-pit mining in several provinces (Svampa/Alvarez 2010). Also, the availability of general skills in social movement organizing can be relevant.

A further factor to which we give special attention is the availability of alternatives—both economic as well as ideological—to the dominant development discourse and the role of mining therein. For the economic alternatives, it has been observed that strong conflicts often arise where mining competes with other crucial income sources for livelihoods; for example, the threat of water shortage and the hampering of market-oriented fruit production in Tambogrande and Quillish Hill in Peru (Arellano-Yanguas 2012: 97). In Turkey, holiday housing development has helped to motivate anti-mining struggles (Hurley and Arı 2011). Such "competing visions of rural capitalism" can easily cause tensions within the population (in Bebbington et al. 2013), between those who fear a hampering of well-established businesses and those who are not benefiting as much and hope to find employment in a newly emerging mining sector.

However, there may also be deeper issues at stake in those cases where people use common-pool resources for livelihoods and refer to their environment as a cultural landscape with ritual and socio-cultural significance, too. Such regions cannot be regarded as "pure" or "unused" nature, but constitute cultural landscape ecosystems that are often inhabited by ancestors and other spiritual beings with whom people interact, as shown by anthropological studies (e.g., Berkes 1999, Fairhead and Leach 1996, Haller et al. 2007, Haller 2010; 2013; Haller et al. 2013). Thereby, the economic and cultural/spiritual significance can hardly be separated and dramatic changes or loss of relation to such an environment will undoubtedly mean a deep and enduring impact to the identity and lifeworld of local people. In such a setting, there might be an "environmentalism of the poor" at stake (Martinez-Alier 2002), or a resistance of "ontological communities" that share "basic presuppositions on 'social and natural' reality" different from a "modern-Western" worldview (Boillat et al. 2008), and therefore less likely to succumb to arguments about the benefits of industrial development and employment.

2 Similar categories are used also by Arellano-Yanguas (2012: 99-100) and De Echave et al. (2009) in their comparative studies of mining conflicts in Peru.

3 Most notably, Free, Prior and Informed Consent (FPIC) as framed in the UNDRIP, and "consultation" as in ILO Convention 169; see Chapter B (1.3), and Part I – 2.

4 For an overview of such referenda in the Americas, see: Kaitlyn Duthie 2012: "Local Votes and Mining in the Americas." Article and Map: <www.miningwatch.ca/article/local-votes-and-mining-americas>; accessed 10.05.2015.

Fig. 2: Poster for a protest against gold mining at the 2013 Gold Forum in Zurich, Switzerland. Source: ch.indymedia.org.

1.3 Means and Methods

Finally, in this section, we discuss the targeted use of different means of protest as a key element of an elaborated strategy. The methods, or tactics, employed in mining conflicts can cover the whole range of possibilities: petitioning, media campaigns, lawsuits, involvement in electoral politics, marches, strike, and direct action such as road blocks, going up to sabotage, kidnapping of company staff, and violent attacks. Of course, they come in combinations and change over time in a dynamics of escalation. The choice of protest tools touches crucial questions such as: who is addressed? Which fora for political expression are used or avoided? With regard to process-related demands, we could assume that local groups might reject to participate in corporate-controlled consultations if they have lost trust due to past action or due to a marked power asymmetry. As from the local perception the question arises whether it is always the best choice to negotiate; non-cooperation or evasion can be strategic means of passive resistance. Contrarily, where a company is not recognizing a certain sector as "stakeholder" at all, the inclusion in "dialogue" might be preferable to "non-existence".

1.4 Timing and the Impact–Resistance Hypothesis

The last aspect of local people's strategies we discuss in this chapter has to do with the "politics of time". Under this concept, Kirsch (2014) relates corporate strategies to spread doubt on negative mining impacts, on the one hand, and the need for critics to pre-emptively confront them at an early stage, on the other hand, in order to prevent damage in advance, rather than trying to hold them accountable for damages already done.[5] The "politics of time" relates well to the initial working hypothesis of the *infoe* book on fossil fuels (Haller et al. 2007), which assumed that increasingly negative impacts of oil extraction over time would be accompanied by increasing resistance. However, this intuitive "impact–resistance" hypothesis needs several qualifications, in line with Kirsch's argument, the findings of Franks et al. (2014) cited below, as well as those already stated in Haller et al. (2007): high resistance at an early stage has more chance of success and may be motivated by the spill-over effects from previous experiences or other resistance movements, which may be transmitted from other settings through alliances on national and global levels. In addition, what counts are less the objective impacts than the expected and subjectively perceived impacts, which relate to a range of other factors.

The stage of the mine-cycle and the related impacts (as discussed in Chapter A) constitute exogenous factors that are commonly perceived by local actors as given, but this does not mean that strategic choices cannot be adapted to them. The right timing of protest actions is crucial to maximize bargaining power. For instance, early direct action can motivate strong CSR and relatively generous compensation agreements.[6] Franks et al. observed an escalation of protest methods, "from campaigns and procedural actions (such as complaints or grievances lodged with governments, companies, tribunals, or courts) through to physical protest." This suggests that companies and authorities did have sufficient opportunities for "dialogue to address issues before escalation", but often failed to use them effectively (2014: 2). Furthermore, the study shows that of the different mine-cycle stages, mining conflicts at the planning and construction stage are most likely to result in the suspension of a project. Possible reasons are that the dramatic changes during this stage encourage mobilization, and that there is a window of opportunity for successful collective action because the company is vulnerable in the national context before the necessary permits are secured and investment decisions taken. Once the massive inflow of capital has happened, economic dependency on mining weakens the potential for fundamental resistance, while demands may centre more on the distribution of benefits and mitigation of impacts (ibid.). Early exploration activities represent the first entry of corporate agents in the field, often under pressure to deliver fast results and with little recognition of ownership rights and customs of local people, which can easily be perceived as the beginning of a major threat.[7]

5 This position is an outcome of Kirsch's own participation in the long-lasting campaign at the major Ok Tedi mine in Papua New Guinea (see Part I – 1).

6 See Welker (2009) for Indonesia and Arellano-Yanguas (2012) for Peru, where road blockades apparently have become a common tool of communication with corporations and are even encouraged by local authorities to strengthen their bargaining position.

7 That this is a problem well known to corporations as well is shown by Welker's description of the effort of Newmont's CSR staff to sensitize the company's geologists on the need to previously inform, consult and negotiate with local people for all their interventions (2014).

2. Case Study Summaries

In this chapter, we provide short summaries of the case studies for the most relevant aspects for comparison as introduced above: from a characterization of actors and coalitions, we move to reasons, demands and means of protest, and their changes over time (see *Table C,* page 477).

Mopani, Zambia

In the 1930s, the Lamba original inhabitants of the Copperbelt region have been displaced to "native reserves". Since then, we do not know of any resistance or attempts to negotiate compensations, nor of any claims based on ethnic minority or indigenous status. Massive labour immigration has made the "mine worker" to be the major common identity in a multi-ethnic "mining town" setting. Historically, the mine workers' union played a major political role, but was weakened by neoliberal reforms and massive lay-offs in the 1980s and 1990s. In recent years, some local NGOs were formed, encouraged by national and international NGOs seeking alliances in the regions. The general impression is that of a highly fragmented population which has few resources to voice demands and put pressure on the mining companies. However, the case study shows an efficient international network of NGOs. Their reports and lobbying led some Swiss mainstream media to cover the case. This information (together with other cases, i.e., El Cerrejón, Colombia) raised public attention among Swiss citizens, also in the communities where some of the managers of the mine owner Glencore live and pay taxes. So far, there seem to be no direct links between Swiss grassroots groups and people in the Copperbelt.

Although there seems to be widespread discontent, the level of resistance and protest in the Copperbelt is low. In recent years, however, the tendency is increasing, with popular demands concerning the reduction of air pollution, compensation for health damages and improvements in social services, healthcare, education etc., and linked to this, increase in mining royalties and taxes for a more just distribution of profits. There is also a resurgence of workers' unions and employment-related demands. Consultation and consent do not seem to be an issue, apart from one incident that concerned the eviction of so-called "squatters" for the construction of a new mine shaft, which led to a complaint of several NGOs to the Canadian OECD contact point.

The population of the Copperbelt has long experience with mining; however, this comes to a limit with regard to newly introduced technologies such as in-situ leaching, bringing new and unknown risks. With regard to movement organizing, the workers' unions provide a strong example. In the 1980s, a series of food-riots took place in reaction to increasing poverty. Also, the region is known to have an important influence on national electoral politics. The involvement of international NGOs in recent years means that more knowledge about environmental issues and tax distribution justice is available. The region has only limited options for subsistence agriculture, and other economic alternatives are hardly available; the large informal sector directly or indirectly depends on mining, too. The main (individual) alternative for some of the inhabitants is to go "back to the village" or migration to other places.

We find little documentation on protest methods, with the most concrete expressions of local people's collective action to be found in the emergence of local NGOs and a recent wave of strikes. Possibly, a lot of protest potential is absorbed in electoral politics, with new governments promising but failing to address the grievances.

Arlit and North Niger

In reaction to the marginalization and lack of opportunities, young Tuareg men especially emigrated to Libya under Ghaddafi, where they often were employed as mercenaries, received military training and established relations with members of other Tuareg groups. This became the base for a series of armed revolts, especially since the 1990s, in reaction to the violent repression of Tuareg autonomy movements by the Niger state. In 2007, the *Mouvement des Nigériens pour la Justice* (MNJ) launched an attack on Areva's camp at the planned Imouraren mining site. In several incidents, employees of Chinese companies and of Areva were kidnapped. With the civil war and subsequent the fall of the Ghaddafi regime in 2010-11, a lot of looted weapons from the Libyan army became available, while the Tuareg mercenaries lost their employment. This supported the emergence of the new *Movement National pour la Liberation de Azawad* (MNLA), which demanded and fought for an independent Tuareg state called Azawad in the north of Niger's neighbouring country Mali. Aided by a political alliance with Jihadi groups, allegedly linked to Al Qaida, the MNLA succeeded with a surprisingly quick defeat of the Malian army. However, the movement was hijacked by the Jihadi groups, which installed sharia law in the north of Mali, until they were defeated by ECOWAS and French troops. The entire area, including the north of Niger, has thus become a "terror zone", heavily supervised and militarized.

A number of civil society organizations are trying to lobby for the Tuareg with non-violent means, both in Niger and abroad. Their space of action has however become very narrow, caught up between the repression of critical voices by Niger's authorities and the "anti-terror" operations supported by the French and US armies. Many Tuareg activists are active in exile, from where they try to forge alliances with international watchdog NGOs, to publicize the situation of the Tuareg and lobby in international fora, such as UN conferences.

On a general level, it can be concluded that the problem in this case is that mining interests intermingle with geo-political interests to such a degree that local collective action is seriously hampered and gets caught up in an actual "resource war", where local people have very low bargaining power in view of a strong state-corporate alliance.

Ranger/Jabiluka, Australia

The larger area surrounding the Ranger and Jabiluka mines is inhabited by different clans of indigenous peoples. The main opposition comes from a very small group, a few dozen members of the Mirarr clan who hold the legally recognized status of the "traditional owners" of the mining area, the mining town of Jabiru and parts of the surrounding Kakadu National Park. Their leadership is apparently strong and they appear very united towards the outside. In the "Stop Jabiluka" campaign, they built alliances with a wide range of support groups: national and international environmental NGOs, student groups, and other indigenous groups in Australia and worldwide became involved at its peak in the late 1990s. The campaign was built from the beginning on a global scale, while remaining under their leadership through the Gundjeihmi Aboriginal Corporation (GAC), a self-constituted representative organization of the Mirarr.

The demands of the "Stop Jabiluka" campaign were both specific (recognition of territorial rights, stopping Jabiluka mine construction, protection of the Kakadu ecosystem, ancient rock art and sacred sites) and very general (end to nuclear power and weapons, respect for cultural and spiritual lifeworlds and self-determination of indigenous peoples). Based on the significance of nature tourism in the Kakadu National Park and the global concern about the dangers of nuclear technologies, this allowed for a broad coalition to identify with the demands.

The indigenous concept of "care for country and people" is crucial to the uncompromising resistance of the Mirarr. The spiritual landscape is perceived as a living being that cannot be disturbed without causing damage to the very existence of humans. As caretakers of the territory inhabited since time immemorial by their ancestors, the Mirarr see themselves responsible for the country's well-being and suffer for not being able to live up to this properly in the current setting. This can even go as far as feeling co-responsibility for the Fukushima disaster, where uranium from Ranger was probably used, due to their failure to stop this mine from operating. The financial benefits they receive from the Ranger mine cannot compensate for the damages, but are conceived as an integral part of a neo-colonial dependency forced upon them. Prior experience certainly played an important role, as indigenous peoples in northern Australia have been affected by mining for a long time, building up legal and organizational expertise to defend their rights for several decades. While certain subsistence activities such as hunting and gathering continue, the focus of the indigenous economy has shifted to related activities as park rangers or tourist guides. The tourist sector is important and largely controlled by indigenous associations, and the Mirarr especially, try to avoid depending directly on mining-related incomes.

The "Stop Jabiluka" campaign used several methods in parallel: petitioning and lobbying, lawsuits, direct action and shaming of the Australian state through international bodies such as the UNESCO World Heritage committee. At its peak, it included a major blockade of the Jabiluka construction site. The campaign was financed with compensation payments of the Ranger mine, together with donations, which allowed to contract permanent staff and external advisers. Also, compensations serve to finance culturally appropriate education for the Mirarr youth, seen as a key element to maintain the resistance in the longer term. The Mirarr use media in a very conscious way, working with leading national and international journalists.

El Cerrejón, Colombia

The population of the Guajira province is 655,000 and consists of indigenous people (48%), Afro-Colombians (14%) and *mestizos* (32%). The Wayuu are by far the largest indigenous group and collective owners of most of the recognized indigenous reserves, which cover close to a third of the Colombian part of the peninsula. The communities most affected by the mine are characterized as either Wayuu or Afro-Colombian. Especially the Afro-Colombian communities have received many immigrants and refugees, hoping for employment or compensation payments from the company. The Wayuu are organized in clans and communities and historically lack any centralized decision-making or leadership institutions. A rather loose cohesion is based on ad-hoc mediation and alliances. There is no indigenous umbrella organization that is recognized by a majority of Wayuu as a legitimate representative of their interests, but there are a number of associations and NGOs run by indigenous people. Especially, the Wayuu women's groups played a central role in the campaign against the diversion of the Ranchería River, which was very successful in cross-sector alliance building. This included directly-affected indigenous and Afro-Colombian communities and sectors of the urban and *mestizo* populations, supported by regional and national indigenous organizations, lawyers' collectives, political parties, music bands and international NGOs.

The typical pattern of interaction between communities and the company was, and still largely is, focussed on negotiation. Open resistance is directed against specific expansion plans, which finds its expression in campaigns at the regional level, temporarily bringing to the open a latent discontent on a variety of issues. These include the protection of the environment and sacred landscape, the indigenous rights for FPIC regarding the territory, as well as improved benefit-sharing, fair compensation for the loss of land, and the recognition and treatment of health problems associated with mining pollution. When mining started in the Guajira in the 1980s, there was no relevant prior experience with mining available in the region. Regarding the increasing resistance, the role of knowledge exchange through indigenous organizations and NGOs was certainly influential in opening up new possibilities for local action. For the possible economic alternatives, subsistence agriculture and animal husbandry are most relevant, but there are little options for cash income that are not directly or indirectly related to mining.

The methods used by local groups critical to the El Cerrejón mine combine petitioning and lobbying, lawsuits, and capacity building. Structured as a network, resistance groups form specific campaign alliances which reach a wide extension in the region, expressed in large protests marches and gatherings. Some Wayuu organizations have built up solid communication skills, and make self-conscious use of own (social) media channels and mainstream media for outreach to a wider public, although this may be restricted by limited access in rural areas.

Ambuja Rawan, Chhattisgarh, India

The local population of the villages surrounding Ambuja's factory in Rawan consists of a mix of rural Hindu castes, dedicated mostly to rice farming. Since the establishment of the cement factory, a significant number of local people have been employed there, mostly as precarious contract workers. In the highly stratified Hindu caste system, socio-economic functions continue to be assigned according to caste membership. However, in the present case, the mobilization was built on an inclusive "peasants/workers" and "Chhattisgarhi" regional identity, against an exploitative "outsider" elite, rather than on caste membership. As most families depend on mixed incomes from farming and contract work, there is little contradiction in this alliance, which finds its organizational expression in the cooperation of a local farmers association with the experienced cement workers' union PCSS. Also, several elected members of the local *panchayats* (village councils) played an important role, although it remains unclear how much these institutions put forward a common strategy. PCSS forms part of a long history of social movement organizing with the *Chhattisgarh Mukti Morcha* (CMM), which helped the local people's voices reach the capitals of Raipur and New Delhi and finally the Holcim headquarters in Switzerland, through international trade union federations as well as activist networks.

The case shows that a general feeling of injustice, accumulated over the years since the start of the operation, has come to the open. The demands concern environmental problems (water usage and pollution), employment and working conditions (low salaries, contract work, discrimination of local workers), and the distribution of benefits and lack of just compensation for land acquisitions, both past and recent. A key demand which links to other factories of Holcim subsidiaries (especially the ACC plant in Jamul) is the recognition of "so-called contract workers" as permanently employed workforce with decent salaries and social security. The movement is however not directed against industrial activity as such, but critical of the way it is carried out. The villagers around Rawan did not have much direct prior experience with industrial development and movement organizing when the factory was built in the 1980s. A first, short-lived agitation took place in the early 1990s under influence of the state-wide movement led by the CMM. When local people were looking for support in movement organizing after 2008, they again contacted CMM for support. The main economic alternative to employment in cement production is local market-oriented agriculture. The lack of land and irrigation water, intensified by industrial use, means that this is under increasing pressure. Other options are seasonal migratory labour in other industries and urban centres.

The protest methods include petitioning, lawsuits, pickets and short strikes. The supporting PCSS union and CMM movement have a history of massive direct actions such as prolonged blockades of factories, though in the Rawan case this did not happen. The actions are not always legal, but strictly non-violent, disciplined and finely attuned to support petitions and lawsuits which aim at reaching favourable conditions for negotiations and collective bargaining. The picture of the Ambuja Rawan case corresponds fully to this strategy. After the local management refused to further negotiate and turned to criminalization of movement leaders instead, the union successfully took the issue to the international level. Supported by a solidarity campaign and an OECD complaint, the international headquarters of Holcim was urged to negotiate. To be recognized as a relevant stakeholder is already considered as an important step, although the conditions of these talks were very unfavourable for the union, as public statements about them are prohibited and it was thus hardly possible to carry on with the international campaign.

Marlin Mine, Guatemala

The local population of the four communities that lost land to the Marlin mine consists foremost of indigenous peoples belonging to the Maya language subgroups of the Mam and Sipacapan. The total population of the two affected municipalities comes to about 50,000 people spread over nearly 100 villages and hamlets, also mostly inhabited by indigenous peoples, with a minority of less than 25 per cent belonging to the dominant ethnic group, the mixed-origin, Spanish-speaking *Ladinos*. The affected communities are described as relatively homogeneous with regard to ethnicity and economic activities, mainly subsistence farming and livestock rearing, together with some cash crop cultivation. Cultivation is done on individual plots on land owned collectively by the community. Since the arrival of the mine, the communities are divided due to the presence of a relatively small but influential "pro-mining" faction, consisting of people who benefit economically from the mine. However, information on that group is scarce. The main actors of the movement against the Marlin mine are the affected communities themselves. Out of the delegitimization of the local authorities that cooperated with the company, an ancient indigenous self-rule institution, the *alcaldía indígena*, has been revived as a permanent forum for mining-related debates and actions. The formation of a national-level coalition against large-scale open-pit mining and the open manifestation of resistance in the region of Marlin coincided around 2004, thereby linking the local and the national level from early on. Later, international NGOs also became important to pressure Goldcorp at its headquarters in Canada.

The local resistance started building up during the land acquisition process due to the lack of adequate compensation, and especially after the failure of the company to inform and consult the people about its plans. The resistance intensified during the construction, accompanied by several incidents of violent repression and human rights abuses. With the operation of the mine, the excessive water use and alleged contamination and health impacts became important, too. Although there are hardly any economic alternatives for cash income except labour migration, the attachment to the territory and its importance for subsistence production as well as cultural and spiritual values seem to be much more important than the expectation of short-term benefits from mining-related jobs and development programs. At the level of the national and international solidarity network, Marlin is seen as a test case for the protection of indigenous peoples' rights and the environment in the context of the fast expansion of extractive industries in Guatemala. Although direct prior experience with modern large-scale mining was not available in the San Marcos province, the general background of colonialization, the civil war experience and continuing exploitation and discrimination may have contributed to the early and massive reaction.

The struggle against the Marlin mine shows a wide variety of methods used by local actors. The most striking collective action is the self-organized referendum in the community of Sipacapa in 2005. The result of 98 per cent objection against the mine gave a very clear signal to the company and the authorities. Since then, similar referenda have been held in 42 municipalities, involving 400,000 citizens who opposed open-pit mining by large majorities, thereby denying the "social licence to operate", both at the local and national level. Another angle is the participation in large protests and gatherings in the capital of Guatemala, coinciding with protests near the mine. Locally, there have been impressive acts of direct action: in 2004, a road blockade was maintained for over 40 days by a large number of people taking turns. In 2008, two acts of sabotage against a power line and an exploration drill rig were justified with the argument that the company had installed them without permission from the landowners. On the legal level, in alliance with various NGOs and sectors of the Catholic Church, all the relevant international mechanisms were tested. Complaints were placed in 2005 to the Compliance Advisor Ombudsman (CAO) of the IFC/World Bank for granting a credit, and in 2009 to the Canadian contact point of the OECD. In 2010, the ILO urged the government to suspend mining at Marlin due to inadequate consultation. This request was followed by a visit and report by the UN Special Rapporteur on Human Rights of Indigenous Peoples, focussing on Marlin. In the same year, the Inter-American Commission on Human Rights decided that the mine should be closed as a precautionary measure to prevent further violations of basic human rights. Though none of this stopped the mine from operating, it certainly had an effect in shaming the company at the international level.

Tampakan, Philippines

The Tampakan mine project is situated in a municipality of the same name in the province of South Cotabato in Mindanao. The local population of the nine directly affected village units (*barangay*), comprise about 17,000 people, consists mostly of indigenous Bla-an and Christian Filipino settlers. Out of them, 870 households or 3500 to 4000 individuals need to be resettled for the mine. With 67%, the majority belongs to the Bla-an, the largest indigenous group of Mindanao with around 240,000 members. They depend mostly on swidden agriculture for subsistence, practised on plots with individual usufruct rights and held in common property, supplemented by foraging in common forest areas. The settlers also mostly practice agriculture, however with a much more market-oriented focus. The Bla-an do not have traditional institutions or leadership beyond the individual communities, but are today represented in Tribal Councils that were introduced since the 1980 as part of the formalization of indigenous land ownership, promoted by state agencies, NGOs and mining companies. Some indigenous leaders are also part of the officially elected *barangay* councils. Negotiations with the company are mostly carried out through these two types of councils. Additionally, the settlers formed five landowners' or farmers' associations to negotiate compensations. Some Bla-an built up informal armed groups, which call upon a traditional concept of a tribal "war of revenge" in opposition to the mine. Also the left-wing New People's Army guerrillas have been active in the area, though it is not known if they receive active support from local people. In the downstream area, the mine could have impacts for about 80,000 farmers practising intensive rice cultivation, which are politically represented through farmers' associations. Finally, there is a diverse civil society and a strong anti-mining movement in Mindanao as well as nationally, with various student groups, sectors of the Catholic Church, indigenous peoples' organizations and other NGOs as major supporters. They form more or less stable coalitions, which are again supported by international NGOs. A major success of that movement—and an indication for its political power—is the ban on open-pit mining passed by the Governor of South Cotabato in 2010, which is a major hindrance to the project.

Both the Bla-an and settlers share concerns over environmental impacts and insufficient compensation, but it seems rather than joining efforts, each group is pursuing its own interests foremost. The relation between the Bla-an indigenous peoples and the Filipino settlers is conflictive due to internal colonization and overlaying land claims. The Bla-an have negative experiences with a gold rush in the late 1980s, bringing many new settlers to the area. Since then, the situation has changed dramatically: while the Bla-an hold recognized collective Ancestral Domain titles, many settlers acquired land through mortgage agreements with Bla-an individuals, for which they have no official titles. Though the settlers are still economically stronger and historically privileged, they now feel side-lined in compensation negotiations. Regarding economic alternatives, small-scale agriculture is the most relevant.

At the time of writing the case study, it seemed that the CSR efforts by the company succeeded in turning the initial opposition from upstream indigenous communities into approval, with only a "silent minority" still resisting. Overall, the position of the Bla-an is described as "actively negotiated dependence". Some Bla-an leaders even sent a letter to the South Cotabato Governor, asking to drop the ban on open-pit mining. Downstream of the planned mine and at the regional level, the opposition remains strong, out of fear that the rich agricultural area could be affected. Also in the capitals of South Cotabato and Mindanao, environmentalist and social justice arguments are at the forefront, based on negative experiences with mining in the country, including a major toxic tailing spill at the Marcopper mine in 1996.

The Tampakan example shows a wide range of collective action methods, both peaceful and violent, part of them embedded in traditional institutions, others more linked to new ways of being active in a globalized world. In the late 1980s, the Bla-an reacted to the lack of consultation and infringement on their territory by arming themselves, and both Bla-an informal armed groups and leftist guerrilla fighters have launched violent attacks on the company staff and infrastructure in recent years. In 2008, Bla-an people blocked an access road to the company's base camp, after Filipino settlers had done the same to demand inclusion in compensation negotiations. At the regional and national level, the Tampakan case is considered a test case for the anti-mining movement, which has deployed diverse means such as mass mobilizations and protests (including at the corporate headquarters in Manila), caravans, fact-finding missions on human rights violations, educational and social media campaigns, and the preparation of an alternative impact assessment presented at public forums.

Las Bambas, Peru

The local population affected by the Las Bambas project in Apurimac is characterized foremost as "peasant communities". These communities are made up by Quechua-speaking subsistence farmers. Situated in a marginalized and hardly accessible mountain region, they depend mostly on livestock and some crop cultivation. Although having all the necessary conditions to claim the status as "indigenous peoples", local people have so far mostly refrained from doing so—a result of ages of cultural discrimination against the formerly so-called "Indios", supposedly backward and uneducated. The communities in Andean Peru are entitled to self-government regarding the common property land-use system and other aspects. There is generally a strong cultural and spiritual attachment to the territory, with mountains considered as seats of ancestor spirits.

The communities are divided by the project: a smaller number situated within the company defined "zone of direct influence", heavily targeted by CSR programs due to the necessity to acquire access to their lands, and a larger group outside of this zone. Furthermore, there is an emerging mining town, attracting immigrants in the hope of employment. Also, locals who had migrated earlier to larger urban centres are returning. Especially, the local youth aspire for employment in the mine. The jobs created already before starting the construction introduced another division in the communities. Collective strategies were not apparent, and regional organizations, most of all the grassroots peasants' association, were unable to act because of internal divisions. A regional NGO with long-standing presence in the communities and a national-level NGO specialized in mining issues provided support and training, though with limited participation from people in the "direct zone of influence".

The main reaction to the project observed at the time of research was acquiescence and individual opportunism. The approval for the project is explained—in accordance with the argumentation of government and corporate officials—with the need for development of a region that has hardly benefited from public services, and opportunities of cash income that were not available before. The doubts however concerned the fast socio-cultural changes, linked to the sudden transformation from subsistence to market economy, and the tensions within and between communities over the access to benefits. Environmental damage is expected to take place, with water availability as the most critical issue for the peasants. There is no direct prior experience with large-scale mining, but knowledge about similar mines in Southern Peru is disseminated by NGOs through info materials and training. A relatively small but motivated number of people participated in such activities, be it out of a critical position towards the project, or to enhance their negotiation skills to be used once they would get in touch with the company. However, apart from this, there was no space for debate independent from the company's influence. Mining is generally being presented as the only viable opportunity for the development of the region, while alternative income sources and development visions are absent.

Due to the lack of collective strategies, there is not much to say on methods used in this case. It looks like the company was the protagonist, while local people were busy finding out how to participate in and benefit from the expected development.

Val Medel, Switzerland

The original population of Val Medel belongs to the small Romansh-speaking minority of Switzerland. Culturally and economically, the valley is characterized by a high percentage of people dedicated to alpine farming and livestock rearing, supplemented by eco-tourism and related activities. A certain tension is described between "original inhabitants" and "outsiders" or "non-locals", the latter often being holiday-home owners. However, there is no clear tendency with regard to mining, with pro- and contra-positions taken by both groups. The main protagonists are the municipal council, the mayor (a "non-local") and an informal support committee on the supporting side, and an ad-hoc campaign group named "Pro Val Medel" on the opposing side. Others, such as various local associations or political parties, did not play any role. National-level environmental organizations assisted on request by Pro Val Medel with information and the release of a press communiqué.

In the debate, both sides emphasized concerns for the future development of the Valley, with little to no reference to experiences elsewhere in the world. The protection of nature and landscape was underscored by the opponents as an asset for sustainable small-scale agricultural and tourism-related activities to counter the proponent's argument of creating employment. The prospect of large numbers of immigrant workers and their integration into the small community was seen as problematic. However, the main argument was about the lack of adequate information and the procedure of decision-making; specifically, the provision that by granting the exploration concession, the authority to cede the exploitation permit later on would be transferred to the municipal council. Prior experience about modern mines is lacking in Switzerland; therefore, the main reference of comparison was drawn with the construction of a hydropower dam in the 1960s and the ongoing train tunnel construction in a nearby locality.

The opinion formation, both within Pro Val Medel and the population in general, seems to have happened mostly in informal exchanges and meetings between residents. The main appearance of collective opposition was a leaflet distributed by Pro Val Medel to all households. The issue attracted considerable interest by national media, which may have increased participation and urgency to the question.

CASE	Mopani Zambia	Arlit / North Niger	Ranger / Jabiluka, Aus.	El Cerrejón Colombia	Ambuja Rawan India	Marlin Guatemala	Tampakan Philippines	Las Bambas Peru	Val Medel Switzerland
a) Local population	Mineworkers, Mixed urban and Lamba orig. pop.	Tuareg nomads; urban pop.	Several indigenous clans; settlers	Indig. Wayuu, Afro-Colombian, *mestizos*	Rural castes, farmers, contract workers	Indig. peasant communities, *ladinos*	Indig. Bla-an communities, Filipino settlers	Peasant communities	Rural, *romansh*, "non-locals"
b) Locally organised actors	Worker unions, local NGOs	Tuareg clans, armed groups, local NGOs	Mirarr clan trad. owners, indig. organisation	communities, local and regional NGOs, women's groups	Workers union, farmers org., *panchayat* leaders	Communities, *alcaldía indigena*	Tribal Councils, informal armed groups, peasant associations	Peasant organisation, regional NGO	Communal council, ad-hoc campaign group
c) Alliances	Int. NGOs and unions	Int. NGOs	Eco NGOs, peace-, anti-nuclear, indig. movements	Nat. and internat. NGOs, indig. org., polit. parties	Internat. trade unions, solidarity groups	Nat. and internat. campaign coalition, Catholic Church	Nat. and internat. campaign coalition, Catholic Church	National NGOs	National Eco NGOs
d) Reasons, demands	Pollution, health impact, benefit sharing (tax, salaries, jobs)	Pollution, health impact, benefits; autonomy	Environment, cultural heritage, indig. rights, end to nuclear energy	Environment, indig. rights, FPIC, sacred landscape; compensation	Working conditions, jobs, compensation, water use, pollution	Environment, water use, consultation and compensation	Environment, resettlement, compensation	Environment, compensation, jobs	Environment, economic development, jobs
e) Prior exp.	Yes	Yes	Yes	No	No	No	Yes	No (Yes: nat.)	No
f) Knowledge ex.	Yes	-	Yes	Yes	Yes	Yes	Yes	Yes	No
g) Alternatives	Few: informal sector, migration	Few: informal sector, migration	Tourism, subsistence	Few; subsistence based	Agriculture, seasonal migration	Subsistence, emigration	Subsistence, market agricult.	Subsistence	Ecotourism, agriculture
h) Means and Methods	Strike, organizing, internat. campaigning	Guerilla warfare; internat. petitioning	Petitioning (UNESCO), legal action, media, marches, direct action	Petitioning, lobbying, legal, protest: gatherings, marches	Petitioning, legal action, pickets, strikes, direct action	Referendum, direct action, petitioning, int. human-rights mechanisms	Direct action, armed attacks, marches, fact-fining, social media a.o.	Participation in info-events	Local referendum; info-flyers, face-to-face interaction

Table C: Overview of local people's strategies from case study summaries.

3. Discussion

In the following sections, we compare and discuss the main aspects identified as challenging collective action: first, the heterogeneity of social settings, actor-groups and coalitions; second, the framing of common motives and demands for action, and third, the employed means and methods. In the conclusion, the changes in the dynamics of resistance/negotiation over time are discussed with a focus on the problems they pose for alliance building.

3.1 Actors and Coalitions

The local population affected by mining in the present examples mostly shows mixture of urban and rural migrated groups in combination with indigenous communities (Australia , India, Colombia, Guatemala, Philippines, and Peru). In Zambia, the picture is of highly heterogeneous urban groups in mining towns with a long history of migration and completely marginalized native communities. None of the cases concerns supposedly "untouched" indigenous groups that have not gone through profound economic, political and institutional changes in contact with the respective dominant society.

The heterogeneity of the local populations is mirrored by the wide range of collective actor-groups and institutions identified in the case studies. On the one side, we have cases of resistance to mining in which actors use elements of their traditional institutions in order to cope with the situation: the Mirarr in Australia have a clearly defined representative organization (the GAC) embedded in a *cosmovision* (worldview) that regulates their relation to the environment and their responsibility to "care for country". In the political context, this is possible as the Mirarr are autonomous in their form of governance, but recognized by the state and corporation. In the case of Marlin, Guatemala, there is a neo-traditional inter-community institution (the "*alcaldía indígena*") that was revived to set up a decision-making space autonomous from the communal bodies, which mining opponents regard as co-opted by the corporation. This seems close to a "classic example" of collective action theory, whereby the revitalization of a traditional institution lowers transaction costs of coordinating resistance and reinforces identity formation as a common base.

In other cases, where neo-traditional institutions play a role, it is rather at the level of individual communities' decision-making than at the inter-community or regional level. Thereby, we observe a tendency that internal divisions and corporate co-optation prompt individual communities and their elites to favour the negotiation of "good deals" for themselves, rather than to organize collective resistance. The Colombian case is illustrative: there are no traditional institutions for centralized or inter-community decision-making, but a community- and clan-based practice of mediation and negotiation, which seems co-opted by "corporate-community relations" to a degree that opponents of the mine rather seek to organize in parallel, non-traditional structures like local NGOs. In Las Bambas Peru, the situation regarding "traditional" institutions of decision-making at the community level is similar; however, there seems to be no collective attempt at creating an alternative forum. In Tampakan Philippines, local institutions play a crucial role, but a critical analysis indicates that they are not only co-opted, but in part even created in the interest of the company in order to obtain the necessary permits and to administer compensation payments. Local resistance is organized rather through informal groups outside of these institutions. In the Swiss case, the authority of the traditional decision-making institution is undisputed, while the opinion formation happens mostly in direct informal exchanges between residents.

In most case studies, neo-traditional leadership roles are hardly mentioned, or they appear as privileged targets for corporate co-optation and alleged corruption, such as the "false leaders" of the Wayuu in Colombia. Again, one of the few examples of strong and legitimate indigenous leadership acting as a moral authority is the case of the Mirarr in Australia. To the contrary, in the example of Tampakan, it seems that at least some of the "traditional" leaders maintain their position thanks to the power gained in negotiations with the company, giving them the possibility to redistribute cash and employment opportunities among their constituency. This corresponds to general observations that roles of indigenous political representation, already transformed during colonial and post-colonial times, are often focussed on negotiating benefits with outside powers. They easily become entry points for corporate consultation and compensation agreements, which might again reinforce their position of power (in line with Haller et al. 2007). Apart from political leadership, the case studies provide little evidence about the position taken by spiritual leaders. However, their actions, based on face-to-face communication or rituals, may not become as visible as those of political leaders, and their role might thus be underestimated. Also, it would be interesting to know more about to what extent the role of activists relates to emic concepts of charismatic leadership (e.g., "war chiefs").

One of the interesting issues is the influence of "traditional warfare" institutions: While in the Philippines this can be seen as one way to organize protest, it is also intermingled with guerrilla issues. In Niger, the Tuareg lead a long-standing struggle against the French colonial and post-colonial power and the state of Niger, which is strongly dependent on France. The history of suppression of this group many times led to upheavals and to armed forms of resistance. Clan and traditional institutions of organizing resistance are well articulated but also overpowered by the external and national state powers. The attempt to create a new Tuareg state in Mali was quickly counteracted. Even worse, it resulted in a terrorism discourse that makes local collective action extremely difficult.

Certainly, there are important gender aspects related to issues of leadership and traditional institutions, as well as to resistance against mining in general (see Lahiri-Dutt ed. 2011; Jenkins 2014). As indicated for instance by the pivotal role of Wayuu women's associations in Colombia, women's subsistence activities and practices such as the gathering of medicinal herbs may be specifically affected, while women may be excluded from, or may participate less in political representation and wage labour. However, we lack sufficient information on this crucial question. Yet, we do have evidence in the case studies on the tendency of young men to embrace the "expectations of modernity" and independence associated with (the hope for) mining-related employment (see case Peru, and Maconachie on Sierra Leone, this volume; generally Ferguson 1999). Additionally, there can also be divergent interests of different age-groups. People within a single household may thus opt for different strategies with regard to mining, which explains the observation that mining conflicts not only divide communities, but often families, too. To uncover such dynamics of competing interests would however need more in-depth ethnographic research.

Although we observe relatively weak involvement of (neo-)traditional leadership and institutions at the regional level, there are other alternative arenas that provide space for knowledge exchange, debate and organization of protest activities: peasants' associations (Peru, India, Philippines), workers unions (Zambia, India), regional NGOs with grassroots participation, and sectors of the Catholic Church (Guatemala, Philippines). In these forms of organization, membership is possible at the local level, but they can also be perceived as external allies. Therefore, a clear distinction between "local actors" and their "allies" is difficult to draw and often it remains unclear as to how far the collective action relies on local initiative alone, or comes in combination with, or is largely encouraged by, outside actors. Certainly, the latter depend on previously established networks of trust to have an impact on the local level. Obvious examples of "external allies" are organizations and NGOs that have their headquarters in the capital or even abroad. Related to this is the question about who is to be considered as "affected". As discussed in Chapter B, it is part of corporate strategies to limit the officially recognized "zone of influence" to the most direct impacts such as land acquisition. This is supported by narrow conceptions of legally established land ownership. However, we see in several cases that self-perceived affectedness can go far beyond that, to include, for instance, downstream villages and the population of a provincial capital.

At the same time, the categorization as either "directly" or "not directly affected" can become a divisive discourse between different communities, even where their interests are otherwise rather homogeneous.

For a simplified positioning of actors regarding impacts and alliance building, we propose a model of three levels of "affectedness". Imagined as concentric circles surrounding a mining site, these levels are: *1) affected by early, direct and strong impacts,* such as land acquisition, resettlement, compensation and employment—the core target for CSR strategies; *2) affected by delayed, less direct impacts*, such as pollution or depletion of water sources—denied and ignored in corporate consultation/compensation; *3) affected by indirect impacts* on the socio-political, economic and environmental conditions in a larger region. In extension, this can include global issues such as climate change, nuclear proliferation and social justice, which potentially concern anyone, even if in a rather abstract way (i.e. as "consumer" or "citizen"). Level one and two can be considered as being local; however, this labelling is subject to different discourses and therefore contested, while the third—subdivided in regional/national/international level—includes actors who may be perceived as possible external allies from a local perspective.

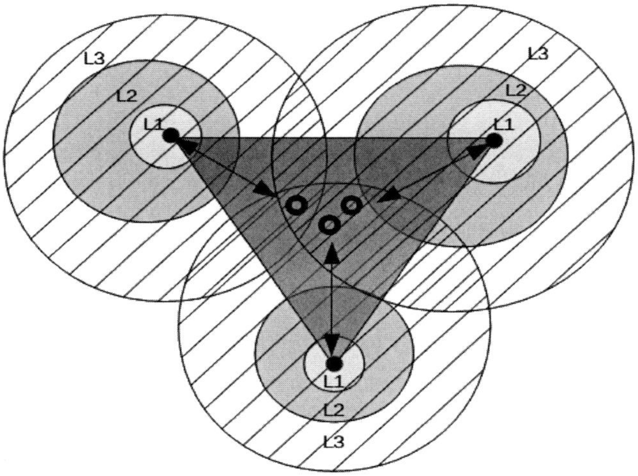

Fig. 3: Three "levels of affectedness" surrounding three mining sites, and alliance building space in between.

Figure 3 illustrates the three levels of affectedness surrounding three hypothetical "mining sites" (*black points*). The latter can as well be exploration areas or other projects against which resistance is building up, possibly directly or indirectly related to mining (e.g., power plants, transport infrastructure). The number of three mining sites is arbitrary, but useful to imagine a "space for alliance building" (*triangle*). Thereby, distance from the core points correlates with decreasing intensity of impacts (i.e., affectedness), while potentially involving much larger numbers of people who may participate only occasionally in any resistance-related activities, but can have important leverage at certain times (e.g., by influencing electoral politics). At the centre of the triangle, solidarity groups and NGOs (*black circles*) function as rallying points for support activities and knowledge exchange. The challenge of framing coalitions is to link all these levels to find a minimal common ground by aligning diverse motivations and priorities, to deal with frictions and distrust by building accountability and discourses that produce legitimacy. These are problems of collective action, too, but the scale and quality are different: we presume high heterogeneity and the need to adapt neo-traditional institutions and to complement them with new ones in order to cope with the challenges. This model also has the advantage of illustrating how resisting minorities at the local level can break up the perceived marginalization through association with other struggles that involve larger numbers of people.

Returning to the case studies, we can now quality of coalitions according to the three levels of affectedness and differentiate actor-groups with diverging tendencies in the resistance/negotiation-dynamics. The separation of local actors in a first and second level (L1 and L2) is most useful where it is created by differential treatment in CSR- and ownership-regimes. In the absence of the latter, one local level (L1) next to the regional/national/international level (L3) is sufficient. For instance, the case of Mopani (Zambia) illustrates that in absence of strong local and regional organizations or grassroots movements (L1/L2), international NGO coalitions (L3) can themselves become the main actors. Thereby, there is a latent possibility that local groups do take up the issue, as the case shows both for Zambia and Switzerland. The Niger case shows relatively weak NGO support (L3), but a strong tradition of warfare—local resistance thereby gets caught up in a discourse of terrorism, which again makes it more difficult to mobilize international support. The campaign against the Jabiluka mine (Australia) is the best example with a clear leadership of an indigenous organization (L1), supported by external allies (L3). Even more, by strategically linking up with affected people at the other end of the nuclear chain (Chernobyl, Fukushima), different local foci of struggle (L1) become related to each other. However, other indigenous groups in the surroundings who do not have the same status as "traditional owners" (L2) appear to be less involved. For El Cerrejón (Colombia), communities that need to be resettled (L1) are clearly treated differently by the company than others (L2), though there is not sufficient information available to differentiate between them in detail. However, the successful campaign to stop a specific expansion project was grounded in the temporary linkage of (sectors of) all three levels into a broad ad-hoc coalition, with strong support from the regional, national and international levels (L3). In the case of Ambuja (India), the division lines do not correspond to the same pattern, but rather to employees of different social and legal status (permanent/contract) and peasants; therefore, a distinction between L1 and 2 seems not useful. In this case, the success for organizing lies in aligning contract workers and peasants and the involvement of international trade union federations (L3). The Guatemala case shows the simultaneous emergence of local resistance (L1/L2) around the Marlin mine and a national coalition against large-scale-mining (L3) in a reciprocal relationship, with international support groups following their lead. For Val Medel, the first level can be defined as the officially registered electorate of the commune, while those of neighbouring villages (L2) were not involved. National environmental organizations (L3) played a minor role. Finally, the cases of Tampakan and Las Bambas are the ones where the differentiation between the first two levels proves most necessary to illuminate the conflict dynamics. The distinction between local communities that benefit from compensation agreements (L1) and those excluded from it (L2) seems strong and internalized to a good degree; this finds expression in divergent local strategies and in limited possibilities for alliance building with external actors (L3).

Summing up the results of this section, we observe that the conventional theories on collective action outlined in the introduction seem to be useful mostly to explain the barriers and hindrances for local organizing in the case studies, which do not show very favourable conditions: mostly, we find that the interests of local actors are highly heterogeneous and that there is a lack of neo-traditional institutions and leadership that could represent local communities autonomous from corporate and state influence. The scale of today's mega-mines and the power of corporations seem so overwhelming that local decision-making institutions have to face serious challenges to cope with them. Under the influence of corporate co-optation strategies, local people may tend to negotiate for benefits; especially, where the quality of common-pool resources and the authority of local institutions have been eroded due to earlier interventions. In such cases, common land ownership appears to be open for corporate appropriation, even if it is legally recognized and protected (e.g.,Las Bambas).[8] However, we argue that campaign coalitions that link up and define common interests of important sectors at several levels are most successful for collective resistance against mining and the way it is carried out. This both in cases where local resistance is rather weak at the level of (neo-)traditional institutions (e.g., El Cerrejón, Tampakan), as well as in cases where the latter are involved in resistance and can be further strengthened though alliance building.

3.2 Reasons and Demands

An overview on the diverse reasons identified for collective action, and the demands formulated thereby, shows that a simple picture of "communities vs. corporations" misses the point. Rather, we find diverging and sometimes contradictory priorities of different sectors, as well as significant changes over time. Outright opposition against mining operations is rather an exception than the rule. In general, the major conflict issues identified by Franks et al. (2014) mentioned above correspond well to the present cases. In *Table C – d*), the main motives are summarized. Frequent complaints concern insufficient compensation (especially, regarding land acquisition and resettlement) and the availability and quality of employment. Often, these come together with environmental concerns; especially, about contamination, related health impacts, and competing water use. Process-related demands, such as lack of prior consultation or the insistence on the right to give or withhold consent (FPIC) on a mining project are strongly relevant in Australia, Colombia and Guatemala. As mentioned, the reasons that motivate collective action have more to do with the emic perception of (potential) impacts or unfulfilled promises and expectations than with objectively measurable facts. For instance, while actual and potential environmental problems are relevant in all the cases, they are not perceived equally by all actors; especially, in advance of the implementation of a project, they are difficult to evaluate for local people as well as for scientists.

In the following, we focus on process-related demands and underlying motivations that have to do with (*C – e*) prior negative experience and (*C – f*) the exchange of knowledge about negative impacts of mining or skills for organizing resistance, and finally with (*C – g*) the availability of alternatives.

With regard to process-related demands and decision-making rights, we can state that the emergence of strong indigenous peoples' movements has brought an empowerment for those who make use of new legal frameworks (see Part I – 2). As rights hereby are limited to specific ethnic categories, this implies an identity politics. The indigenous peoples' claim to their right of self-determination can be used as a powerful strategic option against mining companies. However, this has not happened everywhere, even if the conditions were to be favourable. In the Peruvian case, local people neither self-identify as indigenous Quechua, nor do they strongly insist on the

8 In contrast, mining companies can have serious difficulty to acquire privately owned land if individuals refuse to sell, resulting in a de facto veto right even for a minority, as long as the company cannot resort to the state for requesting forced acquisition (as for e.g., in Rosia Montana, Romania, see Egresi 2011: 63).

far-reaching collective ownership rights they enjoy under the category of "peasant communities". In Chhattisgarh, instead of using the uncertain, weakly recognized and potentially divisive *dalit* and *adivasi* (indigenous) identities, the reference is to a more inclusive "Chhattisgarhi" regional identity. This forms the base for a peasants/workers-alliance with demands that appeal to international working-class solidarity while simultaneously transporting environmental concerns and the protection of peasants' livelihoods. In the African context (Niger, Zambia), the concept of "indigenous peoples" is generally more difficult to apply (see Haller et al. 2007). The call upon an idealized role of "the indigenous" as a "vanguard" for the protection of the environment, biodiversity or against climate change plays a marginal role in most of the cases, though this could be a powerful way to link local problems to global discourses and open up possibilities for global alliances. In Tampakan, indigenous status is used at the local level as an asset to negotiate concrete benefits rather than to relate to global discourses. One could even accuse indigenous Bla-an leaders of "free-riding" on the success of the indigenous peoples' movement in the Philippines, to which they apparently did not contribute a lot; rather, these rights were endowed to them from above. Contrarily, the Mirarr in Australia understand and portray themselves as custodians of territory that contains a dangerous element (uranium), better left in the ground than used for risky nuclear technology. They are consciously forming part of the global indigenous peoples', anti-nuclear and environmentalist movements, in a credible and not opportunistic way. The demand that the company and the state respect their right to veto was at the core of their resistance against the Jabiluka mine. In Colombia and Guatemala, an underlying demand that is transported along with more specific, mining impact related demands, is that FPIC should be mandatory (instead of the weaker "consultation" requested by ILO Convention 169).

Where we do find strong collective resistance at the local level, an "environmentalism of the poor" (i.e., defence of a subsistence base that depends on a healthy environment and access to common resources) is certainly an important factor, often together with a spiritual significance of nature and landscape (e.g., cases Australia, Guatemala, India, and Colombia). The resistance of the Mirarr can be interpreted as an attempt to revive and defend indigenous lifestyles and values which are in a deep crisis due to the overwhelming presence of dominant Western society as expressed in the mining town of Jabiru. However, there are also cases where these conditions are given too, but appear as secondary concerns that do not hinder people from seeking short-term benefits (e.g., Peru, Philippines). As the size of today's mega-mines escapes imagination, it is well possible that people underestimate possible impacts in advance of the construction. The example of Tampakan is instructive, where locals only realized the extent of the planned mine when NGO staff showed them a three-dimensional model. Otherwise, prior negative experiences seem less relevant at the level of the local population than for the mobilization of wider sectors of the public against the project. In Las Bambas, information about mining impacts provided by NGOs is welcomed only by a critical minority. Though the importance of prior negative experience, and/or knowledge transfer about potential impacts for early resistance is undeniable, these may not reach all sectors of local communities equally, and may thus further contribute to social tensions and divisions. At the same time, the proportion of people who prefers one or the other priority may shift quickly when new knowledge or experience is available that seems useful to assess the situation.

The availability of both subsistence and market-oriented economic alternatives is not just a given, but their strengthening and development could be a strategic element of resistance against mining; for example, by promoting cultural heritage and eco-tourism and building up of cooperatives.[9] The Mirarr in Australia have a strategic focus on culturally appropriate and non-discrim-

9 This has been an angle of work of many NGOs, but appears heavily co-opted by corporate "sustainable mining" schemes in several case studies (see Chapter B).

inatory education, which may enhance positive attitudes towards traditional subsistence activities among the young generation. Otherwise, we do not find much evidence that self-organized initiatives for alternative development plays an important role as a long-term strategy for local people here (different to what Walter/Martinez-Alier (2010) describe for Argentina). Likewise, as mentioned above, a spiritual signification of the landscape is not always put forward against mining. Also, with regard to "non-human beings", an emphasis on negotiation is possible.[10] In the Peruvian case for instance, the veneration for "Mother Earth" does not prevent people from selling the land to be destroyed. Contrarily, in Colombia women's groups refer to the protection of "Mother Earth" as a main motivation for resistance against the El Cerrejón mine. This suggests that spirituality needs to be actively mobilized and revived as positive markers for self-identification to become effective in the new context. The role of cultural discrimination and the influence of (e.g., evangelist) missionaries and new religious tendencies in undermining traditional spiritual beliefs and practices, and thus their potential as an ideological base for common action, would be an interesting topic for further research (Haller et al. 2007).

3.3 Means and Methods

The range of methods to express collective action employed in the case studies (*Table C – h*) is quite broad, though certainly not complete. We find a general preference to non-violent means, except for two cases with a tradition of "tribal warfare": the Tuareg of the Sahara desert have been famous warriors for a long time, and in the Philippines case, a militant minority has called upon a tribal war and attacked company staff. Knowledge and experience about the use of violent strategies, linked to a prior conflict history as well as to activities of guerrilla groups, would be available also in Chhattisgarh (India), Colombia, Guatemala and Peru, but have so far not been taken up. Rather, a preference for lawsuits and other forms of legal action can be observed, which may reflect the general strengthening of a human rights discourse over the last decades. The availability and recent extensions of various complaint mechanisms set up by international institutions, such as the World Bank, OECD, the International Labour Organization or the Inter-American Commission on Human Rights are well represented in the case studies. In some cases (e.g., Colombia, Australia), local groups have built up considerable know-how about them. However, these mechanisms are difficult to use without support by specialized NGOs, which also may have their own priorities. Long-lasting lawsuits imply a time-consuming engagement on a territory defined by others, which can cause fatigue, confusion about one's own position and a sense of alienation. Finally, the prevalence of non-violent means is not a one-way road, and some people may take up violent tactics if they are disappointed by the results.

In fact, the improvement of legal instruments and complaint mechanisms does not seem to be sufficient to protect local people against rights violations by corporations, as their implementation is generally deficient. This is the case, for instance, in the Labour Court's decision in Chhattisgarh (delayed by appeals), and the verdict of the Inter-American Human Rights Commission in Guatemala (not implemented by the national government). As mentioned before (Part I – 2), the problem is that most of the international mechanisms are limited by their non-binding character, and some of them by their bias towards economic development. But also the use of binding national law is often hampered by distrust in a judiciary riddled by corruption. Indeed, in all the case studies where coalitions engage in public campaigning, legal action is accompanied by various forms of

10 Many companies have become aware that the issue of sacred sites and ancestor's spirits roaming the land is a sensitive one, and have adopted guidelines for heritage protection and cross-cultural trainings for their staff (see Chapter B).

Fig. 4: Assembly at a protest against the European Gold Forum in Zurich, Switzerland, 17 April 2013. Photo: Philipp Eyer.

protest to publicly "blame and shame" companies and governments. As part of wider campaigns, petitions, public forums, gatherings, protest pickets and marches are commonly used to raise bargaining power of the affected groups.

An important strategy adopted in explicit mining conflicts is direct action taken by the protest groups in the spirit of civil disobedience. As a kind of "third way" between legal action and protests that call upon the authorities, on the one hand, and violent action, on the other, non-violent direct action can be highly effective to mobilize support and create high visibility. The road block seems to be the method of choice in mining conflicts. It is arguably the most direct expression of local people's discontent; especially, if joined by large numbers and maintained over prolonged duration (as in Marlin), and forces the authorities to take position by assuming either a mediating role to de-escalate the conflict, or by forceful intervention, taking side with the corporation's interest in uninterrupted production. As mines and construction sites need continuous flow of materials, the closure of access roads can become very costly for the company. Already the threat of direct action can be a powerful tool to enter into negotiations and to gain leverage.[11] Regarding the conflict dynamics, early direct action is relevant as a negotiation tool in Tampakan. For Jabiluka, blocking the road to the mine construction site came as the culmination point and as an expression of the previous alliance building, to maximize public attention to the irrevocable "no" of the Mirarr to the mine. However, such a strategy can be difficult in contexts where minimal human rights standards are not met and excessive force is used to repress non-violent direct action. In Colombia, the traumatic conflict history is cited as a major reason for being reluctant about direct

11 For instance, the quality of mining agreements between communities and companies in Australia is remarkably higher if the negotiations involved an indigenous organization with a credible track record of direct action (O'Faircheallaigh 2008).

blockades of the El Cerrejón mine.[12] But there are also other forms of direct action, which are less risky. In the Indian case, for instance, spontaneously assembled crowds express their outrage by surrounding and verbally confronting managers and state officials. Similarly, geologists doing exploration in certain areas risk being driven away by angry villagers, as happened in the case of the explorer company Uranio AG in Jujuy, Argentina (see Niederberger, this volume). Such "minor", spontaneous events mostly escape documentation, but certainly have the potential to leave a profound impression on the targeted individuals.

The act of self-organizing in parallel to discredited official structures can be interpreted as strategic refusal to participate in CSR-related schemes under unequal power relations. In Guatemala, for instance, the revival of an indigenous regional institution that does not negotiate with the corporation, the organization of a local referendum and the setting up of a "peoples' tribunal" on health impacts of the mine show this characteristic. Similarly, the resistance in Colombia builds on a campaign coalition, consisting of diverse actors and lacking centralized leadership, which does not enter into negotiations, unlike many community leaders.

Another important development is the availability of new communication channels; especially, the so-called "social media". Again, the Mirarr stick out with a continuous, high quality internet presence and professional media relations. Elsewhere, the impression is that these means are relevant for campaigning in urban settings and at the national and international level (e.g., Colombia, Philippines). However, they seem to be run mostly by urban support groups and not by local people themselves, and are often short-lived and campaign-specific. For local-level organizing, the full potential is not tapped due to lack of access, but this may be changing fast.

It has to be noted that the information available to us is heavily biased, based upon public visibility, and thus the most spectacular and best-documented expressions of dissent or support. More covert forms of everyday resistance, the "weapons of the weak" identified by Scott (1985), mostly escape from our description. Still, one can expect to find subtle forms of subversion; for instance, with regard to development programs, easily to be interpreted by corporate and state actors as proof for local peoples' "backwardness" and "lack of education". For the Mirarr, a certain disinterested and careless attitude about compensation benefits has been observed, possibly as an expression of the fact that they are based on the destruction of the cultural and spiritual landscape that cannot be restored. Another characteristic of passive resistance, as described by Scott, is to wait for a drastic, temporary shift in power relations due to some unusual circumstances, commonly experienced by the ruling elites as a sudden and inexplicable outbreak of rebellion. Such an attitude of passive resistance may be alluded to in the declaration of a peasant leader, cited in the Las Bambas case study, about the need to "wait for the company to make mistakes" for mobilizing people. To uncover expressions of this kind of resistance in everyday life needs long-term field research.

3.4 Local People's Strategies: Conclusion

Most local mining protests appear to be multi-vocal expressions of discontent about its terms and conditions rather than against the activity itself. If there is anything like an underlying common demand, it could be the right to decide on these conditions. In extension, this process-related demand includes the right to refuse mining altogether, if the desired conditions cannot be met. The "tendency toward economic exclusivity" of large-scale mining (Himley 2013) leads to high incentives, or potential benefits, for collective action that aims at reaching compensation agreements,

12 Karmen Ramírez Boscán (this volume), personal communication.

as it may well be the only way to secure at least a minimal participation in mining revenues. Also, the social and environmental costs of failing to organize collectively are very high. Although broad and unified consensus demands are hard to reach in highly heterogeneous settings, this can nevertheless support expressions of shared interests and aspirations of important sectors of the population.

As an assumption, one could suggest that environmental concerns strengthen fundamental resistance (as in Ranger/Jabiluka and Marlin), while demands for better benefit-sharing tend to be sorted out in negotiations (as in Tampakan and Las Bambas). However, this does not hold where the existence of an industry is unquestioned and demands focus on the mitigation of environmental impacts (e.g., Mopani, Niger, Ambuja) and/or a specific expansion project (El Cerrejón). Here, economic dependence on mining revenues restricts too radical demands and the power relations are such that a closure is unthinkable. Rather, it would be the market that forces a company to shut down a mine, if relative prices drop.

At the level of directly affected communities, which benefit from compensations, there may be a preference for negotiations, while communities excluded from negotiations may use resistance to prove that they are relevant stakeholders who cannot be ignored. These tendencies can, in part, be explained by different perceptions and evaluations of the expected impacts prior to construction, and the graded character of different short-term benefits and long-term impacts themselves. Mining typically brings to the fore multiple tensions between collective and individual, as well as between short-term and long-term livelihood strategies. While mining may be a promising option in the short run, especially for individuals well placed to participate in its gains, it often proofs disappointing in the long run and can have devastating impacts on the collective level. Attitudes are thus subject to change according to experience, and local people who first welcomed the expected wealth may shift to an "anti-position" if conditions and promises are not met, or when they become aware of the long-term damages. However, a return to commons-based livelihood strategies may become very difficult once they have been replaced by a mining-driven wage-labour economy. In this regard, it also has to be remembered again that common ownership institutions have been, and continue to be, systematically undermined by colonial and post-colonial states in most parts of the world. In certain local settings, they may thus be eroded to a point that local people perceive that the efforts of upholding them can no longer be justified by the benefits. We view these as relational developments that need to be analysed considering the dynamics of institutional change and local actors' shifting perceptions of costs and benefits.

At the same time, even in cases where local resistance is rather weak, there are inspiring examples of broad coalitions linking the local to the regional, national and international levels, posing fundamental challenges to large-scale mining as carried out under the current capitalist regime—often labelled as having an "anti-mining" position. As this label is used by the mining lobby to push critics into a radical corner, some NGOs insist that they seek to strengthen informed participation in decision-making instead of being "anti-mining" (e.g., Las Bambas); others, however, do not seem to have a problem with it (e.g., Tampakan). Certainly, a clear "anti-position" has the advantage of a simple common denominator for broad alliance building. To participate in such coalitions can help local groups to get leverage in negotiations. However, the possibilities for alliances with local actors may be hampered if the latter tend towards pragmatic negotiations rather than to principled opposition.

The presence of conflicting and shifting attitudes within local populations poses several problems from the perspective of potential external support groups. In view of the often devastating impacts caused by large-scale mining, the temptation is to equate "successful strategy" with resistance that is strong enough to stop a planned mine or even to close down a running one. But if one seriously considers the demand for indigenous self-determination, or the primacy of local people to decide over the resources on which they depend, then one has to accept that they can set other goals: jobs, concrete-block houses, electrification, roads, and on the personal level, to obtain a good compensation for buying a car or a flat-screen TV. Successful strategy then means anything that helps to reach that goal.

The recognition of the right to decide over the use of commons, as expressed in the concept of Free, Prior and Informed Consent (FPIC), is generally understood as empowering people to protect the environment they live in. This is confirmed in the Australian case, although the younger generation might change their ideas, especially under the influence of well-crafted CSR strategies. The Philippine and Peruvian cases, however, show a tendency on seeking short-term benefits and discounting of long-term costs, which creates serious challenges especially for support groups that advocate environmental protection. Proponents of indigenous rights insist that such decisions need to be respected. Consider the comment about an agreement between the Aboriginal Martu community and an uranium mining company in Australia, opposed by conservationists: "Free, Prior, and Informed Consent is applicable to all external entities, not just companies, and communities that choose to develop their natural resources can be strong allies for companies under fire from environmental groups."[13] Although the debate is not new,[14] it has become more pressing with the legal changes and the turn to CSR, especially in the mining sector. The right to self-determination means that indigenous peoples can and should define their own development paths. External actors need to take their material concerns seriously and not try to find the "ecologically noble savage" or assume that resistance is a "natural" or self-evident reaction to mining—as much as companies should accept that "No means No" and that not everyone strives at a "modern" or "Western" lifestyle.

However, the case studies show that even where FPIC is formally maintained, there are good reasons to question how "free" and "informed" decisions can be under the observed conditions of massive corporate "manufacturing of consent". Given that communities are often divided and the companies tend to consult only a few representatives, it remains doubtful if a so-called "consent" reflects a broad consensus, a majority opinion, or just the will of a local elite. Additionally, there are changing attitudes, based on social learning from (often negative) experiences, which escapes a one-time consent. The feeling of being betrayed and the breach of earlier agreements may dramatically change the picture, as the most recent events in the Peruvian case illustrate.[15] Which are the relevant steps that ask for local people's consent, and can they come back to it if a silent minority becomes a loud majority due to broken promises and disappointed expectations? Furthermore, one can point out that the right to decide should involve all the affected people and not be limited to those whose land is required. Considering these objections, we could say that a successful strategy increases the bargaining power of the affected population to make free decisions on an equal basis with the company, fully informed about the short- and long-term benefits and costs, aiming at empowerment and strengthening resilience. This process should not be distorted by dependency, elite-capture, patronage and corporate divide-and-rule, which almost inevitably result from an early distribution of benefits and jobs. Given that such an ideal situation is far from the realities in most parts of the world, there are still ample reasons for legitimate opposition against mining projects.

13 Firstpeoples.org: "Martu Welcome Uranium Development", <firstpeoples.org/wp/martu-welcome-uranium-development> accessed 12.05.2015.

14 See, for instance, the debates in the German support group infoe in the 1980s, with some taking an eco-centric position to limit support only to those indigenous groups who choose to protect the environment, being criticized as paternalism by others (Stüben 1988).

15 See "Update" case study Las Bambas: in September 2015, heavy protests escalated, after the project had been sold and its design was changed by the new owners, causing massive truck traffic and increasing worries about contamination risks.

PART III: Comparison

D) Conclusion and Outlook

In this final conclusion of the case study comparison, we take up the main lines of discussion and identify a number of questions that we think are crucial to follow up in the future. As a main strategic problem, we considered the unequal power relations between local people and corporations in negotiating the conditions of access to mineral ores. In extension, the question is how to prevent their exploitation, if the impacts cannot be justified by the benefits. The turn to CSR and neoliberal resource governance poses further challenges to local communities who find themselves increasingly subject to corporate co-optation and divide-and-rule strategies that exacerbate social divisions, while at the same time creating new but very limited options to participate in decision-making and negotiate mining-related benefits.

An implication of our findings is to stress the limitations of framing the normative debates on resource governance as a choice between voluntary standards and legally binding instruments. Voluntary standards can prompt useful processes to bring different levels, branches and subcontractors of a corporation to comply with CSR policies framed at the top level. The case studies show extensive use of various legal means and grievance mechanisms, as part of mixed strategies, to visualize dissent and build public pressure on corporations and governments. However, they also confirm the criticism about voluntary mechanisms being far from sufficient to address the problems caused by mining, as they typically lack monitoring and enforcement. Indeed, they work best where they are used to ensure compliance with legally binding norms and contractual obligations. And although it is certainly not wrong to call for stricter, legally binding regulations, the cases show that where the latter exist, corporations can capitalize on the fact that their implementation is often vulnerable to economic pressures and corruption.

The shift to enhanced CSR policies emerged in a historical "boom" phase of high price levels and rapidly expanding production. Considering the epilogue of some of the case studies[1] and the current downward trend in commodity market prices, there are doubts in how far our analysis of corporate strategies will still hold for the future. Assuming a long-term lower price level, competitive advantage is secured by radically cutting costs. This includes the closure of mines with relatively high operational costs, irrespective of the CSR claims made earlier. Therefore, Ferguson's (2005) warning that the model of the privately militarized extraction enclave frequently encountered in African conflict zones may become the new norm merits attention, even though we could not confirm it from the case studies in this book. The major TNCs are increasingly challenged by semi-governmental companies from China and smaller companies with investors from newly industrialized countries such as India, Brazil and Russia; companies, which are often willing to take higher risks in social and environmental issues and are more difficult to counter by conventional NGO campaigning.

1 In particular, the post-merger shift of Glencore away from Xstrata's "greenfield" investment strategy and subsequent pull-out from the Las Bambas and Tampakan mine projects (see updates in case studies), as well as the decision to put the Mopani mines on hold for 1.5 years in order to cut operation costs by one third, due to low metal prices (see *Neue Zürcher Zeitung*: "Glencore zieht die Notbremse", 7.9.2015).

In general terms, we consider that CSR discourses, improved environmental management and more efficient technology can only insufficiently address the tendency that mining is becoming increasingly destructive. The fact that the exploited mineral deposits are of decreasing quality means that for the same amount of final product, more land, forests and rock have to be destroyed, using more water, explosives, chemicals and energy, causing more pollution and greenhouse gas emissions;[2] further, the expansion in hitherto unexplored and undeveloped regions brings that mines are more often located in or nearby protected areas, risky geological terrains, conflict regions, and most of all, indigenous peoples' territories. In combination with other factors, this means that the necessary investments in infrastructure and operation costs of increasingly large "mega-mines" are of tremendous sums, further accentuating economic inequality and exclusion, dependencies, and distribution conflicts. These trends reinforce each other in positive feedback loops, indicating that mining conflicts may become more frequent in the future, as long as market prices do not completely collapse due to a global economic crisis. Irrespective of speculations on future developments, to focus exclusively on the improvement of legal mechanisms might be misleading. Likewise, to frame the problem simply as one of "externalized costs" of social and environmental damages that need to be "internalized" in the price of products through market regulations seems insufficient. Although both approaches can certainly provide important elements to check the power imbalances and socio-environmental destruction associated with corporate-driven globalization, we argue that this depends also on deeper, socio-cultural transformations of moral values and attitudes.

In the following, we combine the findings from the case study comparison. In *Figure D*, the conflict trajectories are laid out schematically, roughly in relation to each other, over the mine-cycle stages covered in the case studies.

The four corner-fields are:

- *Exploration and planning/construction stages:* (A) top-left: high resistance, low negotiation; (B) bottom-left: low resistance, high negotiation.
- *Operation and expansion stages:* (C) bottom-right: low resistance, high negotiation; and (D) top-right: high resistance, low negotiation.

Divergent tendencies are included only where they are strongly accentuated (cases Tampakan, Philippines and El Cerrejón, Colombia). The centre line marks the absence of clearly identifiable collective action. From centre to top, "resistance" increases; from centre to bottom, "negotiation" increases. These two dimensions, treated so far rather as an ambiguous pair, are now set as two poles. This does not mean that they represent exclusive positions. Distinct from the indicated predominant tendency, there are still individuals, groups and sectors that follow divergent strategies for shorter or longer periods. "Negotiation" is understood as the active participation in consultation and benefit-sharing; or more generally, a positive attitude towards mainstream notions of development; "resistance" refers to a manifest attitude of opposition, expressed through various forms of collectively organized (protest) activities. We consider resistance as "strong" especially in those cases where it is based on values that are conceived as irreconcilable with mining development, and where it is actively supported by large sectors of the local population.

Because the information given in the case studies is mostly focussing on just one or two stages, the qualifications are incomplete and tentative. Nevertheless, the overall picture is sufficient to identify a range of simplified scenarios, illustrating the predominant tendencies of resistance/negotiation in local people's strategies over time.

2 See Part I – 3) for a general discussion of environmental mining impacts. For the phenomenon of decreasing quality and availability (or "peak minerals"), see Mudd (2007), Mason et al. (2011) and Prior et al. (2012).

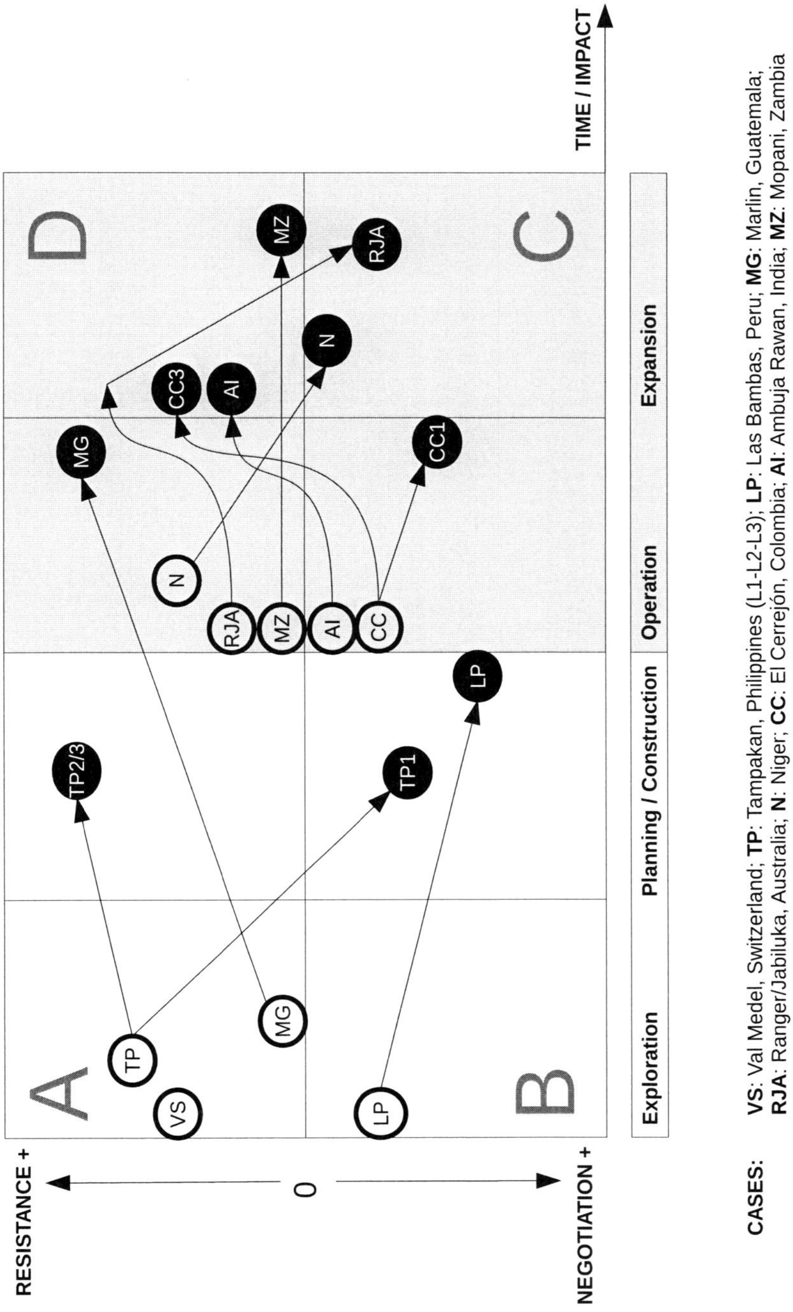

Fig. D: Simplified resistance/negotiation scenarios over mine-cycle stages.

Fields	Scenario	Characterization
"Basic" scenarios:		
0 → D	"Imposition → impact → resistance" (Marlin, Guatemala)	Corresponds to impact-resistance hypothesis, which is, increasing resistance in reaction to corporate imposition and negative impacts, but too late to stop the mine from operating.
0 → 0	"Long-term dependency" (Mopani, Zambia)	Consistent corporate imposition finding weak resistance in operation stage.
B	"CSR works" (Las Bambas, Peru)	Corporate co-optation strategy meets local people's negotiation strategy. Participation in development as main motivation.
"Early resistance" scenarios:		
A	"Spill-over effect triggers early resistance" (Uranio AG, Argentina)	A strong anti-mining movement at the national level fosters early resistance against exploration.
A	"Locals decide" (Val Medel, Switzerland)	The institutional setting (need for local referendum) and initial resistance show the conflict lines and frame the conditions for corporate engagement during early exploration stage.
A → A/B	"Corporate co-optation meets divergent strategies" (Tampakan, Philippines)	Initial resistance at the local level turns to negotiation under the impression of a strong bargaining position and strong corporate co-optation, while resistance at regional/national/international level increases.
"Dependency to resistance" scenarios:		
C → C/D	"Coalition challenges corporate co-optation" (El Cerrejón, Colombia)	Divergent developments, with local actors subject to strong corporate co-optation, while an ad-hoc coalition puts up strong resistance and succeeds to stop an expansion project.
0 → D	"Unexpected outbreak of resistance" (Ambuja, India)	Sudden increase of resistance with the character of a rebellion against long-lasting injustice and accumulated impacts.
D → C	"Strong resistance blocks imposed project" (Ranger/Jabiluka, Australia)	In reaction to strong, value-based resistance, the corporation draws back during construction stage of replacement for an existing mine. After that, both sides change strategy to negotiation.
"Deterioration" scenario:		
D → 0	Decreasing resistance in "resource war" scenario (Niger)	An initially successful rebellion deteriorates to an open "resource war": little room for civil society and non-violent forms of resistance in the context of "anti-terrorism" discourses.

Table 1: Characterization of scenarios as indicated in Figure D.

Several conflict trajectories illustrated in Figure D reflect the emphasis in corporate strategies, and changes thereof, as discussed in Chapter B: in Australia, the shift of the company away from imposing the Jabiluka mine to respecting the lack of consent; in Colombia, Peru and Tampakan, the success of community-relations programs in winning support, and arguably co-opting "directly-affected" local actors through an emphasis on negotiation and participation.

The scenarios illustrated in Figure D are characterized in short form in *Table 1*, emphasizing certain aspects we find most notable out of the complexities and singularities of each case: to begin with, we can state that the simple assumption of increasing impacts being met by increasing resistance can only be confirmed for Marlin (Guatemala), where open resistance starts with land acquisition and intensifies during construction and throughout the operation stage. This is probably the example that is most close to the common notion of a fight of "local communities versus global corporations", whereby the latter gets into some trouble but ultimately prevails. Contrarily, we find that there is little resistance against Mopani (Zambia), the longest operating mine, irrespective of massive negative impacts. This confirms the hypothesis that long-time operating mines are difficult to challenge, mostly due to deeply rooted economic dependency and lack of alternatives. A third "simple" scenario can be identified from the case Las Bambas (Peru). This is a prime example that "CSR works": local people primarily search to participate in mining development and benefit from their relatively good bargaining position to receive the best possible share, rather than to defend their common resources, which seem to offer few perspectives in the short run. This meets a strong CSR strategy, with corporate "expectation management" and the co-optation of public discourse that effectively prevents the establishment of a space of autonomous deliberation from where collective resistance could arise. These three cases correspond well to an intuitive understanding of mining conflicts and can serve as baseline scenarios, from which the others vary for different reasons.

Early resistance and ungovernable conflicts: From a campaigning perspective, the scenarios in field B are especially interesting. High resistance in early stages of the mine-cycle has a good chance to succeed in preventing unwanted developments, using a window of opportunity where local actors have relatively high bargaining power. The question is, why do people react in that manner already during exploration, when negative impacts are still minimal? For Tampakan, Philippines, we can identify a background of colonial and post-colonial conflict history as reason to distrust any activities of external agents. Also in Jujuy, Argentina, this is expressed in immediate direct action at the very first contact with exploration engineers (see chapter on Uranio AG, this volume). Here, exploration activities are not seen as unproblematic, but as the first step in a process of expropriation that stands in continuity with past experiences. In both cases, a "spill-over effect" from strong anti-mining movements at the national level further supports a strong reaction. In the third case in this category (Val Medel, Switzerland), "resistance" means something very different, as it is framed in a formalized institution of democratic deliberation. It is the need for a local referendum, which makes local residents aware about the issue. Resistance is expressed through the mobilization of dissent on a face-to-face level in the community, challenging the proposal of the communal council.

For the early stages of mine development, the comparison between the cases of Las Bambas and Tampakan is especially instructive: both show a similarly strong corporate co-optation strategy; however, with different outcomes. In Las Bambas, this strategy "works well" from a corporate perspective. In Tampakan, early resistance motivated the company to design CSR policies that succeed in turning former opponents into "participants" in negotiating benefits through a corporate-controlled process. However, this is limited to a dominant part of the "privileged" sector of recognized beneficiaries with high bargaining power at the first level of "local affectedness" (L1), while a minority continues to resist. At the second level (L2), there is strong resistance coming from people who fear negative impacts but are largely excluded from potential benefits. They have an incentive to use protest as a strategy to get the negative impacts and related demands recognized, and to be included in the first circle of beneficiaries. This may be hard to impossible to achieve, as it potentially benefits much more people and thus comes with prohibitively high costs for the company. The latter thus has to take an uncompromising position, which may in turn strengthen more principled "anti-mining" positions among local people. The tension between the two local positions (L1 and L2) can easily turn into a shift from community-company to community-community conflicts (see Arellano-Yanguas 2012; visible as emergent tension also in Las Bambas). At the third level, a broad coalition challenges the project on the regional/national/international scales, based on values that can be subsumed as "environmental justice". Overall, the Tampakan situation becomes almost "ungovernable" for the company, to a point where it fails to reach the "social licence to operate" and has to put the project on hold.

Dependency to resistance: Another cluster of cases can be distinguished by a marked increase of resistance after long-term operation of a mine. These cases (Ambuja Chhattisgarh, Ranger/Jabiluka Australia, and El Cerrejón Colombia) contradict the hypothesis that increasing economic dependency and lack of alternatives make it almost impossible to challenge a running mine. Two of the strongest resistance movements concern cases where an existing mine was to be significantly expanded (El Cerrejón) or replaced by a new open-pit nearby (Jabiluka)—both projects were stopped and put on hold for indefinite time. In Australia, this success led to a strategic change of the indigenous organization towards the negotiation of agreements on a variety of issues, while on a deeper level, an attitude of resistance is maintained, based on strong socio-cultural values that are perceived as incompatible with uranium mining. In Colombia, there are divergent developments: in the day-to-day interaction, negotiation and corporate co-optation are prevalent, while open resistance is foremost articulated in the project-specific campaign against the expansion. Here, the campaign coalition constitutes an organizational innovation, which provided a new space to mobilize a "loud minority" powerful enough to break the veil created by dependency and co-optation, at least for a critical period. Thereby, new indigenous actors such as women's groups build upon and at the same time transform established institutional forms, such as the practice of informal mediation. For the third case, Ambuja, institutional innovation based on preexisting networks and the cooperation between various organizational forms (workers' union, farmers' organization, and *panchayat* village councils) opened new possibilities for challenging the well-installed industry, in what appears as an outbreak of rebellion against long-standing, accumulated grievances. Although this coincided with the expansion of the industrial complex, the latter did not provide much leverage, contrary to several other political opportunities (a won court case and the centenary celebrations of the company) that helped that the case was taken up by a wider coalition at the international level.

In all these cases, preexisting and strongly rooted values and institutions evolved through organizational innovation stimulated by coalition building. There was thus a base for collective action, but at the same time a necessity to transform and adapt them to gain leverage by increasing organizational reach and scale, breaking out of corporate-controlled networks of patronage. Furthermore, these cases indicate that mine expansions are especially prone to conflict, as at this stage direct negative experience combines with high economic dependency on mining, possibly resulting in deep social divisions.

Deterioration: The last scenario (Niger) is a kind of "worst case", whereby an initially successful (armed) rebellion against accumulated injustices deteriorates to an open "resource war", followed by military defeat in the context of geo-political power struggles and "terrorism" discourses. This scenario leaves little room for non-violent forms of resistance, and indeed, little hope that the situation could significantly improve in the near future—to prevent a further deterioration can already be seen as a success.

Fig. 1: Protest at the general assembly meeting of Xstrata in Zug, Switzerland, 20.11.2012. Photo: Thomas Niederberger.

Implications for Action and Research

In this section, we draw a number of implications from the above-sketched scenarios, in continuation with the concluding remarks in Chapter C. We address the role and positioning of (potential) allies; that is, people and groups who are or might become involved in mining conflicts, be it as a concerned individual, by working with an NGO, or as an activist researcher.[3]

Starting from an assumption that the inter-linkage of mining (and other) conflicts can create a field of possibilities to challenge power imbalances enshrined in dominant discourses of "development", we see that these possibilities widely vary in the different scenarios. At first sight, from the "basic scenarios", the Marlin case seems most rewarding for anyone seeking an easy confirmation to be on the "right side" by supporting a strong resistance movement, led by the local communities themselves. To the contrary, an involvement in a "long-term dependency" or a "CSR works" scenario such as Mopani or Las Bambas seems less promising; or, even a perfect recipe for cynicism and frustration, in a scenario of generalized powerlessness that apparently provides little room for improvements. More promising are the "early resistance" scenarios. The argument is that, as resources get scarcer and communication and exchange increases, local people confronted with a new mining project are always more likely to have experienced negative with similar projects earlier, or have knowledge of such experiences, and are therefore more likely to resist strongly from the beginning. Thereby, a crucial role is taken by coalitions that involve neo-traditional institutions and succeed in breaking the isolation of local resistance groups, facilitate

3 For different forms of engaged research practices and the trajectory of Action Anthropology, see Part IV.

mutual support, knowledge exchange and linkages with other struggles that challenge political and economic exclusion and inequality. The earlier such coalitions get organized, or even more, if they are already in place due to previous efforts, the higher is the likelihood to stop an unwanted development before corporate co-optation becomes effective.

However, many scenarios, and especially those of mines in operation, are more complex and do not confirm a simple "anti-mining" stance—especially, if large sectors of the local population do not share it, but rather opt for negotiating benefits. Still, there are also good reasons to become involved in these cases: a motivation may be, for instance, to mitigate well-known impacts of mining, and to prevent a deterioration and escalation of a conflict. Furthermore, we showed that even in situations that do not appear very promising for successful collective action, there is a potential that organizational innovation and knowledge transfer, together with political and project-related transformations (e.g., an expansion plan) can suddenly change the scenario. And although this may not lead to shutting down a controversial mine, it can bring significant improvements in the way it is carried out. Finally, challenging long-lasting mines at critical moments may create exactly the momentum that can spill-over, trigger and strengthen other emerging local struggles, and thus set the necessary conditions to not repeat earlier mistakes.

While activist researchers and NGOs do have different criteria and priorities, and are responsive to different constituencies (i.e., the "academic community" for the first, and the members and funders for the second), both can find themselves dealing with similar questions regarding the ethics and methods of engagement. Also, both may tend to fall into a "conflict shopping", jumping on the one case that seems most rewarding at a given time, while side-lining others. In contrast, we argue that rather than privileging one or the other of the mentioned motives (for instance, focussing only on "early resistance" scenarios), the task is to remain attentive to the specific trajectories

Fig. 2: Blockade of an excavator at the open pit coal mine at Hambach, Germany; 01.08.2014.
Photo: <hambacherforst.blogsport.de/media>.

of local struggles in order to uncover and possibly enhance emergent interrelations between them. Such an endeavour could be conceived similarly to human rights observation and civil peace accompaniment in conflict zones, and would need to be sustained over long periods. As this is an excessive demand for any organization with limited resources, it may be best to approach this aim through coalitions of specialized NGOs, research and other institutions in collaboration with social movements and grassroots groups of directly affected people. Thereby, established methods of participative action research, action and advocacy anthropology, and similar forms of engaged research, may be helpful (see Part IV).

Necessarily, such coalitions need to be multi-vocal and embrace diversity and heterogeneity, as well as tensions and contradictions. Furthermore, this could provide entry points for the involvement of people beyond a clearly limited local setting—which seems fair enough given the far-reaching impacts of modern large-scale mines—while remaining attentive to varying degrees of affectedness. Indeed, in the context of today's globalized world it is increasingly difficult to conceive "resource wars" as an insular phenomenon, affecting "failed" or "weak" states only: conflicts of different intensities are happening in so many places, and are so strongly linked to global flows of goods, finances, migration and information, that it is arguably more accurate to conceive them as focal points within a single, global resource war.

Outlook: Mining Conflicts and the "Glocal Commons"

As an outlook, we like to mention some further angles to explore. Something which has largely escaped from this book, due to its exclusive focus on large-scale mining, is the potential of linking different socio-environmental struggles to each other, as well as to other types of social movements, as a driver for profound socio-political transformations. A further question that merits attention is how far the conflicts around the "open cut" of mining can be a breeding ground for the bottom-up emergence and a sense of ownership of new institutional arrangements regarding the governance, ownership, use and distribution of mineral resources. This concerns not only on the "supply side", but also the "demand-side" of resource extraction, and the possibility to conceive mineral resources as a "glocal commons".

To illustrate what we mean by this, the example of Argentina is helpful. There, the *Unión de Asambleas Ciudadanas* (Union of Citizens' Assemblies) has built up an extensive network of local assemblies and support groups, involving almost the entire Argentinian territory and bordering regions, around the core issues of mining, agro-industry and other types of industrial contamination.[4] A special characteristic of the UAC is that it has emerged foremost from the rural periphery, but was able to reach out to the urban population as well, thereby involving diverse social sectors from middle-class professionals to peasants and indigenous communities. In many regions, this movement has become powerful enough to challenge the established nexus of political elites and corporate interests—thereby blocking several mega-mining projects and introducing protective legislation, such as the ban on open-pit mining in some provinces, and the law that prohibits mining in glacier zone on the national level. At a deeper level, the movement transports a positive attitude and respect towards the indigenous peoples of the continent, their lifeworlds and worldviews, and thus becomes an effective agent of de-colonization.[5]

Grown out of the interlinking of diverse local struggles, such movements can thus create a space of interaction for the formation of alternative visions. Thereby, the active revaluation of an unpolluted environment and the mobilization of positive attitudes to small-scale and subsistence

4 For a summary of the genesis of this movement, see: "Construyendo Caminos Colectivos en Defensa de Nuestros Territorios. Sistematiszacion de las Conclusiones de los Encuentros de la UAC", pdf-document availabe from <asambleasciudadanas.org.ar>.

5 Lecture of Maria Victoria Veracierto, 19 March 2012, University of Bern; see also case study "Uranio AG".

economy activities seem to be central elements that act as a counter-discourse to the dominance of neoliberal regimes of resource extraction. In this context, we deal more with conceptions of "resilience" and "sufficiency", or "living well" (*buen vivir/vivir bien*), which includes emic perceptions of what people aim for. Different from the Western concept of "sustainability", these concepts appear to have escaped co-optation through CSR discourse for the time being (which does however not exclude the possibility that this might happen in the future).[6] On a general level, questions for further research could be: what is the potential of mutual inspiration in such "glocal" settings of widely varying challenges and opportunities to consolidate into operative coalitions based on shared values? Can the field of imagination spanned up between diverse local struggles relate to and strengthen the position of local groups involved in mining conflicts such as those described in this book?

This question relates to the concept of "constitutionality", as proposed by Haller/Acciaioli/Rist (2015), and to the debate on the governance of common-pool resources more generally. A crucial problem of formal common property recognition is that, although there have been certain improvements for collective land ownership (in particular, for indigenous peoples), this does not extend to the underground, that is, mineral resources. The latter are still treated almost universally as an exclusive domain of the state, to be privatized according to the discretion of government officials, in non-transparent procedures often riddled by corruption. And although in some regions with a history and practice of artisanal mining, local institutions do regulate the access to mineral ores (see for e.g., Bolay for Guinea, this volume), these are not officially recognized and can hardly cope with large-scale mining. Formal and legal institutions are lacking or inadequate to deal with the massive impacts of mining, but the informal institutions also lag behind.

The issue here is, however, that granting common property rights over cultural landscapes can set the basis for bottom-up institution building processes, creating hence a sense of ownership. This might also be the basis for deliberation on how to conceive the relation of society and mineral resources. Based on our findings, we propose that:

- Only in rare cases people have a sense of collective ownership of mineral resources and thus link their identity and their livelihood to it. Rather, they do so regarding related resources such as land and water, as well as with the institution-building process itself. These are also mostly the cases where strong resistance takes place;
- The expectations of jobs and development usually hold only for a minority and for a limited time-span, while access to the common-pool resources, endangered by mining, could provide a more secure livelihood in the long run;
- Successful local collective action is based on a) well-working local institutions combined with b) innovations and c) extending coalitions and d) the emic perception that mining does not provide a reliable livelihood in the long-term and in addition fuels conflicts that have the potential to severely deteriorate the social conditions.

Although the conflict situations and social divisions described in most of the case studies pose serious challenges to an open and democratic deliberation on how to deal with the sub-soil, we argue that under the above-mentioned conditions, processes of constitutionality can take place. While the opportunities may be restricted in a local setting under highly unequal power-relations, at the larger level it is possible that local people struggling with transnational mining corporations significantly contribute to the debate on where, how and for which purpose mineral resources shall or shall not be extracted. Struggles in the field of mining can thereby trigger a debate on alternatives and the possibility to frame more adequate institutions in a bottom-up process.

6 Certain interpretations of "living well" may be open to co-optation within an extractivist economic policy, as shown by the Governments of Bolivia and Ecuador. See also the introduction of Part I.

Fig. 3: Forest road block near the open pit coal mine at Hambach, Germany 2014. Photo: <hambacherforst.blogsport.de>.

This debate not only concerns the "supply side" (i.e., production), but also the "demand side" (i.e., consumption) of natural resources. Although the focus of this book has been almost exclusively on the first, to gain a full picture necessarily means to address the second, too. There are special challenges to this due to the fact that minerals and metals are no consumer goods and the conventional "consumer politics" model cannot be applied: it is difficult to imagine what an infusion of the market with ethics and value, expressed through consumer choice (see e.g., Fischer 2014 for "German eggs and Guatemalan coffee") could mean for a metal like copper. This might be the reason why most NGOs that campaign for ethical consumption have not addressed mining products so far, with few exceptions. Although there are initiatives that point to this direction, they are still a niche phenomena, only relevant for a limited number of specific local settings.[7]

Apart from the role that solidarity initiatives and NGOs could play in this regard, a further angle that deserves attention is the potential of linking socio-environmental struggles in the "Global South" to similar movements in the "North". In Europe, one can observe an emergent network of inter-related struggles against large-scale development projects, both in rural and urban settings. Several of them concern mines (mentioned in the introduction of Part I), but also an airport in Notre-Dame-des-Landes (France), a high-speed railway-line in Val di Susa (Italy), or early resistance against hydraulic fracturing in Eastern Europe. Many participants in these struggles—both local people and supporters—seem to understand them as expressions of a common, larger movement, as indicated by solidarity declarations and mutual visits.[8] These struggles are qualitatively

7 In the jewellery sector, there is a demand for "fair gold", mostly sourced from artisanal mining cooperatives that work with improved extraction techniques. A good example is the "No Dirty Gold" campaign in the US, see <nodirtygold.earthworksaction.org>; in Switzerland, the Society for Threatened Peoples (GfbV) has launched a similar attempt, albeit focussing more on political lobbying. See Part II – 1 and <www.gfbv.ch/de/kampagnen>

8 See, e.g.: "Collectif de lutte contre l'aéroport de Notre Dame des Landes" <lutteaeroportnddl.com>; "NO TAV Info"

different from the short-term, spectacular confrontation of global conferences that the "global justice movement" put forward around the year 2000,[9] for being more locally rooted long-term engagements. At the same time, there are striking similarities regarding the methods, in particular a preference for direct action, protest camps and blockades (which again relates them to the wave of temporal occupations of urban squares, such as Occupy Wallstreet, Gezi Park in Istanbul or the 15M-movement in Spain).

The transformative power such movements can develop should not be underestimated. A prime example is the anti-nuclear movement in Germany with its yearly blockades of the transports to the nuclear waste disposal site in Gorleben.[10] Drawing broad support from both local residents and solidarity activists throughout the country, this significantly increased the political and economic costs of promoting nuclear energy. After the Fukushima disaster, the German government then took the lead in pulling out from nuclear energy, massively promoting renewable energy instead. The repercussions of such political changes on the uranium price (and the value of nuclear energy companies) can be observed also in (potential) mining regions, where many projects are on hold, not only because of local resistance, but also because they are not viable at current prices (see e.g., the Australian and Niger case studies).

The example of Gorleben highlights the relevance of confronting coal power plants, practised by a range of climate action groups.[11] Furthermore, an increasing number of consumers who opt for renewable energy, excluding nuclear and fossil fuel power, shows that the creation of an "ethical market" also for metals is not a completely hopeless endeavour. In the realm of recycling and product design there have been considerable improvements that need to be further expanded. Most importantly, these are areas where innovations can emerge and be tested. Consumer politics can thus contribute to a more rational use of non-renewable resources, although certainly not sufficient on its own. What is needed, then, is foremost a continuous debate on these issues—not only considering the full picture of costs and benefits of mining products themselves, but also addressing the problem of their unequal distribution.

Moody (2007: 12) suggests to conceive the remaining mineral deposits as a "common heritage of humankind". However, the concept of "global commons", often used for oceans or the atmosphere, has its limitations regarding the interdependence on actions taken on several scales. We propose to conceive mineral deposits rather as a kind of "glocal commons"[12]—they are "local" in the sense that their exploitation affects a number of other common-pool resources such as land, forests, water bodies, wildlife etc., which are subtractable[13] and scattered over larger areas. Therefore, they cannot be controlled by individuals, but only collectively at the local and regional scale and that is the reason why many of these resources are managed by common property institutions (see Ostrom 1990; Haller ed. 2010). A the same time, the exploitation of mineral deposits is "global", as it has impacts on supra-regional, national and global scales as well, including the

<www.notav.info>.

9 See for e.g., Juris (2008), Graeber (2009).

10 See for e.g., "Gorleben Archiv" <www.gorleben-archiv.de> and the "Castor 2010" Infopage <www.castor2010.de>.

11 At the time of writing, the climate campaign groups are focusing on the 2015 United Nations Conference on Climate in Paris (COP21); see for e.g., "Campaign against Climate Change" <www.campaigncc.org>; "Coalition Climat 21" (for Paris COP 21) <coalitionclimat21.org>.

12 Although the literature on (global) common goods is extensive, the term "glocal commons" seems so far little used. A google-scholar search points to Debarbieux/Price (2008) and Shultz (2015). A comparison with the advanced debates on creative commons, the patenting of genetic resources and the protection of traditional knowledge could prove interesting in this regard; see for e.g., Brown (2003).

13 "Subtractibility" means that the amount of resource unit taken from the stock is at the moment no longer available. However, for renewable resources this unit will regenerate or regrow again as long as the stock is not reduced (see Ostrom 1990, Becker/Ostrom 1995, Haller 2007, Haller 2010 ed.).

Fig. 4: "Water is worth more than gold — no to the Mine" Spray-painted slogan on a wall in Argentina. Source: <www.darioaranda.com.ar>.

provision of raw material for indispensable products, far-ranging environmental pollution and the contribution to global climate change. Today, these global aspects are insufficiently regulated through the market economy or through provisions of national and international law, both of which tend to undermine the autonomy of local resource management and ownership institutions.

An adequate institutional setting for the governance of the "glocal commons" should facilitate the arbitration between conflicting interests and aspirations on all levels, always remaining sensitive to the graded affectedness and diverse socio-cultural values along the entire chain of supply and consumption. To enhance distributive and procedural justice, and thereby the legitimacy of resource governance regimes, a shift in power relations is needed, away from the dominance of corporations and private interests toward a predominant position of those who are directly affected by mining. Within such an institutional framework, it should be possible to take differentiated decisions regarding various types of mines and mining products. For instance, from a global perspective there are solid arguments for a "leave it in the ground" policy for coal and uranium, given the enormous risk of their use on a world scale, and the availability of viable alternatives. With other minerals and metals that are indispensable for certain uses (e.g., copper and other industrial metals), extraction should be done according to the best possible standards and be confined to areas where mining is regarded by the local population as a welcome economic activity, and where it is feasible to contain the environmental risks and damages. On the contrary, "no-go areas" should be defined where local people do not give their consent on mining, as well as in regions that are especially vulnerable for environmental, climatic, socio-political and other reasons. For gold, this should be handled in even stricter manner, given that the largest part of its usage (jewellery and saving deposits) is not essential and the remaining part (industrial production) could be covered largely through recycling. The argument is thus similar to Ostrom (1990): to define a general "best case" institution for the management of commons, in the sense of a "one size fits all" solution, seems not useful; rather, what is needed is an open debate on certain basic principles, which may be applied in a variety of ways in different "glocal" settings.

Finally, we argue that the underlying issue at stake in mining is about competing conceptions of values. The ambiguous term "value" can refer to culturally embedded moral standards as well as to quantifiable, monetary worth. A widespread slogan that appears in protests against mines throughout South America sums it up perfectly: *El agua vale mas que el oro* ("Water is worth more than gold!"; see *Figure 4*[14]). Water here is equated with life, while gold stands for dead matter and the cold rationality of capitalism. Mines and other large-scale industrial projects thus constitute value-generating endeavours that bring to the fore underlying deeper conflicts of clashing worldviews, in a geographically highly focussed manner. Mining conflicts thereby act as sand in the gears of the "anti-politics machine" (to paraphrase Ferguson), a de-politicized discourse that presents "development" as a non-negotiable, "natural" fact, associated with whatever capitalist expansion needs or brings. Even more, certain mining conflicts have the potential to grow into veritable "politics machines" that challenge the dominant developmentalist ideology. Or, more precisely, the "politics machine" is the field of relations spanned up through alliances and coalitions wherein mining conflicts and other socio-environmental struggles work as "motors" that drive the debate about diverse visions of development and the "good life". In that sense, the challenge of "healing the open cut" is about how the relationships between humans, and between them and the non-human world, can be structured in a more balanced manner, for the well-being of the present and future generations.

14 Image courtesy of Dario Aranda: "Catamarca movilizada contra la megaminería" (26.01.2015), <www.darioaranda.com.ar>, accessed 10.10.2015.

References

Alimonda, Héctor (ed.) 2011: *La naturaleza colonizada. Ecología Política y Minería en América Latina.* Buenos Aires: CLACSO-CICCUS.

Altman, Jon 2012: Indigenous Rights, Mining Corporations and the Australian State. In: Sawer, Suzana and Terence Gomez (eds.): *The Politics of Resource Extraction. Indigenous Peoples, Corporations and the State.* London: Palgraves.

Arana Zegarra, Marco 2009: *«Cuidar el agua, es cuidar la vida»: Aportes de la cultura andina a la nueva cultura del agua.* In: De Echave, J., R. Hoetmer, and M. Palacios Panéz: *Minería y Territorio en el Perú.* Lima: Democracia y Transformación Global, 219-242.

Arellano-Yanguas, Javier 2012: Mining and Conflict in Peru: Sowing the Minerals, Reaping a Hail of Stones. In: Bebbington, Anthony (ed.): *Social Conflict, Economic Development and Extractive Industry.* London: Routledge.

Auty, Richard 1993: Sustaining Development in Mineral Economies: The Resource Curse Thesis. London: Routledge.

Bakan, Joel 2004: The Corporation: The Pathological Pursuit of Profit. London: Constable.

Ballard, Chris and Glenn Banks 2003: Resource Wars: The Anthropology of Mining. *Annual Review of Anthropology* 32: 287-313.

Bebbington, Anthony (ed.) 2012: Social Conflict, Economic Development and Extractive Industry: Evidence from South America. London: Routledge.

Bebbington, Anthony, Leonith Hinojosa, Denise Bebbington, Maria Burneo and Ximena Warnaars 2008: Contention and Ambiguity: Mining and the Possibilities of Development. *Development and Change* 39(6): 887–914.

Bebbington, Anthony, Teresa Bornschlegl and Adrienne Johnson 2013: Political Economies of Extractive Industry: From Documenting Complexity to Informing Current Debates. *Development and Change* Virtual Issue 2: 1–16.

Becker, Dustin C. and Elinor Ostrom 1995: Human Ecology and Resource Sustainability: The Importance of Institutional Diversity. *Annual Review of Ecology and Systematics* 26: 113-133.

Benson, Peter and Stuart Kirsch 2010a: Capitalism and the Politics of Resignation. *Current Anthropology* 51(4): 459-486.

Benson, Peter and Stuart Kirsch 2010b: Corporate Oxymorons. *Dialectical Anthropology* 34(1): 45–48.

Berkes, Fikret 1999: Sacred Ecology. Traditional Ecological Knowledge and Resource Management. Philadelphia and London: Taylor & Francis.

Boillat, Sebastien; Rist, Stephan; Serrano, Elvira; Ponce Dora; Delgadillo, Jaime 2008: Struggling "Ontological Communities". In: Galvin, Marc and Tobias Haller (eds.): *People, Protected Areas and Global Change.* University of Bern: Geographica Bernensia, 37-80.

Bridge, Gavin 2004: Mapping the Bonanza: Geographies of Mining Investment in an Era of Neoliberal Reform. *The Professional Geographer* 56(3): 406–421.

Brockington, D., R. Duffy and J. Igoe 2008: Nature Unbound: Conservation, Capitalism and the Future of Protected Areas. London: Earthscan.

Brown, Michael F. 2003: Who Owns Native Culture? Cambridge: Harvard University Press.

Bucher, Daniela 2014: Impacts of Uranium Mining in the Kakadu Region: Perceptions of Aboriginal People and Related Stakeholders in a North-Australian Mining Town. University of Bern: Unpublished MA Thesis.

Cambalache (ed.) 2013: *El Oro de Salave: Minería, Especulación y Resistencias.* Oviedo: Local Cambalache.

Campos, Marina T. 2008: From Villains and Victims to Environmental Activists: The Case of the Amazonian Colonos. In: Geiger, Danilo (ed.): *Frontier Encounters: Indigenous Communities and Settlers in Asia and Latin America.* Copenhagen: International Work Group for Indigenous Affairs (IWGIA).

Cleary, David. 1990: Anatomy of the Amazon Gold Rush. Ames, IA: University of Iowa Press.

Coumans, Catherine 2011: Occupying Spaces Created by Conflict: Anthropologists, Development NGOs, Responsible Investment, and Mining. *Current Anthropology* 52(S3): 29-43.

Davis, Rachel and Daniel M. Franks 2014: Costs of Company-Community Conflict in the Extractive Sector, *Corporate Social Responsibility Initiative Report* No. 66. Cambridge, MA: Harvard Kennedy School.

De Echave, José, Alejandro Diez, Ludwig Huber, Bruno Revesz, Xavier Ricard Lanata y Martín Tanaka 2009: *Minería y Conflicto Social.* Lima: CBC, CIPCA, CIES, IEP.

Debarbieux, Bernard and Martin F. Price 2008: Representing Mountains: From Local and National to Global Common Good. *Geopolitics* 13(1): 148-168.

Deegan, Denise 2001: Managing Activism: A Guide to Dealing with Activists and Pressure Groups. London: Kogan Page.

Delgado, Freddy, Stephan Rist, and César Escobar 2010: *El Desarrollo Endógeno Sustentable - Como Interfaz Para Implementar El Vivir Bien En La Gestión Pública Boliviana.* La Paz: Plural.

Dougherty, Michael L. 2011: The Global Gold Mining Industry, Junior Firms, and Civil Society Resistance in Guatemala. *Bulletin of Latin American Research* 30(4): 403-418.

Doyle, Cathal and Andy Whitmore 2014: Indigenous Peoples and the Extractive Sector: Towards a Rights-Respecting Engagement. Baguio: Tebtebba, PIPLinks and Middlesex University.

Doyle, Cathal and Jill Cariño 2013: Making Free, Prior & Informed Consent a Reality: Indigenous Peoples and the Extractive Sector". London: PIPLinks, Middlesex University, ECCR.

Egresi, Istvan 2011: The Curse of the Gold: Discourses Surrounding the Project of the Largest Pit-Mine in Europe. *Human Geographies* 5(2): 57-68.

Ensminger, Jean 1992: Making a Market. The Institutional Transformation of an African Society. Cambridge: Cambridge University Press.

Erklärung von Bern / Berne Declaration (EvB/BD) (ed.) 2012: *Rohstoff: Das gefährlichste Geschäft der Schweiz.* Zürich: Salis. (Also available in English as: "Commodities—Switzerland's most Dangerous Business", and in French as: "Swiss Trading SA", Lausanne: Editions d'en bas).

Escobar, Arturo 1999: After Nature: Steps to an Anti-essentialist Political Ecology. *Current Anthropology* 40(1): 1-30.

Escobar, Arturo 2008: Territories of Difference: Place—Movements—Life—*Redes*. Durham: Duke University Press.

Fairhead, James, and Melissa Leach 1996: Misreading the African Landscape: Society and Ecology in a Forest-Savanna Mosaic. Cambridge University Press.

Ferguson, James 1990: The Anti-Politics Machine: "Development," Depoliticization and Bureaucratic Power in Lesotho. Cambridge: Cambridge University Press.

Ferguson, James 1999: Expectations of Modernity: Myth and Meanings of Urban Life on the Zambian Copperbelt. Berkeley: University of California Press.

Ferguson, James 2005: Seeing Like an Oil Company: Space, Security, and Global Capital in Neoliberal Africa. *American Anthropologist* 107(3): 377-382.

Ferguson, James 2006: Global Shadows: Africa in the Neoliberal World Order. Durham: Duke University Press.

Filer, Colin 1999: The Dialectics of Negation and Negotiation in the Anthropology of Mineral Resource Development in Papua New Guinea. In: Cheater, Angela (ed.): *The Anthropology of Power: Empowerment and Disempowerment in Changing Structures.* London/New York: Routledge: 88-100.

Filer, Colin 2012: The Development Forum in Papua New Guinea: Evaluating Outcomes for Communities. In: M. Langton and J. Longbottom (eds): *Community Futures, Legal Architecture: Foundations for Indigenous Peoples in the Global Mining Boom*, Abingdon: Routledge, 147–61.

Fischer, Edward F. 2014: The Good Life. Aspiration, Dignity, and the Anthropology of Wellbeing. Stanford: Stanford University Press.

Fletcher, Robert 2012: Capitalizing on Chaos: Climate Change and Disaster Capitalism. *Ephemera* 2(1/2): 97-112.

Franks, Daniel M., Rachel Davis, Anthony J. Bebbington, Saleem H. Ali, Deanna Kemp, and Martin Scurrah 2014: Conflict Translates Environmental and Social Risk into Business Costs. *PNAS* 111(21).

Fraser, Nancy 2013: A Triple Movement? *New Left Review* 81: 119–132.

Gedicks Al 1993: The New Resource Wars: Native and Environmental Struggles Against Multinational Corporations. Boston, MA: South End Press.

Galvin, Marc and Tobias Haller (ed.) 2008: People, Protected Areas and Global Change. Participatory Conservation in Latin America, Africa, Asia and Europe. *Perspectives of the Swiss National Centre of Competence in Research NCCR North-South* (Vol. 3). Bern: Geographica Bernensia.

Geiger, Danilo (ed.) 2008: Frontier Encounters: Indigenous Communities and Settlers in Asia and Latin America. Copenhagen: International Work Group for Indigenous Affairs (IWGIA).

Gibson, Clark C. 1999: Politicians and Poachers. The Political Economy of Wildlife Policy in Africa. Cambridge: Cambridge University Press.

Gibson, Ginger and Ciaran O'Faircheallaigh 2010: IBA Community Toolkit. Negotiation and Implementation of Impact and Benefit Agreements. Toronto: The Gordon Foundation. <www.ibacommunitytoolkit.ca>, accessed on 18.5.2015.

Godoy, Ricardo A. 1985: Mining: Anthropological Perspectives. *Annual Review of Anthropology* 14: 199-217.

Goffman, Erving 1959: The Presentation of Self in Everyday Life. New York: Doubleday & Company.

Golub, Alex 2014: Leviathans at the Gold Mine: Creating Indigenous and Corporate Actors in Papua New Guinea. Durham, NC: Duke University Press.

Graeber, David 2009: Direct Action—An Ethnography. Oakland: AK Press.

Gudynas, Eduardo 2011: Buen Vivir: Today's Tomorrow. *Development* 54(4): 441–447.

Gudynas, Eduardo and Alberto Acosta 2011: La Renovación de La Crítica Al Desarrollo Y El Buen Vivir Como Alternativa. *Utopía y Praxis Latinoamericana* 16(53): 71-83.

Haalboom, Bethany 2012: The Intersection of Corporate Social Responsibility Guidelines and Indigenous Rights: Examining Neoliberal Governance of a Proposed Mining Project in Suriname. *Geoforum* 43(5): 969-979.

Haller, Tobias 2013: The Contested Floodplain. Institutional Change of the Commons in the Kafue Flats, Zambia. Lanham USA: Lexington (Rowman & Littlefield).

Haller, Tobias, Annja Blöchlinger, Markus John, Esther Marthaler, Sabine Ziegler (eds.) 2007: Fossil Fuels, Oil Companies, and Indigenous Peoples. Berlin: LIT. (In German: infoe-Studie 12: *Fossile Ressourcen, Erdölkonzerne und indigene Völker.* Giessen: Focus Verlag, 2000.)

Haller, Tobias (ed.) 2010: Disputing the Floodplains: Institutional Change and the Politics of Resource Management in African Floodplains. *African Social Studies Series* 22. Leiden, Boston: Brill.

Haller, Tobias; Merten, Sonja 2010: "We had cattle and did not fish and hunt anyhow!" Institutional Change and Contested Commons in the Kafue Flats Floodplain (Zambia). In: Haller, Tobias (ed.) *Disputing the Floodplains: Institutional Change and the Politics of Resource Management in African Floodplains.* African Social Studies Series 22: 301-359. Leiden, Boston: Brill.

Haller, Tobias and Marc Galvin 2011: Challenges for Participatory Conservation in Times of Global Change: Lessons from a Comparative Analysis and New Developments. In: Wiesmann, Urs Martin; Hurni, Hans (ed.) *Research for Sustainable Development. Foundations, Experiences, and Perspectives.* NCCR North-South (6), University of Bern: Geographica Bernensia, 467-503.

Haller, Tobias; Fokou, Gilbert; Mbeyale, Gimbage; Meroka, Patrick 2013: How Fit Turns into Misfit and Back: Institutional Transformations of Pastoral Commons in African Floodplains. *Ecology and Society* 18(1):1-16.

Haller, Tobias, Greg Acciaioli and Stephan Rist 2015: Constitutionality: Conditions for Crafting Local Ownership of Institution-Building Processes. *Society & Natural Resources* 0: 1–20.

Harvey, David 2003: The New Imperialism. Oxford: Oxford University Press.

Hilson Gavin 2012: Corporate Social Responsibility in the Extractive Industries: Experiences from Developing Countries. *Resources Policy* 37(2): 131-137.

Himley, Matthew 2013: Regularizing Extraction in Andean Peru: Mining and Social Mobilization in an Age of Corporate Social Responsibility. *Antipode* 45(2): 394-416.

Holden, William, Kathleen Nadeau and Daniel Jacobson 2011: Exemplifying Accumulation by Dispossession: Mining and Indigenous Peoples in the Philippines. *Geografiska Annaler: Human Geography* 93(2): 141–161.

Hurley, P.T. and Y. Arı 2011: Mining (Dis)amenity: The Political Ecology of Mining Opposition in the Kaz (Ida) Mountain Region of Western Turkey. *Development and Change* 42(6): 1393–1415.

Hyndman, David. 2001: Academic Responsibility and Representation of the Ok Tedi Crisis in Postcolonial Papua New Guinea. *The Contemporary Pacific* 13(1): 33-54.

Jacks, David S. 2013: From Boom to Bust: A Typology of Real Commodity Prices in the Long Run. *NBER Working Paper* No. 18874.

Jenkins, Katy 2014: Women, Mining and Development: An Emerging Research Agenda. *The Extractive Industries and Society* 1(2): 329–339.

Juris, Jeffrey S. 2008: Networking Futures: The Movements against Corporate Globalization. Durham: Duke University Press.

Kapferer, Bruce 2005: New Formations of Power, the Oligarchic-Corporate State, and Anthropological Ideological Discourse. *Anthropological Theory* 5: 285-299.

Kemp, Deanna, Carol Bond, Daniel Franks, and Claire Cote 2010: Mining, Water and Human Rights: Making the Connection. *Journal of Cleaner Production* 18(15): 1553-1562.

Kirsch, Stuart 2002: Anthropology and Advocacy. A Case Study of the Campaign against the Ok Tedi Mine. *Critique of Anthropology* 22(2): 175-200.

Kirsch, Stuart 2012: Afterword: Extractive Conflicts Compared. In: Bebbington, Anthony (ed.) 2012: *Social Conflict, Economic Development and Extractive Industry.* London: Routledge, 201-213.

Kirsch, Stuart 2014: Mining Capitalism: The Relationship Between Corporations and their Critics. Berkeley: University of California Press.

Klein, Naomi 2007: The Shock Doctrine: The Rise of Disaster Capitalism. Toronto: Alfred A. Knopf.

Kohn, Eduardo 2013: How Forests Think. Berkeley: University of California Press.

Kronenberg, Jakub 2014: Viable Alternatives for Large-Scale Unsustainable Projects in Developing Countries: The Case of the Kumtor Gold Mine in Kyrgyzstan. *Sustainable Development* 22(4): 253–264.

Kuipers, J.R., Maest, A.S., MacHardy, K.A., and Lawson, G. 2006: Comparison of Predicted and Actual Water Quality at Hardrock Mines: The Reliability of Predictions in Environmental Impact Statements. Washington DC: Earthworks.

Kuyek, Joan 2007: Mining Investors. Understanding the Legal Structure of a Mining Company and Identifying its Management, Shareholders and Relationship with the Financial Markets. Ottawa: MiningWatch Canada.

Lahiri-Dutt, Kuntala 2011: Gendering the Field: Towards Sustainable Livelihoods for Mining Communities. Canberra: Australian National University E-Press.

Larsson, V. and M. Ericsson 2014: E&MJ's Annual Survey of Global Mining Investment. *Engineering & Mining Journal* 215(1): 26-31.

Latour, Bruno 1991: *Nous n'avons jamais été modernes - essai d'anthropologie symétrique.* Paris: La Découverte.

Li, Fabiana 2009: Documenting Accountability: Environmental Impact Assessment in a Peruvian Mining Project. *PoLAR: Political and Legal Anthropology Review* 32(2): 218–236.

Lindroth, Marjo 2014: Indigenous Rights as Tactics of Neoliberal Governance Practices of Expertise in the United Nations. *Social & Legal Studies* 23(3): 341-360.

Marfurt, Franziska, Fabian Käser and Samuel Lustenberger 2015: Local Perceptions and Vertical Perspectives of a Large Scale Land Acquisition Project in Northern Sierra Leone. *Homo Oeconomicus,* forthcoming.

Martinez-Alier, Joan 2002: The Environmentalism of the Poor: A Study of Ecological Conflicts and Valuation. Cheltenham: Edward Elgar Publishing.

Mason, Leah, Timothy Prior, Gavin Mudd, and Damien Giurco 2011: Availability, Addiction and Alternatives: Three Criteria for Assessing the Impact of Peak Minerals on Society. *Journal of Cleaner Production* 19(9-10): 958-966.

Mastrángelo, Andrea 2004: *Las Niñas Gutiérrez y la Mina Alumbrera: La Articulación con la Economía Mundial de una Localidad del Noroeste Argentino.* Buenos Aires: Antropofagia.

Moncloa, Alejandra Alayza and Eduardo Gudynas 2012: *Transiciones y Alternativas al Extractivismo en la Región Andina: Una Mirada desde Bolivia, Ecuador y Perú.* Lima Centro Peruano de Estudios Sociales (CEPES).

Moody, Roger 2007: Rocks and Hard Places: The Globalization of Mining. London: Zed Books.

Mudd, Gavin 2007: Global Trends in Gold Mining: Towards Quantifying Environmental and Resource Sustainability. *Resources Policy* 32(1-2): 42-56.

MultiWatch (ed.) 2014: *Milliarden mit Rohstoffen – Der Schweizer Konzern Glencore Xstrata.* Zürich: edition 8 (also available as pdf download in English and Spanish, see www.multiwatch.ch).

Mutti, Diana, Natalia Yakovleva, Diego Vazquez-Brust, and Martín Marco 2012: Corporate Social Responsibility in the Mining Industry: Perspectives from Stakeholder Groups in Argentina. *Resources Policy* 37(2): 212-222.

Nash, June 1979: We Eat the Mines and the Mines Eat Us. Dependency and Exploitation in Bolivian Tin Mines. New York: Columbia University Press.

O'Faircheallaigh, C and Ali, S 2008 (ed.): Earth Matters: Indigenous Peoples, the Extractive Industries and Corporate Social Responsibility. Sheffield: Greenleaf Publishing.

O'Faircheallaigh, Ciaran 2008: Negotiating Cultural Heritage? Aboriginal-Mining Company Agreements in Australia, *Development and Change* 39(1): 25-51.

Ololade, Olusola and Harold Annegarn 2013: Contrasting Community and Corporate Perceptions of Sustainability: A Case Study within the Platinum Mining Region of South Africa. *Resources Policy* 38(4): 568-576.

Olson, Mancur 1965: The Logic of Collective Action. Public Goods and the Theory of Groups. Cambridge Mass.: Harvard University.

Ostrom, Elinor 1990: Governing the Commons. The Evolution of Institutions for Collective Action. Cambridge: Cambridge University Press.

Owen, John, and Deanna Kemp 2013: Social Licence and Mining: A Critical Perspective. *Resources Policy* 38(1): 29-35.

Özkaynak B. et al. 2012: Mining Conflicts Around the World: Common Grounds from Environmental Justice Perspective. *EJOLT Report* No. 7.

Padel, Felix and Samarendra Das 2010: Out of this Earth. East India Adivasis and the Aluminium Cartel. New Delhi: Orient Blackswan.

Parry, Jonathan P. 2008: The Sacrifices of Modernity in a Soviet-Built Steel Town in Central India. In: João de Pina-Cabral and Frances Pine (ed.): *On the Margins of Religion.* New York /Oxford: Berghahn.

Pegg, Scott 2006: Mining and Poverty Reduction: Transforming Rhetoric into Reality. *Journal of Cleaner Production* 14(3-4): 376-387.

Peters, Pauline E. 2013: Conflicts over Land and Threats to Customary Tenure in Africa. *African Affairs* 112(449): 543-562.

Power, Michael 1994: The Audit Explosion. London: Demos.

Prior, T, D Giurco, G Mudd, L Mason, and J Behrisch 2012: Resource Depletion, Peak Minerals and the Implications for Sustainable Resource Management. *Global Environmental Change* 22(3): 577-587.

Prno, Jason 2013: An Analysis of Factors Leading to the Establishment of a Social Licence to Operate in the Mining Industry. *Resources Policy* 38(4): 577-590.

Radcliffe, Sarah 2012: Development for a Postneoliberal Era? Sumak Kawsay, Living Well and the Limits to Decolonisation in Ecuador. *Geoforum* 43(2): 240-249.

Rajak, Dinah 2010: 'HIV/AIDS is our business': the moral economy of treatment in a transnational mining company. *Journal of the Royal Anthropological Institute* 16 (3): 551-571.

Rajak, Dinah 2011: In Good Company: an Anatomy of Corporate Social Responsibility. Stanford: Stanford University Press.

Rajak, Dinah 2014: Corporate memory: historical revisionism, legitimation and the invention of tradition in a multinational mining company. *PoLAR* 37 (2): 259-280.

Rist, Stephan and Farid Dahdouh-Guebas 2006: Ethnosciences–A Step towards the Integration of Scientific and Indigenous Forms of Knowledge in the Management of Natural Resources for the Future. *Environment, Development and Sustainability* 8(4): 467–493.

Sawyer, Suzana and Edmund Terence Gomez (eds.) 2012: The Politics of Resource Extraction: Indigenous Peoples, Multinational Corporations, and the State. London: Palgrave Macmillan.

Scott, James C. 1985: Weapons of the Weak: Everyday Forms of Peasant Resistance. New Haven, Conn.: Yale University Press.

Scott, James C. 1998: Seeing Like a State: How certain schemes to improve the human condition have failed. New Haven, Conn.: Yale University Press.

Shultz, Clifford J. 2015: The Ethical Imperative of Constructive Engagement in a World Confounded by the Commons Dilemma, Social Traps, and Geopolitical Conflicts. In: Alexander Nill (ed.): *Handbook of Marketing Ethics.* London: Edward Elgar Pub., 188-219.

Singer, P.W. 2003: Corporate Warriors: The Rise of the Privatized Military Industry. Ithaca, NY: Cornell University Press.

Strathern, Marilyn 2002: Audit Cultures: Anthropological Studies in Accountability, Ethics and the Academy. New York: Routledge.

Stüben, Peter E. 1988: *Die neuen "Wilden": Umweltschützer unterstützen Stammesvölker – Theorie und Praxis der Ethno-Ökologie.* Giessen: Focus Verlag.

Svampa, Maristella and Mirta A. Antonelli (ed.) 2009: *Minería transnacional, narrativas del desarrollo y resistencias sociales.* Buenos Aires: Ed. Biblos.

Svampa, Maristella y Sola Alvarez, Marian 2010: *Modelo minero, resistencias sociales y estilos de desarrollo: los marcos de la discusión en la Argentina.* In: Ecuador Debate. Conflictos del extractivismo, Quito: Centro Andino de Acción Popular, 105-126.

Taussig, Michael T. 1986 [1980]: The Devil and Commodity Fetishism in South America. Chapel Hill: University of North Carolina Press.

Thorp, R., S. Battistelli, Y. Guichaoua, J.C. Orihuela and M. Paredes (eds) 2012: The Developmental Challenges of Mining and Oil: Lessons from Africa and Latin America. Basingstoke: Palgrave Macmillan.

Tierra Socialista 2011: *Dossier: El Debate sobre el Pachamamismo* (with contributions of Pablo Stefanoni, Hugo Blanco, Arturo Escobar and others). *Tierra Socialista* 2(2): 166-207.

Tsing, Anna L. 2000: Inside the Economy of Appearances. *Public Culture* 12: 115–44.

Tsing, Anna L. 2005: Friction: An Ethnography of Global Connection. Princeton and Oxford: Princeton University Press.

Urkidi, Leire 2010: A Glocal Environmental Movement against Gold Mining: Pascua–Lama in Chile. *Ecological Economics* 70: 219–227.

Voss, Marianne and Emily Greenspan 2012: Community Consent Index: Oil, Gas and Mining Company Public Positions on Free, Prior, and Informed Consent (FPIC). *Oxfam America Research Backgrounder Series.* <www.oxfamamerica.org/publications/community-consent-index> accessed on 18.5.2015.

Walter, Mariana and Joan Martinez-Alier 2010: How to Be Heard When Nobody Wants to Listen: Community Action against Mining in Argentina. *Canadian Journal of Development Studies* 30(1-2): 281-301.

Welker, Marina 2009: "Corporate Security Begins in the Community": Mining, the Corporate Social Responsibility Industry, and Environmental Advocacy in Indonesia. *Cultural Anthropology* 24(1): 142-179.

Welker, Marina 2014: Enacting the Corporation: An American Mining Firm in Post-Authoritarian Indonesia. Berkeley: University of California Press.

Welker, Marina and David Wood 2011: Shareholder Activism and Alienation. *Current Anthropology* 52(S3).

Werthmann, Katja (ed.) 2012: Mining Frontiers in Africa: Anthropological and Historical Perspectives. Köln: Rüdiger Köppe Verlag.

Whitmore, Andy (ed.) 2012: Pitfalls and Pipelines: Indigenous Peoples and Extractive Industries. Baguio City/ Copenhagen: Tebtebba Foundation and IWGIA.

Wilson, Alana and Miguel Cervantes 2013: Annual Survey of Mining Companies 2013, Vancouver: Fraser Institute. <www.fraserinstitute.org> accessed on 18.5.2015.

PART IV

Strategies for Action / Research

Strategies for Action / Research

Germaine F. Spoerri and Thomas Niederberger[1]

Praxis – Theory – Praxis

How to reflect on research processes which aim at unveiling corporations' means and methods of exploiting the earth and its population? How can (academic) forms of knowledge, critical to Transnational Corporations (TNCs), contribute to transformation rather than stabilization of an existing order? How to produce knowledge forcibly framed by institutional structures of neoliberal universities without falling prey to the effects of structural abstraction? Is it at all possible to have people's knowledges flow into universities, 'become academic', and still stand for and maintain the dignity of these knowledges themselves and, more importantly, the voices and (hi)stories behind them? This is just a small selection of doubts that we have to face when we try to speak in the name of others via academic literature.

With these open questions in hand, this chapter proposes a circular process of reflection on the research undertaken for and in this book.[2] Departing from the actual Praxis of research, the praxis of producing this book, specific elements of the knowledge production process are going to be abstracted as and into Theory in order to then be able to feed them back into the actual, concrete Praxis of research. For example, Participative Action Research (PAR) considers 'action' as "the main criterion to validate any theory, prioritizing practical knowledges" (de Molina 2006). It proposes a "procedure that goes from concrete elements (or realities) to the abstract totality, returning afterwards to the concrete", with "the paradigm of objectivity" transformed into reflexivity and dialogical engagement, "embracing two principles similar in ways to feminist epistemology" (ibid.). The goal of this article is thus a reflection on our own research undertakings as an open discussion while walking (as the Zapatista words would say). Alongside the reflection on Praxis, theoretical reflections will touch upon fields which some people have named Action Anthropology, Activist and Militant Research, Participative Action Research as well as 'undisciplined' Geographies. To present and promote these approaches and processes of knowledge production is yet another desire behind our endeavour.

In *Moverse en la incertidumbre: dudas y contradicciones de la investigación activista* ("Moving in incertitude: doubts and contradictions of activist research") the Pantera Rosa collective (Barcelona) reflects on militant research. As guiding questions for reflection on activist and/or militant research/investigation they identify four crucial questions: *with whom / what for / where / and how* is knowledge being produced? (see Pantera Rosa 2004, 191-205). 'With whom?' points to the question of trustful sources and possible allies in the search and production of knowledge(s). 'What for?' poses the question of the direction or aim of the knowledge produced. 'Where?' inquires about the location, on the one hand where is the knowledge to be found, and on the other hand, where it flows to once gathered. 'The how?' is as well a question of method as of the position of the researchers. It reflects upon the relation between research subject and object, the

1 Review by Helen Gambon. Cover: detail from The True Cost of Coal, courtesy by The Beehive Design Collective, creative commons.

2 For a description of this process and the initial intention, see the introduction of Part I.

potential indistinguishability of the two and the methods applied to discuss and exchange ideas and knowledge(s). These four basic questions served to reflect upon the research undertaken in this book at a workshop in Bern in November 2013. As a first step, we sum up some answers to these questions relying on data gathered there, as well as from questionnaires sent out later-on to a range of people involved in the process of making this book.

As the second step, moving away from the concrete, we will outline the history and methodology of the mentioned approaches which we consider promising in contributing to modes of knowledge production sensible to the contradictions which arise whenever we conduct research about people's struggles, transnational corporations and the state. Moving back to the concrete, we will, in a third and last step, discuss three fields of contradiction which emerged from the reflections about the four guiding questions *'with whom? / what for? / where? / and how?'*. The first field of contradiction is an engaged researcher's position in between grassroots movements or so-called "local communities", different kinds of NGOs and Western academia/science. The second is the academic *mode of knowledge production* in the context of partly or fully neoliberalised universities (see De Angelis and Harvie 2009; Olssen and Peters 2005). And, finally, the third field is that of *collective* or *common* knowledge production. Is the knowledge produced in this attempt of making a "collective oeuvre" truly common knowledge? This last section is, above all, directed towards young researchers, be they students or non-academics, who feel that academic knowledge production is often insensitive to basic questions of ethics and involvement.

The Praxis of the *oeuvre*: what for? / where? / how? / with whom?

The four guiding questions 'what for? / where? / how? and with whom?' have been used as a basis for several workshops of reflection with groups of activists and researchers engaged in producing knowledge on TNCs and their means and methods of exploiting earth and its population. Some authors and participants in the production process of this book, together with people from an activist background who had so far not been related to it, took part in one of these collective reflection processes which took place in Bern in November 2013. Further, a more alienated process of reflection took place via Email, with two different questionnaires being answered by authors of the articles and people involved in the production process of the book without having the honour of placing their names above parts of the written product. The following is to be understood as a selection of voices with no illusions regarding its completeness: on the one hand, many of the people involved in gathering the knowledge assembled are scattered around the globe and, on the other hand, the assembled knowledge is, from our point of view, neither quantifiable nor determinable in singular persons but is rather to be grasped as a punctual crystallization of years of lived observation, struggle and suffering by many. Also, rather than describing the actual process that led to the present book, the answers express doubts about its limitations, desires to go beyond them, as well as a critique of its legitimacy from an activist and non-academic perspective. In that sense, they result from a space of reflection outside the normal mode of operation of academic knowledge production, which otherwise has strongly determined the process of writing this book.

Answers to the *'what for?'* of knowledge production brought up comments on personal motivation, affectedness and disquietudes regarding how TNCs conduct their production on a purely growth-based economic perspective. Concerns included aspirations to gather strategic knowledge about this dominant mode of production by studying TNCs' praxis "in order to identify its weaknesses and give direction to action" (X 2014) and "produce knowledge useful to strengthen resistance" (Y 2014); others wanted to develop workable strategies to confront the global "resource wars" as well as to "analyse the mechanics of power and domination elsewhere [than the location of Swiss TNCs headquarters] to sharpen the eye for recognizing the disinformation people are given by the TNCs themselves" (Z 2014). Further, "demystifying the expert guilds' technolanguage" (Y 2014), "immunise ourselves against corporate Orwellian newspeech" in order to read corporate sustainability reports without falling prey to "greenwashing" strategies (V 2014) were also mentioned. Another point brought up was the aim of issuing an open call for the formation of alliances acting towards a "decolonialization of our own hinterlands, inner cities and minds" (W 2014). Shared by many participants of the workshop was the standpoint of *not* wanting to give policy advice or being pushed into the role of policy advisors as many NGOs, legal experts and academic researchers have done and still do, but to clearly and openly break with what has become a norm in the last decades, namely non-confrontational cooperation with TNCs and state representatives.

Derived from the *'what for?'*, 'the *how?*' knowledge is being produced was, on the one hand, understood as a question of methodology, the actual praxis of research conducted, while on the other hand many participants pointed to questions of ethics, of how knowledge should or could be produced, in theory or as desired by those interested in unveiling international corporations' means and methods of exploiting earth and its population. Approved by nodding by most, one contribution to the discussion was that "knowledge production can have no Master [or personalised academic title]" and should be a means of "reflection about our own successes and failures [in struggles, e.g. against mining projects]" (Y 2014). Knowledge should be produced by methods enabling "a learning process from and with each other" in which transparency is perceived as a collective endeavour—transparency as a process and result, because "many eyes can see more than the solitary academic expert" (W 2014). Transparency, non-hierarchy and collectivity were thus largely shared as desires regarding 'the *how?*', while concrete examples of this '*how?*' were get-togethers, open discussions and assemblies involving as many people affected as possible, with modes of discussion sensible to informal hierarchies.

The '*where?*' has mostly been understood as a question of place, namely the locations where knowledge is being produced and locations where it is being diffused, or, where it flows to, once gathered. Most of the participants of the workshop as well as the answers to the questionnaires share the conviction that knowledge on people's struggles is to be published only according to the wishes and decisions of people affected (KS 2014); that knowledge which is produced collectively will be applicable and "useful" to those in struggle and that, finally, knowledge should never be extracted and decoupled from its place of origin. The link between where knowledge(s) lie, namely knowledge sources, and how knowledge(s) are (re)applied should, according to one participant, be maintained as a strong connection. Where and whenever this link is broken, "we can speak of dispossession and extraction" (X 2014).

Who and with whom? (is knowledge being produced) circled around mainly one slogan during the reflection workshop, namely *no somos Greenpeace, somos la ballena* ("we are not Greenpeace, we are the whale") (YG 2013). This slogan and the story behind it was shared by a participant involved in struggles in Argentina: in the context of the anti-mining movement in Esquel, it had emerged as a furious answer to a politician's accusation that Greenpeace (implying foreign activists) had been pulling the strings of the local protest. This slogan of local resistance was thus a clarifying moment on the position of Greenpeace who was *supporting* a movement led by local neighbourhood assemblies who expressed the voice of those directly threatened by the mine project.

The participants of the workshop generally regarded state representatives of countries in which TNCs are given permission to exploit as untrustworthy sources of information, alongside TNCs' communication agents for Corporate Social Responsibility (CSR). The discussion around NGOs brought up more differentiable voices: some participants proclaim that working with (rather small than big) NGOs is necessary to internationalise local struggles, while others, as a result of personal experience, take a very sceptical standpoint towards all forms of paid activism including academic interest in resistance movements. Concluding thoughts shared by all participants of the workshop agreed that cooperative, consensus- and dialogue-based working together of "those who feel affected and recognise each other as such" (X 2014) forms the very basis on which research should be undertaken. Thus, whoever cannot consent to these methods of cooperation, consensus and open dialogue, reveals her or himself as untrustworthy.

After having presented a selection of voices on the four basic questions proposed by the Pantera Rosa collective on "doubts and contradictions of activist research" (Pantera Rosa 2004) as encountered on the concrete level of the production process of this book, we now move on to outline a range of approaches in a more general sense. These approaches are again going to be linked to the questions of '*what for? / where? / how? and with whom?*'. In the following section, we will first focus on the trajectory of Action Anthropology because it constitutes one of the founding lines for the editing group of this book, namely the Swiss section of the "Institute for Ecology and Action Anthropology" (*infoe CH*). Secondly, examples of Activist and Militant Research as well as Participative Action Research will serve to illustrate the diversity of "engaged" forms of knowledge production, finally leading to a call for undisciplined geographies.

Fig. 1: *Con qué fin extraen el oro las minerás a cielo abierto?*; detail of *Cosmovisión Rebelde*, iconoclasistas (2008), creative commons.

A Short History of Action Anthropology

If there is something like a founding myth of Action Anthropology, it could be the humble question of some students at University of Chicago, who were sent to the Tama reservation of the Native American Fox by their Professor Sol Tax in 1948 and found them in a miserable condition. To their question "can we help them?" Tax encouraged them to try it. Of course, this was formulated in more complicated ways (see Schlesier 1980: 38), according to academic customs. The question of Tax' students expresses the same moment of "visceral experience" that many researchers experience when they find their host communities in a situation of injustice and oppression, which drives them to transform their role to active "advocacy" (Seithel 2000: 299).

From this initial exchange between Tax and his students emerged the so-called "Fox project". Schlesier (1980: 36-41) recounts the process as one of trial and error, with emphasis on the second. From 1948 to 1958, in all 19 students lived and worked in the Tama reservation. They enjoyed a lot of freedom, were encouraged by Tax to act as "engaged observers", more than the conventional "participant observer", and tried a variety of interventions including the setting up of cooperatives and fostering cross-cultural awareness. In the process, they came up with some reflections that Tax named "Action Anthropology", described as an attempt to both "help a group of people to solve a problem, and [...] learn something in the process" (Tax 1975: 515). He proposed an ethical position that he clearly distinguished from "pure science", arguing that "people are not rats and ought not to be treated like them [...]. Community research is thus justifiable only to the degree that the results are imminently useful to the community" (ibid.). Also, he proposed a differentiation from Applied Anthropology, which in its praxis applies theoretical knowledge to particular situations according to the priorities of external (funding) agencies, governments and welfare organizations. "The Action Anthropologist", he stated, "can have no master" (ibid.).

Tax insisted that Action Anthropology can produce valuable theoretical knowledge and is not merely a variation of social work or political activism. For example, he defended the conviction that the widespread notion of steady "acculturation" of indigenous groups towards the mainstream up to a point where they would "disappear" was wrong. It simply did not correspond to the facts observed on the ground. Rather, Tax realized that there was a refusal on the part of the indigenous against imposition of any kind of external schemes, particularly if it came from well-intentioned whites who wanted to "help" (Schlesier 1980: 41-42). The roots of many of the problems, he argued, were in the dispossession of territory and unequal power relations that had resulted from conquest and cultural discrimination and thus lay outside the Fox communities. "White" courts and administrative offices, among others, were expressions of unaddressed and ongoing injustice and could therefore hardly resolve any of the Fox' problems, but rather contributed to their perpetuation. Moving ahead from this observation, the role of the Action Anthropologist as characterized by Tax was one of a cross-cultural translator in the service of indigenous community leaders. His/her contribution was "non-directive counselling" without trying to influence the decision itself, in order to maintain the community's traditional decision-making processes (Schlesier 1980: 43).

As formulated by Nader (1972), the principle of "studying up" means that engaged researchers need to have detailed knowledge about the dominant juridical, administrative and corporate system to be able to give advice on different options to act and access necessary resources. Or, as Amborn states, "without the analysis of our own society, Anthropology remains a one-legged monster", because "[f]or an indigenous group that fights for its rights, the enthusiasm of an anthropologist about their shamanic practice is of little use if he does not also know about how corporations assert their interests" (Amborn 1993: 140; own translation). Though the "Fox project" itself did not correspond to the principles of Action Anthropology as outlined by Sol Tax, because it failed to involve the Fox and their traditional authorities in decision making, it nevertheless laid a ground for more successful attempts that followed (Schlesier 1980: 44). For instance, the Chicago conference in 1961, supported by Tax through organizing resources and providing access to university buildings, brought together representatives of indigenous communities from all over the USA and Canada for the first time and constituted an important step towards the formation of the indigenous peoples' "red power" movement (Schlesier 1980: 46).

In the context of a general outburst of rebellious spirits during the late 1960s and 1970s, a number of anthropologists attempted to "jump to the other side of the river" (Sevilla-Casas 1978: 145), by offering to place themselves under the command of indigenous communities, taking their side in struggles for self-determination. Thereby, the decisions around form and content of any publication that would come out of the process should remain with the communities, including the possibility of non-publication of secret knowledge. With this, the new wave of action anthropologists went beyond Tax' postulation that they should not have any "masters". Increasingly, they developed radical anti-civilization and eco-centric perspectives that openly opposed the dominant capitalist and authoritarian communist ideologies (Seithel 2000: 223f.), sometimes linked with at least an implicit affinity to anarchist positions (see Biegert 1981), or a call for a radical "earth first!" style "action ecology" (e.g., Stüben 1988).

One of the many noteworthy action anthropologists of this period was the above cited Karl Schlesier. After an experience in 1968 that he describes as "adoption" and "conversion" (Schlesier 1980: 55), he put his theoretical anthropological knowledge and the privileges of his position as a professor at the Wichita State University to the service of the traditional authorities of the Cheyenne. Over more than 20 years he assisted in reorganizing the institutional setting to push back the order imposed by the US Government, to fight legal battles, defend sacred sites, and on the way convinced a number of students to write their thesis on subjects proposed by the Cheyenne authorities. With time and growing confidence, Schlesier understood himself and apparently was accepted as a member of the Cheyenne. At this point, he went beyond "non-directive counselling" to a "limited active role", as "the action had found him" (ibid.: 63). Schlesier's role is underscored here especially for the proliferation of Action Anthropology in Germany, departing from a guest professorship at Münster University in 1978. In the following years, several German students undertook research in the Cheyenne reservation,[3] and Schlesier's articles were published in German.[4] In 1987, the debates around Action Anthropology then led to the formation of the *Institut für Ökologie und Aktionsethnologie* (*infoe*; "Institute for Ecology and Action Anthropology").

In its early years, infoe supported, for example, the struggle of the Mirarr Aborignial people against the Jabiluka uranium mine[5] by linking it to the anti-nuclear movement in Germany. Later, it became one of the small support NGOs that helped to facilitate access for indigenous delegates at UN conferences, and continues to participate occasionally in larger campaigns.[6] In the mid-1990s, Tobias Haller, then assistant lecturer at University of Zurich, founded the Swiss section (*infoe CH*) together with a small group of students and young researchers. Their first activity was to publish a book on "Fossil Fuels, Oil Companies, and Indigenous Peoples" (Haller et al. 2007) which served as a model for the present book. According to Haller's vision, *infoe CH* can serve as a vehicle close to the academic sphere for students and researchers who want to engage in more value-explicit endeavours, without being bound by the still prevailing notion of "academic neu-

3 Among them Friderike Seithel, who later wrote her PhD study on Action Anthropology from which we cite extensively (2000).

4 For instance, in the journal *Unter dem Pflaster liegt der Strand* (i.e., "underneath the paving stones lies the beach"; edited by Hans Peter Duerr), which was influential in radical circles in and beyond the universities at that time.

5 See case study Ranger/Jabiluka (Australia) in this book.

6 A recent example is the campaign for the Yasuní Amazon forest. See <www.infoe.de> for the German "mother-group" of *infoe CH*. Note that both groups function completely independent from each other, but share a common history and lines of action centring on supporting indigenous peoples affected by resource extraction and environmental destruction.

trality". Out of this, situations can arise where affected communities and their organizations ask for support, e.g., to make their voice heard close to relevant decision-making centres. Infoe can then participate in lobbying and information campaigns. From the "fossil fuels" book resulted, for example, a prolonged exchange with the "Movement for the Survival of Ogoni People" (MOSOP) from 1996 to 2005, around a campaign that demanded to use a part of dictator Sani Abacha's money blocked on Swiss bank accounts to compensate Ogoni people for the damages of oil extraction in their territories.[7] Similarly, the work of *infoe CH* on the present book led to an involvement in ongoing struggles and direct exchanges with members of affected communities which potentially could be a first step for further engagements. Though not directly related, the current association of *infoe CH* with the Institute of Social Anthropology at the University of Bern is not without a historical precedent: based at this otherwise unsuspicious institution, anthropologist Georg Grünberg organized the first Barbados Symposium in 1971, which was highly influential in promoting a more egalitarian exchange between anthropology and indigenous peoples (and in framing the base for Participatory Action Research) in Latin America (Seithel 2000: 247f.; see also Declaration of Barbados I 1971).

It is apparent that the activities of *infoe CH* (including the process of writing this book) are far from the depth and radicalness of Schlesiers' understanding of Action Anthropology as introduced above. Rather, they constitute an attempt towards an "advocacy anthropology" as described by Seithel (2004): a moderate form that mainly aims to offer a port of access for indigenous people and marginalized groups from the "Global South" who try to confront governments and corporations in the "North", be it through information campaigns, making use of international fora or court cases. Seithel observes that this corresponds to a general trend which has led away from the radical positions of the 1970s to more "modest" forms of engagement, which she welcomes as "giving more responsibility for knowing and deciding for themselves to those very people whose lives have been targeted for change" (2004: 39). Apart from "advocacy", there are many other terms that have come up in the 1980s and 90s to describe what can be understood as pragmatic forms of Action Anthropology: collaborative, interactive, proactive, cooperative or community-based research, among others (ibid.: 35). Often, these approaches were developed with marginalized communities in urban contexts, such as slum inhabitants or migrants.

This "pragmatic shift" can be seen as a reaction to a number of serious practical challenges posed by Action Anthropology: most of all, it requires long-term engagement of an "open-ended" character, way beyond the common one-year field research format. Especially if the community in question lies far away from the researcher's base, this is very difficult to maintain. In authoritarian regimes, action anthropologists often had to abandon their work because they were expelled or did not receive visas. To the practical challenges came political ones: Action Anthropology was criticized as non-academic social work, breaching the law of "academic neutrality", or as a covert form of paternalism (Seithel 2004: 26). Schlesier observed that colleagues suspected action anthropologists to be "[...] deserters having withdrawn from the ideal of building monumental theories" (1980: 41). This made it hard to obtain funding and support from university departments (ibid.: 54-55). Although Action Anthropology did have a certain impact on mainstream anthropological theory, it remained marginal even during its "high times", largely dependent on individual professors, and never reached a "critical mass" enough to form its own school within the discipline (Seithel 2004: 26; see also Greenwood 2008).

7 See Max Mader: "Report on the Situation in Ogoni land in the Niger Delta, Rivers State, Nigeria and on the Movement for the Survival of Ogoni People (MOSOP) in January 2002", infoe CH.

Fig. 2: *Pelígro fumigaciones con agrotoxicos*; detail of *Cosmovisión Rebelde,* iconoclasistas (2008), creative commons.

It seems that the debates around Action Anthropology had largely ceased by the mid-1990s. At best, it was treated as a "failed attempt" that formed part of the discipline's history—just at the time when the Zapatista uprising in Chiapas and the emerging "global justice movement" brought new urgency to questions of (academic) engagement (see e.g., Shukaitis/Graeber 2007; Juris 2008). A generation of young researchers marked by the events following the "Battle of Seattle" in 1999 contributed to develop some of the approaches mentioned in the next section, however, the reference to Action Anthropology is largely missing in their texts. It seems that the approach had been shelved precisely at the time when it could have been re-examined in new contexts. Around the same time, the rise of NGOs provided employment for many action-friendly anthropologists. Organizations such as Survival International focused on indigenous peoples' rights and successfully supported legal claims on territory, the set-up of indigenous umbrella organizations and the establishment of a permanent presence of indigenous delegates at different United Nation conferences. These processes were, and still are, studied, accompanied and supported by some academic anthropologists (e.g., Muehlebach 2001). Nevertheless, one can observe a disconnection between anthropologists working in NGOs or engaging in grassroots movements on the one hand, and academic anthropology on the other, whereby a lot of relevant knowledge remains exclusive to specific circles. While the shift to moderate or even covert modes of Action Anthropology in academic research and the professionalization of engaged anthropology in NGOs may be a pragmatic reaction to changed conditions in larger society, it can also be criticized as a de-radicalization, insofar as attention easily gets diverted from the "root of the problem" to the abstractions of the international conference circus, court cases, shareholder activism and media campaigns (e.g., Hardt/Negri 2000, see next section).

Regarding the conflicts around resource extraction and mining, the lessons to be learnt from the history of Action Anthropology appear more pressing then ever in a context where TNCs are increasingly forging alliances with NGOs and universities, and important numbers of anthropologists are being employed in CSR and corporate "community relations" offices (Kirsch 2002: 192; Kirsch 2014). Concluding this section, we think that Action Anthropology can still provide answers to the questions mentioned in the introduction: concerning the 'what for? how? and with whom?' of knowledge production, it calls for an ethically grounded, value-explicit research praxis that aims at the emancipation of everyone involved in the specific case as well as humanity as a whole. With regard to the 'where?' there are inspiring experiences on how to act from an uncertain positioning in-between academia, NGOs and grassroots movements and within fractioned and oppressed communities. Finally, Action Anthropology adds a crucial question to the four: when / for how long? It reminds us of the need for long-term and open-ended engagement, but also the patience and respect for conceptions of time and decision-making that do not adapt to the fast-moving pace of Western activism and the academic machinery. But rather than simple answers, Action Anthropology proposes a "discourse capable of acting" (Amborn 1993; own translation) which results from a consensus finding process about all these questions, always conscious about the involved power inequalities and constantly trying to reduce them (ibid.). If one wants so, this can be expanded even beyond the anthropocentric perspective to include non-human beings, as observed by Duerr: "[y]ou do not exploit human beings you can understand [...]", and "[y]ou do not exploit a nature that speaks to you." (1978: 113; own translation).

Having said all this, what is labelled as Action Anthropology is just one approach capable of contributing to the questions raised at the opening of this chapter—in the next section, we will introduce more recent examples of an engaged research praxis. While we think it is important to recall historical trajectories for a better understanding of current positions, there is also an urgent need to go beyond disciplinary divisions and the tendency of creating exclusive identification with them, which all too fast turn into self-isolation and further marginalization.

From Militant, Activist and Participative Action Research[8] to a Call for Undisciplined Geographies

In this section we are going to depart from concrete examples of what is named Activist Research, Militant Research and Participative Action Research in order to provide a range of possible tools with which research on mining activities of TNCs could be approached. This is not to devalue the methods applied in the case studies of this book, but rather to enlarge the horizon of possibilities for those who wish to conduct radical research in the future by pointing at promising experiments from diverse places driven by diverse aims. To begin with, one has to be aware of the fact that different terminology is used in different linguistic regions and depending on authors' or author collectives' different political vantage points. For example, for some, the term "activist" has a rather negative connotation connected to the development of anti-corporate globalization activists who have, from the late 1990s onwards, taken on paid positions in large NGOs who cooperate with state representatives and 'development' institutions (on the NGO-state-complex, see Hardt/Negri 2000). For others, the terms "activist" and "militant" are being used synonymously.

8 The references regarding Militant, Activist and (Participative) Action Research made in this section are mostly based on teachings and research activities conducted by a group of geographers who built up a Master of Arts in Activism and Social Change at the School of Geography of the University of Leeds, UK. At this point, they should be thanked for their courage (and patience) of attempting to break with the ongoing neoliberalization processes transforming and marginalizing radically critical academic research in the UK and elsewhere.

In this section we propose to focus on praxis rather than etymology of the general terms activist, militant, research, investigation and action. They are going to be presented in the way in which they are proposed and applied in the different, and of course, context-dependent examples. The focus is on the presentation of methods applied, in the hope that they may be used and reapplied in new contexts. Praxis, thus, will here mean methods of creating open and accessible knowledges.

The first example chosen is one of militant research which has been conducted in the city of Madrid, Spain, since 2005. The *Observatorio Metropolitano*, a self-declared militant research group, combines popular education, collective study and militant research. In an open course named *Metrópolis: crisis urbana, periferia y investigación militante* ("Metropolis: urban crisis, periphery and militant investigation"), a team of engaged people developed a 150-hour course composed of workshops, seminars, conferences and tutorials, the target group being "militants and activists of urban movements, artists, university researchers,..." (Observatorio Metropolitano 2013). The requirements of participation entail attendance of the classes, design of a militant investigation as well as conducting research for one of the axis of investigation of the Observatorio Metropolitano, namely 'the political economy of the crisis', 'metropolitan peripheries in crisis', 'the dissolution of the middle class' and 'the urban commons' (ibid.). The method applied in one of the workshops of the course, the *taller de mapeo colectivo: Carabanchelear o como sentarnos juntas a pensar los barrios* ("collective mapping workshop: *Carabanchelear* or how to get together and think about our neighbourhoods"), is one of collective reasoning by localizing places and routes in the neighbourhood which 'to us, appear significant in order to understand what is happening around us' (Universidad Popular de Carabanchel 2014). The aim is to share knowledge(s) and to get to know each other's motives and concerns by mapping the neighbourhood of Carabanchel: "We all know that the crisis is devastating the lives of many, ..., but we would like to know better what we are confronting, recognise signs in space, anticipate social division and point out antidotes which emerge from collective construction of resistance and invention" (Observatorio Metropolitano and Universidad Popular de Carabanchel 2014). This method of self-education by collective mapping is one in which the aim is not "to investigate from above, but the idea of knowing where we come from and what is happening to develop more powerful tools to change the situation" (ibid.). While the theoretical background of the *Observatorio* is one of critical analysis of neoliberal restructuration processes of the city, the method of militant investigation applied is open to all narratives linked to the basic interest of mapping unheard realities in times of crisis. *Contrageografías metropolitanas* ("metropolitan countergeographies") (Observatorio Metropolitano 2007: 150) are thus both method and outcome, in the form of maps, of a militant research process.

Another noteworthy project of collective mapping is that of *iconoclasistas* (see figure 1 and 2), active predominately in Argentina, but also beyond, who since 2008 organize collective mapping workshops "combining graphic art, creative workshops and collective research to produce resources for free circulation, appropriation and use, in order to strengthen communication, forge networks of solidarity and affinity and promote practices of collaboration and resistance" (Ares and Risler 2014). One of the many freely accessible maps of iconoclasistas named *Megaminería en los andes secos* (2010) ("Mega-mining in the dry Andes", see figure 3) contains elements like neighbourhood assemblies, mobilizations against mining, criminalization of and repression against protests, evictions and forced displacements. With a perspective of resistance against large mining projects, the main focus is placed on elements often only marginally mentioned in academic research. Similarly, the mapping projects of the MICLA Research Collective based at the McGill University of Montreal, Canada, aim "to resource public research and debate on Canadian Mining in Latin America" (MICLA 2015). Their maps include not the mere cartography of exploitation by the mining companies, but for e.g. conflicts regarding pollution, water, land rights, biodiversity, community relations, livelihoods, public health and indigenous issues related to the mining activities of Canadian gold mining companies. These maps, further, are updated

Fig. 3: Detail of *Megaminería en los andes secos.* iconoclasistas (2010), creative commons.

and worked upon throughout time and thus, reflect actual developments and struggles taking place today.[9] All the above mentioned maps are published online under *creative commons licenses* making free access and free reproduction possible.

The next example is one of self-declared militant research and was conducted by *Precarias a la Deriva* (Precarious Women Adrift), again based in Madrid, Spain. When the call for a general strike in 2002 mobilized thousands to struggle for better working conditions, a group of self-declared precarious women got together in order to direct public attention to the fact that (predominantly female) domestic workers, per-hour contracted, temporary workers and so-called self-employed women engaged in care work are neither represented nor able to participate in a general strike. The method applied by the collective, *la deriva* ('drifting'), was one of walking through the streets on strike day, visit places of precarious work and pose the question "what is your strike?" to those whose voices were (yet) unheard. *Adrift in the Circuits of Women's Precarious Lives* (A la Deriva 2004) is an example, in the form of a video, of the materialization of this process of producing knowledges about women living in precariousness via the method of *drifting*. The video shows women who share their ways of standing up to oppressive working conditions, far away from all organized labour struggles, therewith shifting the focus from macro-level mobilizations to the micro-level of everyday affective resistance (see also A la Deriva 2004, 2006).

9 See, for e.g., MICLA map Argentina 2015: <micla.ca/countries/argentina>.

The last case we would like to present is one of Participative Action Research (PAR), an academically established approach which "attempts to articulate research and social intervention with local communities' knowledges, know-hows and needs" considering "'action' as the main criterion to validate any theory, prioritizing practical knowledges" (de Molina 2006: 12). A PAR process thus wishes to contribute to the "transformation of reality, generating a new and more just reality" with "social (and transformative) praxis as objects and result of the study" (ibid., see also de Molina 2004). In Latin-American contexts PAR is named *Investigación-Acción Participativa* (IAP) (see Borda 2006). One of many examples of IAP conducted in this region took place in the city of Chaitén, Chile, after the eruption of the volcano Chaitén which buried more than half the city below a lahar in 2008. The method applied here was to conduct three workshop cycles which would bring together youth of Chaitén in order to make them express their situations through drawing, talking and walks through the destroyed city. Their knowledges were, after a first cycle of getting to know each others' stories, later transformed into a community based radio show named *La otra Mirada de Chaitén* ("The other view of Chaitén") in which the participants get to share their experiences and viewpoints with other inhabitants of the city (see Mardones Carrasco 2014).

Of course, the chosen few projects are but a very slim selection of the possible tools we would have wished to present. Also, the superficial presentation limited by space and time cannot do justice to the long processes of involvement that are needed for trust-building and sharing carried on by the people engaged in these research undertakings. Nevertheless, the link between these stories of collective knowledge production are based on specific common grounds: first, all of them put a strong emphasis on active participation of subjects, enabling personal accounts and knowledges to flow into and become a collective narrative of what is going on. This way, a collective subjectivity, a "we in the making" (see Holloway 2010), an identity, is created by doing, materializing *collectively* produced knowledge as maps, a video and a radio show. Thus, all of the examples have a strong emphasis on collectively producing knowledge by means of encounter, exchange of stories and a non-hierarchical research-subject-object relation.

Second, all of the projects focus on producing forms of knowledge as accessible to everyone as possible. To work with graphics, audio and visual information and publish them under *creative commons* licenses is a conscious choice and answer to the copyrighted world of paper-publishing *paypal* academia. In the examples one could speak of knowledge by and knowledge for the people, potentially only restricted by internet censorship, lack of electricity or infrastructure.

Thirdly, especially the mapping projects emphasise not only the sharing of the material outcome, the maps themselves, but also the tools on how to create them. The accessible manuals and handbooks of collective cartography are to be perceived as popular education for the creation of people's own geographies, realities, way beyond the victimization that is being disseminated and taught by schools, mainstream media and other powerful institutions.

What we would like to name 'undisciplined geographies', far from the academic discipline Geography, is geographies as realities, undisciplined with regard to the pressures imposed upon us by borders of faculties and universities, undisciplined with regard to borders of schools of thought, be they radical or liberal, undisciplined with regard to the expertization of our societies leading to inability and incompetence to grasp the complexity which we are exposed to and finally, undisciplined with regard to the limits of the general perception of how knowledge could or should be produced.

How to bring Action back into Research: Concluding Discussion

Starting from the knowledge production process of this book, we explored the more abstract level of approaches in the fields of Anthropology, Geography and other disciplines, in order to now move back to the concrete level of perspective and outlook. The opening questions regarded the possibilities of reflecting on (academic) research processes that aim to produce critical knowledge for the radical transformation of current regimes of resource extraction as promoted by corporations, as well as the effects of neoliberal universities on academic production and the danger of becoming insensitive to the voices of the people directly involved in the struggles we are writing about. From these doubts and the four guiding questions '*what for? / where? / how? / with whom is knowledge being produced?*', we now derive three fields of contradiction which we perceive as emerging from the discussions and reflections with involved activists and researchers.

There would, of course, be many other ways of trying to simplify the complexity in which one finds herself when engaged in critical knowledge production, but for the necessity of breaking down the complex into something which might contribute to prepare young researchers for their endeavours, we chose with best intentions: one, the position of 'researchers' in the 'field' in-between grassroots movements, different kinds of NGOs and academia, two, the academic context from which researchers try to contribute to specific struggles, and three, the question of collectivity in knowledge production. We attempt to delineate these fields of contradiction with a focus on the possibilities on how to bring "action back into research" and to build a tentative framework of reference within which we can reflect upon our praxis in-depth.

The first field of contradiction encompasses an engaged researcher's position in between grassroots movements of indigenous peoples or local communities, different kinds of NGOs and Western academia/science. To illustrate the difficulties of this in-between position, we would like to stereotypically outline three pathways a student in this position can walk down.

The first one is a student as researcher who chooses to travel abroad to get to know a local struggle with a self-appointed expectation to fully grasp the new context. She searches for ways to get involved in and make a valuable contribution to a local struggle. Time goes by and eventually, the pressure to perform at university forces her to shift her focus away from the struggle and return to the individualised academic form of knowledge production. She leaves the local struggle behind and returns to her university. The geographical distance between the struggle and her university makes it very hard, if not impossible to maintain something like direct involvement in the local struggle she got to know.

The second pathway is that of a student who gets carried away by her involvement in a particular struggle. Through her experiences in the new context, she chooses to stay, go as "native" as possible, and dive deeper into grassroots activism, regardless of the risk of dropping out of university. Her focus has shifted away from a university project to a struggle she feels very much part of.

The third pathway tells the story of a student, typically at the end of her studies, who dives into a local context and later decides to use her experiences gained with the grassroots movement as a basis for a professional career in an NGO, in public administration or even the CSR office of a corporation. This pathway has, by the way, critically been mentioned by some activists during the workshop of reflection on what the difficulties are to publish a book like this one. It is a pathway of scientists extracting knowledge from a local struggle to feed it to its counterparts working with authorities, who in the first place try to mute processes of resistance. These three pathways shed light on how the in-between position of engaged students or scientists can create a problem of identity or identification which not only plays into the question of roles one is ascribed to, but of affection and emotion (see Joy and Gordon 2008 for this specific contradiction, or more generally Greenwood 2008).

The second field of contradiction which we tap in as young researchers entails the academic modes of knowledge production in 21st century partly or fully neoliberalised universities.[10] On the one hand, the above mentioned pressure to perform in a specific way, exemplified in the European Credit and Accumulation Transfer System (ECTS), diminishes the possibilities of creating spaces of interaction free of marking, assessment and predetermined periods of engagement. On the other hand, funding applications that somehow link the words "research" with "action" have hardly any chance of being granted at the time, as many of us who intended have experienced in different faculties and departments across the globe. Alongside these institutional restrictions, the individualization of efforts (see Autonomous Geographies Collective 2010) has serious implications on potential collective research or projects below the post-Doc level, not least because undergraduate students are forced to hunt and gather credit points, often regardless of their proper fields of interest, while PhD students are largely involved in administering the bureaucracy behind the marking and assessment schemes.

The academia thus becomes a place of creativity (and innovation) *after* one has elbowed her, but still mostly his, way up the ladder. To be a critical student within the academic modes of knowledge production means having to use your elbows with a smile. This field of contradiction is experienced by most of us and has, we propose, such existential consequences on 'academic freedom', that promising radical thinkers in our (precarious academic) cohort have already left what they perceive to be a sinking boat.

The third field of contradiction is that of *collective* or *common* knowledge production. The "collective oeuvre", how 'collective' is it, respectively, how common is the knowledge produced? And, is it truly common knowledge that is being produced? A differentiation needed to answer these questions is between, on the one hand, knowledge which was exchanged, gained and produced among people involved at different levels and different stages of the project, and, on the other hand, the materialised form of gathered knowledges now printed on paper and put on the shelves.

In the first case, one could state that a common space of interaction has been created insofar as people with focus on the same case study shared experiences and information on an informal basis. Given the limitations of time, the geographical spread and diversity of the contexts of struggle as well as the experiences students brought into the process, collectivity at the level of the whole network of contributors was hardly attainable. Rather, the network had an enabling character in so far as it brought some of the participants together in concrete situations of collectivity, for instance in the mentioned workshop where people involved in very different settings created a space of reflection going beyond 'mere' informal exchange. The search for the reasons behind the relatively low rate of interaction between people involved in different case studies is, we argue, not least to be found in the limitations of the academic frame given to students: as mentioned in the precedent paragraph, the pressure to perform individually and the short terms of engagement in modules or projects do *de facto* weaken the ground for collective knowledge production in general (see De Angelis and Harvie 2009; Olssen and Peters 2005; Olssen 2009).

Regarding the printed version of the book, the materialised form of some of the gathered knowledges now printed on paper, the question of whether 'truly' common knowledge has been produced is one not answerable at this stage. As the impact of a book cannot be foreseen, it might

10 One example of neoliberalized university structures is the right to veto of the Switzerland-based TNC Nestlé AG regarding the election of two professorships financed by the company at the EPFL/ETH Lausanne, Switzerland (see TagesAnzeiger 8 of May 2014). For a campaign against the sponsoring by mining corporations of the Canadian International Resources and Development Institute (CIRDI) in Vancouver, see <www.stoptheinstitute.ca>; for Argentina: REDAJ 2010; and for a general critique of "corporate science" about mining impacts, see Kirsch (2014).

or might not contribute to the local struggles. Clearly, what is contradictory is that the access to the information in the book is restricted through copyright and the use of language, which will, especially for e.g., indigenous communities, make it hard to make any use of it. With focus on the contents, the question of whether the knowledges assembled reproduce the dominant gaze of seeing like a state, a corporation, a large NGO or an all knowing academic, is one to be answered critically by its readers. Summing up, even though the praxis of writing this book did not measure up to the expectations about collectivity of the whole process, moments of collective knowledge production have been triggered and, more importantly, experienced.

Having briefly delineated the three fields of contradiction we identified as emerging from our reflection processes, concluding remarks shall be made on why we think that the outlined examples of Action Anthropology, Activist and Militant Research and Participative Action Research can help to take a stance on how to deal with the mentioned fields of contradiction.

The creation of common spaces of interaction is itself a field of action and at the same time a necessary condition to empower ourselves and others to transform the oppressive and exploitative realities that surround us. The necessity of working collectively in order to capture the complexity of globally interlinked problematics remains, in our view, unquestioned. As mentioned before, one of the aims of the project of writing this book was to create such a collective space of discussion and exchange with a range of people coming from different approaches and backgrounds. Now, this intention of creating and maintaining a common space where knowledges are brought together on the common ground of sharing and channelling them into action comes to a limit from centrifugal and individualising forces surrounding it. This seriously complicates the collective long-term engagement needed as a basis for building relationships of trust.

The more examples and experiments of collective and common knowledge production are being conducted, on all levels and in all disciplines, the larger the pool of methods, experiences and theories from which young researchers can draw inspiration. In the general context of ever more tightened public debates, the conservative-defensive turn and increasing dominance of "control over freedom" (clearly observable in Switzerland, but also elsewhere), we like to think of courageous steps as innovations to what appears to be the aforementioned "sinking ship". If the ship of the 21st century academic world shall turn out to really be sinking regarding its ability to create forms of knowledge non-incorporable into the capitalist mode of knowledge production, we might as well build our own little ships now, along lines of flight and based on methods of emancipation and common grounds. For when old maps expire, new ones appear.

References

A la Deriva, P. 2004: Adrift Through the Circuits of Feminized Precarious Work. *Feminist Review* 77: 157-61.

A la Deriva, P. 2006: A Very Careful Strike—Four Hypotheses. *the commoner* 11: 33-45.

Amborn, H. 1993: Handlungsfähiger Diskurs. In: Schmied-Kowarzik, W./Stagl, J.: *Grundfragen der Ethnologie*. Berlin, 129-150.

Ares, Pablo and Julia Risler 2014: *Iconoclasistas. Critical mapping, collaborative practices and open source graphic resources*. Available from <www.iconoclasistas.net/texts>.

Autonomous Geographies Collective 2010: Beyond Scholar Activism: Making Strategic Interventions Inside and Outside the Neoliberal University. *Acme: An international e-journal for critical geographies* 9(2).

Biegert, C. 1981: Gespräch mit Karl Schlesier über Action Anthropology. *Unter dem Pflaster liegt der Strand* Vol. 8. Berlin: Karin Kramer Verlag.

De Angelis, Massimo and David Harvie 2009: 'Cognitive Capitalism' and the Rat-Race: How Capital Measures Immaterial Labour in British Universities. *Historical Materialism* 17(3): 3-30.

De Molina, Marta Malo (ed.) 2004: *Nociones Comunes. Experiencias y Ensayos entre Investigación y Militancia*. Madrid: Traficantes de Sueños.

De Molina, Marta Malo 2006: Common Notions, Part 2: Institutional Analysis, Participatory Action-Research, Militant Research, available at <http://eipcp.net/transversal/0707/malo/en/>.

Declaration of Barbados I 1971: IWGIA Document 1. Copenhagen: IWGIA.

Duerr, Hans Peter 1978: *Traumzeit. Über die Grenze zwischen Wildnis und Zivilisation*. Frankfurt a. M.: Syndikat.

Fals Borda, O. 2006: Participatory (Action) Research in Social Theory: Origins and Challenges. In: P. Reason and H. Bradbury (eds): *Handbook of Action Research*. London: SAGE.

Greenwood, Davyd 2008: Theoretical Research, Applied Research, and Action Research: The Deinstitutionalization of Activist Research. In: Charles R. Hale (ed.): *Engaging Contradictions: Theory, Politics, and Methods of Activist Scholarship*. Berkeley: University of California Press.

Hardt, Michael and Antonio Negri 2000: Empire. Cambridge: Harvard University Press.

Holloway, J. 2010: Crack Capitalism. London and New York: Pluto Press.

Joy, J. and T. Gordon 2008: Afterword: Activist Scholars or Radical Subjects?. In: C. R. Hales (ed.) *Engaging Contradictions: Theory, Politics, and Methods of Activist Scholarship*. GAIA Books Series. Berkely: University of California Press. (Available at <escholarship.org>).

Juris, Jeffrey S. 2008: Networking Futures: The Movements against Corporate Globalization. Durham: Duke University Press.

Kirsch, Stuart 2002: Anthropology and Advocacy—A Case Study of the Campaign against the Ok Tedi Mine. *Critique of Anthropology* 22(2): 175-200.

Kirsch, Stuart 2014: Mining Capitalism: the Relationship between Corporations and their Critics. California: University of California Press.

Mardones Carrasco, R. 2014: *Sistematización de una experiencia de investigación-acción-participativa (IPA) para el fortalecimiento de la participación comunitaria de jóvenes en el Chaitén post-erupción volcánica a través de la radio local*. Doctoral dissertation, Universidad de Chile.

Muehlebach, Andrea 2001: "Making Place" at the United Nations: Indigenous Cultural Politics at the UN Working Group on Indigenous Populations. *Cultural Anthropology* 16(3): 415–448.

Nader, Laura 1972: Up the Anthropologist – Perspectives Gained from Studying Up. In: Dell Hymes (ed.): *Reinventing Anthropology*. New York: Pantheon.

Observatorio Metropolitano 2007: *Madrid: ¿la suma de todos? Globalización, Territorio, Desigualdad*. Madrid: Traficantes de Sueños.

Olssen, M 2009: Neoliberalism, Education, and the Rise of the Global Common Good. In: Maarten Simons, Mark Olssen, Michael Peters (eds.): *Re-Reading Education Policies: A Handbook for Studying the Policy Agenda of the 21st Century*. Rotterdam, Boston, Taipei: Sense Publishers, 433-57.

Olssen, M and M Peters 2005: Neoliberalism, Higher Education and the Knowledge Economy: From the Free Market to Knowledge Capitalism. *Journal of Education Policy* 20(3): 313-45.

Pantera Rosa 2004: Moverse en la incertidumbre: Dudas y contradicciones de la inverstigacion activista. In: de Molina, Marta Malo (ed.): *Nociones Comunes. Experiencias y Ensayos entre Investigación y Militancia.* Madrid: Traficantes de Sueños, 191-205.

Red de Asistencia Jurídica contra la mega minería (REDAJ) 2010: *Empresas Megamineras y Universidades: YMAD y la Distribución de Fondos al Sistema Universitario Nacional.* Unión de Asambleas Ciudadanas y REDAJ.

Schlesier, Karl H. 1980: Zum Weltbild einer neuen Kulturanthropologie. Erkenntnis und Praxis. Die Rolle der Action Anthropology. *Zeitschrift für Ethnologie* 105(1): 32–66.

Seithel, Friderike 2000: *Von der Kolonialethnologie zur Advocacy Anthropology.* Hamburg/Münster: LIT Verlag.

Seithel, Friderike 2004: Advocacy Anthropology: History and Concepts. *Revista Anthropológicas* 15(1): 5-48.

Sevilla-Casas, Elías 1978: Action Anthropology and Transdisciplinary Perspectives. *Current Anthropology* 17(3): 490-497.

Shukaitis, Stevphen and David Graeber (eds.) 2007: Constituent Imagination: Militant Investigations, Collective Theorization. Oakland: AK Press.

Stüben, Peter E. 1988: Earth First! Ethno-Ökologie: Von der Aktionsethnologie zur Aktionsökologie. In: Peter E. Stüben (ed.): *Die Neuen "Wilden".* Giessen: Focus.

Tax, Sol 1975 [1952]: Action Anthropology. *Current Anthropology* 16(4): 514-517.

Image credits

iconoclasistas 2008: *Cosmovisión rebelde del saqueo neocolonial.* <iconoclasistas.com.ar/wp-content/uploads/2008/05/Argentina01.jpg> (creative commons).

iconoclasistas 2010: *Megaminería en los andes secos.* <www.iconoclasistas.net/2013/2010/05/30/el-grito-de-la-tierra> (creative commons).

Acronyms

ALRA	Aboriginal Land Rights Act of the Northern Territory (Australia)
AMD	Acid Mine Drainage
ASM	Artisanal and Small-scale Mining
BSAC	British South African Company
CAR	Central African Republic
CMM	Chhattisgarh Mukti Morcha
CSR	Corporate Social Responsibility
EIA	Environmental Impact Assessment
EITI	Extractive Industries Transparency Initiative
FPIC	Free, Prior and Informed Consent
GAC	Gundjeihmi Aboriginal Corporation
GRI	Global Reporting Initiative
IACHR	Inter-American Commission on Human Rights
ICMM	International Council on Mining and Metals
IFC	International Finance Corporation
IFIs	International Finance Institutions
ILO	International Labour Organization
IMF	International Monetary Fund
IPRA	Indigenous Peoples Rights Act of 1997 (Philippines)
ISO	International Standard Organization
IUCN	International Union for the Conservation of Nature
MNCs	Multinational Corporations
NGOs	Non-Governmental Organizations
OAS	Organization of American States
OECD	Organization for Economic Co-operation and Development
PCSS	Pragatisheel Cement Shramik Sangh ("progressive cement workers' union")
PNG	Papua New Guniea
SCCJ	Swiss Coalition for Corporate Justice
TNCs	Transnational Corporations
UNDRIP	UN Declaration on the Rights of Indigenous Peoples
UNESCO	United Nations Educational, Scientific and Cultural Organization
WGIP	Working Group on Indigenous Populations
WHO	World Health Organization

Figures, Maps, and Tables

Part I: Introduction

Part II: Case Studies

Index

A

B

D

E

F

G

H

I

J

K

L

M

N

O

P

R

S

T

U

V

W

X

Y

Z